THE BOOK OF
SER MARCO POLO

VOLUME 1

AMS PRESS, INC.
NEW YORK, N.Y.

Library of Congress Cataloging-in-Publication Data

Polo, Marco, 1254–1323?
  The book of Ser Marco Polo, the Venetian, concerning
the kingdoms and the marvels of the East.

  Reprint. Originally published: 3rd ed. London :
J. Murray, 1926.
  Bibliography: p.
  Includes index.
  1. Polo, Marco, 1254–1323? 2. Voyages and travels.
3. Mongols—History. 4. Asia—Description and travel.
I. Yule, Henry, Sir, 1820–1889. II. Yule, Amy Frances.
III. Title.
G370.P9P64   1986   951'.025'0924   74-5240
ISBN 0-404-11540-3 (set)

AMS PRESS, INC.
56 East 13th Street, New York, N.Y. 10003

Reprinted from the edition of 1926, London. Trim size has been
slightly altered. Original trim: 14.5 × 22 cm. The third volume of
this set was originally a separate title, and is a commentary on
the preceding two volumes.

INTERNATIONAL STANDARD BOOK NUMBER
Complete set: 0-404-11540-3
Vol. I: 0-404-11541-1

MANUFACTURED IN
THE UNITED STATES OF AMERICA

H. Yule

THE BOOK OF
SER MARCO POLO

THE VENETIAN CONCERNING THE KINGDOMS AND MARVELS OF THE EAST

TRANSLATED AND EDITED, WITH NOTES, BY
COLONEL SIR HENRY YULE, R.E., C.B., K.C.S.I.,
CORR. INST. FRANCE

THIRD EDITION, REVISED THROUGHOUT IN THE LIGHT OF
RECENT DISCOVERIES BY HENRI CORDIER (OF PARIS)

PROFESSOR OF CHINESE HISTORY AT THE ECOLE DES LANGUES ORIENTALES VIVANTES ; VICE-PRESIDENT
OF THE GEOGRAPHICAL SOCIETY OF PARIS ; MEMBER OF COUNCIL OF THE SOCIÉTÉ ASIATIQUE ; HON.
MEMBER OF THE ROYAL ASIATIC SOCIETY AND OF THE REGIA DEPUTAZIONE VENETA DI STORIA PATRIA

WITH A MEMOIR OF HENRY YULE BY HIS DAUGHTER
AMY FRANCES YULE, L.A.SOC. ANT. SCOT., ETC.

IN TWO VOLUMES—VOL. I.
WITH MAPS AND ILLUSTRATIONS

LONDON
JOHN MURRAY, ALBEMARLE STREET, W.
1926

Ἄνδρα μοι ἔννεπε, Μοῦσα, πολύτροπον, ὃς μάλα πολλὰ
Πλάγχθη . . . .
Πολλῶν δ᾽ ἀνθρώπων ἴδεν ἄστεα καὶ νόον ἔγνω.

Odyssey, I.

————"I AM BECOME A NAME;
FOR ALWAYS ROAMING WITH A HUNGRY HEART
MUCH HAVE I SEEN AND KNOWN; CITIES OF MEN,
AND MANNERS, CLIMATES, COUNCILS, GOVERNMENTS,
MYSELF NOT LEAST, BUT HONOURED OF THEM ALL."

TENNYSON.

"*A SEDER CI PONEMMO IVI AMBODUI*
VÔLTI A LEVANTE, OND' ERAVAM SALITI;
CHÈ SUOLE A RIGUARDAR GIOVARE ALTRUI."

DANTE, *Purgatory*, IV.

Messer Marco Polo, with Messer Nicolo and Messer Maffeo, returned from xxvi years' sojourn in the Orient, is denied entrance to the Ca' Polo. (See *Int.* p. *4.*)

DEDICATION.

TO THE MEMORY OF

SIR RODERICK I. MURCHISON, BART., K.C.B., G.C.St.A., G.C.St.S.
ETC.

THE PERFECT FRIEND

WHO FIRST BROUGHT HENRY YULE AND JOHN MURRAY TOGETHER

(HE ENTERED INTO REST, OCTOBER 22ND, 1871,)

AND TO THAT OF HIS MUCH LOVED NIECE,

HARRIET ISABELLA MURCHISON,

WIFE OF KENNETH ROBERT MURCHISON, D.L., J.P.,

(SHE ENTERED INTO REST, AUGUST 9TH, 1902,)

UNDER WHOSE EVER HOSPITABLE ROOF MANY OF THE PROOF

SHEETS OF THIS EDITION WERE READ BY ME,

I DEDICATE THESE VOLUMES FROM

THE OLD MURCHISON HOME,

IN THANKFUL REMEMBRANCE OF ALL I OWE TO

THE ABIDING AFFECTION, SYMPATHY, AND EXAMPLE OF BOTH.

TARADALE,   AMY FRANCES YULE.
    ROSS-SHIRE,   SEPTEMBER 11th, 1902.
      SCOTLAND.

*   *   *   *

Ed è da noi sì strano,
Che quando ne ragiono
I' non trovo nessuno,
Che l'abbia navicato,

*   *   *   *

Le parti del Levante,
Là dove sono tante
Gemme di gran valute
E di molta salute :
E sono in quello giro
Balsamo, e ambra, e tiro,
E lo pepe, e lo legno
Aloe, ch' è sì degno,
E spigo, e cardamomo,
Giengiovo, e cennamomo ;
E altre molte spezie,
Ciascuna in sua spezie,
E migliore, e più fina,
E sana in medicina.
Appresso in questo loco
Mise in assetto loco
Li tigri, e li grifoni,
Leofanti, e leoni
Cammelli, e dragomene,
Badalischi, e gene,
E pantere, e castoro,
Le formiche dell' oro,
E tanti altri animali,
Ch' io non so ben dir quali,
Che son sì divisati,
E sì dissomigliati
Di corpo e di fazione,
Di sì fera ragione,
E di sì strana taglia,
Ch'io non credo san faglia,
Ch' alcun uomo vivente
Potesse veramente
Per lingua, o per scritture
Recitar le figure
Delle bestie, e gli uccelli . . . . .

—From *Il Tesoretto di Ser Brunetto Latini* (circa MDCCLX.).
(*Florence*, 1824, pp. 83 *seqq.*)

CONTENTS OF VOL. I.

NOTE BY MISS YULE

I DESIRE to take this opportunity of recording my grateful sense of the unsparing labour, learning, and devotion, with which my father's valued friend, Professor Henri Cordier, has performed the difficult and delicate task which I entrusted to his loyal friendship.

Apart from Professor Cordier's very special qualifications for the work, I feel sure that no other Editor could have been more entirely acceptable to my father. I can give him no higher praise than to say that he has laboured in Yule's own spirit.

The slight Memoir which I have contributed (for which I accept all responsibility), attempts no more than a rough sketch of my father's character and career, but it will, I hope, serve to recall pleasantly his remarkable individuality to the few remaining who knew him in his prime, whilst it may also afford some idea of the man, and his work and environment, to those who had not that advantage.

No one can be more conscious than myself of its many shortcomings, which I will not attempt to excuse. I can, however, honestly say that these have not been due to negligence, but are rather the blemishes almost inseparable from the fulfilment under the gloom of bereavement and amidst the pressure of other duties, of a task undertaken in more favourable circumstances.

Nevertheless, in spite of all defects, I believe this sketch to be such a record as my father would himself have approved, and I know also that he would have chosen my hand to write it.

In conclusion, I may note that the first edition of this work was dedicated to that very noble lady, the Queen (then Crown Princess) Margherita of Italy. In the second edition the Dedication was reproduced within brackets (as also the original preface), but not renewed. That precedent is again followed.

I have, therefore, felt at liberty to associate the present edition of my father's work with the Name MURCHISON, which for more than a generation was the name most generally representative of British Science in Foreign Lands, as of Foreign Science in Britain.

A. F. YULE.

PREFACE TO THIRD EDITION.

LITTLE did I think, some thirty years ago, when I received a copy of the first edition of this grand work, that I should be one day entrusted with the difficult but glorious task of supervising the third edition. When the first edition of the *Book of Ser Marco Polo* reached "Far Cathay," it created quite a stir in the small circle of the learned foreigners, who then resided there, and became a starting-point for many researches, of which the results have been made use of partly in the second edition, and partly in the present. The Archimandrite PALLADIUS and Dr. E. BRETSCHNEIDER, at Peking, ALEX. WYLIE, at Shang-hai—friends of mine who have, alas! passed away, with the exception of the Right Rev. Bishop G. E. MOULE, of Hang-chau, the only survivor of this little group of hard-working scholars,—were the first to explore the Chinese sources of information which were to yield a rich harvest into their hands.

When I returned home from China in 1876, I was introduced to Colonel HENRY YULE, at the India Office, by our common friend, Dr. REINHOLD ROST, and from that time we met frequently and kept up a correspondence which terminated only with the life of the great geographer, whose friend I had become. A new edition of the travels of Friar Odoric of Pordenone,

our "mutual friend," in which Yule had taken the
greatest interest, was dedicated by me to his memory.
I knew that Yule contemplated a third edition of his
Marco Polo, and all will regret that time was not
allowed to him to complete this labour of love, to see
it published.   If the duty of bringing out the new
edition of *Marco Polo* has fallen on one who considers
himself but an unworthy successor of the first illustrious
commentator, it is fair to add that the work could not
have been entrusted to a more respectful disciple.
Many of our tastes were similar; we had the same
desire to seek the truth, the same earnest wish to
be exact, perhaps the same sense of humour, and,
what is necessary when writing on Marco Polo,
certainly the same love for Venice and its history.
Not only am I, with the late CHARLES SCHEFER, the
founder and the editor of the *Recueil de Voyages et de
Documents pour servir à l'Histoire de la Géographie
depuis le XIIIe jusqu'à la fin du XVIe siècle*, but I
am also the successor, at the Ecole des langues
Orientales Vivantes, of G. PAUTHIER, whose book on
the Venetian Traveller is still valuable, so the mantle
of the last two editors fell upon my shoulders.

I therefore, gladly and thankfully, accepted Miss
AMY FRANCES YULE's kind proposal to undertake the
editorship of the third edition of the *Book of Ser Marco
Polo*, and I wish to express here my gratitude to her
for the great honour she has thus done me.*

Unfortunately for his successor, Sir Henry
Yule, evidently trusting to his own good memory,
left but few notes. These are contained in an inter-
leaved copy obligingly placed at my disposal by Miss
Yule, but I luckily found assistance from various other

---

* Miss Yule has written the Memoir of her father and the new Dedication.

quarters. The following works have proved of the greatest assistance to me :—The articles of General HOUTUM-SCHINDLER in the *Journal of the Royal Asiatic Society*, and the excellent books of Lord CURZON and of Major P. MOLESWORTH SYKES on Persia, M. GRENARD'S account of DUTREUIL de RHINS' Mission to Central Asia, BRETSCHNEIDER'S and PALLADIUS' remarkable papers on Mediæval Travellers and Geography, and above all, the valuable books of the Hon. W. W. ROCKHILL on Tibet and Rubruck, to which the distinguished diplomatist, traveller, and scholar kindly added a list of notes of the greatest importance to me, for which I offer him my hearty thanks.

My thanks are also due to H.H. Prince ROLAND BONAPARTE, who kindly gave me permission to reproduce some of the plates of his *Recueil de Documents de l'Epoque Mongole*, to M. LÉOPOLD DELISLE, the learned Principal Librarian of the Bibliothèque Nationale, who gave me the opportunity to study the inventory made after the death of the Doge Marino Faliero, to the Count de SEMALLÉ, formerly French Chargé d'Affaires at Peking, who gave me for reproduction a number of photographs from his valuable personal collection, and last, not least, my old friend Comm. NICOLÒ BAROZZI, who continued to lend me the assistance which he had formerly rendered to Sir Henry Yule at Venice.

Since the last edition was published, more than twenty-five years ago, Persia has been more thoroughly studied ; new routes have been explored in Central Asia, Karakorum has been fully described, and Western and South-Western China have been opened up to our knowledge in many directions. The results of these investigations form the main features of this new edition of *Marco Polo*. I have suppressed hardly any of Sir

Henry Yule's notes and altered but few, doing so only when the light of recent information has proved him to be in error, but I have supplemented them by what, I hope, will be found useful, new information.*

Before I take leave of the kind reader, I wish to thank sincerely Mr. JOHN MURRAY for the courtesy and the care he has displayed while this edition was going through the press.

HENRI CORDIER.

PARIS, 1st of October, 1902.

---

* Paragraphs which have been altered are marked thus +; my own additions are placed between brackets [ ].—H. C.

"Now strike your Sailes yee jolly Mariners,
  For we be come into a quiet Rode" . . . .

—THE FAERIE QUEENE, I. xii. 42.

PREFACE TO SECOND EDITION.

—◇◇◇—

THE unexpected amount of favour bestowed on the former edition of this Work has been a great encouragement to the Editor in preparing this second one.

Not a few of the kind friends and correspondents who lent their aid before have continued it to the present revision. The contributions of Mr. A. WYLIE of Shang-hai, whether as regards the amount of labour which they must have cost him, or the value of the result, demand above all others a grateful record here. Nor can I omit to name again with hearty acknowledgment Signor Comm. G. BERCHET of Venice, the Rev. Dr. CALDWELL, Colonel (now Major-General) R. MACLAGAN, R.E., Mr. D. HANBURY, F.R.S., Mr. EDWARD THOMAS, F.R.S. (Corresponding Member of the Institute), and Mr. R. H. MAJOR.

But besides these old names, not a few new ones claim my thanks.

The Baron F. VON RICHTHOFEN, now President of the Geographical Society of Berlin, a traveller who not only has trodden many hundreds of miles in the footsteps of our Marco, but has perhaps travelled over more of the Interior of China than Marco ever did, and who carried to that survey high scientific accomplish-

ments of which the Venetian had not even a rudimentary
conception, has spontaneously opened his bountiful stores
of new knowledge in my behalf.   Mr. NEY ELIAS,
who in 1872 traversed and mapped a line of upwards
of 2000 miles through the almost unknown tracts of
Western Mongolia, from the Gate in the Great Wall
at Kalghan to the Russian frontier in the Altai, has
done likewise.*   To the Rev. G. MOULE, of the Church
Mission at Hang-chau, I owe a mass of interesting
matter regarding that once great and splendid city,
the KINSAY of our Traveller, which has enabled me,
I trust, to effect great improvement both in the Notes
and in the Map, which illustrate that subject.   And to
the Rev. CARSTAIRS DOUGLAS, LL.D., of the English
Presbyterian Mission at Amoy, I am scarcely less
indebted.   The learned Professor BRUUN, of Odessa,
whom I never have seen, and have little likelihood
of ever seeing in this world, has aided me with zeal
and cordiality like that of old friendship.   To Mr.
ARTHUR BURNELL, Ph.D., of the Madras Civil Service,
I am grateful for many valuable notes bearing on these
and other geographical studies, and particularly for
his generous communication of the drawing and photograph of the ancient Cross at St. Thomas's Mount,
long before any publication of that subject was made

---

* It would be ingratitude if this Preface contained no acknowledgment of the
medals awarded to the writer, mainly for this work, by the Royal Geographical
Society, and by the Geographical Society of Italy, the former under the Presidence of
Sir Henry Rawlinson, the latter under that of the Commendatore C. Negri.   Strongly
as I feel the too generous appreciation of these labours implied in such awards, I
confess to have been yet more deeply touched and gratified by practical evidence
of the approval of the two distinguished Travellers mentioned above ; as shown by
Baron von Richthofen in his spontaneous proposal to publish a German version of
the book under his own immediate supervision (a project in abeyance, owing to
circumstances beyond his or my control) ; by Mr. Ney Elias in the fact of his having
carried these ponderous volumes with him on his solitary journey across the
Mongolian wilds !

on his own account. My brother officer, Major OLIVER ST. JOHN, R.E., has favoured me with a variety of interesting remarks regarding the Persian chapters, and has assisted me with new data, very materially correcting the Itinerary Map in Kerman.

Mr. BLOCHMANN of the Calcutta Madrasa, Sir DOUGLAS FORSYTH, C.B., lately Envoy to Kashgar, M. de MAS LATRIE, the Historian of Cyprus, Mr. ARTHUR GROTE, Mr. EUGENE SCHUYLER of the U.S. Legation at St. Petersburg, Dr. BUSHELL and Mr. W. F. MAYERS, of H.M.'s Legation at Peking, Mr. G. PHILLIPS of Fuchau, Madame OLGA FEDTCHENKO, the widow of a great traveller too early lost to the world, Colonel KEATINGE, V.C., C.S.I., Major-General KEYES, C.B., Dr. GEORGE BIRDWOOD, Mr. BURGESS, of Bombay, my old and valued friend Colonel W. H. GREATHED, C.B., and the Master of Mediæval Geography, M. D'AVEZAC himself, with others besides, have kindly lent assistance of one kind or another, several of them spontaneously, and the rest in prompt answer to my requests.

Having always attached much importance to the matter of illustrations,* I feel greatly indebted to the liberal action of Mr. Murray in enabling me largely to increase their number in this edition. Though many are original, we have also borrowed a good many;† a proceeding which seems to me entirely unobjectionable when the engravings are truly illustrative of the text, and not hackneyed.

I regret the augmented bulk of the volumes. There

---

* I am grateful to Mr. de Khanikoff for his especial recognition of these in a kindly review of the first edition in the *Academy*.

† Especially from Lieutenant Garnier's book, mentioned further on ; the only existing source of illustration for many chapters of Polo.

has been some excision, but the additions visibly and palpably preponderate. The truth is that since the completion of the first edition, just four years ago, large additions have been made to the stock of our knowledge bearing on the subjects of this Book; and how these additions have continued to come in up to the last moment, may be seen in Appendix L,* which has had to undergo repeated interpolation after being put in type. KARAKORUM, for a brief space the seat of the widest empire the world has known, has been visited; the ruins of SHANG-TU, the "Xanadu of Cublay Khan," have been explored; PAMIR and TANGUT have been penetrated from side to side; the famous mountain Road of SHEN-SI has been traversed and described; the mysterious CAINDU has been unveiled; the publication of my lamented friend Lieutenant Garnier's great work on the French Exploration of Indo-China has provided a mass of illustration of that YUN-NAN for which but the other day Marco Polo was well-nigh the most recent authority. Nay, the last two years have thrown a promise of light even on what seemed the wildest of Marco's stories, and the bones of a veritable RUC from New Zealand lie on the table of Professor Owen's Cabinet!

M. VIVIEN de St. MARTIN, during the interval of which we have been speaking, has published a History of Geography. In treating of Marco Polo, he alludes to the first edition of this work, most evidently with no intention of disparagement, but speaks of it as merely a revision of Marsden's Book. The last thing I should allow myself to do would be to apply to a

---

* [Merged into the notes of the present edition.—H. C.]

Geographer, whose works I hold in so much esteem, the disrespectful definition which the adage quoted in my former Preface * gives of the *vir qui docet quod non sapit;* but I feel bound to say that on this occasion M. Vivien de St. Martin has permitted himself to pronounce on a matter with which he had not made himself acquainted; for the perusal of the very first lines of the Preface (I will say nothing of the Book) would have shown him that such a notion was utterly unfounded.

In concluding these "forewords" I am probably taking leave of Marco Polo,† the companion of many pleasant and some laborious hours, whilst I have been contemplating with him ("*vôlti a levante*") that Orient in which I also had spent years not a few.

*   *   *   *   *   *

And as the writer lingered over this conclusion, his thoughts wandered back in reverie to those many venerable libraries in which he had formerly made search for mediæval copies of the Traveller's story; and it seemed to him as if he sate in a recess of one of these with a manuscript before him which had never till then been examined with any care, and which he found with delight to contain passages that appear in no version of the Book hitherto known. It was written in clear Gothic text, and in the Old French tongue of the early 14th century. Was it possible that he had lighted on the long-

---

* See page xxix.

† Writing in Italy, perhaps I ought to write, according to too prevalent modern Italian custom, *Polo Marco.* I have already *seen*, and in the work of a writer of reputation, the Alexandrian geographer styled *Tolomeo Claudio!* and if this preposterous fashion should continue to spread, we shall in time have *Tasso Torquato, Jonson Ben*, Africa explored by *Park Mungo*, Asia conquered by *Lane Tamer*, Copperfield David by *Dickens Charles*, Homer Englished by *Pope Alexander*, and the Roman history done into French from the original of *Live Tite!*

lost original of Ramusio's Version? No; it proved to be different. Instead of the tedious story of the northern wars, which occupies much of our Fourth Book, there were passages occurring in the later history of Ser Marco, some years after his release from the Genoese captivity. They appeared to contain strange anachronisms certainly; but we have often had occasion to remark on puzzles in the chronology of Marco's story! * And in some respects they tended to justify our intimated suspicion that he was a man of deeper feelings and wider sympathies than the book of Rusticiano had allowed to appear.† Perhaps this time the Traveller had found an amanuensis whose faculties had not been stiffened by fifteen years of Malapaga?‡ One of the most important passages ran thus:—

"*Bien est voirs que, après ce que* Messires Marc Pol *avoit pris fame et si estoit demouré plusours ans de sa vie a* Venysse, *il avint que mourut* Messires Mafés *qui oncles* Monseignour Marc *estoit:* (*et mourut ausi ses granz chiens mastins qu'avoit amenei dou Catai,§ et qui avoit non* Bayan *pour l'amour au bon chievetain* Bayan Cent-iex); *adonc n'avoit. oncques puis* Messires Marc *nullui, fors son esclave* Piere le Tartar, *avecques lequel pouvoit penre soulas à s'entretenir de ses voiages et des choses dou Levant. Car la gent de* Venysse *si avoit de grant piesce moult anuy pris des loncs contes* Monseignour Marc; *et quand ledit* Messires Marc *issoit de l'uys sa meson ou* Sain Grisostome, *souloient li petit marmot es voies dariere-li courir en cryant* Messer Marco Miliòn! *cont' a nu un busiòn! que veult dire en François '*Messires Marcs des millions di-nous un de vos gros mensonges.' *En oultre, la Dame* Donate *fame anuyouse estoit, et de trop estroit esprit, et plainne de couvoitise.‖ Ansi avint que* Messires Marc *desiroit es voiages rantrer durement.*

"*Si se partist de* Venisse *et chevaucha aux parties d'occident. Et demoura mainz jours es contrées de* Provence *et de* France *et puys fist passaige aux* Ysles *de la tremontaingne et s'en retourna par* la Magne, *si comme vous orrez cy-après. Et fist-il escripre son voiage atout les devisements les contrées; mes de la* France *n'y parloit mie grantment pour ce que maintes genz la scevent apertement. Et pour ce en lairons atant, et commencerons d'autres choses, assavoir, de* BRETAINGNE LA GRANT.

---

* Introduction p. *24*, and *passim* in the notes.   † *Ibid.*, p. *112.*
‡ See Introduction, pp. *51, 57.*   § See Title of present volumes.
‖ Which quite agrees with the story of the document quoted at p. 77 of Introduction.

Cy devyse dou roiaume de Bretaingne la grant.

"*Et sachiés que quand l'en se part de* Calés, *et l'en nage* XX *ou* XXX *milles à trop grant mesaise, si treuve l'en une grandisme Ysle qui s'apelle* Bretaingne la Grant. *Elle est à une grant royne et n'en fait treuage à nulluy. Et ensevelissent lor mors, et ont monnoye de chartres et d'or et d'argent, et ardent pierres noyres, et vivent de marchandises et d'ars, et ont toutes choses de vivre en grant habondance mais non pas à bon marchié. Et c'est une Ysle de trop grant richesce, et li marinier de celle partie dient que c'est li plus riches royaumes qui soit ou monde, et qu'il y a li mieudre marinier dou monde et li mieudre coursier et li mieudre chevalier (ains ne chevauchent mais lonc com* François). *Ausi ont-il trop bons homes d'armes et vaillans durement (bien que maint n'y ait), et les dames et damoseles bonnes et loialles, et belles com lys souef florant. Et quoi vous en diroie-je? Il y a citez et chasteau assez, et tant de marchéanz et si riches qui font venir tant d'avoir-depoiz et de toute espece de marchandise qu'il n'est hons qui la verité en sceust dire. Font venir* d'Ynde *et d'autres parties coton a grant planté, et font venir soye de* Manzi *et de* Bangala, *et font venir laine des ysles de la Mer Occeane et de toutes parties. Et si labourent maintz bouquerans et touailles et autres draps de coton et de laine et de soye. Encores sachiés que ont vaines d'acier assez, et si en labourent trop soubtivement de tous hernois de chevalier, et de toutes choses besoignables à ost; ce sont espées et glaive et esperon et heaume et haches, et toute espèce d'arteillerie et de coutelerie, et en font grant gaaigne et grant marchandise. Et en font si grant habondance que tout li mondes en y puet avoir et à bon marchié.*

Encores cy devise dou dyt roiaume, et de ce qu'en dist Messires Marcs.

"*Et sachiés que tient icelle Royne la seigneurie de l'*Ynde *majeure et de* Mutfili *et de* Bangala, *et d'une moitié de* Mien. *Et moult est saige et noble dame et pourvéans, si que est elle amée de chascun. Et avoit jadis mari; et depuys qu'il mourut bien* XIV *ans avoit; adonc la royne sa fame l'ama tant que oncques puis ne se voult marier a nullui, pour l'amour le prince son baron, ançois moult maine quoye vie. Et tient son royaume ausi bien ou miex que oncques le tindrent li roy si aioul. Mes ores en ce royaume li roy n'ont guieres pooir, ains la poissance commence a trespasser à la menue gent. Et distrent aucun marinier de celes parties à* Monseignour Marc *que hui-et-le jour li royaumes soit auques abastardi come je vous diroy. Car bien est voirs que ci-arrières estoit ciz pueple de* Bretaingne la Grant *bonne et granz et loialle gent qui servoit Diex moult volontiers selonc lor usaige; et tuit li labour qu'il labouroient et portoient a vendre estoient honnestement labouré, et dou greigneur vaillance, et chose pardurable; et se vendoient à jouste pris sanz barguignier. En tant que se aucuns labours portoit l'estanpille* Bretaingne la Grant *c'estoit regardei com pleges de bonne estoffe. Mes orendroit li labours n'est mie tousjourz si bons; et quand l'en achate pour un quintal pesant de toiles de coton, adonc, par trop souvent, si treuve l'en de chascun* C *pois de coton, bien* XXX *ou* XL *pois de plastre de gifs, ou de blanc* *d'Espaigne, ou de choses semblables. Et se l'en achate de cammeloz ou de tireteinne ou d'autre dras de laine, cist ne durent mie, ains sont plain d'empoise, ou de glu et de balieures.*

"*Et bien qu'il est voirs que chascuns hons egalement doit de son cors servir son seigneur ou sa commune, pour aler en ost en tens de besoingne; et bien que trestuit li autre royaume d'occident tieingnent ce pour ordenance, ciz pueple de* Bretaingne la Grant *n'en veult nullement, ains si dient: 'Veez-là: n'avons nous pas la* Manche *pour fossé de nostre pourpris, et pourquoy nous penerons-nous pour nous faire homes d'armes, en lessiant nos gaaignes et nos soulaz? Cela lairons aus soudaiers.' Or li preudhome entre eulx moult scevent bien com tiex paroles sont nyaises; mes si ont paour de lour en dire la verité pour ce que cuident desplaire as bourjois et à la menue gent.*

"*Or je vous di sanz faille que, quand* Messires Marcs Pols *sceust ces choses, moult en ot pitié de cestui pueple, et il li vint à remembrance ce que avenu estoit, ou tens* Monseignour Nicolas *et* Monseignour Mafé, *à l'ore quand* Alau, *frère charnel dou Grant Sire* Cublay, *ala en ost seur* Baudas, *et print le* Calife *et sa maistre cité, atout son vaste tresor d'or et d'argent, et l'amère parolle que dist ledit Alau au Calife, com l'a escripte li Maistres Rusticiens ou chief de cestui livre.**

"*Car sachiés tout voirement que* Messires Marc *moult se deleitoit à faire appert combien sont pareilles au font les condicions des diverses regions dou monde, et soloit-il clorre son discours si disant en son language de* Venisse: '*Sto mondo xe fato tondo, com uzoit dire mes oncles Mafés.*'

"*Ore vous lairons à conter de ceste matière et retournerons à parler de la Loy des genz de* Bretaingne la Grant.

Cy devise des diverses créances de la gent Bretaingne la Grant et de ce qu'en cuidoit Messires Marcs.

"*Il est voirs que li pueples est Crestiens, mes non pour le plus selonc la foy de l'Apostoille Rommain, ains tiennent le en mautalent assez. Seulement il y en a aucun qui sont féoil du dit Apostoille et encore plus forment que li nostre prudhome de* Venisse. *Car quand dit li Papes: 'Telle ou telle chose est noyre,' toute ladite gent si en jure: 'Noyre est com poivre.' Et puis se dira li Papes de la dite chose: 'Elle est blanche,' si en jurera toute ladite gent: 'Il est voirs qu'elle est blanche; blanche est com noifs.' Et dist* Messires Marc Pol: '*Nous n'avons nullement tant de foy à* Venyse, *ne li prudhome de* Florence *non plus, com l'en puet savoir bien apertement dou livre* Monseignour Dantès Aldiguiere, *que j'ay congneu a* Padoe *le meisme an que* Messires Thibault de Cepoy à Venisse *estoit.† Mes c'est joustement ce que j'ay veu autre foiz près le Grant* Bacsi *qui est com li Papes des Ydres.*'

"*Encore y a une autre manière de gent; ce sont de celz qui s'appellent filsoufes;‡ et si il disent: 'S'il y a Diex n'en scavons nul, mes il est voirs*

---

* Vol. i. p. 64, and p. 67.

† *I.e.* 1306; see Introduction, pp. 68-69.

‡ The form which Marco gives to this word was probably a reminiscence of the Oriental corruption *failsúf.* It recalls to my mind a Hindu who was very fond of the word, and especially of applying it to certain of his fellow-servants. But as he used it, *bara failsúf*—"great philosopher"—meant exactly the same as the modern slang "*Artful Dodger*"!

qu'il est une certeinne courance des choses laquex court devers le bien.' Et fist Messires Marcs : *'Encore la créance des* Bacsi *qui dysent que n'y a ne Diex Eternel ne Juge des homes, ains il est une certeinne chose laquex s'apelle* Kerma.' *

"*Une autre foiz avint que disoit un des filsoufes à* Monseignour Marc : *'Diex n'existe mie jeusqu'ores, ainçois il se fait desorendroit.' Et fist encore* Messires Marcs : *' Veez-là une autre foiz la créance des ydres, car dient que li seuz Diex est icil hons qui par force de ses vertuz et de son savoir tant pourchace que d'home il se face Diex presentement. Et li Tartar l'appelent* Borcan. *Tiex Diex* Sagamoni Borcan *estoit, dou quel parle li livres Maistre* Rusticien.'†

"*Encore ont une autre manière de filsoufes, et dient-il: 'Il n'est mie ne Diex ne* Kerma *ne courance vers le bien, ne Providence, ne Créerres, ne Sauvours, ne sainteté ne pechiés ne conscience de pechié, ne proyère ne response à proyère, il n'est nulle riens fors que trop minime grain ou paillettes qui ont à nom* atosmes, *et de tiex grains devient chose qui vive, et chose qui vive devient une certeinne creature qui demoure au rivaige de la Mer : et ceste creature devient poissons, et poissons devient lezars, et lezars devient blayriaus, et blayriaus devient gat-maimons, et gat-maimons devient hons sauvaiges qui menjue char d'homes, et hons sauvaiges devient hons crestien.'*

"*Et dist* Messires Marc : *'Encore une foiz, biaus sires, li* Bacsi *de* Tebet *et de* Kescemir *et li prestre de* Seilan, *qui si dient que l'arme vivant doie trespasser par tous cez changes de vestemens; si com se treuve escript ou livre* Maistre Rusticien *que* Sagamoni Borcan *mourut iiij vint et iiij foiz et tousjourz resuscita, et à chascune foiz d'une diverse manière de beste, et à la derreniere foyz mourut hons et devint diex, selonc ce qu'il dient.'‡ Et fist encore* Messires Marc : *'A moy pert-il trop estrange chose se juesques à toutes les créances des ydolastres deust dechéoir ceste grantz et saige nation. Ainsi peuent jouer Misire li filsoufe atout lour propre perte, mes à l'ore quand tiex fantaisies se respanderont es joenes bacheliers et parmy la menue gent, celz averont pour toute Loy* manducemus et bibamus, cras enim moriemur ; *et trop isnellement l'en raccomencera la descente de l'eschiele, et d'home crestien deviendra hons sauvaiges, et d'home sauvaige gat-maimons, et de gat-maimon blayriaus.' Et fist encores* Messires Marc : *'Maintes contrées et provinces et ysles et citéz je* Marc Pol *ay veues et de maintes genz de maintes manières ay les condicionz congneues, et je croy bien que il est plus assez dedens l'univers que ce que li nostre prestre n'y songent. Et puet bien estre, biaus sires, que li mondes n'a estés creés à tous poinz com nous creiens, ains d'une sorte encore plus merveillouse. Mes cil n' amenuise nullement nostre pensée de Diex et de sa majesté, ains la fait greingnour. Et contrée n'ay veue ou Dame Diex ne manifeste apertement les granz euvres de sa tout-poissante saigesse; gent n'ay congneue esquiex ne se fait sentir li fardels de pechié, et la besoingne de Phisicien des maladies de l'arme tiex com est nostre Seignours Ihesus Crist, Beni soyt son Non. Pensez doncques à cel qu'a dit uns de ses*

---

* See for the explanation of *Karma*, "the power that controls the universe," in the doctrine of atheistic Buddhism, HARDY's *Eastern Monachism*, p. 5.

† Vol. ii. p. 316 (see also i. 348).

‡ Vol. ii. pp. 318–319.

Apostres: Nolite esse prudentes apud bosmet ipsos; *et uns autres:* Quoniam multi pseudo-prophetae exierint; *et uns autres:* Quod benient in nobissimis diebus illusores . . . dicentes, Ubi est promissio? *et encores aus parolles que dist li Signours meismes:* Bide ergo ne lumen quod in te est tenebrae sint.)

Commant Messires Marcs se partist de l'ysle de Bretaingne et de la propère que fist.

" *Et pourquoy vous en feroie-je lonc conte? Si print nef* Messires Marcs *et se partist en nageant vers la terre ferme. Or* Messires Marc Pol *moult ama cel roiaume de* Bretaingne *la* grant *pour son viex renon et s'ancienne franchise, et pour sa saige et bonne Royne* (*que Diex gart*), *et pour les mainz homes de vaillance et bons chaceours et les maintes bonnes et honnestes dames qui y estoient. Et sachiés tout voirement que en estant delez le bort la nef, et en esgardant aus roches blanches que l'en par dariere-li lessoit,* Messires Marc *prieoit Diex, et disoit-il:* ' *Ha Sires Diex ay merci de cestuy vieix et noble royaume; fay-en pardurable forteresse de liberté et de joustice, et garde-le de tout meschief de dedens et de dehors; donne à sa gent droit esprit pour ne pas Diex guerroyer de ses dons, ne de richesce ne de savoir; et conforte-les fermement en ta foy'* . . ."

A loud *Amen* seemed to peal from without, and the awakened reader started to his feet. And lo! it was the thunder of the winter-storm crashing among the many-tinted crags of Monte Pellegrino,—with the wind raging as it knows how to rage here in sight of the Isles of Æolus, and the rain dashing on the glass as ruthlessly as it well could have done, if, instead of Æolic Isles and many-tinted crags, the window had fronted a dearer shore beneath a northern sky, and looked across the grey Firth to the rain-blurred outline of the Lomond Hills.

But I end, saying to Messer Marco's prayer, Amen.

PALERMO, 31*st December*, 1874

ORIGINAL PREFACE.

THE amount of appropriate material, and of acquaintance with the mediæval geography of some parts of Asia, which was acquired during the compilation of a work of kindred character for the Hakluyt Society,* could hardly fail to suggest as a fresh labour in the same field the preparation of a new English edition of Marco Polo. Indeed one kindly critic (in the *Examiner*) laid it upon the writer as a duty to undertake that task.

Though at least one respectable English edition has appeared since Marsden's,† the latter has continued to be the standard edition, and maintains not only its reputation but its market value. It is indeed the work of a sagacious, learned, and right-minded man, which can never be spoken of otherwise than with respect. But since Marsden published his quarto (1818) vast stores of new knowledge have become available in elucidation both of the contents of Marco Polo's book and of its literary history. The works of writers such as Klaproth, Abel Rémusat, D'Avezac, Reinaud, Quatremère, Julien, I. J. Schmidt, Gildemeister, Ritter, Hammer-Purgstall, Erdmann, D'Ohsson, Defrémery, Elliot, Erskine, and many more, which throw light directly or incidentally on Marco Polo, have, for the most part, appeared since then. Nor, as regards the literary history of the book, were any just views possible at a time when what may be called the *Fontal* MSS. (in French) were unpublished and unexamined.

Besides the works which have thus occasionally or inci-

---

* *Cathay and The Way Thither, being a Collection of Minor Medieval Notices of China.* London, 1866. The necessities of the case have required the repetition in the present work of the substance of some notes already printed (but hardly published) in the other.

† Viz. Mr. Hugh Murray's. I mean no disrespect to Mr. T. Wright's edition, but it is, and professes to be, scarcely other than a reproduction of Marsden's, with abridgment of his notes.

dentally thrown light upon the Traveller's book, various editions of the book itself have since Marsden's time been published in foreign countries, accompanied by comments of more or less value. All have contributed something to the illustration of the book or its history ; the last and most learned of the editors, M. Pauthier, has so contributed in large measure. I had occasion some years ago* to speak freely my opinion of the merits and demerits of M. Pauthier's work ; and to the latter at least I have no desire to recur here.

Another of his critics, a much more accomplished as well as more favourable one,† seems to intimate the opinion that there would scarcely be room in future for new commentaries. Something of the kind was said of Marsden's at the time of its publication. I imagine, however, that whilst our libraries endure the *Iliad* will continue to find new translators, and Marco Polo—though one hopes not so plentifully—new editors.

The justification of the book's existence must however be looked for, and it is hoped may be found, in the book itself, and not in the Preface. The work claims to be judged as a whole, but it may be allowable, in these days of scanty leisure, to indicate below a few instances of what is believed to be new matter in an edition of Marco Polo ; by which however it is by no means intended that all such matter is claimed by the editor as his own.‡

---

* In the *Quarterly Review* for July, 1868.   † M. Nicolas Khanikoff.

‡ In the Preliminary Notices will be found new matter on the Personal and Family History of the Traveller, illustrated by Documents ; and a more elaborate attempt than I have seen elsewhere to classify and account for the different texts of the work, and to trace their mutual relation.

As regards geographical elucidations, I may point to the explanation of the name *Gheluchelan* (i. p. 58), to the discussion of the route from Kerman to Hormuz, and the identification of the sites of Old Hormuz, of *Cobinan* and *Dogana*, the establishment of the position and continued existence of *Keshm*, the note on *Pein* and *Charchan*, on *Gog* and *Magog*, on the geography of the route from *Sindafu* to *Carajan*, on *Anin* and *Coloman*, on *Mutafili, Cail*, and *Ely*.

As regards historical illustrations, I would cite the notes regarding the Queens

From the commencement of the work it was felt that the task was one which no man, though he were far better equipped and much more conveniently situated than the present writer, could satisfactorily accomplish from his own resources, and help was sought on special points wherever it seemed likely to be found. In scarcely any quarter was the application made in vain. Some who have aided most materially are indeed very old and valued friends; but to many others who have done the same the applicant was unknown; and some of these again, with whom the editor began correspondence on this subject as a stranger, he is happy to think that he may now call friends.

To none am I more indebted than to the Comm. GUGLIELMO BERCHET, of Venice, for his ample, accurate, and generous assistance in furnishing me with Venetian documents, and in many other ways. Especial thanks are also due to Dr. WILLIAM LOCKHART, who has supplied the materials for some of the most valuable illustrations; to Lieutenant FRANCIS GARNIER, of the French Navy, the gallant and accomplished leader (after the death of Captain Doudart de la Grée) of the memorable expedi-

---

Bolgana and *Cocachin*, on the *Karaunahs*, etc., on the title of King of *Bengal* applied to the K. of Burma, and those bearing upon the Malay and Abyssinian chronologies.

In the interpretation of outlandish phrases, I may refer to the notes on *Ondanique, Nono, Barguerlac, Argon, Sensin, Keshican, Toscaol, Bularguchi, Gat-paul*, etc.

Among miscellaneous elucidations, to the disquisition on the *Arbre Sol* or *Sec* in vol. i., and to that on Mediæval Military Engines in vol. ii.

In a variety of cases it has been necessary to refer to Eastern languages for pertinent elucidations or etymologies. The editor would, however, be sorry to fall under the ban of the mediæval adage:

> " *Vir qui docet quod non sapit*
> *Definitur Bestia !*"

and may as well reprint here what was written in the Preface to *Cathay:*

" I am painfully sensible that in regard to many subjects dealt with in the following pages, nothing can make up for the want of genuine Oriental learning. A fair familiarity with Hindustani for many years, and some reminiscences of elementary Persian, have been useful in their degree; but it is probable that they may sometimes also have led me astray, as such slender lights are apt to do."

tion up the Mekong to Yun-nan; to the Rev. Dr. CALDWELL, of the S. P. G. Mission in Tinnevelly, for copious and valuable notes on Southern India; to my friends Colonel ROBERT MACLAGAN, R.E., Sir ARTHUR PHAYRE, and Colonel HENRY MAN, for very valuable notes and other aid; to Professor A. SCHIEFNER, of St. Petersburg, for his courteous communication of very interesting illustrations not otherwise accessible; to Major-General ALEXANDER CUNNINGHAM, of my own corps, for several valuable letters; to my friends Dr. THOMAS OLDHAM, Director of the Geological Survey of India, Mr. DANIEL HANBURY, F.R.S., Mr. EDWARD THOMAS, Mr. JAMES FERGUSSON, F.R.S., Sir BARTLE FRERE, and Dr. HUGH CLEGHORN, for constant interest in the work and readiness to assist its progress; to Mr. A. WYLIE, the learned Agent of the B. and F. Bible Society at Shang-hai, for valuable help; to the Hon. G. P. MARSH, U.S. Minister at the Court of Italy, for untiring kindness in the communication of his ample stores of knowledge, and of books. I have also to express my obligations to Comm. NICOLÒ BAROZZI, Director of the City Museum at Venice, and to Professor A. S. MINOTTO, of the same city; to Professor ARMINIUS VÁMBÉRY, the eminent traveller; to Professor FLÜCKIGER of Bern; to the Rev. H. A. JAESCHKE, of the Moravian Mission in British Tibet; to Colonel LEWIS PELLY, British Resident in the Persian Gulf; to Pandit MANPHUL, C.S.I. (for a most interesting communication on Badakhshan); to my brother officer, Major T. G. MONTGOMERIE, R.E., of the Indian Trigonometrical Survey; to Commendatore NEGRI, the indefatigable President of the Italian Geo-graphical Society; to Dr. ZOTENBERG, of the Great Paris Library, and to M. CH. MAUNOIR, Secretary-General of the Société de Géographie; to Professor HENRY

GIGLIOLI, at Florence ; to my old friend Major-General ALBERT FYTCHE, Chief Commissioner of British Burma ; to Dr. ROST and Dr. FORBES-WATSON, of the India Office Library and Museum ; to Mr. R. H. MAJOR, and Mr. R. K. DOUGLAS, of the British Museum ; to Mr. N. B. DENNYS, of Hong-kong ; and to Mr. C. GARDNER, of the Consular Establishment in China.   There are not a few others to whom my thanks are equally due ; but it is feared that the number of names already mentioned may seem ridiculous, compared with the result, to those who do not appreciate from how many quarters the facts needful for a work which in its course intersects so many fields required to be collected, one by one.   I must not, however, omit acknowledgments to the present Earl of DERBY for his courteous permission, when at the head of the Foreign Office, to inspect Mr. Abbott's valuable unpublished Report upon some of the Interior Provinces of Persia ; and to Mr. T. T. COOPER, one of the most adventurous travellers of modern times, for leave to quote some passages from his unpublished diary.

PALERMO, 31st December, 1870.

[Original Dedication.]

TO

HER ROYAL HIGHNESS,

MARGHERITA,

Princess of Piedmont,

THIS ENDEAVOUR TO ILLUSTRATE THE LIFE AND WORK
OF A RENOWNED ITALIAN
IS
BY HER ROYAL HIGHNESS'S GRACIOUS PERMISSION
Dedicated
WITH THE DEEPEST RESPECT
BY
H. YULE.

TO

HENRY YULE.

UNTIL you raised dead monarchs from the mould
 And built again the domes of Xanadu,
 I lay in evil case, and never knew
The glamour of that ancient story told
By good Ser Marco in his prison-hold.
 But now I sit upon a throne and view
 The Orient at my feet, and take of you
And Marco tribute from the realms of old.

If I am joyous, deem me not o'er bold ;
 If I am grateful, deem me not untrue ;
For you have given me beauties to behold,
 Delight to win, and fancies to pursue,
Fairer than all the jewelry and gold
 Of Kublaï on his throne in Cambalu.

E. C. BABER.

20th July, 1884.

MEMOIR OF SIR HENRY YULE.

HENRY YULE was the youngest son of Major William Yule, by his first wife, Elizabeth Paterson, and was born at Inveresk, in Midlothian, on 1st May, 1820. He was named after an *aunt* who, like Miss Ferrier's immortal heroine, owned a man's name.

On his father's side he came of a hardy agricultural stock,[1] improved by a graft from that highly-cultured tree, Rose of Kilravock.[2] Through his mother, a somewhat prosaic person herself, he inherited strains from Huguenot and Highland ancestry. There were recognisable traces of all these elements

---

[1] There is a vague tradition that these Yules descend from the same stock as the Scandinavian family of the same name, which gave Denmark several men of note, including the great naval hero Niels Juel. The portraits of these old Danes offer a certain resemblance of type to those of their Scots namesakes, and Henry Yule liked to play with the idea, much in the same way that he took humorous pleasure in his reputed descent from Michael Scott, the Wizard! (This tradition was more historical, however, and stood thus : Yule's great grandmother was a Scott of Ancrum, and the Scotts of Ancrum had established their descent from Sir Michael Scott of Balwearie, reputed to be the Wizard.) Be their origin what it may, Yule's forefathers had been already settled on the Border hills for many generations, when in the time of James VI. they migrated to the lower lands of East Lothian, where in the following reign they held the old fortalice of Fentoun Tower of Nisbet of Dirleton. When Charles II. empowered his Lord Lyon to issue certificates of arms (in place of the Lyon records removed and lost at sea by the Cromwellian Government), these Yules were among those who took out confirmation of arms, and the original document is still in the possession of the head of the family.

Though Yules of sorts are still to be found in Scotland, the present writer is the only member of the Fentoun Tower family now left in the country, and of the few remaining out of it most are to be found in the Army List.

[2] The literary taste which marked William Yule probably came to him from his grandfather, the Rev. James Rose, Episcopal Minister of Udny, in Aberdeenshire. James Rose, a non-jurant (*i.e.* one who refused to acknowledge allegiance to the Hanoverian King), was a man of devout, large, and tolerant mind, as shown by writings still extant. His father, John Rose, was the younger son of the 14th Hugh of Kilravock. He married Margaret Udny of Udny, and was induced by her to sell his pleasant Ross-shire property and invest the proceeds in her own bleak Buchan. When George Yule (about 1759) brought home Elizabeth Rose as his wife, the popular feeling against the Episcopal Church was so strong and bitter in Lothian, that all the men of the family—themselves Presbyterians—accompanied Mrs. Yule as a bodyguard on the occasion of her first attendance at the Episcopal place of worship. Years after, when dissensions had arisen in the Church of Scotland, Elizabeth Yule succoured and protected some of the dissident Presbyterian ministers from their persecutors.

in Henry Yule, and as was well said by one of his oldest friends :
" He was one of those curious racial compounds one finds on
the east side of Scotland, in whom the hard Teutonic grit is
sweetened by the artistic spirit of the more genial Celt."[3]   His
father, an officer of the Bengal army (born 1764, died 1839),
was a man of cultivated tastes and enlightened mind, a good
Persian and Arabic scholar, and possessed of much miscellaneous
Oriental learning.   During the latter years of his career in India,
he served successively as Assistant Resident at the (then
independent) courts of Lucknow[4] and Delhi.   In the latter
office his chief was the noble Ouchterlony.   William Yule,
together with his younger brother Udny,[5] returned home in
1806.   " A recollection of their voyage was that they hailed an
outward bound ship, somewhere off the Cape, through the
trumpet: ' What news?'   Answer: 'The King's mad, and
Humfrey's beat Mendoza' (two celebrated prize-fighters and
often matched).   'Nothing more?'   'Yes, Bonaparty's made
his *Mother* King of Holland!'

" Before his retirement, William Yule was offered the Lieut.-
Governorship of St. Helena.   Two of the detailed privileges of
the office were residence at Longwood (afterwards the house of
Napoleon), and the use of a certain number of the Company's
slaves.   Major Yule, who was a strong supporter of the anti-
slavery cause till its triumph in 1834, often recalled both of these
offers with amusement."[6]

---

[3] General Collinson in *Royal Engineers' Journal*, 1st Feb. 1890.   The gifted author
of this excellent sketch himself passed away on 22nd April 1902.

[4] The grave thoughtful face of William Yule was conspicuous in the picture of a
Durbar (by an Italian artist, but *not* Zoffany), which long hung on the walls of the
Nawab's palace at Lucknow.   This picture disappeared during the Mutiny of 1857.

[5] Colonel Udny Yule, C.B.   " When he joined, his usual *nomen* and *cognomen*
puzzled the staff-sergeant at Fort-William, and after much boggling on the cadet
parade, the name was called out *Whirly Wheel*, which produced no reply, till some
one at a venture shouted, ' sick in hospital.' " (*Athenæum*, 24th Sept. 1881.)   The ship
which took Udny Yule to India was burnt at sea.   After keeping himself afloat for several
hours in the water, he was rescued by a passing ship and taken back to the Mauritius,
whence, having lost everything but his cadetship, he made a fresh start for India,
where he and William for many years had a common purse.   Colonel Udny Yule commanded a brigade at the Siege of Cornelis (1811), which gave us Java, and afterwards
acted as Resident under Sir Stamford Raffles.   Forty-five years after the retrocession
of Java, Henry Yule found the memory of his uncle still cherished there.

[6] Article on the Oriental Section of the British Museum Library in *Athenæum*,
24th Sept. 1881.   Major Yule's Oriental Library was presented by his sons to the
British Museum a few years after his death.

William Yule was a man of generous chivalrous nature, who took large views of life, apt to be unfairly stigmatised as Radical in the narrow Tory reaction that prevailed in Scotland during the early years of the 19th century.[7] Devoid of literary ambition, he wrote much for his private pleasure, and his knowledge and library (rich in ·Persian and Arabic MSS.) were always placed freely at the service of his friends and correspondents, some of whom, such as Major C. Stewart and Mr. William Erskine, were more given to publication than himself. He never travelled without a little 8vo MS. of Hafiz, which often lay under his pillow. Major Yule's only printed work was a lithographed edition of the *Apothegms* of 'Ali, the son of Abu Talib, in the Arabic, with an old Persian version and an English translation interpolated by himself. "This was privately issued in 1832, when the Duchesse d'Angoulême was living at Edinburgh, and the little work was inscribed to her, with whom an accident of neighbourhood and her kindness to the Major's youngest child had brought him into relations of goodwill."[8]

Henry Yule's childhood was mainly spent at Inveresk. He used to say that his earliest recollection was sitting with the little cousin, who long after became his wife, on the doorstep of her father's house in George Street, Edinburgh (now the Northern Club), listening to the performance of a passing piper. There was another episode which he recalled with humorous satisfaction. Fired by his father's tales of the jungle, Yule (then about six years old) proceeded to improvise an elephant pit in the back garden, only too successfully, for soon, with mingled terror and delight, he saw his uncle John[9] fall headlong into the snare. He lost his mother before he was eight, and almost his only remembrance of her was the circumstance of her having given him a little lantern to light him home on winter nights from his first school. On Sundays it was the Major's custom

---

[7] It may be amusing to note that he was considered an almost dangerous person because he read the *Scotsman* newspaper !

[8] *Athenæum*, 24th Sept. 1881. A gold chain given by the last Dauphiness is in the writer's possession.

[9] Dr. John Yule (b. 176– d. 1827), a kindly old *savant*. He was one of the earliest corresponding members of the Society of Antiquaries of Scotland, and the author of some botanical tracts.

to lend his children, as a picture-book, a folio Arabic translation of the Four Gospels, printed at Rome in 1591, which contained excellent illustrations from Italian originals.[10] Of the pictures in this volume Yule seems never to have tired. The last page bore a MS. note in Latin to the effect that the volume had been read in the Chaldæan Desert by *Georgius Strachanus, Milnensis, Scotus,* who long remained unidentified, not to say mythical, in Yule's mind. But George Strachan never passed from his memory, and having ultimately run him to earth, Yule, sixty years later, published the results in an interesting article.[11]

Two or three years after his wife's death, Major Yule removed to Edinburgh, and established himself in Regent's Terrace, on the face of the Calton Hill.[12] This continued to be Yule's home until his father's death, shortly before he went to India. "Here he learned to love the wide scenes of sea and land spread out around that hill—a love he never lost, at home or far away. And long years after, with beautiful Sicilian hills before him and a lovely sea, he writes words of fond recollection of the bleak Fife hills, and the grey Firth of Forth." [13]

Yule now followed his elder brother, Robert, to the famous High School, and in the summer holidays the two made ex-

---

[10] According to Brunet, by Lucas Pennis after Antonio Tempesta.

[11] *Concerning some little-known Travellers in the East.* ASIATIC QUARTERLY, vol. v. (1888).

[12] William Yule died in 1839, and rests with his parents, brothers, and many others of his kindred, in the ruined chancel of the ancient Norman Church of St. Andrew, at Gulane, which had been granted to the Yule family as a place of burial by the Nisbets of Dirleton, in remembrance of the old kindly feeling subsisting for genera-tions between them and their tacksmen in Fentoun Tower. Though few know its history, a fragrant memorial of this wise and kindly scholar is still conspicuous in Edinburgh. The magnificent wall-flower that has, for seventy summers, been a glory of the Castle rock, was originally all sown by the patient hand of Major Yule, the self-sowing of each subsequent year, of course, increasing the extent of bloom. Lest the extraordinarily severe spring of 1895 should have killed off much of the old stock, another (but much more limited) sowing on the northern face of the rock was in that year made by his grand-daughter, the present writer, with the sanction and active personal help of the lamented General (then Colonel) Andrew Wauchope of Niddrie Marischal. In Scotland, where the memory of this noble soldier is so greatly revered, some may like to know this little fact. May the wall-flower of the Castle rock long flourish a fragrant memorial of two faithful soldiers and true-hearted Scots.

[13] Obituary notice of Yule, by Gen. R. Maclagan, R.E. *Proceedings, R.G.S.* 1890.

peditions to the West Highlands, the Lakes of Cumberland, and elsewhere. Major Yule chose his boys to have every reasonable indulgence and advantage, and when the British Association, in 1834, held its first Edinburgh meeting, Henry received a member's ticket. So, too, when the passing of the Reform Bill was celebrated in the same year by a great banquet, at which Lord Grey and other prominent politicians were present, Henry was sent to the dinner, probably the youngest guest there.[14]

At this time the intention was that Henry should go to Cambridge (where his name was, indeed, entered), and after taking his degree study for the Bar. With this view he was, in 1833, sent to Waith, near Ripon, to be coached by the Rev. H. P. Hamilton, author of a well-known treatise, *On Conic Sections*, and afterwards Dean of Salisbury. At his tutor's hospitable rectory Yule met many notabilities of the day. One of them was Professor Sedgwick.

There was rumoured at this time the discovery of the first known (?) fossil monkey, but its tail was missing. " Depend upon it, Daniel O'Conell's got hold of it ! " said ' Adam ' briskly.[15] Yule was very happy with Mr. Hamilton and his kind wife, but on his tutor's removal to Cambridge other arrangements became necessary, and in 1835 he was transferred to the care of the Rev. James Challis, rector of Papworth St. Everard, a place which " had little to recommend it except a dulness which made reading almost a necessity." [16] Mr. Challis had at this time two other resident pupils, who both, in most diverse ways, attained distinction in the Church. These were John Mason Neale, the future eminent ecclesiologist and founder of the devoted Anglican Sisterhood of St. Margaret, and Harvey Goodwin, long afterwards the studious and large-minded Bishop of Carlisle. With the latter, Yule remained on terms of cordial friendship to the end of his life. Looking back through more than fifty years to these boyish days, Bishop Goodwin wrote that Yule then " showed much more liking for Greek plays and for German than for mathematics, though he had considerable geometrical

---

[14] This was the famous " Grey Dinner," of which The Shepherd made grim fun in the *Noctes*.

[15] Probably the specimen from South America, of which an account was published in 1833.

[16] Rawnsley, *Memoir of Harvey Goodwin, Bishop of Carlisle.*

ingenuity."[17] On one occasion, having solved a problem that puzzled Goodwin, Yule thus discriminated the attainments of the three pupils : "The difference between you and me is this : You like it and can't do it ; I don't like it and can do it. Neale neither likes it nor can do it." Not bad criticism for a boy of fifteen.[18]

On Mr. Challis being appointed Plumerian Professor at Cambridge, in the spring of 1836, Yule had to leave him, owing to want of room at the Observatory, and he became for a time, a most dreary time, he said, a student at University College, London.

By this time Yule had made up his mind that not London and the Law, but India and the Army should be his choice, and accordingly in Feb. 1837 he joined the East India Company's Military College at Addiscombe. From Addiscombe he passed out, in December 1838, at the head of the cadets of his term (taking the prize sword[19]), and having been duly appointed to the Bengal Engineers, proceeded early in 1839 to the Headquarters of the Royal Engineers at Chatham, where, according to custom, he was enrolled as a " local and temporary Ensign." For such was then the invidious designation at Chatham of the young Engineer officers of the Indian army, who ranked as full lieutenants in their own Service, from the time of leaving Addiscombe.[20] Yule once audaciously tackled the formidable Pasley on this very grievance. The venerable Director, after a minute's pondering, replied : " Well, I don't remember what the reason was, but I have *no* doubt (*staccato*) it . . . was . . . a very . . . *good* reason." [21]

" When Yule appeared among us at Chatham in 1839," said his friend Collinson, " he at once took a prominent place in our little Society by his slightly advanced age [he was then 18½], but more by his strong character. . . . His earlier education . . . gave him a better classical knowledge than most of us possessed ;

---

[17] [18] Biog. Sketch of Yule, by C. Trotter, *Proceedings*, *R.S.E.* vol. xvii.

[19] After leaving the army, Yule always used this sword when wearing uniform.

[20] The Engineer cadets remained at Addiscombe a term (=6 months) longer than the Artillery cadets, and as the latter were ordinarily gazetted full lieutenants six months after passing out, unfair seniority was obviated by the Engineers receiving the same rank on passing out of Addiscombe.

[21] Yule, in *Memoir of General Becher.*

then he had the reserve and self-possession characteristic of his race; but though he took small part in the games and other recreations of our time, his knowledge, his native humour, and his good comradeship, and especially his strong sense of right and wrong, made him both admired and respected. . . . Yule was not a scientific engineer, though he had a good general knowledge of the different branches of his profession; his natural capacity lay rather in varied knowledge, combined with a strong understanding and an excellent memory, and also a peculiar power as a draughtsman, which proved of great value in after life. . . . Those were nearly the last days of the old *régime*, of the orthodox double sap and cylindrical pontoons, when Pasley's genius had been leading to new ideas, and when Lintorn Simmons' power, G. Leach's energy, W. Jervois' skill, and R. Tylden's talent were developing under the wise example of Henry Harness." [22]

In the Royal Engineer mess of those days (the present anteroom), the portrait of Henry Yule now faces that of his first chief, Sir Henry Harness. General Collinson said that the pictures appeared to eye each other as if the subjects were continuing one of those friendly disputes in which they so often engaged. [23]

It was in this room that Yule, Becher, Collinson, and other young R.E.'s, profiting by the temporary absence of the austere Colonel Pasley, acted some plays, including *Pizarro*. Yule bore the humble part of one of the Peruvian Mob in this performance, of which he has left a droll account. [24]

On the completion of his year at Chatham, Yule prepared to sail for India, but first went to take leave of his relative, General White. An accident prolonged his stay, and before he left he had proposed to and been refused by his cousin Annie. This occurrence, his first check, seems to have cast rather a gloom over his start for India. He went by the then newly-opened Overland Route, visiting Portugal, stopping at Gibraltar to see

---

[22] Collinson's *Memoir of Yule* in *R.E. Journal*.

[23] The picture was subscribed for by his brother officers in the corps, and painted in 1880 by T. B. Wirgman. It was exhibited at the Royal Academy in 1881. A reproduction of the artist's etching from it forms the frontispiece of this volume.

[24] In *Memoir of Gen. John Becher*.

his cousin, Major (afterwards General) Patrick Yule, R.E.[25] He was under orders "to stop at Aden (then recently acquired), to report on the water supply, and to deliver a set of meteorological and magnetic instruments for starting an observatory there. The overland journey then really meant so ; tramping across the desert to Suez with camels and Arabs, a proceeding not conducive to the preservation of delicate instruments ; and on arriving at Aden he found that the intended observer was dead, the observatory not commenced, and the instruments all broken. There was thus nothing left for him but to go on at once" to Calcutta,[26] where he arrived at the end of 1840.

His first service lay in the then wild Khasia Hills, whither he was detached for the purpose of devising means for the transport of the local coal to the plains. In spite of the depressing character of the climate (Cherrapunjee boasts the highest rainfall on record), Yule thoroughly enjoyed himself, and always looked back with special pleasure on the time he spent here. He was unsuccessful in the object of his mission, the obstacles to cheap transport offered by the dense forests and mighty precipices proving insurmountable, but he gathered a wealth of interesting observations on the country and people, a very primitive Mongolian race, which he subsequently embodied in two excellent and most interesting papers (the first he ever published).[27]

In the following year, 1842, Yule was transferred to the

---

[25] General Patrick Yule (b. 1795, d. 1873) was a thorough soldier, with the repute of being a rigid disciplinarian. He was a man of distinguished presence, and great charm of manner to those whom he liked, which were by no means all. The present writer holds him in affectionate remembrance, and owes to early correspondence with him much of the information embodied in preceding notes. He served on the Canadian Boundary Commission of 1817, and on the Commission of National Defence of 1859, was prominent in the Ordnance Survey, and successively Commanding R.E. in Malta and Scotland. He was Engineer to Sir C. Fellows' Expedition, which gave the nation the Lycian Marbles, and while Commanding R.E. in Edinburgh, was largely instrumental in rescuing St. Margaret's Chapel in the Castle from desecration and oblivion. He was a thorough Scot, and never willingly tolerated the designation N.B. on even a letter. He had cultivated tastes, and under a somewhat austere exterior he had a most tender heart. When already past sixty, he made a singularly happy marriage to a truly good woman, who thoroughly appreciated him. He was the author of several Memoirs on professional subjects. He rests in St. Andrew's, Gulane.

[26] Collinson's *Memoir of Yule.*

[27] Notes on the Iron of the Khasia Hills and Notes on the Khasia Hills and People, both in Journal of the R. Asiatic Society of Bengal, vols. xi. and xiii.

irrigation canals of the north-west with head-quarters at Kurnaul. Here he had for chief Captain (afterwards General Sir William) Baker, who became his dearest and most steadfast friend. Early in 1843 Yule had his first experience of field service. The death without heir of the Khytul Rajah, followed by the refusal of his family to surrender the place to the native troops sent to receive it, obliged Government to send a larger force against it, and the canal officers were ordered to join this. Yule was detailed to serve under Captain Robert Napier (afterwards F.-M. Lord Napier of Magdala). Their immediate duty was to mark out the route for a night march of the troops, barring access to all side roads, and neither officer having then had any experience of war, they performed the duty " with all the elaborate care of novices." Suddenly there was an alarm, a light detected, and a night attack awaited, when the danger resolved itself into Clerk Sahib's *khansamah* with welcome hot coffee ! [28] Their hopes were disappointed, there was no fighting, and the Fort of Khytul was found deserted by the enemy. It " was a strange scene of confusion—all the paraphernalia and accumulation of odds and ends of a wealthy native family lying about and inviting loot. I remember one beautiful crutch-stick of ebony with two rams' heads in jade. I took it and sent it in to the political authority, intending to buy it when sold. There was a sale, but my stick never appeared. Somebody had a more developed taste in jade. . . . Amid the general rummage that was going on, an officer of British Infantry had been put over a part of the palace supposed to contain treasure, and they —officers and all—were helping themselves. Henry Lawrence was one of the politicals under George Clerk. When the news of this affair came to him I was present. It was in a white marble loggia in the palace, where was a white marble chair or throne on a basement. Lawrence was sitting on this throne in great excitement. He wore an Afghan *choga*, a sort of dressing-gown garment, and this, and his thin locks, and thin beard were

---

[28] Mr. (afterwards Sir) George Clerk, Political Officer with the expedition. Was twice Governor of Bombay and once Governor of the Cape : " A diplomatist of the true English stamp—undaunted in difficulties and resolute to maintain the honour of his country." (Sir H. B. Edwardes, *Life of Henry Lawrence*, i. 267). He died in 1889.

streaming in the wind. He always dwells in my memory as a sort of pythoness on her tripod under the afflatus." [29]

During his Indian service, Yule had renewed and continued by letters his suit to Miss White, and persistency prevailing at last, he soon after the conclusion of the Khytul affair applied for leave to go home to be married. He sailed from Bombay in May, 1843, and in September of the same year was married, at Bath, to the gifted and large-hearted woman who, to the end, remained the strongest and happiest influence in his life. [30]

Yule sailed for India with his wife in November 1843. The next two years were employed chiefly in irrigation work, and do not call for special note. They were very happy years, except in the one circumstance that the climate having seriously affected his wife's health, and she having been brought to death's door, partly by illness, but still more by the drastic medical treatment of those days, she was imperatively ordered back to England by the doctors, who forbade her return to India.

Having seen her on board ship, Yule returned to duty on the canals. The close of that year, December, 1845, brought some variety to his work, as the outbreak of the first Sikh War called nearly all the canal officers into the field. "They went up to the front by long marches, passing through no stations, and quite unable to obtain any news of what had occurred, though on the 21st December the guns of Ferozshah were distinctly heard in their camp at Pehoa, at a distance of 115 miles south-east from the field, and some days later they came successively on the fields of Moodkee and of Ferozshah itself, with all the recent traces of battle. When the party of irrigation officers reached head-quarters, the arrangements for attacking the Sikh army in its entrenchments at Sobraon were beginning (though suspended till weeks later for the arrival of the tardy siege guns), and the opposed forces were lying in sight of each other." [31]

Yule's share in this campaign was limited to the sufficiently arduous task of bridging the Sutlej for the advance of the British army. It is characteristic of the man that for this

---

[29] Note by Yule, communicated by him to Mr R. B. Smith and printed by the latter in *Life of Lord Lawrence*.

[30] And when nearing his own end, it was to her that his thoughts turned most constantly.

[31] Yule and Maclagan's *Memoir of Sir W. Baker*.

reason he always abstained from wearing his medal for the Sutlej campaign.

His elder brother, Robert Yule, then in the 16th Lancers, took part in that magnificent charge of his regiment at the battle of Aliwal (Jan. 28, 1846) which the Great Duke is said to have pronounced unsurpassed in history.   From particulars gleaned from his brother and others present in the action, Henry Yule prepared a spirited sketch of the episode, which was afterwards published as a coloured lithograph by M'Lean (Haymarket).

At the close of the war, Yule succeeded his friend Strachey as Executive Engineer of the northern division of the Ganges Canal, with his head-quarters at Roorkee, "the division which, being nearest the hills and crossed by intermittent torrents of great breadth and great volume when in flood, includes the most important and interesting engineering works." [32]

At Roorkee were the extensive engineering workshops connected with the canal.   Yule soon became so accustomed to the din as to be undisturbed by the noise, but the un-punctuality and carelessness of the native workmen sorely tried his patience, of which Nature had endowed him with but a small reserve.   Vexed with himself for letting temper so often get the better of him, Yule's conscientious mind devised a characteristic remedy.   Each time that he lost his temper, he transferred a fine of two rupees (then about five shillings) from his right to his left pocket.   When about to leave Roorkee, he devoted this accumulation of self-imposed fines to the erection of a sun-dial, to teach the natives the value of time.   The late Sir James Caird, who told this legend of Roorkee as he heard it there in 1880, used to add, with a humorous twinkle of his kindly eyes, " It was a *very* handsome dial." [33]

From September, 1845, to March, 1847, Yule was much occupied intermittently, in addition to his professional work, by service on a Committee appointed by Government "to investi-gate the causes of the unhealthiness which has existed at Kurnal, and other portions of the country along the line of the Delhi Canal," and further, to report "whether an injurious

---

[32] Maclagan's *Memoir of Yule*, *P.R.G.S.*, Feb. 1890.

[33] On hearing this, Yule said to him, " Your story is quite correct except in one particular ; you understated the *amount* of the fine."

effect on the health of the people of the Doab is, or is not, likely to be produced by the contemplated Ganges Canal."

"A very elaborate investigation was made by the Committee, directed principally to ascertaining what relation subsisted between certain physical conditions of the different districts, and the liability of their inhabitants to miasmatic fevers." The principal conclusion of the Committee was, "that in the extensive epidemic of 1843, when Kurnaul suffered so seriously . . . the greater part of the evils observed had not been the necessary and unavoidable results of canal irrigation, but were due to interference with the natural drainage of the country, to the saturation of stiff and retentive soils, and to natural disadvantages of site, enhanced by excess of moisture. As regarded the Ganges Canal, they were of opinion that, with due attention to drainage, improvement rather than injury to the general health might be expected to follow the introduction of canal irrigation."[34] In an unpublished note written about 1889, Yule records his ultimate opinion as follows: "At this day, and after the large experience afforded by the Ganges Canal, I feel sure that a verdict so favourable to the sanitary results of canal irrigation would not be given." Still the fact remains that the Ganges Canal has been the source of unspeakable blessings to an immense population.

The Second Sikh War saw Yule again with the army in the field, and on 13th Jan. 1849, he was present at the dismal 'Victory' of Chillianwallah, of which his most vivid recollection seemed to be the sudden apparition of Henry Lawrence, fresh from London, but still clad in the legendary Afghan cloak.

On the conclusion of the Punjab campaign, Yule, whose health had suffered, took furlough and went home to his wife. For the next three years they resided chiefly in Scotland, though paying occasional visits to the Continent, and about 1850 Yule bought a house in Edinburgh. There he wrote "The African Squadron vindicated" (a pamphlet which was afterwards re-published in French), translated Schiller's *Kampf mit dem Drachen* into English verse, delivered Lectures on Fortification at the, now long defunct, Scottish Naval and Military Academy, wrote on Tibet for his friend Blackwood's

---

[34] Yule and Maclagan's *Memoir of Baker.*

Magazine, attended the 1850 Edinburgh Meeting of the British Association, wrote his excellent lines, "On the Loss of the *Birkenhead*," and commenced his first serious study of Marco Polo (by whose wondrous tale, however, he had already been captivated as a boy in his father's library—in Marsden's edition probably). But the most noteworthy literary result of these happy years was that really fascinating volume, entitled *Fortification for Officers of the Army and Students of Military History*, a work that has remained unique of its kind. This was published by Blackwood in 1851, and seven years later received the honour of (unauthorised) translation into French. Yule also occupied himself a good deal at this time with the practice of photography, a pursuit to which he never after reverted.

In the spring of 1852, Yule made an interesting little semi-professional tour in company with a brother officer, his accomplished friend, Major R. B. Smith. Beginning with Kelso, "the only one of the Teviotdale Abbeys which I had not as yet seen," they made their way leisurely through the north of England, examining with impartial care abbeys and cathedrals, factories, brick-yards, foundries, timber-yards, docks, and railway works. On this occasion Yule, contrary to his custom, kept a journal, and a few excerpts may be given here, as affording some notion of his casual talk to those who did not know him.

At Berwick-on-Tweed he notes the old ramparts of the town : "These, erected in Elizabeth's time, are interesting as being, I believe, the only existing sample in England of the bastioned system of the 16th century. . . . The outline of the works seems perfect enough, though both earth and stone work are in great disrepair. The bastions are large with obtuse angles, square orillons, and double flanks originally casemated, and most of them crowned with cavaliers." On the way to Durham, "much amused by the discussions of two passengers, one a smooth-spoken, semi-clerical looking person ; the other a brusque well-to-do attorney with a Northumbrian burr. Subject, among others, Protection. The Attorney all for 'cheap bread'—'You wouldn't rob the poor man of his loaf,' and so forth. 'You must go with the *stgheam*, sir, you must go with the stgheam.' 'I never did, Mr Thompson, and I never will,' said the other in an oily manner, singularly inconsistent with the sentiment." At Durham they dined with a dignitary of the Church, and Yule was roasted by being placed with his back to an enormous fire. "Coals are cheap at Durham," he notes feelingly, adding, "The party we found as heavy as any Edinburgh one. Smith, indeed, evidently has had little experience of really stupid Edinburgh parties, for he had never met with anything approaching to this before." (Happy Smith!) But thanks to the kindness and hospitality of the astronomer, Mr. Chevalier, and his gifted daughter, they had a delightful visit to beautiful Durham, and came away full of admiration for the (then newly established) University, and its grand *locale*. They went on to stay with an uncle by marriage of Yule's, in Yorkshire. At dinner he was asked by his host to explain Foucault's pendulum experiment. "I endeavoured to explain it somewhat, I hope, to the satisfaction of his doubts, but not at all to that of Mr G. M., who most resolutely declined to take in *any* elucidation, coming at last to the conclusion that he entirely differed with me as to what North meant, and that it was useless to argue until we could agree about that!" They went next to Leeds, to visit Kirkstall Abbey, "a mediæval fossil, curiously embedded among the squalid brickwork and chimney stalks of a manufacturing suburb. Having established ourselves at the hotel, we went to deliver a letter to Mr. Hope, the official assignee, a very handsome, aristocratic-looking gentleman, who seemed as much out of place at Leeds as the Abbey." At Leeds they visited the flax mills of Messrs. Marshall, "a firm noted for the conscientious care they take of their workpeople . . . We mounted on the roof of the building, which is covered with grass, and formerly was actually grazed by a few sheep, until the repeated inconvenience of their tumbling through the glass domes put a stop to this." They next visited some tile and brickworks on land belonging to a friend. "The owner of the tile works, a well-to-do burgher, and the apparent model of a West Riding Radical, received us in rather a dubious way: 'There are a many people has come and brought introductions, and looked at all my works, and then gone and set up for themselves close by. Now des you mean to say that you be really come all the way from Beng*ul*?' 'Yes, indeed we have, and we are going all the way back again, though we didn't exactly come from there to look at your brickworks.' 'Then you're not in

the brick-making line, are you?' 'Why we've had a good deal
to do with making bricks, and may have again ; but we'll engage
that if we set up for ourselves, it shall be ten thousand miles
from you.' This seemed in some degree to set his mind
at rest. . . ."

"A dismal day, with occasional showers, prevented our
seeing Sheffield to advantage. On the whole, however, it is
more cheerful and has more of a country-town look than Leeds
—a place utterly without beauty of aspect. At Leeds you have
vast barrack-like factories, with their usual suburbs of squalid
rows of brick cottages, and everywhere the tall spiracles of the
steam, which seems the pervading power of the place. Everything there is machinery—the machine is the intelligent agent, it
would seem, the man its slave, standing by to tend it and pick
up a broken thread now and then. At Sheffield . . . you might
go through most of the streets without knowing anything of the
kind was going on. And steam here, instead of being a ruler,
is a drudge, turning a grindstone or rolling out a bar of steel,
but all the accuracy and skill of hand is the Man's. And consequently there was, we thought, a healthier aspect about the
men engaged. None of the Rodgers remain who founded the
firm in my father's time. I saw some pairs of his scissors in the
show-room still kept under the name of *Persian* scissors." [35]

From Sheffield Yule and his friend proceeded to Boston,
"where there is the most exquisite church tower I have ever
seen," and thence to Lincoln, Peterborough, and Ely, ending
their tour at Cambridge, where Yule spent a few delightful
days.

In the autumn the great Duke of Wellington died, and
Yule witnessed the historic pageant of his funeral. His furlough
was now nearly expired, and early in December he again
embarked for India, leaving his wife and only child, of a few

---

[35] It would appear that Major Yule had presented the Rodgers with some specimens of Indian scissors, probably as suggestions in developing that field of export.
Scissors of elaborate design, usually damascened or gilt, used to form a most important
item in every set of Oriental writing implements. Even long after adhesive envelopes
had become common in European Turkey, their use was considered over familiar, if
not actually disrespectful, for formal letters, and there was a particular traditional
knack in cutting and folding the special envelope for each missive, which was included
in the instruction given by every competent *Khoja*, as the present writer well remem-
bers in the quiet years that ended with the disasters of 1877.

weeks old, behind him. Some verses dated "Christmas Day near the Equator," show how much he felt the separation.

Shortly after his return to Bengal, Yule received orders to proceed to Aracan, and to examine and report upon the passes between Aracan and Burma, as also to improve communications and select suitable sites for fortified posts to hold the same. These orders came to Yule quite unexpectedly late one Saturday evening, but he completed all preparations and started at daybreak on the following Monday, 24th Jan. 1853.

From Calcutta to Khyook Phyoo, Yule proceeded by steamer, and thence up the river in the *Tickler* gunboat to Krenggyuen. "Our course lay through a wilderness of wooded islands (50 to 200 feet high) and bays, sailing when we could, anchoring when neither wind nor tide served . . . slow progress up the river. More and more like the creeks and lagoons of the Niger or a Guiana river rather than anything I looked for in India. The densest tree jungle covers the shore down into the water. For miles no sign of human habitation, but now and then at rare intervals one sees a patch of hillside rudely cleared, with the bare stems of the burnt trees still standing. . . . Sometimes, too, a dark tunnel-like creek runs back beneath the thick vault of jungle, and from it silently steals out a slim canoe, manned by two or three wild-looking Mugs or Kyens (people of the Hills), driving it rapidly along with their short paddles held vertically, exactly like those of the Red men on the American rivers."

At the military post of Bokhyong, near Krenggyuen, he notes (5th Feb.) that "Captain Munro, the adjutant, can scarcely believe that I was present at the Duke of Wellington's funeral, of which he read but a few days ago in the newspapers, and here am I, one of the spectators, a guest in this wild spot among the mountains—2½ months since I left England."

Yule's journal of his arduous wanderings in these border wilds is full of interest, but want of space forbids further quotation. From a note on the fly-leaf it appears that from the time of quitting the gun-boat at Krenggyuen to his arrival at Toungoop he covered about 240 miles on foot, and that under immense difficulties, even as to food. He commemorated his tribulations in some cheery humorous verse, but ultimately fell seriously ill of the local fever, aided doubtless by previous exposure and privation. His servants successively fell ill,

some died and others had to be sent back, food supplies
failed, and the route through those dense forests was uncertain ;
yet under all difficulties he seems never to have grumbled or
lost heart. And when things were nearly at the worst, Yule
restored the spirits of his local escort by improvising a
wappenshaw, with a Sheffield gardener's knife, which he happened
to have with him, for prize! When at last Yule emerged
from the wilds and on 25th March marched into Prome, he
was taken for his own ghost! "Found Fraser (of the Engineers)
in a rambling phoongyee house, just under the great gilt pagoda.
I went up to him announcing myself, and his astonishment was
so great that he would scarcely shake hands!" It was on this
occasion at Prome that Yule first met his future chief Captain
Phayre—"a very young-looking man—very cordial," a description no less applicable to General Sir Arthur Phayre at the age
of seventy !

After some further wanderings, Yule embarked at Sandong,
and returned by water, touching at Kyook Phyoo and Akyab, to
Calcutta, which he reached on 1st May—his birthday.

The next four months were spent in hard work at Calcutta.
In August, Yule received orders to proceed to Singapore, and
embarked on the 29th. His duty was to report on the defences
of the Straits Settlements, with a view to their improvement.
Yule's recommendations were sanctioned by Government, but
his journal bears witness to the prevalence then, as since, of the
penny-wise-pound-foolish system in our administration. On all
sides he was met by difficulties in obtaining sites for batteries,
etc., for which heavy compensation was demanded, when by the
exercise of reasonable foresight, the same might have been
secured earlier at a nominal price.

Yule's journal contains a very bright and pleasing picture of
Singapore, where he found that the majority of the European
population "were evidently, from their tongues, from benorth
the Tweed, a circumstance which seems to be true of four-fifths
of the Singaporeans. Indeed, if I taught geography, I should
be inclined to class Edinburgh, Glasgow, Dundee, and Singapore
together as the four chief towns of Scotland."

Work on the defences kept Yule in Singapore and its
neighbourhood until the end of November, when he embarked
for Bengal. On his return to Calcutta, Yule was appointed

Deputy Consulting Engineer for Railways at Head-quarters. In this post he had for chief his old friend Baker, who had in 1851 been appointed by the Governor-General, Lord Dalhousie, Consulting Engineer for Railways to Government. The office owed its existence to the recently initiated great experiment of railway construction under Government guarantee.

The subject was new to Yule, " and therefore called for hard and anxious labour. He, however, turned his strong sense and unbiased view to the general question of railway communication in India, with the result that he became a vigorous supporter of the idea of narrow gauge and cheap lines in the parts of that country outside of the main trunk lines of traffic." [36]

The influence of Yule, and that of his intimate friends and ultimate successors in office, Colonels R. Strachey and Dickens, led to the adoption of the narrow (metre) gauge over a great part of India. Of this matter more will be said further on ; it is sufficient at this stage to note that it was occupying Yule's thoughts, and that he had already taken up the position in this question that he thereafter maintained through life. The office of Consulting Engineer to Government for Railways ultimately developed into the great Department of Public Works.

As related by Yule, whilst Baker "held this appointment, Lord Dalhousie was in the habit of making use of his advice in a great variety of matters connected with Public Works projects and questions, but which had nothing to do with guaranteed railways, there being at that time no officer attached to the Government of India, whose proper duty it was to deal with such questions. In August, 1854, the Government of India sent home to the Court of Directors a despatch and a series of minutes by the Governor-General and his Council, in which the constitution of the Public Works Department as a separate branch of administration, both in the local governments and the government of India itself, was urged on a detailed plan."

In this communication Lord Dalhousie stated his desire to appoint Major Baker to the projected office of Secretary for the Department of Public Works. In the spring of 1855 these recommendations were carried out by the creation of the Depart-

---

[36] Collinson's *Memoir of Yule, Royal Engineer Journal.*

ment, with Baker as Secretary and Yule as Under Secretary for Public Works.

Meanwhile Yule's services were called to a very different field, but without his vacating his new appointment, which he was allowed to retain. Not long after the conclusion of the second Burmese War, the King of Burma sent a friendly mission to the Governor-General, and in 1855 a return Embassy was despatched to the Court of Ava, under Colonel Arthur Phayre, with Henry Yule as Secretary, an appointment the latter owed as much to Lord Dalhousie's personal wish as to Phayre's good-will. The result of this employment was Yule's first geographical book, a large volume entitled *Mission to the Court of Ava in* 1855, originally printed in India, but subsequently re-issued in an embellished form at home (see over leaf). To the end of his life, Yule looked back to this " social progress up the Irawady, with its many quaint and pleasant memories, as to a bright and joyous holiday." [37]   It was a delight to him to work under Phayre, whose noble and lovable character he had already learned to appreciate two years before in Pegu. Then, too, Yule has spoken of the intense relief it was to escape from the monotonous scenery and depressing conditions of official life in Bengal (Resort to Simla was the exception, not the rule, in these days!) to the cheerfulness and unconstraint of Burma, with its fine landscapes and merry-hearted population. " It was such a relief to find natives who would laugh at a joke," he once remarked in the writer's presence to the lamented E. C. Baber, who replied that he had experienced exactly the same sense of relief in passing from India to China.

Yule's work on Burma was largely illustrated by his own sketches. One of these represents the King's reception of the Embassy, and another, the King on his throne. The originals were executed by Yule's ready pencil, surreptitiously within his cocked hat, during the audience.

From the latter sketch Yule had a small oil-painting executed under his direction by a German artist, then resident in Calcutta, which he gave to Lord Dalhousie. [38]

---

[37] Extract from Preface to *Ava*, edition of 1858.

[38] The present whereabouts of this picture is unknown to the writer. It was lent to Yule in 1889 by Lord Dalhousie's surviving daughter (for whom he had strong

The Government of India marked their approval of the Embassy by an unusual concession. Each of the members of the mission received a souvenir of the expedition. To Yule was given a very beautiful and elaborately chased small bowl, of nearly pure gold, bearing the signs of the Zodiac in relief.[39]

On his return to Calcutta, Yule threw himself heart and soul into the work of his new appointment in the Public Works Department. The nature of his work, the novelty and variety of the projects and problems with which this new branch of the service had to deal, brought Yule into constant, and eventually very intimate association with Lord Dalhousie, whom he accompanied on some of his tours of inspection. The two men thoroughly appreciated each other, and, from first to last, Yule experienced the greatest kindness from Lord Dalhousie. In this intimacy, no doubt the fact of being what French soldiers call *pays* added something to the warmth of their mutual regard: their forefathers came from the same *airt,* and neither was unmindful of the circumstance. It is much to be regretted that Yule preserved no sketch of Lord Dalhousie, nor written record of his intercourse with him, but the following lines show some part of what he thought:

At this time [1849] there appears upon the scene that vigorous and masterful spirit, whose arrival to take up the government of India had been greeted by events so inauspicious. No doubt from the beginning the Governor-General was desirous to let it be understood that although new to India he was, and meant to be, master; . . . Lord Dalhousie was by no means averse to frank dissent, provided *in the manner* it was never forgotten that he was Governor-General. Like his great predecessor Lord Wellesley, he was jealous of all familiarity and resented it. . . . The general sentiment of those who worked under that ἄναξ ανδρῶν was one of strong and admiring affection . . . and we doubt if a Governor-General ever embarked on the Hoogly amid deeper feeling than attended him who, shattered by sorrow and

---

regard and much sympathy), and was returned to her early in 1890, but is not named in the catalogue of Lady Susan's effects, sold at Edinburgh in 1898 after her death. At that sale the present writer had the satisfaction of securing for reverent preservation the watch used throughout his career by the great Marquess.

[39] Now in the writer's possession. It was for many years on exhibition in the Edinburgh and South Kensington Museums,

physical suffering, but erect and undaunted, quitted Calcutta on the 6th March 1856." [40]

His successor was Lord Canning, whose confidence in Yule and personal regard for him became as marked as his predecessor's.

In the autumn of 1856, Yule took leave and came home. Much of his time while in England was occupied with making arrangements for the production of an improved edition of his book on Burma, which so far had been a mere government report. These were completed to his satisfaction, and on the eve of returning to India, he wrote to his publishers [41] that the correction of the proof sheets and general supervision of the publication had been undertaken by his friend the Rev. W. D. Maclagan, formerly an officer of the Madras army (and now Archbishop of York).

Whilst in England, Yule had renewed his intimacy with his old friend Colonel Robert Napier, then also on furlough, a visitor whose kindly sympathetic presence always brought special pleasure also to Yule's wife and child. One result of this intercourse was that the friends decided to return together to India. Accordingly they sailed from Marseilles towards the end of April, and at Aden were met by the astounding news of the outbreak of the Mutiny.

On his arrival in Calcutta Yule, who retained his appointment of Under Secretary to Government, found his work indefinitely increased. Every available officer was called into the field, and Yule's principal centre of activity was shifted to the great fortress of Allahabad, forming the principal base of operations against the rebels. Not only had he to strengthen or create defences at Allahabad and elsewhere, but on Yule devolved the principal burden of improvising accommodation for the European troops then pouring into India, which ultimately meant providing for an army of 100,000 men. His task was made the more difficult by the long-standing chronic friction, then and long after, existing between the officers of the Queen's and the Company's services. But in a far more important matter he was always fortunate. As he subsequently recorded in a Note for Government: " Through all consciousness of mistakes and short-

---

[40] Article by Yule on Lord Lawrence, *Quarterly Review* for April, 1883.
[41] Messrs. Smith & Elder.

comings, I have felt that I had the confidence of those whom I served, a feeling which has lightened many a weight."

It was at Allahabad that Yule, in the intervals of more serious work, put the last touches to his Burma book. The preface of the English edition is dated, " Fortress of Allahabad, Oct. 3, 1857," and contains a passage instinct with the emotions of the time. After recalling the "joyous holiday" on the Irawady, he goes on : "But for ourselves, standing here on the margin of these rivers, which a few weeks ago were red with the blood of our murdered brothers and sisters, and straining the ear to catch the echo of our avenging artillery, it is difficult to turn the mind to what seem dreams of past days of peace and security ; and memory itself grows dim in the attempt to repass the gulf which the last few months has interposed between the present and the time to which this narrative refers."[42]

When he wrote these lines, the first relief had just taken place, and the second defence of Lucknow was beginning. The end of the month saw Sir Colin Campbell's advance to the second—the real—relief of Lucknow. Of Sir Colin, Yule wrote and spoke with warm regard : " Sir Colin was delightful, and when in a good humour and at his best, always reminded me very much, both in manner and talk, of the General (*i.e.* General White, his wife's father). The voice was just the same and the

---

[42] Preface to *Narrative of a Mission to the Court of Ava.* Before these words were written, Yule had had the sorrow of losing his elder brother Robert, who had fallen in action before Delhi (19th June, 1857), whilst in command of his regiment, the 9th Lancers. Robert Abercromby Yule (born 1817) was a very noble character and a fine soldier. He had served with distinction in the campaigns in Afghanistan and the Sikh Wars, and was the author of an excellent brief treatise on Cavalry Tactics. He had a ready pencil and a happy turn for graceful verse. In prose his charming little allegorical tale for children, entitled *The White Rhododendron,* is as pure and graceful as the flower whose name it bears. Like both his brothers, he was at once chivalrous and devout, modest, impulsive, and impetuous. No officer was more beloved by his men than Robert Yule, and when some one met them carrying back his covered body from the field and enquired of the sergeant : " Who have you got there ? " the reply was : " Colonel Yule, and better have lost half the regiment, sir." It was in the chivalrous effort to extricate some exposed guns that he fell. Some one told afterwards that when asked to go to the rescue, he turned in the saddle, looked back wistfully on his regiment, well knowing the cost of such an enterprise, then gave the order to advance and charge. " No stone marks the spot where Yule went down, but no stone is needed to commemorate his valour " (Archibald Forbes, in *Daily News,* 8th Feb. 1876). At the time of his death Colonel R. A. Yule had been recommended for the C.B. His eldest son, Colonel J. H. Yule, C.B., distinguished himself in several recent campaigns (on the Burma-Chinese frontier, in Tirah, and South Africa).

quiet gentle manner, with its underlying keen dry humour. But then if you did happen to offend Sir Colin, it was like treading on crackers, which was not our General's way."

When Lucknow had been relieved, besieged, reduced, and finally remodelled by the grand Roads and Demolitions Scheme of his friend Napier, the latter came down to Allahabad, and he and Yule sought diversion in playing quoits and skittles, the only occasion on which either of them is known to have evinced any liking for games.

Before this time Yule had succeeded his friend Baker as *de facto* Secretary to Government for Public Works, and on Baker's retirement in 1858, Yule was formally appointed his successor.[43] Baker and Yule had, throughout their association, worked in perfect unison, and the very differences in their characters enhanced the value of their co-operation ; the special qualities of each friend mutually strengthened and completed each other. Yule's was by far the more original and creative mind, Baker's the more precise and, at least in a professional sense, the more highly-trained organ. In chivalrous sense of honour, devotion to duty, and natural generosity, the men stood equal ; but while Yule was by nature impatient and irritable, and liable, until long past middle age, to occasional sudden bursts of uncontrollable anger, generally followed by periods of black depression and almost absolute silence,[44] Baker was the very reverse. Partly by natural temperament, but also certainly by severe self-discipline, his manner was invincibly placid and his temper imperturbable.[45] Yet none was more tenacious in maintaining whatever he judged right.

Baker, whilst large-minded in great matters, was extremely conventional in small ones, and Yule must sometimes have tried his feelings in this respect. The particulars of one such tragic occurrence have survived. Yule, who was colour-blind,[46] and in

---

[43] Baker went home in November, 1857, but did not retire until the following year.

[44] Nothing was more worthy of respect in Yule's fine character than the energy and success with which he mastered his natural temperament in the last ten years of his life, when few would have guessed his original fiery disposition.

[45] Not without cause did Sir J. P. Grant officially record that "to his imperturbable temper the Government of India owed much."

[46] Yule's colour-blindness was one of the cases in which Dalton, the original investigator of this optical defect, took special interest. At a later date (1859) he sent Yule, through Professor Wilson, skeins of coloured silks to name. Yule's elder brother Robert had the same peculiarity of sight, and it was also present in two earlier

early life whimsically obstinate in maintaining his own view of colours, had selected some cloth for trousers undeterred by his tailor's timid remonstrance of "Not *quite* your usual taste, sir." The result was that the Under-Secretary to Government startled official Calcutta by appearing in brilliant claret - coloured raiment.   Baker remonstrated: "Claret-colour!   Nonsense, my trousers are silver grey," said Yule, and entirely declined to be convinced.   "I think I *did* convince him at last," said Baker with some pride, when long after telling the story to the present writer. "And *then* he gave them up?"   "Oh, no," said Sir William ruefully, "he wore those claret - coloured trousers to the very end."   That episode probably belonged to the Dalhousie period.

When Yule resumed work in the Secretariat at Calcutta at the close of the Mutiny, the inevitable arrears of work were enormous.   This may be the proper place to notice more fully his action with respect to the choice of gauge for Indian rail-ways already adverted to in brief.   As we have seen, his own convictions led to the adoption of the metre gauge over a great part of India.   This policy had great disadvantages not at first foreseen, and has since been greatly modified.   In justice to Yule, however, it should be remembered that the con-ditions and requirements of India have largely altered, alike through the extraordinary growth of the Indian export, especially the grain, trade, and the development of new necessities for Imperial defence.   These new features, however, did but accentuate defects inherent in the system, but which only prolonged practical experience made fully apparent.

At the outset the supporters of the narrow gauge seemed to have the stronger position, as they were able to show that the cost was much less, the rails employed being only about ⅔rds the weight of those required by the broad gauge, and many other subsidiary expenses also proportionally less.   On the other

---

and two later generations of their mother's family—making five generations in all. But in no case did it pass from parent to child, always passing in these examples, by a sort of Knight's move, from uncle to nephew.   Another peculiarity of Yule's more difficult to describe was the instinctive association of certain architectural forms or images with the days of the week.   He once, and once only (in 1843), met another person, a lady who was a perfect stranger, with the same peculiarity.   About 1878-79 he contributed some notes on this obscure subject to one of the newspapers, in connec-tion with the researches of Mr. Francis Galton, on Visualisation, but the particulars are not now accessible.

hand, as time passed and practical experience was gained, its opponents were able to make an even stronger case against the narrow gauge. The initial expenses were undoubtedly less, but the durability was also less. Thus much of the original saving was lost in the greater cost of maintenance, whilst the small carrying capacity of the rolling stock and loss of time and labour in shifting goods at every break of gauge, were further serious causes of waste, which the internal commercial development of India daily made more apparent. Strategic needs also were clamant against the dangers of the narrow gauge in any general scheme of Indian defence. Yule's connection with the Public Works Department had long ceased ere the question of the gauges reached its most acute stage, but his interest and indirect participation in the conflict survived. In this matter a certain parental tenderness for a scheme which he had helped to originate, combined with his warm friendship for some of the principal supporters of the narrow gauge, seem to have influenced his views more than he himself was aware. Certainly his judgment in this matter was not impartial, although, as always in his case, it was absolutely sincere and not consciously biased.

In reference to Yule's services in the period following the Mutiny, Lord Canning's subsequent Minute of 1862 may here be fitly quoted. In this the Governor-General writes : " I have long ago recorded my opinion of the value of his services in 1858 and 1859, when with a crippled and overtaxed staff of Engineer officers, many of them young and inexperienced, the G.-G. had to provide rapidly for the accommodation of a vast English army, often in districts hitherto little known, and in which the authority of the Government was barely established, and always under circumstances of difficulty and urgency. I desire to repeat that the Queen's army in India was then greatly indebted to Lieut.-Colonel Yule's judgment, earnestness, and ability ; and this to an extent very imperfectly understood by many of the officers who held commands in that army.

" Of the manner in which the more usual duties of his office have been discharged it is unnecessary for me to speak. It is, I believe, known and appreciated as well by the Home Government as by the Governor-General in Council."

In the spring of 1859 Yule felt the urgent need of a rest, and took the, at that time, most unusual step of coming home on three months' leave, which as the voyage then occupied a month each way, left him only one month at home. He was accompanied by his elder brother George, who had not been out of India for thirty years. The visit home of the two brothers was as bright and pleasant as it was brief, but does not call for further notice.

In 1860, Yule's health having again suffered, he took short leave to Java. His journal of this tour is very interesting, but space does not admit of quotation here. He embodied some of the results of his observations in a lecture he delivered on his return to Calcutta.

During these latter years of his service in India, Yule owed much happiness to the appreciative friendship of Lord Canning and the ready sympathy of Lady Canning. If he shared their tours in an official capacity, the intercourse was much more than official. The noble character of Lady Canning won from Yule such whole-hearted chivalrous devotion as, probably, he felt for no other friend save, perhaps in after days, Sir Bartle Frere. And when her health failed, it was to Yule's special care that Lord Canning entrusted his wife during a tour in the Hills. Lady Canning was known to be very homesick, and one day as the party came in sight of some ilexes (the evergreen oak), Yule sought to cheer her by calling out pleasantly: "Look, Lady Canning! There are *oaks!*" "No, no, Yule, *not* oaks," cried Sir C. B. "They are (solemnly) IBEXES." "No, *not* Ibexes, Sir C., you mean SILEXES," cried Capt. ——, the A.D.C.; Lady Canning and Yule the while almost choking with laughter.

On another and later occasion, when the Governor-General's camp was peculiarly dull and stagnant, every one yawning and grumbling, Yule effected a temporary diversion by pretending to tap the telegraph wires, and circulating through camp, what purported to be, the usual telegraphic abstract of news brought to Bombay by the latest English mail. The news was of the most astounding character, with just enough air of probability, in minor details, to pass muster with a dull reader. The effect was all he could wish—or rather more—and there was a general flutter in the camp. Of course the Governor-General and one or two others were in the secret, and mightily relished the diversion. But this pleasant and cheering intercourse was drawing to its mournful close.   On her way back from Darjeeling, in November, 1861, Lady Canning (not then in Yule's care) was unavoidably exposed to the malaria of a specially unhealthy season.   A few days' illness followed, and on 18th November, 1861, she passed calmly to

"That remaining rest where night and tears are o'er." [47]

It was to Yule that Lord Canning turned in the first anguish of his loss, and on this faithful friend devolved the sad privilege of preparing her last resting-place.   This may be told in the touching words of Lord Canning's letter to his only sister, written on the day of Lady Canning's burial, in the private garden at Barrackpoor [48] :—

"The funeral is over, and my own darling lies buried in a spot which I am sure she would have chosen of all others. . . . From the grave can be seen the embanked walk leading from the house to the river's edge, which she made as a landing-place three years ago, and from within 3 or 4 paces of the grave there is a glimpse of the terrace-garden and its balustrades, which she made near the house, and of the part of the grounds with which she most occupied herself. . . .   I left Calcutta yesterday . . . and on arriving here, went to look at the precise spot chosen for the grave.   I could see by the clear full moon . . . that it was exactly right.   Yule was there superintending the workmen, and before daylight this morning a solid masonry vault had been completely finished.

" Bowie [Military Secretary] and Yule have done all this for me.   It has all been settled since my poor darling died.   She liked Yule.   They used to discuss together her projects of improvement for this place, architecture, gardening, the Cawnpore monument, etc., and they generally agreed.   He knew her tastes well. . . ."

The coffin, brought on a gun-carriage from Calcutta, " was carried by twelve soldiers of the 6th Regiment (Queen's), the A.D.C.'s bearing the pall.   There were no hired men or ordinary funeral attendants of any kind at any part of the ceremony, and no lookers-on. . . . Yule was the only person not of the house-

---

[47] From Yule's verses on her grave.

[48] Lord Canning to Lady Clanricarde : Letter dated Barrackpoor, 19th **Nov.** 1861, 7 A.M., printed n *Two Noble Lives,* by A. J. C. Hare, and here reproduced by M**r.** Hare's permission.

hold staff. Had others who had asked" to attend "been allowed to do so, the numbers would have been far too large.

"On coming near the end of the terrace walk I saw that the turf between the walk and the grave, and for several yards all round the grave, was strewed thick with palm branches and bright fresh-gathered flowers—quite a thick carpet. It was a little matter, but so exactly what she would have thought of." [49]

And, therefore, Yule thought of this for her! He also recorded the scene two days later in some graceful and touching lines, privately printed, from which the following may be quoted :

> "When night lowered black, and the circling shroud
> Of storm rolled near, and stout hearts learned dismay ;
> Not Hers ! To her tried Lord a Light and Stay
> Even in the Earthquake and the palpable cloud
> Of those dark months ; and when a fickle crowd
> Panted for blood and pelted wrath and scorn
> On him she loved, her courage never stooped :
> But when the clouds were driven, and the day
> Poured Hope and glorious Sunshine, she who had borne,
> The night with such strong Heart, withered and drooped,
> Our queenly lily, and smiling passed away.
> Now ! let no fouling touch profane her clay,
> Nor odious pomps and funeral tinsels mar
> Our grief. But from our England's cannon car
> Let England's soldiers bear her to the tomb
> Prepared by loving hands. Before her bier
> Scatter victorious palms ; let Rose's bloom
> Carpet its passage . . . ."

Yule's deep sympathy in this time of sorrow strengthened the friendship Lord Canning had long felt for him, and when the time approached for the Governor-General to vacate his high office, he invited Yule, who was very weary of India, to accompany him home, where his influence would secure Yule congenial employment. Yule's weariness of India at this time was extreme. Moreover, after serving under such leaders as Lord Dalhousie and Lord Canning, and winning their full confidence and friendship, it was almost repugnant to him to begin afresh with new men and probably new measures, with which he might

---

[49] Lord Canning's letter to Lady Clanricarde. He gave to Yule Lady Canning's own silver drinking-cup, which she had constantly used. It is carefully treasured, with other Canning and Dalhousie relics, by the present writer.

not be in accord. Indeed, some little clouds were already visible on the horizon. In these circumstances, it is not surprising that Yule, under an impulse of lassitude and impatience, when accepting Lord Canning's offer, also 'burnt his boats' by sending in his resignation of the service. This decision Yule took against the earnest advice of his anxious and devoted wife, and for a time the results justified all her misgivings. She knew well, from past experience, how soon Yule wearied in the absence of compulsory employment. And in the event of the life in England not suiting him, for even Lord Canning's good-will might not secure perfectly congenial employment for his talents, she knew well that his health and spirits would be seriously affected. She, therefore, with affectionate solicitude, urged that he should adopt the course previously followed by his friend Baker, that is, come home on furlough, and only send in his resignation after he saw clearly what his prospects of home employment were, and what he himself wished in the matter.

Lord Canning and Yule left Calcutta late in March, 1862 ; at Malta they parted never to meet again in this world. Lord Canning proceeded to England, and Yule joined his wife and child in Rome. Only a few weeks later, at Florence, came as a thunderclap the announcement of Lord Canning's unexpected death in London, on 17th June. Well does the present writer remember the day that fatal news came, and Yule's deep anguish, not assuredly for the loss of his prospects, but for the loss of a most noble and magnanimous friend, a statesman whose true greatness was, both then and since, most imperfectly realised by the country for which he had worn himself out.[50] Shortly after Yule went to England,[51] where he was cordially received by Lord Canning's representatives, who gave him a touching re-

---

[50] Many years later Yule wrote of Lord Canning as follows : " He had his defects, no doubt. He had not at first that entire grasp of the situation that was wanted at such a time of crisis. But there is a virtue which in these days seems unknown to Parliamentary statesmen in England—Magnanimity. Lord Canning was an English statesman, and he was surpassingly magnanimous. There is another virtue which in Holy Writ is taken as the type and sum of all righteousness—Justice—and he was eminently just. The misuse of special powers granted early in the Mutiny called for Lord Canning's interference, and the consequence was a flood of savage abuse ; the violence and bitterness of which it is now hard to realise." (*Quarterly Review*, April, 1883, p. 306.)

[51] During the next ten years Yule continued to visit London annually for two or three months in the spring or early summer.

membrance of his lost friend, in the shape of the silver travelling candlesticks, which had habitually stood on Lord Canning's writing-table.[52]   But his offer to write Lord Canning's *Life* had no result, as the relatives, following the then recent example of the Hastings family, in the case of another great Governor-General, refused to revive discussion by the publication of any Memoir.

Nor did Yule find any suitable opening for employment in England, so after two or three months spent in visiting old friends, he rejoined his family in the Black Forest, where he sought occupation in renewing his knowledge of German.   But it must be confessed that his mood both then and for long after was neither happy nor wholesome. The winter of 1862 was spent somewhat listlessly, partly in Germany and partly at the Hôtel des Bergues, Geneva, where his old acquaintance Colonel Tronchin was hospitably ready to open all doors.   The picturesque figure of John Ruskin also flits across the scene at this time. But Yule was unoccupied and restless, and could neither enjoy Mr. Ruskin's criticism of his sketches nor the kindly hospitality of his Genevan hosts. Early in 1863 he made another fruitless visit to London, where he remained four or five months, but found no opening.   Though unproductive of work, this year brought Yule official recognition of his services in the shape of the C.B., for which Lord Canning had long before recommended him.[53]

On rejoining his wife and child at Mornex in Savoy, Yule found the health of the former seriously impaired.   During his absence, the kind and able English Doctor at Geneva had felt obliged to inform Mrs. Yule that she was suffering from disease of the heart, and that her life might end suddenly at any moment.   Unwilling to add to Yule's anxieties, she made all necessary arrangements, but did not communicate this intelligence until he had done all he wished and returned, when she broke it to him very gently. Up to this year Mrs. Yule, though not strong and often ailing, had not allowed herself to be considered

---

[52] Now in the writer's possession.   They appear in the well-known portrait of Lord Canning reading a despatch.

[53] Lord Canning's recommendation had been mislaid, and the India Office was disposed to ignore it.   It was Lord Canning's old friend and Eton chum, Lord Granville, who obtained this tardy justice for Yule, instigated thereto by that most faithful friend, Sir Roderick Murchison.

an invalid, but from this date doctor's orders left her no choice in the matter.[54]

About this time, Yule took in hand the first of his studies of mediæval travellers. His translation of the *Travels of Friar Jordanus* was probably commenced earlier ; it was completed during the leisurely journey by carriage between Chambéry and Turin, and the Dedication to Sir Bartle Frere written during a brief halt at Genoa, from which place it is dated. Travelling slowly and pleasantly by *vetturino* along the Riviera di Levante, the family came to Spezzia, then little more than a quiet village. A chance encounter with agreeable residents disposed Yule favourably towards the place, and a few days later he opened negotiations for land to build a house ! Most fortunately for himself and all concerned these fell through, and the family continued their journey to Tuscany, and settled for the winter in a long rambling house, with pleasant garden, at Pisa, where Yule was able to continue with advantage his researches into mediæval travel in the East. He paid frequent visits to Florence, where he had many pleasant acquaintances, not least among them Charles Lever ("Harry Lorrequer"), with whom acquaintance ripened into warm and enduring friendship. At Florence he also made the acquaintance of the celebrated Marchese Gino Capponi, and of many other Italian men of letters. To this winter of 1863-64 belongs also the commencement of a lasting friendship with the illustrious Italian historian, Villari, at that time holding an appointment at Pisa. Another agreeable acquaintance, though less intimate, was formed with John Ball, the well-known President of the Alpine Club, then resident at Pisa, and with many others, among whom the name of a very cultivated German scholar, H. Meyer, specially recurs to memory.

---

[54] I cannot let the mention of this time of lonely sickness and trial pass without recording here my deep gratitude to our dear and honoured friend, John Ruskin. As my dear mother stood on the threshold between life and death at Mornex that sad spring, he was untiring in all kindly offices of friendship. It was her old friend, Principal A. J. Scott (then eminent, now forgotten), who sent him to call. He came to see us daily when possible, sometimes bringing MSS. of Rossetti and others to read aloud (and who could equal his reading ?), and when she was too ill for this, or himself absent, he would send not only books and flowers to brighten the bare rooms of the hillside inn (then very primitive), but his own best treasures of Turner and W. Hunt, drawings and illuminated missals. It was an anxious solace ; and though most gratefully enjoyed, these treasures were never long retained.

In the spring of 1864, Yule took a spacious and delightful old villa, situated in the highest part of the Bagni di Lucca,[55] and commanding lovely views over the surrounding chestnut-clad hills and winding river.

Here he wrote much of what ultimately took form in *Cathay, and the Way Thither*. It was this summer, too, that Yule commenced his investigations among the Venetian archives, and also visited the province of Friuli in pursuit of materials for the history of one of his old travellers, the *Beato Odorico*. At Verona—then still Austrian—he had the amusing experience of being arrested for sketching too near the fortifications. However, his captors had all the usual Austrian *bonhomie* and courtesy, and Yule experienced no real inconvenience. He was much more disturbed when, a day or two later, the old mother of one of his Venetian acquaintances insisted on embracing him on account of his supposed likeness to Garibaldi!

As winter approached, a warmer climate became necessary for Mrs. Yule, and the family proceeded to Sicily, landing at Messina in October, 1864. From this point, Yule made a very interesting excursion to the then little known group of the Lipari Islands, in the company of that eminent geologist, the late Robert Mallet, F.R.S., a most agreeable companion.

On Martinmas Day, the Yules reached the beautiful capital of Sicily, Palermo, which, though they knew it not, was to be their home—a very happy one—for nearly eleven years.

During the ensuing winter and spring, Yule continued the preparation of *Cathay*, but his appetite for work not being satisfied by this, he, when in London in 1865, volunteered to make an Index to the third decade of the *Journal of the Royal Geographical Society*, in exchange for a set of such volumes as he did not possess. That was long before any Index Society existed; but Yule had special and very strong views of his own as to what an Index should be, and he spared no labour to realise his ideal.[56] This proved a heavier task than he had anticipated, and he got very weary before the Index was completed.

---

[55] Villa Mansi, nearly opposite the old Ducal Palace. With its private chapel, it formed three sides of a small *place* or court.

[56] He also at all times spared no pains to enforce that ideal on other index-makers, who were not always grateful for his sound doctrine !

In the spring of 1866, *Cathay and the Way Thither* appeared, and at once took the high place which it has ever since retained. In the autumn of the same year Yule's attention was momentarily turned in a very different direction by a local insurrection, followed by severe reprisals, and the bombardment of Palermo by the Italian Fleet. His sick wife was for some time under rifle as well as shell fire ; but cheerfully remarking that "every bullet has its billet," she remained perfectly serene and undisturbed. It was the year of the last war with Austria, and also of the suppression of the Monastic Orders in Sicily ; two events which probably helped to produce the outbreak, of which Yule contributed an account to *The Times*, and subsequently a more detailed one to the *Quarterly Review*.[57]

Yule had no more predilection for the Monastic Orders than most of his countrymen, but his sense of justice was shocked by the cruel incidence of the measure in many cases, and also by the harshness with which both it and the punishment of suspected insurgents was carried out. Cholera was prevalent in Italy that year, but Sicily, which had maintained stringent quarantine, entirely escaped until large bodies of troops were landed to quell the insurrection, when a devastating epidemic immediately ensued, and re-appeared in 1867. In after years, when serving on the Army Sanitary Committee at the India Office, Yule more than once quoted this experience as indicating that quarantine restrictions may, in some cases, have more value than British medical authority is usually willing to admit.

In 1867, on his return from London, Yule commenced systematic work on his long projected new edition of the *Travels of Marco Polo*. It was apparently in this year that the scheme first took definite form, but it had long been latent in his mind. The Public Libraries of Palermo afforded him much good material, whilst occasional visits to the Libraries of Venice, Florence, Paris, and London, opened other sources. But his most important channel of supply came from his very extensive private correspondence, extending to nearly all parts of Europe and many centres in Asia. His work brought him many new and valued friends, indeed too many to mention, but amongst whom, as

---

[57] He saw a good deal of the outbreak when taking small comforts to a friend, the Commandant of the Military School, who was captured and imprisoned by the insurgents.

belonging specially to this period, three honoured names must be recalled here : Commendatore (afterwards Baron) CRISTO-FORO NEGRI, the large-hearted Founder and First President of the Geographical Society of Italy, from whom Yule received his first public recognition as a geographer, Commendatore GUGLIELMO BERCHET (affectionately nicknamed *il Bello e Buono*), ever generous in learned help, who became a most dear and honoured friend, and the Hon. GEORGE P. MARSH, U.S. Envoy to the Court of Italy, a man, both as scholar and friend, unequalled in his nation, perhaps almost unique anywhere.

Those who only knew Yule in later years, may like some account of his daily life at this time. It was his custom to rise fairly early ; in summer he sometimes went to bathe in the sea,[58] or for a walk before breakfast ; more usually he would write until break-fast, which he preferred to have alone. After breakfast he looked through his notebooks, and before ten o'clock was usually walking rapidly to the library where his work lay. He would work there until two or three o'clock, when he returned home, read the *Times*, answered letters, received or paid visits, and then resumed work on his book, which he often continued long after the rest of the household were sleeping. Of course his family saw but little of him under these circumstances, but when he had got a chapter of *Marco* into shape, or struck out some new discovery of interest, he would carry it to his wife to read. She always took great interest in his work, and he had great faith in her literary instinct as a sound as well as sympathetic critic.

The first fruits of Yule's Polo studies took the form of a review of Pauthier's edition of *Marco Polo*, contributed to the *Quarterly Review* in 1868.

In 1870 the great work itself appeared, and received prompt generous recognition by the grant of the very beautiful gold medal of the Geographical Society of Italy,[59] followed in 1872 by the award of the Founder's Medal of the Royal Geographical Society, while the Geographical and Asiatic Societies of Paris, the Geographical Societies of Italy and Berlin, the Academy of Bologna, and other learned bodies, enrolled him as an Honorary Member.

---

[58] After 1869 he discontinued sea-bathing.

[59] This was Yule's first geographical honour, but he had been elected into the Athenæum Club, under " Rule II.," in January, 1867.

Reverting to 1869, we may note that Yule, when passing through Paris early in the spring, became acquainted, through his friend M. Charles Maunoir, with the admirable work of exploration lately performed by Lieut. Francis Garnier of the French Navy. It was a time of much political excitement in France, the eve of the famous *Plébiscite*, and the importance of Garnier's work was not then recognised by his countrymen. Yule saw its value, and on arrival in London went straight to Sir Roderick Murchison, laid the facts before him, and suggested that no other traveller of the year had so good a claim to one of the two gold medals of the R.G.S. as this French naval Lieutenant. Sir Roderick was propitious, and accordingly in May the Patron's medal was assigned to Garnier, who was touchingly grateful to Yule; whilst the French Minister of Marine marked his appreciation of Yule's good offices by presenting him with the magnificent volumes commemorating the expedition.[60]

Yule was in Paris in 1871, immediately after the suppression of the Commune, and his letters gave interesting accounts of the extraordinary state of affairs then prevailing. In August, he served as President of the Geographical Section of the British Association at its Edinburgh meeting.

On his return to Palermo, he devoted himself specially to the geography of the Oxus region, and the result appeared next year in his introduction and notes to Wood's *Journey*. Soon after his return to Palermo, he became greatly interested in the plans, about which he was consulted, of an English church, the gift to the English community of two of its oldest members, Messrs Ingham and Whitaker. Yule's share in the enterprise gradually expanded, until he became a sort of volunteer clerk of the works, to the great benefit of his health, as this occupation during the next three years, whilst adding to his interests, also kept him longer in the open air than would otherwise have been the case. It was a real misfortune to Yule (and one of which he was himself at times conscious) that he had no taste for any out-of-door pursuits, neither for any form of natural science, nor for gardening, nor for

---

[60] Garnier took a distinguished part in the Defence of Paris in 1870-71, after which he resumed his naval service in the East, where he was killed in action. His last letter to Yule contained the simple announcement "*J ai pris Hanoï*," a modest terseness of statement worthy of the best naval traditions.

any kind of sport nor games. Nor did he willingly ride.[61] He was always restless away from his books. There can be no doubt that want of sufficient air and exercise, reacting on an impaired liver, had much to do with Yule's unsatisfactory state of health and frequent extreme depression. There was no lack of agreeable and intelligent society at Palermo (society that the present writer recalls with cordial regard), to which every winter brought pleasant temporary additions, both English and foreign, the best of whom generally sought Yule's acquaintance. Old friends too were not wanting ; many found their way to Palermo, and when such came, he was willing to show them hospitality and to take them excursions, and occasionally enjoyed these. But though the beautiful city and surrounding country were full of charm and interest, Yule was too much pre-occupied by his own special engrossing pursuits ever really to get the good of his surroundings, of which indeed he often seemed only half conscious.

By this time Yule had obtained, without ever having sought it, a distinct and, in some respects, quite unique position in geographical science. Although his *Essay on the Geography of the Oxus Region* (1872) received comparatively little public attention at home, it had yet made its mark once for all,[62] and from this time, if not earlier, Yule's high authority in all questions of Central Asian geography was generally recognised. He had long ere this, almost unconsciously, laid the broad foundations of that "Yule method," of which Baron von Richthofen has written so eloquently, declaring that not only in his own land, " but also in the literatures of France, Italy, Germany, and other countries, the powerful stimulating influence of the Yule method is visible."[63] More than one writer has indeed boldly com-

---

[61] One year the present writer, at her mother's desire, induced him to take walks of 10 to 12 miles with her, but interesting and lovely as the scenery was, he soon wearied for his writing-table (even bringing his work with him), and thus little permanent good was effected. And it was just the same afterwards in Scotland, where an old High-land gillie, describing his experience of the Yule brothers, said : " I was liking to take out Sir George, for *he* takes the time to enjoy the hills, but (plaintively), the Kornel is no good, for he's just as restless as a water-wagtail ! " If there be any *mal de l'écritoire* corresponding to *mal du pays*, Yule certainly had it.

[62] The Russian Government in 1873 paid the same work the very practical compliment of circulating it largely amongst their officers in Central Asia.

[63] " Auch in den Literaturen von Frankreich, Italien, Deutschland und andere Ländern ist der mächtig treibende Einfluss der Yuleschen Methode, welche

pared Central Asia before Yule to Central Africa before Livingstone!

Yule had wrought from sheer love of the work and without expectation of public recognition, and it was therefore a great surprise as well as gratification to him, to find that the demand for his *Marco Polo* was such as to justify the appearance of a second edition only a few years after the first. The preparation of this enlarged edition, with much other miscellaneous work (see subjoined bibliography), and the superintendence of the building of the church already named, kept him fully occupied for the next three years.

Amongst the parerga and miscellaneous occupations of Yule's leisure hours in the period 1869-74, may be mentioned an interesting correspondence with Professor W. W. Skeat on the subject of *William of Palerne* and Sicilian examples of the Werwolf; the skilful analysis and exposure of Klaproth's false geography;[64] the purchase and despatch of Sicilian seeds and young trees for use in the Punjab, at the request of the Indian Forestry Department; translations (prepared for friends) of tracts on the cultivation of Sumach and the collection of Manna as practised in Sicily; also a number of small services rendered to the South Kensington Museum, at the request of the late Sir Henry Cole. These latter included obtaining Italian and Sicilian bibliographic contributions to the Science and Art Department's *Catalogue of Books on Art*, selecting architectural subjects to be photographed;[65] negotiating the purchase of the original drawings illustrative of Padre B. Gravina's great work on the Cathedral of Monreale; and superintending the execution of a copy in mosaic of the large mosaic picture (in the Norman Palatine Chapel, Palermo,) of the Entry of our Lord into Jerusalem.

In the spring of 1875, just after the publication of the second

---

wissenschaftliche Grundlichkeit mit anmuthender Form verbindet, bemerkbar." (*Verhandlungen der Gesellschaft für Erdkunde zu Berlin*, Band XVII. No. 2.)

[64] This subject is too lengthy for more than cursory allusion here, but the patient analytic skill and keen venatic instinct with which Yule not only proved the forgery of the alleged *Travels of Georg Ludwig von* ——— (that had been already established by Lord Strangford, whose last effort it was, and Sir Henry Rawlinson), but step by step traced it home to the arch-culprit Klaproth, was nothing less than masterly.

[65] This is probably the origin of the odd misstatement as to Yule occupying himself at Palermo with photography, made in the delightful *Reminiscences* of the late Colonel Balcarres Ramsay. Yule never attempted photography after 1852.

edition of *Marco Polo*, Yule had to mourn the loss of his noble wife. He was absent from Sicily at the time, but returned a few hours after her death on 30th April. She had suffered for many years from a severe form of heart disease, but her end was perfect peace. She was laid to rest, amid touching tokens of both public and private sympathy, in the beautiful camposanto on Monte Pellegrino. What her loss was to Yule only his oldest and closest friends were in a position to realise. Long years of suffering had impaired neither the soundness of her judgment nor the sweetness, and even gaiety, of her happy, unselfish disposition. And in spirit, as even in appearance, she retained to the very last much of the radiance of her youth. Nor were her intellectual gifts less remarkable. Few who had once conversed with her ever forgot her, and certainly no one who had once known her intimately ever ceased to love her.[66]

Shortly after this calamity, Yule removed to London, and on the retirement of his old friend, Sir William Baker, from the India Council early that autumn, Lord Salisbury at once selected him for the vacant seat. Nothing would ever have made him a party-man, but he always followed Lord Salisbury with conviction, and worked under him with steady confidence.

In 1877 Yule married, as his second wife, the daughter of an old friend,[67] a very amiable woman twenty years his junior, who made him very happy until her untimely death in 1881. From the time of his joining the India Council, his duties at the India Office of course occupied a great part of his time, but he also continued to do an immense amount of miscellaneous literary work, as may be seen by reference to the subjoined bibliography,

---

[66] She was a woman of fine intellect and wide reading ; a skilful musician, who also sang well, and a good amateur artist in the style of Aug. Delacroix (of whom she was a favourite pupil). Of French and Italian she had a thorough and literary mastery, and how well she knew her own language is shown by the sound and pure English of a story she published in early life, under the pseudonym of Max Lyle (*Fair Oaks, or The Experiences of Arnold Osborne, M.D.*, 2 vols., 1856). My mother was partly of Highland descent on both sides, and many of her fine qualities were very characteristic of that race. Before her marriage she took an active part in many good works, and herself originated the useful School for the Blind at Bath, in a room which she hired with her pocket-money, where she and her friend Miss Elwin taught such of the blind poor as they could gather together.

In the tablet which he erected to her memory in the family burial-place of St. Andrew's, Gulane, her husband described her thus :—" A woman singular in endowments, in suffering, and in faith ; to whom to live was Christ, to die was gain."

[67] Mary Wilhelmina, daughter of F. Skipwith, Esq., B.C.S.

(itself probably incomplete).   In Council he invariably " showed his strong determination to endeavour to deal with questions on their own merits and not only by custom and precedent." [68] Amongst subjects in which he took a strong line of his own in the discussions of the Council, may be specially instanced his action in the matter of the cotton duties (in which he defended native Indian manufactures as against hostile Manchester interests); the Vernacular Press Act, the necessity for which he fully recognised; and the retention of Kandahar, for which he recorded his vote in a strong minute.   In all these three cases, which are typical of many others, his opinion was overruled, but having been carefully and deliberately formed, it remained unaffected by defeat.

In all matters connected with Central Asian affairs, Yule's opinion always carried great weight; some of his most competent colleagues indeed preferred his authority in this field to that of even Sir Henry Rawlinson, possibly for the reason given by Sir M. Grant Duff, who has epigrammatically described the latter as good in Council but dangerous in counsel.[69]

Yule's courageous independence and habit of looking at all public questions by the simple light of what appeared to him right, yet without fads or doctrinairism, earned for him the respect of the successive Secretaries of State under whom he served, and the warm regard and confidence of his other colleagues.   The value attached to his services in Council was sufficiently shown by the fact that when the period of ten years (for which members are usually appointed), was about to expire, Lord Hartington (now Duke of Devonshire), caused Yule's appointment to be renewed for life, under a special Act of Parliament passed for this purpose in 1885.

His work as a member of the Army Sanitary Committee, brought him into communication with Miss Florence Nightingale, a privilege which he greatly valued and enjoyed, though he used to say : " She is worse than a Royal Commission to answer, and, in the most gracious charming manner possible, immediately finds out all I don't know!"   Indeed his devotion to the " Lady-in-Chief" was scarcely less complete than Kinglake's.

---

[68] Collinson's *Memoir of Yule*.
[69] See *Notes from a Diary*, 1888-91.

In 1880, Yule was appointed to the Board of Visitors of the Government Indian Engineering College at Cooper's Hill, a post which added to his sphere of interests without materially increasing his work. In 1882, he was much gratified by being named an Honorary Fellow of the Society of Antiquaries of Scotland, more especially as it was to fill one of the two vacancies created by the deaths of Thomas Carlyle and Dean Stanley.

Yule had been President of the Hakluyt Society from 1877, and in 1885 was elected President also of the Royal Asiatic Society. He would probably also have been President of the Royal Geographical Society, but for an untoward incident. Mention has already been made of his constant determination to judge all questions by the simple touchstone of what he believed to be right, irrespective of personal considerations. It was in pursuance of these principles that, at the cost of great pain to himself and some misrepresentation, he in 1878 sundered his long connection with the Royal Geographical Society, by resigning his seat on their Council, solely in consequence of their adoption of what he considered a wrong policy. This severance occurred just when it was intended to propose him as President. Some years later, at the personal request of the late Lord Aberdare, a President in all respects worthy of the best traditions of that great Society, Yule consented to rejoin the Council, which he re-entered as a Vice-President.

In 1883, the University of Edinburgh celebrated its Tercentenary, when Yule was selected as one of the recipients of the honorary degree of LL.D. His letters from Edinburgh, on this occasion, give a very pleasant and amusing account of the festivity and of the celebrities he met. Nor did he omit to chronicle the envious glances cast, as he alleged, by some British men of science on the splendours of foreign Academic attire, on the yellow robes of the Sorbonne, and the Palms of the Institute of France! Pasteur was, he wrote, the one most enthusiastically acclaimed of all who received degrees.

I think it was about the same time that M. Renan was in England, and called upon Sir Henry Maine, Yule, and others at the India Office. On meeting just after, the colleagues compared notes as to their distinguished but unwieldy visitor. "It seems that *le style n'est pas l'homme même* in *this* instance," quoth " Ancient Law" to " Marco Polo."   And here it may be remarked that Yule so completely identified himself with his favourite traveller that he frequently signed contributions to the public press as MARCUS PAULUS VENETUS or M.P.V.   His more intimate friends also gave him the same *sobriquet*, and once, when calling on his old friend, Dr. John Brown (the beloved chronicler of *Rab and his Friends*), he was introduced by Dr. John to some lion-hunting American visitors as " our Marco Polo."   The visitors evidently took the statement in a literal sense, and scrutinised Yule closely.[70]

In 1886 Yule published his delightful *Anglo-Indian Glossary*, with the whimsical but felicitous sub-title of *Hobson-Jobson* (the name given by the rank and file of the British Army in India to the religious festival in celebration of Hassan and Husaïn).

This *Glossary* was an abiding interest to both Yule and the present writer.   Contributions of illustrative quotations came from most diverse and unexpected sources, and the arrival of each new word or happy quotation was quite an event, and gave such pleasure to the recipients as can only be fully understood by those who have shared in such pursuits.   The volume was dedicated in affecting terms to his elder brother, Sir George Yule, who, unhappily, did not survive to see it completed.

In July 1885, the two brothers had taken the last of many happy journeys together, proceeding to Cornwall and the Scilly Isles.   A few months later, on 13th January 1886, the end came suddenly to the elder, from the effects of an accident at his own door.[71]

It may be doubted if Yule ever really got over the shock of this loss, though he went on with his work as usual, and served that year as a Royal Commissioner on the occasion of the Indian and Colonial Exhibition of 1886.

From 1878, when an accidental chill laid the foundations of an exhausting, though happily quite painless, malady, Yule's strength had gradually failed, although for several years longer his general health and energies still appeared unimpaired to a casual observer.   The condition of public affairs also, in some

---

[70] The identification was not limited to Yule, for when travelling in Russia many years ago, the present writer was introduced by an absent-minded Russian *savant* to his colleagues as *Mademoiselle Marco Paulovna!*

[71] See Note on Sir George Yule's career at the end of this Memoir.

degree, affected his health injuriously. The general trend of political events from 1880 to 1886 caused him deep anxiety and distress, and his righteous wrath at what he considered the betrayal of his country's honour in the cases of Frere, of Gordon, and of Ireland, found strong, and, in a noble sense, passionate expression in both prose and verse. He was never in any sense a party man, but he often called himself "one of Mr. Gladstone's converts," *i.e.* one whom Gladstonian methods had compelled to break with liberal tradition and prepossessions.

Nothing better expresses Yule's feeling in the period referred to than the following letter, written in reference to the R. E. Gordon Memorial,[72] but of much wider application: "Will you allow me an inch or two of space to say to my brother officers, 'Have nothing to do with the proposed Gordon Memorial.'

"That glorious memory is in no danger of perishing and needs no memorial. Sackcloth and silence are what it suggests to those who have guided the action of England; and Englishmen must bear the responsibility for that action and share its shame. It is too early for atoning memorials; nor is it possible for those who take part in them to dissociate themselves from a repulsive hypocrisy.

"Let every one who would fain bestow something in honour of the great victim, do, in silence, some act of help to our soldiers or their families, or to others who are poor and suffering.

"In later days our survivors or successors may look back with softened sorrow and pride to the part which men of our corps have played in these passing events, and Charles Gordon far in the front of all; and then they may set up our little tablets, or what not—not to preserve the memory of our heroes, but to maintain the integrity of our own record of the illustrious dead."

Happily Yule lived to see the beginning of better times for his country. One of the first indications of that national awakening was the right spirit in which the public, for the most part, received Lord Wolseley's stirring appeal at the close of 1888, and Yule was so much struck by the parallelism between Lord Wolseley's warning and some words of his own contained

---

[72] Addressed to the Editor, *Royal Engineers' Journal,* who did not, however, publish it.

in the pseudo-Polo fragment (see above, end of Preface), that he sent Lord Wolseley the very last copy of the 1875 edition of *Marco Polo*, with a vigorous expression of his sentiments.

That was probably Yule's last utterance on a public question. The sands of life were now running low, and in the spring of 1889, he felt it right to resign his seat on the India Council, to which he had been appointed for life. On this occasion Lord Cross, then Secretary of State for India, successfully urged his acceptance of the K.C.S.I., which Yule had refused several years before.

In the House of Lords, Viscount Cross subsequently referred to his resignation in the following terms. He said : " A vacancy on the Council had unfortunately occurred through the resignation from ill-health of Sir Henry Yule, whose presence on the Council had been of enormous advantage to the natives of the country. A man of more kindly disposition, thorough intelligence, high-minded, upright, honourable character, he believed did not exist ; and he would like to bear testimony to the estimation in which he was held, and to the services which he had rendered in the office he had so long filled." [73]

This year the Hakluyt Society published the concluding volume of Yule's last work of importance, the *Diary of Sir William Hedges*. He had for several years been collecting materials for a full memoir of his great predecessor in the domain of historical geography, the illustrious Rennell.[74] This work was well advanced as to preliminaries, but was not sufficiently developed for early publication at the time of Yule's death, and ere it could be completed its place had been taken by a later enterprise.

During the summer of 1889, Yule occupied much of his leisure by collecting and revising for re-issue many of his miscellaneous writings. Although not able to do much at a time, this desultory work kept him occupied and interésted, and gave him much pleasure during many months. It was, however, never completed. Yule went to the seaside for a few weeks

---

[73] Debate of 27th August, 1889, as reported in *The Times* of 28th August.

[74] Yule had published a brief but very interesting Memoir of Major Rennell in the *R. E. Journal* in 1881. He was extremely proud of the circumstance that Rennell's surviving grand-daughter presented to him a beautiful wax medallion portrait of the great geographer. This wonderfully life-like presentment was bequeathed by Yule to his friend Sir Joseph Hooker, who presented it to the Royal Society.

in the early summer, and subsequently many pleasant days were spent by him among the Surrey hills, as the guest of his old friends Sir Joseph and Lady Hooker. Of their constant and unwearied kindness, he always spoke with most affectionate gratitude. That autumn he took a great dislike to the English climate; he hankered after sunshine, and formed many plans, eager though indefinite, for wintering at Cintra, a place whose perfect beauty had fascinated him in early youth. But increasing weakness made a journey to Portugal, or even the South of France, an alternative of which he also spoke, very inexpedient, if not absolutely impracticable. Moreover, he would certainly have missed abroad the many friends and multifarious interests which still surrounded him at home. He continued to take drives, and occasionally called on friends, up to the end of November, and it was not until the middle of December that increasing weakness obliged him to take to his bed. He was still, however, able to enjoy seeing his friends—some to the very end, and he had a constant stream of visitors, mostly old friends, but also a few newer ones, who were scarcely less welcome. He also kept up his correspondence to the last, three attached brother R.E.'s, General Collinson, General Maclagan, and Major W. Broadfoot, taking it in turn with the present writer to act as his amanuensis.

On Friday, 27th December, Yule received a telegram from Paris, announcing his nomination that day as Corresponding Member of the Institute of France (Académie des Inscriptions), one of the few distinctions of any kind of which it can still be said that it has at no time lost any of its exalted dignity.

An honour of a different kind that came about the same time, and was scarcely less prized by him, was a very beautiful letter of farewell and benediction from Miss Florence Nightingale,[75] which he kept under his pillow and read many times. On the 28th, he dictated to the present writer his acknowledgment, also by telegraph, of the great honour done him by the Institute. The message was in the following words: " Reddo gratias,

---

[75] Knowing his veneration for that noble lady, I had written to tell her of his condition, and to ask her to give him this last pleasure of a few words. The response was such as few but herself could write. This letter was not to be found after my father's death, and I can only conjecture that it must either have been given away by himself (which is most improbable), or was appropriated by some unauthorised outsider.

Illustrissimi Domini, ob honores tanto nimios quanto immeritos! Mihi robora deficiunt, vita collabitur, accipiatis voluntatem pro facto.   Cum corde pleno et gratissimo moriturus vos, Illustrissimi Domini, saluto.   YULE."

Sunday, 29th December, was a day of the most dense black fog, and he felt its oppression, but was much cheered by a visit from his ever faithful friend, Collinson, who, with his usual unselfishness, came to him that day at very great personal inconvenience.

On Monday, 30th December, the day was clearer, and Henry Yule awoke much refreshed, and in a peculiarly happy and even cheerful frame of mind.   He said he felt so comfortable.   He spoke of his intended book, and bade his daughter write about the inevitable delay to his publisher : " Go and write to John Murray," were indeed his last words to her.   During the morning he saw some friends and relations, but as noon approached his strength flagged, and after a period of unconsciousness, he passed peacefully away in the presence of his daughter and of an old friend, who had come from Edinburgh to see him, but arrived too late for recognition.   Almost at the same time that Yule fell asleep, his "stately message,"[76] was being read under the great Dome in Paris.   Some two hours after Yule had passed away, F.-M. Lord Napier of Magdala, called on an errand of friendship, and at his desire was admitted to see the last of his early friend.   When Lord Napier came out, he said to the present writer, in his own reflective way : " He looks as if he had just settled to some great work."   With these suggestive words of the great soldier, who was so soon, alas, to follow his old friend to the work of another world, this sketch may fitly close.

The following excellent verses (of unknown authorship) on Yule's death, subsequently appeared in the *Academy* : [77]

> "' Moriturus vos saluto'
> Breathes his last the dying scholar—
> Tireless student, brilliant writer ;
> He ' salutes his age' and journeys
> To the Undiscovered Country.

[76] So Sir M. E. Grant Duff well calls it.
[77] *Academy*, 29th March, 1890.

> " There await him with warm welcome
> All the heroes of old Story—
> The Venetians, the Cà Polo,
> Marco, Nicolo, Maffeo,
> Odoric of Pordenone,
> Ibn Batuta, Marignolli,
> Benedict de Goës—' Seeking
> Lost Cathay and finding Heaven.'
> Many more whose lives he cherished
> With the piety of learning ;
> Fading records, buried pages,
> Failing lights and fires forgotten,
> By his energy recovered,
> By his eloquence re-kindled.
> ' Moriturus vos saluto '
> Breathes his last the dying scholar,
> And the far off ages answer :
> *Immortales te salutant.*   D. M."

The same idea had been previously embodied, in very felicitous language, by the late General Sir William Lockhart, in a letter which that noble soldier addressed to the present writer a few days after Yule's death.   And Yule himself would have taken pleasure in the idea of those meetings with his old travellers, which seemed so certain to his surviving friends.[78]

He rests in the old cemetery at Tunbridge Wells, with his second wife, as he had directed.   A great gathering of friends attended the first part of the burial service which was held in London on 3rd January, 1890.   Amongst those present were witnesses of every stage of his career, from his boyish days at the High School of Edinburgh downwards.   His daughter, of course, was there, led by the faithful, peerless friend who was so soon to follow him into the Undiscovered Country.[79]   She and his youngest nephew, with two cousins and a few old friends, followed his remains over the snow to the graveside.   The epitaph subsequently inscribed on the tomb was penned by Yule himself, but is by no means representative of his powers in a kind of composition in which he had so often excelled in the service of others.   As a composer of epitaphs and other monumental inscriptions few of our time have surpassed, if any have equalled him, in his best efforts.

---

[78] He was much pleased, I remember, by a letter he once received from a kindly Franciscan friar, who wrote : " You may rest assured that the Beato Odorico will not forget all you have done for him."

[79] F.-M. Lord Napier of Magdala, died 14th January, 1890.

SIR GEORGE UDNY YULE, C.B., K.C.S.I.*

GEORGE UDNY YULE, born at Inveresk in 1813, passed through Haileybury into the Bengal Civil Service, which he entered at the age of 18 years. For twenty-five years his work lay in Eastern Bengal. He gradually became known to the Government for his activity and good sense, but won a far wider reputation as a mighty hunter, alike with hog-spear and double barrel. By 1856 the roll of his slain tigers exceeded four hundred, some of them of special fame; after that he continued slaying his tigers, but ceased to count them. For some years he and a few friends used annually to visit the plains of the Brahmaputra, near the Garrow Hills—an entirely virgin country then, and swarming with large game. Yule used to describe his once seeing seven rhinoceroses at once on the great plain, besides herds of wild buffalo and deer of several kinds. One of the party started the theory that Noah's Ark had been shipwrecked there ! In those days George Yule was the only man to whom the Maharajah of Nepaul, Sir Jung Bahadur, conceded leave to shoot within his frontier.

Yule was first called from his useful obscurity in 1856. The year before, the Sonthals in insurrection disturbed the long unbroken peace of the Delta. These were a numerous non-Aryan, uncivilised, but industrious race, driven wild by local mismanagement, and the oppressions of Hindoo usurers acting through the regulation courts. After the suppression of their rising, Yule was selected by Sir F. Halliday, who knew his man, to be Commissioner of the Bhagulpoor Division, containing some six million souls, and embracing the hill country of the Sonthals. He obtained sanction to a code for the latter, which removed these people entirely from the Court system, and its tribe of leeches, and abolished all intermediaries between the Sahib and the Sonthal peasant. Through these measures, and his personal influence, aided by picked assistants, he was able to effect, with extraordinary rapidity, not only their entire pacification, but such a beneficial change in their material condition, that they have risen from a state of barbarous penury to comparative prosperity and comfort.

George Yule was thus engaged when the Mutiny broke out, and it soon made itself felt in the districts under him. To its suppression within his limits, he addressed himself with characteristic vigour. Thoroughly trusted by every class—by his Government, by those under him, by planters and by Zemindars—he organised a little force, comprising a small detachment of the 5th Regiment, a party of British sailors, mounted volunteers from the districts, etc., and of this he became practically the captain. Elephants were collected from all quarters to spare the legs of his infantry and sailors ; while dog-carts were turned into limbers for the small three-pounders of the seamen. And with this little army George Yule scoured the Trans-Gangetic districts, leading it against bodies of the Mutineers, routing them upon more than one occasion, and out-manœuvring them by

---

* This notice includes the greater part of an article written by my father, and published in the *St. James' Gazette* of 18th January, 1886, but I have added other details from personal recollection and other sources.—A. F. Y.

his astonishing marches, till he succeeded in driving them across the Nepaul frontier. No part of Bengal was at any time in such danger, and nowhere was the danger more speedily and completely averted.

After this Yule served for two or three years as Chief Commissioner of Oudh, where in 1862 he married Miss Pemberton, the daughter of a very able father, and the niece of Sir Donald MacLeod, of honoured and beloved memory. Then for four or five years he was Resident at Hyderabad, where he won the enduring friendship of Sir Salar Jung. "Everywhere he showed the same characteristic firm but benignant justice. Everywhere he gained the lasting attachment of all with whom he had intimate dealings—except tigers and scoundrels."

Many years later, indignant at the then apparently supine attitude of the British Government in the matter of the Abyssinian captives, George Yule wrote a letter (necessarily published without his name, as he was then on the Governor-General's Council), to the editor of an influential Indian paper, proposing a private expedition should be organised for their delivery from King Theodore, and inviting the editor (Dr. George Smith) to open a list of subscriptions in his paper for this purpose, to which Yule offered to contribute £2000 by way of beginning. Although impracticable in itself, it is probable that, as in other cases, the existence of such a project may have helped to force the Government into action. The particulars of the above incident were printed by Dr. Smith in his *Memoir of the Rev. John Wilson,* but are given here from memory.

From Hyderabad he was promoted in 1867 to the Governor-General's Council, but his health broke down under the sedentary life, and he retired and came home in 1869.

After some years of country life in Scotland, where he bought a small property, he settled near his brother in London, where he was a principal instrument in enabling Sir George Birdwood to establish the celebration of Primrose Day (for he also was " one of Mr. Gladstone's converts "). Sir George Yule never sought ' London Society ' or public employment, but in 1877 he was offered and refused the post of Financial Adviser to the Khedive under the Dual control. When his feelings were stirred he made useful contributions to the public press, which, after his escape from official trammels, were always signed. The very last of these (*St. James' Gazette,* 24th February 1885) was a spirited protest against the snub administered by the late Lord Derby, as Secretary of State, to the Colonies, when they had generously offered assistance in the Soudan campaign. He lived a quiet, happy, and useful life in London, where he was the friend and unwearied helper of all who needed help. He found his chief interests in books and flowers, and in giving others pleasure. Of rare unselfishness and sweet nature, single in mind and motive, fearing God and knowing no other fear, he was regarded by a large number of people with admiring affection. He met his death by a fall on the frosty pavement at his door, in the very act of doing a kindness. An interesting sketch of Sir George Yule's Indian career, by one who knew him thoroughly, is to be found in Sir Edward Braddon's *Thirty Years of Shikar.* An account of his share in the origin of Primrose Day appeared in the *St. James' Gazette* during 1891.

A BIBLIOGRAPHY OF SIR HENRY YULE'S WRITINGS

COMPILED BY H. CORDIER AND A. F. YULE *

1842 Notes on the Iron of the Kasia Hills. (*Jour. Asiatic Soc. Bengal*, XI., Part II. July-Dec. 1842, pp. 853-857.)
Reprinted in *Proceedings of the Museum of Economic Geology*, 1852.

1844 Notes on the Kasia Hills and People. By Lieut. H. Yule. (*Jour. Asiatic Soc. Bengal*, XII. Part II. July-Dec. 1844, pp. 612-631.)

1846 A Canal Act of the Emperor Akbar, with some notes and remarks on the History of the Western Jumna Canals. By Lieut. Yule. (*Jour. Asiatic Society Bengal*, XV. 1846, pp. 213-223.)

1850 The African Squadron vindicated. By Lieut. H. **Yule**. Second Edition. London, J. Ridgway, 1850, 8vo, pp. 41.
Had several editions. Reprinted in the Colonial Magazine of March, 1850.

—— L'Escadre Africaine vengée. Par le lieutenant H. Yule. Traduit du *Colonial Magazine* de Mars, 1850. (*Revue Coloniale*, Mai, 1850.)

1851 Fortification for Officers of the Army and Students of Military History, with Illustrations and Notes. By Lieut. H. Yule, Blackwood, MDCCCLI. 8vo, pp. xxii.-210. (There had been a previous edition privately printed.)

—— La Fortification mise à la portée des Officiers de l'Armée et des personnes qui se livrent à l'étude de l'histoire militaire (avec Atlas). Par H. Yule. Traduit de l'Anglais par M. Sapia, Chef de Bataillon d'Artillerie de Marine et M. Masselin, Capitaine du Génie. Paris, J. Corréard, 1858, 8vo, pp. iii.-263, and Atlas.

1851 The Loss of the *Birkenhead* (Verses). (*Edinburgh Courant*, Dec. 1851.)
Republished in Henley's *Lyra Heroica*, a Book of Verse for Boys. London, D. Nutt, 1890.

1852 Tibet. (*Blackwood's Edinburgh Magazine*, 1852.)

1856 Narrative of Major Phayre's Mission to the Court of Ava, with Notices of the Country, Government, and People. Compiled by Capt. H. Yule. Printed for submission to the Government of India. Calcutta, J. Thomas, ... 1856, 4to, pp. xxix. + 1 f. n. ch. p. l. er. + pp. 315 + pp. cxiv. + pp. iv. and pp. 70.
The last pp. iv.-70 contain: Notes on the Geological features of the banks of the River Irawadee and on the Country north of the Amarapoora, by Thomas Oldham. .... Calcutta, 1856.

—— A Narrative of the Mission sent by the Governor-General of India to the Court of Ava in 1855, with Notices of the Country, Government, and People. By Capt. H. Yule. With Numerous Illustrations. London, Smith, Elder & Co., 1858, 4to.

1857 On the Geography of Burma and its Tributary States, in illustration of a New Map of those Regions. (*Journal, R.G.S.*, XXVII. 1857, pp. 54-108.)

—— Notes on the Geography of Burma, in illustration of a Map of that

* This list is based on the excellent preliminary List compiled by E. Delmar Morgan, published in the *Scottish Geographical Magazine*, vol. vi., pp. 97-98, but the present compilers have much more than doubled the number of entries. It is, however, known to be still incomplete, and any one able to add to the list, will greatly oblige the compilers by sending additions to the Publisher.—A. F. Y.

Country. (*Proceedings R. G. S.*,vol.i. 1857, pp. 269-273.)

1857 An Account of the Ancient Buddhist Remains at Pagân on the Iráwadi. By Capt. H. Yule. (*Jour. Asiatic Society, Bengal*, XXVI. 1857, pp. 1-51.)

1861 A few notes on Antiquities near Jubbulpoor. By Lieut.-Col. H. Yule. (*Journal Asiatic Society, Bengal*, XXX. 1861, pp. 211-215.)

—— Memorandum on the Countries between Thibet, Yunân, and Burmah. By the Very Rev. Thomine D'Mazure (*sic*), communicated by Lieut.-Col. A. P. Phayre (with notes and a comment by Lieut.-Col. H. Yule). With a Map of the N. E. Frontier, prepared in the Office of the Surveyor-Gen. of India, Calcutta, Aug. 1861. (*Jour. Asiatic Soc. Bengal*, XXX. 1861, pp. 367-383.)

1862 Notes of a brief Visit to some of the Indian Remains in Java. By Lieut.-Col. H. Yule. (*Jour. Asiatic Society, Bengal*, XXXI. 1862, pp. 16-31.)

—— Sketches of Java. A Lecture delivered at the Meeting of the Bethune Society, Calcutta, 13th Feb. 1862.

—— Fragments of Unprofessional Papers gathered from an Engineer's portfolio after twenty-three years of service. Calcutta, 1862. Ten copies printed for private circulation.

1863 *Mirabilia descripta*. The Wonders of the East. By Friar Jordanus, of the Order of Preachers and Bishop of Columbum in India the Greater (*circa* 1330). Translated from the Latin original, as published at Paris in 1839, in the *Recueil de Voyages et de Mémoires*, of the Society of Geography, with the addition of a Commentary, by Col. H. Yule, London. Printed for the Hakluyt Society, M.DCCC.LXIII, 8vo, p. iv.-xvii.-68.

—— Report on the Passes between Arakan and Burma [written in 1853]. (*Papers on Indian Civil Engineering*, vol. i. Roorkee.)

1866 Notices of Cathay. (*Proceedings, R. G. S.*, X. 1866, pp. 270-278.)

—— Cathay and the Way Thither, being a Collection of Mediæval Notices of China. Translated and Edited by Col. H. Yule. With a Preliminary Essay on the Intercourse between China and the Western Nations previous to the Discovery of the Cape route. London, printed for the Hakluyt Society. M.DCCC.LXVI. 2 vols. 8vo.

1866 The Insurrection at Palermo. (*Times*, 29th Sep., 1866.)

—— Lake People. (*The Athenæum*, No. 2042, 15th Dec. 1866, p. 804.) Letter dated Palermo, 3rd Dec. 1866.

1867 General Index to the third ten Volumes of the Journal of the Royal Geographical Society. Compiled by Col. H. Yule. London, John Murray, M.DCCCLXVII, 8vo, pp. 228.

—— A Week's Republic at Palermo. (*Quarterly Review*, Jan. 1867.)

—— On the Cultivation of Sumach (*Rhus coriaria*), in the Vicinity of Colli, near Palermo. By Prof. Inzenga. Translated by Col. H. Yule. Communicated by Dr. Cleghorn. *From the Trans. Bot. Society*, vol. ix., 1867-68, ppt. 8vo, p. 15. Original first published in the *Annali di Agricoltura Siciliana, redatti per l'Istituzione del Principe di Castelnuovo*. Palermo, 1852.

1868 Marco Polo and his Recent Editors. (*Quarterly Review*, vol. 125, July and Oct. 1868, pp. 133 and 166.)

1870 An Endeavour to Elucidate Rashi-duddin's Geographical Notices of India. (*Journal R. Asiatic Society*, N.S. iv. 1870, pp. 340-356.)

—— Some Account of the Senbyú Pagoda at Mengún, near the Burmese Capital, in a Memorandum by Capt. E. H. Sladen, Political Agent at Mandalé; with Remarks on the Subject, by Col. H. Yule. (*Ibid.* pp. 406-429.)

—— Notes on Analogies of Manners between the Indo-Chinese and the Races of the Malay Archipelago. (*Report Fortieth Meeting British Association, Liverpool*, Sept. 1870, p. 178.)

1871 The Book of Ser Marco Polo, the Venetian, Concerning the Kingdoms and Marvels of the East. Newly translated and edited with notes. By Col. H. Yule. In two volumes. With Maps and other Illustrations. London, John Murray, 1871, 2 vols. 8vo.

—— The Book of Ser Marco Polo, the Venetian, concerning the Kingdoms and Marvels of the East. Newly translated and edited, with Notes, Maps, and other Illustrations. By

Col. H. Yule. Second edition. London, John Murray, 1875, 2 vols. 8vo.

1871 Address by Col. H. Yule. (*Report Forty-First Meeting British Association, Edinburgh*, Aug. 1871, pp. 162-174.)

1872 A Journey to the Source of the River Oxus. By Captain John Wood, Indian Navy. New edition, edited by his Son. With an Essay on the Geography of the Valley of the Oxus. By Col. H. Yule. With maps. London, John Murray, 1872. In-8, pp. xc.-280.

—— Papers connected with the Upper Oxus Regions. (*Journal*, xlii. 1872, pp. 438-481.)

—— Letter [on Yule's edition of Wood's *Oxus*]. (*Ocean Highways*, Feb. 1874, p. 475.)
Palermo, 9th Jan. 1874.

1873 Letter [about the route of M. Polo through Southern Kerman]. (*Ocean Highways*, March, 1873, p. 385.)
Palermo, 11th Jan. 1873.

—— On Northern Sumatra and especially Achin. (*Ocean Highways*, Aug. 1873, pp. 177-183.)

—— Notes on Hwen Thsang's Account of the Principalities of Tokharistan, in which some previous Geographical Identifications are reconsidered. (*Jour. Royal Asiatic Society*, N.S. vi. 1873, pp. 92-120 and p. 278.)

1874 Francis Garnier (In Memoriam). (*Ocean Highways*, pp. 487-491.) March, 1874.

—— Remarks on Mr. Phillips's Paper [*Notices of Southern Mangi*]. (*Journal*, XLIV. 1874, pp. 103-112.)
Palermo, 22nd Feb. 1874.

—— [Sir Frederic Goldsmid's] "Telegraph and Travel." (*Geographical Magazine*, April, 1874, p. 34 ; Oct. 1874, pp. 300-303.)

—— Geographical Notes on the Basins of the Oxus and the Zarafshán. By the late Alexis Fedchenko. (*Geog. Mag.*, May, 1874, pp. 46-54.)

—— [Mr. Ashton Dilke on the Valley of the Ili.] (*Geog. Mag.*, June, 1874, p. 123.)
Palermo, 16th May, 1874.

—— The *Atlas Sinensis* and other Sinensiana. (*Geog. Mag.*, 1st July, 1847, pp. 147-148.)

—— Letter [on Belasaghun]. (*Geog. Mag.*, 1st July, 1874, p. 167 ; *Ibid.* 1st Sept. 1874, p. 254.)
Palermo, 17th June, 1874 ; 8th Aug. 1874.

1874 Bala Sagun and Karakorum. By Eugene Schuyler. With note by Col. Yule. (*Geog. Mag.*, 1st Dec. 1874, p. 389.)

—— M. Khanikoff's Identifications of Names in Clavijo. (*Ibid.* pp. 389-390.)

1875 Notes [to the translation by Eugene Schuyler of Palladius's version of *The Journey of the Chinese Traveller, Chang Fe-hui*]. (*Geog. Mag.*, 1st Jan. 1875, pp. 7-11).

—— Some Unscientific Notes on the History of Plants. (*Geog. Mag.*, 1st Feb. 1875, pp. 49-51.)

—— Trade Routes to Western China. (*Geog. Mag.*, April, 1875, pp. 97-101.)

—— Garden of Transmigrated Souls [Friar Odoric]. (*Geog. Mag.*, 1st May, 1875, pp. 137-138.)

—— A Glance at the Results of the Expedition to Hissar. By Herr P. Lerch. (*Geog. Mag.*, 1st Nov. 1875, pp. 334-339.)

—— Kathay or Cathay. (*Johnson's American Cyclopædia.*)

—— Achín. (*Encycl. Brit.* 9th edition, 1875, I. pp. 95-97.)

—— Afghánistán. (*Ibid.* pp. 227-241.)

—— Andaman Islands. (*Ibid.* II. 1875, pp. 11-13.)

—— India [Ancient]. (Map No. 31, 1874, in *An Atlas of Ancient Geography, edited by William Smith and George Grove*. London, John Murray, 1875.)

1876 Mongolia, the Tangut Country, and the Solitudes of Northern Tibet, being a Narrative of Three Years' Travel in Eastern High Asia. By Lieut.-Col. N. Prejevalsky, of the Russian Staff Corps ; Mem. of the Imp. Russ. Geog. Soc. Translated by E. Delmar Morgan, F.R.G.S. With Introduction and Notes by Col. H. Yule. With Maps and Illustrations. London, Sampson Low, 1876, 8vo.

—— Tibet . . . Edited by C. R. Markham. Notice of. (*Times*, 1876, —— ?)

—— Eastern Persia. Letter. (*The Athenæum*, No. 2559, 11th Nov. 1876.)

—— Review of *H. Howorth's History of the Mongols*, Part I. (*The Athenæum*, No. 2560, 18th Nov. 1876, pp. 654-656.) Correspondence. (*Ibid.* No. 2561, 25th Nov. 1876.)

—— Review of *T. E. Gordon's Roof of the World*. (*The Academy*, 15th July, 1876, pp. 49-50.)

1876 Cambodia. (*Encycl. Brit.* IV. 1876, pp. 723-726.)
1877 Champa. (*Geog. Mag.*, 1st March, 1877, pp. 66-67.)
Article written for the *Encycl. Brit.* 9th edition, but omitted for reasons which the writer did not clearly understand.
—— *Quid, si Mundus evolvatur?* (*Spectator*, 24th March, 1877.)
Written in 1875.—Signed MARCUS PAULUS VENETUS.
—— On Louis de Backer's *L'Extrême-Orient au Moyen - Age.* (*The Athenæum*, No. 2598, 11th Aug. 1877, pp. 174-175.)
—— On P. Dabry de Thiersant's *Catholicisme en Chine.* (*The Athenæum*, No. 2599, 18th Aug. 1877, pp. 209-210.)
—— Review of *Thomas de Quincey, His Life and Writings. By H. A. Page.* (*Times*, 27th Aug. 1877.)
—— Companions of Faust. Letter on the Claims of P. Castaldi. (*Times*, Sept. 1877.)
1878 The late Col. T. G. Montgomerie, R.E. (Bengal). (*R. E. Journal*, April, 1878.) 8vo, pp. 8.
—— Mr. Henry M. Stanley and the Royal Geographical Society; being the Record of a Protest. By Col. H. Yule and H. M. Hyndman B.A., F.R.G.S. London: Bickers and Son, 1878, 8vo, pp. 48
—— Review of *Burma, Past and Present; with Personal Reminiscences of the Country.* By Lieut.-Gen. Albert Fytche. (*The Athenæum*, No. 2634, 20th April, 1878, pp. 499-500.)
—— Kayal. (*The Athenæum*, No. 2634, 20th April, 1878, p. 515.)
Letter dated April, 1878.
—— Missions in Southern India. (Letter to *Pall Mall Gazette*, 20th June, 1878.)
—— Mr. Stanley and his Letters of 1875. (Letter to *Pall Mall Gazette*, 30th Jan. 1878.)
—— Review of *Richthofen's China*, Bd. I. (*The Academy*, 13th April, 1878, pp. 315-316.)
—— [A foreshadowing of the Phonograph.] (*The Athenæum*, No. 2636, 4th May, 1878.)
1879 A Memorial of the Life and Services of Maj.-Gen. W. W. H. Greathed, C.B., Royal Engineers (Bengal), (1826-1878). Compiled by a Friend and Brother Officer. London, printed for private circulation, 1879, 8vo, pp. 57.

1879 Review of *Gaur: its Ruins and Inscriptions.* By John Henry Ravenshaw. (*The Athenæum*, No. 2672, 11th Jan. 1879, pp. 42-44.)
—— Wellington College. (Letter to *Pall Mall Gazette*, 14th April, 1879.)
—— Dr. Holub's Travels. (*The Athenæum*, No. 2710, 4th Oct. 1879, pp. 436-437.)
—— Letter to Comm. Berchet, dated 2nd Dec. 1878. (*Archivio Veneto* XVII. 1879, pp. 360-362.)
Regarding some documents discovered by the Ab. Cav. V. Zanetti.
—— Gaur. (*Encyclop. Brit.* X. 1879, pp. 112-116.)
—— Ghazni. (*Ibid.* pp. 559-562.)
—— Gilgit. (*Ibid.* pp. 596-599.)
—— Singular Coincidences. (*The Athenæum*, No. 2719, 6th Dec. 1879.)
1880 [Brief Obituary Notice of] General W. C. Macleod. (*Pall Mall Gazette*, 10th April, 1880.)
—— [Obituary Notice of] Gen. W. C. Macleod. (*Proc. R. Geog. Soc.*, June, 1880.)
—— An Ode in Brown Pig. Suggested by reading Mr. Lang's *Ballades in Blue China.* [Signed MARCUS PAULUS VENETUS.] (*St. James' Gazette*, 17th July, 1880.)
—— Notes on Analogies of Manners between the Indo-Chinese Races and the Races of the Indian Archipelago. By Col. Yule (*Journ. Anthrop. Inst. of Great Britain and Ireland*, vol. ix., 1880, pp. 290-301.)
—— Sketches of Asia in the Thirteenth Century and of Marco Polo's Travels, delivered at Royal Engineer Institute, 18th Nov. 1880.
[This Lecture, with slight modification, was also delivered on other occasions both before and after. Doubtful if ever fully reported.
—— Dr. Holub's Collections. (*The Athenæum*, No. 2724, 10th Jan. 1880.)
—— Prof. Max Müller's Paper at the Royal Asiatic Society. (*The Athenæum*, No. 2731, 28th Feb. 1880, p. 285.)
—— The Temple of Buddha Gaya. (Review of *Dr. Rajendralála Mitra's Buddha Gaya.*) (*Sat. Rev.*, 27th March, 1870.)
—— Mr. Gladstone and Count Karoiyi. (Letter to *The Examiner*, 22nd May, 1880, signed TRISTRAM SHANDY.)

1880 Stupa of Barhut. [Review of Cunningham's work.] (*Sat. Rev.*, 5th June, 1880.)
—— From Africa: Southampton, Fifth October, 1880. [Verses to Sir Bartle Frere.] (*Blackwood's Edinburgh Magazine*, Nov. 1880.)
—— Review of *H. Howorth's History of the Mongols*, Part II. (*The Athenæum*, No. 2762, 2nd Oct. 1880, pp. 425-427.)
—— *Verboten ist*, a Rhineland Rhapsody. (Printed for private circulation only.)
—— Hindú-Kúsh. (*Encyclop. Brit.* XI. 1880, pp. 837-839.)
—— The River of Golden Sand, the Narrative of a Journey through China and Eastern Tibet to Burmah, With Illustrations and ten Maps from Original Surveys. By Capt. W. Gill, Royal Engineers. With an Introductory Essay. By Col. H. Yule, London, John Murray, . . . 1880, 2 vols. 8vo, pp. 95-420, 11-453.
—— The River of Golden Sand : Being the Narrative of a Journey through China and Eastern Tibet to Burmah. By the late Capt. W. Gill, R.E. Condensed by Edward Colborne Baber, Chinese Secretary to H.M.'s Legation at Peking. Edited, with a Memoir and Introductory Essay, by Col. H. Yule. With Portrait, Map, and Woodcuts. London, John Murray, 1883, 8vo., pp. 141-332.
—— Memoir of Captain W. Gill, R.E., and Introductory Essay as prefixed to the New Edition of the " River of Golden Sand." By Col. H. Yule. London, John Murray, . . . 1884, 8vo. [Paged 19-141.]
1881 [Notice on William Yule] in Persian Manuscripts in the British Museum. By Sir F. J. Goldsmid. (*The Athenæum*, No. 2813, 24th Sept. 1881, pp. 401-403.)
—— Il Beato Odorico di Pordenone, ed i suoi Viaggi : Cenni dettati dal Col. Enrico Yule, quando s'inaugurava in Pordenone il Busto di Odorico il giorno, 23° Settembre, MDCCC-LXXXI, 8vo. pp. 8.
—— Hwen T'sang. (*Encyclop. Brit.* XII. 1881, pp. 418-419.)
—— Ibn Batuta. (*Ibid.* pp. 607-609.)
—— Kâfiristân. (*Ibid.* XIII. 1881, pp. 820-823.)
—— Major James Rennell, F.R.S., of the Bengal Engineers. [Reprinted from the *Royal Engineers' Journal*], 8vo., pp. 16. (Dated 7th Dec. 1881.)
1881 Notice of Sir William E. Baker. (*St. James' Gazette*, 27th Dec. 1881.)
—— Parallels [Matthew Arnold and de Barros]. (*The Athenæum*, No. 2790, 16th April, 1881, pp. 536.)
1882 Memoir of Gen. Sir William Erskine Baker, K.C.B., Royal Engineers (Bengal). Compiled by two old friends, brother officers and pupils. London. Printed for private circulation, 1882, 8vo., pp. 67. By H. Y [ule] and R. M. [Gen. R. Maclagan].
—— Etymological Notes. (*The Athenæum*, No. 2837, 11th March, 1882; No. 2840, 1st April, 1882, p. 413.)
—— Lhása. (*Encyclop. Brit.* XIV. 1882, pp. 496-503.)
—— *Wadono*. (*The Athenæum*, No. 2846, 13th May, 1882, p. 602.)
—— Dr. John Brown. (*The Athenæum*, No. 2847, 20th May, 1882, pp. 635-636.)
—— A Manuscript of Marco Polo. (*The Athenæum*, No. 2851, 17th June, 1882, pp. 765-766.) [About Baron Nordenskiöld's Facsimile Edition.]
—— Review of *Ancient India as described by Ktesias the Knidian*, etc. By J. W. M'Crindle. (*The Athenæum*, No. 2860, 19th Aug. 1882, pp. 237-238.)
—— The Silver Coinage of Thibet. (Review of Terrien de Lacouperie's Paper.) (*The Academy*, 19th Aug. 1882, pp. 140-141.)
—— Review of *The Indian Balhara and the Arabian Intercourse with India*. By Edward Thomas. (*The Athenæum*, No. 2866, 30th Sept. 1882, pp. 428-429.)
—— The Expedition of Professor Palmer, Capt. Gill, and Lieut. Charrington. (Letter in *The Times*, 16th Oct. 1882.)
—— Obituary Notice of Dr. Arthur Burnell. (*Times*, 20th Oct. 1882.)
—— Capt. William Gill, R.E. [Notice of]. (*The Times*, 31st Oct. 1882.) See *supra*, first col. of this page.
—— Notes on the Oldest Records of the Sea Route to China from Western Asia. By Col. Yule. *Proc. of the Royal Geographical Society, and Monthly Record of Geography*, Nov. No. 1882, 8vo. *Proceedings*, N.S. IV. 1882, pp.

649-660. Read at the Geographical Section, Brit. Assoc., Southampton Meeting, augmented and revised by the author.

1883 Lord Lawrence. [Review of *Life of Lord Lawrence*. By R. Bosworth Smith.] (*Quarterly Review*, vol. 155, April, 1883, pp. 289-326.)

—— Review of *Across Chrysé*. By A. R. Colquhoun. (*The Athenæum*, No. 2900, 26th May, 1883, pp. 663-665.)

—— La Terra del Fuoco e Carlo Darwin. (Extract from Letter published by the *Fanfulla*, Rome 2nd June, 1883.)

—— How was the Trireme rowed? (*The Academy*, 6th Oct. 1883, p. 237.)

—— *Across Chrysé*. (*The Athenæum*, No. 2922, 27th Oct. 1883.)

—— Political Fellowship in the India Council. (Letter in *The Times*, 15th Dec. 1883.) [Heading was not Yule's.]

—— Maldive Islands. (*Encyclop. Brit.* XV. 1883, pp. 327-332.)

—— Mandeville. (*Ibid.* pp. 473-475.)

1884 A Sketch of the Career of Gen. John Reid Becher, C.B., Royal Engineers (Bengal). By an old friend and brother officer. Printed for private circulation, 1884, 8vo, pp. 40.

—— Ruc Quills. (*The Academy*, No. 620, 22nd March, 1884, pp. 204-205.) Reprinted in present ed. of Marco Polo, vol. ii. p. 596.

—— Lord Canning. (Letter in *The Times*, 2nd April, 1884.)

—— Sir Bartle Frere [Letter respecting Memorial of]. (*St. James' Gazette*, 27th July, 1884.)

—— Odoric. (*Encyclop. Brit.* XVIII. 1884, pp. 728-729.)

—— Ormus. (*Ibid.* pp. 856-858.)

1885 Memorials of Gen. Sir Edward Harris Greathed, K.C.B. Compiled by the late Lieut.-Gen. Alex. Cunningham Robertson, C.B. Printed for private circulation. (With a prefatory notice of the compiler.) London, Harrison & Sons, . . . 1885, 8vo, pp. 95.

The Prefatory Notice of Gen. A. C. Robertson is by H. Yule, June, 1885, p. iii.-viii.

—— Anglo-Indianisms. (Letter in the *St. James' Gazette*, 30th July, 1885.)

—— Obituary Notice of Col. Grant Allan, Madras Army. (*From the Army and Navy Gazette*, 22nd Aug. 1885.)

—— Shameless Advertisements. (Letter in *The Times*, 28th Oct. 1885.)

1886 Marco Polo. (*Encyclop. Brit.* XIX. 1885, pp. 404-409.)

—— Prester John. (*Ibid.* pp. 714-718.)

—— Brief Notice of Sir Edward Clive Bayley. Pages ix.-xiv. [Prefixed to *The History of India as told by its own Historians: Gujarát*. By the late Sir Edward Clive Bayley.] London, Allen, 1886, 8vo.

—— Sir George Udny Yule. In Memoriam. (*St. James' Gazette*, 18th Jan. 1886.)

—— Cacothanasia. [Political Verse, Signed Μηνυ 'ΑΕΙΔΕ (*St. James' Gazette*, 1st Feb. 1886.)

—— William Kay, D.D. [Notice of]. (Letter to *The Guardian*, 3rd Feb. 1886.)

—— Col. George Thomson, C.B., R.E. (*Royal Engineers' Journal*, 1886.)

—— Col. George Thomson, C.B. [Note]. (*St. James' Gazette*, 16th Feb. 1886.)

—— Hidden Virtues [A Satire on W. E. Gladstone]. (Letter to the *St. James' Gazette*, 21st March, 1886. Signed M. P. V.)

—— Burma, Past and Present. (*Quart. Rev.* vol. 162, Jan. and April, 1886, pp. 210-238.)

——— Errors of Facts, in two well-known Pictures. (*The Athenæum*, No. 3059, 12th June, 1886, p. 788.)

—— [Obituary Notice of] Lieut.-Gen. Sir Arthur Phayre, C.B., K.C.S.I., G.C.M.G. (*Proc. R.G.S.*, N.S. 1886, VIII. pp. 103-112.)

—— " Lines suggested by a Portrait in the Millais Exhibition."
Privately printed and (though never published) widely circulated. These powerful verses on Gladstone are those several times referred to by Sir Mountstuart Grant Duff, in his published Diaries.

—— Introductory Remarks on *The Rock-Cut Caves and Statues of Bamian*. By Capt. the Hon. M. G. Talbot. (*Journ. R. As. Soc.* N.S. XVIII. 1886, pp. 323-329.)

—— Opening Address. (*Ibid.* pp. i.-v.)

—— Opening Address. (*Ibid.* xix. pp. i.-iii.)

—— Hobson-Jobsoniana. By H. Yule (*Asiatic Quarterly Review*, vol. i. 1886, pp. 119-140.)

—— HOBSON-JOBSON : Being a Glossary of Anglo-Indian Colloquial Words and Phrases, and of Kindred Terms ; etymological, historical, geographical, and discursive. By Col. H. Yule, and the late Arthur Coke

Burnell, Ph.D., C.I.E., author of "The Elements of South Indian Palaeography," etc., London, John Murray, 1886. (All rights reserved), 8vo, p. xliii.-870. Preface, etc. A new edition is in preparation under the editorship of Mr. William Crooke (1902).

1886 John Bunyan. (Letter in *St. James' Gazette*, *circa* 31st Dec. 1886. Signed M. P. V.)

—— Rennell. (*Encyclop. Brit.* XX. 1886, pp. 398-401.)

—— Rubruquis (*Ibid.* XXI. 1886, pp. 46-47.)

1887 Lieut.-Gen. W. A. Crommelen, C.B., R.E. (*Royal Engineers' Journal*, 1887.)

—— [Obituary Notice] Col. Sir J. U. Bateman Champain. (*Times*, 2nd Feb. 1887).

—— "Pulping Public Records." (*Notes and Queries*, 19th March, 1887.)

—— A Filial Remonstrance (Political Verses). Signed M. P. V. (*St. James' Gazette*, 8th Aug. 1887.)

—— Memoir of Major-Gen. J. T. Boileau, R.E., F.R.S. By C. R. Low, I.N., F.R.G.S. With a Preface by Col. H. Yule, C.B., London, Allen, 1887.

—— The Diary of William Hedges, Esq. (afterwards Sir William Hedges), during his Agency in Bengal; as well as on his voyage out and return overland (1681-1687). Transcribed for the Press, with Introductory Notes, etc., by R. Barlow, Esq., and illustrated by copious extracts from unpublished records, etc., by Col. H. Yule. Pub. for Hakluyt Society. London, 1887-1889, 3 vols. 8vo.

1888 Concerning some little known Travellers in the East. (*Asiatic Quarterly Review*, V. 1888, pp. 312-335.) No. I.—George Strachan.

—— Concerning some little known Travellers in the East. (*Asiatic Quarterly Review*, VI. 1888, pp. 382-398.) No. II.—William, Earl of Denbigh; Sir Henry Skipwith; and others.

—— Notes on the St. James's of the 6th Jan. [A Budget of Miscellaneous interesting criticism.] (Letter to *St. James' Gazette*, 9th Jan. 1888.)

—— Deflections of the Nile. (Letter in *The Times*, 15th Oct. 1888.)

—— The History of the Pitt Diamond, being an excerpt from Documentary Contributions to a Biography of Thomas Pitt, prepared for issue [in Hedges' Diary] by the Hakluyt Society. London, 1888, 8vo. pp. 23. Fifty Copies printed for private circulation.

1889 The Remains of Pagan. By H. Yule. (*Trübner's Record*, 3rd ser. vol. i. pt. i. 1889, p. 2.) To introduce notes by Dr. E. Forchammer.

—— A Coincident Idiom. By H. Yule. (*Trübner's Record*, 3rd ser. vol. i. pt. iii. pp. 84-85.)

—— The Indian Congress [a Disclaimer]. (Letter to *The Times*, 1st Jan. 1889.)

—— Arrowsmith, the Friend of Thomas Poole. (Letter in *The Academy*, 9th Feb. 1889, p. 96.)

BIOGRAPHIES OF SIR HENRY YULE.

—— Colonel Sir Henry Yule, K.C.S.I., C.B., LL.D., R.E. By General Robert Maclagan, R.E. (*Proceed. Roy. Geog. Soc.* XII. 1890, pp. 108-113.)

—— Colonel Sir Henry Yule, K.C.S.I., C.B., LL.D., R.E., etc. (With a Portrait). By E. Delmar Morgan. (*Scottish Geographical Magazine*, VI. 1890, pp. 93-98.) Contains a very good Bibliography.

—— Col. Sir H. Yule, R.E., C.B., K.C.S.I., by Maj.-Gen. T. B. Collinson, R.E., *Royal Engineers' Journal*, March, 1890. [This is the best of the Notices of Yule which appeared at the time of his death.]

—— Sir Henry Yule, K.C.S.I., C.B., LL.D., R.E., by E. H. Giglioli. Roma, 1890, ppt. 8vo, pp. 8. Estratto dal *Bollettino della Società Geografica Italiana*, Marzo, 1890.

—— Sir Henry Yule. By J. S. C[otton]. (*The Academy*, 11th Jan. 1890, No. 923, pp. 26-27.)

—— Sir Henry Yule. (*The Athenæum*, No. 3245, 4th Jan. 1900, p. 17; No. 3246, 11th Jan. p. 53; No. 3247, 18th Jan. p. 88.)

—— *In Memoriam*. Sir Henry Yule. By H. M. (*The Academy*, 29th March, 1890, p. 222.) See end of *Memoir* in present work.

—— Le Colonel Sir Henry Yule. Par M. Henri Cordier. Extrait du *Journal Asiatique*. Paris, Imprimerie nationale, MDCCCXC, in-8, pp. 26.

—— The same, *Bulletin de la Société de Géographie*. Par M. Henri Cordier. 1890, 8vo, pp. 4. Meeting 17th Jan. 1890.

1889 Baron F. von Richthofen. (*Verhandlungen der Gesellschaft für Erdkunde zu Berlin*, xvii. 2.)

—— Colonel Sir Henry Yule, R.E., C.B., K.C.S.I. Memoir by General R. Maclagan, *Journ. R. Asiatic Society*, 1890.

—— Memoir of Colonel Sir Henry Yule, R.E., C.B., K.C.S.I., LL.D., etc. By Coutts Trotter. (*Proceedings of the Royal Society of Edinburgh*, 1891, p. xliii. to p. lvi.)

1889 Sir Henry Yule (1820-1889). By Coutts Trotter. (*Dict. of National Biography*, lxiii. pp. 405-407.)

1903 Memoir of Colonel Sir Henry Yule, R.E., C.B., K.C.S.I., Corr. Inst. France, by his daughter, Amy Frances Yule, L.A.Soc. Ant. Scot., etc. Written for third edition of Yule's Marco Polo. Reprinted for private circulation only.

SYNOPSIS OF CONTENTS.

MARCO POLO AND HIS BOOK.

INTRODUCTORY NOTICES.

THE BOOK OF MARCO POLO.

—◇—

PROLOGUE.

BOOK FIRST.

*Account of Regions Visited or heard of on the Journey from the
Lesser Armenia to the Court of the Great Kaan at Chandu.*

BOOK SECOND.

—◇—

PART I.

EXPLANATORY LIST OF ILLUSTRATIONS
TO VOLUME I.

—◇—

INSERTED PLATES AND MAPS.

To face Title . . PORTRAIT of Sir HENRY YULE. From the Painting by Mr. T. B. Wirgman, in the Royal Engineers' Mess House at Chatham.

Illuminated Title, with Medallion representing the POLOS ARRIVING AT VENICE after 26 years' absence, and being refused admittance to the Family Mansion; as related by Ramusio, p. *4* of Introductory Essay. Drawn by Signor QUINTO CENNI, No. 7 Via Solferino, Milan; from a Design by the Editor.

  ,,   *page*   *1.* DOORWAY of the HOUSE of MARCO POLO in the Corte Sabbionera at Venice (see p. *27*). Woodcut from a drawing by Signor L. ROSSO, Venice.

  ,,   ,,   *26. Corte del Milione,* Venice.

  ,,   ,,   *28. Malibran Theatre,* Venice.

  ,,   ,,   *30.* Entrance to the Corte del Milione, Venice. From photographs taken for the present editor, by Signor NAYA.

  ,,   ,,   *42.* Figures from St. Sabba's, sent to Venice. From a photograph of Signor NAYA.

  ,,   ,,   *50.* Church of SAN MATTEO, at Genoa.

  ,,   ,,   *62. Palazzo di S. Giorgio,* at Genoa.

  ,,   ,,   *68. Miracle of S. Lorenzo.* From the Painting by V. CARPACCIO.

  ,,   ,,   *70.* FACSIMILE of the WILL of MARCO POLO, preserved in St. Mark's Library. Lithographed from a photograph specially taken by Bertani at Venice.

  ,,   ,,   *74.* Pavement in front of S. Lorenzo.

  ,,   ,,   *76.* Mosaic Portrait of Marco Polo, at Genoa.

  ,,   ,,   *78.* The Pseudo Marco Polo at Canton.

  ,,   ,,   *80.* Porcelain Incense-Burner, from the Louvre.

  ,,   ,,   *82.* Temple of 500 Genii, at Canton, after a drawing by FÉLIX RÉGAMEY.

  ,,   ,,   *108.* Probable view of MARCO POLO'S OWN GEOGRAPHY: a Map of the World, formed as far as possible from the Traveller's own data. Drawn by the Editor.

  ,,   ,,   *134.* Part of the *Catalan Map* of 1375.

WOODCUTS PRINTED WITH THE TEXT.

INTRODUCTORY NOTICES.

BOOK SECOND.—PART FIRST.

MARCO POLO AND HIS BOOK.

INTRODUCTORY NOTICES.

Doorway of the House of Marco Polo in the Corte Sabbionera, at Venice.

(To face p. 1

MARCO POLO AND HIS BOOK.

INTRODUCTORY NOTICES.

———◇———

I. Obscurities in the History of his Life and Book.
Ramusio's Statements.

1. WITH all the intrinsic interest of Marco Polo's Book it may perhaps be doubted if it would have continued to exercise such fascination on many minds through succes- Obscurities of Polo's sive generations were it not for the difficult questions Book, and which it suggests. It is a great book of puzzles, History. whilst our confidence in the man's veracity is such that we feel certain every puzzle has a solution.

And such difficulties have not attached merely to the identification of places, the interpretation of outlandish terms, or the illustration of obscure customs; for strange entanglements have perplexed also the chief circumstances of the Traveller's life and authorship. The time of the dictation of his Book and of the execution of his Last Will have been almost the only undisputed epochs in his biography. The year of his birth has been contested, and the date of his death has not been recorded; the critical occasion of his capture by the Genoese, to which we seem to owe the happy fact that he did not go down mute to the tomb of his fathers, has been made the subject of chronological difficulties; there are in the various texts of his story variations hard to account for; the very tongue in which it was written down has furnished a question, solved only in our own age, and in a most unexpected manner.

2. The first person who attempted to gather and string
the facts of Marco Polo's personal history was his
countryman, the celebrated John Baptist Ramusio.
His essay abounds in what we now know to be errors
of detail, but, prepared as it was when traditions of the Traveller were still rife in Venice, a genuine thread runs through
it which could never have been spun in later days, and its
presentation seems to me an essential element in any full
discourse upon the subject.

Ramusio's preface to the Book of Marco Polo, which opens
the second volume of his famous Collection of Voyages and
Travels, and is addressed to his learned friend Jerome Fra-
castoro, after referring to some of the most noted geographers
of antiquity, proceeds : *—

"Of all that I have named, Ptolemy, as the latest, possessed the greatest
extent of knowledge. Thus, towards the North, his knowledge carries
him beyond the Caspian, and he is aware of its being shut in all round
like a lake,—a fact which was unknown in the days of Strabo and Pliny,
though the Romans were already lords of the world. But though his knowledge extends so far, a tract of 15 degrees beyond that sea he can describe
only as Terra Incognita ; and towards the South he is fain to apply the
same character to all beyond the Equinoxial. In these unknown regions,
as regards the South, the first to make discoveries have been the Portuguese captains of our own age ; but as regards the North and North-
East the discoverer was the Magnifico Messer Marco Polo, an honoured
nobleman of Venice, nearly 300 years since, as may be read more fully in
his own Book. And in truth it makes one marvel to consider the immense
extent of the journeys made, first by the Father and Uncle of the said
Messer Marco, when they proceeded continually towards the East-North-
East, all the way to the Court of the Great Can and the Emperor of the
Tartars ; and afterwards again by the three of them when, on their return
homeward, they traversed the Eastern and Indian Seas. Nor is that all,
for one marvels also how the aforesaid gentleman was able to give such
an orderly description of all that he had seen ; seeing that such an accomplishment was possessed by very few in his day, and he had had a large
part of his nurture among those uncultivated Tartars, without any regular
training in the art of composition. His Book indeed, owing to the endless
errors and inaccuracies that had crept into it, had come for many years
to be regarded as fabulous ; and the opinion prevailed that the names of
cities and provinces contained therein were all fictitious and imaginary,
without any ground in fact, or were (I might rather say) mere dreams.

---

* The Preface is dated Venice, 7th July, 1553. Fracastorius died in the same
year, and Ramusio erected a statue of him at Padua. Ramusio himself died in
July, 1557.

Ramusio,
his earliest
biographer.
His account
of Polo.

3. "Howbeit, during the last hundred years, persons acquainted with
Persia have begun to recognise the existence of Cathay. The
voyages of the Portuguese also towards the North-East, beyond
the Golden Chersonese, have brought to knowledge many cities
and provinces of India, and many islands likewise, with those
very names which our Author applies to them ; and again, on reaching
the Land of China, they have ascertained from the people of that region
(as we are told by Sign. John de Barros, a Portuguese gentleman, in his
Geography) that Canton, one of the chief cities of that kingdom, is in $30\frac{2}{3}°$
of latitude, with the coast running N.E. and S.W. ; that after a distance of
275 leagues the said coast turns towards the N.W. ; and that there are
three provinces along the sea-board, Mangi, Zanton, and Quinzai, the last
of which is the principal city and the King's Residence, standing in 46° of
latitude. And proceeding yet further the coast attains to 50°.* Seeing
then how many particulars are in our day becoming known of that part
of the world concerning which Messer Marco has written, I have deemed
it reasonable to publish his book, with the aid of several copies written
(as I judge) more than 200 years ago, in a perfectly accurate form, and
one vastly more faithful than that in which it has been heretofore read.
And thus the world shall not lose the fruit that may be gathered from so
much diligence and industry expended upon so honourable a branch of
knowledge."

Ramusio vindicates Polo's Geography.

4. Ramusio, then, after a brief apologetic parallel of the
marvels related by Polo with those related by the Ancients
and by the modern discoverers in the West, such as Columbus
and Cortes, proceeds :—

"And often in my own mind, comparing the land explorations of these
our Venetian gentlemen with the sea explorations of the aforesaid Signor
Don Christopher, I have asked myself which of the two were
really the more marvellous. And if patriotic prejudice delude
me not, methinks good reason might be adduced for setting the
land journey above the sea voyage. Consider only what a
height of courage was needed to undertake and carry through so difficult
an enterprise, over a route of such desperate length and hardship,
whereon it was sometimes necessary to carry food for the supply of man
and beast, not for days only but for months together. Columbus, on the
other hand, going by sea, readily carried with him all necessary provision ;
and after a voyage of some 30 or 40 days was conveyed by the wind
whither he desired to go, whilst the Venetians again took a whole year's
time to pass all those great deserts and mighty rivers. Indeed that the
difficulty of travelling to Cathay was so much greater than that of reaching the New World, and the route so much longer and more perilous, may
be gathered from the fact that, since those gentlemen twice made this

Ramusio compares Polo with Columbus.

---

* The Geography of De Barros, from which this is quoted, has never been
printed. I can find nothing corresponding to this passage in the Decades.

journey, no one from Europe has dared to repeat it,* whereas in the very
year following the discovery of the Western Indies many ships immediately retraced the voyage thither, and up to the present day continue to
do so, habitually and in countless numbers. Indeed those regions are
now so well known, and so thronged by commerce, that the traffic between
Italy, Spain, and England is not greater."

5. Ramusio goes on to explain the light regarding the first
part or prologue of Marco Polo's book that he had derived
from a recent piece of luck which had made him
partially acquainted with the geography of Abulfeda,
and to make a running commentary on the whole
of the preliminary narrative until the final return of the
travellers to Venice :—

Recounts a
tradition of
the travel-
lers' return
to Venice.

"And when they got thither the same fate befel them as befel Ulysses,
who, when he returned, after his twenty years' wanderings, to his native
Ithaca, was recognized by nobody. Thus also those three gentlemen
who had been so many years absent from their native city were recognized by none of their kinsfolk, who were under the firm belief that they
had all been dead for many a year past, as indeed had been reported.
Through the long duration and the hardships of their journeys, and
through the many worries and anxieties that they had undergone, they
were quite changed in aspect, and had got a certain indescribable smack
of the Tartar both in air and accent, having indeed all but forgotten their
Venetian tongue. Their clothes too were coarse and shabby, and of a
Tartar cut. They proceeded on their arrival to their house in this city in
the confine of St. John Chrysostom, where you may see it to this day.
The house, which was in those days a very lofty and handsome palazzo,
is now known by the name of the *Corte del Millioni* for a reason that I
will tell you presently. Going thither they found it occupied by some of
their relatives, and they had the greatest difficulty in making the latter
understand who they should be. For these good people, seeing them to
be in countenance so unlike what they used to be, and in dress so shabby,
flatly refused to believe that they were those very gentlemen of the Ca'
Polo whom they had been looking upon for ever so many years as among
the dead.† So these three gentlemen,—this is a story I have often heard
when I was a youngster from the illustrious Messer GASPARO MALPIERO,
a gentleman of very great age, and a Senator of eminent virtue and
integrity, whose house was on the Canal of Santa Marina, exactly at the
corner over the mouth of the Rio di S. Giovanni Chrisostomo, and just
midway among the buildings of the aforesaid Corte del Millioni, and he
said he had heard the story from his own father and grandfather, and
from other old men among the neighbours,—the three gentlemen, I say,
devised a scheme by which they should at once bring about their recog-

* A grievous error of Ramusio's.
† See the decorated title-page of this volume for an attempt to realise the scene.

nition by their relatives, and secure the honourable notice of the whole city ; and this was it :—

"They invited a number of their kindred to an entertainment, which they took care to have prepared with great state and splendour in that house of theirs ; and when the hour arrived for sitting down to table they came forth of their chamber all three clothed in crimson satin, fashioned in long robes reaching to the ground such as people in those days wore within doors. And when water for the hands had been served, and the guests were set, they took off those robes and put on others of crimson damask, whilst the first suits were by their orders cut up and divided among the servants. Then after partaking of some of the dishes they went out again and came back in robes of crimson velvet, and when they had again taken their seats, the second suits were divided as before. When dinner was over they did the like with the robes of velvet, after they had put on dresses of the ordinary fashion worn by the rest of the company.* These proceedings caused much wonder and amazement among the guests. But when the cloth had been drawn, and all the servants had been ordered to retire from the dining hall, Messer Marco, as the youngest of the three, rose from table, and, going into another chamber, brought forth the three shabby dresses of coarse stuff which they had worn when they first arrived. Straightway they took sharp knives and began to rip up some of the seams and welts, and to take out of them jewels of the greatest value in vast quantities, such as rubies, sapphires, carbuncles, diamonds and emeralds, which had all been stitched up in those dresses in so artful a fashion that nobody could have suspected the fact. For when they took leave of the Great Can they had changed all the wealth that he had bestowed upon them into this mass of rubies, emeralds, and other jewels, being well aware of the impossibility of carrying with them so great an amount in gold over a journey of such extreme length and difficulty. Now this exhibition of such a huge treasure of jewels and precious stones, all tumbled out upon the table, threw the guests into fresh amazement, insomuch that they seemed quite bewildered and dumbfounded. And now they recognized that in spite of all former doubts these were in truth those honoured and worthy gentlemen of the Ca' Polo that they claimed to be ; and so all paid them the greatest honour and reverence. And when the story got wind in Venice, straightway the whole city, gentle and simple, flocked to the house to embrace them, and to make much of them, with every conceivable demonstration of affection and respect. On Messer Maffio, who was the eldest, they conferred the honours of an office that was of great dignity in those days ; whilst the young men came daily to visit and converse with the ever polite and gracious Messer Marco, and to ask him questions about Cathay and the Great Can, all which he answered with such kindly courtesy that every man felt himself in a manner his debtor. And as it happened that in the story, which he was constantly called on to repeat, of the magnificence of the Great Can, he would speak of his revenues as

---

* At first sight this fantastic tradition seems to have little verisimilitude ; but when we regard it in the light of genuine Mongol custom, such as is quoted from Rubruquis, at p. 389 of this volume, we shall be disposed to look on the whole story with respect.

amounting to ten or fifteen *millions* of gold ; and in like manner, when recounting other instances of great wealth in those parts, would always make use of the term *millions*, so they gave him the nickname of MESSER MARCO MILLIONI : a thing which I have noted also in the Public Books of this Republic where mention is made of him.* The Court of his House, too, at S. Giovanni Chrisostomo, has always from that time been popularly known as the Court of the Millioni.

6. " Not many months after the arrival of the travellers at Venice, news came that LAMPA DORIA, Captain of the Genoese Fleet, had advanced with 70 galleys to the Island of Curzola, upon which orders were issued by the Prince of the Most Illustrious Signory for the arming of 90 galleys with

Recounts all the expedition possible, and Messer Marco Polo for his valour
Marco's cap- was put in charge of one of these. So he with the others, under
ture by the the command of the Most Illustrious MESSER ANDREA DAN-
Genoese.

DOLO, Procurator of St. Mark's, as Captain General, a very brave and worthy gentleman, set out in search of the Genoese Fleet. They fought on the September feast of Our Lady, and, as is the common hazard of war, our fleet was beaten, and Polo was made prisoner. For, having pressed on in the vanguard of the attack, and fighting with high and worthy courage in defence of his country and his kindred, he did not receive due support, and being wounded, he was taken, along with Dandolo, and immediately put in irons and sent to Genoa.

" When his rare qualities and marvellous travels became known there, the whole city gathered to see him and to speak with him, and he was no longer entreated as a prisoner but as a dear friend and honoured gentleman. Indeed they showed him such honour and affection that at all hours of the day he was visited by the noblest gentlemen of the city, and was continually receiving presents of every useful kind. Messer Marco finding himself in this position, and witnessing the general eagerness to hear all about Cathay and the Great Can, which indeed compelled him daily to repeat his story till he was weary, was advised to put the matter in writing. So having found means to get a letter written to his father here at Venice, in which he desired the latter to send the notes and memoranda which he had brought home with him, after the receipt of these, and assisted by a Genoese gentleman, who was a great friend of his, and who took great delight in learning about the various regions of the world, and used on that account to spend many hours daily in the prison with him, he wrote this present book (to please him) in the Latin tongue.

"To this day the Genoese for the most part write what they have to write in that language, for there is no possibility of expressing their natural dialect with the pen.† Thus then it came to pass that the Book was put forth at first by Messer Marco in Latin ; but as many copies were taken, and as it was rendered into our vulgar tongue, all Italy became filled with it, so much was this story desired and run after.

---

* This curious statement is confirmed by a passage in the records of the Great Council, which, on a late visit to Venice, I was enabled to extract, through an obliging communication from Professor Minotto. (See below, p. 67.)

† This rather preposterous skit at the Genoese dialect naturally excites a remon-strance from the Abate Spotorno. (*Storia Letteraria della Liguria*, II. 217.)

7. "The captivity of Messer Marco greatly disturbed the minds of Messer Maffio and his father Messer Nicolo. They had decided, whilst still on their travels, that Marco should marry as soon as they should get to Venice ; but now they found themselves in this unlucky pass, with so much wealth and nobody to inherit it. Fearing that Marco's imprisonment might endure for many years, or, worse still, that he might not live to quit it (for many assured them that numbers of Venetian prisoners had been kept in Genoa a score of years before obtaining liberty) ; seeing too no prospect of being able to ransom him,—a thing which they had attempted often and by various channels,—they took counsel together, and came to the conclusion that Messer Nicolo, who, old as he was, was still hale and vigorous, should take to himself a new wife. This he did ; and at the end of four years he found himself the father of three sons, Stefano, Maffio, and Giovanni. Not many years after, Messer Marco aforesaid, through the great favour that he had acquired in the eyes of the first gentlemen of Genoa, and indeed of the whole city, was discharged from prison and set free. Returning home he found that his father had in the meantime had those three other sons. Instead of taking this amiss, wise and discreet man that he was, he agreed also to take a wife of his own. He did so accordingly, but he never had any son, only two girls, one called Moreta and the other Fantina.

"When at a later date his father died, like a good and dutiful son he caused to be erected for him a tomb of very honourable kind for those days, being a great sarcophagus cut from the solid stone, which to this day may be seen under the portico before the Church of S. Lorenzo in this city, on the right hand as you enter, with an inscription denoting it to be the tomb of Messer Nicolo Polo of the contrada of S. Gio. Chrisostomo. The arms of his family consist of a *Bend* with three birds on it, and the colours, according to certain books of old histories in which you see all the coats of the gentlemen of this city emblazoned, are the field *azure*, the bend *argent*, and the three birds *sable*. These last are birds of that kind vulgarly termed *Pole*,* or, as the Latins call them, *Gracculi*.

8. "As regards the after duration of this noble and worthy family, I

---

* *Jackdaws*, I believe, in spite of some doubt from the imbecility of ordinary dictionaries in such matters.

They are under this name made the object of a similitude by Dante (surely a most unhappy one) in reference to the resplendent spirits flitting on the celestial stairs in the sphere of Saturn :—

<blockquote>
" E come per lo natural costume
   Le *Pole* insieme, al cominciar del giorno,
   Si muovono a scaldar le fredde piume :
Poi altre vanno vià senza ritorno,
   Altre rivolgon sè, onde son mosse,
   Ed altre roteando fan soggiorno."—*Parad.* XXI. 34.
</blockquote>

There is some difference among authorities as to the details of the Polo blazon. According to a MS. concerning the genealogies of Venetian families written by Marco Barbaro in 1566, and of which there is a copy in the Museo Civico, the field is *gules*, the bend *or*. And this I have followed in the cut. But a note by S. Stefani

Arms of the Polo.[1]

---

[1] [This coat of arms is reproduced from the Genealogies of Priuli, Archivio di Stato, Venice.—H. C.]

find that Messer Andrea Polo of San Felice had three sons, the first of
whom was Messer Marco, the second Maffio, the third Nicolo.
The two last were those who went to Constantinople first, and
afterwards to Cathay, as has been seen.   Messer Marco the elder
being dead, the wife of Messer Nicolo who had been left at home
with child, gave birth to a son, to whom she gave the name of
Marco in memory of the deceased, and this is the Author of our Book.   Of
the brothers who were born from his father's second marriage, viz. Stephen,
John, and Matthew, I do not find that any of them had children, except
Matthew.   He had five sons and one daughter called Maria ; and she, after
the death of her brothers without offspring, inherited in 1417 all the property of her father and her brothers.   She was honourably married to
Messer AZZO TREVISANO of the parish of Santo Stazio in this city, and
from her sprung the fortunate and honoured stock of the Illustrious Messer
DOMENICO TREVISANO, Procurator of St. Mark's, and valorous Captain
General of the Sea Forces of the Republic, whose virtue and singular good
qualities are represented with augmentation in the person of the Most
Illustrious Prince Ser MARC' ANTONIO TREVISANO, his son.*

(margin) Ramusio's account of the Family Polo and its termination.

" Such has been the history of this noble family of the Ca' Polo, which
lasted as we see till the year of our Redemption 1417, in which year died
childless Marco Polo, the last of the five sons of Maffeo, and so it came to
an end.   Such be the chances and changes of human affairs ! "

Arms of the Ca' Polo.

II. SKETCH OF THE STATE OF THE EAST AT THE TIME OF THE JOURNEYS OF THE POLO FAMILY.

9. The story of the travels of the Polo family opens in
1260.

Christendom had recovered from the alarm into which it had

_____

of Venice, with which I have been favoured since the cut was made, informs me that
a fine 15th-century MS. in his possession gives the field as *argent*, with no *bend*,
and the three birds *sable* with beaks *gules*, disposed thus ****.

* Marco Antonio Trevisano was elected Doge, 4th June, 1553, but died on the
31st of May following.   We do not here notice Ramusio's numerous errors, which will
be corrected in the sequel.   [See p. *78.*]

been thrown some 18 years before when the Tartar cataclysm had threatened to engulph it. The Tartars State of the
themselves were already becoming an object of curi- Levant.
osity rather than of fear, and soon became an object of hope, as
a possible help against the old Mahomedan foe. The frail
Latin throne in Constantinople was still standing, but tottering
to its fall. The successors of the Crusaders still held the Coast
of Syria from Antioch to Jaffa, though a deadlier brood of
enemies than they had yet encountered was now coming to
maturity in the Dynasty of the Mamelukes, which had one
foot firmly planted in Cairo, the other in Damascus. The
jealousies of the commercial republics of Italy were daily waxing
greater. The position of Genoese trade on the coasts of the
Aegean was greatly depressed, through the predominance which
Venice had acquired there by her part in the expulsion of the
Greek Emperors, and which won for the Doge the lofty style of
Lord of Three-Eighths of the Empire of Romania. But Genoa
was biding her time for an early revenge, and year by year her
naval strength and skill were increasing. Both these republics
held possessions and establishments in the ports of Syria, which
were often the scene of sanguinary conflicts between their
citizens. Alexandria was still largely frequented in the
intervals of war as the great emporium of Indian wares, but the
facilities afforded by the Mongol conquerors who now held the
whole tract from the Persian Gulf to the shores of the Caspian
and of the Black Sea, or nearly so, were beginning to give a
great advantage to the caravan routes which debouched at the
ports of Cilician Armenia in the Mediterranean and at Trebizond
on the Euxine. Tana (or Azov) had not as yet become the
outlet of a similar traffic; the Venetians had apparently
frequented to some extent the coast of the Crimea for local
trade, but their rivals appear to have been in great measure
excluded from this commerce, and the Genoese establishments
which so long flourished on that coast, are first heard of some
years after a Greek dynasty was again in possession of
Constantinople.*

10. In Asia and Eastern Europe scarcely a dog might bark
without Mongol leave, from the borders of Poland and the Gulf

---

* See Heyd, *Le Colonie Commerciali degli Italiani*, etc., passim.

of Scanderoon to the Amur and the Yellow Sea. The

The various Mongol Sovereignties in Asia and Eastern Europe. vast empire which Chinghiz had conquered still owned a nominally supreme head in the Great Kaan,* but practically it was splitting up into several great monarchies under the descendants of the four sons of Chinghiz, Juji, Chaghatai, Okkodai, and Tuli; and wars on a vast scale were already brewing between them. Hulaku, third son of Tuli, and brother of two Great Kaans, Mangku and Kúblái, had become practically independent as ruler of Persia, Babylonia, Mesopotamia, and Armenia, though he and his sons, and his sons' sons, continued to stamp the name of the Great Kaan upon their coins, and to use the Chinese seals of state which he bestowed upon them. The Seljukian Sultans of Iconium, whose dominion bore the proud title of Rúm (Rome), were now but the struggling bondsmen of the Ilkhans. The Armenian

---

* We endeavour to preserve throughout the book the distinction   at was made in the age of the Mongol Empire between *Khán* and *Kaán* (ﺧﺎﻥ and ﻗﺎﺁﻥ, as written by Arabic and Persian authors). The former may be rendered *Lord*, and was applied generally to Tartar chiefs whether sovereign or not; it has since become in Persia, and especially in Afghanistan, a sort of "Esq.," and in India is now a common affix in the names of (Musulman) Hindustanis of all classes; in Turkey alone it has been reserved for the Sultan. *Kaán*, again, appears to be a form of *Khákán*, the Χαγάνος of the Byzantine historians, and was the peculiar title of the supreme sovereign of the Mongols; the Mongol princes of Persia, Chaghatai, etc., were entitled only to the former affix (Khán), though *Kaán* and *Khakán* are sometimes applied to them in adulation. Polo always writes *Kaan* as applied to the Great Khan, and does not, I think, use *Khan* in any form, styling the subordinate princes by their name only, as *Argon, Alau,* etc. *Ilkhan* was a special title assumed by Hulátu and his successors in Persia; it is said to be compounded from a word *Il*, signifying tribe or nation. The relation between *Khán* and *Khakán* seems to be probably that the latter signifies "*Khán of Kháns*," Lord of Lords. Chinghiz, it is said, did not take the higher title; it was first assumed by his son Okkodai. But there are doubts about this. (See *Quatremère's Rashid*, pp. 10 *seqq.*, and *Pavet de Courteille, Dict. Turk-Oriental.*) The tendency of swelling titles is always to degenerate, and when the value of Khan had sunk, a new form, *Khán-khánán*, was devised at the Court of Delhi, and applied to one of the high officers of state.

[Mr. Rockhill writes (*Rubruck*, p. 108, note): "The title *Khan*, though of very great antiquity, was only used by the Turks after A.D. 560, at which time the use of the word *Khatun* came in use for the wives of the Khan, who himself was termed *Ilkhan*. The older title of *Shan-yü* did not, however, completely disappear among them, for Albiruni says that in his time the chief of the Ghuz Turks, or Turkomans, still bore the title of *Jenuyeh*, which Sir Henry Rawlinson (*Proc. R. G. S.*, v. 15) takes to be the same word as that transcribed *Shan-yü* by the Chinese (see *Ch'ien Han shu*, Bk. 94, and *Chou shu*, Bk. 50, 2). Although the word *Khakhan* occurs in Menander's account of the embassy of Zemarchus, the earliest mention I have found of it in a Western writer is in the *Chronicon* of Albericus Trium Fontium, where (571), under the year 1239, he uses it in the form *Cacanus*."—Cf. *Terrien de Lacouperie, Khan, Khakan, and other Tartar Titles.* Lond., Dec. 1888.—H. C.]

Hayton in his Cilician Kingdom had pledged a more frank allegiance to the Tartar, the enemy of his Moslem enemies.

Barka, son of Juji, the first ruling prince of the House of Chinghiz to turn Mahomedan, reigned on the steppes of the Volga, where a standing camp, which eventually became a great city under the name of Sarai, had been established by his brother and predecessor Batu.

The House of Chaghatai had settled upon the pastures of the Ili and the valley of the Jaxartes, and ruled the wealthy cities of Sogdiana.

Kaidu, the grandson of Okkodai who had been the successor of Chinghiz in the Kaanship, refused to acknowledge the transfer of the supreme authority to the House of Tuli, and was through the long life of Kúblái a thorn in his side, perpetually keeping his north-western frontier in alarm. His immediate authority was exercised over some part of what we should now call Eastern Turkestan and Southern Central Siberia; whilst his hordes of horsemen, force of character, and close neighbourhood brought the Khans of Chaghatai under his influence, and they generally acted in concert with him.

The chief throne of the Mongol Empire had just been ascended by Kúblái, the most able of its occupants after the Founder. Before the death of his brother and predecessor Mangku, who died in 1259 before an obscure fortress of Western China, it had been intended to remove the seat of government from Kara Korum on the northern verge of the Mongolian Desert to the more populous regions that had been conquered in the further East, and this step, which in the end converted the Mongol Kaan into a Chinese Emperor,* was carried out by Kúblái.

11. For about three centuries the Northern provinces of China had been detached from native rule, and subject to foreign dynasties; first to the *Khitan*, a people from China. the basin of the Sungari River, and supposed (but doubtfully) to have been akin to the Tunguses, whose rule subsisted for 200 years, and originated the name of KHITAI, Khata, or CATHAY, by which for nearly 1000 years China has been known to the nations of Inner Asia, and to those

---

* "China is a sea that salts all the rivers that flow into it."—*P. Parrenin* in *Lett. Édif.* XXIV. 58.

whose acquaintance with it was got by that channel.* The Khitan, whose dynasty is known in Chinese history as the *Liao* or " Iron," had been displaced in 1123 by the Chúrchés or Niu-chen, another race of Eastern Tartary, of the same blood as the modern Manchus, whose Emperors in their brief period of prosperity were known by the Chinese name of Tai-*Kin,* by the Mongol name of the *Altun* Kaans, both signifying " Golden." Already in the lifetime of Chinghiz himself the northern Provinces of China Proper, including their capital, known as Chung-tu or Yen-King, now Peking, had been wrenched from them, and the conquest of the dynasty was completed by Chinghiz's successor Okkodai in 1234.

Southern China still remained in the hands of the native dynasty of the Sung, who had their capital at the great city now well known as Hang-chau fu. Their dominion was still substantially untouched, but its subjugation was a task to which Kúblái before many years turned his attention, and which became the most prominent event of his reign.

12. In India the most powerful sovereign was the Sultan of Delhi, Nassir-uddin Mahmud of the Turki House of Iltit-

India, and Indo-China.

mish ;† but, though both Sind and Bengal acknowledged his supremacy, no part of Peninsular India had yet been invaded, and throughout the long period of our Traveller's residence in the East the Kings of Delhi had their hands too full, owing to the incessant incursions of the Mongols across the Indus, to venture on extensive campaigning in the south. Hence the Dravidian Kingdoms of Southern India were as yet untouched by foreign conquest, and the accumulated gold of ages lay in their temples and treasuries, an easy prey for the coming invader.

In the Indo-Chinese Peninsula and the Eastern Islands a variety of kingdoms and dynasties were expanding and contracting, of which we have at best but dim and shifting glimpses. That they were advanced in wealth and art, far

---

* *E.g.*, the Russians still call it Khitai. The pair of names, *Khitai* and *Machin*, or Cathay and China, is analogous to the other pair, *Seres* and *Sinae. Seres* was the name of the great nation in the far East as known by land, *Sinae* as known by sea ; and they were often supposed to be diverse, just as Cathay and China were afterwards.

† There has been much doubt about the true form of this name. *Iltitmish* is that sanctioned by Mr. Blochmann (see *Proc. As. Soc. Bengal*, 1870, p. 181).

beyond what the present state of those regions would suggest, is attested by vast and magnificent remains of Architecture, nearly all dating, so far as dates can be ascertained, from the 12th to the 14th centuries (that epoch during which an architectural afflatus seems to have descended on the human race), and which are found at intervals over both the Indo-Chinese continent and the Islands, as at Pagán in Burma, at Ayuthia in Siam, at Angkor in Kamboja, at Borobodor and Brambánan in Java. All these remains are deeply marked by Hindu influence, and, at the same time, by strong peculiarities, both generic and individual.

Autograph of Hayton, King of Armenia, *circa* A.D. 1243.

". . . e por so qui cestes lettres soient fermes e establis ci abuns escrit l'escrit de notre main bermoil e sapelé de notre ceau pendant . . . ."

III. THE POLO FAMILY. PERSONAL HISTORY OF THE TRAVELLERS DOWN TO THEIR FINAL RETURN FROM THE EAST.

13. In days when History and Genealogy were allowed to draw largely on the imagination for the *origines* of states and families, it was set down by one Venetian Antiquary that among the companions of King Venetus, or of Prince Antenor of Troy, when they settled on the northern shores of the Adriatic, there was one LUCIUS POLUS, who became the progenitor of our Traveller's Family ;* whilst another deduces it from PAOLO the first Doge † (Paulus Lucas Anafestus of Heraclea, A.D. 696).

Alleged origin of the Polos.

---

* *Zurla*, I. 42, quoting a MS. entitled *Petrus Ciera S. R. E. Card. de Origine Venetorum et de Civitate Venetiarum.* Cicogna says he could not find this MS. as it had been carried to England ; and then breaks into a diatribe against foreigners who purchase and carry away such treasures, " not to make a serious study of them, but for mere vain-glory . . . . or in order to write books contradicting the very MSS. that they have bought, and with that dishonesty and untruth which are so notorious ! " (IV. 227.)

† *Campidoglio Veneto* of Cappellari (MS. in St. Mark's Lib.), quoting "the Venetian Annals of Giulio Faroldi."

More trustworthy traditions, recorded among the Family Histories of Venice, but still no more it is believed than traditions, represent the Family of Polo as having come from Sebenico in Dalmatia, in the 11th century.* Before the end of the century they had taken seats in the Great Council of the Republic; for the name of Domenico Polo is said to be subscribed to a grant of 1094, that of Pietro Polo to an act of the time of the Doge Domenico Michiele in 1122, and that of a Domenico Polo to an acquittance granted by the Doge Domenico Morosini and his Council in 1153.†

The ascertained genealogy of the Traveller, however, begins only with his grandfather, who lived in the early part of the 13th century.

Two branches of the Polo Family were then recognized, distinguished by the *confini* or Parishes in which they lived, as Polo of S. Geremia, and Polo of S. Felice. ANDREA POLO of S. Felice was the father of three sons, MARCO, NICOLO, and MAFFEO. And Nicolo was the Father of our Marco.

14. Till quite recently it had never been precisely ascertained whether the immediate family of our Traveller belonged to the *Nobles* of Venice properly so called, who had seats in the Great Council and were enrolled in the Libro d'Oro. Ramusio indeed styles our Marco *Nobile* and *Magnifico*, and Rusticiano, the actual scribe of the Traveller's recollections, calls him "*sajes et noble citaiens de Venece*," but Ramusio's accuracy and Rusticiano's precision were scarcely to be depended on. Very recently, however, since the subject has been discussed with accomplished students of the Venice Archives, proofs have been found establishing Marco's personal claim to nobility, inasmuch as both in judicial decisions and in official resolutions of the Great Council, he is designated *Nobilis Vir*, a formula which would never have been used in such documents (I am assured) had he not been technically noble.‡

Claims to be styled noble.

---

* The *Genealogies* of Marco Barbaro specify 1033 as the year of the migration to Venice ; on what authority does not appear (MS. copy in *Museo Civico* at Venice).

† *Cappellari*, u.s., and *Barbaro*. In the same century we find (1125, 1195) indications of Polos at Torcello, and of others (1160) at Equileo, and (1179, 1206) Lido Maggiore ; in 1154 a Marco Polo of Rialto. Contemporary with these is a family of Polos (1139, 1183, 1193, 1201) at Chioggia (*Documents and Lists of Documents from various Archives at* Venice).

‡ See Appendix C, Nos. 4, 5, and 16. It was supposed that an autograph of Marco as member of the Great Council had been discovered, but this proves to be a

15. Of the three sons of Andrea Polo of S. Felice, Marco seems to have been the eldest, and Maffeo the youngest.* They were all engaged in commerce, and apparently Marco the
Elder. in a partnership, which to some extent held good even when the two younger had been many years absent in the Far East.† Marco seems to have been established for a time at Constantinople,‡ and also to have had a house (no doubt of business) at Soldaia, in the Crimea, where his son and daughter, Nicolo and Maroca by name, were living in 1280. This year is the date of the Elder Marco's Will, executed at Venice, and when he was "weighed down by bodily ailment." Whether he survived for any length of time we do not know.

16. Nicolo Polo, the second of the Brothers, had two legitimate sons, MARCO, the Author of our Book, born in 1254,§ and MAFFEO, of whose place in the family we shall Nicolo and
Maffeo com-
mence their
travels. have a few words to say presently. The story opens, as we have said, in 1260, when we find the two brothers, Nicolo and Maffeo the Elder, at Constantinople. How long they had been absent from Venice we are not distinctly told. Nicolo had left his wife there behind him; Maffeo apparently was a bachelor. In the year named they started on a trading venture to the Crimea, whence a succession of openings and chances, recounted in the Introductory chapters of Marco's work, carried them far north along the Volga, and thence first to Bokhara, and then to the Court of the Great Kaan Kúblái in the Far East, on or within the borders of CATHAY. That a great and civilized country so called existed in the extremity of Asia had already been reported in Europe by the Friars Plano Carpini (1246) and William Rubruquis (1253), who had not indeed reached its

---

mistake, as will be explained further on (see p. *74*, note). In those days the demarcation between Patrician and non-Patrician at Venice, where all classes shared in commerce, all were (generally speaking) of one race, and where there were neither castles, domains, nor trains of horsemen, formed no wide gulf. Still it is interesting to establish the verity of the old tradition of Marco's technical nobility.

* Marco's seniority rests only on the assertion of Ramusio, who also calls Maffeo older than Nicolo. But in Marco the Elder's Will these two are always (3 times) specified as "*Nicolaus et Matheus.*"

† This seems implied in the Elder Marco's Will (1280): "*Item de bonis quæ me habere contingunt* de fraternâ Compagniâ *a suprascriptis Nicolao et Matheo Paulo,*" etc.

‡ In his Will he terms himself "Ego Marcus Polo quondam de Constantinopoli."

§ There is no real ground for doubt as to this. All the extant MSS. agree in making Marco fifteen years old when his father returned to Venice in 1269.

frontiers, but had met with its people at the Court of the Great Kaan in Mongolia; whilst the latter of the two with characteristic acumen had seen that they were identical with the Seres of classic fame.

17. Kúblái had never before fallen in with European gentlemen. He was delighted with these Venetians, listened

Their inter-
course with
Kúblái
Kaan. with strong interest to all that they had to tell him of the Latin world, and determined to send them back as his ambassadors to the Pope, accompanied by an officer of his own Court. His letters to the Pope, as the Polos represent them, were mainly to desire the despatch of a large body of educated missionaries to convert his people to Christianity. It is not likely that religious motives influenced Kúblái in this, but he probably desired religious aid in softening and civilizing his rude kinsmen of the Steppes, and judged, from what he saw in the Venetians and heard from them, that Europe could afford such aid of a higher quality than the degenerate Oriental Christians with whom he was familiar, or the Tibetan Lamas on whom his patronage eventually devolved when Rome so deplorably failed to meet his advances.

18. The Brothers arrived at Acre in April,* 1269, and found that no Pope existed, for Clement IV. was dead the

Their return
home, and
Marco's ap-
pearance on
the scene. year before, and no new election had taken place. So they went home to Venice to see how things stood there after their absence of so many years.

The wife of Nicolo was no longer among the living, but he found his son Marco a fine lad of fifteen.

The best and most authentic MSS. tell us no more than this. But one class of copies, consisting of the Latin version made by our Traveller's contemporary, Francesco Pipino, and of the numerous editions based indirectly upon it, represents that Nicolo had left Venice when Marco was as yet unborn, and consequently had never seen him till his return from the East in 1269.†

---

* Baldelli and Lazari say that the Bern MS. specifies 30th April; but this is a mistake.

† Pipino's version runs : "Invenit Dominus Nicolaus Paulus uxorem suam esse defunctam, quae in recessu suo fuit praegnans. Invenitque filium, Marcum nomine, qui jam annos xv. habebat aetatis, qui post discessum ipsius de Venetiis natus fuerat de uxore

We have mentioned that Nicolo Polo had another legitimate son, by name Maffeo, and him we infer to have been younger than Marco, because he is named last (*Marcus et Matheus*) in the Testament of their uncle Marco the Elder. We do not know if they were by the same mother. They could not have been so if we are right in supposing Maffeo to have been the younger, and if Pipino's version of the history be genuine. If however we reject the latter, as I incline to do, no ground remains for supposing that Nicolo went to the East much before we find him there viz., in 1260, and Maffeo may have been born of the same mother during the interval between 1254 and 1260. If on the other hand Pipino's version be held to, we must suppose that Maffeo (who is named by his uncle in 1280, during his father's second absence in the East) was born of a marriage contracted during Nicolo's residence at home after his first journey, a residence which lasted from 1269 to 1271.*

---

sua praefatâ." To this Ramusio adds the further particular that the mother died in giving birth to Mark.

The interpolation is older even than Pipino's version, for we find in the rude Latin published by the Société de Géographie "quam cum Venetiis primo recessit praegnantem dimiserat." But the statement is certainly an *interpolation*, for it does not exist in any of the older texts; nor have we any good reason for believing that it was an *authorised* interpolation. I suspect it to have been introduced to harmonise with an erroneous date for the commencement of the travels of the two brothers.

Lazari prints: "Messer Nicolò trovò che la sua donna era morta, e n'era rimasto un fanciullo di *dodici* anni per nome Marco, *che il padre non avea veduto mai, perchè non era ancor nato quando egli partì.*" These words have no equivalent in the French Texts, but are taken from one of the Italian MSS. in the Magliabecchian Library, and are I suspect also interpolated. The *dodici* is pure error (see p. 21 *infra*).

* The last view is in substance, I find, suggested by Cicogna (ii. 389).

The matter is of some interest, because in the Will of the younger Maffeo, which is extant, he makes a bequest to his uncle (*Avunculus*) Jordan Trevisan. This seems an indication that his mother's name may have been Trevisan. The same Maffeo had a daughter *Fiordelisa*. And Marco the Elder, in his Will (1280), appoints as his executors, during the absence of his brothers, the same Jordan Trevisan and his own sister-in-law *Fiordelisa* ("Jordanum Trivisanum de confinio S. Antonini: et Flordelisam cognatam meam"). Hence I conjecture that this *cognata Fiordelisa* (Trevisan?) was the wife of the absent Nicolo, and the mother of Maffeo. In that case of course Maffeo and Marco were the sons of different mothers. With reference to the above suggestion of Nicolo's second marriage in 1269 there is a curious variation in a fragmentary Venetian Polo in the Barberini Library at Rome. It runs, in the passage corresponding to the latter part of ch. ix. of Prologue: "i qual do fratelli steteno do anni in Veniezia aspettando la elletion de nuovo Papa, *nel qual tempo Mess. Nicolo si tolse moier et si la lasò graveda.*" I believe, however, that it is only a careless misrendering of Pipino's statement about Marco's birth.

19. The Papal interregnum was the longest known, at
least since the dark ages. Those two years passed, and yet

The Piazzetta at Venice. (From the Bodleian MS. of Polo.)

the Cardinals at Viterbo had come to no agreement. The
brothers were unwilling to let the Great Kaan think
them faithless, and perhaps they hankered after the
virgin field of speculation that they had discovered;
so they started again for the East, taking young
Mark with them. At Acre they took counsel with an
eminent churchman, TEDALDO (or Tebaldo) VISCONTI, Arch-

Second
Journey of
the Polo
Brothers,
accompanied
by Marco.

deacon of Liège, whom the Book represents to have been Legate in Syria, and who in any case was a personage of much gravity and influence. From him they got letters to authenticate the causes of the miscarriage of their mission, and started for the further East. But they were still at the port of Ayas on the Gulf of Scanderoon, which was then becoming one of the chief points of arrival and departure for the inland trade of Asia, when they were overtaken by the news that a Pope was at last elected, and that the choice had fallen upon their friend Archdeacon Tedaldo. They immediately returned to Acre, and at last were able to execute the Kaan's commission, and to obtain a reply. But instead of the hundred able teachers of science and religion whom Kúblái is said to have asked for, the new Pope, Gregory X., could supply but two Dominicans; and these lost heart and drew back when they had barely taken the first step of the journey.

Judging from certain indications we conceive it probable that the three Venetians, whose second start from Acre took place about November 1271, proceeded by Ayas and Sivas, and then by Mardin, Mosul, and Baghdad, to Hormuz at the mouth of the Persian Gulf, with the view of going on by sea, but that some obstacle arose which compelled them to abandon this project. and turn north again from Hormuz.* They then

* [Major Sykes, in his remarkable book on *Persia*, ch. xxiii. pp. 262-263, does not share Sir Henry Yule's opinion regarding this itinerary, and he writes :

"To return to our travellers, who started on their second great journey in 1271, Sir Henry Yule, in his introduction,[1] makes them travel *via* Sivas to Mosul and Baghdád, and thence by sea to Hormuz, and this is the itinerary shown on his sketch map. This view I am unwilling to accept for more than one reason. In the first place, if, with Colonel Yule, we suppose that Ser Marco visited Baghdád, is it not unlikely that he should term the River Volga the Tigris,[2] and yet leave the river of Baghdád nameless? It may be urged that Marco believed the legend of the reappearance of the Volga in Kurdistán, but yet, if the text be read with care and the character of the traveller be taken into account, this error is scarcely explicable in any other way, than that he was never there.

"Again, he gives no description of the striking buildings of Baudas, as he terms it, but this is nothing to the inaccuracy of his supposed onward journey. To quote the text, 'A very great river flows through the city, ... and merchants descend some eighteen days from Baudas, and then come to a certain city called Kisi,[3] where they enter the Sea of India.' Surely Marco, had he travelled down the Persian Gulf, would never have given this description of the route, which is so untrue as to point

[1] Page 19.
[2] *Vide Yule*, vol. i. p. 5. It is noticeable that John of Pian de Carpine, who travelled 1245 to 1247, names it correctly.
[3] The modern name is Keis, an island lying off Linga.

traversed successively Kerman and Khorasan, Balkh and
Badakhshan, whence they ascended the Panja or upper Oxus to
the Plateau of Pamir, a route not known to have been since
followed by any European traveller except Benedict Goës, till
the spirited expedition of Lieutenant John Wood of the Indian
Navy in 1838.* Crossing the Pamir highlands the travellers
descended upon Kashgar, whence they proceeded by Yarkand
and Khotan, and the vicinity of Lake Lob, and eventually
across the Great Gobi Desert to Tangut, the name then applied
by Mongols and Persians to territory at the extreme North-west
of China, both within and without the Wall. Skirting the

---

to the conclusion that it was vague information given by some merchant whom he
met in the course of his wanderings.

"Finally, apart from the fact that Baghdád, since its fall, was rather off the main
caravan route, Marco so evidently travels east from Yezd and thence south to
Hormuz, that unless his journey be described backwards, which is highly improbable,
it is only possible to arrive at one conclusion, namely, that the Venetians entered
Persia near Tabriz, and travelled to Sultania, Kashán, and Yezd. Thence they proceeded to Kermán and Hormuz, where, probably fearing the sea voyage, owing to
the manifest unseaworthiness of the ships, which he describes as 'wretched
affairs,' the Khorasán route was finally adopted. Hormuz, in this case, was not
visited again until the return from China, when it seems probable tha the same route
was retraced to Tabriz, where their charge, the Lady Kokachin, 'moult bele dame
et avenant,' was married to Gházan Khán, the son of her fiancé Arghun. It remains
to add that Sir Henry Yule may have finally accepted this view in part, as in the
plate showing *Probable View of Marco Polo's own Geography*,[1] the itinerary is not
shown as running to Baghdád."

I may be allowed to answer that when Marco Polo *started* for the East, Baghdád
was not rather off the main caravan route. The fall of Baghdád was not immediately
followed by its decay, and we have proof of its prosperity at the beginning of the
14th century. Tauris had not yet the importance it had reached when the Polos
visited it on their *return* journey. We have the will of the Venetian Pietro Viglioni,
dated from Tauris, 10th December, 1264 (*Archiv. Veneto*, xxvi. 161-165), which
shows that he was but a pioneer. It was only under Arghún Khan (1284-1291) that
Tauris became the great market for foreign, especially Genoese, merchants, as Marco
Polo remarks on his return journey; with Gházan and the new city built by that
prince, Tauris reached a very high degree of prosperity, and was then really the chief
emporium on the route from Europe to Persia and the far East. Sir Henry Yule
had not changed his views, and if in the plate showing *Probable View of Marco Polo's
own Geography*, the itinerary is not shown as running to Baghdád, it is mere neglect
on the part of the draughtsman.—H. C.]

* It is stated by Neumann that this most estimable traveller once intended to have
devoted a special work to the elucidation of Marco's chapters on the Oxus Provinces,
and it is much to be regretted that this intention was never fulfilled. Pamir has
been explored more extensively and deliberately, whilst this book was going through
the press, by Colonel Gordon, and other officers, detached from Sir Douglas Forsyth's
Mission. [We have made use of the information given by these officers and by more
recent travellers.—H. C.]

---

[1] Vol. i. p. *110* (Introduction).

northern frontier of China they at last reached the presence of
the Kaan, who was at his usual summer retreat at Kai-ping fu,
near the base of the Khingan Mountains, and nearly 100 miles
north of the Great Wall at Kalgan. If there be no mistake in
the time (three years and a half) ascribed to this journey in all
the existing texts, the travellers did not reach the Court till
about May of 1275.*

20. Kúblái received the Venetians with great cordiality,
and took kindly to young Mark, who must have been by this
time one-and-twenty. The *Joenne Bacheler*, as the
story calls him, applied himself to the acquisition of
the languages and written characters in chief use
among the multifarious nationalities included in the
Kaan's Court and administration ; and Kúblái after a time,
seeing his discretion and ability, began to employ him in the
public service. M. Pauthier has found a record in the Chinese
Annals of the Mongol Dynasty, which states that in the year
1277, a certain POLO was nominated a second-class commissioner or agent attached to the Privy Council, a passage
which we are happy to believe to refer to our young traveller.†

His first mission apparently was that which carried him
through the provinces of Shan-si, Shen-si, and Sze-ch'wan, and
the wild country on the East of Tibet, to the remote province of
Yun-nan, called by the Mongols Karájàng, and which had been
partially conquered by an army under Kúblái himself in 1253,
before his accession to the throne.‡ Mark, during his stay at
court, had observed the Kaan's delight in hearing of strange
countries, their marvels, manners, and oddities, and had heard

Marco's employment by Kúblái Kaan ; and his journeys.

---

* Half a year earlier, if we suppose the three years and a half to count from
Venice rather than Acre. But at that season (November) Kúblái would not have
been at Kai-ping fu (otherwise Shang-tu).

† *Pauthier*, p. ix., and p. 361.

‡ That this was Marco's first mission is positively stated in the Ramusian edition ;
and though this may be only an editor's gloss it seems well-founded. The French
texts say only that the Great Kaan, "l'envoia en un message en une terre ou bien
avoit vj. mois de chemin." The traveller's actual Itinerary affords to Vochan
(Yung-ch'ang), on the frontier of Burma, 147 days' journey, which with halts might
well be reckoned six months in round estimate. And we are enabled by various
circumstances to fix the date of the Yun-nan journey between 1277 and 1280. The
former limit is determined by Polo's account of the battle with the Burmese, near
Vochan, which took place according to the Chinese Annals in 1277. The latter is
fixed by his mention of Kúblái's son, Mangalai, as governing at Kenjanfu (Si-ngan fu),
a prince who died in 1280. (See vol. ii. pp. 24, 31, also 64, 80.)

his Majesty's frank expressions of disgust at the stupidity of his
commissioners when they could speak of nothing but the official
business on which they had been sent. Profiting by these
observations, he took care to store his memory or his note-books
with all curious facts that were likely to interest Kúblái, and
related them with vivacity on his return to Court. This first
journey, which led him through a region which is still very
nearly a *terra incognita*, and in which there existed and still
exists, among the deep valleys of the Great Rivers flowing down
from Eastern Tibet, and in the rugged mountain ranges
bordering Yun-nan and Kwei-chau, a vast Ethnological Garden,
as it were, of tribes of various race and in every stage of
uncivilisation, afforded him an acquaintance with many strange
products and eccentric traits of manners, wherewith to delight
the Emperor.

Mark rose rapidly in favour, and often served Kúblái again
on distant missions, as well as in domestic administration, but
we gather few details as to his employments. At one time we
know that he held for three years the government of the great
city of Yang-chau, though we need not try to magnify this office,
as some commentators have done, into the viceroyalty of one of
the great provinces of the Empire; on another occasion we
find him with his uncle Maffeo, passing a year at Kan-chau in
Tangut; again, it would appear, visiting Kara Korum, the old
capital of the Kaans in Mongolia; on another occasion in
Champa or Southern Cochin China; and again, or perhaps as a
part of the last expedition, on a mission to the Indian Seas,
when he appears to have visited several of the southern states of
India. We are not informed whether his father and uncle
shared in such employments;* and the story of their services
rendered to the Kaan in promoting the capture of the city of
Siang-yang, by the construction of powerful engines of attack, is
too much perplexed by difficulties of chronology to be cited
with confidence. Anyhow they were gathering wealth, and
after years of exile they began to dread what might follow old
Kúblái's death, and longed to carry their gear and their own
grey heads safe home to the Lagoons. The aged Emperor

---

* Excepting in the doubtful case of Kan-chau, where one reading says that the
three Polos were there on business of their own not necessary to mention, and
another, that only Maffeo and Marco were there, "*en légation.*"

growled refusal to all their hints, and but for a happy chance we should have lost our mediæval Herodotus.

21. Arghún Khan of Persia, Kúblái's great-nephew, had in 1286 lost his favourite wife the Khatun Bulughán; and, mourning her sorely, took steps to fulfil her dying injunction that her place should be filled only by a lady of her own kin, the Mongol Tribe of Bayaut. Ambassadors were despatched to the Court of Kaan- balígh to seek such a bride. The message was courteously received, and the choice fell on the lady Kokáchin, a maiden of 17, "*moult bele dame et avenant.*" The overland road from Peking to Tabriz was not only of portentous length for such a tender charge, but was imperilled by war, so the envoys desired to return by sea. Tartars in general were strangers to all navigation; and the envoys, much taken with the Venetians, and eager to profit by their experience, especially as Marco had just then returned from his Indian mission, begged the Kaan as a favour to send the three *Firinghis* in their company. He consented with reluctance, but, having done so, fitted the party out nobly for the voyage, charging the Polos with friendly messages for the potentates of Europe, including the King of England. They appear to have sailed from the port of Zayton (as the Westerns called T'swan-chau or Chin-cheu in Fo-kien) in the beginning of 1292. It was an ill-starred voyage, involving long detentions on the coast of Sumatra, and in the South of India, to which, however, we are indebted for some of the best chapters in the book; and two years or upwards passed before they arrived at their destination in Persia.* The three hardy

Circum-stances of the Depar-ture of the Polos from the Kaan's Court.

---

* Persian history seems to fix the arrival of the lady Kokáchin in the North of Persia to the winter of 1293-1294. The voyage to Sumatra occupied three months (vol. i. p. 34); they were five months detained there (ii. 292); and the remainder of the voyage extended to eighteen more (i. 35),—twenty-six months in all.

The data are too slight for unexceptional precision, but the following adjustment will fairly meet the facts. Say that they sailed from Fo-kien in January 1292. In April they would be in Sumatra, and find the S.W. Monsoon too near to admit of their crossing the Bay of Bengal. They remain in port till September (five months), and then proceed, touching (perhaps) at Ceylon, at Kayal, and at several ports of Western India. In one of these, *e.g.* Kayal or Tana, they pass the S.W. Monsoon of 1293, and then proceed to the Gulf. They reach Hormuz in the winter, and the camp of the Persian Prince Gházán, the son of Arghún, in March, twenty-six months from their departure.

I have been unable to trace Hammer's authority (not Wassáf I find), which

Venetians survived all perils, and so did the lady, who had come
to look on them with filial regard ; but two of the three envoys,
and a vast proportion of the suite, had perished by the way.*
Arghún Khan too had been dead even before they quitted
China ; † his brother Kaikhátú reigned in his stead ; and his son
Gházán succeeded to the lady's hand. We are told by one who
knew both the princes well that Arghún was one of the handsomest men of his time, whilst Gházán was, among all his host,
one of the most insignificant in appearance. But in other
respects the lady's change was for the better. Gházán had some
of the highest qualities of a soldier, a legislator and a king,
adorned by many and varied accomplishments; though his reign
was too short for the full development of his fame.

22. The princess, whose enjoyment of her royalty was brief,
wept as she took leave of the kindly and noble Venetians.
They went on to Tabriz, and after a long halt there proceeded
They pass homewards, reaching Venice, according to all the texts
by Persia
to Venice. some time in 1295.‡
Their rela-
tions there. We have related Ramusio's interesting tradition,
like a bit out of the Arabian Nights, of the reception that the
Travellers met with from their relations, and of the means that
they took to establish their position with those relations, and

---

perhaps gives the precise date of the Lady's arrival in Persia (see *infra*, p. 38).
From his narrative, however (*Gesch. der Ilchane*, ii. 20), March 1294 is perhaps too
late a date. But the five months' stoppage in Sumatra *must* have been in the
S.W. Monsoon ; and if the arrival in Persia is put earlier, Polo's numbers can
scarcely be held to. Or, the eighteen months mentioned at vol. i. p. 35, must *include*
the five months' stoppage. We may then suppose that they reached Hormuz about
November 1293, and Gházán's camp a month or two later.

* The French text which forms the *basis* of my translation says that, excluding
mariners, there were 600 souls, out of whom only 8 survived. The older MS. which
I quote as G. T., makes the number 18, a fact that I had overlooked till the sheets
were printed off.

† Died 12th March, 1291.

‡ All dates are found so corrupt that even in this one I do not feel absolute confidence. Marco in dictating the book is aware that Gházán had attained the throne of
Persia (see vol. i. p. 36, and ii. pp. 50 and 477), an event which did not occur till
October, 1295. The date assigned to it, however, by Marco (ii. 477) is 1294, or the
year *before* that assigned to the return home.

The travellers may have stopped some time at Constantinople on their way, or even
may have visited the northern shores of the Black Sea ; otherwise, indeed, how did
Marco acquire his knowledge of that Sea (ii. 486-488) and of events in Kipchak (ii. 496
seqq.)? If 1296 was the date of return, moreover, the six-and-twenty years assigned
in the preamble as the period of Marco's absence (p. 2) would be nearer accuracy.
For he left Venice in the spring or summer of 1271.

with Venetian society.*   Of the relations, Marco the Elder had probably been long dead ; † Maffeo the brother of our Marco was alive, and we hear also of a cousin (*consanguineus*) Felice Polo, and his wife Fiordelisa, without being able to fix their precise position in the family.   We know also that Nicolo, who died before the end of the century, left behind him two illegitimate sons, Stefano and Zannino.   It is not unlikely that these were born from some connection entered into during the long

---

* Marco Barbaro, in his account of the Polo family, tells what seems to be the same tradition in a different and more mythical version :—

"From ear to ear the story has past till it reached mine, that when the three Kinsmen arrived at their home they were dressed in the most shabby and sordid manner, insomuch that the wife of one of them gave away to a beggar that came to the door one of those garments of his, all torn, patched, and dirty as it was. The next day he asked his wife for that mantle of his, in order to put away the jewels that were sewn up in it ; but she told him she had given it away to a poor man, whom she did not know. Now, the stratagem he employed to recover it was this. He went to the Bridge of Rialto, and stood there turning a wheel, to no apparent purpose, but as if he were a madman, and to all those who crowded round to see what prank was this, and asked him why he did it, he answered : 'He'll come if God pleases.' So after two or three days he recognised his old coat on the back of one of those who came to stare at his mad proceedings, and got it back again. Then, indeed, he was judged to be quite the reverse of a madman ! And from those jewels he built in the contrada of S. Giovanni Grisostomo a very fine palace for those days ; and the family got among the vulgar the name of the *Ca' Million*, because the report was that they had jewels to the value of a million of ducats ; and the palace has kept that name to the present day—*viz.*, 1566." (*Genealogies*, MS. copy in *Museo Civico ;* quoted also by *Baldelli Boni, Vita*, p. xxxi.)

† The Will of the Elder Marco, to which we have several times referred, is dated at Rialto 5th August, 1280.

The testator describes himself as formerly of Constantinople, but now dwelling in the confine of S. Severo.

His brothers *Nicolo* and *Maffeo*, if at Venice, are to be his sole trustees and executors, but in case of their continued absence he nominates *Jordano Trevisano*, and his sister-in-law *Fiordelisa* of the confine of S. Severo.

The proper tithe to be paid. All his clothes and furniture to be sold, and from the proceeds his funeral to be defrayed, and the balance to purchase masses for his soul at the discretion of his trustees.

Particulars of money due to him from his partnership with Donato Grasso, now of Justinople (Capo d'Istria), 1200 *lire* in all. (Fifty-two lire due by said partnership to Angelo di Tumba of S. Severo.)

The above money bequeathed to his son *Nicolo*, living at *Soldachia*, or failing him, to his beloved brothers *Nicolo* and *Maffeo*. Failing them, to the sons of his said brothers (*sic*) *Marco* and *Maffeo*. Failing them, to be spent for the good of his soul at the discretion of his trustees.

To his son Nicolo he bequeaths a silver-wrought girdle of vermilion silk, two silver spoons, a silver cup without cover (or saucer? *sine cembalo*), his desk, two pairs of sheets, a velvet quilt, a counterpane, a feather-bed—all on the same conditions as above, and to remain with the trustees till his son returns to Venice.

Meanwhile the trustees are to invest the money at his son's risk and benefit, but only here in Venice (*investiant seu investire faciant*).

residence of the Polos in Cathay, though naturally their presence in the travelling company is not commemorated in Marco's Prologue.*

IV. DIGRESSION CONCERNING THE MANSION OF THE POLO FAMILY AT VENICE.

23. We have seen that Ramusio places the scene of the story recently alluded to at the mansion in the parish of S. Giovanni Grisostomo, the court of which was known in his time as the Corte del Millioni; and indeed he speaks of

Probable period of their establishment at S. Giovanni Grisostomo. the Travellers as at once on their arrival resorting to that mansion as their family residence. Ramusio's details have so often proved erroneous that I should not be surprised if this also should be a mistake. At least we find (so far as I can learn) no previous intimation that the family were connected with that locality. The grand-father Andrea is styled of *San Felice*. The will of Maffeo Polo the younger, made in 1300, which we shall give hereafter in abstract, appears to be the first document that connects the family with S. Giovanni Grisostomo. It indeed styles the testator's father "the late Nicolo Paulo of the confine of St. John Chrysostom," but that only shows what is not disputed, that the Travellers after their return from the East settled in this locality. And the same will appears to indicate a surviving connexion with S. Felice, for the priests and clerks who drew it up and witness it are all of the church of S. Felice, and it is to the parson of S. Felice and his successor that Maffeo bequeaths an annuity to procure their prayers for the souls of

---

From the proceeds to come in from his partnership with his brothers Nicolo and Maffeo, he bequeaths 200 lire to his daughter Maroca.

From same source 100 lire to his natural son Antony.

Has in his desk (*capsella*) two hyperperae (Byzantine gold coins), and three golden florins, which he bequeaths to the sister-in-law *Fiordelisa*.

Gives freedom to all his slaves and handmaidens.

Leaves his house in Soldachia to the Minor Friars of that place, reserving life-occupancy to his son Nicolo and daughter Maroca.

The rest of his goods to his son Nicolo.

* The terms in which the younger Maffeo mentions these half-brothers in his Will (1300) seem to indicate that they were still young.

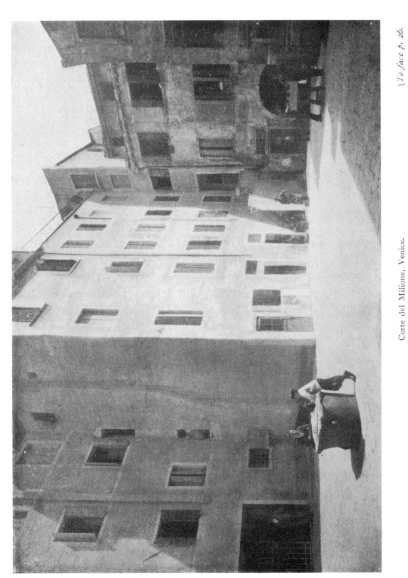

Corte del Milione, Venice.

Vo face p. 26.

his father, his mother, and himself, though after the successor
the annuity is to pass on the same condition to the senior
priest of S. Giovanni Grisostomo. Marco Polo the Elder is
in his will described as of *S. Severo*, as is also his sister-in-
law Fiordelisa, and the document contains no reference to
S. Giovanni. On the whole therefore it seems probable that
the Palazzo in the latter parish was purchased by the Travellers after their return from the East.*

24. The Court which was known in the 16th century as the
Corte del Millioni has been generally understood to be that now
known as the Corte Sabbionera, and here is still pointed
out a relic of Marco Polo's mansion. [Indeed it is
called now (1899) *Corte del Milione;* see p. *30.*—H. C.] *Relic of the Casa Polo in the Corte Sabbionera.*

M. Pauthier's edition is embellished with a good engraving
which purports to represent the House of Marco Polo. But
he has been misled. His engraving in fact exhibits, at
least as the prominent feature, an embellished representation
of a small house which exists on the *west side* of the Sabbionera,
and which had at one time perhaps that pointed style of
architecture which his engraving shows, though its present
decoration is paltry and unreal. But it is on the *north side*
of the Court, and on the foundations now occupied by the
Malibran theatre, that Venetian tradition and the investigations
of Venetian antiquaries concur in indicating the site of the
Casa Polo. At the end of the 16th century a great fire
destroyed the Palazzo,† and under the description of "an old

---

* Marco Barbaro's story related at p. *25* speaks of the Ca' Million as *built* by the travellers.

From a list of parchments existing in the archives of the *Casa di Ricovero,* or Great Poor House, at Venice, Comm. Berchet obtained the following indication :—

" *No.* 94. *Marco Galetti invests* Marco Polo *S. of* Nicolo *with the ownership of his possessions* (beni) *in* S. Giovanni Grisostomo ; 10 *September,* 1319 ; *drawn up by the Notary Nicolo, priest of S. Canciano.*"

This document would perhaps have thrown light on the matter, but unfortunately recent search by several parties has failed to trace it. [The document has been dis-covered since : see vol. ii., *Calendar,* No. 6.—H. C.]

† ——-" Sua casa che era posta nel confin di S. Giovanni Chrisostomo, *che hor fà l'anno s'abbrugiò totalmente,* con gran danno di molti." (*Doglioni, Hist. Venetiana,* Ven. 1598, pp. 161-162.)

" 1596. 7 *Nov. Senato* (Arsenal . . . . ix c. 159 t).

" Essendo conveniente usar qualche ricognizione a quelli della maestranza del-l'Arsenal nostro, che prontamente sono concorsi all' incendio occorso ultimamente a S. Zuane Grizostomo nelli stabeli detti di CA' MILION dove per la relazion fatta nell collegio nostro dalli patroni di esso Arsenal hanno nell' estinguere il foco prestato ogni buon servitio. . . ."—(Comm. by Cav. Cecchetti through Comm. Berchet.)

mansion ruined from the foundation" it passed into the hands
of one Stefano Vecchia, who sold it in 1678 to Giovanni
Carlo Grimani. He built on the site of the ruins a theatre
which was in its day one of the largest in Italy, and was
called the Theatre of S. Giovanni Grisostomo; afterwards
the *Teatro Emeronitio*. When modernized in our own day the
proprietors gave it the name of Malibran, in honour of that
famous singer, and this it still bears.*

[In 1881, the year of the Venice International Geographical
Congress, a Tablet was put up on the Theatre with the
following inscription :—

QVI FURONO LE CASE

DI

MARCO POLO

CHE VIAGGIÒ LE PIÙ LONTANE REGIONI DELL' ASIA

E LE DESCRISSE

———

PER DECRETO DEL COMUNE

MDCCCLXXXI].

There is still to be seen on the north side of the Court an
arched doorway in Italo-Byzantine style, richly sculptured
with scrolls, disks, and symbolical animals, and on the wall
above the doorway is a cross similarly ornamented.† The
style and the decorations are those which were usual in
Venice in the 13th century. The arch opens into a passage
from which a similar doorway at the other end, also retaining
some scantier relics of decoration, leads to the entrance of the
Malibran Theatre. Over the archway in the Corte Sabbionera
the building rises into a kind of tower. This, as well as the
sculptured arches and cross, Signor Casoni, who gave a good
deal of consideration to the subject, believed to be a relic of
the old Polo House. But the tower (which Pauthier's view
does show) is now entirely modernized.‡

Other remains of Byzantine sculpture, which are probably

---

* See a paper by G. C. (the Engineer Giovanni Casoni) in *Teatro Emeronitio,
Almanacco per l'Anno* 1835.

† This Cross is engraved by Mr. Ruskin in vol. ii. of the *Stones of Venice* : see
p. 139, and Pl. xi. Fig. 4.

‡ Casoni's only doubt was whether the *Corte del Millioni* was what is now the
Sabbionera, or the interior area of the theatre. The latter seems most probable.

One Illustration of this volume, p. *1*, shows the archway in the Corte Sabbionera,
and also the decorations of the soffit.

Malibran Theatre, Venice.

[To face p. 25.

The site of the
CA' POLO.

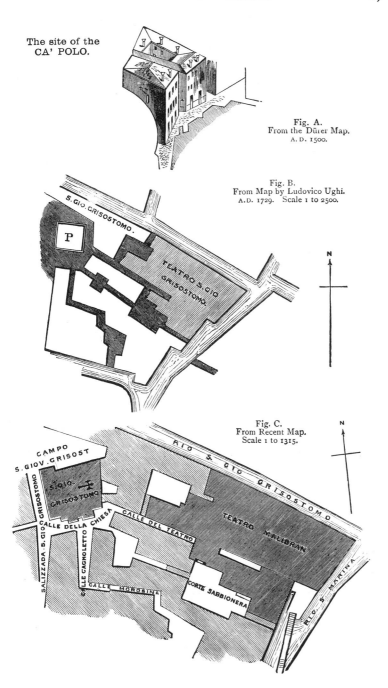

Fig. A.
From the Dürer Map.
A.D. 1500.

Fig. B.
From Map by Ludovico Ughi.
A.D. 1729.   Scale 1 to 2500.

Fig. C.
From Recent Map.
Scale 1 to 1315.

fragments of the decoration of the same mansion, are found imbedded in the walls of neighbouring houses.* It is impossible to determine anything further as to the form or extent of the house of the time of the Polos, but some slight idea of its appearance about the year 1500 may be seen in the extract (fig A) which we give from the famous pictorial map of Venice attributed erroneously to Albert Dürer. The state of the buildings in the last century is shown in (fig. B) an extract from the fine Map of Ughi; and their present condition in one (fig. C) reduced from the Modern Official Map of the Municipality.

[Coming from the Church of S. G. Grisostomo to enter the calle del Teatro on the left and the passage (*Sottoportico*) leading to the *Corte del Milione*, one has in front of him a building with a door of the epoch of the Renaissance; it was the office of the *provveditori* of silk; on the architrave are engraved the words:

PROVISORES SERICI

and below, above the door, is the Tablet which] in the year 1827 the Abate Zenier caused to be put up with this inscription :—

AEDES PROXIMA THALIAE CVLTVI MODO ADDICTA
MARCI POLO P. V. ITINERVM FAMA PRAECLARI
JAM HABITATIO FVIT.

24a. I believe that of late years some doubts have been thrown on the tradition of the site indicated as that of the

Recent corroboration as to the traditional site of the Casa Polo.

Casa Polo, though I am not aware of the grounds of such doubts. But a document recently discovered at Venice by Comm. Barozzi, one of a series relating to the testamentary estate of Marco Polo, goes far to confirm the tradition. This is the copy of a technical definition of two pieces of house property adjoining the property of Marco Polo and his brother Stephen, which were sold to Marco Polo by his wife Donata † in June 1321. Though the definition is not decisive, from the rarity of topographical references and absence of points of the compass, the description

---

* See *Ruskin*, iii. 320.

† Comm. Barozzi writes : "Among us, contracts between husband and wife are and were very common, and recognized by law. The wife sells to the husband property not included in dowry, or that she may have inherited, just as any third person might."

Entrance to the Corte del Milione, Venice.

[To face p. 30.

of Donata's tenements as standing on the Rio (presumably that of S. Giovanni Grisostomo) on one side, opening by certain porticoes and stairs on the other to the Court and common alley leading to the Church of S. Giovanni Grisostomo, and abutting in two places on the CA' POLO, the property of her husband and Stefano, will apply perfectly to a building occupying the western portion of the area on which now stands the Theatre, and perhaps forming the western side of a Court of which Casa Polo formed the other three sides.*

We know nothing more of Polo till we find him appearing a year or two later in rapid succession as the Captain of a Venetian Galley, as a prisoner of war, and as an author.

V. DIGRESSION CONCERNING THE WAR-GALLEYS OF THE MEDITERRANEAN STATES IN THE MIDDLE AGES.

25. And before entering on this new phase of the Traveller's biography it may not be without interest that we say something regarding the equipment of those galleys which are so prominent in the mediæval history of the Mediterranean.† *Arrangement of the Rowers in Mediæval Galleys: a separate oar to every man.*

Eschewing that "Serbonian Bog, where armies whole have sunk" of Books and Commentators, the theory of the classification of the Biremes and Triremes of the Ancients, we can at least assert on secure grounds that in *mediæval* armament, up to the middle of the 16th century or thereabouts, the characteristic distinction of galleys of different calibres, so far as such differences existed, was based *on the number of rowers that sat on one bench pulling each his separate oar, but through one* portella *or rowlock-port.‡* And to the classes

---

* See Appendix C, No. 16.

† I regret not to have had access to Jal's learned memoirs (*Archéologie Navale*, Paris, 1839) whilst writing this section, nor since, except for a hasty look at his Essay on the difficult subject of the oar arrangements. I see that he rejects so great a number of oars as I deduce from the statements of Sanudo and others, and that he regards a large number of the rowers as supplementary.

‡ It seems the more desirable to elucidate this, because writers on mediæval subjects so accomplished as Buchon and Capmany have (it would seem) entirely misconceived the matter, assuming that all the men on one bench pulled at one oar.

of galleys so distinguished the Italians, of the later Middle Age at least, did certainly apply, rightly or wrongly, the classical terms of *Bireme, Trireme,* and *Quinquereme,* in the sense of galleys having two men and two oars to a bench, three men and three oars to a bench, and five men and five oars to a bench.*

That this was the mediæval arrangement is very certain from the details afforded by Marino Sanudo the Elder, confirmed by later writers and by works of art. Previous to 1290, Sanudo tells us, almost all the galleys that went to the Levant had but two oars and men to a bench; but as it had been found that three oars and men to a bench could be employed with great advantage, after that date nearly all galleys adopted this arrangement, which was called *ai Terzaruoli.*†

Moreover experiments made by the Venetians in 1316 had shown that four rowers to a bench could be employed still more advantageously. And where the galleys could be used on inland waters, and could be made more bulky, Sanudo would even recommend five to a bench, or have gangs of rowers on two decks with either three or four men to the bench on each deck.

26. This system of grouping the oars, and putting only one man to an oar, continued down to the 16th century, during the first half of which came in the more modern system of using great oars, equally spaced, and requiring from four to seven men each to ply them, in the manner which endured till late in the last century, when galleys became altogether obsolete. Captain Pantero Pantera, the author of a work on Naval Tactics (1616), says he had heard, from veterans

Change of System in the 16th century.

---

* See *Coronelli, Atlante Veneto,* I. 139, 140. Marino Sanudo the Elder, though not using the term *trireme,* says it was well understood from ancient authors that the Romans employed their rowers *three to a bench* (p. 59).

† "*Ad terzarolos*" (*Secreta Fidelium Crucis,* p. 57). The Catalan Worthy, Ramon de Muntaner, indeed constantly denounces the practice of manning *all* the galleys with *terzaruoli,* or *tersols,* as his term is. But his reason is that these thirds-men were taken from the oar when crossbowmen were wanted, to act in that capacity, and as such they were good for nothing ; the crossbowmen, he insists, should be men specially enlisted for that service and kept to that. He would have some 10 or 20 per cent. only of the fleet built very light and manned in threes. He does not seem to have contemplated oars three-banked, and crossbowmen *besides,* as Sanudo does. (See below ; and *Muntaner,* pp. 288, 323, 525, etc.)

In Sanudo we have a glimpse worth noting of the word *soldiers* advancing towards the modern sense ; he expresses a strong preference for *soldati* (viz. *paid* soldiers) over *crusaders* (viz. volunteers), p. 74.

who had commanded galleys equipped in the antiquated fashion, that *three* men to a bench, with separate oars, answered better than three men to one great oar, but four men to one great oar (he says) were certainly more efficient than four men with separate oars. The new-fashioned great oars, he tells us, were styled *Remi di Scaloccio*, the old grouped oars *Remi a Zenzile*,— terms the etymology of which I cannot explain.*

It may be doubted whether the four-banked and five-banked galleys, of which Marino Sanudo speaks, really then came into practical use. A great five-banked galley on this system, built in 1529 in the Venice Arsenal by Vettor Fausto, was the subject of so much talk and excitement, that it must evidently have been something quite new and unheard of.† So late as 1567 indeed the King of Spain built at Barcelona a galley of thirty-six benches to the side, and seven men to the bench, with a separate oar to each in the old fashion. But it proved a failure. ‡

Down to the introduction of the great oars the usual system appears to have been three oars to a bench for the larger galleys, and two oars for lighter ones. The *fuste* or lighter galleys of the Venetians, even to about the middle of the 16th century, had their oars in pairs from the stern to the mast, and single oars only from the mast forward.§

27. Returning then to the three-banked and two-banked galleys of the latter part of the 13th century, the number of benches on each side seems to have run from twenty- Some details five to twenty-eight, at least as I interpret Sanudo's of the 13th century calculations. The 100-oared vessels often mentioned Galleys. (*e.g.* by *Muntaner*, p. 419) were probably two-banked vessels with twenty-five benches to a side.

The galleys were very narrow, only $15\frac{1}{2}$ feet in beam.||

---

* *L'Armata Navale*, Roma, 1616, pp. 150-151.

† See a work to which I am indebted for a good deal of light and information, the Engineer Giovanni Casoni's Essay: " *Dei Navigli Poliremi usati nella Marina dagli Antichi Veneziani*," in " *Esercitazioni dell' Ateneo Veneto*," vol. ii. p. 338. This great *Quinquereme*, as it was styled, is stated to have been struck by a fire-arrow, and blown up, in January 1570.

‡ *Pantera*, p. 22.

§ *Lazarus Bayfius de Re Navali Veterum*, in *Gronovii Thesaurus*, Ven. 1737, vol. xi. p. 581. This writer also speaks of the Quinquereme mentioned above (p. 577).

|| *Marinus Sanutius*, p. 65.

But to give room for the play of the oars and the passage of the fighting-men, &c., this width was largely augmented by an *opera-morta*, or outrigger deck, projecting much beyond the ship's sides and supported by timber brackets.* I do not find it stated how great this projection was in the mediæval galleys, but in those of the 17th century it was *on each side* as much as ⅔ths of the true beam. And if it was as great in the 13th-century galleys the total width between the false gunnels would be about 22¼ feet.

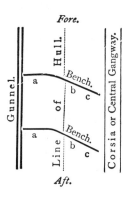

Fore.

Aft.

In the centre line of the deck ran, the whole length of the vessel, a raised gangway called the *corsia*, for passage clear of the oars.

The benches were arranged as in this diagram. The part of the bench next the gunnel was at right angles to it, but the other two-thirds of the bench were thrown forward obliquely. *a, b, c*, indicate the position of the three rowers. The shortest oar *a* was called *Terlicchio*, the middle one *b Posticcio*, the long oar *c Piamero.*†

I do not find any information as to how the oars worked on the gunnels. The Siena fresco (see p. *35*) appears to show them attached by loops and pins, which is the usual practice in boats of the Mediterranean now. In the cut from D. Tintoretto (p. *37*) the groups of oars protrude through regular ports in the bulwarks, but this probably represents the use of a later day. In any case the oars of each bench must have worked in very close proximity. Sanudo states the length of the galleys of his time (1300-1320) as 117 feet. This was doubtless length of *keel*, for that is specified ("*da ruoda a ruoda*") in other Venetian measurements, but the whole oar space could scarcely have been so much, and with twenty-eight benches to a side there could not have been more than 4 feet

---

* See the woodcuts opposite and at p. *37*; also *Pantera*, p. 46 (who is here, however, speaking of the great-oared galleys), and *Coronelli*, i. 140.

† *Casoni*, p. 324. He obtains these particulars from a manuscript work of the 16th century by Cristoforo Canale.

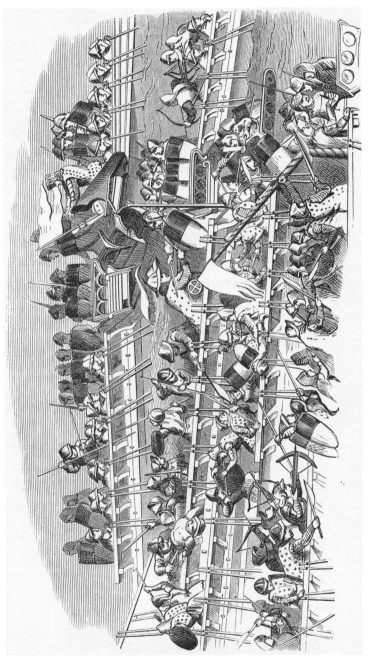

Galley-Fight, from a Mediæval Fresco at Siena. (See p. 36.)

gunnel-space to each bench. And as one of the objects of the grouping of the oars was to allow room between the benches for the action of cross-bowmen, &c., it is plain that the rowlock space for the three oars must have been very much compressed.*

The rowers were divided into three classes, with graduated pay. The highest class, who pulled the poop or stroke oars, were called *Portolati;* those at the bow, called *Prodieri*, formed the second class.†

Some elucidation of the arrangements that we have tried to describe will be found in our cuts. That at p. *35* is from a drawing, by the aid of a very imperfect photograph, of part of one of the frescoes of Spinello Aretini in the Municipal Palace at Siena, representing a victory of the Venetians over the Emperor Frederick Barbarossa's fleet, commanded by his son Otho, in 1176; but no doubt the galleys, &c., are of the artist's own age, the

---

* Signor Casoni (p. 324) expresses his belief that no galley of the 14th century had more than 100 oars. I differ from him with hesitation, and still more as I find M. Jal agrees in this view. I will state the grounds on which I came to a different conclusion. (1) Marino Sanudo assigns 180 rowers for a galley equipped *ai Terzaruoli* (p. 75). This seemed to imply something near 180 oars, for I do not find any allusion to reliefs being provided. In the French galleys of the 18th century there were no reliefs except in this way, that in long runs without urgency only half the oars were pulled. (See *Mém. d'un Protestant condamné aux Galères*, etc., Réimprimés, Paris, 1865, p. 447.) If four men to a bench were to be employed, then Sanudo seems to calculate for his smaller galleys 220 men actually rowing (see pp. 75-78). This seems to assume 55 benches, *i.e.*, 28 on one side and 27 on the other, which with 3-banked oars would give 165 rowers. (2) Casoni himself refers to Pietro Martire d'Anghieria's account of a Great Galley of Venice in which he was sent ambassador to Egypt from the Spanish Court in 1503. The crew amounted to 200, of whom 150 were for working the sails and oars, *that being the number of oars in each galley*, one man to each oar and three to each bench. Casoni assumes that this vessel must have been much larger than the galleys of the 14th century ; but, however that may have been, Sanudo to his galley assigns the larger crew of 250, of whom almost exactly the same proportion (180) were rowers. And in the *galeazza* described by Pietro Martire the oars were used only as an occasional auxiliary. (See his *Legationis Babylonicæ Libri Tres*, appended to his 3 Decads concerning the New World; *Basil.* 1533, f. 77 *ver.*) (3) The galleys of the 18th century, with their great oars 50 feet long pulled by six or seven men each, had 25 benches to the side, and only 4' 6" (French) gunnel-space to each oar. (See *Mém. d'un Protest.*, p. 434.) I imagine that a smaller space would suffice for the 3 light oars of the mediæval system, so that this need scarcely be a difficulty in the face of the preceding evidence. Note also the *three hundred rowers* in Joinville's description quoted at p. *40*. The great galleys of the Malay Sultan of Achin in 1621 had, according to Beaulieu, from 700 to 800 rowers, but I do not know on what system.

† *Marinus Sanutius*, p. 78. These titles occur also in the *Documenti d'Amore* of Fr. Barberino referred to at p. 117 of this volume :—

" Convienti qui manieri
Portolatti e prodieri
E presti galeotti
Aver, e forti e dotti.

middle of the 14th century.* In this we see plainly the projecting *opera-morta*, and the rowers sitting two to a bench, each with his oar, for these are two-banked. We can also discern the Latin rudder on the quarter. (See this volume, p. 119.) In a picture in the Uffizj, at Florence, of about the same date, by Pietro Laurato (it is in the corridor near the entrance), may be seen a small figure of a galley with the oars also very distinctly coupled.† Casoni has engraved, after Cristoforo Canale, a pictorial plan of a Venetian trireme of the 16th century, which shows the arrangement of the oars in *triplets* very plainly.

The following cut has been sketched from an engraving of a

Part of a Sea Fight, after Dom. Tintoretto.

picture by Domenico Tintoretto in the Doge's palace, representing, I believe, the same action (real or imaginary) as Spinello's fresco, but with the costume and construction of a later date. It shows, however, very plainly, the projecting *opera-morta*, and the arrangement of the oars in fours, issuing through row-ports in high bulwarks.

28. Midships in the mediæval galley a castle was erected, of

---

* Spinello's works, according to Vasari, extended from 1334 till late in the century. A religious picture of his at Siena is assigned to 1385, so the frescoes may probably be of about the same period. Of the battle represented I can find no record.

† Engraved in Jal, i. 330 ; with other mediæval illustrations of the same points.

the width of the ship, and some 20 feet in length; its platform
being elevated sufficiently to allow of free passage
under it and over the benches. At the bow was the
battery, consisting of mangonels (see vol. ii. p.
161 *seqq.*) and great cross-bows with winding gear, * whilst
there were shot-ports† for smaller cross-bows along the gunnels
in the intervals between the benches. Some of the larger galleys
had openings to admit horses at the stern, which were closed
and caulked for the voyage, being under water when the vessel
was at sea.‡

Fighting arrange-ments.

It seems to have been a very usual piece of tactics, in attacking as well as in awaiting attack, to connect a large number of
galleys by hawsers, and sometimes also to link the oars together,
so as to render it difficult for the enemy to break the line or run
aboard. We find this practised by the Genoese on the defensive
at the battle of Ayas (*infra*, p. *43*), and it is constantly resorted to
by the Catalans in the battles described by Ramon de
Muntaner.§

Sanudo says the toil of rowing in the galleys was excessive,
almost unendurable. Yet it seems to have been performed by
freely-enlisted men, and therefore it was probably less severe
than that of the great-oared galleys of more recent times,

---

* To these Casoni adds *Sifoni* for discharging Greek fire; but this he seems to
take from the Greek treatise of the Emperor Leo. Though I have introduced Greek
fire in the cut at p. *49*, I doubt if there is evidence of its use by the Italians in the
thirteenth century. Joinville describes it like something strange and new.

In after days the artillery occupied the same position, at the bow of the
galley.

Great beams, hung like battering rams, are mentioned by Sanudo, as well as iron
crow's-feet with fire attached, to shoot among the rigging, and jars of quick-lime and
soft soap to fling in the eyes of the enemy. The lime is said to have been used by
Doria against the Venetians at Curzola (*infra*, p. *48*), and seems to have been a
usual provision. Francesco Barberini specifies among the stores for his galley :—
" *Calcina*, con lancioni, Pece, pietre, e ronconi " (p. 259.) And Christine de Pisan,
in her *Faiz du Sage Roy Charles* (V. of France), explains also the use of the soap :
" *Item*, on doit avoir pluseurs vaisseaulx legiers à rompre, comme *poz plains de chauls*
ou pouldre, et gecter dedens ; et, par ce, seront comme avuglez, au brisier des poz.
Item, on doit avoir autres *poz de mol savon* et gecter ẽs nefzs des adversaires, et quant
les vaisseaulx brisent, le savon est glissant, si ne se peuent en piez soustenir et
chiéent en l'eaue " (pt. ii. ch. 38).

† *Balistariæ*, whence no doubt *Balistrada* and our *Balustrade*. Wedgwood's
etymology is far-fetched. And in his new edition (1872), though he has shifted his
ground, he has not got nearer the truth.

‡ *Sanutius*, p. 53 ; *Joinville*, p. 40 ; *Muntaner*, 316, 403.

§ See pp. 270, 288, 324, and especially 346.

which it was found impracticable to work by free enlistment, or
otherwise than by slaves under the most cruel driving.*   I
am not well enough read to say that war-galleys were never
rowed by slaves in the Middle Ages, but the only doubtful
allusion to such a class that I have met with is in one passage of
Muntaner, where he says, describing the Neapolitan and Catalan
fleets drawing together for action, that the gangs of the galleys
had to toil *like* " forçats" (p. 313). Indeed, as regards Venice
at least, convict rowers are stated to have been first introduced
in 1549, previous to which the gangs were of *galeotti
assoldati.*†

29. We have already mentioned that Sanudo requires for his
three-banked galley a ship's company of 250 men. Crew of a Galley and
They are distributed as follows :— Staff of a Fleet.

Comito or Master . . . 1	Orderlies 2
Quartermasters . . . 8	Cook 1
Carpenters 2	Arblasteers 50
Caulkers 2	Rowers 180
In charge of stores and arms . 4	
	250‡

This does not include the *Sopracomito,* or Gentleman-Commander,
who was expected to be *valens homo et probus,* a soldier and a
gentleman, fit to be consulted on occasion by the captain-
general. In the Venetian fleet he was generally a
noble.§

The aggregate pay of such a crew, not including the sopracomito, amounted monthly to 60 *lire de' grossi,* or 600 florins,
equivalent to 280*l.* at modern gold value; and the cost for a
year to nearly 3160*l.,* exclusive of the victualling of the vessel
and the pay of the gentleman-commander. The build or
purchase of a galley complete is estimated by the same author
at 15,000 florins, or 7012*l.*

We see that war cost a good deal in money even then.

Besides the ship's own complement Sanudo gives an estimate
for the general staff of a fleet of 60 galleys. This consists of a
captain-general, two (vice) admirals, and the following :—

---

* See the *Protestant,* cited above, p. 441, *et seqq.*
† *Venezia e le sue Lagune,* ii. 52.   ‡ *Mar. Sanut.* p. 75.
§ *Mar. Sanut.,* p. 30.

6 *Probi homines*, or gentlemen of character, forming a council to the Captain-General ;	15 Master Smiths ;
	12 Master Fletchers ;
	5 Cuirass men and Helmet-makers ;
4 Commissaries of Stores ;	15 Oar-makers and Shaft-makers ;
2 Commissaries over the Arms ;	10 Stone cutters for stone shot ;
3 Physicians ;	10 Master Arblast-makers ;
3 Surgeons ;	20 Musicians ;
5 Master Engineers and Carpenters ;	20 Orderlies, &c.

30. The musicians formed an important part of the equipment. Sanudo says that in going into action every vessel should make the greatest possible display of colours ; gonfalons and broad banners should float from stem to stern, and gay pennons all along the bulwarks ; whilst it was impossible to have too much of noisy music, of pipes, trumpets, kettle-drums, and what not, to put heart into the crew and strike fear into the enemy.*

Music; and other particulars.

So Joinville, in a glorious passage, describes the galley of his kinsman, the Count of Jaffa, at the landing of St. Lewis in Egypt:—

"That galley made the most gallant figure of them all, for it was painted all over, above water and below, with scutcheons of the count's arms, the field of which was *or* with a cross *patée gules*.† He had a good 300 rowers in his galley, and every man of them had a target blazoned with his arms in beaten gold. And, as they came on, the galley looked to be some flying creature, with such spirit did the rowers spin it along ;—or rather, with the rustle of its flags, and the roar of its nacaires and drums and Saracen horns, you might have taken it for a rushing bolt of heaven."‡

The galleys, which were very low in the water,§ could not keep the sea in rough weather, and in winter they never willingly kept the sea at night, however fair the weather might

---

* The Catalan Admiral Roger de Loria, advancing at daybreak to attack the Provençal Fleet of Charles of Naples (1283) in the harbour of Malta, " did a thing which should be reckoned to him rather as an act of madness," says Muntaner, "than of reason. He said, ' God forbid that I should attack them, all asleep as they are ! Let the trumpets and nacaires sound to awaken them, and I will tarry till they be ready for action. No man shall have it to say, if I beat them, that it was by catching them asleep.'" (*Munt.* p. 287.) It is what Nelson might have done !

The Turkish admiral Sidi 'Ali, about to engage a Portuguese squadron in the Straits of Hormuz, in 1553, describes the Franks as " dressing their vessels with flags and coming on." (*J. As.* ix. 70.)

† A cross *patée*, is one with the extremities broadened out into *feet* as it were.

‡ Page 50.

§ The galley at p. *49* is somewhat too high ; and I believe it should have had no *shrouds*.

be. Yet Sanudo mentions that he had been with armed galleys to Sluys in Flanders.

I will mention two more particulars before concluding this digression. When captured galleys were towed into port it was stern foremost, and with their colours dragging on the surface of the sea.* And the custom of saluting at sunset (probably by music) was in vogue on board the galleys of the 13th century.†

We shall now sketch the circumstances that led to the appearance of our Traveller in the command of a war-galley.

VI. The Jealousies and Naval Wars of Venice and Genoa. Lamba Doria's Expedition to the Adriatic; Battle of Curzola; and Imprisonment of Marco Polo by the Genoese.

31. Jealousies, too characteristic of the Italian communities, were, in the case of the three great trading republics of Venice, Genoa, and Pisa, aggravated by commercial rivalries, whilst, between the two first of those states, and also between the two last, the bitterness of such feelings had been augmenting during the whole course of the 13th century.‡

Growing jealousies and outbreaks between the Republics.

The brilliant part played by Venice in the conquest of Constantinople (1204), and the preponderance she thus acquired on the Greek shores, stimulated her arrogance and the resentment of her rivals. The three states no longer stood on a level as bidders for the shifting favour of the Emperor of the East. By treaty, not only was Venice established as the most important ally of the empire and as mistress of a large fraction of its territory, but all members of nations at war with her were prohibited from entering its limits. Though the Genoese colonies continued to exist, they stood at a great

---

* See *Muntaner*, passim, *e.g.* 271, 286, 315, 349.   † *Ibid.* 346.

‡ In this part of these notices I am repeatedly indebted to *Heyd*. (See *supra*, p. *9*.)

disadvantage, where their rivals were so predominant and enjoyed exemption from duties, to which the Genoese remained subject. Hence jealousies and resentments reached a climax in the Levantine settlements, and this colonial exacerbation reacted on the mother States.

A dispute which broke out at Acre in 1255 came to a head in a war which lasted for years, and was felt all over Syria. It began in a quarrel about a very old church called St. Sabba's, which stood on the common boundary of the Venetian and Genoese estates in Acre,* and this flame was blown by other unlucky occurrences. Acre suffered grievously.† Venice at this time generally kept the upper hand, beating Genoa by land and sea, and driving her from Acre altogether.‡ Four ancient porphyry figures from St. Sabba's were sent in triumph to Venice, and with their strange devices still stand at the exterior corner of St. Mark's, towards the Ducal Palace.‡

But no number of defeats could extinguish the spirit of Genoa, and the tables were turned when in her wrath she allied herself with Michael Palaeologus to upset the feeble and tottering Latin Dynasty, and with it the preponderance of Venice on the Bosphorus. The new emperor handed over to his allies the castle of their foes, which they tore down with jubilations, and now it was their turn to send its stones as trophies to Genoa. Mutual hate waxed fiercer than ever ; no merchant fleet of either state could go to sea without convoy, and wherever their ships met they fought.§ It was something like the state of things between Spain and England in the days of Drake.

The energy and capacity of the Genoese seemed to rise with

---

* On or close to the Hill called *Monjoie ;* see the plan from Marino Sanudo at p. 18.

† " Throughout that year there were not less than 40 machines all at work upon the city of Acre, battering its houses and its towers, and smashing and overthrowing everything within their range. There were at least ten of those engines that shot stones so big and heavy that they weighed a good 1500 lbs. by the weight of Champagne ; insomuch that nearly all the towers and forts of Acre were destroyed, and only the religious houses were left. And there were slain in this same war good 20,000 men on the two sides, but chiefly of Genoese and Spaniards." (*Lettre de Jean Pierre Sarrasin,* in *Michel's Joinville,* p. 308.)

‡ The origin of these columns is, however, somewhat uncertain. [See *Cicogna,* I. p. 379.]

§ In 1262, when a Venetian squadron was taken by the Greek fleet in alliance with the Genoese, the whole of the survivors of the captive crews were *blinded* by order of Palaeologus. (*Roman.* ii. 272.)

Figures from St. Sabba's, sent to Venice. [*To face p.* 42.

their success, and both in seamanship and in splendour they began almost to surpass their old rivals. The fall of Acre (1291), and the total expulsion of the Franks from Syria, in great measure barred the southern routes of Indian trade, whilst the predominance of Genoa in the Euxine more or less obstructed the free access of her rival to the northern routes by Trebizond and Tana.

32. Truces were made and renewed, but the old fire still smouldered. In the spring of 1294 it broke into flame, in consequence of the seizure in the Grecian seas of three Genoese vessels by a Venetian fleet. This led to an action with a Genoese convoy which sought redress.
Battle in Bay of Ayas in 1294.
The fight took place off Ayas in the Gulf of Scanderoon,* and though the Genoese were inferior in strength by one-third they gained a signal victory, capturing all but three of the Venetian galleys, with rich cargoes, including that of Marco Basilio (or Basegio), the commodore.

This victory over their haughty foe was in its completeness evidently a surprise to the Genoese, as well as a source of immense exultation, which is vigorously expressed in a ballad of the day, written in a stirring salt-water rhythm.† It represents the Venetians, as they enter the bay, in arrogant mirth reviling the Genoese with very unsavoury epithets as having deserted their ships to skulk on shore. They are described as saying :—

> " 'Off they've slunk ! and left us nothing ;
> We shall get nor prize nor praise ;
> Nothing save those crazy timbers
> Only fit to make a blaze.' "

So they advance carelessly—

> "On they come ! But lo their blunder !
> When our lads start up anon,
> Breaking out like unchained lions,
> With a roar, 'Fall on ! Fall on !' " ‡

---

* See pp. 16, 41, and Plan of Ayas at beginning of Bk. I.
† See *Archivio Storico Italiano*, Appendice, tom. iv.

‡ *Niente ne resta a prender*   *Como li fom aproximai*
   Se no li corpi de li legni :   *Queli si levan lantor*
   Preixi som senza difender ;   *Como leon descaenai*
    De bruxar som tute degni !   *Tuti criando* " Alor ! Alor !"
    *   *   *   *

This *Alor ! Alor !* (" Up, Boys, and at 'em "), or something similar, appears to have been the usual war-cry of both parties. So a trumpet-like poem of the

After relating the battle and the thoroughness of the victory, ending in the conflagration of five-and-twenty captured galleys, the poet concludes by an admonition to the enemy to moderate his pride and curb his arrogant tongue, harping on the obnoxious epithet *porci leproxi*, which seems to have galled the Genoese.* He concludes :—

> " Nor can I at all remember
> Ever to have heard the story
> Of a fight wherein the Victors
> Reaped so rich a meed of glory ! " †

The community of Genoa decreed that the victory should be commemorated by the annual presentation of a golden pall to the monastery of St. German's, the saint on whose feast (28th May) it had been won.‡

The startling news was received at Venice with wrath and grief, for the flower of their navy had perished, and all energies were bent at once to raise an overwhelming force.§ The Pope (Boniface VIII.) interfered as arbiter, calling for plenipotentiaries from both sides. But spirits were too much inflamed, and this mediation came to nought.

Troubadour warrior Bertram de Born, whom Dante found in such evil plight below (xxviii. 118 *seqq.*), in which he sings with extraordinary spirit the joys of war :—

> " Ie us dic que tan no m'a sabor
> Manjars, ni beure, ni dormir,
> Cum a quant aug cridar, Alor!
> D'ambas la partz ; et aug agnir
> Cabals boitz per l'ombratge. . . ."

> " I tell you a zest far before
> Aught of slumber, or drink, or of food,
> I snatch when the shouts of Alor
> Ring from both sides : and out of the wood
> Comes the neighing of steeds dimly seen. . . ."

In a galley fight at Tyre in 1258, according to a Latin narrative, the Genoese shout " Ad arma, ad arma ! *ad ipsos, ad ipsos !* " The cry of the Venetians before engaging the Greeks is represented by Martino da Canale, in his old French, as " *or à yaus ! or à yaus !* " that of the Genoese on another occasion as *Aur ! Aur !* and this last is the shout of the Catalans also in Ramon de Muntaner. (*Villemain, Litt. du Moyen Age*, i. 99 ; *Archiv. Stor. Ital.* viii. 364, 506 ; *Pertz, Script.* xviii. 239 ; *Muntaner*, 269, 287.) Recently in a Sicilian newspaper, narrating an act of gallant and successful reprisal (only too rare) by country folk on a body of the brigands who are such a scourge to parts of the island, I read that the honest men in charging the villains raised a shout of " *Ad iddi ! Ad iddi !* "

* A phrase curiously identical, with a similar sequence, is attributed to an Austrian General at the battle of Skalitz in 1866. (*Stoffel's Letters.*)

†

> *E no me posso aregordar*
> *Dalcuno romanzo vertadi*
> *Donde oyse uncha cointar*
> *Alcun triumfo si sobré !*

‡ *Stella* in *Muratori*, xvii. 984.   § *Dandulo*, Ibid. xii. 404-405.

Further outrages on both sides occurred in 1296. The Genoese residences at Pera were fired, their great alum works on the coast of Anatolia were devastated, and Caffa was stormed and sacked; whilst on the other hand a number of the Venetians at Constantinople were massacred by the Genoese, and Marco Bembo, their Bailo, was flung from a house-top. Amid such events the fire of enmity between the cities waxed hotter and hotter.

33. In 1298 the Genoese made elaborate preparations for a great blow at the enemy, and fitted out a powerful fleet which they placed under the command of LAMBA DORIA, a younger brother of Uberto of that illustrious house, under whom he had served fourteen years before in the great rout of the Pisans at Meloria.

Lamba Doria's Expedition to the Adriatic.

The rendezvous of the fleet was in the Gulf of Spezia, as we learn from the same pithy Genoese poet who celebrated Ayas. This time the Genoese were bent on bearding St. Mark's Lion in his own den; and after touching at Messina they steered straight for the Adriatic:—

> "Now, as astern Otranto bears,
>   Pull with a will! and, please the Lord,
>   Let them who bragged, with fire and sword,
> To waste our homesteads, look to theirs!"*

On their entering the gulf a great storm dispersed the fleet. The admiral with twenty of his galleys got into port at Antivari on the Albanian coast, and next day was rejoined by fifty-eight more, with which he scoured the Dalmatian shore, plundering all Venetian property. Some sixteen of his galleys were still missing when he reached the island of Curzola, or Scurzola as the more popular name seems to have been, the Black Corcyra of the Ancients—the chief town of which, a rich and flourishing

---

*

> *Or entram con gran vigor,*
> *En De sperando aver triumpho,*
> *Queli zerchando inter lo Gorfo*
> *Chi menazeram zercha lor!*

And in the next verse note the pure Scotch use of the word *bra* :—

> *Sichè da Otranto se partim*
> *Quella bra compagnia,*
> *Per assar in Ihavonia,*
> *D'Avosto a vinte nove dì.*

place, the Genoese took and burned.* Thus they were engaged when word came that the Venetian fleet was in sight.

Venice, on first hearing of the Genoese armament, sent Andrea Dandolo with a large force to join and supersede Maffeo Quirini, who was already cruising with a squadron in the Ionian sea; and, on receiving further information of the strength of the hostile expedition, the Signory hastily equipped thirty-two more galleys in Chioggia and the ports of Dalmatia, and despatched them to join Dandolo, making the whole number under his command up to something like ninety-five. Recent drafts had apparently told heavily upon the Venetian sources of enlistment, and it is stated that many of the complements were made up of rustics swept in haste from the Euganean hills. To this the Genoese poet seems to allude, alleging that the Venetians, in spite of their haughty language, had to go begging for men and money up and down Lombardy. "Did *we* do like that, think you?" he adds :—

> "Beat up for aliens? *We* indeed?
> When lacked we homeborn Genoese?
> Search all the seas, no salts like these,
> For Courage, Seacraft, Wit at need."†

Of one of the Venetian galleys, probably in the fleet which sailed under Dandolo's immediate command, went Marco Polo as *Sopracomito* or Gentleman-Commander.‡

---

* The island of Curzola now counts about 4000 inhabitants; the town half the number. It was probably reckoned a dependency of Venice at this time. The King of Hungary had renounced his claims on the Dalmatian coasts by treaty in 1244. (*Romanin*, ii. 235.) The gallant defence of the place against the Algerines in 1571 won for Curzola from the Venetian Senate the honourable title in all documents of *fedelissima*. (*Paton's Adriatic*, I. 47.)

† *Ma sè si gran colmo avea*
Perchè andava mendigando

Per terra de Lombardia
Peccunia, gente a sodi?
Pone mente tu che l'odi
Se noi tegnamo questa via?

No, ma' più! ajamo omi nostrar
Destri, valenti, e avisti,
Che mai par de lor n' o visti
In tuti officj de mar.

‡ In July 1294, a Council of Thirty decreed that galleys should be equipped by the richest families in proportion to their wealth. Among the families held to equip one galley each, or one galley among two or more, in this list, is the CA' POLO. But this was before the return of the travellers from the East, and just after the battle of Ayas. (*Romanin*, ii. 332; this author misdates Ayas, however.) When a levy was required in Venice for any expedition the heads of each *contrada* divided the male inhabitants, between the ages of twenty and sixty, into groups of twelve each, called *duodene*. The dice were thrown to decide who should go first on service. He who went received five *lire* a month from the State, and one *lira* from each of his colleagues in

34. It was on the afternoon of Saturday the 6th September that the Genoese saw the Venetian fleet approaching, but, as sunset was not far off, both sides tacitly agreed to defer the engagement.* _{The Fleets come in sight of each other at Curzola.}

The Genoese would appear to have occupied a position near the eastern end of the Island of Curzola, with the Peninsula of Sabbioncello behind them, and Meleda on their left, whilst the Venetians advanced along the south side of Curzola. (See map on p. *50*).

According to Venetian accounts the Genoese were staggered at the sight of the Venetian armaments, and sent more than once to seek terms, offering finally to surrender galleys and munitions of war, if the crews were allowed to depart. This is an improbable story, and that of the Genoese ballad seems more like truth. Doria, it says, held a council of his captains in the evening at which they all voted for attack, whilst the Venetians, with that overweening sense of superiority which at this time is reflected in their own annals as distinctly as in those of their enemies, kept scout-vessels out to watch that the Genoese fleet, which they looked on as already their own, did not steal away in the darkness. A vain imagination, says the poet :—

> " Blind error of vainglorious men
> To dream that we should seek to flee
> After those weary leagues of sea
> Crossed, but to hunt them in their den ! " †

the *duodena.* Hence his pay was sixteen *lire* a month, about 2*s.* a day in silver value, if these were *lire ai grossi,* or 1*s.* 4*d.* if *lire dei piccoli.* (See *Romanin,* ii. 393-394.)

Money on such occasions was frequently raised by what was called an *Estimo* or *Facion,* which was a forced loan levied on the citizens in proportion to their estimated wealth ; and for which they were entitled to interest from the State.

* Several of the Italian chroniclers, as Ferreto of Vicenza and Navagiero, whom Muratori has followed in his " Annals," say the battle was fought on the 8th September. the so-called Birthday of the Madonna. But the inscription on the Church of St. Matthew at Genoa, cited further on, says the 7th, and with this agree both Stella and the Genoese poet. For the latter, though not specifying the day of the month, says it was on a Sunday :—

> " Lo di de Domenga era
> Passa prima en l'ora bona
> Stormezam fin provo nona
> Con bataio forte e fera."

Now the 7th September, 1298, fell on a Sunday.

† *Ma li pensavam grande error*
Che in fuga se fussem tuti metui
Che de si lonzi eram vegnui
Per *cerchali a casa lor.*

35. The battle began early on Sunday and lasted till the afternoon. The Venetians had the wind in their favour, but

The Vene-
tians de-
feated, and
Marco Polo
a prisoner. the morning sun in their eyes. They made the attack, and with great impetuosity, capturing ten Genoese galleys; but they pressed on too wildly, and some of their vessels ran aground. One of their galleys too, being taken, was cleared of her crew and turned against the Venetians. These incidents caused confusion among the assailants; the Genoese, who had begun to give way, took fresh heart, formed a close column, and advanced boldly through the Venetian line, already in disorder. The sun had begun to decline when there appeared on the Venetian flank the fifteen or sixteen missing galleys of Doria's fleet, and fell upon it with fresh force. This decided the action. The Genoese gained a complete victory, capturing all but a few of the Venetian galleys, and including the flagship with Dandolo. The Genoese themselves lost heavily, especially in the early part of the action, and Lamba Doria's eldest son Octavian is said to have fallen on board his father's vessel.* The number of prisoners taken was over 7000, and among these was Marco Polo.†

The prisoners, even of the highest rank, appear to have been chained. Dandolo, in despair at his defeat, and at the prospect of being carried captive into Genoa, refused food, and ended by dashing his head against a bench.‡ A Genoese account asserts

---

* "Note here that the Genoese generally, commonly, and by nature, are the most covetous of Men, and the Love of Gain spurs them to every Crime. Yet are they deemed also the most valiant Men in the World. Such an one was Lampa, of that very Doria family, a man of an high Courage truly. For when he was engaged in a Sea-Fight against the Venetians, and was standing on the Poop of his Galley, his Son, fighting valiantly at the Forecastle, was shot by an Arrow in the Breast, and fell wounded to the Death; a Mishap whereat his Comrades were sorely shaken, and Fear came upon the whole Ship's Company. But Lampa, hot with the Spirit of Battle, and more mindful of his Country's Service and his own Glory than of his Son, ran forward to the spot, loftily rebuked the agitated Crowd, and ordered his Son's Body to be cast into the Deep, telling them for their Comfort that the Land could never have afforded his Boy a nobler Tomb. And then, renewing the Fight more fiercely than ever, he achieved the Victory." (*Benvenuto of Imola*, in *Comment. on Dante. in Muratori, Antiq.* i. 1146.)

("Yet like an English General will I die,
And all the Ocean make my spacious Grave:
Women and Cowards on the Land may lie,
The Sea's the Tomb that's proper for the Brave!"—Annus Mirabilis.)

† The particulars of the battle are gathered from *Ferretus Vicentinus*, in *Murat.* ix. 985 *seqq.; And. Dandulo*, in xii. 407-408; *Navagiero*, in xxiii. 1009-1010; and the Genoese Poem as before.

‡ *Navagiero, u. s.* Dandulo says, "after a few days he died of grief"; Ferretus, that he was killed in the action and buried at Curzola.

Marco Polo's Galley going into action at Curzola.

" . . . . . "il sembloit que la galie bolast, par les nageurs qui la contreingnoient aux avirons, et sembloit que fondre cheist des ciex, au bruit que les pennonciaus menoient ; et que les nacaires les tabours et les cors sarrazinnois menoient, qui estoient en sa galie."

(Joinville, vide ante, p. 40.)

that a noble funeral was given him after the arrival of the fleet at Genoa, which took place on the evening of the 16th October.* It was received with great rejoicing, and the City voted the annual presentation of a pallium of gold brocade to the altar of the Virgin in the Church of St. Matthew, on every 8th of September, the Madonna's day, on the eve of which the Battle had been won. To the admiral himself a Palace was decreed. It still stands, opposite the Church of St. Matthew, though it has passed from the possession of the Family. On the striped marble façades, both of the Church and of the Palace, inscriptions of that age, in excellent preservation, still commemorate Lamba's

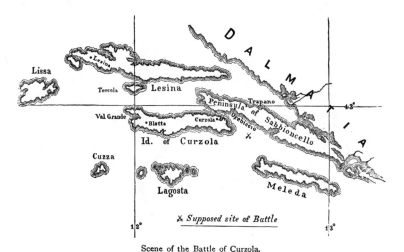

Scene of the Battle of Curzola.

achievement.† Malik al Mansúr, the Mameluke Sultan of Egypt,

---

* For the funeral, a MS. of Cibo Recco quoted by *Jacopo Doria* in *La Chiesa di San Matteo descritta*, etc., Genova, 1860, p. 26. For the date of arrival the poem so often quoted :—

> "*De Oitover*, a zoia, *a seze di*
> Lo nostro ostel, con gran festa
> En nostro porto, a or di sesta
> Domine De restitui.*"

† S. Matteo was built by Martin Doria in 1125, but pulled down and rebuilt by the family in a slightly different position in 1278. On this occasion is recorded a remarkable anticipation of the feats of American engineering : "As there was an ancient and very fine picture of Christ upon the apse of the Church, it was thought a great pity that so fine a work should be destroyed. And so they contrived an ingenious method by which the apse bodily was transported without injury, picture and all, for a distance of 25 ells, and firmly set upon the foundations where it now exists." (*Jacopo de Varagine* in *Muratori*, vol. ix. 36.)

Church of San Matteo, Genoa.

[*To face p.* 50.

as an enemy of Venice, sent a complimentary letter to Doria accompanied by costly presents.*

The latter died at Savona 17th October, 1323, a few months before the most illustrious of his prisoners, and his bones were laid in a sarcophagus which may still be seen forming the sill of one of the windows of S. Matteo (on the right as you enter). Over this sarcophagus stood the Bust of Lamba till 1797, when the mob of Genoa, in idiotic imitation of the French proceedings of that age, threw it down. All of Lamba's six sons had fought with him at Meloria. In 1291 one of them, Tedisio, went forth into the Atlantic in company with Ugolino Vivaldi on a voyage of discovery, and never returned. Through Cæsar, the youngest, this branch of the Family still survives, bearing the distinctive surname of *Lamba-Doria*.†

As to the treatment of the prisoners, accounts differ; a thing usual in such cases. The Genoese Poet asserts that the hearts of his countrymen were touched, and that the captives were treated with compassionate courtesy. Navagiero the Venetian, on the other hand, declares that most of them died of hunger.‡

The inscription on S. Matteo regarding the battle is as follows :—" *Ad Honorem Dei et Beate Virginis Marie Anno MCCLXXXXVIII Die Dominico VII Septembris iste Angelus captus fuit in Gulfo Venetiarum in Civitate Scursole et ibidem fuit prelium Galearum LXXVI Januensium cum Galeis LXXXXVI Veneciarum. Capte fuerunt LXXXIIII per Nobilem Virum Dominum Lambam Aurie Capitaneum et Armiratum tunc Comunis et Populi Janue cum omnibus existentibus in eisdem, de quibus conduxit Janue homines vivos carceratos VII cccc et Galeas XVIII, reliquas LXVI fecit cumbu̯ri in dicto Gulfo Veneciarum. Qui obiit Sagone I. MCCCXXIII.*" It is not clear to what the *Angelus* refers.

* *Rampoldi, Ann. Musulm.* ix. 217.　　† *Jacopo Doria*, p. 280.

‡ *Murat.* xxiii. 1010. I learn from a Genoese gentleman, through my friend Professor Henry Giglioli (to whose kindness I owe the transcript of the inscription just given), that a faint tradition exists as to the place of our traveller's imprisonment. It is alleged to have been a massive building, standing between the *Grazie* and the Mole, and bearing the name of the *Malapaga*, which is now a barrack for Doganieri, but continued till comparatively recent times to be used as a civil prison. "It is certain," says my informant, "that men of fame in arms who had fallen into the power of the Genoese *were* imprisoned there, and among others is recorded the name of the Corsican Giudice dalla Rocca and Lord of Cinarca, who died there in 1312;" a date so near that of Marco's imprisonment as to give some interest to the hypothesis, slender as are its grounds. Another Genoese, however, indicates as the scene of Marco's captivity certain old prisons near the Old Arsenal, in a site still known as the *Vico degli Schiavi*. (*Celesia, Dante in Liguria*, 1865, p. 43.) [Was not the place of Polo's captivity the basement of the *Palazzo del Capitan del Popolo*, afterwards *Palazzo del Comune al Mare*, where the Customs (*Dogana*) had their office, and from the 15th century the *Casa* or *Palazzo di S. Giorgio*?—H. C.]

36. Howsoever they may have been treated, here was Marco Polo one of those many thousand prisoners in Genoa ; and here, Marco Polo in prison dictates his book to Rusticiano of Pisa. before long, he appears to have made acquaintance with a man of literary propensities, whose destiny had brought him into the like plight, by name RUSTICIANO Release of Venetian prisoners. or RUSTICHELLO of Pisa. It was this person perhaps who persuaded the Traveller to defer no longer the reduction to writing of his notable experiences ; but in any case it was he who wrote down those experiences at Marco's dictation ; it is he therefore to whom we owe the preservation of this record, and possibly even that of the Traveller's very memory. This makes the Genoese imprisonment so important an episode in Polo's biography.

To Rusticiano we shall presently recur. But let us first bring to a conclusion what may be gathered as to the duration of Polo's imprisonment.

It does not appear whether Pope Boniface made any new effort for accommodation between the Republics ; but other Italian princes did interpose, and Matteo Visconti, Captain-General of Milan, styling himself Vicar-General of the Holy Roman Empire in Lombardy, was accepted as Mediator, along with the community of Milan. Ambassadors from both States presented themselves at that city, and on the 25th May, 1299, they signed the terms of a Peace.

These terms were perfectly honourable to Venice, being absolutely equal and reciprocal ; from which one is apt to conclude that the damage to the City of the Sea was rather to her pride than to her power ; the success of Genoa, in fact, having been followed up by no systematic attack upon Venetian commerce.* Among the terms was the mutual release of prisoners on a day to be fixed by Visconti after the completion of all formalities. This day is not recorded, but as the Treaty was ratified by the Doge of Venice on the 1st July, and the latest extant document connected with the formalities appears to be dated 18th July, we may believe that before the end of August

---

* The Treaty and some subsidiary documents are printed in the Genoese *Liber Jurium*, forming a part of the *Monumenta Historiae Patriae*, published at Turin. (See *Lib. Jur.* II. 344, *seqq.*) Muratori in his Annals has followed John Villani (Bk. VIII. ch. 27) in representing the terms as highly unfavourable to Venice. But for this there is no foundation in the documents. And the terms are stated with substantial accuracy in Navagiero. (*Murat. Script.* xxiii. 1011.)

Marco Polo was restored to the family mansion in S. Giovanni
Grisostomo.

37. Something further requires to be said before quitting this
event in our Traveller's life.   For we confess that a critical reader
may have some justification in asking what evidence Grounds on
there is that Marco Polo ever fought at Curzola, and story of
ever was carried a prisoner to Genoa from that unfor- capture at
tunate action ? rests.

A learned Frenchman, whom we shall have to quote freely
in the immediately ensuing pages, does not venture to be more
precise in reference to the meeting of Polo and Rusticiano than
to say of the latter : " In 1298, being in durance in the Prison of
Genoa, he there became acquainted with Marco Polo, whom the
Genoese had deprived of his liberty *from motives equally
unknown.*"*

To those who have no relish for biographies that round the
meagre skeleton of authentic facts with a plump padding of
what *might have been*, this sentence of Paulin Paris is quite
refreshing in its stern limitation to positive knowledge.   And
certainly no contemporary authority has yet been found for the
capture of our Traveller at Curzola.   Still I think that the fact
is beyond reasonable doubt.

Ramusio's biographical notices certainly contain many errors
of detail ; and some, such as the many years' interval which he
sets between the Battle of Curzola and Marco's return, are errors
which a very little trouble would have enabled him to eschew.
But still it does seem reasonable to believe that the main fact of
Marco's command of a galley at Curzola, and capture there, was
derived from a genuine tradition, if not from documents.

Let us then turn to the words which close Rusticiano's
preamble (see *post*, p. 2) :—" Lequel (Messire Marc) puis demo-
rant en le charthre de Jene, fist retraire toutes cestes chouses à
Messire Rustacians de Pise que en celle meissme charthre estoit,
au tens qu'il avoit 1298 anz que Jezu eut vesqui."   These words
are at least thoroughly consistent with Marco's capture at
Curzola, as regards both the position in which they present him,
and the year in which he is thus presented.

There is however another piece of evidence, though it is
curiously indirect.

* *Paulin Paris, Les Manuscrits François de la Bibliothèque du Roi*, ii. 355.

The Dominican Friar Jacopo of Acqui was a contemporary of Polo's, and was the author of a somewhat obscure Chronicle called *Imago Mundi.** Now this Chronicle does contain mention of Marco's capture in action by the Genoese, but attributes it to a different action from Curzola, and one fought at a time when Polo could not have been present. The passage runs as follows in a manuscript of the Ambrosian Library, according to an extract given by Baldelli Boni :—

" In the year of Christ MCCLXXXXVI, in the time of Pope Boniface VI., of whom we have spoken above, a battle was fought in Arminia, at the place called Layaz, between XV. galleys of Genoese merchants and XXV. of Venetian merchants ; and after a great fight the galleys of the Venetians were beaten, and (the crews) all slain or taken ; and among them was taken Messer Marco the Venetian, who was in company with those merchants, and who was called *Milono*, which is as much as to say ' a thousand thousand pounds,' for so goes the phrase in Venice. So this Messer Marco Milono the Venetian, with the other Venetian prisoners, is carried off to the prison of Genoa, and there kept for a long time. This Messer Marco was a long time with his father and uncle in Tartary, and he there saw many things, and made much wealth, and also learned many things, for he was a man of ability. And so, being in prison at Genoa, he made a Book concerning the great wonders of the World, *i.e.*, concerning such of them as he had seen. And what he told in the Book was not as much as he had really seen, because of the tongues of detractors, who, being ready to impose their own lies on others, are over hasty to set down as lies what they in their perversity disbelieve, or do not understand. And because there are many great and strange things in that Book, which are reckoned past all credence, he was asked by his friends on his death-bed to correct the Book by removing everything that went beyond the facts. To which his reply was that he had not told *one-half* of what he had really seen ! " †

This statement regarding the capture of Marco *at the Battle of Ayas* is one which cannot be true, for we know that he did not reach Venice till 1295, travelling from Persia by way of Trebizond and the Bosphorus, whilst the Battle of Ayas of which we have purposely given some detail, was fought in May, 1294.

---

* Though there is no precise information as to the birth or death of this writer, who belonged to a noble family of Lombardy, the Bellingeri, he can be traced with tolerable certainty as in life in 1289, 1320, and 1334. (See the Introduction to his Chronicle in the Turin *Monumentà, Scriptores* III.)

† There is another MS. of the *Imago Mundi* at Turin, which has been printed in the *Monumenta*. The passage about Polo in that copy differs widely in wording, is much shorter, and contains no date. But it relates his capture as having taken place at *Là Glazà*, which I think there can be no doubt is also intended for Ayas (sometimes called *Giàzza*), a place which in fact is called *Glaza* in three of the MSS. of which various readings are given in the edition of the Société de Géographie (p. 535).

The date MCCLXXXXVI assigned to it in the preceding extract has given rise to some unprofitable discussion. Could that date be accepted, no doubt it would enable us also to accept this, the sole statement from the Traveller's own age of the circumstances which brought him into a Genoese prison ; it would enable us to place that imprisonment within a few months of his return from the East, and to extend its duration to three years, points which would thus accord better with the general tenor of Ramusio's tradition than the capture of Curzola. But the matter is not open to such a solution. The date of the Battle of Ayas is not more doubtful than that of the Battle of the Nile. It is clearly stated by several independent chroniclers, and is carefully established in the Ballad that we have quoted above.* We shall see repeatedly in the course of this Book how uncertain are the transcriptions of dates in Roman numerals, and in the present case the LXXXXVI is as certainly a mistake for LXXXXIV as is Boniface VI. in the same quotation a mistake for Boniface VIII.

But though we cannot accept the statement that Polo was taken prisoner at *Ayas, in the spring of* 1294, we may accept the passage as evidence from a contemporary source that he was *taken prisoner in some sea-fight with the Genoese*, and thus admit it in corroboration of the Ramusian Tradition of his capture in a sea-fight at Curzola in 1298, which is perfectly consistent with all other facts in our possession.

VII. RUSTICIANO OR RUSTICHELLO OF PISA, MARCO POLO'S FELLOW-PRISONER AT GENOA, THE SCRIBE WHO WROTE DOWN THE TRAVELS.

38. We have now to say something of that Rusticiano to whom all who value Polo's book are so much indebted.

The relations between Genoa and Pisa had long been so

---

*
> " *E per meio esse aregordenti*
> *De si grande scacho mato*
> *Correa mille duxenti*
> *Zonto ge novanta e quatro.*"

The Armenian Prince Hayton or Héthum has put it under 1293. (See *Langlois, Mém sur les Relations de Gênes avec la Petite-Arménie.*)

hostile that it was only too natural in 1298 to find a Pisan in
the gaol of Genoa.   An unhappy multitude of such
prisoners had been carried thither fourteen years before,
and the survivors still lingered there in vastly dwindled
numbers.   In the summer of 1284 was fought the battle from
which Pisa had to date the commencement of her long decay.   In
July of that year the Pisans, at a time when the Genoese had no
fleet in their own immediate waters, had advanced to the very
port of Genoa and shot their defiance into the proud city in the
form of silver-headed arrows, and stones belted with scarlet.*
They had to pay dearly for this insult.   The Genoese, recalling
their cruisers, speedily mustered a fleet of eighty-eight galleys,
which were placed under the command of another of that
illustrious House of Doria, the Scipios of Genoa as they have
been called, Uberto, the elder brother of Lamba.   Lamba himself with his six sons, and another brother, was in the fleet,
whilst the whole number of Dorias who fought in the ensuing
action amounted to 250, most of them on board one great galley
bearing the name of the family patron, St. Matthew. †

(margin note: Rusticiano, perhaps a prisoner from Meloria.)

The Pisans, more than one-fourth inferior in strength, came
out boldly, and the battle was fought off the Porto Pisano, in
fact close in front of Leghorn, where a lighthouse on a remarkable arched basement still marks the islet of MELORIA, whence
the battle got its name.   The day was the 6th of August, the
feast of St. Sixtus, a day memorable in the Pisan Fasti for
several great victories.   But on this occasion the defeat of Pisa
was overwhelming.   Forty of their galleys were taken or sunk,
and upwards of 9000 prisoners carried to Genoa.   In fact so
vast a sweep was made of the flower of Pisan manhood that it
was a common saying then: " *Che vuol veder Pisa, vada a*

---

* B. Marangone, *Croniche della C. di Pisa*, in *Rerum Ital. Script.* of *Tartini*,
Florence, 1748, i. 563 ; *Dal Borgo, Dissert. sopra l'Istoria Pisana*, ii. 287.

† The list of the whole number is preserved in the Doria archives, and has been
published by Sign. Jacopo D'Oria.   Many of the Baptismal names are curious, and
show how far sponsors wandered from the Church Calendar. *Assan, Aiton, Turco,
Soldan* seem to come of the constant interest in the East. *Alaone*, a name which
remained in the family for several generations, I had thought certainly borrowed from
the fierce conqueror of the Khalif (*infra*, p. 63).   But as one Alaone, present at this
battle, had a son also there, he must surely have been christened before the fame of
Hulaku could have reached Genoa.   (See *La Chiesa di S. Matteo*, pp. 250, *seqq.*)

In documents of the kingdom of Jerusalem there are names still more anomalous,
e.g., Gualterius Baffumeth, Joannes Mahomet.   (See *Cod. Dipl. del Sac. Milit. Ord.
Gerosol.* I. 2-3, 62.)

Genova!" Many noble ladies of Pisa went in large companies on foot to Genoa to seek their husbands or kinsmen : "And when they made enquiry of the Keepers of the Prisons, the reply would be, 'Yesterday there died thirty of them, to-day there have died forty ; all of whom we have cast into the sea ; and so it is daily.'"*

A body of prisoners so numerous and important naturally exerted themselves in the cause of peace, and through their efforts, after many months of ne-gotiation, a formal peace was signed (15th April, 1288). But through the influence, as was alleged, of Count Ugo-lino (Dante's) who was then in power at Pisa, the peace became abortive ; war almost immediately recommenced, and the prisoners had no re-lease.† And, when the 6000 or 7000 Venetians were thrown into the prisons of Genoa in October 1298, they

Seal of the Pisan Prisoners.

would find there the scanty surviving remnant of the Pisan Prisoners of Meloria, and would gather from them dismal forebodings of the fate before them.

It is a fair conjecture that to that remnant Rusticiano of Pisa may have belonged.

We have seen Ramusio's representation of the kindness shown to Marco during his imprisonment by a certain Genoese gentleman who also assisted him to reduce his travels to writing. We may be certain that this Genoese gentleman is only a dis-torted image of Rusticiano, the Pisan prisoner in the gaol of

---

* *Memorial. Potestat. Regiens.* in *Muratori*, viii. 1162.

† See *Fragm. Hist. Pisan.* in *Muratori*, xxiv. 651, *seqq.* ; and *Caffaro, id.* vi. 588, 594-595. The cut in the text represents a striking memorial of those Pisan Prisoners, which perhaps still survives, but which at any rate existed last century in a collection at Lucca. It is the seal of the prisoners as a body corporate : SIGILLUM UNIVERSITATIS CARCERATORUM PISANORUM JANUE DETENTORUM, and was doubtless used in their negotiations for peace with the Genoese Commissioners. It represents two of the prisoners imploring the Madonna, Patron of the Duomo at Pisa. It is from *Manni, Osserv. Stor. sopra Sigilli Antichi,* etc., Firenze, 1739, tom. xii. The seal is also engraved in *Dal Borgo, op. cit.* ii. 316.

Genoa, whose name and part in the history of his hero's book
Ramusio so strangely ignores. Yet patriotic Genoese writers in
our own times have striven to determine the identity of this
their imaginary countryman! *

39. Who, then, was Rusticiano, or, as the name actually is
read in the oldest type of MS., " Messire Rustacians de Pise "?

Rusticiano,   Our knowledge of him is but scanty. Still some-
a person
known from thing is known of him besides the few words con-
other
sources.   cluding his preamble to our Traveller's Book, which
you may read at pp. 1-2 of the body of this volume.

In Sir Walter Scott's " Essay on Romance," when he speaks
of the new mould in which the subjects of the old metrical
stories were cast by the school of prose romancers which arose
in the 13th century, we find the following words :—

"Whatever fragments or shadows of true history may yet remain hidden
under the mass of accumulated fable which had been heaped upon them
during successive ages, must undoubtedly be sought in the metrical romances
. . . . . But those prose authors who wrote under the imaginary names of
RUSTICIEN DE PISE, Robert de Borron, and the like, usually seized upon the
subject of some old minstrel ; and recomposing the whole narrative after
their own fashion, with additional character and adventure, totally obliterated
in that operation any shades which remained of the original and probably
authentic tradition," &c.†

Evidently, therefore, Sir Walter regarded Rustician of Pisa
as a person belonging to the same ghostly company as his own
Cleishbothams and Dryasdusts. But in this we see that he was
wrong.

In the great Paris Library and elsewhere there are manuscript
volumes containing the stories of the Round Table abridged and
somewhat clumsily combined from the various Prose Romances
of that cycle, such as *Sir Tristan, Lancelot, Palamedes, Giron le
Courtois,* &c., which had been composed, it would seem, by
various Anglo-French gentlemen at the court of Henry III.,
styled, or styling themselves, Gasses le Blunt, Luces du Gast,

---

* The Abate Spotorno in his *Storia Letteraria della Liguria*, II. 219, fixes on
a Genoese philosopher called Andalo del Negro, mentioned by Boccaccio.
† I quote from Galignani's ed. of Prose Works, v. 712. This has "Rusticien de
Puise." In this view of the fictitious character of the names of Rusticien and the
rest, Sir Walter seems to have been following Ritson, as I gather from a quotation in
Dunlop's H. of Fiction. (*Liebrecht's* German Version, p. 63.)

Robert de Borron, and Hélis de Borron.   And these abridgments or recasts are professedly the work of *Le Maistre Rusticien
de Pise.* Several of them were printed at Paris in the end of
the 15th and beginning of the 16th centuries as the works of
Rusticien de Pise; and as the preambles and the like, especially
in the form presented in those printed editions, appear to be due
sometimes to the original composers (as Robert and Hélis de
Borron) and sometimes to Rusticien de Pise the recaster, there
would seem to have been a good deal of confusion made in
regard to their respective personalities.

From a preamble to one of those compilations which undoubtedly belongs to Rustician, and which we shall quote at
length by and bye, we learn that Master Rustician "translated"
(or perhaps *transferred?*) his compilation from a book belonging
to King Edward of England, at the time when that prince went
beyond seas to recover the Holy Sepulchre.   Now Prince
Edward started for the Holy Land in 1270, spent the winter of
that year in Sicily, and arrived in Palestine in May 1271.   He
quitted it again in August, 1272, and passed again by Sicily,
where in January, 1273, he heard of his father's death and his own
consequent accession.   Paulin Paris supposes that Rustician
was attached to the Sicilian Court of Charles of Anjou, and that
Edward "may have deposited with that king the Romances
of the Round Table, of which all the world was talking, but the
manuscripts of which were still very rare, especially those of the
work of Helye de Borron * . . . . whether by order, or only
with permission of the King of Sicily, our Rustician made
haste to read, abridge, and re-arrange the whole, and when
Edward returned to Sicily he recovered possession of the
book from which the indefatigable Pisan had extracted the
contents."

But this I believe is, in so far as it passes the facts stated in
Rustician's own preamble, pure hypothesis, for nothing is cited
that connects Rustician with the King of Sicily.   And if there
be not some such confusion of personality as we have alluded to,
in another of the preambles, which is quoted by Dunlop as an
utterance of Rustician's, that personage would seem to claim to
have been a comrade in arms of the two de Borrons.   We

---

* *Giron le Courtois,* and the conclusion of *Tristan.*

might, therefore, conjecture that Rustician himself had accompanied Prince Edward to Syria.*

40. Rustician's literary work appears from the extracts and remarks of Paulin Paris to be that of an industrious simple Character of man, without method or much judgment. "The haste Rustician's Romance with which he worked is too perceptible; the advencompila- tions. tures are told without connection; you find long stories of Tristan followed by adventures of his father Meliadus." For the latter derangement of historical sequence we find a quaint and ingenuous apology offered in Rustician's epilogue to Giron le Courtois :—

"Cy fine le Maistre Rusticien de Pise son conte en louant et regraciant le Père le Filz et le Saint Esperit, et ung mesme Dieu, Filz de la Benoiste Vierge Marie, de ce qu'il m'a doné grace, sens, force, et mémoire, temps et lieu, de me mener à fin de si haulte et si noble matière come ceste-cy dont j'ay traicté les faiz et proesses recitez et recordez à mon livre. Et se aucun me demandoit pour quoy j'ay parlé de Tristan avant que de son père le Roy Meliadus, le respons que ma matière n'estoist pas congneue. Car je ne puis pas scavoir tout, ne mettre toutes mes paroles par ordre. Et ainsi fine mon conte. Amen."†

In a passage of these compilations the Emperor Charlemagne is asked whether in his judgment King Meliadus or his son Tristan were the better man? The Emperor's answer is: "I should say that the King Meliadus was the better man, and I will tell you why I say so. As far as I can see, everything that Tristan did was done for Love, and his great feats would never have been done but under the constraint of Love, which was his

---

* The passage runs thus as quoted (from the preamble of the *Meliadus*—I suspect in one of the old printed editions) :—

"Aussi Luces du Gau (Gas) translata en langue Françoise une partie de l'Hystoire de Monseigneur Tristan, et moins assez qu'il ne deust. Moult commença bien son livre et si ny mist tout les faicts de Tristan, ains la greigneur partie. Après s'en entremist Messire Gasse le Blond, qui estoit parent au Roy Henry, et divisa l'Hystoire de Lancelot du Lac, et d'autre chose ne parla il mye grandement en son livre. Messire Robert de Borron s'en entremist et Helye de Borron, par la prière du dit Robert de Borron, *et pource que compaignons feusmes d'armes longuement*, je commencay mon livre," etc. (*Liebrecht's Dunlop*, p. 80.) If this passage be authentic it would set beyond doubt the age of the de Borrons and the other writers of Anglo-French Round Table Romances, who are placed by the *Hist. Littéraire de la France*, and apparently by Fr. Michel, under Henry II. I have no means of pursuing the matter, and have preferred to follow Paulin Paris, who places them under Henry III. I notice, moreover, that the *Hist. Litt.* (xv. p. 498) puts not only the de Borrons but Rustician himself under Henry II.; and, as the last view is certainly an error, the first is probably so too.

† Transc. from MS. 6975 (now Fr. 355) of Paris Library.

spur and goad. Now that never can be said of King Meliadus! For what deeds he did, he did them not by dint of Love, but by dint of his strong right arm. Purely out of his own goodness he did good, and not by constraint of Love." " It will be seen," remarks on this Paulin Paris, " that we are here a long way removed from the ordinary principles of Round Table Romances. And one thing besides will be manifest, viz., that Rusticien de Pise was no Frenchman!" *

The same discretion is shown even more prominently in a passage of one of his compilations, which contains the romances of Arthur, Gyron, and Meliadus (No. 6975—see last note but one) :—

" No doubt," Rustician says, " other books tell the story of the Queen Ginevra and Lancelot differently from this ; and there were certain passages between them of which the Master, in his concern for the honour of both those personages, will say not a word." Alas, says the French Bibliographer, that the copy of Lancelot, which fell into the hands of poor Francesca of Rimini, was not one of those *expurgated* by our worthy friend Rustician ! †

41. A question may still occur to an attentive reader as to the identity of this Romance-compiler Rusticien de Pise with the Messire *Rustacians de Pise*, of a solitary MS. of Polo's work (though the oldest and most authentic), a name which appears in other copies as *Rusta Pisan, Rasta Pysan, Rustichelus Civis Pisanus, Rustico, Restazio da Pisa, Stazio da Pisa*, and who is stated in the preamble to have acted as the Traveller's scribe at Genoa.

Identity of the Romance Compiler with Polo's fellow-prisoner.

M. Pauthier indeed ‡ asserts that the French of the MS. Romances of Rusticien de Pise is of the same barbarous character as that of the early French MS. of Polo's Book to which we have just alluded, and which we shall show to be the nearest presentation of the work as originally dictated by the Traveller. The language of the latter MS. is so peculiar that this would be almost perfect evidence of the identity of the writers, if it were really the fact. A cursory inspection which I have made of two of those MSS. in Paris, and the extracts which I have given

---

* *MSS. François*, iii. 60-61. † *Ibid.* 56-59.
‡ *Introd.* pp. lxxxvi.-vii. note.

and am about to give, do not, however, by any means support M. Pauthier's view. Nor would that view be consistent with the judgment of so competent an authority as Paulin Paris, implied in his calling Rustician a *nom recommandable* in old French literature, and his speaking of him as "versed in the secrets of the French Romance Tongue." * In fact the difference of language in the two cases would really be a difficulty in the way of identification, if there were room for doubt. This, however, Paulin Paris seems to have excluded finally, by calling attention to the peculiar formula of preamble which is common to the Book of Marco Polo and to one of the Romance compilations of Rusticien de Pise.

The former will be found in English at pp. 1, 2, of our Translation ; but we give a part of the original below † for comparison with the preamble to the Romances of Meliadus, Tristan, and Lancelot, as taken from MS. 6961 (Fr. 340) of the Paris Library :—

"*Seigneurs Empereurs et Princes, Ducs et Contes et Barons et Chevaliers et Vavasseurs et Bourgeois, et tous les preudommes de cestui monde qui avez talent de vous deliter en rommans, si prenez cestui (livre) et le faites lire de chief en chief, si orrez toutes les grans aventure* qui advindrent entre les Chevaliers errans du temps au Roy Uter Pendragon, jusques à le temps au Roy Artus son fils, et des compaignons de la Table Ronde. Et sachiez tout vraiment que cist livres fust translatez du livre Monseigneur Edouart le Roy d'Engleterre en cellui temps qu'il passa oultre la mer au service nostre Seigneur Damedieu pour conquester le Sant Sepulcre, et Maistre Rusticiens de Pise, lequel est ymaginez yci dessus,‡ compila ce rommant, car il en translata toutes les merveilleuses nouvelles et aventures qu'il trouva en celle livre et traita tout certainement de toutes les aventures du monde, et si sachiez qu'il traitera plus de Monseigneur Lancelot du Lac, et Monsʳ Tristan le fils au Roy Meliadus de Leonnoie que d'autres, porcequ'ilz furent sans faille les meilleurs chevaliers qui à ce temps furent en terre ; et li Maistres en dira de ces deux pluseurs choses et pluseurs nouvelles que l'en treuvera escript en tous les autres livres ; et porce que le Maistres les trouva escript au Livre d'Engleterre."

"Certainly," Paulin Paris observes, "there is a singular

---

* See *Jour. As.* sér. II. tom. xii. p. 251.

† *Seignors Enperaor, ⁊ Rois, Dux ⁊ Marquois, Cuens, Chevaliers ⁊ Bargions* [for Borgiois] ⁊ *toutes gens qe uoles sauoir les deuerses jenerasions des homes, ⁊* les deuersités des deuerses region dou monde, *si prennés cestui liure* ⁊ *le feites lire* ⁊ *chi trouerés toutes les grandismes meruoilles,*" etc.

‡ The portrait of Rustician here referred to would have been a precious illustration for our book. But unfortunately it has not been transferred to MS. 6961, nor apparently to any other noticed by Paulin Paris.

Palazzo di S. Giorgio, Genoa. [*To face p. 62.*

analogy between these two prefaces. And it must be remarked that the formula is not an ordinary one with translators, compilers, or authors of the 13th and 14th centuries. Perhaps you would not find a single other example of it." *

This seems to place beyond question the identity of the Romance-compiler of Prince Edward's suite in 1270, and the Prisoner of Genoa in 1298.

42. In Dunlop's History of Fiction a passage is quoted from the preamble of *Meliadus*, as set forth in the Paris printed edition of 1528, which gives us to understand that Rusticien de Pise had received as a reward for some of his compositions from King Henry III. the prodigal gift of two *chateaux*. I gather, however, from passages in the work of Paulin Paris that this must certainly be one of those confusions of persons to which I have referred before, and that the recipient of the chateaux was in reality Helye de Borron, the author of some of the originals which Rustician manipulated.† This supposed incident in Rustician's scanty history must therefore be given up.

Further particulars concerning Rustician.

We call this worthy *Rustician* or *Rusticiano*, as the nearest probable representation in Italian form of the *Rusticien* of the Round-Table MSS. and the *Rustacians* of the old text of Polo. But it is highly probable that his real name was *Rustichello*, as is suggested by the form *Rustichelus* in the early Latin version published by the *Société de Géographie*. The change of one liquid for another never goes for much in Italy,‡ and Rustichello might easily Gallicize himself as Rusticien. In a very long list of Pisan officials during the Middle Ages I find several bearing the name of *Rustichello* or *Rustichelli*, but no *Rusticiano* or *Rustigiano*.§

Respecting him we have only to add that the peace between Genoa and Venice was speedily followed by a treaty between Genoa and Pisa. On the 31st July, 1299, a truce for twenty-five years was signed between those two

---

* *Jour. As.* as above.

† See *Liebrecht's Dunlop*, p. 77 ; and *MSS. François*, II. 349, 353. The alleged gift to Rustician is also put forth by D'Israeli the Elder in his *Amenities of Literature*, 1841, I. p. 103.

‡ *E.g.* Geronimo, *Girolamo;* and garofalo, *garofano;* Cristoforo, *Cristovalo;* gonfalone, *gonfanone*, etc.

§ See the List in *Archivio Stor. Ital.* VI. p. 64, *seqq.*

Republics. It was a very different matter from that between Genoa and Venice, and contained much that was humiliating and detrimental to Pisa. But it embraced the release of prisoners; and those of Meloria, reduced it is said to less than one tithe of their original number, had their liberty at last. Among the prisoners then released no doubt Rustician was one. But we hear of him no more.

VIII. Notices of Marco Polo's History, after the Termination of his Imprisonment at Genoa.

43. A few very disconnected notices are all that can be collected of matter properly biographical in relation to the quarter

Death of
Marco's
Father
before 1300. century during which Marco Polo survived the Genoese captivity.

Will of his
brother
Maffeo. We have seen that he would probably reach Venice in the course of August, 1299. Whether he found his aged father alive is not known; but we know at least that a year later (31st August, 1300) Messer Nicolo was no longer in life.

This we learn from the Will of the younger Maffeo, Marco's brother, which bears the date just named, and of which we give an abstract below.* It seems to imply strong regard for the

---

* 1. The Will is made in prospect of his voyage to Crete.

2. He had drafted his will with his own hand, sealed the draft, and made it over to Pietro Pagano, priest of S. Felice and Notary, to draw out a formal testament in faithful accordance therewith in case of the Testator's death ; and that which follows is the substance of the said draft rendered from the vernacular into Latin. ("Ego Matheus Paulo . . . volens ire in Cretam, ne repentinus casus hujus vite fragilis me subreperet intestatum, mea propria manu meum scripsi et condidi testamentum, rogans Petrum Paganum ecclesie Scti. Felicis presbiterum et Notarium, sana mente et integro consilio, ut, secundum ipsius scripturam quam sibi tunc dedi meo sigillo munitam, meum scriberet testamentum, si me de hoc seculo contigeret pertransire ; cujus scripture tenor translato vulgari in latinum per omnia talis est.")

3. Appoints as Trustees Messer Maffeo Polo his uncle, Marco Polo his brother, Messer Nicolo Secreto (or Sagredo) his father-in-law, and Felix Polo his cousin (consanguineum).

4. Leaves 20 soldi to each of the Monasteries from Grado to Capo d'Argine ; and 150 lire to all the congregations of Rialto, on condition that the priests of these maintain an annual service in behalf of the souls of his father, mother, and self.

5. To his daughter Fiordelisa 2000 lire to marry her withal. To be invested in safe mortgages in Venice, and the interest to go to her.

Also leaves her the interest from 1000 lire of his funds in Public Debt (? de meis imprestitis) to provide for her till she marries. After her marriage this 1000 lire and its interest shall go to his male heir if he has one, and failing that to his brother Marco.

testator's brother Marco, who is made inheritor of the bulk of
the property, failing the possible birth of a son. I have already
indicated some conjectural deductions from this document. I
may add that the terms of the second clause, as quoted in the
note, seem to me to throw considerable doubt on the genealogy
which bestows a large family of sons upon this brother Maffeo.
If he lived to have such a family it seems improbable that the
draft which he thus left in the hands of a notary, to be converted
into a Will in the event of his death (a curious example of the
validity attaching to all acts of notaries in those days), should
never have been superseded, but should actually have been so
converted after his death, as the existence of the parchment

---

6. To his wife Catharine 400 *lire* and all her clothes as they stand now. To the
Lady Maroca 100 *lire*.

7. To his natural daughter Pasqua 400 *lire* to marry her withal. Or, if she likes
to be a nun, 200 *lire* shall go to her convent and the other 200 shall purchase securities
for her benefit. After her death these shall come to his male heir, or failing that be sold,
and the proceeds distributed for the good of the souls of his father, mother, and self.

8. To his natural brothers Stephen and Giovannino he leaves 500 *lire*. If one
dies the whole to go to the other. If both die before marrying, to go to his male heir ;
failing such, to his brother Marco or *his* male heir.

9. To his uncle Giordano Trevisano 200 *lire*. To Marco de Tumba 100.
To Fiordelisa, wife of Felix Polo, 100. To Maroca, the daughter of the late
Pietro Trevisano, living at Negropont, 100. To Agnes, wife of Pietro Lion, 100 ;
and to Francis, son of the late Pietro Trevisano, in Negropont, 100.

10. To buy Public Debt producing an annual 20 *lire ai grossi* to be paid yearly to
Pietro Pagano, Priest of S. Felice, who shall pray for the souls aforesaid : on death of
said Pietro the income to go to Pietro's cousin Lionardo, Clerk of S. Felice ; and after
him always to the senior priest of S. Giovanni Grisostomo with the same obligation.

11. Should his wife prove with child and bear a son or sons they shall have his
whole property not disposed of. If a daughter, she shall have the same as Fiordelisa.

12. If he have no male heir his Brother Marco shall have the Testator's share of
his Father's bequest, and 2000 *lire* besides. Cousin Nicolo shall have 500 *lire*, and
Uncle Maffeo 500.

13. Should Daughter Fiordelisa die unmarried her 2000 *lire* and interest to go
to his male heir, and failing such to Brother Marco and his male heir. But in that
case Marco shall pay 500 *lire* to Cousin Nicolo or his male heir.

14. Should his wife bear him a male heir or heirs, but these should die under age,
the whole of his undisposed property shall go to Brother Marco or his male heir.
But in that case 500 *lire* shall be paid to Cousin Nicolo.

15. Should his wife bear a daughter and she die unmarried, her 2000 *lire* and
interest shall go to Brother Marco, with the same stipulation in behalf of Cousin Nicolo.

16. Should the whole amount of his property between cash and goods not amount
to 10,000 *lire* (though he believes he has fully as much), his bequests are to be ratably
diminished, except those to his own children which he does not wish diminished.
Should any legatee die before receiving the bequest, its amount shall fall to the
Testator's heir male, and failing such, the half to go to Marco or his male heir, and
the other half to be distributed for the good of the souls aforesaid.

The witnesses are Lionardo priest of S. Felice, Lionardo clerk of the same, and
the Notary Pietro Pagano priest of the same.

seems to prove. But for this circumstance we might suppose the Marcolino mentioned in the ensuing paragraph to have been a son of the younger Maffeo.

Messer Maffeo, the uncle, was, we see, alive at this time. We do not know the year of his death. But it is alluded to by Friar Pipino in the Preamble to his Translation of the Book, supposed to have been executed about 1315-1320; and we learn from a document in the Venetian archives (see p. 77) that it must have been previous to 1318, and subsequent to February 1309, the date of his last Will. The Will itself is not known to be extant, but from the reference to it in this document we learn that he left 1000 *lire* of public debt * (? *imprestitorum*) to a certain Marco Polo, called *Marcolino*. The relationship of this Marco to old Maffeo is not stated, but we may suspect him to have been an illegitimate son. [Marcolino was a son of Nicolo, son of Marco the Elder; see vol. ii., *Calendar*, No. 6.—H. C.]

44. In 1302 occurs what was at first supposed to be a glimpse of Marco as a citizen, slight and quaint enough ; being a resolution on the Books of the Great Council to exempt the respectable Marco Polo from the penalty incurred by him on account of the omission to have his water-pipe duly inspected. But since our Marco's claims to the designation of *Nobilis Vir* have been established, there is a doubt whether the *providus vir* or *prud'-homme* here spoken of may not have been rather his namesake Marco Polo of Cannareggio or S. Geremia, of whose existence we learn from another entry of the same year.† It is, however, possible

Side note: Documentary notices of Polo at this time. The sobriquet of Milione.

---

* According to Romanin (I. 321) the *lira dei grossi* was also called *Lira d'imprestidi*, and if the *lire* here are to be so taken, the sum will be 10,000 ducats, the largest amount by far that occurs in any of these Polo documents, unless, indeed, the 1000 *lire* in § 5 of Maffeo Junior's Will be the like ; but I have some doubt if such lire are intended in either case.

† "(Resolved) That grace be granted to the respectable MARCO PAULO, relieving him of the penalty he has incurred for neglecting to have his water-pipe examined, seeing that he was ignorant of the order on that subject." (See *Appendix C.* No. 3.) The other reference, to M. Polo, of S. Geremia, runs as follows :—

[*MCCCII. indic. XV. die VIII. Macii q̃ fiat grã Güillõ aurifici q̃ ipe absolvat a pena ĩ qua dicit icurisse p̃ uno spõtono sibi iũeto veuiẽdo de Mestre p̃pe domũ Macĩ Pauli de Canareglo ũi descenderat ad bibendũ.*]

"That grace be granted to William the Goldsmith, relieving him of the penalty which he is stated to have incurred on account of a spontoon (*spontono*, a loaded bludgeon) found upon him near the house of MARCO PAULO of Cannareggio, where he had landed to drink on his way from Mestre." (See *Cicogna*, V. p. 606.)

that Marco the Traveller was called to the Great Council *after* the date of the document in question.

We have seen that the Traveller, and after him his House and his Book, acquired from his contemporaries the surname, or nickname rather, of *Il Milione*. Different writers have given different explanations of the origin of this name ; some, beginning with his contemporary Fra Jacopo d'Acqui (*supra*, p. *54*), ascribing it to the family's having brought home a fortune of a million of *lire*, in fact to their being *millionaires*. This is the explanation followed by Sansovino, Marco Barbaro, Coronelli, and others.* More far-fetched is that of Fontanini, who supposes the name to have been given to the Book as containing a great number of stories, like the *Cento Novelle* or the *Thousand and One Nights!* But there can be no doubt that Ramusio's is the true, as it is the natural, explanation ; and that the name was bestowed on Marco by the young wits of his native city, because of his frequent use of a word which appears to have been then unusual, in his attempts to convey an idea of the vast wealth and magnificence of the Kaan's Treasury and Court.† Ramusio has told us (*supra*, p. *6*) that he had seen Marco styled by this sobriquet in the Books of the Signory ; and it is pleasant to be able to confirm this by the next document which we cite. This is an extract from the Books of the Great Council under 10th April, 1305, condoning the offence of a certain Bonocio of Mestre in smuggling wine, for whose penalty one of the sureties had been the NOBILIS VIR MARCHUS PAULO MILIONI.‡

It is alleged that long after our Traveller's death there was always, in the Venetian Masques, one individual who assumed the character of Marco Milioni, and told Munchausenlike stories

---

* *Sansovino, Venezia, Città Nobilissima e Singolare, Descritta*, etc., Ven. 1581, f. 236 *v.* ; *Barbaro, Alberi ; Coronelli, Atlante Veneto*, I. 19.

† The word *Millio* occurs several times in the Chronicle of the Doge Andrea Dandolo, who wrote about 1342 ; and *Milion* occurs at least once (besides the application of the term to Polo) in the History of Giovanni Villani ; viz. when he speaks of the Treasury of Avignon :—" *diciotto* milioni *di fiorini d'oro* ec. *che ogni* milione *è mille migliaja di fiorini d' oro la valuta.*" (xi. 20, § 1 ; *Ducange*, and *Vocab. Univ. Ital.*). But the definition, thought necessary by Villani, in itself points to the use of the word as rare. *Domilion* occurs in the estimated value of houses at Venice in 1367, recorded in the *Cronaca Magna* in St. Mark's Library. (*Romanin*, III. 385).

‡ " Also ; that Pardon be granted to Bonocio of Mestre for that 152 *lire* in which he stood condemned by the Captains of the Posts, on account of wine smuggled by him, in such wise : to wit, that he was to pay the said fine in 4 years by annual

to divert the vulgar. Such, if this be true, was the honour of our prophet among the populace of his own country.*

45. A little later we hear of Marco once more, as presenting a copy of his Book to a noble Frenchman in the service of Charles of Valois.

This Prince, brother of Philip the Fair, in 1301 had married Catharine, daughter and heiress of Philip de Courtenay, titular Polo's relations with Thibault de Cepoy. Emperor of Constantinople, and on the strength of this marriage had at a later date set up his own claim to the Empire of the East. To this he was prompted by Pope Clement V., who in the beginning of 1306 wrote to Venice, stimulating that Government to take part in the enterprise. In the same year, Charles and his wife sent as their envoys to Venice, in connection with this matter, a noble knight called THIBAULT DE CEPOY, along with an ecclesiastic of Chartres called Pierre le Riche, and these two succeeded in executing a treaty of alliance with Venice, of which the original, dated 14th December, 1306, exists at Paris. Thibault de Cepoy eventually went on to Greece with a squadron of Venetian Galleys, but accomplished nothing of moment, and returned to his master in 1310.†

During the stay of Thibault at Venice he seems to have made acquaintance with Marco Polo, and to have received from him a copy of his Book. This is recorded in a curious note which appears on two existing MSS. of Polo's Book, viz., that

---

instalments of one fourth, to be retrenched from the pay due to him on his journey in the suite of our ambassadors, with assurance that anything then remaining deficient of his instalments should be made good by himself or his securities. And his securities are the Nobles Pietro Morosini and MARCO PAULO MILIOÑ." Under *Milioñ* is written in an ancient hand "*mortuus.*" (See *Appendix C*, No. 4.)

* Humboldt tells this (*Examen*, II. 221), alleging *Jacopo d'Acqui* as authority; and Libri (*H. des Sciences Mathématiques*, II. 149), quoting *Doglioni, Historia Veneziana.* But neither authority bears out the citations. The story seems really to come from Amoretti's commentary on the *Voyage du Cap. L. F. Maldonado*, Plaisance, 1812, p. 67. Amoretti quotes as authority *Pignoria, Degli Dei Antichi.*

An odd revival of this old libel was mentioned to me recently by Mr. George Moffatt. When he was at school it was common among the boys to express incredulity by the phrase: "Oh, what a Marco Polo!"

† Thibault, according to Ducange, was in 1307 named Grand Master of the Arblasteers of France; and Buchon says his portrait is at Versailles among the Admirals (No. 1170). Ramon de Muntaner fell in with the Seigneur de Cepoy in Greece, and speaks of him as "but a Captain of the Wind, as his Master was King of the Wind." (See *Ducange, H. de l'Empire de Const. sous les Emp. François*, Venice ed. 1729, pp. 109, 110; *Buchon, Chroniques Etrangères*, pp. lv. 467-470.)

Miracle of S. Lorenzo.

of the Paris Library (10,270 or Fr. 5649), and that of Bern, which is substantially identical in its text with the former, and is, as I believe, a copy of it.* The note runs as follows :—

"Here you have the Book of which My Lord THIEBAULT, Knight and LORD OF CEPOY, (whom may God assoil!) requested a copy from SIRE MARC POL, Burgess and Resident of the City of Venice. And the said Sire Marc Pol, being a very honourable Person, of high character and respect in many countries, because of his desire that what he had witnessed should be known throughout the World, and also for the honour and reverence he bore to the most excellent and puissant Prince my Lord CHARLES, Son of the King of France and COUNT OF VALOIS, gave and presented to the aforesaid Lord of Cepoy the first copy (that was taken) of his said Book after he had made the same. And very pleasing it was to him that his Book should be carried to the noble country of France and there made known by so worthy a gentleman. And from that copy which the said Messire Thibault, Sire de Cepoy above-named, did carry into France, Messire John, who was his eldest son and is the present Sire de Cepoy,† after his Father's decease did have a copy made, and that very first copy that was made of the Book after its being carried into France he did present to his very dear and dread Lord Monseigneur de Valois. Thereafter he gave copies of it to such of his friends as asked for them.

"And the copy above-mentioned was presented by the said Sire Marc Pol to the said Lord de Cepoy when the latter went to Venice, on the part of Monseigneur de Valois and of Madame the Empress his wife, as Vicar General for them both in all the Territories of the Empire of Constantinople. And this happened in the year of the Incarnation of our Lord Jesus Christ one thousand three hundred and seven, and in the month of August."

Of the bearings of this memorandum on the literary history of Polo's Book we shall speak in a following section.

46. When Marco married we have not been able to ascertain, but it was no doubt early in the 14th century, for in 1324, we find that he had two married daughters besides one unmarried. His wife's Christian name was *Donata*, but of her family we have as yet found no assurance. I suspect, however, that her name may have been Loredano (*vide infra*, p. 77).

His marriage and his daughters. Marco as a merchant.

Under 1311 we find a document which is of considerable in-

---

* The note is not found in the Bodleian MS., which is the third known one of this precise type.

† Messire Jean, the son of Thibault, is mentioned in the accounts of the latter in the *Chambre des Comptes* at Paris, as having been with his Father in Romania. And in 1344 he commanded a confederate Christian armament sent to check the rising power of the Turks, and beat a great Turkish fleet in the Greek seas. (*Heyd*. I. 377 ; *Buchon*, 468.)

terest, because it is the only one yet discovered which exhibits Marco under the aspect of a practical trader. It is the judgment of the Court of Requests upon a suit brought by the NOBLE MARCO POLO of the parish of S. Giovanni Grisostomo against one Paulo Girardo of S. Apollinare. It appears that Marco had entrusted to the latter as a commission agent for sale, on an agreement for half profits, a pound and a half of musk, priced at six *lire of grossi* (about *22l. 10s.* in value of silver) the pound. Girardo had sold half-a-pound at that rate, and the remaining pound which he brought back was deficient of a *saggio*, or, one-sixth of an ounce, but he had accounted for neither the sale nor the deficiency. Hence Marco sues him for three *lire of Grossi*, the price of the half-pound sold, and for twenty *grossi* as the value of the saggio. And the Judges cast the defendant in the amount with costs, and the penalty of imprisonment in the common gaol of Venice if the amounts were not paid within a suitable term.*

Again in May, 1323, probably within a year of his death, Ser Marco appears (perhaps only by attorney), before the Doge and his judicial examiners, to obtain a decision respecting a question touching the rights to certain stairs and porticoes in contact with his own house property, and that obtained from his wife, in S. Giovanni Grisostomo. To this allusion has been already made (*supra*, p. *31*).

47. We catch sight of our Traveller only once more. It is
Marco Polo's Last Will and Death. on the 9th of January, 1324 ; he is labouring with disease, under which he is sinking day by day ; and he has sent for Giovanni Giustiniani, Priest of S. Proculo and Notary, to make his Last Will and Testament. It runs thus :—

"IN THE NAME OF THE ETERNAL GOD AMEN !

"In the year from the Incarnation of our Lord Jesus Christ 1323, on the

---

* The document is given in *Appendix C*, No. 5. It was found by Comm. Barozzi, the Director of the Museo Civico, when he had most kindly accompanied me to aid in the search for certain other documents in the archives of the *Casa di Ricovero*, or Poor House of Venice. These archives contain a great mass of testamentary and other documents, which probably have come into that singular depository in connection with bequests to public charities.

The document next mentioned was found in as strange a site, viz., the *Casa degli Esposti* or Foundling Hospital, which possesses similar muniments. This also I owe to Comm. Barozzi, who had noted it some years before, when commencing an arrangement of the archives of the Institution.

(Dimensions of Original, 26·4 inches by 9·4 inches).

to face page 70

SLIGHTLY REDUCED FROM A PHOTOGRAPH SPECIALLY TAKEN IN
ST. MARK'S LIBRARY BY SIGNOR BERTANI.

9th day of the month of January, in the first half of the 7th Indiction,* at Rialto.

" It is the counsel of Divine Inspiration as well as the judgment of a provident mind that every man should take thought to make a disposition of his property before death become imminent, lest in the end it should remain without any disposition :

"Wherefore I MARCUS PAULO of the parish of St. John Chrysostom, finding myself to grow daily feebler through bodily ailment, but being by the grace of God of a sound mind, and of senses and judgment unimpaired, have sent for JOHN GIUSTINIANI, Priest of S. Proculo and Notary, and have instructed him to draw out in complete form this my Testament :

"Whereby I constitute as my Trustees DONATA my beloved wife, and my dear daughters FANTINA, BELLELA, and MORETA,† in order that after my decease they may execute the dispositions and bequests which I am about to make herein.

" First of all : I will and direct that the proper Tithe be paid.‡ And over and above the said tithe I direct that 2000 *lire* of Venice denari be distributed as follows : §

" *Viz.*, 20 *soldi* of Venice *grossi* to the Monastery of St. Lawrence where I desire to be buried.

---

* The Legal Year at Venice began on the 1st of March. And 1324 was 7th of the Indiction. Hence the date is, according to the modern Calendar, 1324.

† Marsden says of Moreta and Fantina, the only daughters named by Ramusio, that these may be thought rather familiar terms of endearment than baptismal names. This is a mistake however. *Fantina* is from one of the parochial saints of Venice, S. Fantino, and the male name was borne by sundry Venetians, among others by a son of Henry Dandolo's. Moreta is perhaps a variation of Maroca, which seems to have been a family name among the Polos. We find also the male name of Bellela, written *Bellello, Bellero, Belletto*.

‡ The *Decima* went to the Bishop of Castello (eventually converted into Patriarch of Venice) to divide between himself, the Clergy, the Church, and the Poor. It became a source of much bad feeling, which came to a head after the plague of 1348, when some families had to pay the tenth three times within a very short space. The existing Bishop agreed to a composition, but his successor Paolo Foscari (1367) claimed that on the death of every citizen an exact inventory should be made, and a full tithe levied. The Signory fought hard with the Bishop, but he fled to the Papal Court and refused all concession. After his death in 1376 a composition was made for 5500 ducats yearly. (*Romanin*, II. 406 ; III. 161, 165.)

§ There is a difficulty about estimating the value of these sums from the variety of Venice pounds or *lire*. Thus the *Lira dei piccoli* was reckoned 3 to the ducat or zecchin, the *Lira ai grossi* 2 to the ducat, but the *Lira* dei *grossi* or *Lira d'imprestidi* was equal to 10 ducats, or (allowing for higher value of silver then) about 3*l*. 15*s*. ; a little more than the equivalent of the then Pound sterling. This last money is *specified* in some of the bequests, as in the 20 soldi (or 1 lira) to St. Lorenzo, and in the annuity of 8 lire to Polo's wife ; but it seems doubtful what money is meant when *libra* only or *libra denariorum venetorum* is used. And this doubt is not new. Galicciolli relates that in 1232 Giacomo Menotto left to the Church of S. Cassiano as an annuity *libras denariorum venetorum quatuor*. Till 1427 the church received the income as of *lire dei piccoli*, but on bringing a suit on the subject it was adjudged that *lire ai grossi* were to be understood. (*Delle Mem. Venet. Ant.* II. 18.) This story, however, cuts both ways, and does not decide our doubt.

"Also 300 *lire* of Venice denari to my sister-in-law YSABETA QUIRINO,* that she owes me.

"Also 40 *soldi* to each of the Monasteries and Hospitals all the way from Grado to Capo d'Argine.†

"Also I bequeath to the Convent of SS. Giovanni and Paolo, of the Order of Preachers, that which it owes me, and also 10 *lire* to Friar RENIER, and 5 *lire* to Friar BENVENUTO the Venetian, of the Order of Preachers, in addition to the amount of his debt to me.

"I also bequeath 5 *lire* to every Congregation in Rialto, and 4 *lire* to every Guild or Fraternity of which I am a member.‡

"Also I bequeath 20 *soldi* of Venetian grossi to the Priest Giovanni Giustiniani the Notary, for his trouble about this my Will, and in order that he may pray the Lord in my behalf.

"Also I release PETER the Tartar, my servant, from all bondage, as completely as I pray God to release mine own soul from all sin and guilt. And I also remit him whatever he may have gained by work at his own house ; and over and above I bequeath him 100 *lire* of Venice denari.§

---

* The form of the name *Ysabeta* aptly illustrates the transition that seems so strange from *Elizabeth* into the *Isabel* that the Spaniards made of it.

† *I.e.* the extent of what was properly called the Dogado, all along the Lagoons from Grado on the extreme east to Capo d'Argine (Cavarzere at the mouth of the Adige) on the extreme west.

‡ The word rendered *Guilds* is "*Scholarum.*" The crafts at Venice were united in corporations called *Fraglie* or *Scholae*, each of which had its statutes, its head called the *Gastald*, and its place of meeting under the patronage of some saint. These acted as societies of mutual aid, gave dowries to poor girls, caused masses to be celebrated for deceased members, joined in public religious processions, etc., nor could any craft be exercised except by members of such a guild. (*Romanin*, I. 390.)

§ A few years after Ser Marco's death (1328) we find the Great Council granting to this Peter the rights of a natural Venetian, as having been a long time at Venice, and well-conducted. (See App. C, *Calendar of Documents*, No. 13.) This might give some additional colour to M. Pauthier's supposition that this Peter the Tartar was a faithful servant who had accompanied Messer Marco from the East 30 years before. But yet the supposition is probably unfounded. Slavery and slave-trade were very prevalent at Venice in the Middle Ages, and V. Lazari, a writer who examined a great many records connected therewith, found that by far the greater number of slaves were described as *Tartars*. There does not seem to be any clear information as to how they were imported, but probably from the factories on the Black Sea, especially Tana after its establishment.

A tax of 5 ducats per head was set on the export of slaves in 1379, and as the revenue so received under the Doge Tommaso Mocenigo (1414-1423) amounted (so says Lazari) to 50,000 ducats, the startling conclusion is that 10,000 slaves yearly were exported ! This it is difficult to accept. The slaves were chiefly employed in domestic service, and the records indicate the women to have been about twice as numerous as the men. The highest price recorded is 87 ducats paid for a Russian girl sold in 1429. All the higher prices are for young women ; a significant circumstance. With the existence of this system we may safely connect the extraordinary frequence of mention of illegitimate children in Venetian wills and genealogies. (See *Lazari, Del Traffico degli Schiavi in Venezia*, etc., in *Miscellanea di Storia Italiana*, I. 463 *seqq.*) In 1308 the Khan Toktai of Kipchak (see Polo, II. 496), hearing that the Genoese and other Franks were in the habit of carrying off Tartar children to sell,

"And the residue of the said 2000 *lire*, free of tithe, I direct to be distributed for the good of my soul, according to the discretion of my trustees.

"Out of my remaining property I bequeath to the aforesaid Donata, my Wife and Trustee, 8 *lire* of Venetian grossi annually during her life, for her own use, over and above her settlement, and the linen and all the household utensils,* with 3 beds garnished.

"And all my other property movable and immovable that has not been disposed of [here follow some lines of mere technicality] I specially and expressly bequeath to my aforesaid Daughters Fantina, Bellela, and Moreta, freely and absolutely, to be divided equally among them. And I constitute them my heirs as regards all and sundry my property movable and immovable, and as regards all rights and contingencies tacit and expressed, of whatsoever kind as hereinbefore detailed, that belong to me or may fall to me. Save and except that before division my said daughter Moreta shall receive the same as each of my other daughters hath received for dowry and outfit [here follow many lines of technicalities, ending]

"And if any one shall presume to infringe or violate this Will, may he incui the malediction of God Almighty, and abide bound under the anathema of the 318 Fathers; and farthermore he shall forfeit to my Trustees aforesaid five pounds of gold;† and so let this my Testament abide in force. The signature of the above named Messer Marco Paulo who gave instructions for this deed.

"‡ I Peter Grifon, Priest, Witness.
"* I Humfrey Barberi, Witness.
"† I John Giustiniani, Priest of S. Proculo, and Notary,
        have completed and authenticated (this testament)."‡

---

sent a force against Caffa, which was occupied without resistance, the people taking refuge in their ships. The Khan also seized the Genoese property in Sarai. (*Heyd.* II. 27.)

* " *Stracium et omne capud massariciorum* "; in Scotch phrase "*napery and plenishing.*" A Venetian statute of 1242 prescribes that a bequest of *massariticum* shall be held to carry to the legatee all articles of common family use except those of gold and silver plate or jeweller's work. (See *Ducange, sub voce.*) *Stracci* is still used technically in Venice for "household linen."

† In the original *aureas libras quinque.* According to Marino Sanudo the Younger (*Vite dei Dogi* in *Muratori*, xxii. 521) this should be pounds or *lire* of *aureole*, the name of a silver coin struck by and named after the Doge *Aurio* Mastropietro (1178-1192): " Ancora fu fatta una Moneta d'argento che si chiamava *Aureola* per la casata del Doge; *è quella Moneta che i Notai de Venezia mettevano di pena sotto i loro instrumenti.*" But this was a vulgar error. An example of the penalty of 5 pounds of gold is quoted from a decree of 960; and the penalty is sometimes expressed " *auri purissimi librae* 5." A coin called the *lira d'oro* or *redonda* is alleged to have been in use before the ducat was introduced. (See *Gallicciolli*, II. 16.) But another authority seems to identify the *lira a oro* with the *lira dei grossi.* (See *Zanetti, Nuova Racc. delle Monete &c. d'Italia*, 1775. I. 308.)

‡ We give a photographic reduction of the original document. This, and the other two Polo Wills already quoted, had come into the possession of the Noble Filippo Balbi, and were by him presented in our own time to the St. Mark's Library. They are all on parchment, in writing of that age, and have been officially examined and declared to be originals. They were first published by

We do not know, as has been said, how long Marco survived the making of this will, but we know, from a scanty series of documents commencing in June of the following year (1325), that he had *then* been some time dead.*

48. He was buried, no doubt, according to his declared wish,

Place of Sepulture.

Professed Portraits of Polo.

in the Church of S. Lorenzo; and indeed Sansovino bears testimony to the fact in a confused notice of our Traveller.† But there does not seem to have been any monument to Marco, though the sarcophagus which had been erected to his father Nicolo, by his own filial care, existed till near the end of the 16th century in the porch or corridor leading to the old Church of S. Lorenzo, and bore the inscription: "SEPULTURA DOMINI NICOLAI PAULO DE CONTRATA S. IOANNIS GRISOSTEMI." The church was renewed from its foundations in 1592, and then, probably, the sarcophagus was cast aside and lost, and with it all certainty as to the position of the tomb.‡

Cicogna, Iscrizioni Veneziane, III. 489-493. We give Marco's in the original language, line for line with the facsimile, in *Appendix C.*

There is no signature, as may be seen, except those of the Witnesses and the Notary. The sole presence of a Notary was held to make a deed valid, and from about the middle of the 13th century in Italy it is common to find no actual signature (even of witnesses) except that of the Notary. The peculiar flourish before the Notary's name is what is called the *Tabellionato,* a fanciful distinctive monogram which each Notary adopted. Marco's Will is unfortunately written in a very cramp hand with many contractions. The other two Wills (of Marco the Elder and Maffeo) are in beautiful and clear Gothic penmanship.

* We have noticed formerly (pp. *14-15, note*) the recent discovery of a document bearing what was supposed to be the autograph signature of our Traveller. The document in question is the Minute of a Resolution of the Great Council, attested by the signatures of three members, of whom the last is MARCUS PAULLO. But the date alone, 11th March, 1324, is sufficient to raise the gravest doubts as to this signature being that of our Marco. And further examination, as I learn from a friend at Venice, has shown that the same name occurs in connection with analogous entries on several subsequent occasions up to the middle of the century. I presume that this Marco Polo is the same that is noticed in our *Appendix B*, II. as a voter in the elections of the Doges Marino Faliero and Giovanni Gradenigo. I have not been able to ascertain his relation to either branch of the Polo family; but I suspect that he belonged to that of S. Geremia, of which there *was* certainly a Marco about the middle of the century.

† "Under the *angiporta* (of S. Lorenzo) [see plate] is buried that Marco Polo surnamed Milione, who wrote the Travels in the New World, and who was the first before Christopher Columbus to discover new countries. No faith was put in him because of the extravagant things that he recounted; but in the days of our Fathers Columbus augmented belief in him, by discovering that part of the world which eminent men had heretofore judged to be uninhabited." (*Venezia . . . . Descritta,* etc., f. *23 v.*) Marco Barbaro attests the same inscription in his Genealogies (copy in Museo Civico at Venice).

‡ *Cicogna,* II. 385.

[To face p. 74.

Pavement in front of San Lorenzo, Venice.

There is no portrait of Marco Polo in existence with any claim to authenticity. The quaint figure which we give in the *Bibliography*, vol. ii. p. 555, extracted from the earliest printed edition of his book, can certainly make no such pretension. The oldest one after this is probably a picture in the collection of Monsignor Badia at Rome, of which I am now able, by the owner's courtesy, to give a copy. It is set down in the catalogue to Titian, but is probably a work of 1600, or thereabouts, to which the aspect and costume belong. It is inscribed "*Marcus Polvs Venetvs Totivs Orbis et Indie Peregrator Primus.*" Its history unfortunately cannot be traced, but I believe it came from a collection at

S. Lorenzo as it was in the 15th century.

Urbino. A marble statue was erected in his honour by a family at Venice in the 17th century, and is still to be seen in the Palazzo Morosini-Gattemburg in the Campo S. Stefano in that city. The medallion portrait on the wall of the *Sala dello Scudo* in the ducal palace, and which was engraved in Bettoni's "Collection of Portraits of Illustrious Italians," is a work of imagination painted by Francesco Griselini in 1761.* From this, however, was taken the medal by Fabris, which was struck in 1847 in honour of the last meeting of the Italian Congresso Scientifico; and from the medal again is copied, I believe, the elegant woodcut which adorns the introduction to M. Pauthier's

---

* *Lazari*, xxxi.

edition, though without any information as to its history.   A
handsome bust, by Augusto Gamba, has lately been placed
among the illustrious Venetians in the inner arcade of the Ducal
Palace.*   There is also a mosaic portrait of Polo, opposite the
similar portrait of Columbus in the Municipio at Genoa.

49. From the short series of documents recently alluded to,†
we gather all that we know of the remaining history of Marco

Further
History of
the Polo
Family.

Polo's immediate family.   We have seen in his will an
indication that the two elder daughters, Fantina and
Bellela, were married before his death.   In 1333 we
find the youngest, Moreta, also a married woman, and Bellela
deceased.   In 1336 we find that their mother Donata had died
in the interval.   We learn, too, that Fantina's husband was
MARCO BRAGADINO, and Moreta's, RANUZZO DOLFINO.‡   The
name of Bellela's husband does not appear.

Fantina's husband is probably the Marco Bragadino, son of
Pietro, who in 1346 is mentioned to have been sent as
Provveditore-Generale to act against the Patriarch of Acqui-
leia.§   And in 1379 we find Donna Fantina herself, presumably in widowhood, assessed as a resident of S. Giovanni
Grisostomo, on the *Estimo* or forced loan for the Genoese war,
at 1300 *lire*, whilst Pietro Bragadino of the same parish—her son
as I imagine—is assessed at 1500 *lire*. ‖   [See vol. ii., *Calendar*.]

The documents show a few other incidents which may be
briefly noted.   In 1326 we have the record of a charge against
one Zanino Grioni for insulting Donna Moreta in the Campo
of San Vitale ; a misdemeanour punished by the Council of
Forty with two months' imprisonment.

---

* In the first edition I noticed briefly a statement that had reached me from China
that, in the Temple at Canton vulgarly called " of the 500 gods," there is a foreign
figure which from the name attached had been supposed to represent Marco Polo !
From what I have heard from Mr. Wylie, a very competent authority, this is
nonsense.   The temple contains 500 figures of *Arhans* or Buddhist saints, and one of
these attracts attention from having a hat like a sailor's straw hat.   Mr. Wylie had
not remarked the name.   [A model of this figure was exhibited at Venice at the
international Geographical Congress, in 1881.   I give a reproduction of this figure
and of the Temple of 500 Genii (*Fa Lum Sze*) at Canton, from drawings by Félix
Régamey made after photographs sent to me by my late friend, M. Camille Imbault
Huart, French Consul at Canton.—H. C.]

† These documents are noted in Appendix C, Nos. 9-12, 14, 17, 18.

‡ I can find no *Ranuzzo* Dolfino among the Venetian genealogies, but several
Reniers.   And I suspect Ranuzzo may be a form of the latter name.

§ *Cappellari* (see p. 77, ‡) under *Bragadino*.   ‖ *Ibid.* and *Gallicciolli*, II. 146.

Mosaic Portrait of Marco Polo at Genoa.   [*To face p.* 76.

In March, 1328, Marco Polo, called Marcolino, of St. John Chrysostom (see p. *66*), represents before the *Domini Advocatores* of the Republic that certain *imprestita* that had belonged to the late Maffeo Polo the Elder, had been alienated and transferred in May, 1318, by the late Marco Polo of St. John Chrysostom and since his death by his heirs, without regard to the rights of the said Marcolino, to whom the said Messer Maffeo had bequeathed 1000 *lire* by his will executed on 6th February, 1308 (*i.e.* 1309). The Advocatores find that the transfer was to that extent unjust and improper, and they order that to the same extent it should be revoked and annulled. Two months later the Lady Donata makes rather an unpleasant figure before the Council of Forty. It would seem that on the claim of Messer Bertuccio Quirino a mandate of sequestration had been issued by the Court of Requests affecting certain articles in the Ca' Polo; including two bags of money which had been tied and sealed, but left in custody of the Lady Donata. The sum so sealed was about 80 *lire* of grossi (300*l.* in silver value), but when opened only 45 *lire* and 22 *grossi* (about 170*l.*) were found therein, and the Lady was accused of abstracting the balance *non bono modo*. Probably she acted, as ladies sometimes do, on a strong sense of her own rights, and a weak sense of the claims of law. But the Council pronounced against her, ordering restitution, and a fine of 200 *lire* over and above "*ut ceteris transeat in exemplum.*" *

It will have been seen that there is nothing in the amounts mentioned in Marco's will to bear out the large reports as to his wealth, though at the same time there is no positive ground for a deduction to the contrary.†

The mention in two of the documents of Agnes Loredano as the sister of the Lady Donata suggests that the latter may have belonged to the Loredano family, but as it does not appear whether Agnes was maid or wife this remains uncertain.‡

---

* The *lire* of the fine are not specified ; but probably *ai grossi*, which would be = 37*l.* 10*s.*; not, we hope, *dei* grossi !

† Yet, if the family were so wealthy as tradition represents, it is strange that Marco's brother Maffeo, *after* receiving a share of his father's property, should have possessed barely 10,000 *lire*, probably equivalent to 5000 ducats at most. (See p. 65, *supra*.)

‡ An Agnes Loredano, Abbess of S. Maria delle Vergini, died in 1397. (*Cicogna*, V. 91 and 629.) The interval of 61 years makes it somewhat improbable that it should be the same.

Respecting the further history of the family there is nothing certain, nor can we give unhesitating faith to Ramusio's statement that the last male descendant of the Polos of S. Giovanni Grisostomo was Marco, who died Castellano of Verona in 1417 (according to others, 1418, or 1425),* and that the family property then passed to Maria (or *Anna,* as she is styled in a MS. statement furnished to me from Venice), who was married in 1401 to Benedetto Cornaro, and again in 1414 to Azzo Trevisan. Her descendant in the fourth generation by the latter was Marc Antonio Trevisano,† who was chosen Doge in 1553.

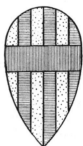

Arms of the Trevisan family.

The genealogy recorded by Marco Barbaro, as drawn up from documents by Ramusio, makes the Castellano of Verona a grandson of our Marco by a son Maffeo, whom we may safely pronounce not to have existed, and makes Maria the daughter of Maffeo, Marco's brother—that is to say, makes a lady marry in 1414 and have children, whose father was born in 1271 at the very latest! The genealogy is given in several other ways, but as I have satisfied myself that they all (except perhaps this of Barbaro's, which we see to be otherwise erroneous) confound together the two distinct families of Polo of S. Geremia and Polo of S. Giov. Grisostomo, I reserve my faith, and abstain from presenting them. Assuming that the Marco or Marcolino Polo, spoken of in the preceding page, was a near relation (as is

---

* In the *Museo Civico* (No. 2271 of the Cicogna collection) there is a commission addressed by the Doge Michiel Steno in 1408, "*Nobili Viro Marcho Paulo,*" nominating him Podestà of Arostica (a Castello of the Vicentino). This is probably the same Marco.

† The descent runs: (1) Azzo = Maria Polo; (2) Febo, Captain at Padua; (3) Zaccaria, Senator; (4) Domenico, Procurator of St. Mark's; (5) Marc' Antonio, Doge (*Cappellari, Campidoglio Veneto,* MS. St. Mark's Lib.).

Marc' Antonio *nolebat ducari* and after election desired to renounce. His friends persuaded him to retain office, but he lived scarcely a year after. (*Cicogna,* IV. 566.) [See p. *8.*]

The Pseudo Marco Polo at Canton.   (To face p. 78.

probable, though perhaps an illegitimate one), he is the only male descendant of old Andrea of San Felice whom we can indicate as having survived Marco himself; and from a study of the links in the professed genealogies I think it not unlikely that both Marco the Castellano of Verona and Maria Trevisan belonged to the branch of S. Geremia.* [See vol. ii., *App. C*, p. 510.]

[49. *bis.*—It is interesting to note some of the *reliques* left by our traveller.

1. The unfortunate Doge of Venice, Marino Faliero, seems to have possessed many souvenirs of Marco Polo, and among them two manuscripts, one in the handwriting of his celebrated fellow-citizen(?), and one adorned with miniatures. M. Julius von Schlosser has reprinted (*Die ältesten Medaillen und die Antike*, Bd. XVIII., *Jahrb. d. Kunsthist. Samml. d. Allerhöchsten Kaiserhauses*, Vienna, 1897, pp. 42-43) from the *Bulletino di arti, industrie e curiosità veneziane*, III., 1880-81, p. 101,† the inventory of the curiosities kept in the "Red Chamber" of Marino Faliero's palace in the Parish of the SS. Apostles; we give the following abstract of it :—

Anno ab incarnacione domini nostri Jesu Christi 1351° indictione sexta

---

* In Appendix B will be found tabulated all the facts that seem to be positively ascertained as to the Polo genealogies.

In the Venetian archives occurs a procuration executed by the Doge in favour of the *Nobilis Vir* SER MARCO PAULO that he may present himself before the king of Sicily; under date, Venice 9th November, 1342. And some years later we have in the Sicilian Archives an order by King Lewis of Sicily, directed to the Maestri Procuratori of Messina, which grants to MARCO POLO of Venice, on account of services rendered to the king's court, the privilege of free import and export at the port of Messina, without payment of customs of goods to the amount annually of 20 ounces. Dated in Catania 13th January, 1346 (1347?).

For the former notice I am indebted to the courtesy of Signor B. Cecchetti of the Venetian Archives, who cites it as "transcribed in the *Commemor.* IV. p. 5"; for the latter to that of the Abate Carini of the *Reale Archivio* at Palermo; it is in *Archivio della Regia Cancellaria* 1343-1357, f. 58.

The mission of this MARCO POLO is mentioned also in a rescript of the Sicilian king Peter II., dated Messina, 14th November, 1340, in reference to certain claims of Venice, about which the said Marco appeared as the Doge's ambassador. This is printed in F. TESTA, *De Vitâ et Rebus Gestis Federici II., Siciliæ Regis*, Panormi, 1775, pp. 267 *seqq.* The Sicilian Antiquary Rosario Gregorio identifies the Envoy with our Marco, dead long before. (See *Opere scelte del Canon Ros. Gregorio*, Palermo, 1845, 3za ediz., p. 352.)

It is possible that this Marco, who from the latter notice seems to have been engaged in mercantile affairs, may have been the Marcolino above mentioned, but it is perhaps on the whole more probable that this *nobilis vir* is the Marco spoken of in the note at p. 74.

† *La Collezione del Doge Marin Faliero e i Tesori di Marco Polo*, pp. 98-103. I have seen this article.—H. C.

mensis aprilis.   Inuentarium rerum qui sunt in camera rubea domi habitationis clarissimi domini MARINI FALETRO de confinio SS. Apostolorum, scriptum per me Johannem, presbiterum, dicte ecclesie.

Item alia capsaleta cum ogiis auri et argenti, inter quos unum anulum con inscriptione que dicit : *Ciuble Can Marco Polo*, et unum torques cum multis animalibus Tartarorum sculptis, que res donum dedit predictus MARCUS cuidam Faletrorum.

Item 2 capsalete de corio albo cum variis rebus auri et argenti, quas habuit praedictus MARCUS a Barbarorum rege.

Item 1 ensem mirabilem, qui habet 3 enses simul, quem habuit in suis itineribus praedictus MARCUS.

Item 1 tenturam de pannis indicis, quam habuit praedictus MARCUS.
Item de itineribus MARCI praedicti liber in corio albo cum multis figuris.
Item aliud volumen quod vocatur *de locis mirabilibus Tartarorum*, *scriptum manu praedicti* MARCI.

11. There is kept at the Louvre, in the very valuable collection of China Ware given by M. Ernest Grandidier, a white porcelain incense-burner said to come from Marco Polo. This incense-burner, which belonged to Baron Davillier, who received it, as a present, from one of the keepers of the Treasury of St. Mark's at Venice, is an octagonal *ting* from the Fo-kien province, and of the time of the Sung Dynasty. By the kind permission of M. P. Grandidier, we reproduce it from Pl. II. 6, of the *Céramique chinoise*, Paris, 1894, published by this learned amateur.—H. C.]

IX. MARCO POLO'S BOOK; AND THE LANGUAGE IN WHICH IT WAS
FIRST WRITTEN.

50. The Book itself consists essentially of Two Parts. *First*, of a Prologue, as it is termed, the only part which is

General statement of what the Book contains.

actual personal narrative, and which relates, in a very interesting but far too brief manner, the circumstances which led the two elder Polos to the Kaan's Court, and those of their second journey with Mark, and of their return to Persia through the Indian Seas. *Secondly*, of a long series of chapters of very unequal length, descriptive of notable sights and products, of curious manners and remarkable events, relating to the different nations and states of Asia, but, above all, to the

Porcelain Incense-Burner, from the Louvre.

[To face p. 80.

Emperor Kúbláí, his court, wars, and administration. A series of chapters near the close treats in a verbose and monotonous manner of sundry wars that took place between the various branches of the House of Chinghiz in the latter half of the 13th century. This last series is either omitted or greatly curtailed in all the copies and versions except one; a circumstance perfectly accounted for by the absence of interest as well as value in the bulk of these chapters. Indeed, desirous though I have been to give the Traveller's work complete, and sharing the dislike that every man who *uses* books must bear to abridgments, I have felt that it would be sheer waste and dead-weight to print these chapters in full.

This second and main portion of the Work is in its oldest forms undivided, the chapters running on consecutively to the end.* In some very early Italian or Venetian version, which Friar Pipino translated into Latin, it was divided into three Books, and this convenient division has generally been adhered to. We have adopted M. Pauthier's suggestion in making the final series of chapters, chiefly historical, into a Fourth.

51. As regards the language in which Marco's Book was first committed to writing, we have seen that Ramusio assumed, somewhat arbitrarily, that it was *Latin;* Marsden supposed it to have been the *Venetian* dialect; Baldelli Boni first showed, in his elaborate edition (Florence, 1827), by arguments that have been illustrated and corroborated by learned men since, that it was *French.*

Language of the original Work.

That the work was originally written in *some* Italian dialect was a natural presumption, and slight contemporary evidence can be alleged in its favour; for Fra Pipino, in the Latin version of the work, executed whilst Marco still lived, describes his task as a translation *de vulgari*. And in one MS. copy of the same Friar Pipino's Chronicle, existing in the library at Modena, he refers to the said version as made " *ex vulgari idiomate* Lombardico." But though it may seem improbable that at so early a date a Latin version should have been made at second hand, I believe this to have been the case, and that some internal evidence also is traceable that Pipino translated *not* from the original but from an Italian *version* of the original.

---

* 232 chapters in the oldest French which we quote as the *Geographic Text* (or G. T.), 200 in Pauthier's Text, 183 in the Crusca Italian.

The oldest MS. (it is supposed) in any Italian dialect is one in the Magliabecchian Library at Florence, which is known in Italy as *L'Ottima*, on account of the purity of its Tuscan, and as *Della Crusca* from its being one of the authorities cited by that body in their Vocabulary.* It bears on its face the following note in Italian :—

"This Book called the Navigation of Messer Marco Polo, a noble Citizen of Venice, was written in Florence by Michael Ormanni my great grandfather by the Mother's side, who died in the Year of Grace One Thousand Three Hundred and Nine ; and my mother brought it into our Family of Del Riccio, and it belongs to me Pier del Riccio and to my Brother; 1452."

As far as I can learn, the age which this note implies is considered to be supported by the character of the MS. itself.† If it be accepted, the latter is a performance going back to within eleven years *at most* of the first dictation of the Travels. At first sight, therefore, this would rather argue that the original had been written in pure Tuscan. But when Baldelli came to prepare it for the press he found manifest indications of its being a Translation from the *French*. Some of these he has noted; others have followed up the same line of comparison. We give some detailed examples in a note.‡

---

* The MS. has been printed by Baldelli as above, and again by Bartoli in 1863.

† This is somewhat peculiar. I traced a few lines of it, which with Del Riccio's note were given in facsimile in the First Edition.

‡ The Crusca is cited from Bartoli's edition.

French idioms are frequent, as *l'uomo* for the French *on ; quattro-vinti* instead of *ottanta ;* etc.

We have at p. 35, "*Questo piano è molto* cavo," which is nonsense, but is explained by reference to the French (G. T.) "*Voz di qu' il est celle plaingne mout* chaue" (*chaude*).

The bread in Kerman is bitter, says the G. T. "*por ce que l'eive hi est* amer," because the water there is bitter. The Crusca mistakes the last word and renders (p. 40) "*e questi è per lo* mare *che vi viene.*"

"*Sachiés de voir qe* endementiers," know for a truth that whilst——, by some misunderstanding of the last word becomes (p. 129) "*Sappiate di vero* sanza mentire."

"*Mès de sel* font-il monoie "—"They make money of salt," becomes (p. 168) "*ma fannole* da loro," *sel* being taken for a pronoun, whilst in another place *sel* is transferred bodily without translation.

"*Chevoil,*" "hair" of the old French, appears in the Tuscan (p. 20) as *cavagli,* "horses."—"*La Grant Provence* Jereraus," the great general province, appears (p. 68) as a province whose proper name is *Ienaraus.* In describing Kúblái's expedition against Mien or Burma, Polo has a story of his calling on the Jugglers at his court to undertake the job, promising them a Captain and other help, "*Cheveitain*

Temple of 500 Genii, at Canton, *after a Drawing by* FÉLIX RÉGAMEY

[*To face p. 82.*

52. The French Text that we have been quoting, published by the Geographical Society of Paris in 1824, affords on the other hand the strongest corresponding proof that it is an original and not a Translation. Rude as is the language of the manuscript (Fr. 1116, formerly No. 7367, of Paris Library), it is, in the correctness of the proper names, and the intelligible exhibition of the itineraries, much superior to any form of the Work previously published. Old French Text published by the Société de Géographie.

The language is very peculiar. We are obliged to call it French, but it is not "Frenche of Paris." "Its style," says Paulin Paris, " is about as like that of good French authors of the age, as in our day the natural accent of a German, an Englishman, or an Italian, is like that of a citizen of Paris or Blois." The author is at war with all the practices of French grammar; subject and object, numbers, moods, and tenses, are in consummate confusion. Even readers of his own day must at times have been fain to guess his meaning. Italian words are constantly introduced, either quite in the crude or rudely Gallicized.* And words

---

et aide." This has fairly puzzled the Tuscan, who converts these (p. 186) into two Tartar tribes, "*quegli d'* Aide *e quegli di* Caveità."

So also we have *lievre* for hare transferred without change; *lait*, milk, appearing as *laido* instead of *latte*; *très*, rendered as "three"; *bue*, "mud," Italianised as *buoi*, "oxen," and so forth. Finally, in various places when Polo is explaining Oriental terms we find in the Tuscan MS. "*cioè a dire in* Francesco."

The blunders m ntioned are intelligible enough as in a version *from the French;* but in the description of the Indian pearl-fishery we have a startling one not so easy to account for. The French says, "the divers gather the sea-oysters (*hostrige de Mer*), and in these the pearls are found." This appears in the Tuscan in the extraordinary form that the divers catch those fishes called *Herrings* (Aringhe), and in those Herrings are found the Pearls !

* As examples of these Italianisms : " *Et ont del* olio *de la lanpe dou* sepolchro *de Crist*"; "*L'Angel ven en vision pour mesajes de Deu à un* Veschevo *qe mout estoient home de* sante vite"; "*E certes il estoit bien* beizongno"; "*ne trop caut ne trop fredo*"; "*la* crense" (*credei za*); "remort" for noise (*rumore*); "inverno"; "jorno"; "dementiqué" (*dimenticato*); "enferme" for sickly; "leign" (*legno*); "devisce" (*dovizia*); "ammalaide" (*ammalato*), etc. etc.

Professor Bianconi points out that there are also traces of *Venetian* dialect, as *Pare* for *père; Mojer* for wife; *Zabater*, cobbler; *cazaor*, huntsman, etc.

I have not been able to learn to what extent books in this kind of mixed language are extant. I have observed one, a romance in verse called *Macaire (Altfranzösische Gedichte aus Venez. Handschriften*, von *Adolf Mussafia*, Wien, 1864), the language of which is not unlike this jargon of Rustician's, *e.g.* :—

> "' Dama,' fait-il, ' molto me poso merviler
> De ves enfant quant le fi batecer
> De un signo qe le vi sor la spal'a droiturer
> Qe non ait nul se no filz d'inperer.'"—(p. 41)

also, we may add, sometimes slip in which appear to be purely Oriental, just as is apt to happen with Anglo-Indians in these days.* All this is perfectly consistent with the supposition that we have in this MS. a copy at least of the original words as written down by Rusticiano a Tuscan, from the dictation of Marco an Orientalized Venetian, in French, a language foreign to both.

But the character of the language *as French* is not its only peculiarity. There is in the style, apart from grammar or vocabulary, a rude angularity, a rough dramatism like that of oral narrative ; there is a want of proportion in the style of different parts, now over curt, now diffuse and wordy, with at times even a hammering reiteration ; a constant recurrence of pet colloquial phrases (in which, however, other literary works of the age partake) ; a frequent change in the spelling of the same proper names, even when recurring within a few lines, as if caught by ear only ; a literal following to and fro of the hesitations of the narrator ; a more general use of the third person in speaking of the Traveller, but an occasional lapse into the first. All these characteristics are strikingly indicative of the unrevised product of dictation, and many of them would *necessarily* disappear either in translation or in a revised copy.

Of changes in representing the same proper name, take as an example that of the Kaan of Persia whom Polo calls *Quiacatu* (Kaikhátú), but also *Acatu*, *Catu*, and the like.

As an example of the literal following of dictation take the following :—

"Let us leave Rosia, and I will tell you about the Great Sea (the Euxine), and what provinces and nations lie round about it, all in detail ; and we will begin with Constantinople——First, however, I should tell you about a province, etc. . . . There is nothing more worth mentioning, so I will speak of other subjects,—but there is one thing more to tell you about Rosia that I had forgotten. . . . Now then let us speak of the Great Sea as I was about to do. To be sure many merchants and others have

---

* As examples of such Orientalisms : *Bonus*, "ebony," and *calamanz*, "pencases," seem to represent the Persian *abnús* and *ḳalạmdàn ;* the dead are mourned by *les mères et les* Araines, the *Harems ;* in speaking of the land of the Ismaelites or Assassins, called *Mulhete, i.e.* the Arabic *Muláhidah*, "Heretics," he explains this term as meaning "des *Aram*" (*Ḥarám*, "the reprobate "). Speaking of the Viceroys of Chinese Provinces, we are told that they rendered their accounts yearly to the *Safators* of the Great Kaan. This is certainly an Oriental word. Sir H. Rawlinson has suggested that it stands for *dafátir* ("registers or public books "), pl. of *daftar*. This seems probable, and in that case the true reading may have been *dafators.*

been here, but still there are many again who know nothing about it, so it
will be well to include it in our Book. We will do so then, and let us begin
first with the Strait of Constantinople.

"At the Straits leading into the Great Sea, on the West Side, there is a
hill called the Faro.——But since beginning on this matter I have changed
my mind, because so many people know all about it, so we will not put it in
our description but go on to something else." (See vol. ii. p. 487 *seqq.*)

And so on.

As a specimen of tautology and hammering reiteration the
following can scarcely be surpassed. The Traveller is speaking
of the *Chughi, i.e.* the Indian Jogis :—

" And there are among them certain devotees, called *Chughi ;* these are
longer-lived than the other people, for they live from 150 to 200 years ; and
yet they are so hale of body that they can go and come wheresoever they
please, and do all the service needed for their monastery or their idols, and
do it just as well as if they were younger ; and that comes of the great
abstinence that they practise, in eating little food and only what is wholesome ; for they use to eat rice and milk more than anything else. And
again I tell you that these Chughi who live such a long time as I have told
you, do also eat what I am going to tell you, and you will think it a great
matter. For I tell you that they take quicksilver and sulphur, and mix them
together, and make a drink of them, and then they drink this, and they
say that it adds to their life ; and in fact they do live much longer for it ;
and I tell you that they do this twice every month. And let me tell you
that these people use this drink from their infancy in order to live longer,
and without fail those who live so long as I have told you use this drink
of sulphur and quicksilver." (See G. T. p. 213.)

Such talk as this does not survive the solvent of translation ;
and we may be certain that we have here the nearest approach
to the Traveller's reminiscences as they were taken down from his
lips in the prison of Genoa.

53. Another circumstance, heretofore I believe unnoticed, is
in itself enough to demonstrate the Geographic Text to be the
source of all other versions of the Work. It is this. Conclusive
proof that
In reviewing the various classes or types of texts the Old
French Text
of Polo's Book, which we shall hereafter attempt to dis- is the source
of all the
criminate, there are certain proper names which we find others.
in the different texts to take very different forms, each class
adhering in the main to one particular form.

Thus the names of the Mongol ladies introduced at pp. 32 and
36 of this volume, which are in proper Oriental form *Bulughán*
and *Kukáchin*, appear in the class of MSS. which Pauthier has
followed as *Bolgara* and *Cogatra ;* in the MSS. of Pipino's

version, and those founded on it, including Ramusio, the names appear in the correcter forms *Bolgana* or *Balgana* and *Cogacin.* Now *all the forms* Bolgana, Balgana, Bolgara, *and* Cogatra, Cocacin *appear in the Geographic Text.*

Kaikhátú Kaan appears in the Pauthier MSS. as *Chiato*, in the Pipinian as *Acatu*, in the Ramusian as *Chiacato. All three forms*, Chiato, Achatu, and Quiacatu *are found in the Geographic Text.*

The city of Koh-banan appears in the Pauthier MSS. as *Cabanant*, in the Pipinian and Ramusian editions as *Cobinam* or *Cobinan. Both forms are found in the Geographic Text.*

The city of the Great Kaan (Khanbalig) is called in the Pauthier MSS. *Cambaluc*, in the Pipinian and Ramusian less correctly *Cambalu. Both forms appear in the Geographic Text.*

The aboriginal People on the Burmese Frontier who received from the Western officers of the Mongols the Persian name (translated from that applied by the Chinese) of *Zardandán*, or Gold-Teeth, appear in the Pauthier MSS. most accurately as Zardandan, but in the Pipinian as *Ardandan* (still further corrupted in some copies into *Arcladam*). Now *both forms are found in the Geographic Text.* Other examples might be given, but these I think may suffice to prove that this Text was the common source of both classes.

In considering the question of the French original too we must remember what has been already said regarding Rusticien de Pise and his other French writings; and we shall find hereafter an express testimony borne in the next generation that Marco's Book was composed *in vulgari Gallico.*

54. But, after all, the circumstantial evidence that has been adduced from the texts themselves is the most conclusive. We Greatly diffused employment of French in that age. have then every reason to believe both that the work was written in French, and that an existing French Text is a close representation of it as originally committed to paper. And that being so we may cite some circumstances to show that the use of French or quasi-French for the purpose was not a fact of a very unusual or surprising nature. The French language had at that time almost as wide, perhaps relatively a wider, diffusion than it has now. It was still spoken at the Court of England, and still used by many English writers, of whom the authors or translators of the Round Table

Romances at Henry III.'s Court are examples.* In 1249 Alexander III. King of Scotland, at his coronation spoke in Latin and French ; and in 1291 the English Chancellor addressing the Scotch Parliament did so in French. At certain of the Oxford Colleges as late as 1328 it was an order that the students should converse *colloquio latino vel saltem gallico*.† Late in the same century Gower had not ceased to use French, composing many poems in it, though apologizing for his want of skill therein :—

> " Et si jeo nai de Francois la faconde
> *   *   *   *
> Jeo suis Englois ; si quier par tiele voie
> Estre excusé." ‡

Indeed down to nearly 1385, boys in the English grammar-schools were taught to construe their Latin lessons into French.§ St. Francis of Assisi is said by some of his biographers to have had his original name changed to Francesco because of his early mastery of that language as a qualification for commerce. French had been the prevalent tongue of the Crusaders, and was that of the numerous Frank Courts which they established in the East, including Jerusalem and the states of the Syrian coast, Cyprus, Constantinople during the reign of the Courtenays, and the principalities of the Morea. The Catalan soldier and chronicler Ramon de Muntaner tells us that it was commonly said of the Morean chivalry that they spoke as good French as at Paris.‖ Quasi-French at least was still spoken half a century later by the numerous Christians settled at Aleppo, as John Marignolli testifies ; ¶ and if we may trust Sir John Maundevile the Soldan of Egypt himself and four of his chief Lords "*spak Frensche righte wel!*" ** Gházán Kaan, the accomplished Mongol Sovereign of Persia, to whom our Traveller conveyed a

---

* Luces du Gast, one of the first of these, introduces himself thus :—" Je Luces, Chevaliers et Sires du Chastel du Gast, voisins prochain de Salebieres, comme chevaliers amoureus enprens à translater du Latin en François une partie de cette estoire, non mie pour ce que je sache gramment de François, ainz apartient plus ma langue et ma parleure à la manière de l'Engleterre que à celle de France, comme cel qui fu en Engleterre nez, mais tele est ma volentez et mon proposement, que je en langue françoise le translaterai." (*Hist. Litt. de La France*, xv. 494.)

† *Hist. Litt. de la France*, xv. 500.   ‡ *Ibid.* 508.
§ *Tyrwhitt's Essay on Lang., etc., of Chaucer*, p. xxii. (Moxon's Ed. 1852.)
‖ *Chroniques Etrangères*, p. 502.
¶ " *Loquuntur linguam quasi Gallicam, scilicet quasi de Cipro.*" (See *Cathay*, p. 332.)   ** Page 138.

bride from Cambaluc, is said by the historian Rashiduddin to have known something of the Frank tongue, probably French.* Nay, if we may trust the author of the Romance of Richard Cœur-de-Lion, French was in his day the language of still higher spheres! †

Nor was Polo's case an exceptional one even among writers on the East who were not Frenchmen. Maundevile himself tells us that he put his book first "out of Latyn into Frensche," and then out of French into English.‡ The History of the East which the Armenian Prince and Monk Hayton dictated to Nicolas Faulcon at Poictiers in 1307 was taken down in French. There are many other instances of the employment of French by foreign, and especially by Italian authors of that age. The Latin chronicle of the Benedictine Amato of Monte Cassino was translated into French early in the 13th century by another monk of the same abbey, at the particular desire of the Count of Militrée (or Malta), "*Pour ce qu'il set lire et entendre fransoize et s'en delitte.*" § Martino da Canale, a countryman and contemporary of Polo's, during the absence of the latter in the East wrote a Chronicle of Venice in the same language, as a reason for which he alleges its general popularity.‖ The like does the most notable example of all, Brunetto Latini, Dante's master, who wrote in French his encyclopædic and once highly popular work *Li Tresor.*¶ Other examples might be given, but in fact

---

* *Hammer's Ilchan*, II. 148.

† After the capture of Acre, Richard orders 60,000 Saracen prisoners to be executed :—

"They wer brought out off the toun,	*They sayde*: 'SEVNYORS, TUEZ, TUEZ!
Save twenty, he heeld to raunsoun.	'Spares hem nought ! Behedith these!'
They wer led into the place ful evene :	Kyng Rychard herde the Aungelys voys,
Ther they herden Aungeles off Hevene:	And thankyd God, and the Holy Croys."
	—*Weber*, II. 144.

Note that, from the rhyme, the Angelic French was apparently pronounced "*Too-eese! Too-eese!*"

‡ [Refer to the edition of Mr. George F. Warner, 1889, for the Roxburghe Club, and to my own paper in the *T'oung Pao*, Vol. II., No. 4, regarding the compilation published under the name of Maundeville. Also *App. L.* 13—H. C.]

§ *L'Ystoire de li Normand*, etc., edited by M. Champollion-Figeac, Paris, 1835, p. v.

‖ "*Porce que lengue Frenceise cort parmi le monde, et est la plus delitable à lire et à oir que nule autre, me sui-je entremis de translater l'ancien estoire des Veneciens de Latin en Franceis.*" (Archiv. Stor. Ital. viii. 268.)

¶ "*Et se aucuns demandoit por quoi cist livres est escriz en Romans, selonc le langage des Francois, puisque nos somes Ytaliens, je diroie que ce est por. ij. raisons: l'une, car nos somes en France ; et l'autre porce que la parleure est plus delitable et plus commune à toutes gens.*" (Li Livres dou Tresor, p. 3.)

such illustration is superfluous when we consider that Rusticiano himself was a compiler of French Romances.

But why the language of the Book as we see it in the Geographic Text should be so much more rude, inaccurate, and Italianized than that of Rusticiano's other writings, is a question to which I can suggest no reply quite satisfactory to myself. Is it possible that we have in it a literal representation of Polo's own language in dictating the story,—a rough draft which it was intended afterwards to reduce to better form, and which was so reduced (after a fashion) in French copies of another type, regarding which we shall have to speak presently ? * And, if this be the true answer, why should Polo have used a French jargon in which to tell his story ? Is it possible that his own mother Venetian, such as he had carried to the East with him and brought back again, was so little intelligible to Rusticiano that French of some kind was the handiest medium of communication between the two ? I have known an Englishman and a Hollander driven to converse in Malay ; Chinese Christians of different provinces are said sometimes to take to English as the readiest means of intercommunication ; and the same is said even of Irish-speaking Irishmen from remote parts of the Island.

It is worthy of remark how many notable narratives of the Middle Ages have been dictated instead of being written by their authors, and that in cases where it is impossible to ascribe this to ignorance of writing. The Armenian Hayton, though evidently a well-read man, possibly could not write in Roman characters. But Joinville is an illustrious example. And the narratives of four of the most famous Mediæval Travellers †️ seem to have been drawn from them by a kind of pressure, and committed to paper by other hands. I have elsewhere remarked this as indicating how little diffused was literary ambition or vanity ; but it would perhaps be more correct to ascribe it to that intense dislike which is still seen on the shores of the Mediter-

---

* It is, however, not improbable that Rusticiano's hasty and abbreviated original was extended by a scribe who knew next to nothing of French ; otherwise it is hard to account for such forms as *perlinage* (pelerinage), *peseries* (espiceries), *proque* (see vol. ii. p. 370), *oisi* (G. T. p. 208), *thochere* (toucher), etc. (See *Bianconi*, 2nd Mem. pp. 30-32.)

† Polo, Friar Odoric, Nicolo Conti, Ibn Batuta.

ranean to the use of pen and ink. On certain of those shores at least there is scarcely any inconvenience that the majority of respectable and good-natured people will not tolerate—inconvenience to their neighbours be it understood—rather than put pen to paper for the purpose of preventing it.

X. Various Types of Text of Marco Polo's Book.

55. In treating of the various Texts of Polo's Book we must

Four Principal Types of Text.
First, that of the Geographic, or oldest French.

necessarily go into some irksome detail.

Those Texts that have come down to us may be classified under Four principal Types.

I. The First Type is that of the Geographic Text of which we have already said so much. This is found nowhere *complete* except in the unique MS. of the Paris Library, to which it is stated to have come from the old Library of the French Kings at Blois. But the Italian *Crusca*, and the old Latin version (No. 3195 of the Paris Library) published with the Geographic Text, are evidently derived entirely from it, though both are considerably abridged. It is also demonstrable that neither of these copies has been translated from the other, for each has passages which the other omits, but that both have been taken, the one as a copy more or less loose, the other as a translation, from an intermediate *Italian* copy.* A special

---

* In the following citations, the Geographic Text (G. T.) is quoted by page from the printed edition (1824) ; the Latin published in the same volume (G. L.) also by page ; the Crusca, as before, from Bartoli's edition of 1863. References in parentheses are to the present translation :—

A. *Passages showing the G. L. to be a translation from the Italian, and derived from the same Italian text as the* Crusca.

		Page			
(1).	G.T.	17	(I.	43).	Il hi se laborent *le souran tapis* dou monde.
	Crusca,	17	. .		E quivi si fanno *i sovrani tappeti* del mondo.
	G.L.	311	. .		Et ibi fiunt *soriani et tapeti* pulcriores de mundo.
(2).	G.T.	23	(I.	69).	Et adonc le calif mande par tuit les cristiez . . . *que en sa tere estoient.*
	Crusca,	27	. .		*Ora mandò* lo aliffo per tutti gli Cristiani *ch' erano di là.*
	G.L.	316	. .		*Or misit* califus pro Christianis *qui erant ultra fluvium* (the last words being clearly a misunderstanding of the Italian *di là*).

difference lies in the fact that the Latin version is divided into three Books, whilst the Crusca has no such division. I shall show in a tabular form the *filiation* of the texts which these facts seem to demonstrate (see Appendix G).

There are other Italian MSS. of this type, some of which show signs of having been derived independently from the French ;* but I have not been able to examine any of them with the care needful to make specific deductions regarding them.

		Page		
(3).	G.T.	198	(II. 313).	Ont *sosimain* (sesamum) de coi il font le olio.
	Crusca,	253	. .	Hanno *sosimai* onde fanno l' olio.
	G.L.	448	. .	Habent *turpes manus* (taking *sosimani* for *sozze mani* " Dirty hands " !).
(4).	Crusca,	52	(I. 158).	*Cacciare e uccellare* v' è lo migliore del mondo.
	G.L.	332	. .	Et est ibi optimum *caciare et ucellare*.
(5).	G.T.	124	(II. 36).	Adonc treuve . . . . une Provence *qe est encore* de le confin dou Mangi.
	Crusca,	162-3	. .	L' uomo truova una Provincia *ch' è chiamata ancora* delle confine de' Mangi.
	G.L.	396	. .	Invenit unam Provinciam *quae vocatur Anchota* de confinibus Mangi.
(6).	G.T.	146	(II. 119.)	Les dames portent as jambes et es braces, braciaus d'or et d'arjent de grandisme vailance.
	Crusca,	189	. .	Le donne *portano alle braccia e alle gambe bracciali d'oro* e d'ariento di gran valuta.
	G.L.	411	. .	Dominæ eorum *portant ad brachia et ad gambas brazalia de auro* et de argento magni valoris.

B. *Passages showing additionally the errors, or other peculiarities of a translation from a French original, common to the Italian and the Latin.*

(7).	G.T.	32	(I. 97.)	Est celle plaingne mout *chaue* (chaude).
	Crusca,	35	. .	Questo piano è molto *cavo*.
	G.L.	322	. .	Ista planities est multum *cava*.
(8).	G.T.	36	(I. 110).	Avent por ce que l'eive *hi est amer*.
	Crusca,	40	. .	E questo è *per lo mare* che vi viene.
	G.L.	324	. .	Istud est *propter mare* quod est ibi.
(9).	G.T.	18	(I. 50).	Un roi qi est apelés par tout tens Davit Melic, que veut à dir *en fransois* Davit Roi.
	Crusca,	20	. .	Uno re il quale si chiama *sempre* David Melic, ciò è a dire *in francesco* David Re.
	G.L.	312	. .	Rex qui *semper* vocatur David Mellic, quod sonat *in gallico* David Rex.

These passages, and many more that might be quoted, seem to me to demonstrate (1) that the Latin and the Crusca have had a common original, and (2) that this original was an Italian version from the French.

* Thus the *Pucci* MS. at Florence, in the passage regarding the Golden King (vol. ii. p. 17) which begins in G. T. " *Lequel fist faire* jadis *un rois qe fu apellés le Roi Dor*," renders " *Lo quale fa fare* Jaddis *uno re*," a mistake which is not in the Crusca nor in the Latin, and seems to imply derivation from the French directly, or by some other channel (*Baldelli Boni*),

56. II. The next Type is that of the French MSS. on which
M. Pauthier's Text is based, and for which he claims the highest
authority, as having had the mature revision and
sanction of the Traveller. There are, as far as I know,
five MSS. which may be classed together under this
type, three in the Great Paris Library, one at Bern, and
one in the Bodleian.

The high claims made by Pauthier on behalf of this class of
MSS. (on the first three of which his Text is formed) rest mainly
upon the kind of certificate which two of them bear regarding
the presentation of a copy by Marco Polo to Thibault de Cepoy,
which we have already quoted (*supra*, p. *69*). This certificate is
held by Pauthier to imply that the original of the copies which
bear it, and of those having a general correspondence with them,
had the special seal of Marco's revision and approval. To
some considerable extent their character is corroborative of such
a claim, but they are far from having the perfection which
Pauthier attributes to them, and which leads him into many
paradoxes.

It is not possible to interpret rigidly the bearing of this so-
called certificate, as if no copies had previously been taken of
any form of the Book ; nor can we allow it to impugn the
authenticity of the Geographic Text, which demonstratively
represents an older original, and has been (as we have seen) the
parent of all other versions, including some very old ones,
Italian and Latin, which certainly owe nothing to this revision.

The first idea apparently entertained by d'Avezac and
Paulin Paris was that the Geographic Text was *itself* the
copy given to the Sieur de Cepoy, and that the differences in
the copies of the class which we describe as Type II. merely
resulted from the modifications which would naturally arise in
the process of transcription into purer French. But closer
examination showed the differences to be too great and too
marked to admit of this explanation. These differences consist
not only in the conversion of the rude, obscure, and half Italian
language of the original into good French of the period. There
is also very considerable curtailment, generally of tautology, but
also extending often to circumstances of substantial interest ;
whilst we observe the omission of a few notably erroneous
statements or expressions ; and a few insertions of small im-

portance. None of the MSS. of this class contain more than a few of the historical chapters which we have formed into Book IV.

The only *addition* of any magnitude is that chapter which in our translation forms chapter xxi. of Book II. It will be seen that it contains no new facts, but is only a tedious recapitulation of circumstances already stated, though scattered over several chapters. There are a few minor additions. I have not thought it worth while to collect them systematically here, but two or three examples are given in a note.*

There are also one or two corrections of erroneous statements in the G. T. which seem not to be accidental and to indicate some attempt at revision. Thus a notable error in the account of Aden, which seems to conceive of the Red Sea as a *river,* disappears in Pauthier's MSS. A and B.† And we find in these MSS. one or two interesting names preserved which are not found in the older Text.‡

But on the other hand this class of MSS. contains many erroneous readings of names, either adopting the worse of two forms occurring in the G. T. or originating blunders of its own.§

---

* In the Prologue (vol. i. p. 34) this class of MSS. alone names the King of England.

In the account of the Battle with Nayan (i. p. 337) this class alone speaks of the two-stringed instruments which the Tartars played whilst awaiting the signal for battle. But the circumstance appears elsewhere in the G. T. (p. 250).

In the chapter on *Malabar* (vol. ii. p. 390), it is said that the ships which go with cargoes towards Alexandria are not one-tenth of those that go to the further East. This is not in the older French.

In the chapter on *Coilun* (ii. p. 375), we have a notice of the Columbine ginger so celebrated in the Middle Ages, which is also absent from the older text.

† See vol. ii. p. 439. It is, however, remarkable that a like mistake is made about the Persian Gulf (see i. 63, 64). Perhaps Polo *thought* in Persian, in which the word *darya* means either *sea* or a *large river.* The same habit and the ambiguity of the Persian *sher* led him probably to his confusion of lions and tigers (see i. 397).

‡ Such are Pasciai-*Dir* and *Ariora* Kesciemur (i. p. 98.)

§ Thus the MSS. of this type have elected the erroneous readings *Bolgara, Cogatra, Chiato, Cabanant,* etc., instead of the correcter *Bolgana, Cocacin, Quiacatu, Cobinan,* where the G. T. presents both (*supra,* p. 86). They read *Esanar* for the correct *Etzina* ; *Chascun* for *Casvin ; Achalet* for *Acbalec ; Sardansu* for *Sindafu , Kayteu, Kayton, Sarcon* for *Zaiton* or *Caiton ; Soucat* for *Locac ; Falec* for *Ferlec,* and so on, the worse instead of the better. They make the *Mer Occeane* into *Mer Occident ;* the wild asses (*asnes*) of the Kerman Desert into wild geese (*oes*) ; the *escoillez* of Bengal (ii. p. 115) into *escoliers ;* the *giraffes* of Africa into *girofles,* or cloves, etc., etc.

M. Pauthier lays great stress on the character of these MSS. as the sole authentic form of the work, from their claim to have been specially revised by Marco Polo. It is evident, however, from what has been said, that this revision can have been only a very careless and superficial one, and must have been done in great measure by deputy, being almost entirely confined to curtailment and to the improvement of the expression, and that it is by no means such as to allow an editor to dispense with a careful study of the Older Text.

57. There is another curious circumstance about the MSS. of this type, viz., that they clearly divide into two distinct recensions, The Bern MS. and two others form a sub-class of this Type. of which both have so many peculiarities and errors in common that they must necessarily have been both derived from *one* modification of the original text, whilst at the same time there are such differences between the two as cannot be set down to the accidents of transcription. Pauthier's MSS. A and B (Nos. 16 and 15 of the List in App. F) form one of these subdivisions: his C (No. 17 of List), Bern (No. 56), and Oxford (No. 6), the other. Between A and B the differences are only such as seem constantly to have arisen from the whims of transcribers or their dialectic peculiarities. But between A and B on the one side, and C on the other, the differences are much greater. The readings of proper names in C are often superior, sometimes worse; but in the latter half of the work especially it contains a number of substantial passages * which are to be found in the G. T., but are altogether absent from the MSS. A and B; whilst in one case at least (the history of the Siege of Saianfu, vol. ii. p. 159) it diverges considerably from the G. T. *as well* as from A and B.†

I gather from the facts that the MS. C represents an older form of the work than A and B. I should judge that the latter had been derived from that older form, but intentionally modified from it. And as it is the MS. C, with its copy at Bern, that alone presents the certificate of derivation from the Book given

---

* There are about five-and-thirty such passages altogether.

† The Bern MS. I have satisfied myself is an actual *copy* of the Paris MS. C.

The Oxford MS. closely resembles both, but I have not made the comparison minutely enough to say if it is an exact copy of either.

to the Sieur de Cepoy, there can be no doubt that it is the true representative of that recension.

58. III. The next Type of Text is that found in Friar Pipino's Latin version. It is the type of which MSS. are by far the most numerous. In it condensation and curtail- Third; ment are carried a good deal further than in Type II. The work is also divided into three Books. But this division does not seem to have originated with Pipino, as we find it in the ruder and perhaps older Latin version of which we have already spoken under Type I. And we have demonstrated that this ruder Latin is a translation from an Italian copy. It is probable therefore that an Italian version similarly divided was the common source of what we call the Geographic Latin and of Pipino's more condensed version.*

Pipino's version appears to have been executed in the later years of Polo's life.† But I can see no ground for the idea entertained by Baldelli-Boni and Professor Bianconi that it was executed with Polo's cognizance and retouched by him,

59. The absence of effective publication in the Middle Ages led to a curious complication of translation and retranslation. Thus the Latin version published by Grynæus in the *Novus Orbis* (Basle, 1532) is different from Pipino's, and yet clearly traceable to it as a base. In fact it

Marginal notes: Third; Friar Pipino's Latin. The Latin of Grynæus a translation at fifth hand.

---

* The following comparison will also show that these two Latin versions have probably had a common source, such as is here suggested.

At the end of the Prologue the Geographic Text reads simply :—

" Or puis que je voz ai contez tot le fat dou prologue ensi con voz avés oï, adonc (commencerai) le Livre."

Whilst the Geographic Latin has :—

" *Postquam recitavimus et diximus facta et condictiones morum, itinerum* et ea quae nobis contigerunt per vias, *incipiemus dicere ea quae vidimus. Et primo dicemus de Minore Hermenia.*"

And Pipino :—

" *Narratione facta nostri itineris, nunc ad ea narranda quae vidimus accedamus. Primo autem Armeniam Minorem describemus breviter.*"

† Friar Francesco Pipino of Bologna, a Dominican, is known also as the author of a lengthy chronicle from the time of the Frank Kings down to 1314 ; of a Latin Translation of the French History of the Conquest of the Holy Land, by Bernard the Treasurer ; and of a short Itinerary of a Pilgrimage to Palestine in 1320. Extracts from the Chronicle, and the version of Bernard, are printed in Muratori's Collection. As Pipino states himself to have executed the translation of Polo by order of his Superiors, it is probable that the task was set him at a general chapter of the order which was held at Bologna in 1315. (See *Muratori*, IX. 583; and *Quétif, Script. Ord. Praed.* I. 539). We do not know why Ramusio assigned the translation specifically to 1320, but he may have had grounds.

is a retranslation into Latin from some version (Marsden thinks the printed Portuguese one) of Pipino. It introduces many minor modifications, omitting specific statements of numbers and values, generalizing the names and descriptions of specific animals, exhibiting frequent sciolism and self-sufficiency in modifying statements which the Editor disbelieved.* It is therefore utterly worthless as a Text, and it is curious that Andreas Müller, who in the 17th century devoted himself to the careful editing of Polo, should have made so unfortunate a choice as to reproduce this fifth-hand Translation. I may add that the French editions published in the middle of the 16th century are *translations* from Grynæus. Hence they complete this curious and vicious circle of translation: French—Italian—Pipino's Latin—Portuguese? —Grynæus's Latin—French! †

60. IV. We now come to a Type of Text which deviates largely from any of those hitherto spoken of, and the history and true character of which are involved in a cloud of difficulty. We mean that Italian version prepared for the press by G. B. Ramusio, with most interesting, though, as we have seen, not always accurate preliminary dissertations, and published at Venice two years after his death, in the second volume of the *Navigationi e Viaggi*.‡

Fourth; Ramusio's Italian.

The peculiarities of this version are very remarkable. Ramusio seems to imply that he used as one basis at least the Latin of Pipino; and many circumstances, such as the division into Books, the absence of the terminal historical chapters and of

* See *Bianconi*, 1st Mem. 29 *seqq.*

† C. Dickens somewhere narrates the history of the equivalents for a sovereign as changed and rechanged at every frontier on a continental tour. The final equivalent received at Dover on his return was some 12 or 13 shillings; a fair parallel to the comparative value of the first and last copies in the circle of translation.

‡ The Ramusios were a family of note in literature for several generations. Paolo, the father of Gian Battista, came originally from Rimini to Venice in 1458, and had a great repute as a jurist, besides being a littérateur of some eminence, as was also his younger brother Girolamo. G. B. Ramusio was born at Treviso in 1485, and early entered the public service. In 1533 he became one of the Secretaries of the Council of X. He was especially devoted to geographical studies, and had a school for such studies in his house. He retired eventually from public duties, and lived at his Villa Ramusia, near Padua. He died in the latter city, 10th July, 1557, but was buried at Venice in the Church of S. Maria dell' Orto. There was a portrait of him by Paul Veronese in the Hall of the Great Council, but it perished in the fire of 1577; and that which is now seen in the Sala dello Scudo is, like the companion portrait of Marco Polo, imaginary. Paolo Ramusio, his son, was the author of the well-known History of the Capture of Constantinople. (*Cicogna*, II. 310 *seqq.*)

those about the Magi, and the form of many proper names, confirm this. But also many additional circumstances and anecdotes are introduced, many of the names assume a new shape, and the whole style is more copious and literary in character than in any other form of the work.

Whilst some of the changes or interpolations seem to carry us further from the truth, others contain facts of Asiatic nature or history, as well as of Polo's own experiences, which it is extremely difficult to ascribe to any hand but the Traveller's own. This was the view taken by Baldelli, Klaproth, and Neumann; * but Hugh Murray, Lazari, and Bartoli regard the changes as interpolations by another hand; and Lazari is rash enough to ascribe the whole to a *rifacimento* of Ramusio's own age, asserting it to contain interpolations not merely from Polo's own contemporary Hayton, but also from travellers of later centuries, such as Conti, Barbosa, and Pigafetta. The grounds for these last assertions have not been cited, nor can I trace them. But I admit *to a certain extent* indications of modern tampering with the text, especially in cases where proper names seem to have been identified and more modern forms substituted. In days, however, where an Editor's duties were ill understood, this was natural.

61. Thus we find substituted for the *Bastra* (or *Bascra*) of the older texts the more modern and incorrect *Balsora*, dear to memories of the Arabian Nights; among the provinces of Persia we have *Spaan* (Ispahan) where older texts read *Istanit;* for *Cormos* we have *Ormus;* for *Herminia* and *Laias*, *Armenia* and *Giazza; Coulàm* for the older *Coilum; Socotera* for *Scotra*. With these changes may be classed the chapter-headings, which are undisguisedly modern, and probably Ramusio's own. In some other cases this editorial spirit has been over-meddlesome and has gone astray. Thus *Malabar* is substituted wrongly for *Maabar* in one place, and by a grosser error for *Dàlivar* in another. The age of young Marco, at the time of his father's first return to Venice, has been arbitrarily altered from 15 to 19, in order to correspond with a date which is itself erroneous. Thus also Polo is made to describe Ormus

Injudicious tamperings in Ramusio.

---

* The old French texts were unknown in Marsden's time. Hence this question did not present itself to him.

as on an Island, contrary to the old texts and to the fact; for the city of Hormuz was not transferred to the island, afterwards so famous, till some years after Polo's return from the East. It is probably also the editor who in the notice of the oil-springs of Caucasus (i. p. 46) has substituted *camel-loads* for *ship-loads*, in ignorance that the site of those alluded to was probably Baku on the Caspian.

Other erroneous statements, such as the introduction of window-glass as one of the embellishments of the palace at Cambaluc, are probably due only to accidental misunderstanding.

62. Of circumstances certainly genuine, which are peculiar to this edition of Polo's work, and which it is difficult to assign to any one but himself, we may note the specification of the woods east of Yezd as composed of *date trees* (vol. i. pp. 88-89); the unmistakable allusion to the subterranean irrigation channels of Persia (p. 123); the accurate explanation of the term *Mulehet* applied to the sect of Assassins (pp. 139-142); the mention of the Lake (Sirikul?) on the plateau of Pamer, of the wolves that prey on the wild sheep, and of the piles of wild rams' horns used as landmarks in the snow (pp. 171-177). To the description of the Tibetan Yak, which is in all the texts, Ramusio's version alone adds a fact probably not recorded again till the present century, viz., that it is the practice to cross the Yak with the common cow (p. 274). Ramusio alone notices the prevalence of *goître* at Yarkand, confirmed by recent travellers (i. p. 187); the vermilion seal of the Great Kaan imprinted on the paper-currency, which may be seen in our plate of a Chinese note (p. 426); the variation in Chinese dialects (ii. p. 236); the division of the hulls of junks into water-tight compartments (ii. p. 249); the introduction into China from Egypt of the art of refining sugar (ii. p. 226). Ramusio's account of the position of the city of Sindafu (Ch'êng-tu fu) encompassed and intersected by many branches of a great river (ii. p. 40), is much more just than that in the old text, which speaks of but one river through the middle of the city. The intelligent notices of the Kaan's charities as originated by his adoption of "idolatry" or Buddhism; of the astrological superstitions of the Chinese, and of the manners and character of the latter nation, are found in Ramusio alone. To whom but Marco himself, or one of his party, can we refer the brief but vivid picture of the delicious

(marginal note:) Genuine statements peculiar to Ramusio.

atmosphere and scenery of the Badakhshan plateaux (i. p. 158), and of the benefit that Messer Marco's health derived from a visit to them? In this version alone again we have an account of the oppressions exercised by Kúblái's Mahomedan Minister Ahmad, telling how the Cathayans rose against him and murdered him, with the addition that Messer Marco was on the spot when all this happened. Now not only is the whole story in substantial accordance with the Chinese Annals, even to the name of the chief conspirator,* but those annals also tell of the courageous frankness of "Polo, assessor of the Privy Council," in opening the Kaan's eyes to the truth.

Many more such examples might be adduced, but these will suffice. It is true that many of the passages peculiar to the Ramusian version, and indeed the whole version, show a freer utterance and more of a literary faculty than we should attribute to Polo, judging from the earlier texts. It is possible, however, that this may be almost, if not entirely, due to the fact that the version is the result of a double translation, and probably of an editorial fusion of several documents; processes in which angularities of expression would be dissolved.†

---

* *Wangcheu* in the Chinese Annals; *Vanchu* in Ramusio. I assume that Polo's *Vanchu* was pronounced as in English; for in Venetian the *ch* very often has that sound. But I confess that I can adduce no other instance in Ramusio where I suppose it to have this sound, except in the initial sound of *Chinchitalas* and twice in *Choiach* (see II. 364).

Professor Bianconi, who has treated the questions connected with the Texts of Polo with honest enthusiasm and laborious detail, will admit nothing genuine in the Ramusian interpolations beyond the preservation of some *oral traditions* of Polo's supplementary recollections. But such a theory is out of the question in face of a chapter like that on Ahmad.

† Old Purchas appears to have greatly relished Ramusio's comparative lucidity: "I found (says he) this Booke translated by Master Hakluyt out of the Latine (*i.e.* among Hakluyt's MS. collections). But where the blind leade the blind both fall: as here the corrupt *Latine* could not but yeeld a corruption of truth in *English*. Ramusio, Secretarie to the *Decemviri* in *Venice*, found a better Copie and published the same, whence you have the worke in manner new : so renewed, that I have found the Proverbe true, that it is better to pull downe an old house and to build it anew, then to repaire it ; as I also should have done, had I knowne that which in the event I found. The *Latine* is Latten, compared to *Ramusio's* Gold. And hee which hath the *Latine* hath but *Marco Polo's* carkasse or not so much, but a few bones, yea, sometimes stones rather then bones ; things divers, averse, adverse, perverted in manner, disjoynted in manner, beyond beliefe. I have seene some Authors maymed, but never any so mangled and so mingled, so present and so absent, as this vulgar *Latine* of *Marco Polo;* not so like himselfe, as the Three *Polo's* were at their returne to *Venice*, where none knew them. . . . . Much are wee beholden to *Ramusio*, for restoring this *Pole* and Load-starre of *Asia*, out of that mirie poole or puddle in which he lay drouned." (III. p. 65.)

63. Though difficulties will certainly remain,* the most probable explanation of the origin of this text seems to me to be some such hypothesis as the following :—I suppose that Polo in his latter years added with his own hand supplementary notes and reminiscences, marginally or otherwise, to a copy of his book ; that these, perhaps in his lifetime, more probably after his death, were digested and translated into Latin ;† and that Ramusio, or some friend of his, in retranslating and fusing them with Pipino's version for the *Navigationi*, made those minor modifications in names and other matters which we have already noticed. The mere facts of digestion from memoranda and double translation would account for a good deal of unintentional corruption.

Hypothesis of the sources of the Ramusian Version.

That more than one version was employed in the composition of Ramusio's edition we have curious proof in at least one passage of the latter. We have pointed out at p. 410 of this volume a curious example of misunderstanding of the old French

---

* Of these difficulties the following are some of the more prominent :—

1. The mention of the death of Kúblái (see note 7, p. 38 of this volume), whilst throughout the book Polo speaks of Kúblái as if still reigning.

2. Mr Hugh Murray objects that whilst in the old texts Polo appears to look on Kúblái with reverence as a faultless Prince, in the Ramusian we find passages of an opposite tendency, as in the chapter about Ahmad.

3. The same editor points to the manner in which one of the Ramusian additions represents the traveller to have visited the Palace of the Chinese Kings at Kinsay, which he conceives to be inconsistent with Marco's position as an official of the Mongol Government. (See vol. ii. p. 208.)

If we could conceive the Ramusian additions to have been originally notes written by old Maffeo Polo on his nephew's book, this hypothesis would remove almost all difficulty.

One passage in Ramusio seems to bear a reference to the date at which these interpolated notes were amalgamated with the original. In the chapter on Samarkand (i. p. 191) the conversion of the Prince Chagatai is said in the old texts to have occurred " not a great while ago " (*il ne a encore grament de tens*). But in Ramusio the supposed event is fixed at " one hundred and twenty-five years since." This number could not have been uttered with reference to 1298, the year of the dictation at Genoa, nor to any year of Polo's own life. Hence it is probable that the original note contained a date or definite term which was altered by the compiler to suit the date of his own compilation, some time in the 14th century.

† In the first edition of Ramusio the preface contained the following passage, which is omitted from the succeeding editions ; but as even the first edition was issued after Ramusio's own death, I do not see that any stress can be laid on this :

" A copy of the Book of Marco Polo, as it was originally written in Latin, marvel- lously old, and perhaps directly copied from the original as it came from M. Marco s own hand, has been often consulted by me and compared with that which we now publish, having been lent me by a nobleman of this city, belonging to the Ca' Ghisi."

Text, a passage in which the term *Roi des Pelaines*, or "King of Furs," is applied to the Sable, and which in the Crusca has been converted into an imaginary Tartar phrase *Leroide pelame*, or as Pipino makes it *Rondes* (another indication that Pipino's Version and the Crusca passed through a common medium). But Ramusio exhibits *both* the true reading and the perversion : "*E li Tartari la chiamano* Regina delle pelli" (there is the true reading), "*E gli animali si chiamano* Rondes" (and there the perverted one).

We may further remark that Ramusio's version betrays indications that one of its bases either was in the Venetian dialect, or had passed through that dialect ; for a good many of the names appear in Venetian forms, *e.g.*, substituting the *z* for the sound of *ch*, *j*, or soft *g*, as in *Goza, Zorzania, Zagatay, Gonza* (for Giogiu), *Quenzanfu, Coiganzu, Tapinzu, Zipangu, Ziamba*.

64. To sum up.   It is, I think, beyond reasonable dispute that we have, in what we call the Geographic Text, as nearly as may be an exact transcript of the Traveller's words as ₛ Summary in originally taken down in the prison of Genoa.   We regard to Text of have again in the MSS. of the second type an edition Polo. pruned and refined, probably under instructions from Marco Polo, but not with any critical exactness.   And lastly, I believe, that we have, imbedded in the Ramusian edition, the supplementary recollections of the Traveller, noted down at a later period of his life, but perplexed by repeated translation, compilation, and editorial mishandling.

And the most important remaining problem in regard to the text of Polo's work is the discovery of the supplemental manuscript from which Ramusio derived those passages which are found only in his edition.   It is possible that it may still exist, but no trace of it in anything like completeness has yet been found ; though when my task was all but done I discovered a small part of the Ramusian peculiarities in a MS. at Venice.*

---

* For a moment I thought I had been lucky enough to light on a part of the missing original of Ramusio in the Barberini Library at Rome.   A fragment of a Venetian version in that library (No. 56 in our list of MSS.) bore on the fly-leaf the title "*Alcuni primi capi del Libro di S. Marco Polo, copiati dall esemplare manoscritto di PAOLO RANNUSIO.*"   But it proved to be of no importance.   One brief passage of those which have been thought peculiar to Ramusio ; viz., the

65. Whilst upon this subject of manuscripts of our Author, I will give some particulars regarding a very curious one, containing a version in the *Irish* language.

This remarkable document is found in the *Book of Lismore*, belonging to the Duke of Devonshire. That magnificent book, finely written on vellum of the largest size, was discovered in 1814, enclosed in a wooden box, along with a superb crozier, on opening a closed doorway in the castle of Lismore. It contained Lives of the Saints, the (Romance) History of Charlemagne, the History of the Lombards, histories and tales of Irish wars, etc., etc., and among the other matter this version of Marco Polo. A full account of the Book and its mutilations will be found in *O'Curry's Lectures on the MS. Materials of Ancient Irish History*, p. 196 *seqq.*, Dublin, 1861. The *Book of Lismore* was written about 1460 for

Notice of a curious Irish Version of Polo.

---

reference to the Martyrdom of St. Blaize at Sebaste (see p. 43 of this volume), is found also in the Geographic Latin.

It was pointed out by Lazari, that another passage (vol. i. p. 60) of those otherwise peculiar to Ramusio, is found in a somewhat abridged Latin version in a MS. which belonged to the late eminent antiquary Emanuel Cicogna. (See List in Appendix F, No. 35.) This fact induced me when at Venice in 1870 to examine the MS. throughout, and, though I could give little time to it, the result was very curious.

I find that this MS. contains, not one only, but at least *seven* of the passages otherwise peculiar to Ramusio, and must have been one of the elements that went to the formation of his text. Yet of his more important interpolations, such as the chapter on Ahmad's oppressions and the additional matter on the City of Kinsay, there is no indication. The seven passages alluded to are as follows; the words corresponding to Ramusian peculiarities are in italics, the references are to my own volumes.

1. In the chapter on Georgia :
"Mare quod dicitur Gheluchelan *vel ABACU* " . . .
"Est ejus stricta via et dubia. Ab una parte est mare *quod dixi de ABACU* et ab aliâ nemora invia," etc. (See I. p. 59, note 8.)

2. "Et ibi optimi austures *dicti AVIGI*" (1. 50).

3. After the chapter on Mosul is another short chapter, already alluded to :
"*Prope hanc civitatem* (est) *alia provincia dicta MUS e MEREDIEN in quâ nascitur magna quantitas bombacis, et hic fiunt bocharini et alia multa, et sunt mercatores homines et artiste.*" (See i. p. 60.)

4. In the chapter on *Tarcan* (for Carcan, *i.e.* Yarkand) :
"*Et maior pars horum habent unum ex pedibus grossum et habent gosum in gulâ ;* et est hic fertilis contracta." (See i. p. 187.)

5. In the Desert of Lop :
"*Homines trasseuntes appendunt bestiis suis capanullas* [*i.e.* campanellas] *ut ipsas senciant et ne deviare possint*" (i. p. 197.)

6. "Ciagannor, *quod sonat in Latino STAGNUM ALBUM.*" (i. p. 296.)

7. "Et in medio hujus viridarii est palacium sive logia, *tota super columpnas. Et in summitate cujuslibet columnæ est draco magnus circundans totam columpnam, et hic substinet eorum cohoperturam cum ore et pedibus ;* et est cohopertura tota de cannis hoc modo," etc. (See i. p. 299.)

Finghin MacCarthy and his wife Catharine Fitzgerald, daughter of Gerald, Eighth Earl of Desmond.

The date of the Translation of Polo is not known, but it may be supposed to have been executed about the above date, probably in the Monastery of Lismore (county of Waterford).

From the extracts that have been translated for me, it is obvious that the version was made, with an astounding freedom certainly, from Friar Francesco Pipino's Latin.

Both beginning and end are missing. But what remains opens thus; compare it with Friar Pipino's real prologue as we give it in the Appendix! *

"rigaib ⁊ tairsch na cathr̄ s̄. bai bᵉtꝗ rigui anaibit san frises is̄ cathr intans̄. ba eoluċ dā is̄ nahilberlaib fransiscī aaim̄. buir iar̄ du ambat̄ na maite ucut ⁊cumgit fꝗ inleabor toċloḋ ḟcula otlngaiḋ natꝗtairlḋ cg intlng laitanda." &c.

——"Kings and chieftains of that city. There was then in the city a princely Friar in the habit of St. Francis, named Franciscus, who was versed in many languages. He was brought to the place where those nobles were, and they requested of him to translate the book from the Tartar (!) into the Latin language. 'It is an abomination to me,' said he, 'to devote my mind or labour to works of Idolatry and Irreligion.' They entreated him again. 'It shall be done,' said he; 'for though it be an irreligious narrative that is related therein, yet the things are miracles of the True God; and every one who hears this much against the Holy Faith shall pray fervently for their conversion. And he who will not pray shall waste the vigour of his body to convert them.' I am not in dread of this Book of Marcus, for there is no lie in it. My eyes beheld him bringing the relics of the holy Church with him, and he left [his testimony], whilst tasting of death, that it was true. And Marcus was a devout man. What is there in it, then, but that Franciscus translated this Book of Marcus from the Tartar into Latin; and the years of the Lord at that time were fifteen years, two score, two hundred, and one thousand" (1255).

It then describes *Armein Bec* (Little Armenia), *Armein Mor* (Great Armenia), *Musul*, *Taurisius*, *Persida*, *Camandi*, and so forth. The last chapter is that on *Abaschia*:—

"ABASCHIA also is an extensive country, under the government of Seven

* My valued friend Sir Arthur Phayre made known to me the passage in *O'Curry's Lectures*. I then procured the extracts and further particulars from Mr. J. Long, Irish Transcriber and Translator in Dublin, who took them from the Transcript of the *Book of Lismore*, in the possession of the Royal Irish Academy. [Cf. *Anecdota Oxoniensia. Lives of the Saints from the Book of Lismore, edited with a translation . . . . by* Whitley Stokes, Oxford, 1890.—*Marco Polo* forms fo. 79 a, 1—fo. 89 b, 2, of the MS., and is described pp. xxii.-xxiv. of Mr. Whitley Stokes' Book, who has since published the Text in the *Zeit. f. Celtische Philol.* (See *Bibliography*, vol. ii. p. 573.)—H. C.]

Kings, four of whom worship the true God, and each of them wears a golden cross on the forehead ; and they are valiant in battle, having been brought up fighting against the Gentiles of the other three kings, who are Unbelievers and Idolaters. And the kingdom of ADEN ; a Soudan rules over them.

" The king of Abaschia once took a notion to make a pilgrimage to the Sepulchre of Jesus. 'Not at all,' said his nobles and warriors to him, 'for we should be afraid lest the infidels through whose territories you would have to pass, should kill you. There is a Holy Bishop with you,' said they ; 'send him to the Sepulchre of Jesus, and much gold with him'"———

The rest is wanting.

XI. SOME ESTIMATE OF THE CHARACTER OF POLO AND HIS BOOK.

66. That Marco Polo has been so universally recognised as the King of Mediæval Travellers is due rather to the width of his experience, the vast compass of his journeys, and the romantic nature of his personal history, than to transcendent superiority of character or capacity.

Grounds of Polo's pre-eminence among mediæval travellers.

The generation immediately preceding his own has bequeathed to us, in the Report of the Franciscan Friar William de Rubruquis,* on the Mission with which

---

* M. d'Avezac has refuted the common supposition that this admirable traveller was a native of Brabant.

The form *Rubruquis* of the name of the traveller William de Rubruk has been habitually used in this book, perhaps without sufficient consideration, but it is the most familiar in England, from its use by Hakluyt and Purchas. The former, who first published the narrative, professedly printed from an imperfect MS. belonging to the Lord Lumley, which does not seem to be now known. But all the MSS. collated by Messrs. Francisque-Michel and Wright, in preparing their edition of the Traveller, call him simply Willelmus de Rubruc or Rubruk.

Some old authors, apparently without the slightest ground, having called him *Risbroucke* and the like, it came to be assumed that he was a native of Ruysbroeck, a place in South Brabant.

But there is a place still called *Rubrouck* in French Flanders. This is a commune containing about 1500 inhabitants, belonging to the Canton of Cassel and *arrondissement* of Hazebrouck, in the Department du Nord. And we may take for granted, till facts are alleged against it, that *this* was the place from which the envoy of St. Lewis drew his origin. Many documents of the Middle Ages, referring expressly to this place Rubrouck, exist in the Library of St. Omer, and a detailed notice of them has been published by M. Edm. Coussemaker, of Lille. Several of these documents refer to persons bearing the same name as the Traveller, *e.g.*, in 1190, Thierry de Rubrouc ; in 1202 and 1221, Gauthier du Rubrouc ; in 1250, Jean du Rubrouc ; and in 1258, Woutermann de Rubrouc. It is reasonable to suppose that Friar William was of the same stock. See *Bulletin de la Soc. de Géographie*, 2nd vol. for 1868, pp. 569-570, in which there are some remarks on the subject by M. d'Avezac ; and

St. Lewis charged him to the Tartar Courts, the narrative of one great journey, which, in its rich detail, its vivid pictures, its acuteness of observation and strong good sense, seems to me to form a Book of Travels of much higher claims than *any one series* of Polo's chapters; a book, indeed, which has never had justice done to it, for it has few superiors in the whole Library of Travel.

Enthusiastic Biographers, beginning with Ramusio, have placed Polo on the same platform with Columbus. But where has our Venetian Traveller left behind him any trace of the genius and lofty enthusiasm, the ardent and justified previsions which mark the great Admiral as one of the lights of the human race?* It is a juster praise that the spur which his Book eventually gave to geographical studies, and the beacons which it hung out at the Eastern extremities of the Earth helped to guide the aims, though scarcely to kindle the fire, of the greater son of the rival Republic. His work was at

---

I am indebted to the kind courtesy of that eminent geographer himself for the indication of this reference and the main facts, as I had lost a note of my own on the subject.

It seems a somewhat complex question whether a native even of *French* Flanders at that time should be necessarily claimable as a Frenchman;* but no doubt on this point is alluded to by M. d'Avezac, so he probably had good ground for that assumption. [See also *Yule's* article in the *Encyclopædia Britannica*, and *Rockhill's Rubruck*, Int., p. xxxv.—H. C.]

That cross-grained Orientalist, I. J. Schmidt, on several occasions speaks contemptuously of this veracious and delightful traveller, whose evidence goes in the teeth of some of his crotchets. But I am glad to find that Professor Peschel takes a view similar to that expressed in the text: "The narrative of Ruysbroek [Rubruquis], almost immaculate in its freedom from fabulous insertions, may be indicated on account of its truth to nature as the greatest geographical masterpiece of the Middle Ages." (*Gesch. der Erdkunde*, 1865, p. 151.)

* High as Marco's name deserves to be set, his place is not beside the writer of such burning words as these addressed to Ferdinand and Isabella: "From the most tender age I went to sea, and to this day I have continued to do so. Whosoever devotes himself to this craft must desire to know the secrets of Nature here below. For 40 years now have I thus been engaged, and wherever man has sailed hitherto on the face of the sea, thither have I sailed also. I have been in constant relation with men of learning, whether ecclesiastic or secular, Latins and Greeks, Jews and Moors, and men of many a sect besides. To accomplish this my longing (to know the Secrets of the World) I found the Lord favourable to my purposes; it is He who hath given me the needful disposition and understanding. He bestowed upon me abundantly the knowledge of seamanship: and of Astronomy He gave me enough to work withal, and so with Geometry and Arithmetic. . . . . In the days of my youth I studied

---

* The County of Flanders was at this time in large part a fief of the French Crown. (See *Natalis de Wailly*, notes to Joinville, p. 576.) But that would not much affect the question either one way or the other.

least a link in the Providential chain which at last dragged the New World to light.*

67. Surely Marco's real, indisputable, and, in their kind, unique claims to glory may suffice! *He was the first Tra-*
His true claims to glory. *veller to trace a route across the whole longitude of* ASIA, *naming and describing kingdom after kingdom which he had seen with his own eyes; the Deserts of* PERSIA, *the flowering plateaux and wild gorges of* BADAKH-SHAN, *the jade-bearing rivers of* KHOTAN, *the* MONGOLIAN *Steppes, cradle of the power that had so lately threatened to swallow up Christendom, the new and brilliant Court that had been established at* CAMBALUC: *The first Traveller to reveal*

---

works of all kinds, history, chronicles, philosophy, and other arts, and to apprehend these the Lord opened my understanding. Under His manifest guidance I navigated hence to the Indies; for it was the Lord who gave me the will to accomplish that task, and it was in the ardour of that will that I came before your Highnesses. All those who heard of my project scouted and derided it; all the acquirements I have mentioned stood me in no stead; and if in your Highnesses, and in you alone, Faith and Constancy endured, to Whom are due the Lights that have enlightened you as well as me, but to the Holy Spirit?" (Quoted in *Humboldt's Examen Critique*, I. 17, 18.)

* Libri, however, speaks too strongly when he says: "The finest of all the results due to the influence of Marco Polo is that of having stirred Columbus to the discovery of the New World. Columbus, jealous of Polo's laurels, spent his life in preparing means to get to that Zipangu of which the Venetian traveller had told such great things; his desire was to reach China by sailing westward, and in his way he fell in with America." (*H. des Sciences Mathém.* etc. II. 150.)

The fact seems to be that Columbus knew of Polo's revelations only at second hand, from the letters of the Florentine Paolo Toscanelli and the like; and I cannot find that he *ever* refers to Polo by name. [How deep was the interest taken by Colombus in Marco Polo's travels is shown by the numerous marginal notes of the Admiral in the printed copy of the latin version of Pipino kept at the Bib. Colombina at Seville. See *Appendix H.* p. 558.—H. C.] Though to the day of his death he was full of imaginations about Zipangu and the land of the Great Kaan as being in immediate proximity to his discoveries, these were but accidents of his great theory. It was the intense conviction he had acquired of the absolute smallness of the Earth, of the vast extension of Asia eastward, and of the consequent narrowness of the Western Ocean, on which his life's project was based. This conviction he seems to have derived chiefly from the works of Cardinal Pierre d'Ailly. But the latter borrowed his collected arguments from Roger Bacon, who has stated them, erroneous as they are, very forcibly in his *Opus Majus* (p. 137), as Humboldt has noticed in his *Examen* (vol. i. p. 64). The Spanish historian Mariana makes a strange jumble of the alleged guides of Columbus, saying that some ascribed his convictions to "the information given by *one Marco Polo, a Florentine Physician!*" ("como otros dizen, por aviso que le dio *un cierto Marco Polo, Medico Florentin;*" *Hist. de España*, lib. xxvi. cap 3). Toscanelli is called by Columbus *Maestro Paulo*, which seems to have led to this mistake; see Sign. G. *Uzielli*, in *Boll. della Soc. Geog. Ital.* IX. p. 119. [Also by the same: *Paolo dal Pozzo Toscanelli iniziatore della scoperta d'America*, Florence, 1892; *Toscanelli*, No. 1; *Toscanelli*, Vol. V. of the *Raccolta Colombiana*, 1894.—H. C.]

CHINA *in all its wealth and vastness, its mighty rivers, its huge cities, its rich manufactures, its swarming population, the inconceivably vast fleets that quickened its seas and its inland waters; to tell us of the nations on its borders with all their eccentricities of manners and worship; of* TIBET *with its sordid devotees; of* BURMA *with its golden pagodas and their tinkling crowns; of* LAOS, *of* SIAM, *of* COCHIN CHINA, *of* JAPAN, *the Eastern Thule, with its rosy pearls and golden-roofed palaces; the first to speak of that Museum of Beauty and Wonder, still so imperfectly ransacked, the* INDIAN ARCHIPELAGO, *source of those aromatics then so highly prized and whose origin was so dark; of* JAVA *the Pearl of Islands; of* SUMATRA *with its many kings, its strange costly products, and its cannibal races; of the naked savages of* NICOBAR *and* ANDAMAN; *of* CEYLON *the Isle of Gems with its Sacred Mountain and its Tomb of Adam; of* INDIA THE GREAT, *not as a dream-land of Alexandrian fables, but as a country seen and partially explored, with its virtuous Brahmans, its obscene ascetics, its diamonds and the strange tales of their acquisition, its sea-beds of pearl, and its powerful sun; the first in mediæval times to give any distinct account of the secluded Christian Empire of* ABYSSINIA, *and the semi-Christian Island of* SOCOTRA; *to speak, though indeed dimly, of* ZANGIBAR *with its negroes and its ivory, and of the vast and distant* MADAGASCAR, *bordering on the Dark Ocean of the South, with its Ruc and other monstrosities; and, in a remotely opposite region, of* SIBERIA *and the* ARCTIC OCEAN, *of dog-sledges, white bears, and reindeer-riding Tunguses.*

That all this rich catalogue of discoveries should belong to the revelations of one Man and one Book is surely ample ground enough to account for and to justify the Author's high place in the roll of Fame, and there can be no need to exaggerate his greatness, or to invest him with imaginary attributes.[*]

68. What manner of man was Ser Marco? It is a question hard to answer. Some critics cry out against personal detail in books of Travel; but as regards him who would not welcome a little more egotism! In his Book impersonality is carried to excess; and we are often

His personal attributes seen but dimly.

---

[*] " C'est diminuer l'expression d'un éloge que de l'exagérer." (*Humboldt, Examen,* III. 13.)

driven to discern by indirect and doubtful indications alone, whether he is speaking of a place from personal knowledge or only from hearsay. In truth, though there are delightful exceptions, and nearly every part of the book suggests interesting questions, a desperate meagreness and baldness does extend over considerable tracts of the story. In fact his book reminds us sometimes of his own description of Khorasan :—
" *On chevauche par beaus plains et belles costieres, là où il a moult beaus herbages et bonne pasture et fruis assez . . . . . Et aucune fois y treuve l'en un desert de soixante milles ou de mains, esquels desers ne treuve l'en point d'eaue; mais la convient porter o lui !* "

Still, some shadowy image of the man may be seen in the Book; a practical man, brave, shrewd, prudent, keen in affairs, and never losing his interest in mercantile details, very fond of the chase, sparing of speech; with a deep wondering respect for Saints, even though they be Pagan Saints, and their asceticism, but a contempt for Patarins and such like, whose consciences would not run in customary grooves, and on his own part a keen appreciation of the World's pomps and vanities. See, on the one hand, his undisguised admiration of the hard life and long fastings of Sakya Muni; and on the other how enthusiastic he gets in speaking of the great Kaan's command of the good things of the world, but above all of his matchless opportunities of sport ! *

Of humour there are hardly any signs in his Book. His almost solitary joke (I know but one more, and it pertains to the οὐκ ἀνήκοντα) occurs in speaking of the Kaan's paper-money when he observes that Kúblái might be said to have the true Philosopher's Stone, for he made his money at pleasure out of the bark of Trees.† Even the oddest eccentricities of outlandish tribes scarcely seem to disturb his gravity; as when he relates in his brief way of the people called Gold-Teeth on the frontier of Burma, that ludicrous custom which Mr. Tylor has so well illustrated under the name of the *Couvade*. There is more savour of laughter in the few lines of a Greek Epic, which relate precisely the same custom of a people on the Euxine :—

---

* See vol. ii. p. 318, and vol. i. p. 404.   † Vol. i. p. 423.

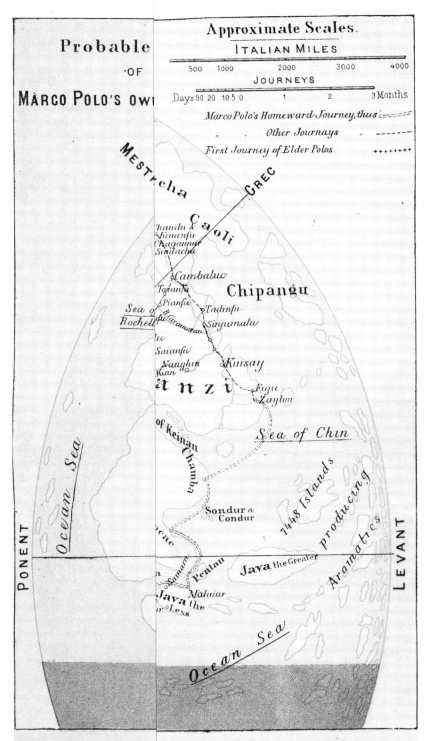

Probable

·OF·

MARCO POLO'S OW[N]

Approximate Scales.

ITALIAN MILES

500   1000   2000   3000   4000

JOURNEYS

Days 30 20 10 5 0   1   2   3 Months

Marco Polo's Homeward Journey, thus ‡‡‡‡‡‡

"   Other Journeys   — — — —

First Journey of Elder Polos.   ++++++++

MESTrcha   CREC

Caoli

Chandu &
Kimensu
Chaganmu
Sindachu

Cambaluc

Tadanfu

Pianfu   Chipangu
        Tadinfu
Sea of   Caramoran
Rochell   Sinjumatu
lee
Saianfu
Nanghin   Kinsay
Kian
ânzi   Fuju
        Zayton

of Keinau   Sea of Chin
Chamba

Ocean Sea   Sondur &
        Condur

        7448 Islands
            producing
        Aromatics

ac   Samara
        Pentan   Java the Greater

PONENT   Java the
        r Less   Malaiur   LEVANT

Ocean Sea

—— "In the Tibarenian Land
When some good woman bears her lord a babe,
'Tis *he* is swathed and groaning put to bed ;
Whilst *she*, arising, tends his baths, and serves
Nice possets for her husband in the straw." *

69. Of scientific notions, such as we find in the unvera-
cious Maundevile, we have no trace in truthful Marco. The
former, "lying with a circumstance," tells us boldly
that he was in 33° of South Latitude ; the latter is
full of wonder that some of the Indian Islands
Absence of
scientific
notions.
where he had been lay so far to the south that you lost sight
of the Pole-star. When it rises again on his horizon he estimates the Latitude by the Pole-star's being so many *cubits*
high. So the gallant Baber speaks of the sun having mounted
spear-high when the onset of battle began at Paniput. Such
expressions convey no notion at all to such as have had their
ideas sophisticated by angular perceptions of altitude, but
similar expressions are common among Orientals,† and indeed
I have heard them from educated Englishmen. In another
place Marco states regarding certain islands in the Northern
Ocean that they lie so very far to the north that in going
thither one actually leaves the Pole-star a trifle behind towards
the south ; a statement to which we know only one parallel,
to wit, in the voyage of that adventurous Dutch skipper who
told Master Moxon, King Charles II.'s Hydrographer, that he
had sailed two degrees beyond the Pole !

70. The Book, however, is full of bearings and distances,
and I have thought it worth while to construct a map from its
indications, in order to get some approximation to
Polo's own idea of the face of that world which
he had traversed so extensively. There are three
Map con-
structed on
Polo's data.
allusions to maps in the course of his work (II. 245, 312, 424).

In his own bearings, at least on land journeys, he usually
carries us along a great general traverse line, without much
caring about small changes of direction. Thus on the great
outward journey from the frontier of Persia to that of China
the line runs almost continuously "*entre Levant et Grec*" or
E.N.E. In his journey from Cambaluc or Peking to Mien or

---

* Vol. ii. p. 85, and *Apollonius Rhodius, Argonaut.* II. 1012.
† Chinese Observers record the length of Comets' tails by *cubits !*

Burma, it is always *Ponent* or W.; and in that from Peking to
Zayton in Fo-kien, the port of embarkation for India, it is
Sceloc or S.E.   The line of bearings in which he deviates most
widely from truth is that of the cities on the Arabian Coast
from Aden to Hormuz, which he makes to run steadily *vers
Maistre* or N.W., a conception which it has not been very easy
to realise on the map.*

71. In the early part of the Book we are told that Marco
acquired several of the languages current in the Mongol

Singular
omissions
of Polo in
regard to
China ; His-
torical inac-
curacies. Empire, and no less than four written characters.
We have discussed what these are likely to have
been (i. pp. 28-29), and have given a decided opinion
that Chinese was not one of them.   Besides intrinsic
improbability, and positive indications of Marco's ignorance
of Chinese, in no respect is his book so defective as in regard
to Chinese manners and peculiarities.   The Great Wall is
never mentioned, though we have shown reason for believing
that it was in his mind when one passage of his book was
dictated.†   The use of Tea, though he travelled through the

---

* The map, perhaps, gives too favourable an idea of Marco's geographical con-
ceptions.   For in such a construction much has to be supplied for which there are no
data, and that is apt to take mould from modern knowledge.   Just as in the book
illustrations of ninety years ago we find that Princesses of Abyssinia, damsels of Otaheite,
and Beauties of Mary Stuart's Court have all somehow a savour of the high waists,
low foreheads, and tight garments of 1810.

We are told that Prince Pedro of Portugal in 1426 received from the Signory of
Venice a map which was supposed to be either an original or a copy of one by Marco
Polo's own hand.   (*Major's P. Henry*, p. 62.)   There is no evidence to justify any
absolute expression of disbelief ; and if any map-maker with the spirit of the author
of the Carta Catalana then dwelt in Venice, Polo certainly could not have gone to
his grave uncatechised.   But I should suspect the map to have been a copy of the
old one that existed in the Sala dello Scudo of the Ducal Palace.

The maps now to be seen painted on the walls of that Hall, and on which Polo's
route is marked, are not of any great interest.   But in the middle of the 15th
century there was an old *Descriptio Orbis sive Mappamundus* in the Hall, and when
the apartment was renewed n 1459 a decree of the Senate ordered that such a map
should be repainted on the new walls.   This also perished by a fire in 1483.   On the
motion of Ramusio, in the next century, four new maps were painted.   These had
become dingy and ragged, when, in 1762, the Doge Marco Foscarini caused them to
be renewed by the painter Francesco Grisellini.   He professed to have adhered
closely to the old maps, but he certainly did not, as Morelli testifies.   Eastern Asia
looks as if based on a work of Ramusio's age, but Western Asia is of undoubtedly
modern character. (See *Operetti di Iacopo Morelli*, Ven. 1820, I. 299.)

† "Humboldt confirms the opinion I have more than once expressed that too
much must not be inferred from the silence of authors.   He adduces three important
and perfectly undeniable matters of fact, as to which no evidence is to be found where
it would be most anticipated : In the archives of Barcelona no trace of the triumphal

Tea districts of Fo-kien, is never mentioned ; * the compressed feet of the women and the employment of the fishing cormorant (both mentioned by Friar Odoric, the contemporary of his later years), artificial egg-hatching, printing of books (though the notice of this art seems positively challenged in his account of paper-money), besides a score of remarkable arts and customs which one would have expected to recur to his memory, are never alluded to. Neither does he speak of the great characteristic of the Chinese writing. It is difficult to account for these omissions, especially considering the comparative fulness with which he treats the manners of the Tartars and of the Southern Hindoos ; but the impression remains that his associations in China were chiefly with foreigners. Wherever the place he speaks of had a Tartar or Persian name he uses that rather than the Chinese one. Thus *Cathay, Cambaluc, Pulisanghin, Tangut, Chagannor, Saianfu, Kenjanfu, Tenduc, Acbalec, Carajan, Zardandan, Zayton, Kemenfu, Brius, Caramoran, Chorcha, Juju,* are all Mongol, Turki, or Persian forms, though all have Chinese equivalents.†

In reference to the then recent history of Asia, Marco is often inaccurate, *e.g.* in his account of the death of Chinghiz, in the list of his successors, and in his statement of the relation-

---

entry of Columbus into that city ; *in Marco Polo no allusion to the Chinese Wall;* in the archives of Portugal nothing about the voyages of Amerigo Vespucci in the service of that crown." (*Varnhagen* v. *Ense,* quoted by Hayward, *Essays,* 2nd Ser. I. 36.) See regarding the Chinese Wall the remarks referred to above, at p. 292 of this volume.

* [It is a strange fact that Polo never mentions the use of *Tea* in China, although he travelled through the Tea districts in Fu Kien, and tea was then as generally drunk by the Chinese as it is now. It is mentioned more than four centuries earlier by the Mohammedan merchant Soleyman, who visited China about the middle of the 9th century. He states (*Reinaud, Relation des Voyages faits par les Arabes et les Persans dans l'Inde et à la Chine,* 1845, I. 40) : "The people of China are accustomed to use as a beverage an infusion of a plant, which they call *sakh,* and the leaves of which are aromatic and of a bitter taste. It is considered very wholesome. This plant (the leaves) is sold in all the cities of the empire." (*Bretschneider, Hist. Bot. Disc.* I. p. 5.)—H. C.]

† It is probable that Persian, which had long been the language of Turanian courts, was also the common tongue of foreigners at that of the Mongols. *Pulisanghin* and *Zardandán,* in the preceding list, are pure Persian. So are several of the Oriental phrases noted at p. *84.* See also notes on *Ondanique* and *Vernique* at pp. 93 and 384 of this volume, on *Tacuin* at p. 448, and a note at p. *93 supra.* The narratives of Odoric, and others of the early travellers to Cathay, afford corroborative examples. Lord Stanley of Alderley, in one of his contributions to the Hakluyt Series, has given evidence from experience that Chinese Mahomedans still preserve the knowledge of numerous Persian words.

ship between notable members of that House.* But the most perplexing knot in the whole book lies in the interesting account which he gives of the Siege of Sayanfu or Siang-yang, during the subjugation of Southern China by Kúblái. I have entered on this matter in the notes (vol. ii. p. 167), and will only say here that M. Pauthier's solution of the difficulty is no solution, being absolutely inconsistent with the story as told by Marco himself, and that I see none; though I have so much faith in Marco's veracity that I am loath to believe that the facts admit of no reconciliation.

Our faint attempt to appreciate some of Marco's qualities, as gathered from his work, will seem far below the very high estimates that have been pronounced, not only by some who have delighted rather to enlarge upon his frame than to make themselves acquainted with his work,† but also by persons whose studies and opinions have been worthy of all respect. Our estimate, however, does not abate a jot of our intense interest in his Book and affection for his memory. And we have a strong feeling that, owing partly to his reticence, and partly to the great disadvantages under which the Book was committed to writing, we have in it a singularly imperfect image of the Man.

72. A question naturally suggests itself, how far Polo's narrative, at least in its expression, was modified by passing under the pen of a professed littérateur of somewhat humble claims, such as Rusticiano was. The case is not a singular one, and in our own day the ill-judged use of such assistance has been fatal to the reputation of an adventurous Traveller.

Was Polo's Book materially affected by the Scribe Rusticiano?

---

* Compare these errors with like errors of Herodotus, *e.g.*, regarding the conspiracy of the False Smerdis. (See Rawlinson's Introduction, p. 55.) There is a curious parallel between the two also in the supposed occasional use of Oriental state records, as in Herodotus's accounts of the revenues of the satrapies, and of the army of Xerxes, and in Marco Polo's account of Kinsay, and of the Kaan's revenues. (Vol. ii pp. 185, 216.)

† An example is seen in the voluminous *Annali Musulmani* of *G. B. Rampoldi,* Milan, 1825. This writer speaks of the Travels of Marco Polo with his *brother* and uncle; declares that he visited *Tipango* (*sic*), Java, Ceylon, and the *Maldives,* collected all the geographical notions of his age, traversed the two peninsulas of the Indies, examined the islands of *Socotra, Madagascar, Sofala,* and traversed with *philosophic eye* the regions of Zanguebar, Abyssinia, Nubia, and Egypt! and so forth (ix. 174). And whilst Malte-Brun bestows on Marco the sounding and ridiculous title of *"the Humboldt of the 13th century,"* he shows little real acquaintance with his Book. (See his *Précis,* ed. of 1836, I. 551 *seqq.*)

We have, however, already expressed our own view that in the Geographic Text we have the nearest possible approach to a photographic impression of Marco's oral narrative. If there be an exception to this we should seek it in the descriptions of battles, in which we find the narrator to fall constantly into a certain vein of bombastic commonplaces, which look like the stock phrases of a professed romancer, and which indeed have a strong resemblance to the actual phraseology of certain metrical romances.* Whether this feature be due to Rusticiano I cannot say, but I have not been able to trace anything of the same character in a cursory inspection of some of his romance-compilations. Still one finds it impossible to conceive of our sober and reticent Messer Marco pacing the floor of his Genoese dungeon, and seven times over rolling out this magniloquent bombast, with sufficient deliberation to be overtaken by the pen of the faithful amanuensis!

73. On the other hand, though Marco, who had left home at fifteen years of age, naturally shows very few signs of reading, there are indications that he had read romances, especially those dealing with the fabulous adventures of Alexander.

Marco's reading embraced the Alexandrian Romances. Examples.

To these he refers explicitly or tacitly in his notices of the Irongate and of Gog and Magog, in his allusions to the marriage of Alexander with Darius's daughter, and to the battle between those two heroes, and in his repeated mention of the *Arbre Sol* or *Arbre Sec* on the Khorasan frontier.

The key to these allusions is to be found in that Legendary History of Alexander, entirely distinct from the true history of the Macedonian Conqueror, which in great measure took the place of the latter in the imagination of East and West for more than a thousand years. This fabulous history is believed to be of Græco-Egyptian origin, and in its earliest extant compiled form, in the Greek of the Pseudo-Callisthenes, can be traced back to at least about A.D. 200. From the Greek its marvels spread eastward at an early date; some part at least of their matter was known to Moses of Chorene, in the

---

* See for example vol. i. p. 338, and note 4 at p. 341 ; also vol. ii. p. 103. The descriptions in the style referred to recur in all seven times ; but most of them (which are in Book IV.) have been omitted in this translation.

5th century ;* they were translated into Armenian, Arabic, Hebrew, and Syriac; and were reproduced in the verses of Firdusi and various other Persian Poets; spreading eventually even to the Indian Archipelago, and finding utterance in Malay and Siamese. At an early date they had been rendered into Latin by Julius Valerius; but this work had probably been lost sight of, and it was in the 10th century that they were re-imported from Byzantium to Italy by the Archpriest Leo, who had gone as Envoy to the Eastern Capital from John Duke of Campania.† Romantic histories on this foundation, in verse and prose, became diffused in all the languages of Western Europe, from Spain to Scandinavia, rivalling in popularity the romantic cycles of the Round Table or of Charlemagne. Nor did this popularity cease till the 16th century was well advanced.

The heads of most of the Mediæval Travellers were crammed with these fables as genuine history.‡ And by the help of that community of legend on this subject which they found wherever Mahomedan literature had spread, Alexander Magnus was to be traced everywhere in Asia. Friar Odoric found Tana, near Bombay, to be the veritable City of King Porus; John Marignolli's vainglory led him to imitate King Alexander in setting up a marble column "in the corner of the world over against Paradise," *i.e.* somewhere on the coast of Travancore; whilst Sir John Maundevile, with a cheaper ambition, borrowed wonders from the Travels of Alexander to adorn his own. Nay, even in after days, when the Portuguese stumbled with amazement on those vast ruins in Camboja, which have so lately become familiar to us through the works of Mouhot, Thomson, and Garnier, they ascribed them to Alexander.§

Prominent in all these stories is the tale of Alexander's shutting up a score of impure nations, at the head of which were Gog and Magog, within a barrier of impassable moun-

---

* [On the subject of Moses of Chorene and his works, I must refer to the clever researches of the late Auguste Carrière, Professor of Armenian at the École des Langues Orientales.—H.C.]

† *Zacher, Forschungen zur Critik, &c., der Alexandersage,* Halle, 1867, p. 108.

‡ Even so sagacious a man as Roger Bacon quotes the fabulous letter of Alexander to Aristotle as authentic. (*Opus Majus,* p. 137.)

§ *J. As.* sér. VI. tom. xviii. p. 352.

tains, there to await the latter days ; a legend with which the disturbed mind of Europe not unnaturally connected that cataclysm of unheard-of Pagans that seemed about to deluge Christendom in the first half of the 13th century. In these stories also the beautiful Roxana, who becomes the bride of Alexander, is *Darius's* daughter, bequeathed to his arms by the dying monarch. Conspicuous among them again is the Legend of the Oracular Trees of the Sun and Moon, which with audible voice foretell the place and manner of Alexander's death. With this Alexandrian legend some of the later forms of the story had mixed up one of Christian origin about the Dry Tree, *L'Arbre Sec.* And they had also adopted the Oriental story of the Land of Darkness and the mode of escape from it, which Polo relates at p. 484 of vol. ii.

74. We have seen in the most probable interpretation of the nickname *Milioni* that Polo's popular reputation in his lifetime was of a questionable kind ; and a contem- Injustice long done to porary chronicler, already quoted, has told us how on Polo. Singular modern his death-bed the Traveller was begged by anxious instance. friends to retract his extraordinary stories.* A little later one who copied the Book "*per passare tempo e malinconia*" says frankly that he puts no faith in it.† Sir Thomas Brown is content "to carry a wary eye" in reading "Paulus Venetus"; but others of our countrymen in the last century express strong doubts whether he ever was in Tartary or China.‡ Marden's edition might well have extinguished the last sparks of scepticism.§ Hammer meant praise in calling Polo "*der Vater orientalischer Hodogetik,*" in spite of the uncouthness of

---

* See passage from Jacopo d'Acqui, *supra*, p. *54.*

† It is the transcriber of one of the Florence MSS. who appends this terminal note, worthy of Mrs. Nickleby :—"Here ends the Book of Messer M. P. of Venice, written with mine own hand by me Amalio Bonaguisi when Podestà of Cierreto Guidi, to get rid of time and *ennui.* The contents seem to me incredible things, not lies so much as miracles ; and it may be all very true what he says, but I don't believe it ; though to be sure throughout the world very different things are found in different countries. But these things, it has seemed to me in copying, are entertaining enough, but not things to believe or put any faith in ; that at least is my opinion. And I finished copying this at Cierreto aforesaid, 12th November, A.D. 1392."

‡ *Vulgar Errors*, Bk. I. ch. viii. ; *Astley's Voyages*, IV. 583.

§ A few years before Marsden's publication, the Historical branch of the R. S. of Science at Göttingen appears to have put forth as the subject of a prize Essay the Geography of the Travels of Carpini, Rubruquis, and especially of Marco Polo. (See *L. of M. Polo*, by *Zurla*, in *Collezione di Vite e Ritratti d'Illustri Italiani.* Pad. 1816.)

the eulogy. But another grave German writer, ten years after Marsden's publication, put forth in a serious book that the whole story was a clumsy imposture!*

XII. CONTEMPORARY RECOGNITION OF POLO AND HIS BOOK.

75. But we must return for a little to Polo's own times. Ramusio states, as we have seen, that immediately after the first commission of Polo's narrative to writing (in Latin as he imagined), many copies of it were made, it was translated into the vulgar tongue, and in a few months all Italy was full of it.

How far was there diffusion of his Book in his own day?

The few facts that we can collect do not justify this view of the rapid and diffused renown of the Traveller and his Book. The number of MSS. of the latter dating from the 14th century is no doubt considerable, but a large proportion of these are of Pipino's condensed Latin Translation, which was not put forth, if we can trust Ramusio, till 1320, and certainly not much earlier. The whole number of MSS. in various languages that we have been able to register, amounts to about eighty. I find it difficult to obtain statistical data as to the comparative number of copies of different works existing in manuscript. With

---

* See *Städtewesen des Mittelalters*, by *K. D. Hüllmann*, Bonn, 1829, vol. iv.

After speaking of the Missions of Pope Innocent IV. and St. Lewis, this author sketches the Travels of the Polos, and then proceeds :—"Such are the clumsily compiled contents of this ecclesiastical fiction (*Kirchengeschichtlichen Dichtung*) disguised as a Book of Travels, a thing devised generally in the spirit of the age, but specially in the interests of the Clergy and of Trade. . . . This compiler's aim was analogous to that of the inventor of the Song of Roland, to kindle enthusiasm for the conversion of the Mongols, and so to facilitate commerce through their dominions. . . . Assuredly the Poli never got further than Great Bucharia, which was then reached by many Italian Travellers. What they have related of the regions of the Mongol Empire lying further east consists merely of recollections of the bazaar and travel-talk of traders from those countries; whilst the notices of India, Persia, Arabia, and Ethiopia, are borrowed from Arabic Works. The compiler no doubt carries his audacity in fiction a long way, when he makes his hero Marcus assert that he had been seventeen years in Kúbláí's service," etc. etc. (pp. 360-362).

In the French edition of *Malcolm's History of Persia* (II. 141), Marco is styled "*prêtre Venetien*"! I do not know whether this is due to Sir John or to the translator.

[Polo is also called "a Venetian Priest," in a note, vol. i., p. 409, of the original edition of London, 1815, 2 vols., 4to.—H. C.]

Dante's great Poem, of which there are reckoned close upon 500 MSS.,* comparison would be inappropriate. But of the Travels of Friar Odoric, a poor work indeed beside Marco Polo's, I reckoned thirty-nine MSS., and could now add at least three more to the list. [I described seventy-three in my edition of *Odoric.*—H. C.] Also I find that of the nearly contemporary work of Brunetto Latini, the *Tresor*, a sort of condensed Encyclopædia of knowledge, but a work which one would scarcely have expected to approach the popularity of Polo's Book, the Editor enumerates some fifty MSS. And from the great frequency with which one encounters in Catalogues both MSS. and early printed editions of Sir John Maundevile, I should suppose that the lying wonders of our English Knight had a far greater popularity and more extensive diffusion than the veracious and more sober marvels of Polo.† To Southern Italy Polo's popularity certainly does not seem at any time to have extended. I cannot learn that any MS. of his Book exists in any Library of the late Kingdom of Naples or in Sicily.‡

Dante, who lived for twenty-three years after Marco's work

---

* See *Ferrazzi, Manuele Dantesca*, Bassano, 1865, p. 729.

† In Quaritch's catalogue for Nov. 1870 there is only one *old* edition of Polo ; there are *nine* of Maundevile. In 1839 there were nineteen MSS. of the latter author *catalogued* in the British Museum Library. There are *now* only six of Marco Polo. At least twenty-five editions of Maundevile and only five of Polo were printed in the 15th century.

‡ I have made personal enquiry at the National Libraries of Naples and Palermo, at the Communal Library in the latter city, and at the Benedictine Libraries of Monte Cassino, Monreale, S. Martino, and Catania.

In the 15th century, when Polo's book had become more generally diffused we find three copies of it in the Catalogue of the Library of Charles VI. of France, made at the Louvre in 1423, by order of the Duke of Bedford.

The estimates of value are curious. They are in *sols parisis*, which we shall not estimate very wrongly at a shilling each :—

"No. 295. *Item.* Marcus Paulus ; *en ung cahier escript de lettre formée, en françois, à deux coulombes. Commt. ou iie fo.* ' deux frères prescheurs,' *et ou derrenier* ' que sa arrières.' *X. s. p.*
   *   *   *

"No. 334. *Item.* Marcus Paulus. *Couvert de drap d'or, bien escript & enluminé, de lettre de forme en françois, à deux coulombes. Commt. ou iie fol. ;* ' il fut Roys,' *& ou derrenier* ' propremen,' *à deux fermouers de laton. XV. s. p.*
   *   *   *

"No. 336. *Item.* Marcus Paulus ; *non enluminé, escript en françois, de lettre de forme. Commt. ou iie fo.* 'vocata moult grant,' *& ou derrenier* 'ilec dist il.' *Couvert de cuir blanc, à deux fermouers de laton. XII. s. p.*"
                 (*Inventaire de la Bibliothèque du Roi Charles VI.*, etc.,
                   Paris, Société des Bibliophiles, 1867.)

was written, and who touches so many things in the seen and unseen Worlds, never alludes to Polo, nor I think to anything that can be connected with his Book. I believe that no mention of *Cathay* occurs in the *Divina Commedia*. That distant region is indeed mentioned more than once in the poems of a humbler contemporary, Francesco da Barberino, but there is nothing in his allusions besides this name to suggest any knowledge of Polo's work.*

Neither can I discover any trace of Polo or his work in that of his contemporary and countryman, Marino Sanudo the Elder, though this worthy is well acquainted with the somewhat later work of Hayton, and many of the subjects which he touches in his own book would seem to challenge a reference to Marco's labours.

76. Of contemporary or nearly contemporary references to our Traveller by name, the following are all that I can produce, and none of them are new.

Contemporary references to Polo.

First there is the notice regarding his presentation of his book to Thibault de Cepoy, of which we need say no more (*supra*, p. 68).

Next there is the Preface to Friar Pipino's Translation, which we give at length in the Appendix (E) to these notices. The phraseology of this appears to imply that Marco was still alive, and this agrees with the date assigned to the work by Ramusio.

---

* See *Del Reggimento e de' Costumi delle donne di Messer Francesco da Barberino*, Roma, 1815, pp. 166 and 271. The latter passage runs thus, on *Slavery:*—

> " E fu indutta prima da Noé,
> E fu cagion lo vin, perchè si egge:
> Ch' egli è un paese, dove
> Son molti servi in parte di Cathay:
> Che per questa cagione
> Hanno a nimico il vino,
> E non ne beon, nè voglion vedere."

The author was born the year before Dante (1264), and though he lived to 1348 it is probable that the poems in question were written in his earlier years. *Cathay* was no doubt known by dim repute long before the final return of the Polos, both through the original journey of Nicolo and Maffeo, and by information gathered by the Missionary Friars. Indeed, in 1278 Pope Nicolas III., in consequence of information said to have come from Abaka Khan of Persia, that Kúblái was a baptised Christian, sent a party of Franciscans with a long letter to the Kaan *Quobley*, as he is termed. They never seem to have reached their destination. And in 1289 Nicolas IV. entrusted a similar mission to Friar John of Monte Corvino, which eventually led to very tangible results. Neither of the Papal letters, however, mentions *Cathay*. (See *Mosheim*, App. pp. 76 and 94.)

Pipino was also the author of a Chronicle, of which a part was printed by Muratori, and this contains chapters on the Tartar wars, the destruction of the Old Man of the Mountain, etc., derived from Polo. A passage not printed by Muratori has been extracted by Prof. Bianconi from a MS. of this Chronicle in the Modena Library, and runs as follows :—

"The matters which follow, concerning the magnificence of the Tartar Emperors, whom in their language they call *Cham* as we have said, are related by Marcus Paulus the Venetian in a certain Book of his which has been translated by me into Latin out of the Lombardic Vernacular. Having gained the notice of the Emperor himself and become attached to his service, he passed nearly 27 years in the Tartar countries."*

Again we have that mention of Marco by Friar Jacopo d'Acqui, which we have quoted in connection with his capture by the Genoese, at p. *54*.† And the Florentine historian GIOVANNI VILLANI,‡ when alluding to the Tartars, says :—

" Let him who would make full acquaintance with their history examine the book of Friar Hayton, Lord of Colcos in Armenia, which he made at the instance of Pope Clement V., and also the Book called *Milione* which was made by Messer Marco Polo of Venice, who tells much about their power and dominion, having spent a long time among them. And so let us quit the Tartars and return to our subject, the History of Florence."§

77. Lastly, we learn from a curious passage in a medical work by PIETRO OF ABANO, a celebrated physician and philosopher, and a man of Polo's own generation, that he was personally acquainted with the Traveller. In a discussion on the old notion of the non-habitability of the Equatorial regions, which Pietro controverts, he says :‖

Further contemporary references.

---

* See *Muratori*, IX. 583, *seqq.* ; *Bianconi*, Mem. I. p. 37.

† This Friar makes a strange hotch-potch of what he had read, *e.g.* : " The Tartars, when they came out of the mountains, made them a king, viz., the son of Prester John, who is thus vulgarly termed *Vetulus de la Montagna!* " (*Mon. Hist. Patr.* Script. III. 1557.)

‡ G. Villani died in the great plague of 1348. But his book was begun soon after Marco's was written, for he states that it was the sight of the memorials of greatness which he witnessed at Rome, during the Jubilee of 1300, that put it into his head to write the history of the rising glories of Florence, and that he began the work after his return home. (Bk. VIII. ch. 36.)   § Book V. ch. 29.

‖ *Petri Aponensis Medici ac Philosophi Celeberrimi, Conciliator,* Venice, 1521, fol. 97. Peter was born in 1250 at Abano, near Padua, and was Professor of Medicine at the University in the latter city. He twice fell into the claws of the Unholy Office, and only escaped them by death in 1316.

"In the country of the ZINGHI there is seen a star as big as a sack. I know a man who has seen it, and he told me it had a faint light like a piece of a cloud, and is always in the south.*

Star at the Antarctic as sketched
by Marco Polo (†).

I have been told of this and other matters by MARCO the Venetian, the most extensive traveller and the most diligent inquirer whom I have ever known. He saw this same star under the Antarctic ; he described it as having a great tail, and drew a figure of it *thus.* He also told me that he saw the Antarctic Pole at an altitude above the earth apparently equal to the length of a soldier's lance, whilst the Arctic Pole was as much below the horizon. 'Tis from that place, he says, that they export to us camphor, lign-aloes, and brazil. He says the heat there is intense, and the habitations few. And these things he witnessed in a certain island at which he arrived by Sea. He tells me also that there are (wild ?) men there, and also certain very great rams that have very coarse and stiff wool just like the bristles of our pigs."‡

In addition to these five I know no other contemporary references to Polo, nor indeed any other within the 14th century, though such there must surely be, excepting in a Chronicle written after the middle of that century by JOHN of

---

* The great Magellanic cloud ?   In the account of Vincent Yanez Pinzon's Voyage to the S.W. in 1499 as given in Ramusio (III. 15) after Pietro Martire d'Anghieria, it is said :—" Taking the astrolabe in hand, and ascertaining the Antarctic Pole, they did not see any star like our Pole Star ; but they related that they saw another manner of stars very different from ours, and which they could not clearly discern because of a certain dimness which diffused itself about those stars, and obstructed the view of them."   Also the Kachh mariners told Lieutenant Leech that midway to Zanzibar there was a town (?) called Marethee, where the North Pole Star sinks below the horizon, and they steer by *a fixed cloud in the heavens.*   (Bombay Govt. Selections, No. XV. N.S. p. 215.)

The great Magellan cloud is mentioned by an old Arab writer as a white blotch at the foot of Canopus, visible in the Tehama along the Red Sea, but not in Nejd or 'Irák.   Humboldt, in quoting this, calculates that in A.D. 1000 the Great Magellan would have been visible at Aden some degrees above the horizon. (*Examen,* V. 235.)

† [It is curious that this figure is almost exactly that which among oriental carpets is called a "cloud."   I have heard the term so applied by Vincent Robinson.   It often appears in old Persian carpets, and also in Chinese designs.   Mr. Purdon Clarke tells me it is called *nebula* in heraldry; it is also called in Chinese by a term signifying cloud ; in Persian, by a term which he called *silen-i-khitai,* but of this I can make nothing.—*MS. Note by Yule.*]

‡ This passage contains points that are omitted in Polo's book, besides the drawing implied to be from Marco's own hand !   The island is of course Sumatra.   The animal is perhaps the peculiar Sumatran wild-goat, figured by Marsden, the hair of which on the back is "coarse and strong, almost like bristles."   (*Sumatra,* p. 115.)

YPRES, Abbot of St. Bertin, otherwise known as Friar John the Long, and himself a person of very high merit in the history of Travel, as a precursor of the Ramusios, Hakluyts and Purchases, for he collected together and translated (when needful) into French all of the most valuable works of Eastern Travel and Geography produced in the age immediately preceding his own.* In his Chronicle the Abbot speaks at some length of the adventures of the Polo Family, concluding with a passage to which we have already had occasion to refer :

"And so Messers Nicolaus and Maffeus, with certain Tartars, were sent a second time to these parts ; but Marcus Pauli was retained by the Emperor and employed in his military service, abiding with him for a space of 27 years. And the Cham, on account of his ability despatched him upon affairs of his to various parts of Tartary and India and the Islands, on which journeys he beheld many of the marvels of those regions. And concerning these he afterwards composed a book in the French vernacular, which said Book of Marvels, with others of the same kind, we do possess." (*Thesaur. Nov. Anecdot.* III. 747.)

78. There is, however, a notable work which is ascribed to a rather early date in the 14th century, and which, though it contains no reference to Polo by name, shows a thorough acquaintance with his book, and borrows themes largely from it. This is the poetical Romance of Bauduin de Sebourc, Curious borrowings an exceedingly clever and vivacious production, par- from Polo in the taking largely of that bantering, half-mocking Romance of Bauduin de spirit which is, I believe, characteristic of many of the Sebourc.

---

* A splendid example of Abbot John's Collection is the *Livre des Merveilles* of the Great French Library (No. 18 in our *App. F.*). This contains Polo, Odoric, William of Boldensel, the Book of the Estate of the Great Kaan by the Archbishop of Soltania, Maundevile, Hayton, and Ricold of Montecroce, of which all but Polo and Maundevile are French versions by this excellent Long John. A list of the Polo miniatures is given in *App. F.* of this Edition, p. 527.

It is a question for which there is sufficient ground, whether the Persian Historians Rashiduddin and Wassáf, one or other or both, did not derive certain information that appears in their histories, from Marco Polo personally, he having spent many months in Persia, and at the Court of Tabriz, when either or both may have been there. Such passages as that about the Cotton-trees of Guzerat (vol. ii. p. 393, and note), those about the horse trade with Maabar (id. p. 340, and note), about the brother-kings of that country (id. p. 331), about the naked savages of Necuveram (id. p. 306), about the wild people of Sumatra calling themselves subjects of the Great Kaan (id. pp. 285, 292, 293, 299), have so strong a resemblance to parallel passages in one or both of the above historians, as given in the first and third volumes of Elliot, that the probability, at least, of the Persian writers having derived their information from Polo might be fairly maintained.

later mediæval French Romances.* Bauduin is a knight who,
after a very wild and loose youth, goes through an extraordinary
series of adventures, displaying great faith and courage, and
eventually becomes King of Jerusalem. I will cite some of the
traits evidently derived from our Traveller, which I have met
with in a short examination of this curious work.

Bauduin, embarked on a dromond in the Indian Sea, is
wrecked in the territory of Baudas, and near a city called Falise,
which stands on the River of Baudas. The people of this city
were an unbelieving race.

> " Il ne créoient Dieu, Mahon, né Tervogant,
> Ydole, cruchéfis, déable, né tirant." P. 300.

Their only belief was this, that when a man died a great fire
should be made beside his tomb, in which should be burned all
his clothes, arms, and necessary furniture, whilst his horse
and servant should be put to death, and then the dead man
would have the benefit of all these useful properties in the other
world.† Moreover, if it was the king that died—

> " Sé li rois de la terre i aloit trespassant,
>   *   *   *   *   *
> Si fasoit-on tuer, .viij. jour en un tenant,
> Tout chiaus c'on encontroit par la chité passant,
> Pour tenir compaingnie leur ségnor soffisant.
> Telle estoit le créanche ou païs dont je cant ! "‡ P. 301.

Baudin arrives when the king has been dead three days, and
through dread of this custom all the people of the city are shut
up in their houses. He enters an inn, and helps himself to a
vast repast, having been fasting for three days. He is then
seized and carried before the king, Polibans by name. We
might have quoted this prince at p. *87* as an instance of the
diffusion of the French tongue :

> " Polibans sot Fransois, car on le doctrina :
> j. renoiés de Franche. vij. ans i demora,
> Qui li aprist Fransois, si que bel en parla." P. 309.

---

* *Li Romans de Bauduin de Sebourc IIIᵉ Roy de Jhérusalem ;* Poëme du
XIVᵉ Siècle ; Valenciennes, 1841. 2 vols. 8vo. I was indebted to two references
of M. Pauthier's for knowledge of the existence of this work. He cites the legends
of the Mountain, and of the Stone of the Saracens from an abstract, but does not
seem to have consulted the work itself, nor to have been aware of the extent of its
borrowings from Marco Polo. M. Génin, from whose account Pauthier quotes,
ascribes the poem to an early date after the death of Philip the Fair (1314). See
Pauthier, pp. 57, 58, and 140.

† See Polo, vol. i. p. 204, and vol. ii. p. 191. ‡ See Polo, vol. i. p. 246.

Bauduin exclaims against their barbarous belief, and declares the Christian doctrine to the king, who acknowledges good points in it, but concludes :

> "Vassaus, dist Polibans, à le chière hardie,
> Jà ne crerrai vou Dieux, à nul jour de ma vie ;
> Né vostre Loy ne vaut une pomme pourie !" P. 311.

Bauduin proposes to prove his Faith by fighting the prince, himself unarmed, the latter with all his arms. The prince agrees, but is rather dismayed at Bauduin's confidence, and desires his followers, in case of his own death, to burn with him horses, armour, etc., asking at the same time which of them would consent to burn along with him, in order to be his companions in the other world :

> "Là en i ot. ij⁰. dont cascuns s'escria :
> "Nous morons volentiers, quant vo corps mort sara !"* P. 313.

Bauduin's prayer for help is miraculously granted ; Polibans is beaten, and converted by a vision. He tells Bauduin that in his neighbourhood, beyond Baudas—

> "ou. v. liewes, ou. vi.
> Ché un felles prinches, orgoellieus et despis ;
> De la Rouge-Montaingne est Prinches et Marchis.
> Or vous dirai comment il a ses gens nouris :
> Je vous di que chius Roys a fait un Paradis
> Tant noble et gratieus, et plain de tels déliis,
> * * * * *
> Car en che Paradis est un riex establis,
> Qui se partist en trois, en che noble pourpris :
> En l'un coert li clarés, d'espises bien garnis ;
> Et en l'autre li miés, qui les a resouffis ;
> Et li vins di pieument i queurt par droit avis—
> * * * * *
> Il n'i vente, né gèle. Che liés est de samis,
> De riches dras de soie, bien ouvrés à devis.
> Et aveukes tout che que je chi vous devis,
> I a. ij⁰ puchelles qui moult ont cler les vis,
> Carolans et tresquans, menans gales et ris.
> Et si est li dieuesse, dame et suppellatis,
> Qui doctrine les autres et en fais et en dis,
> Celle est la fille au Roy c'on dist des *Haus-Assis.*"† Pp. 319-320.

---

* See Polo, vol. ii. p. 339.
† See Polo, vol. i. p. 140. *Hashishi* has got altered into *Haus Assis.*
VOL. I.   *p* 2

This Lady Ivorine, the Old Man's daughter, is described among other points as having—

> "Les iex vairs com faucons, nobles et agentis."* P. 320.

The King of the Mountain collects all the young male children of the country, and has them brought up for nine or ten years :

> "Dedens un lieu oscur : là les met-on toudis
> Aveukes males bestes ; kiens, et cas, et soris,
> Culoères, et lisaerdes, escorpions petis.
> Là endroit ne peut nuls avoir joie, né ris." Pp. 320-321.

And after this dreary life they are shown the Paradise, and told that such shall be their portion if they do their Lord's behest.

> "S'il disoit à son homme : 'Va-t-ent droit à Paris ;
> Si me fier d'un coutel le Roy de Saint Denis,
> Jamais n'aresteroit, né par nuit né par dis,
> S'aroit tué le Roy, voïant tous ches marchis ;
> Et déuist estre à fources traïnés et mal mis.'" P. 321.

Bauduin determines to see this Paradise and the lovely Ivorine. The road led by Baudas :

> "Or avoit à che tamps, sé l'istoire ne ment,
> En le chit de Baudas Kristiens jusqu' à cent ;
> Qui manonent illoec par tréu d'argent,
> Que cascuns cristiens au Roy-Calife rent.
>   Li pères du Calife, qui régna longement,
> Ama les Crestiens, et Dieu primièrement :
>   *   *   *   *   *
> Et lor fist establir. j. monstier noble et gent,
> Où Crestien faisoient faire lor sacrement.
> Une mout noble pière lor donna proprement,
> Où on avoit posé Mahon moult longement." † P. 322.

The story is, in fact, that which Marco relates of Samarkand.‡ The Caliph dies. His son hates the Christians. His people complain of the toleration of the Christians and their minister ; but he says his father had pledged him not to interfere, and he dared not forswear himself. If, without

---

* See vol. i. p. 358, note.   † See vol. i. p. 189, note 2.
‡ Vol. i. pp. 183-186.

doing so, he could do them an ill turn, he would gladly. The people then suggest their claim to the stone:

> " Or leur donna vos pères, dont che fu mesprisons.
> Ceste pierre, biaus Sire, Crestiens demandons :
> Il ne le porront rendre, pour vrai le vous disons,
> Si li monstiers n'est mis et par pièches et par mons ;
> Et s'il estoit desfais, jamais ne le larons
> Refaire chi-endroit. Ensément averons
> Faites et acomplies nostres ententions." P. 324.

The Caliph accordingly sends for Maistre Thumas, the Priest of the Christians, and tells him the stone must be given up:

> " Il a. c. ans ut plus c'on i mist à solas
> Mahon, le nostre Dieu : dont che n'est mie estas
> Que li vous monstiers soit fais de nostre harnas ! " P. 324.

Master Thomas, in great trouble, collects his flock, mounts the pulpit, and announces the calamity. Bauduin and his convert Polibans then arrive. Bauduin recommends confession, fasting, and prayer. They follow his advice, and on the third day the miracle occurs:

> " L'escripture le dist, qui nous achertéfie
> Que le pierre Mahon,qui ou mur fut fiquie,
> Sali hors du piler, coi que nul vous en die,
> Droit enmi le monstier, c'onques ne fut brisie.
> Et demoura li traus, dont le pière ert widie,
> Sans pière est sans quailliel, à cascune partie ;
> Chou deseure soustient, par divine maistrie,
> Tout en air proprement, n'el tenés à falie.
>   Encore le voit-on en ichelle partie :
> Qui croire ne m'en voelt, si voist ; car je l'en prie ! " P. 327.

The Caliph comes to see, and declares it to be the Devil's doing. Seeing Polibans, who is his cousin, he hails him, but Polibans draws back, avowing his Christian faith. The Caliph in a rage has him off to prison. Bauduin becomes very ill, and has to sell his horse and arms. His disease is so offensive that he is thrust out of his hostel, and in his wretchedness sitting on a stone he still avows his faith, and confesses that even then he has not received his deserts. He goes to beg in the Christian

quarter, and no one gives to him; but still his faith and love to
God hold out:

> "Ensément Bauduins chelle rue cherqua,
> Tant qu'à .j. chavetier Bauduins s'arresta,
> Qui chavates cousoit; son pain en garigna :
> Jones fu et plaisans, apertement ouvra.
> Bauduins le regarde, c'onques mot ne parla." P. 334.

The cobler is charitable, gives him bread, shoes, and a grey coat
that was a foot too short. He then asks Bauduin if he will not
learn his trade; but that is too much for the knightly stomach :

> "Et Bauduins respont, li preus et li membrus :
> J'ameroie trop miex que je fuisse pendus !" P. 335.

The Caliph now in his Council expresses his vexation about the
miracle, and says he does not know how to disprove the faith of
the Christians. A very sage old Saracen who knew Hebrew,
and Latin, and some thirty languages, makes a suggestion,
which is, in fact, that about the moving of the Mountain, as
related by Marco Polo.* Master Thomas is sent for again,
and told that they must transport the high mountain of *Thir*
to the valley of *Joaquin*, which lies to the westward. He goes
away in new despair and causes his clerk to *sonner le clocke*
for his people. Whilst they are weeping and wailing in the
church, a voice is heard desiring them to seek a certain holy
man who is at the good cobler's, and to do him honour. God
at his prayer will do a miracle. They go in procession to
Bauduin, who thinks they are mocking him. They treat him
as a saint, and strive to touch his old coat. At last he consents
to pray along with the whole congregation.

The Caliph is in his palace with his princes, taking his ease
at a window. Suddenly he starts up exclaiming:

> "'Seignour, par Mahoumet que j'aoure et tieng chier,
> Le Mont de Thir enportent le déable d'enfeir !'
> Li Califes s'écrie : 'Seignour, franc palasin,
> Voïés le Mont de Thir qui ch'est mis au chemin !
> Vés-le-là tout en air, par mon Dieu Apolin ;
> Jà bientost le verrons ens ou val Joaquin !'" P. 345.

The Caliph is converted, releases Polibans, and is baptised,

---

* Vol. i. pp. 68 *seqq.* The virtuous cobler is not left out, but is made to play
second fiddle to the hero Bauduin.

taking the name of Bauduin, to whom he expresses his fear of the Viex de la Montagne with his *Hauts-Assis*, telling anew the story of the Assassin's Paradise, and so enlarges on the beauty of Ivorine that Bauduin is smitten, and his love heals his malady.   Toleration is not learned however :

> " Bauduins, li Califes, fist baptisier sa gent,
> Et qui ne voilt Dieu crore, li teste on li pourfent ! "   P. 350.

The Caliph gives up his kingdom to Bauduin, proposing to follow him to the Wars of Syria.   And Bauduin presents the Kingdom to the Cobler.

Bauduin, the Caliph, and Prince Polibans then proceed to visit the Mountain of the Old Man.   The Caliph professes to him that they want help against Godfrey of Bouillon.   The Viex says he does not give a *bouton* for Godfrey ; he will send one of his *Hauts-Assis* straight to his tent, and give him a great knife of steel between *fie et poumon !*

After dinner they go out and witness the feat of devotion which we have quoted elsewhere.*   They then see the Paradise and the lovely Ivorine, with whose beauty Bauduin is struck dumb.   The lady had never smiled before ; now she declares that he for whom she had long waited was come.   Bauduin exclaims :

> " ' Madame, fu-jou chou qui sui le vous soubgis ? '
> Quant la puchelle l'ot, lors li geta. j. ris ;
> Et li dist : ' Bauduins, vous estes mes amis ! ' "   Pp. 362-363.

The Old One is vexed, but speaks pleasantly to his daughter, who replies with frightfully bad language, and declares herself to be a Christian.   The father calls out to the Caliph to kill her.   The Caliph pulls out a big knife and gives him a blow that nearly cuts him in two.   The amiable Ivorine says she will go with Bauduin :

> " ' Sé mes pères est mors, n'en donne. j. paresis ! ' "   P. 364.

We need not follow the story further, as I did not trace beyond this point any distinct derivation from our Traveller, with the exception of that allusion to the incombustible cover-

---

* Vol. i. p. 144.

ing of the napkin of St. Veronica, which I have quoted at
p. 216 of this volume. But including this, here are at least
seven different themes borrowed from Marco Polo's book, on
which to be sure his poetical contemporary plays the most
extraordinary variations.

[78 *bis*.—In the third volume of *The Complete Works of
Geoffrey Chaucer*, Oxford, 1894, the Rev. Walter W. Skeat gives
Chaucer
and Marco
Polo. (pp. 372 *seqq.*) an *Account of the Sources of the
Canterbury Tales*. Regarding *The Squieres Tales*, he
says that one of his sources was the Travels of Marco ;
Mr. Keighley in his *Tales and Popular Fictions*, published in
1834, at p. 76, distinctly derives Chaucer's Tale from the
travels of Marco Polo. (*Skeat, l. c.*, p. 463, note.) I cannot quote
all the arguments given by the Rev. W. W. Skeat to support his
theory, pp. 463-477.

Regarding the opinion of Professor Skeat of Chaucer's in-
debtedness to Marco Polo, cf. *Marco Polo and the Squire's Tale*,
by Professor John Matthews Manly, vol. xi. of the *Publications
of the Modern Language Association of America*, 1896, pp. 349-
362. Mr. Manly says (p. 360) : " It seems clear, upon reviewing
the whole problem, that if Chaucer used Marco Polo's narrative,
he either carelessly or intentionally confused all the features of
the setting that could possibly be confused, and retained not a
single really characteristic trait of any person, place or event.
It is only by twisting everything that any part of Chaucer's
story can be brought into relation with any part of Polo's. To
do this might be allowable, if any rational explanation could
be given for Chaucer's supposed treatment of his ' author,' or
if there were any scarcity of sources from which Chaucer might
have obtained as much information about Tartary as he seems
really to have possessed ; but such an explanation would be
difficult to devise, and there is no such scarcity. Any one of
half a dozen accessible accounts could be distorted into almost
if not quite as great resemblance to the *Squire's Tale* as Marco
Polo's can."

Mr. A. W. Pollard, in his edition of *The Squire's Tale*
(Lond., 1899) writes : " A very able paper, by Prof. J. M. Manly,
demonstrates the needlessness of Prof. Skeat's theory, which
has introduced fresh complications into an already complicated
story. My own belief is that, though we may illustrate the

Squire's Tale from these old accounts of Tartary, and especially
from Marco Polo, because he has been so well edited by Colonel
Yule, there is very little probability that Chaucer consulted any
of them. It is much more likely that he found these details
where he found more important parts of his story, *i.e.* in some
lost romance. But if we must suppose that he provided his
own local colour, we have no right to pin him down to using
Marco Polo to the exclusion of other accessible authorities."
Mr. Pollard adds in a note (p. xiii.): "There are some features
in these narratives, *e.g.* the account of the gorgeous dresses worn
at the Kaan's feast, which Chaucer with his love of colour could
hardly have helped reproducing if he had known them."—H. C.]

XIII. Nature of Polo's Influence on Geographical Knowledge.

79. Marco Polo contributed such a vast amount of new
facts to the knowledge of the Earth's surface, that *Tardy opera*-
one might have expected his book to have had a *tion, and causes*
sudden effect upon the Science of Geography : but *thereof.*
no such result occurred speedily, nor was its beneficial effect
of any long duration.

No doubt several causes contributed to the slowness of its
action upon the notions of Cosmographers, of which the unreal
character attributed to the Book, as a collection of romantic
marvels rather than of geographical and historical facts, may
have been one, as Santarem urges. But the essential causes
were no doubt the imperfect nature of publication before the
invention of the press ; the traditional character which clogged
geography as well as all other branches of knowledge in the
Middle Ages ; and the entire absence of scientific principle in
what passed for geography, so that there was no organ competent to the assimilation of a large mass of new knowledge.

Of the action of the first cause no examples can be more
striking than we find in the false conception of the Caspian
as a gulf of the Ocean, entertained by Strabo, and the opposite
error in regard to the Indian Sea held by Ptolemy, who regards
it as an enclosed basin, when we contrast these with the correct

ideas on both subjects possessed by Herodotus. The later Geographers no doubt knew his statements, but did not appreciate them, probably from not possessing the evidence on which they were based.

80. As regards the second cause alleged, we may say that down nearly to the middle of the 15th century cosmographers, General character-acteristics of Mediæval Cosmography. as a rule, made scarcely any attempt to reform their maps by any elaborate search for new matter, or by lights that might be collected from recent travellers. Their world was in its outline that handed down by the traditions of their craft, as sanctioned by some Father of the Church, such as Orosius or Isidore, as sprinkled with a combination of classical and mediæval legend; Solinus being the great authority for the former. Almost universally the earth's surface is represented as filling the greater part of a circular disk, rounded by the ocean; a fashion that already existed in the time of Aristotle and was ridiculed by him.* No dogma of false geography was more persistent or more pernicious than this. Jerusalem occupies the central point, because it was found written in the Prophet Ezekiel: "*Haec dicit Dominus Deus: Ista est Jerusalem*, in medio gentium *posui eam, et in circuitu ejus terras;*"† a declaration supposed to be corroborated by the Psalmist's expression, regarded as prophetic of the death of Our Lord: "*Deus autem, Rex noster, ante secula operatus est salutem* in medio Terrae" (Ps. lxxiii. 12).‡ The Terrestrial

---

* " They draw nowadays the map of the world in a laughable manner, for they draw the inhabited earth as a circle; but this is impossible, both from what we see and from reason." (*Meteorolog. Lib.* II. cap. 5.) Cf. *Herodotus*, iv. 36.

† In Dante's Cosmography, Jerusalem is the centre of our οἰκουμένη, whilst the Mount of Purgatory occupies the middle of the Antipodal hemisphere :—

> " Come ciò sia, se'l vuoi poter pensare,
> Dentro raccolto immagina Sion
> Con questo monte in su la terra stare,
> Sì, ch' ambodue hann' un solo orrizon
> E diversi emisperi " . . . . . .
> —*Purg.* IV. 67.

‡ The belief, with this latter ground of it, is alluded to in curious verses by Jacopo Alighieri, Dante's son :—

> " E molti gran Profeti   E per la Santa fede
> Filosofi e Poeti   Cristiana ancor si vede
> Fanno il colco dell' Emme   Che' l' suo principio Cristo
> Dov' è Gerusalemme ;   Nel suo mezzo conquisto
> Se le loro scritture   Per cui prese morte
> Hanno vere figure :   E vi pose la sorte."
> —(*Rime Antiche Toscane, III.* 9.)

Though the general meaning of the second couplet is obvious, the expression *il*

Paradise was represented as occupying the extreme East, because it was found in Genesis that the Lord planted a garden east ward in Eden.* *Gog and Magog* were set in the far north or north-east, because it was said again in Ezekiel : "*Ecce Ego super te Gog Principem capitis Mosoch et Thubal . . . et ascendere te faciam de lateribus Aquilonis,*" whilst probably the topography of those mysterious nationalities was completed by a girdle of mountains out of the Alexandrian Fables. The loose and scanty nomenclature was mainly borrowed from Pliny or Mela through such Fathers as we have named ; whilst vacant spaces were occupied by Amazons, Arimaspians, and the realm. of Prester John. A favourite representation of the inhabited earth was this (T) ; a great O enclosing a T, which thus divides the circle in three parts ; the greater or half-circle being Asia, the two quarter circles Europe and Africa.† These Maps were known to St. Augustine.‡

81. Even Ptolemy seems to have been almost unknown ; and indeed had his Geography been studied it might, with all its errors, have tended to some greater endeavours after accuracy. Roger Bacon, whilst lamenting the exceeding deficiency of geographical knowledge in the Latin world, and purposing to essay an exacter distribution of countries, says he will not attempt to do so by latitude and longitude, for that is a system of which the Latins have learned

Roger Bacon as a geographer.

---

colco dell' Emme, "the couch of the M," is puzzling. The best solution that occurs to me is this : In looking at the world map of Marino Sanudo, noticed on p. *133,* as engraved by Bongars in the *Gesta Dei per Francos,* you find geometrical lines laid down, connecting the N.E., N.W., S.E., and S.W. points, and thus forming a square inscribed in the circular disk of the Earth, with its diagonals passing through the Central Zion. The eye easily discerns in these a great M inscribed in the circle, with its middle angular point at Jerusalem. Gervasius of Tilbury (with some confusion in his mind between tropic and equinoxial, like that which Pliny makes in speaking of the Indian Mons Malleus) says that "some are of opinion that the Centre is in the place where the Lord spoke to the woman of Samaria at the well, for there, at the summer solstice, the noonday sun descends perpendicularly into the water of the well, casting no shadow ; a thing which the philosophers say occurs at Syene" ! (*Otia Imperialia,* by Liebrecht, p. 1.)

* This circumstance does not, however, show in the Vulgate.

† "Veggiamo in prima in general la terra
Come risiede e come il mar la serra.

Un T dentro ad un O mostra il disegno
Come in tre parti fu diviso il Mondo,
E la superiore è il maggior regno
Che quasi piglia la metà del tondo.

ASIA chiamata : il gambo ritto è segno
Che parte il terzo nome dal secondo
AFFRICA dico da EUROPA : il mare
Mediterran tra esse in mezzo appare."
—*La Sfera,* di F. Leonardo di Stagio Dati, Lib. iii. st. 11.

‡ *De Civ. Dei,* xvi. 17, quoted by *Peschel,* 92.

nothing. He himself, whilst still somewhat burdened by the authoritative dicta of "saints and sages" of past times, ventures at least to criticise some of the latter, such as Pliny and Ptolemy, and declares his intention to have recourse to the information of those who have travelled most extensively over the Earth's surface. And judging from the good use he makes, in his description of the northern parts of the world, of the Travels of Rubruquis, whom he had known and questioned, besides diligently studying his narrative,* we might have expected much in Geography from this great man, had similar materials been available to him for other parts of the earth. He did attempt a map with mathematical determination of places, but it has not been preserved.†

It may be said with general truth that the world-maps current up to the end of the 13th century had more analogy to the mythical cosmography of the Hindus than to any thing properly geographical. Both, no doubt, were originally based in the main on real features. In the Hindu cosmography these genuine features are symmetrised as in a kaleidoscope; in the European cartography they are squeezed together in a manner that one can only compare to a pig in brawn. Here and there some feature strangely compressed and distorted is just recognisable. A splendid example of this kind of map is that famous one at Hereford, executed about A.D. 1275, of which a facsimile has lately been published, accompanied by a highly meritorious illustrative Essay.‡

82. Among the Arabs many able men, from the early days of Islám, took an interest in Geography, and devoted labour to geographical compilations, in which they often made use of their own observations, of the itineraries of travellers, and of other fresh knowledge. But somehow or other their maps were always far behind their books. Though they appear to have had an early translation of Ptolemy, and elaborate Tables of Latitudes and Longitudes form a prominent feature in many of their geographical treatises, there appears to be no Arabic map in

---

* *Opus Majus,* Venice ed. pp. 142, *seqq.*

† *Peschel,* p. 195. This had escaped me.

‡ By the Rev. W. L. Bevan, M.A., and the Rev. H. W. Phillott, M.A. In Asia, they point out, the only name showing any recognition of modern knowledge is Samarcand.

existence, laid down with meridians and parallels ; whilst *all* of their best known maps are on the old system of the circular disk. This apparent incapacity for map-making appears to have acted as a heavy drag and bar upon progress in Geography among the Arabs, notwithstanding its early promise among them, and in spite of the application to its furtherance of the great intellects of some (such as Abu Rihán al-Biruni), and of the indefatigable spirit of travel and omnivorous curiosity of others (such as Mas'údi).

83. Some distinct trace of acquaintance with the Arabian Geography is to be found in the World-Map of Marino Sanudo the Elder, constructed between 1300 and 1320 ; and this may be regarded as an exceptionally favourable specimen of the cosmography in vogue, for the author

Marino Sanudo the Elder.

was a diligent investigator and compiler, who evidently took a considerable interest in geographical questions, and had a strong enjoyment and appreciation of a map.* Nor is the map in question without some result of these characteristics. His representation of Europe, Northern Africa, Syria, Asia Minor, Arabia and its two gulfs, is a fair approximation to general facts ; his collected knowledge has enabled him to locate, with more or less of general truth, Georgia, the Iron Gates, Cathay, the Plain of Moghan, Euphrates and Tigris, Persia, Bagdad, Kais, Aden (though on the wrong side of the Red Sea), Abyssinia (*Habesh*), Zangibar (*Zinz*), Jidda (Zede), etc. But after all the traditional forms are too strong for him. Jerusalem is still the centre of the disk of the habitable earth, so that the distance is as great from Syria to Gades in the extreme West, as from Syria to the India Interior of Prester John which terminates the extreme East. And Africa beyond the Arabian Gulf is carried, according to the Arabian modification of Ptolemy's misconception, far to the eastward until it almost meets the prominent shores of India.

84. The first genuine mediæval attempt at a geographical construction that I know of, absolutely free from the traditional *idola*, is the Map of the known World from the Portulano

---

* His work, *Liber Secretorum Fidelium Crucis*, intended to stimulate a new Crusade, has three capital maps, besides that of the World, one of which, translated, but otherwise in facsimile, is given at p. 18 of this volume. But besides these maps, he gives, in a tabular form of parallel columns, the reigning sovereigns in Europe and Asia connected with his historical retrospect, just on the plan presented in Sir Harris Nicolas's Chronology of History.

Mediceo (in the Laurentian Library), of which an extract is
engraved in the atlas of Baldelli-Boni's Polo. I need not
The Catalan describe it, however, because I cannot satisfy myself
Map of 1375, that it makes much use of Polo's contributions, and
the most
complete its facts have been embodied in a more ambitious
mediæval
embodiment work of the next generation, the celebrated Catalan
of Polo's
Geography. Map of 1375 in the great Library of Paris. This also,
but on a larger scale and in a more comprehensive manner, is
an honest endeavour to represent the known world on the basis
of collected facts, casting aside all theories pseudo-scientific or
pseudo-theological; and a very remarkable work it is. In this
map it seems to me Marco Polo's influence, I will not say on
geography, but on map-making, is seen to the greatest advantage. His Book is the basis of the Map as regards Central
and Further Asia, and partially as regards India. His names
are often sadly perverted, and it is not always easy to understand the view that the compiler took of his itineraries. Still
we have Cathay admirably placed in the true position of China,
as a great Empire filling the south-east of Asia. The Eastern
Peninsula of India is indeed absent altogether, but the Peninsula of Hither India is for the first time in the History of
Geography represented with a fair approximation to its correct
form and position,* and Sumatra also (*Jaua*) is not badly
placed. Carajan, Vocian, Mien, and Bangala, are located with
a happy conception of their relation to Cathay and to India.
Many details in India foreign to Polo's book,† and some in
Cathay (as well as in Turkestan and Siberia, which have been
entirely derived from other sources) have been embodied
in the Map. But the study of his Book has, I conceive, been
essentially the basis of those great portions which I have
specified, and the additional matter has not been in mass
sufficient to perplex the compiler. Hence we really see

---

* I do not see that al-Birúni deserves the credit in this respect assigned to him
by Professor Peschel, so far as one can judge from the data given by Sprenger
(*Peschel*, p. 128 ; *Post und Reise-Routen*, 81-82.)

† For example, *Delli*, which Polo does not name ; *Diogil* (Deogír) ; on the
Coromandel coast *Setemelti*, which I take to be a clerical error for *Sette-Templi*, the
Seven Pagodas ; round the Gulf of Cambay we have *Cambetum* (Kambayat),
Cocintaya (Kokan-Tana, see vol. ii. p. 396), *Goga, Baroche, Neruala* (Anharwala),
and to the north *Moltan*. Below Multan are *Hocibelch* and *Bargelidoa*, two puzzles.
The former is, I think, *Uch-baligh*, showing that part of the information was from
Perso-Mongol sources.

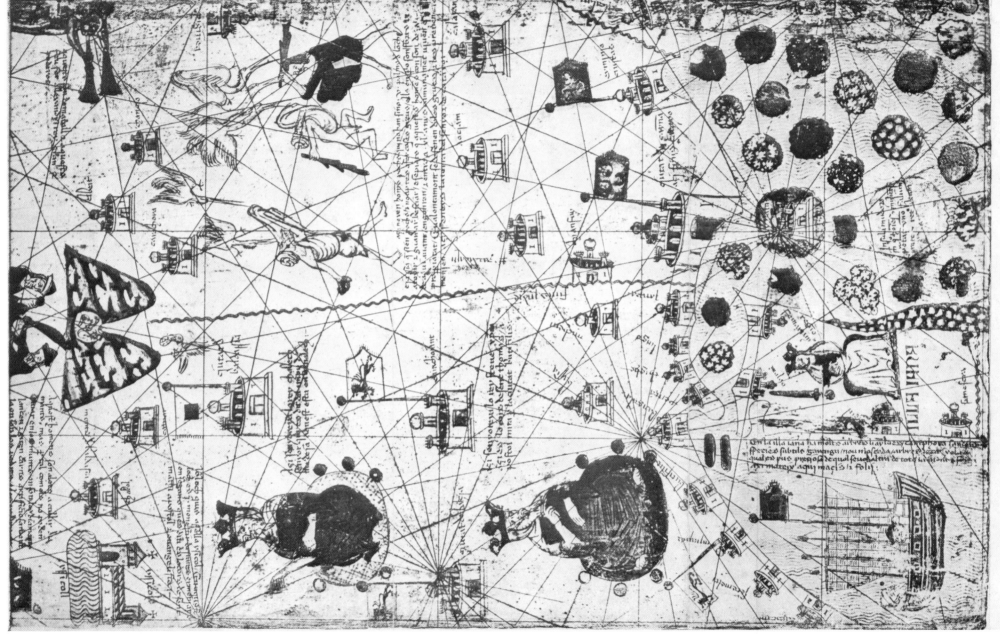

Part of the Catalan Map (1375).

in this Map something like the idea of Asia that the Traveller himself would have presented, had he bequeathed a Map to us.

[Some years ago, I made a special study of the Far East in the Catalan Map. (*L'Extrême-Orient dans l'Atlas catalan de Charles V.*, Paris, 1895), and I have come to the conclusion that the cartographer's knowledge of Eastern Asia is drawn almost entirely from Marco Polo. We give a reproduction of part of the Catalan Map.—H. C.]

85. In the following age we find more frequent indications that Polo's book was diffused and read. And now that the spirit of discovery began to stir, it was apparently regarded in a juster light as a Book of Facts, and not as a mere *Romman du Grant Kaan.** But in fact this age produced new supplies of crude information in greater abundance than the knowledge of geographers was prepared to digest or co-ordinate, and the consequence is that the magnificent Work of Fra Mauro (1459), though the result of immense labour in the collection of facts and the endeavour to combine them, really gives a considerably less accurate idea of Asia than that which the Catalan Map had afforded.†

Confusions in Cartography of the 16th century, from the endeavour to combine new and old information.

And when at a still later date the great burst of discovery eastward and westward took effect, the results of all attempts to combine the new knowledge with the old was most unhappy. The first and crudest forms of such combinations attempted to realise the ideas of Columbus regarding the identity of his discoveries with the regions of the Great Kaan's dominion ;‡ but even after AMERICA had vindicated its independent position on the surface of the globe, and the new

---

* I see it stated by competent authority that *Romman* is often applied to any prose composition in a Romance language.

In or about 1426, Prince Pedro of Portugal, the elder brother of the illustrious Prince Henry, being on a visit to Venice, was presented by the Signory with a copy of Marco Polo's book, together with a map already alluded to. (*Major's P. Henry*, pp. 61, 62.)

† This is partly due also to Fra Mauro's reversion to the fancy of the circular disk limiting the inhabited portion of the earth.

‡ An early graphic instance of this is Ruysch's famous map (1508). The following extract of a work printed as late as 1533 is an example of the like confusion in verbal description : " The Territories which are beyond the limits of Ptolemy's Tables have not yet been described on certain authority. Behind the Sinae and the Seres, and beyond 180° of East Longitude, many countries were discovered by one [*quendam*] Marco

knowledge of the Portuguese had introduced CHINA where the Catalan Map of the 14th century had presented CATHAY, the latter country, with the whole of Polo's nomenclature, was shoved away to the north, forming a separate system.* Henceforward the influence of Polo's work on maps was simply injurious; and when to his nomenclature was added a sprinkling of Ptolemy's, as was usual throughout the 16th century, the result was a most extraordinary hotch-potch, conveying no approximation to any consistent representation of facts.

Thus, in a map of 1522,† running the eye along the north of Europe and Asia from West to East, we find the following succession of names: Groenlandia, or Greenland, as a great peninsula overlapping that of Norvegia and Suecia; Livonia, Plescovia and Moscovia, Tartaria bounded on the South by *Scithia extra Imaum*, and on the East, by the Rivers *Ochardes* and *Bautisis* (out of Ptolemy), which are made to flow into the Arctic Sea. South of these are *Aureacithis* and *Asmirea* (Ptolemy's *Auxacitis* and *Asmiræa*), and *Serica Regio*. Then following the northern coast *Balor Regio*,‡ *Judei Clausi, i.e.* the Ten Tribes who are constantly associated or confounded with the Shut-up Nations of Gog and Magog. These impinge upon the River *Polisacus*, flowing into the Northern Ocean in Lat. 75°, but which is in fact no other than Polo's *Pulisanghin!* § Immediately south of this is *Tholomon Provincia* (Polo's again), and on the coast *Tangut, Cathaya*, the Rivers

---

Polo a Venetian and others, and the sea-coasts of those countries have now recently again been explored by Columbus the Genoese and Amerigo Vespucci in navigating the Western Ocean. . . . To this part (of Asia) belong the territory called that of the *Bachalaos* [or Codfish, Newfoundland], *Florida, the Desert of Lop, Tangut, Cathay,* the realm of *Mexico* (wherein is the vast city of *Temistitan*, built in the middle of a great lake, but which the older travellers styled QUINSAY), besides *Paria, Uraba,* and the countries of the *Canibals."* (*Joannis Schoneri Carolostadtii Opusculum Geogr.*, quoted by Humboldt, *Examen*, V. 171, 172.)

* In Robert Parke's Dedication of his Translation of Mendoza's, London, 1st of January, 1589, he identifies China and Japan with the regions of which *Paulus Venetus* and *Sir John Mandeuill* "wrote long agoe."—*MS. Note by Yule.*

† "*Totius Europae et Asiae Tabula Geographica, Auctore Thoma D. Aucupario. Edita Argentorati,* MDXXII." Copied in Witsen.

‡ This strange association of *Balor (i.e.,* Bolor, that name of so many odd vicissitudes, see pp. 178-179 *infra*) with the shut-up Israelites must be traced to a passage which Athanasius Kircher quotes from *R. Abraham Pizol* (qu. Peritsol?): "*Regnum,* inquit, Belor *magnum et excelsum nimis, juxta omnes illos qui scripserunt Historicos.* Sunt in eo Judaei *plurimi inclusi, et illud in latere Orientali et Boreali,*" etc. (*China Illustrata,* p. 49.) § Vol. ii. p. 1.

Caramoran and *Oman* (a misreading of Polo's *Quian*), *Quinsay* and *Mangi*.

86. The Maps of Mercator (1587) and Magini (1597) are similar in character, but more elaborate, introducing China as a separate system. Such indeed also is Blaeu's Map (1663) excepting that Ptolemy's contributions are reduced to one or two.

Gradual disappearance of Polo's nomenclature.

In Sanson's Map (1659) the data of Polo and the mediæval Travellers are more cautiously handled, but a new element of confusion is introduced in the form of numerous features derived from Edrisi.

It is scarcely worth while to follow the matter further. With the increase of knowledge of Northern Asia from the Russian side, and that of China from the Maps of Martini, followed by the surveys of the Jesuits, and with the real science brought to bear on Asiatic Geography by such men as De l'Isle and D'Anville, mere traditional nomenclature gradually disappeared. And the task which the study of Polo has provided for the geographers of later days has been chiefly that of determining the true localities that his book describes under obsolete or corrupted names.

[My late illustrious friend, Baron *A. E. Nordenskiöld*, who has devoted much time and labour to the study of Marco Polo (see his *Periplus*, Stockholm, 1897), and published a facsimile edition of one of the French MSS. kept in the Stockholm Royal Library (see vol. ii. *Bibliography*, p. 570), has given to *The Geographical Journal* for April, 1899, pp. 396-406, a paper on *The Influence of the " Travels of Marco Polo" on Jacobo Gastaldi's Maps of Asia.* He writes (p. 398) that as far as he knows, none "of the many learned men who have devoted their attention to the discoveries of Marco Polo, have been able to refer to any maps in which all or almost all those places mentioned by Marco Polo are given. All friends of the history of geography will therefore be glad to hear that such an atlas from the middle of the sixteenth century really does exist, viz. Gastaldi's ' Prima, seconda e terza parte dell Asia.'" All the names of places in Ramusio's Marco Polo are introduced in the maps of Asia of Jacobo Gastaldi (1561). Cf. *Periplus,* liv., lv., and lvi.

I may refer to what both Yule and myself say *supra* of the Catalan Map.—H. C.]

87. Before concluding, it may be desirable to say a few words on the subject of important knowledge other than Alleged introduction of Block-printed Books into Europe by Marco Polo. geographical, which various persons have supposed that Marco Polo must have introduced from Eastern Asia to Europe.

Respecting the mariner's compass and gunpowder I shall say nothing, as no one now, I believe, imagines Marco to have had anything to do with their introduction. But from a highly respectable source in recent years we have seen the introduction of Block-printing into Europe connected with the name of our Traveller. The circumstances are stated as follows : *

" In the beginning of the 15th century a man named Pamphilo Castaldi, of Feltre . . . . was employed by the Seignory or Government of the Republic, to engross deeds and public edicts of various kinds . . . . the initial letters at the commencement of the writing being usually ornamented with red ink, or illuminated in gold and colours

" According to Sansovino, certain stamps or types had been invented some time previously by Pietro di Natali, Bishop of Aquilœa.† These were made at Murano of glass, and were used to stamp or print the outline of the large initial letters of public documents, which were afterwards filled up by hand. . . . Pamphilo Castaldi improved on these glass types, by having others made of wood or metal, and having seen several Chinese books which the famous traveller Marco Polo had brought from China, and of which the entire text was printed with wooden blocks, he caused moveable wooden types to be made, each type containing a single letter ; and with these he printed several broadsides and single leaves, at Venice, in the year 1426. Some of these single sheets are said to be preserved among the archives at Feltre. . . .

" The tradition continues that John Faust, of Mayence . . . . became acquainted with Castaldi, and passed some time with him, at his *Scriptorium*, . . . at Feltre ; "

and in short developed from the knowledge so acquired the great invention of printing. Mr. Curzon goes on to say that

---

* *A short Account of Libraries of Italy*, by the Hon. R. Curzon (the late Lord de la Zouche) ; in *Bibliog. and Hist. Miscellanies ; Philobiblon Society*, vol. i, 1854, pp. 6. *seqq.*

† P. dei Natali was Bishop of Equilio, a city of the Venetian Lagoons, in the latter part of the 14th century. (See *Ughelli, Italia Sacra*, X. 87.) There is no ground whatever for connecting him with these inventions. The story of the glass types appears to rest entirely and solely on one obscure passage of Sansovino, who says that under the Doge Marco Corner (1365-1367) : " *certe Natale Veneto lasciò un libro della materie delle forme da giustar intorno alle lettere, ed il modo di formarle di vetro.*" There is absolutely nothing more. Some kind of stencilling seems indicated,

Panfilo Castaldi was born in 1398, and died in 1490, and that he gives the story as he found it in an article written by Dr. Jacopo Facen, of Feltre, in a (Venetian?) newspaper called *Il Gondoliere*, No. 103, of 27th December, 1843.

In a later paper Mr. Curzon thus recurs to the subject :*

"Though none of the early block-books have dates affixed to them, many of them are with reason supposed to be more ancient than any books printed with moveable types. Their resemblance to Chinese block-books is so exact, that they would almost seem to be copied from the books commonly used in China. *The impressions are taken off on one side of the paper only, and in binding, both the Chinese, and ancient German, or Dutch block-books, the blank sides of the pages are placed opposite each other*, and sometimes pasted together . . . . The impressions are not taken off with printer's ink, but *with a brown paint or colour, of a much thinner description, more in the nature of Indian ink, as we call it, which is used in printing Chinese books*. Altogether the German and Oriental block-books are so precisely alike, in almost every respect, that . . we must suppose that the process of printing then must have been copied from ancient Chinese specimens, brought from thatcountry by some early travellers, whose names have not been handed down to our times."

The writer then refers to the tradition about *Guttemberg* (so it is stated on this occasion, not Faust) having learned Castaldi's art, etc., mentioning a circumstance which he supposes to indicate that Guttemberg had relations with Venice ; and appears to assent to the probability of the story of the art having been founded on specimens brought home by Marco Polo.

This story was in recent years diligently propagated in Northern Italy, and resulted in the erection at Feltre of a public statue of Panfilo Castaldi, bearing this inscription (besides others of like tenor) :—

"*To Panfilo Castaldi the illustrious Inventor of Movable Printing Types, Italy renders this Tribute of Honour, too long deferred.*"

In the first edition of this book I devoted a special note to the exposure of the worthlessness of the evidence for this story.†
This note was, with the present Essay, translated and published at Venice by Comm. Berchet, but this challenge to the supporters

---

* *History of Printing in China and Europe*, in *Philobiblon*, vol. vi. p. **23.**
† See *Appendix L.* in First Edition.

of the patriotic romance, so far as I have heard, brought none of them into the lists in its defence.

But since Castaldi has got his statue from the printers of Lombardy, would it not be mere equity that the mariners of Spain should set up a statue at Huelva to the Pilot Alonzo Sanchez of that port, who, according to Spanish historians, after discovering the New World, died in the house of Columbus at Terceira, and left the crafty Genoese to appropriate his journals, and rob him of his fame?

Seriously; if anybody in Feltre cares for the real reputation of his native city, let him do his best to have that preposterous and discreditable fiction removed from the base of the statue. If Castaldi has deserved a statue on other and truer grounds let *him* stand; if not, let him be burnt into honest lime! I imagine that the original story that attracted Mr. Curzon was more *jeu d'esprit* than anything else; but that the author, finding what a stone he had set rolling, did not venture to retract.

88. Mr. Curzon's own observations, which I have italicised about the resemblance of the two systems are, however, very

Frequent opportunities for such introduction in the age following Polo's. striking, and seem clearly to indicate the derivation of the art from China. But I should suppose that in the tradition, if there ever was any genuine tradition of the kind at Feltre (a circumstance worthy of all doubt), the name of Marco Polo was introduced merely because it was so prominent a name in Eastern Travel. The fact has been generally overlooked and forgotten * that, for many years in the course of the 14th century, not only were missionaries of the Roman Church and Houses of the Franciscan Order established in the chief cities of China, but a regular trade was carried on overland between Italy and China, by way of Tana (or Azov), Astracan, Otrar and Kamul, insomuch that instructions for the Italian merchant following that route form the two first chapters in the Mercantile Handbook of Balducci Pegolotti (*circa* 1340).† Many a traveller besides Marco Polo might therefore have brought home the block-books. And this is the less to be ascribed to him because

---

* Ramusio himself appears to have been entirely unconscious of it, *vide supra*, p. *3*.

† This subject has been fully treated in *Cathay and the Way Thither.*

he so curiously omits to speak of the art of printing, when his subject seems absolutely to challenge its description.

XIV. Explanations regarding the Basis adopted for the present Translation.

89. It remains to say a few words regarding the basis adopted for our English version of the Traveller's record.

Ramusio's recension was that which Marsden selected for translation. But at the date of his most meritorious publication nothing was known of the real literary history of Polo's Book, and no one was aware of the peculiar value and originality of the French manuscript texts, nor had Marsden seen any of them. A translation *Text followed by Marsden and by Pauthier* from one of those texts is a translation at first hand ; a translation from Ramusio's Italian is, as far as I can judge, the translation of a translated compilation from two or more translations, and therefore, whatever be the merits of its matter, inevitably carries us far away from the spirit and style of the original narrator. M. Pauthier, I think, did well in adopting for the text of his edition the MSS. which I have classed as of the second Type, the more as there had hitherto been no publication from those texts. But editing a text in the original language, and translating, are tasks substantially different in their demands.

90. It will be clear from what has been said in the preceding pages that I should not regard as a fair or full representation of Polo's Work, a version on which the Geographic Text did not exercise a material influence. But to adopt *Eclectic formation of the English Text of this Translation.* that Text, with all its awkwardnesses and tautologies, as the absolute subject of translation, would have been a mistake. What I have done has been, in the first instance, to translate from Pauthier's Text. The process of abridgment in this text, however it came about, has been on the whole judiciously executed, getting rid of the intolerable prolixities of manner which belong to many parts of the Original Dictation, but *as a general rule* preserving the matter. Having translated this,—not always from the Text adopted by Pauthier himself,

but with the exercise of my own judgment on the various readings which that Editor lays before us,—I then compared the translation with the Geographic Text, and transferred from the latter not only all items of real substance that had been omitted, but also all expressions of special interest and character, and occasionally a greater fulness of phraseology where condensation in Pauthier's text seemed to have been carried too far. And finally I introduced *between brackets* everything peculiar to Ramusio's version that seemed to me to have a just claim to be reckoned authentic, and that could be so introduced without harshness or mutilation. Many passages from the same source which were of interest in themselves, but failed to meet one or other of these conditions, have been given in the notes.*

91. As regards the reading of proper names and foreign words, in which there is so much variation in the different MSS.

Mode of
rendering
proper
names.
and editions, I have done my best to select what seemed to be the true reading from the G. T. and Pauthier's three MSS., only in some rare instances transgressing this limit.

Where the MSS. in the repetition of a name afforded a choice of forms, I have selected that which came nearest the real name when known. Thus the G. T. affords *Baldasciain, Badascian, Badasciam, Badausiam, Balasian.* I adopt BADASCIAN, or in English spelling BADASHAN, because it is closest to the real name *Badakhshan.* Another place appears as COBINAN, *Cabanat, Cobian.* I adopt the first because it is the truest expression of the real name *Koh-benán.* In chapters 23, 24 of Book I., we have in the G. T. *Asisim, Asciscin, Asescin,* and in Pauthier's MSS. *Hasisins, Harsisins,* etc. I adopt ASCISCIN, or in English spelling ASHISHIN, for the same reason as before.

---

* This " eclectic formation of the English text," as I have called it for brevity in the marginal rubric, has been disapproved by Mr. de Khanikoff, a critic worthy of high respect. But I must repeat that the duties of a translator, and of the Editor of an original text, at least where the various recensions bear so peculiar a relation to each other as in this case, are essentially different; and that, on reconsidering the matter after an interval of four or five years, the plan which I have adopted, whatever be the faults of execution, still commends itself to me as the only appropriate one.

Let Mr. de Khanikoff consider what course he would adopt if he were about to publish Marco Polo in Russian. I feel certain that with whatever theory he might set out, before his task should be concluded he would have arrived practically at the same system that I have adopted.

So with *Creman, Crerman, Crermain,* QUERMAN, Anglicè
KERMAN ; Cormos, HORMOS, and many more.*

In two or three cases I have adopted a reading which I cannot show *literatim* in any authority, but because such a form
appears to be the just resultant from the variety of readings
which are presented ; as in surveying one takes the mean of a
number of observations when no one can claim an absolute
preference.

Polo's proper names, even in the French Texts, are *in the
main* formed on an Italian fashion of spelling.† I see no object
in preserving such spelling in an English book, so after selecting
the best reading of the name I express it in English spelling,
printing *Badashan, Pashai, Kerman,* instead of *Badascian, Pasciai,
Querman,* and so on.

And when a little trouble has been taken to ascertain the
true form and force of Polo's spelling of Oriental names and
technical expressions, it will be found that they are in the main
as accurate as Italian lips and orthography will admit, and not
justly liable either to those disparaging epithets‡ or to those
exegetical distortions which have been too often applied to them.
Thus, for example, *Cocacin, Ghel* or *Ghelan, Tonocain, Cobinan,
Ondanique, Barguerlac, Argon, Sensin, Quescican, Toscaol,
Bularguci, Zardandan, Anin, Caugigu, Coloman, Gauenispola,
Mutfili, Avarian, Choiach,* are not, it will be seen, the ignorant
blunderings which the interpretations affixed by some commentators would imply them to be, but are, on the contrary, all but
perfectly accurate utterances of the names and words intended.

---

* In Polo's diction C frequently represents H., *e.g., Cormos* = Hormuz ; *Camadi*
probably = Hamadi ; *Caagiu* probably = Hochau ; *Cacianfu* = Hochangfu, and so on.
This is perhaps attributable to Rusticiano's Tuscan ear. A true Pisan will absolutely
contort his features in the intensity of his efforts to aspirate sufficiently the letter C.
Filippo Villani, speaking of the famous Aguto (Sir J. Hawkwood), says his name in
English was *Kauchouvole.* (*Murat. Script.* xiv. 746.)

† In the Venetian dialect *ch* and *j* are often sounded as in English, not as in
Italian. Some traces of such pronunciation I think there are, as in *Coja, Carajan,*
and in the Chinese name *Vanchu* (occurring only in Ramusio, *supra,* p. 99). But the
scribe of the original work being a Tuscan, the spelling is in the main Tuscan. The
sound of the *Qu* is, however, French, as in *Quescican, Quinsai,* except perhaps in the
case of *Quenianfu,* for a reason given in vol. ii. p. 29.

‡ For example, that enthusiastic student of mediæval Geography, Joachim
Lelewel, speaks of Polo's "gibberish" (*le baragouinage du Venitien*) with special
reference to such names as *Zayton* and *Kinsay,* whilst we now know that these names
were in universal use by all foreigners in China, and no more deserve to be called
gibberish than *Bocca-Tigris, Leghorn, Ratisbon,* or *Buda.*

The -*tchéou* (of French writers), -*choo*, -*chow*, or -*chau* * of English writers, which so frequently forms the terminal part in the names of Chinese cities, is almost invariably rendered by Polo as -*giu*. This has frequently in the MSS., and constantly in the printed editions, been converted into -*gui*, and thence into -*guy*. This is on the whole the most constant canon of Polo's geographical orthography, and holds in *Caagiu* (Ho-chau), *Singiu* (Sining-chau), *Cui-giu* (Kwei-chau), *Sin-giu* (T'sining-chau), *Pi-giu* (Pei-chau), *Coigangiu* (Hwaingan-chau), *Si-giu* (Si-chau), *Ti-giu* (Tai-chau), *Tin-giu* (Tung-chau), *Yan-giu* (Yang-chau), *Sin-giu* (Chin-chau), *Cai-giu* (Kwa-chau), *Chinghi-giu* (Chang-chau), *Su-giu* (Su-chau), *Vu-giu* (Wu-chau), and perhaps a few more. In one or two instances only (as *Sinda-ciu*, *Caiciu*) he has -*ciu* instead of -*giu*.

The chapter-headings I have generally taken from Panthier's Text, but they are no essential part of the original work, and they have been slightly modified or enlarged where it seemed desirable.

> "Behold! I see the Haven nigh at Hand,
> To which I meane my wearie Course to bend;
> Vere the maine Shete, and beare up with the Land,
> The which afore is fayrly to be kend,
> And seemeth safe from Storms that may offend.
>   *   *   *   *   *
> There eke my feeble Barke a while may stay,
> Till mery Wynd and Weather call her thence away."
>   —THE FAERIE QUEENE, I. xii. 1.

---

* I am quite sensible of the diffidence with which any outsider should touch any question of Chinese language or orthography. A Chinese scholar and missionary (Mr. Moule) objects to my spelling *chau*, whilst he, I see, uses *chow*. I imagine we mean the same sound, according to the spelling which I try to use throughout the book. Dr. C. Douglas, another missionary scholar, writes *chau*

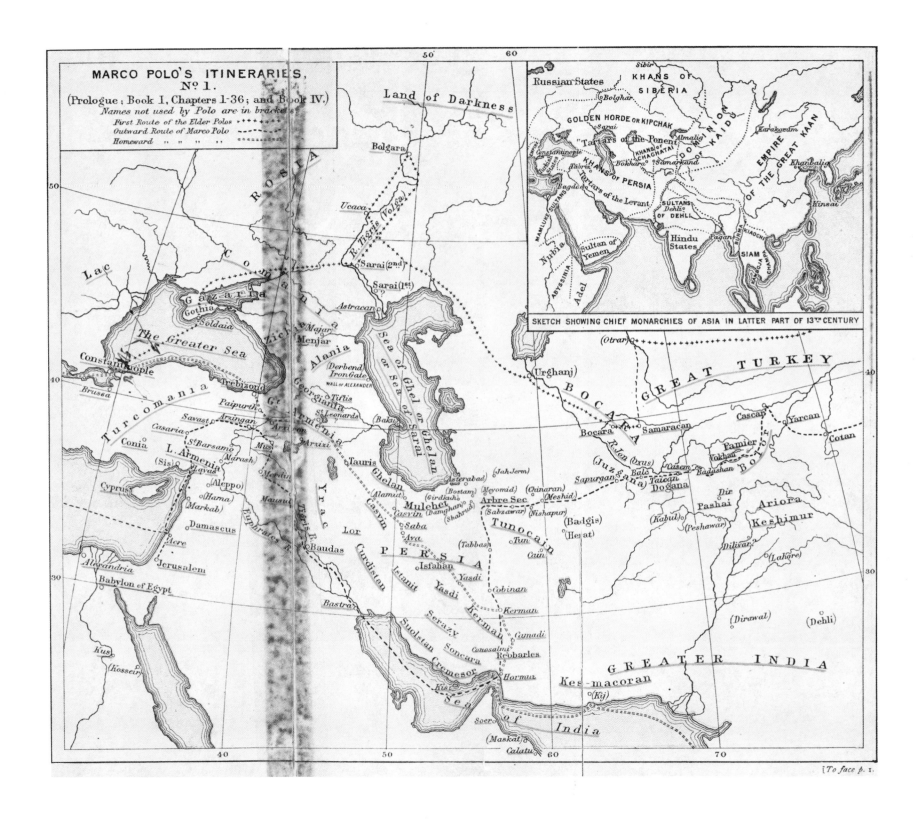

MARCO POLO'S ITINERARIES,
Nᵒ 1.

(Prologue; Book I, Chapters 1-36; and Book IV.)

Names not used by Polo are in brackets.

First Route of the Elder Polos ++++++
Outward Route of Marco Polo ------
Homeward   "   "   "

Land of Darkness

Bolgara

Ucaca

R. Tigris (Volga)

Sarai (2nd)

Sarai (1st)

Astracan

R O S I A

C o m a n i a

Lac

Gazaria

Gothia

Soldaia

The Greater Sea

Lichaostania

Majar

Menjar

Alania

Constantinople

Trebizond

Brussa

Paipurth

Arzingan

Savast

Casaria

Conia

Arziron

St Barsamo

L. Armenia

(Marash)

(Sis)

Ayas

(Aleppo)

Cyprus

(Hama)

Merdin

Markab

Mus

Arzizi

Georgiania

St Leonards

Tiflis

(Derbend
Iron Gate)
WALL of ALEXANDER

(Baku)

Sea of Ghel or Sea of Sarai

Tauris

Ghelan

(Asferabad)

(JahJerm)

(Bostam)

(Girdkuh)

Meyomid)

(Chinaran)

Muleket

(Alamut)

(Damghan)

(Shahrud)

Arbre Sec

(Sabzawar)

(Meshid)

Casvin

(Nishapur)

Casan

Saba

Ava

(Tabbas)

Tun

Cain

Tunocain

(Balgis)

(Herat)

Damascus

Acre

Euphrates R.

Tigris R.

Mausul

Lor

Kurdistan

Baudas

Yrac

PERSIA

Isfahan

Yasdi

Istanit

Yasdi

Cobinan

Kerman

Alexandria

Jerusalem

Babylon of Egypt

Bastra

Serazy

Soncara

Cremesor

Suolstan

Kerman

Canadi

Cenosalmi

Reobarles

Kisi

Hormuz

Kus

(Kosseir)

Sea of India

Soer

(Maskat)

Calatu

BOCARA

GREAT TURKEY

(Urghanj)

Bocara

Samaracan

Cascar

Yarcan

R. Jon (Oxus)

Pamier

Cotan

(Juzganaj)

Balc

Vokhan

Sapurgan

Taican

Casem

Badashan

Bolor

Dogana

Dir

Pashai

Ariora

Keshimur

(Kabul)

(Peshawar)

Dilivar?

(Lahore)

(Dirawal)

(Dehli)

GREATER INDIA

Kes-macoran

(Kij)

SKETCH SHOWING CHIEF MONARCHIES OF ASIA IN LATTER PART OF 13ᵀᴴ CENTURY

Sibir

KHANS OF SIBERIA

Russian States

Bolghar

GOLDEN HORDE OR KIPCHAK

Sarai

"Tartars of the Ponent"

KHANS OF CHAGHATAI

Almalig

Karakorum

DOMINION OF KAIDU

Constantinople

Rum Seljuks

Tabriz

Bokhara

Samarkand

EMPIRE OF THE GREAT KAAN

Khanbalig

KHANS OF PERSIA

Bagdad

Tartars of the Levant

SULTANS DEHLI OF DEHLI

Kinsai

MAMLUKE SULTANS

Nubia

Sultan of Yemen

Hindu States

Pagan

BURMA

KIAOCHI

SIAM

CAMBOJA CHAMPA

ABYSSINIA

Adel

[To face p. 1.

THE

BOOK OF MARCO POLO.

PROLOGUE.

GREAT PRINCES, Emperors, and Kings, Dukes and Marquises, Counts, Knights, and Burgesses! and People of all degrees who desire to get knowledge of the various races of mankind and of the diversities of the sundry regions of the World, take this Book and cause it to be read to you. For ye shall find therein all kinds of wonderful things, and the divers histories of the Great Hermenia, and of Persia, and of the Land of the Tartars, and of India, and of many another country of which our Book doth speak, particularly and in regular succession, according to the description of Messer Marco Polo, a wise and noble citizen of Venice, as he saw them with his own eyes. Some things indeed there be therein which he beheld not ; but these he heard from men of credit and veracity. And we shall set down things seen as seen, and things heard as heard only, so that no jot of falsehood may mar the truth of our Book, and that all who shall read it or hear it read may put full faith in the truth of all its contents.

For let me tell you that since our Lord God did mould with his hands our first Father Adam, even until this day, never hath there been Christian, or Pagan, or

Tartar, or Indian, or any man of any nation, who in
his own person hath had so much knowledge and
experience of the divers parts of the World and its
Wonders as hath had this Messer Marco! And for
that reason he bethought himself that it would be a
very great pity did he not cause to be put in writing
all the great marvels that he had seen, or on sure
information heard of, so that other people who had not
these advantages might, by his Book, get such knowledge. And I may tell you that in acquiring this knowledge he spent in those various parts of the World good
six-and-twenty years. Now, being thereafter an inmate
of the Prison at Genoa, he caused Messer Rusticiano
of Pisa, who was in the said Prison likewise, to reduce
the whole to writing ; and this befell in the year 1298
from the birth of Jesus.

CHAPTER I.

HOW THE TWO BROTHERS POLO SET FORTH FROM CONSTANTINOPLE TO TRAVERSE THE WORLD.

IT came to pass in the year of Christ 1260, when
Baldwin was reigning at Constantinople,[1] that Messer
Nicolas Polo, the father of my lord Mark, and Messer
Maffeo Polo, the brother of Messer Nicolas, were at
the said city of CONSTANTINOPLE, whither they had gone
from Venice with their merchants' wares. Now these
two Brethren, men singularly noble, wise, and provident,
took counsel together to cross the GREATER SEA on a
venture of trade ; so they laid in a store of jewels and
set forth from Constantinople, crossing the Sea to
SOLDAIA.[2]

Note 1.—Baldwin II. (de Courtenay), the last Latin Emperor of Constantinople, reigned from 1237 to 1261, when he was expelled by Michael Palaeologus.

The date in the text is, as we see, that of the Brothers' voyage across the Black Sea. It stands 1250 in all the chief texts. But the figure is certainly wrong. We shall see that, when the Brothers return to Venice in 1269, they find Mark, who, according to Ramusio's version, was *born after their departure*, a lad of fifteen. Hence, if we rely on Ramusio, they must have left Venice about 1253-54. And we shall see also that they reached the Volga in 1261. Hence their start from Constantinople may well have occurred in 1260, and this I have adopted as the most probable correction. Where they spent the interval between 1254 (if they really left Venice so early) and 1260, nowhere appears. But as their brother, Mark the Elder, in his Will styles himself *"whilom of Constantinople,"* their headquarters were probably there.

Castle of Soldaia or Sudak.

Note 2.—In the Middle Ages the Euxine was frequently called *Mare Magnum* or *Majus*. Thus Chaucer :—

"In the Grete See,
At many a noble Armee hadde he be."

The term Black Sea (*Mare Maurum* v. *Nigrum*) was, however, in use, and Abulfeda says it was general in his day. That name has been alleged to appear as early as the 10th century, in the form Σκοτεινή, "The Dark Sea"; but an examination of the passage cited, from Constantine Porphyrogenitus, shows that it refers rather to the Baltic, whilst that author elsewhere calls the Euxine simply Pontus. (*Reinaud's Abulf.* I. 38 ; *Const. Porph. De Adm. Imp.* c. 31, c. 42.)

-:- *Sodaya, Soldaia,* or *Soldachia,* called by Orientals *Súdák,* stands on the S.E.

coast of the Crimea, west of Kaffa. It had belonged to the Greek Empire, and had a considerable Greek population. After the Frank conquest of 1204 it apparently fell to Trebizond. It was taken by the Mongols in 1223 for the first time, and a second time in 1239, and during that century was the great port of intercourse with what is now Russia. At an uncertain date, but about the middle of the century, the Venetians established a factory there, which in 1287 became the seat of a consul. In 1323 we find Pope John XXII. complaining to Uzbek Khan of Sarai that the Christians had been ejected from Soldaia and their churches turned into mosques. Ibn Batuta, who alludes to this strife, counts Sudak as one of the four great ports of the World. The Genoese got Soldaia in 1365 and built strong defences, still to be seen. Kaffa, with a good anchorage, in the 14th century, and later on Tana, took the place of Soldaia as chief emporium in South Russia. Some of the Arab Geographers call the Sea of Azov the Sea of Sudak.

The Elder Marco Polo in his Will (1280) bequeaths to the Franciscan Friars of the place a house of his in *Soldachia*, reserving life occupation to his own son and daughter, then residing in it. Probably this establishment already existed when the two Brothers went thither. (*Elie de Laprimaudaie*, passim; *Gold. Horde*, 87; *Mosheim*, App. 148; *Ibn Bat.* I. 28, II. 414; *Cathay*, 231-33; *Heyd*, II. passim.)

CHAPTER II.

How the Two Brothers went on beyond Soldaia.

Having stayed a while at Soldaia, they considered the matter, and thought it well to extend their journey further. So they set forth from Soldaia and travelled till they came to the Court of a certain Tartar Prince, Barca Kaan by name, whose residences were at Sara[1] and at Bolgara [and who was esteemed one of the most liberal and courteous Princes that ever was among the Tartars.][2] This Barca was delighted at the arrival of the Two Brothers, and treated them with great honour; so they presented to him the whole of the jewels that they had brought with them. The Prince was highly pleased with these, and accepted the offering most graciously, causing the Brothers to receive at least twice its value.

After they had spent a twelvemonth at the court of this Prince there broke out a great war between Barca

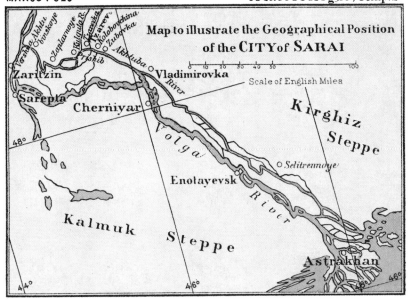

Map to illustrate the Geographical Position of the CITY of SARAI

Scale of English Miles

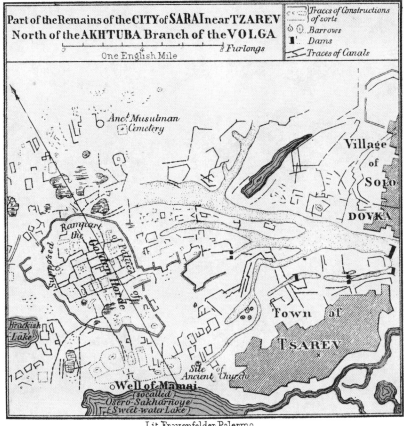

Part of the Remains of the CITY of SARAI near TZAREV North of the AKHTUBA Branch of the VOLGA

One English Mile

Traces of Constructions of sorts
Barrows
Dams
Traces of Canals

Lit. Frauenfelder, Palermo

[To face p. 4.

and Aláu, the Lord of the Tartars of the Levant, and great hosts were mustered on either side.[3]

But in the end Barca, the Lord of the Tartars of the Ponent, was defeated, though on both sides there was great slaughter.   And by reason of this war no one could travel without peril of being taken ; thus it was at least on the road by which the Brothers had come, though there was no obstacle to their travelling forward. So the Brothers, finding they could not retrace their steps, determined to go forward.   Quitting Bolgara, therefore, they proceeded to a city called UCACA, which was at the extremity of the kingdom of the Lord of the Ponent ; [4] and thence departing again, and passing the great River Tigris, they travelled across a Desert which extended for seventeen days' journey, and wherein they found neither town nor village, falling in only with the tents of Tartars occupied with their cattle at pasture.[5]

Note I.—|- Barka Khan, third son of Jújí, the first-born of Chingniz, ruled the *Ulús* of Juji and Empire of Kipchak (Southern Russia) from 1257 to 1265. He was the first Musulman sovereign of his race. His chief residence was at SARAI (Sara of the text), a city founded by his brother and predecessor Bátú, on the banks of the Akhtuba branch of the Volga. In the next century Ibn Batuta describes Sarai as a very handsome and populous city, so large that it made half a day's journey to ride through it. The inhabitants were Mongols, Aás (or Alans), Kipchaks, Circassians, Russians, and Greeks, besides the foreign Moslem merchants, who had a walled quarter. Another Mahomedan traveller of the same century says the city itself was not walled, but, "The Khan's Palace was a great edifice surmounted by a golden crescent weighing two *kantars* of Egypt, and encompassed by a wall flanked with towers," etc. Pope John XXII., on the 26th February 1322, defined the limits of the new Bishopric of Kaffa, which were Sarai to the east and Varna to the west.

Sarai became the seat of both a Latin and a Russian metropolitan, and of more than one Franciscan convent. It was destroyed by Timur on his second invasion of Kipchak (1395-6), and extinguished by the Russians a century later. It is the scene of Chaucer's half-told tale of Cambuscan :—

"At *Sarra*, in the Londe of Tartarie,
There dwelt a King that werriëd Russie."

[" *Mesalek-al-absar* (285, 287), says Sarai, meaning 'the Palace,' was founded by Bereké, brother of Batu. It stood in a salty plain, and was without walls, though the palace had walls flanked by towers. The town was large, had markets, *madrasas* —and baths. It is usually identified with Selitrennoyé Gorodok, about 70 miles above Astrakhan." (*Rockhill, Rubruck*, p. 260, note.)—H. C.]

Several sites exhibiting extensive ruins near the banks of the Akhtuba have been identified with Sarai ; two in particular. One of these is not far from the great elbow of the Volga at Tzaritzyn : the other much lower down, at Selitrennoyé Gorodok or Saltpetre-Town, not far above Astrakhan.

The upper site exhibits by far the most extensive traces of former population, and, is declared unhesitatingly to be the sole site of Sarai by M. Gregorieff, who carried on excavations among the remains for four years, though with what precise results I have not been able to learn. The most dense part of the remains, consisting of mounds and earth-works, traces of walls, buildings, cisterns, dams, and innumerable canals, extends for about 7½ miles in the vicinity of the town of Tzarev, but a tract of 66 miles in length and 300 miles in circuit, commencing from near the head of the Akhtuba, presents remains of like character, though of less density, marking the ground occupied by the villages which encircled the capital. About 2½ miles to the N.W. of Tzarev a vast mass of such remains, surrounded by the traces of a brick rampart, points out the presumable position of the Imperial Palace.

M. Gregorieff appears to admit no alternative. Yet it seems certain that the indications of Abulfeda, Pegolotti, and others, with regard to the position of the capital in the early part of the 14th century, are not consistent with a site so far from the Caspian. Moreover, F. H. Müller states that the site near Tzarev is known to the Tartars as the "Sarai of Janibek Khan" (1341-1357). Now it is worthy of note that in the coinage of Janibek we repeatedly find as the place of mintage, *New Sarai*. Arabsháh in his History of Timur states that 63 years had elapsed from the foundation to the destruction of Sarai. But it must have been at least 140 years since the foundation of Batu's city. Is it not possible, therefore, that both the sites which we have mentioned were successively occupied by the Mongol capital ; that the original Sarai of Batu was at Selitrennoyé Gorodok, and that the *New Sarai* of Janibek was established by him, or by his father Uzbeg in his latter days, on the upper Akhtuba ? Pegolotti having carried his merchant from Tana (Azov) to Gittarchan (Astrakhan), takes him *one day* by river to Sara, and from Sara to *Saracanco*, also by river, eight days more. (*Cathay*, p. 287.) In the work quoted I have taken Saracanco for Saraichik, on the Yaik. But it was possibly the Upper or New Sarai on the Akhtuba. Ibn Batuta, marching on the frozen river, reached Sarai in three days from Astrakhan. This could not have been at Tzarev, 200 miles off.

In corroboration (*quantum valeat*) of my suggestion that there must have been two Sarais near the Volga, Professor Bruun of Odessa points to the fact that Fra Mauro's map presents *two* cities of Sarai on the Akhtuba ; only the Sarai of Janibeg is with him no longer *New* Sarai, but *Great* Sarai.

The use of the latter name suggests the possibility that in the *Saracanco* of Pegolotti the latter half of the name may be the Mongol *Kúnk* "Great." (See *Pavet de Courteille*, p. 439.)

Professor Bruun also draws attention to the impossibility of Ibn Batuta's travelling from Astrakhan to Tzarev in three days, an argument which had already occurred to me and been inserted above.

[The Empire of Kipchak founded after the Mongol Conquest of 1224, included also parts of Siberia and Khwarizm ; it survived nominally until 1502.—H. C.]

(*Four Years of Archæological Researches among the Ruins of Sarai* [in Russian]. by M. Gregorieff [who appears to have also published a pamphlet specially on the site, but this has not been available]; *Historisch-geographische Darstellung des Strom-systems der Wolga, von Ferd. Heinr. Müller*, Berlin, 1839, 568-577 ; *Ibn. Bat.* II. 447 ; *Not. et Extraits*, XIII. i. 286 ; *Pallas, Voyages* ; *Cathay*, 231, etc. ; *Erdmann, Numi Asiatici*, pp. 362 *seqq* ; *Arabs.* I. p. 381.)

NOTE 2.—BOLGHAR, our author's Bolgara, was the capital of the region some-times called Great Bulgaria, by Abulfeda *Inner Bulgaria*, and stood a few miles from the left bank of the Volga, in latitude about 54° 54', and 90 miles below Kazan. The old Arab writers regarded it as nearly the limit of the habitable world, and told wonders of the cold, the brief summer nights, and the fossil ivory that was found in its vicinity. This was exported, and with peltry, wax, honey, hazel-nuts, and Russia leather, formed the staple articles of trade. The last item derived from Bolghar the name which it still bears all over Asia. (See Bk. II. ch. xvi., and Note.) Bolghar seems to have been the northern limit of Arab travel, and was visited by the curious (by Ibn Batuta among others) in order to witness the phenomena of the short summer night, as tourists now visit Hammerfest to witness its entire absence.

Russian chroniclers speak of an earlier capital of the Bulgarian kingdom, Brakhimof, near the mouth of the Kama, destroyed by Andrew, Grand Duke of Rostof and Susdal, about 1160 ; and this may have been the city referred to in the earlier Arabic accounts. The fullest of these is by Ibn Fozlán, who accompanied an embassy from the Court of Baghdad to Bolghar, in A.D. 921. The King and people had about this time been converted to Islam, having previously, as it would seem, pro-fessed Christianity. Nevertheless, a Mahomedan writer of the 14th century says the people had then long renounced Islam for the worship of the Cross. (*Not. et Extr.* XIII. i. 270.)

Ruins of Bolghar.

Bolghar was first captured by the Mongols in 1225. It seems to have perished early in the 15th century, after which Kazan practically took its place. Its position is still marked by a village called Bolgari, where ruins of Mahomedan character remain, and where coins and inscriptions have been found. Coins of the Kings of Bolghar, struck in the 10th century, have been described by Fraehn, as well as coins of the Mongol period struck at Bolghar. Its latest known coin is of A.H. 818 (A.D. 1415-16). A history of Bolghar was written in the first half of the 12th century by Yakub Ibn Noman, Kadhi of the city, but this is not known to be extant.

Fraehn shows ground for believing the people to have been a mixture of Fins, Slavs, and Turks. Nicephorus Gregoras supposes that they took their name from the great river on which they dwelt (Βούλγα).

["The ruins [of Bolghar]," says Bretschneider, in his *Mediæval Researches,* published in 1888, vol. ii. p. 82, "still exist, and have been the subject of learned investigation by several Russian scholars. These remains are found on the spot where now the village *Uspenskoye,* called also *Bolgarskoye* (Bolgari), stands, in the district of Spask, province of Kazan. This village is about 4 English miles distant from the Volga, east of it, and 83 miles from Kazan." Part of the Bulgars removed to the Balkans ; others remained in their native country on the shores of the Azov Sea, and were subjugated by the Khazars. At the beginning of the 9th century, they marched northwards to the Volga and the Kama, and established the kingdom of Great Bulgaria. Their chief city, Bolghar, was on the bank of the Volga, but the river runs now to the west ; as the Kama also underwent a change in its course, it is possible that formerly Bolghar was built at the junction of the two rivers. (Cf. *Reclus,*

Europe russe, p. 761.) The Bulgars were converted to Islam in 922. Their country was first invaded by the Mongols under Subutai in 1223 ; this General conquered it in 1236, the capital was destroyed the following year, and the country annexed to the kingdom of Kipchak. Bolghar was again destroyed in 1391 by Tamerlan. In 1438, Ulugh Mohammed, cousin of Toka Timur, younger son of Juji, transformed this country into the khanate of Kazan, which survived till 1552. It had probably been the capital of the Golden Horde before Sarai.

With reference to the early Christianity of the Bulgarians, to which Yule refers in his note, the *Laurentian Chronicle* (A.D. 1229), quoted by Shpilevsky, adduces evidence to show that in the Great City, *i.e. Bulgar*, there were Russian Christians and a Christian cemetery, and the death of a Bulgarian Christian martyr is related in the same chronicle as well as in the Nikon, Tver, and Tatischef annals in which his name is given. (Cf. Shpilevsky, *Anc. towns and other Bulgaro-Tartar monuments*, Kazan, 1877, p. 158 *seq. ; Rockhill's Rubruck*, Hakl. Soc. p. 121, note.)—H. C.]

The severe and lasting winter is spoken of by Ibn Fozlán and other old writers in terms that seem to point to a modern mitigation of climate. It is remarkable, too, that Ibn Fozlán speaks of the aurora as of very frequent occurrence, which is not now the case in that latitude. We may suspect this frequency to have been connected with the greater cold indicated, and perhaps with a different position of the magnetic pole. Ibn Fozlán's account of the aurora is very striking :—" Shortly before sunset the horizon became all very ruddy, and at the same time I heard sounds in the upper air, with a dull rustling. I looked up and beheld sweeping over me a fire-red cloud, from which these sounds issued, and in it movements, as it were, of men and horses ; the men grasping bows, lances, and swords. This I saw, or thought I saw. Then there appeared a white cloud of like aspect ; in it also I beheld armed horsemen, and these rushed against the former as one squadron of horse charges another. We were so terrified at this that we turned with humble prayer to the Almighty, whereupon the natives about us wondered and broke into loud laughter. We, however, continued to gaze, seeing how one cloud charged the other, remained confused with it a while, and then sundered again. These movements lasted deep into the night, and then all vanished."

(*Fraehn, Ueber die Wolga Bulgaren*, Petersb. 1832 ; *Gold. Horde*, 8, 9, 423-424 ; *Not. et Extr.* II. 541 ; *Ibn Bat.* II. 398 ; *Büschings Mag.* V. 492 ; *Erdmann, Numi Asiat.* I. 315-318, 333-334, 520-535 ; *Niceph. Gregoras*, II. 2, 2.)

NOTE 3.—ALAU is Polo's representation of the name of Hulákú, brother of the Great Kaans Mangu and Kublai and founder of the Mongol dynasty in Persia. In the Mongol pronunciation guttural and palatal consonants are apt to be elided, hence this spelling. The same name is written by Pope Alexander IV., in addressing the Khan, *Olao*, by Pachymeres and Gregoras Χαλαὺ and Χαλαοῦ, by Hayton *Haolon*, by Ibn Batuta *Huláún*, as well as in a letter of Hulaku's own, as given by Makrizi.

The war in question is related in Rashíduddín's history, and by Polo himself towards the end of the work. It began in the summer of 1262, and ended about eight months later. Hence the Polos must have reached Barka's Court in 1261.

Marco always applies to the Mongol Khans of Persia the title of " Lords of the East " (*Levant*), and to the Khans of Kipchak that of " Lords of the West " (*Ponent*). We use the term *Levant* still with a similar specific application, and in another form *Anatolia*. I think it best to preserve the terms *Levant* and *Ponent* when used in this way.

[Robert Parke in his translation out of Spanish of Mendoza, *The Historie of the great and mightie kingdome of China* . . . London, printed by I. Wolfe for Edward White, 1588, uses the word *Ponent :* " You shall understande that this mightie kingdome is the Orientalest part of all Asia, and his next neighbour towards the *Ponent* is the kingdome of *Quachinchina* . . . (p. 2)."—H. C.]

NOTE 4.—UCACA or UKEK was a town on the right bank of the Volga, nearly equidistant between Sarai and Bolghar, and about six miles south of the modern Saratov, where a village called *Uwek* still exists. Ukek is not mentioned before the Mongol domination, and is supposed to have been of Mongol foundation, as the name Ukek is said in Mongol to signify a dam of hurdles. The city is mentioned by Abulfeda as marking the extremity of "the empire of the Barka Tartars," and Ibn Batuta speaks of it as "one day distant from the hills of the Russians." Polo therefore means that it was the frontier of the Ponent towards Russia. Ukek was the site of a Franciscan convent in the 14th century ; it is mentioned several times in the campaigns of Timur, and was destroyed by his army. It is not mentioned under the form Ukek after this, but appears as *Uwek* and *Uwesh* in Russian documents of the 16th century. Perhaps this was always the Slavonic form, for it already is written *Uguech* (= Uwek) in Wadding's 14th century catalogue of convents. Anthony Jenkinson, in Hakluyt, gives an observation of its latitude, as *Oweke* (51° 40'), and Christopher Burrough, in the same collection, gives a description of it as *Oueak*, and the latitude as 51° 30' (some 7' too much). In his time (1579) there were the remains of a "very faire stone castle" and city, with old tombs exhibiting sculptures and inscriptions. All these have long vanished. Burrough was told by the Russians that the town "was swallowed into the earth by the justice of God, for the wickednesse of the people that inhabited the same." Lepechin in 1769 found nothing remaining but part of an earthen rampart and some underground vaults of larger bricks, which the people dug out for use. He speaks of coins and other relics as frequent, and the like have been found more recently. Coins with Mongol-Arab inscriptions. struck at Ukek by Tuktugai Khan in 1306, have been described by Fraehn and Erdmann.

(*Fraehn, Ueber die ehemalige Mong. Stadt Ukek, etc.*, Petersb. 1835; *Gold. Horde; Ibn Bat.* II.. 414; *Abulfeda, in Büsching,* V. 365; *Ann. Minorum,* sub anno 1400; *Petis de la Croix,* II. 355, 383, 388; *Hakluyt,* ed. 1809, I. 375 and 472; *Lepechin, Tagebuch der Reise, etc.,* I. 235-237; *Rockhill, Rubruck,* 120-121, note 2.)

NOTE 5.—The great River Tigeri or Tigris is the Volga, as Pauthier rightly shows. It receives the same name from the Monk Pascal of Vittoria in 1338. (*Cathay,* p. 234.) Perhaps this arose out of some legend that the Tigris was a reappearance of the same river. The ecclesiastical historian, Nicephorus Callistus, appears to imply that the Tigris coming from Paradise flows under the Caspian to emerge in Kurdistan. (See IX. 19.)

The "17 days" applies to one stretch of desert. The whole journey from Ukek Bokhara would take some 60 days at least. Ibn Batuta is 58 days from Sarai to Bokhara, and of the last section he says, "we entered the desert which extends between Khwarizm and Bokhara, and *which has an extent of 18 days' journey.*" (III. 19.)

CHAPTER III.

How the Two Brothers, after crossing a Desert, came to the City of Bocara, and fell in with certain Envoys there.

AFTER they had passed the desert, they arrived at a very great and noble city called BOCARA, the territory of which belonged to a king whose name was Barac,

and is also called Bocara. The city is the best in all
Persia.[1] And when they had got thither, they found
they could neither proceed further forward nor yet turn
back again; wherefore they abode in that city of Bocara
for three years. And whilst they were sojourning in
that city, there came from Alau, Lord of the Levant,
Envoys on their way to the Court of the Great Kaan,
the Lord of all the Tartars in the world. And when
the Envoys beheld the Two Brothers they were amazed,
for they had never before seen Latins in that part of
the world. And they said to the Brothers: "Gentlemen, if ye will take our counsel, ye will find great
honour and profit shall come thereof." So they replied
that they would be right glad to learn how. "In
truth," said the Envoys, "the Great Kaan hath never
seen any Latins, and he hath a great desire so to do.
Wherefore, if ye will keep us company to his Court, ye
may depend upon it that he will be right glad to see
you, and will treat you with great honour and liberality;
whilst in our company ye shall travel with perfect
security, and need fear to be molested by nobody."[2]

NOTE 1.—Hayton also calls Bokhara a city of Persia, and I see Vámbéry says
that, up till the conquest by Chinghiz, Bokhara, Samarkand, Balkh, etc., were considered to belong to Persia. (*Travels*, p. 377.) The first Mongolian governor of
Bokhara was Buka Bosha.

King Barac is Borrak Khan, great-grandson of Chagatai, and sovereign of the
Ulús of Chagatai, from 1264 to 1270. The Polos, no doubt, reached Bokhara
before 1264, but Borrak must have been sovereign some time before they left it.

NOTE 2.—The language of the envoys seems rather to imply that they were
the Great Kaan's own people returning from the Court of Hulaku. And Rashid
mentions that Sartak, the Kaan's ambassador to Hulaku, returned from Persia in the
year that the latter prince died. It may have been his party that the Venetians
joined, for the year almost certainly was the same, viz. 1265. If so, another of the
party was Bayan, afterwards the greatest of Kublai's captains, and much celebrated
in the sequel of this book. (See *Erdmann's Temudschin*, p. 214.)

Marsden justly notes that Marco habitually speaks of *Latins*, never of *Franks*.
Yet I suspect his own mental expression was *Farangi.*

CHAPTER IV.

How the Two Brothers took the Envoys' counsel, and went
to the Court of the Great Kaan.

So when the Two Brothers had made their arrangements,
they set out on their travels, in company with the Envoys,
and journeyed for a whole year, going northward and
north-eastward, before they reached the Court of that
Prince.   And on their journey they saw many marvels of
divers and sundry kinds, but of these we shall say nothing
at present, because Messer Mark, who has likewise seen
them all, will give you a full account of them in the Book
which follows.

CHAPTER V.

How the Two Brothers arrived at the Court of the
Great Kaan.

When the Two Brothers got to the Great Kaan, he received them with great honour and hospitality, and showed
much pleasure at their visit, asking them a great number
of questions.   First, he asked about the emperors, how
they maintained their dignity, and administered justice in
their dominions ; and how they went forth to battle, and
so forth.   And then he asked the like questions about the
kings and princes and other potentates.

CHAPTER VI.

How the Great Kaan asked all about the manners of the
Christians, and particularly about the Pope of Rome.

And then he inquired about the Pope and the Church,
and about all that is done at Rome, and all the customs of
the Latins.   And the Two Brothers told him the truth in
all its particulars, with order and good sense, like sensible
men as they were ; and this they were able to do as they
knew the Tartar language well.[1]

Note 1.—The word generally used for Pope in the original is *Apostoille*
(*Apostolicus*), the usual French expression of that age.

It is remarkable that for the most part the text edited by Pauthier has the
correcter Oriental form *Tatar*, instead of the usual *Tartar*.   *Tattar* is the word used
by Yvo of Narbonne, in the curious letter given by Matthew Paris under 1243.

We are often told that *Tartar* is a vulgar European error.   It is in any case a
very old one ; nor does it seem to be of European origin, but rather Armenian ; *
though the suggestion of Tartarus may have given it readier currency in Europe.
Russian writers, or rather writers who have been in Russia, sometimes try to force on
us a specific limitation of the word *Tartar* to a certain class of Oriental Turkish race,
to whom the Russians appropriate the name.   But there is no just ground for this.
Tátár is used by Oriental writers of Polo's age exactly as Tartar was then, and is
still, used in Western Europe, as a generic title for the Turanian hosts who followed
Chinghiz and his successors.   But I believe the name in this sense was unknown to
Western Asia before the time of Chinghiz.   And General Cunningham must over-
look this when he connects the *Tátaríya* coins, mentioned by Arab geographers of
the 9th century, with "the Scythic or Tátár princes who ruled in Kabul" in the
beginning of our era.   Tartars on the Indian frontier in those centuries are surely
to be classed with the Frenchmen whom Brennus led to Rome, or the Scotchmen
who fought against Agricola.

---

* See *J. As.* sér. V. tom. xi. p. 204.

CHAPTER VII.

How the Great Kaan sent the Two Brothers as his Envoys to the Pope.

When that Prince, whose name was Cublay Kaan, Lord of the Tartars all over the earth, and of all the kingdoms and provinces and territories of that vast quarter of the world, had heard all that the Brothers had to tell him about the ways of the Latins, he was greatly pleased, and he took it into his head that he would send them on an Embassy to the Pope.   So he urgently desired them to undertake this mission along with one of his Barons; and they replied that they would gladly execute all his commands as those of their Sovereign Lord. Then the Prince sent to summon to his presence one of his Barons whose name was Cogatal, and desired him to get ready, for it was proposed to send him to the Pope along with the Two Brothers.   The Baron replied that he would execute the Lord's commands to the best of his ability.

After this the Prince caused letters from himself to the Pope to be indited in the Tartar tongue,[1] and committed them to the Two Brothers and to that Baron of his own, and charged them with what he wished them to say to the Pope.   Now the contents of the letter were to this purport : He begged that the Pope would send as many as an hundred persons of our Christian faith ; intelligent men, acquainted with the Seven Arts,[2] well qualified to enter into controversy, and able clearly to prove by force of argument to idolaters and other kinds of folk, that the Law of Christ was best, and that all other religions were false and naught ; and that if they would prove this, he and all under him would become Christians and the

Church's liegemen. Finally he charged his Envoys to bring back to him some Oil of the Lamp which burns on the Sepulchre of our Lord at Jerusalem.[3]

NOTE 1.—·|· The appearance of the Great Kaan's letter may be illustrated by two letters on so-called Corean paper preserved in the French archives; one from Arghún Khan of Persia (1289), brought by Buscarel, and the other from his son Oljaitu (May, 1305), to Philip the Fair. These are both in the Mongol language, and according to Abel Rémusat and other authorities, in the Uighúr character, the parent of the present Mongol writing. Facsimiles of the letters are given in Rémusat's paper on intercourse with Mongol Princes, in *Mém. de l'Acad. des Inscript.* vols. vii. and viii., reproductions in J. B. Chabot's *Hist. de Mar Jabalaha III.*, Paris, 1895, and preferably in Prince Roland Bonaparte's beautiful *Documents Mongols*, Pl. XIV., and we give samples of the two in vol. ii.*

NOTE 2.—"The Seven Arts," from a date reaching back nearly to classical times, and down through the Middle Ages, expressed the whole circle of a liberal education, and it is to these Seven Arts that the degrees in arts were understood to apply. They were divided into the *Trivium* of Rhetoric, Logic, and Grammar, and the *Quadrivium* of Arithmetic, Astronomy, Music, and Geometry. The 38th epistle of Seneca was in many MSS. (according to Lipsius) entitled "*L. Annaei Senecae Liber de Septèm Artibus liberalibus.*" I do not find, however, that Seneca there mentions categorically more than five, viz., Grammar, Geometry, Music, Astronomy, and Arithmetic. In the 5th century we find the Seven Arts to form the successive subjects of the last seven books of the work of Martianus Capella, much used in the schools during the early Middle Ages. The Seven Arts will be found enumerated in the verses of Tzetzes (*Chil. XI.* 525), and allusions to them in the mediæval romances are endless. Thus, in one of the "Gestes d'Alexandre," a chapter is headed "*Comment Aristotle aprent à Alixandre les Sept Arts.*" In the tale of the Seven Wise Masters, Diocletian selects that number of tutors for his son, each to instruct him in one of the Seven Arts. In the romance of *Erec and Eneide* we have a dress on which the fairies had portrayed the Seven Arts (*Franc. Michel, Recherches, etc.* II. 82); in the *Roman de Mahommet* the young impostor is master of all the seven. There is one mediæval poem called the *Marriage of the Seven Arts*, and another called the *Battle of the Seven Arts.* (See also Dante, *Convito*, Trat. II. c. 14; *Not. et Ex.* V., 491 *seqq.*)

NOTE 3.—The Chinghizide Princes were eminently liberal—or indifferent—in religion; and even after they became Mahomedan, which, however, the Eastern branch never did, they were rarely and only by brief fits persecutors. Hence there was scarcely one of the non-Mahomedan Khans of whose conversion to Christianity there were not stories spread. The first rumours of Chinghiz in the West were as of a Christian conqueror; tales may be found of the Christianity of Chagatai, Hulaku, Abaka, Arghun, Baidu, Ghazan, Sartak, Kuyuk, Mangu, Kublai, and one or two of the latter's successors in China, all probably false, with one or two doubtful exceptions.

---

* See plates with ch. xvii. of Bk. IV. See also the Uighúr character in the second *Païza*, Bk. II. ch. vii.

The Great Kaan delivering a Golden Tablet to the Brothers.   From a miniature of the 14th century.

CHAPTER VIII.

HOW THE GREAT KAAN GAVE THEM A TABLET OF GOLD, BEARING HIS ORDERS IN THEIR BEHALF.

WHEN the Prince had charged them with all his commission, he caused to be given them a Tablet of Gold, on which was inscribed that the three Ambassadors should be supplied with everything needful in all the countries through which they should pass—with horses, with escorts, and, in short, with whatever they should require.   And when they had made all needful preparations, the three Ambassadors took their leave of the Emperor and set out.

When they had travelled I know not how many days, the Tartar Baron fell sick, so that he could not ride, and being very ill, and unable to proceed further, he halted at a certain city.   So the Two Brothers judged it best that they should leave him behind and proceed to carry out their commission; and, as he was well content that they should do so, they continued their journey. And I can assure you, that whithersoever they went they were honourably provided with whatever they stood in need of, or chose to command. And this was owing to that Tablet of Authority from the Lord which they carried with them.[1]

So they travelled on and on until they arrived at Layas in Hermenia, a journey which occupied them, I assure you, for three years.[2] It took them so long because they could not always proceed, being stopped sometimes by snow, or by heavy rains falling, or by great torrents which they found in an impassable state.

NOTE 1.—On these Tablets, see a note under Bk. II. ch. vii.

NOTE 2.—AYAS, called also Ayacio, Aiazzo, Giazza, Glaza, La Jazza, and *Layas*, occupied the site of ancient Aegae, and was the chief port of Cilician Armenia, on the Gulf of Scanderoon. *Aegae* had been in the 5th century a place of trade with the West, and the seat of a bishopric, as we learn from the romantic but incomplete

Castle of Avas.

story of Mary, the noble slave-girl, told by Gibbon (ch. 33). As Ayas it became in the latter part of the 13th century one of the chief places for the shipment of Asiatic wares arriving through Tabriz, and was much frequented by the vessels of the Italian Republics. The Venetians had a *Bailo* resident there.

Ayas is the *Leyes* of Chaucer's Knight,—

("At Leyes was he and at Satalie")—

and the Layas of Froissart. (Bk. III. ch. xxii.) The Gulf of Layas is described in the xix. Canto of Ariosto, where Mafisa and Astolfo find on its shores a country of barbarous Amazons :—

" Fatto è 'l porto a sembranza d' una luna," etc.

Marino Sanuto says of it : " Laiacio has a haven, and a shoal in front of it that we might rather call a reef, and to this shoal the hawsers of vessels are moored whilst the anchors are laid out towards the land." (II. IV. ch. xxvi.)

The present Ayas is a wretched village of some 15 huts, occupied by about 600 Turkmans, and standing inside the ruined walls of the castle. This castle, which is still in good condition, was built by the Armenian kings, and restored by Sultan Suleiman ; it was constructed from the remains of the ancient city ; fragments of old columns are embedded in its walls of cut stone. It formerly communicated by a causeway with an advanced work on an island before the harbour. The ruins of the city occupy a large space. (*Langlois, V. en Cilicie,* pp. 429-31 ; see also *Beaufort's Karamania,* near the end.) A plan of Ayas will be found at the beginning of Bk. I. —H. Y. and H. C.

CHAPTER IX.

How the Two Brothers came to the city of Acre.

THEY departed from Layas and came to ACRE, arriving there in the month of April, in the year of Christ 1269, and then they learned that the Pope was dead. And when they found that the Pope was dead (his name was Pope * *),[1] they went to a certain wise Churchman who was Legate for the whole kingdom of Egypt, and a man of great authority, by name THEOBALD OF PIACENZA, and told him of the mission on which they were come. When the Legate heard their story, he was greatly surprised, and deemed the thing to be of great honour and advantage for the whole of Christendom. So his answer to the two Ambassador Brothers was this : " Gentlemen, ye see that

the Pope is dead ; wherefore ye must needs have patience
until a new Pope be made, and then shall ye be able to
execute your charge." Seeing well enough that what the
Legate said was just, they observed : "But while the
Pope is a-making, we may as well go to Venice and visit
our households." So they departed from Acre and went

CIVITAS ACON SIVE PTOLOMAYDA.

ACRE AS IT WAS WHEN LOST (A.D. 1291).
FROM THE PLAN GIVEN BY
MARINO SANUTO.

to Negropont, and from Negropont they continued their
voyage to Venice.[2]   On their arrival there, Messer
Nicolas found that his wife was dead, and that she
had left behind her a son of fifteen years of age, whose
name was MARCO; and 'tis of him that this Book tells.[3]
The Two Brothers abode at Venice a couple of years,
tarrying until a Pope should be made.

NOTE 1.—The deceased Pope's name is omitted both in the Geog. Text and in
Pauthier's, clearly because neither Rusticiano nor Polo remembered it. It is supplied
correctly in the Crusca Italian as *Clement*, and in Ramusio as *Clement IV.*

It is not clear that *Theobald*, though generally adopted, is the ecclesiastic's proper
name. It appears in different MSS. as *Teald* (G. T.), *Ceabo* for *Teabo* (Pauthier),
Odoaldo (Crusca), and in the Riccardian as *Thebaldus de Vice-comitibus de Placentia,*

which corresponds to Ramusio's version.   Most of the ecclesiastical chroniclers call him *Tedaldus*, some *Thealdus*.   *Tedaldo* is a real name, occurring in Boccaccio.   (Day iii. Novel 7.)

NOTE 2.—After the expulsion of the Venetians from Constantinople, Negropont was the centre of their influence in Romania.   On the final return of the travellers they again take Negropont on their way.   [It was one of the ports on the route from Venice to Constantinople, Tana, Trebizond.—H. C.]

NOTE 3.—The *edition* of the Soc. de Géographie makes Mark's age *twelve*, but I have verified from inspection the fact noticed by Pauthier that the *manuscript* has distinctly xv. like all the other old texts.   In Ramusio it is *nineteen*, but this is doubtless an arbitrary correction to suit the mistaken date (1250) assigned for the departure of the father from Constantinople.

There is nothing in the old French texts to justify the usual statement that Marco was born after the departure of his father from Venice.   All that the G. T. says is : "Meser Nicolau treuve que sa fame estoit morte, et les remès un filz de xv. anz que avoit à nom Marc," and Pauthier's text is to the same effect.   Ramusio, indeed, has : " M. Nicolò trovò, che sua moglie era morta, la quale nella sua partita haveva partorito un figliuolo," and the other versions that are based on Pipino's seem all to have like statements.

CHAPTER X.

HOW THE TWO BROTHERS AGAIN DEPARTED FROM VENICE, ON THEIR WAY BACK TO THE GREAT KAAN, AND TOOK WITH THEM MARK, THE SON OF MESSER NICOLAS.

WHEN the Two Brothers had tarried as long as I have told you, and saw that never a Pope was made, they said that their return to the Great Kaan must be put off no longer. So they set out from Venice, taking Mark along with them, and went straight back to Acre, where they found the Legate of whom we have spoken.   They had a good deal of discourse with him concerning the matter, and asked his permission to go to JERUSALEM to get some Oil from the Lamp on the Sepulchre, to carry with them to the Great Kaan, as he had enjoined.[1]   The Legate giving them leave, they went from Acre to Jerusalem and got some of the Oil, and then returned to Acre, and went to the Legate and said to him : "As we see no sign of a

Pope's being made, we desire to return to the Great
Kaan; for we have already tarried long, and there has
been more than enough delay." To which the Legate
replied: " Since 'tis your wish to go back, I am well content." Wherefore he caused letters to be written for
delivery to the Great Kaan, bearing testimony that the
Two Brothers had come in all good faith to accomplish
his charge, but that as there was no Pope they had been
unable to do so.

NOTE I.—In a Pilgrimage of date apparently earlier than this, the Pilgrim says of
the Sepulchre: " The Lamp which had been placed by His head (when He lay there)
still burns on the same spot day and night. *We took a blessing from it* (*i.e.* apparently took some of the oil as a beneficent memorial), and replaced it." (*Itinerarium
Antonini Placentini* in *Bollandists*, May, vol. ii. p. xx.)

["Five great oil lamps," says Daniel, the Russian Hégoumène, 1106-1107
(*Itinéraires russes en Orient*, trad. pour la Soc. de l'Orient Latin, par Mme. B. de
Khitrowo, Geneva, 1889, p. 13), "burning continually night and day, are hung in the
Sepulchre of Our Lord."—H. C.]

CHAPTER XI.

How the Two Brothers set out from Acre, and Mark along
with them.

WHEN the Two Brothers had received the Legate's
letters, they set forth from Acre to return to the Grand
Kaan, and got as far as Layas. But shortly after their
arrival there they had news that the Legate aforesaid was
chosen Pope, taking the name of Pope Gregory of
Piacenza; news which the Two Brothers were very glad
indeed to hear. And presently there reached them at
Layas a message from the Legate, now the Pope, desiring
them, on the part of the Apostolic See, not to proceed
further on their journey, but to return to him incontinently. And what shall I tell you? The King of

Hermenia caused a galley to be got ready for the Two Ambassador Brothers, and despatched them to the Pope at Acre.[1]

NOTE 1.—The death of Pope Clement IV. occurred on St Andrew's Day (29th November), 1268 ; the election of Tedaldo or Tebaldo of Piacenza, a member of the Visconti family, and Archdeacon of Liège, did not take place till 1st September, 1271, owing to the factions among the cardinals. And it is said that some of them, anxious only to get away, voted for Theobald in full belief that he was dead. The conclave, in its inability to agree, had named a committee of six with full powers which the same day elected Theobald, on the recommendation of the Cardinal Bishop of Portus (John de Toleto, said, in spite of his name, to have been an Englishman). This facetious dignitary had suggested that the roof should be taken off the Palace at Viterbo where they sat, to allow the divine influences to descend more freely on their counsels (*quia nequeunt ad nos per tot tecta ingredi*). According to some, these doggerel verses, current on the occasion, were extemporised by Cardinal John in the pious exuberance of his glee :—

Portrait of Pope Gregory X.

" Papatûs munus tulit Archidiaconus unus
Quem Patrem Patrum fecit discordia Fratrum."

The Archdeacon, a man of great weight of character, in consequence of differences with his Bishop (of Liège), who was a disorderly liver, had gone to the Holy Land, and during his stay there he contracted great intimacy with Prince Edward of England (Edward I.). Some authors, *e.g.* John Villani (VIII. 39), say that he was Legate in Syria ; others, as Rainaldus, deny this ; but Polo's statement, and the authority which the Archdeacon took on himself in writing to the Kaan, seem to show that he had some such position.

He took the name of Gregory X., and before his departure from Acre, preached a moving sermon on the text, " *If I forget thee, O Jerusalem,*" etc. Prince Edward fitted him out for his voyage.

Gregory reigned barely four years, dying at Arezzo 10th January, 1276. His character stood high to the last, and some of the Northern Martyrologies enrolled him among the saints, but there has never been canonisation by Rome. The people of Arezzo used to celebrate his anniversary with torch-light gatherings at his tomb, and plenty of miracles were alleged to have occurred there. The tomb still stands in the

Duomo at Arezzo, a handsome work by Margaritone, an artist in all branches, who was the Pope's contemporary.   There is an engraving of it in *Gonnelli, Mon. Sepolc. di Toscana.*

(*Fra Pipino* in *Muratori*, IX. 700; *Rainaldi Annal.* III. 252 *seqq.* ; *Wadding,* sub. an. 1217 : *Bollandists,* 10th January ; *Palatii, Gesta Pontif. Roman.* vol. iii., and *Fasti Cardinalium,* I. 463, etc.)

CHAPTER XII.

How the Two Brothers presented themselves before the new Pope.

And when they had been thus honourably conducted to Acre they proceeded to the presence of the Pope, and paid their respects to him with humble reverence.   He received them with great honour and satisfaction, and gave them his blessing.   He then appointed two Friars of the Order of Preachers to accompany them to the Great Kaan, and to do whatever might be required of them. These were unquestionably as learned Churchmen as were to be found in the Province at that day—one being called Friar Nicolas of Vicenza, and the other Friar William of Tripoli.[1]   He delivered to them also proper credentials, and letters in reply to the Great Kaan's messages [and gave them authority to ordain priests and bishops, and to bestow every kind of absolution, as if given by himself in proper person ; sending by them also many fine vessels of crystal as presents to the Great Kaan].[2]   So when they had got all that was needful, they took leave of the Pope, receiving his benediction ; and the four set out together from Acre, and went to Layas, accompanied always by Messer Nicolas's son Marco.

Now, about the time that they reached Layas, Bendocquedar, the Soldan of Babylon, invaded Hermenia with a great host of Saracens, and ravaged the country, so that our Envoys ran a great peril of being taken or slain.[3] And when the Preaching Friars saw this they were greatly frightened, and said that go they never would. So they made over to Messer Nicolas and Messer Maffeo all their credentials and documents, and took their leave, departing in company with the Master of the Temple.[4]

NOTE I.—Friar William, of Tripoli, of the Dominican convent at Acre, appears to have served there as early as 1250. [He was born *circa* 1220, at Tripoli, in Syria, whence his name.—H. C.] He is known as the author of a book, *De Statu Saracenorum post Ludovici Regis de Syriâ reditum*, dedicated to Theoldus, Archdeacon of Liège (*i.e.* Pope Gregory). Of this some extracts are printed in Duchesne's *Hist. Francorum Scriptores*. There are two MSS. of it, with different titles, in the Paris Library, and a French version in that of Berne. A MS. in Cambridge Univ. Library, which contains among other things a copy of Pipino's Polo, has also the work of Friar William :—" *Willelmus Tripolitanus, Aconensis Conventus, de Egressu Machometi et Saracenorum, atque progressu eorumdem, de Statu Saracenorum*," etc. It is imperfect ; it is addressed THEOBALDO *Ecclesiarcho digno Sancte Terre Peregrino Sancto*. And from a cursory inspection I imagine that the Tract appended to one of the Polo MSS. in the British Museum (Addl. MSS., No. 19,952) is the same work or part of it. To the same author is ascribed a tract called *Clades Damiatae*. (*Duchesne*, V. 432 ; *D'Avezac* in *Rec. de Voyages*, IV. 406 ; *Quétif, Script. Ord. Praed.* I. 264-5 ; *Catal. of MSS. in Camb. Univ. Library*, I. 22.)

NOTE 2.—I presume that the powers, stated in this passage from Ramusio to have been conferred on the Friars, are exaggerated. In letters of authority granted in like cases by Pope Gregory's successors, Nicolas III. (in 1278) and Boniface VIII. (in 1299), the missionary friars to remote regions are empowered to absolve from excommunication and release from vows, to settle matrimonial questions, to found churches and appoint *idoneos rectores*, to authorise Oriental clergy who should publicly submit to the Apostolic See to enjoy the *privilegium clericale*, whilst in the absence of bishops those among the missionaries who were priests might consecrate cemeteries, altars, palls, etc., admit to the Order of Acolytes, but nothing beyond. (See *Mosheim, Hist. Tartar. Eccles.* App. Nos. 23 and 42.)

NOTE 3.—The statement here about Bundúkdár's invasion of Cilician Armenia is a difficulty. He had invaded it in 1266, and his second devastating invasion, during which he burnt both Layas and Sis, the king's residence, took place in 1275, a point on which Marino Sanuto is at one with the Oriental Historians. Now we know from Rainaldus that Pope Gregory left Acre in November or December, 1271, and the text appears to imply that our travellers left Acre before him. The utmost corroboration that I can find lies in the following facts stated by Makrizi :—

On the 13th Safar, A.H. 670 (20th September 1271), Bundúkdár arrived unexpectedly at Damascus, and after a brief raid against the Ismaelians he returned to that city. In the middle of Rabi I. (about 20-25 October) the Tartars made an incursion in northern Syria, and the troops of Aleppo retired towards Hamah. There was great alarm at Damascus ; the Sultan sent orders to Cairo for reinforcements, and these arrived at Damascus on the 9th November. The Sultan then advanced on Aleppo, sending corps likewise towards Marash (which was within the Armenian frontier) and Harran. At the latter place the Tartars were attacked and those in the town slaughtered ; the rest retreated. The Sultan was back at

Damascus, and off on a different expedition, by 7th December. Hence, if the travellers arrived at Ayas towards the latter part of November they would probably find alarm existing at the advance of Bundúkdár, though matters did not turn out so serious as they imply.

"Babylon," of which Bundúkdár is here styled Sultan, means Cairo, commonly so styled (*Bambellonia d'Egitto*) in that age. Babylon of Egypt is mentioned by Diodorus quoting Ctesias, by Strabo, and by Ptolemy; it was the station of a Roman Legion in the days of Augustus, and still survives in the name of *Babul*, close to old Cairo.

Malik Dáhir Ruknuddín Bíbars Bundúkdári, a native of Kipchak, was originally sold at Damascus for 800 dirhems (about 18*l.*), and returned by his purchaser because of a blemish. He was then bought by the Amir Aláuddín Aidekín *Bundúkdár* ("The Arblasteer") whose surname he afterwards adopted. He became the fourth of the Mameluke Sultans, and reigned from 1259 to 1276. The two great objects of his life were the repression of the Tartars and the expulsion of the Christians from Syria, so that his reign was one of constant war and enormous activity. William of Tripoli, in the work above mentioned, says: "Bondogar, as a soldier, was not inferior to Julius Caesar, nor in malignity to Nero." He admits, however, that the Sultan was sober, chaste, just to his own people, and even kind to his Christian subjects; whilst Makrizi calls him one of the best princes that ever reigned over Musulmans. Yet if we take Bibars as painted by this admiring historian and by other Arabic documents, the second of Friar William's comparisons is justified, for he seems almost a devil in malignity as well as in activity. More than once he played tennis at Damascus and Cairo within the same week. A strange sample of the man is the letter which he wrote to Boemond, Prince of Antioch and Tripoli, to announce to him the capture of the former city. After an ironically polite address to Boemond as having by the loss of his great city had his title changed from Princeship (*Al-Brensíyah*) to Countship (*Al-Komasíyah*), and describing his own devastations round Tripoli, he comes to the attack of Antioch: "We carried the place, sword in hand, at the 4th hour of Saturday, the 4th day of Ramadhán, ... Hadst thou but seen thy Knights trodden under the hoofs of the horses! thy palaces invaded by plunderers and ransacked for booty! thy treasures weighed out by the hundredweight! thy ladies (*Dámátaka*, 'tes DAMES') bought and sold with thine own gear, at four for a dinár! hadst thou but seen thy churches demolished, thy crosses sawn in sunder, thy garbled Gospels hawked about before the sun, the tombs of thy nobles cast to the ground; thy foe the Moslem treading thy Holy of the Holies; the monk, the priest, the deacon slaughtered on the Altar; the rich given up to misery; princes of royal blood reduced to slavery! Couldst thou but have seen the flames devouring thy halls; thy dead cast into the fires temporal with the fires eternal hard at hand; the churches of Paul and of Cosmas rocking and going down——, then wouldst thou have said, 'Would God that I were dust!' . . . . As not a man hath escaped to tell thee the tale, I TELL IT THEE!"

A little later, when a mission went to treat with Boemond, Bibars himself accompanied it in disguise, to have a look at the defences of Tripoli. In drawing out the terms, the Envoys styled Boemond *Count*, not *Prince*, as in the letter just quoted. He lost patience at their persistence, and made a movement which alarmed them. Bibars nudged the Envoy Mohiuddin (who tells the story) with his foot to give up the point, and the treaty was made. On their way back the Sultan laughed heartily at their narrow escape, "sending to the devil all the counts and princes on the face of the earth."

(*Quatremère's Makrizi*, II. 92-101, and 190 *seqq.*; *J. As.* sér. I. tom. xi. p. 89; *D'Ohsson*, III. 459-474; *Marino Sanuto* in Bongars, 224-226, etc.)

NOTE 4.—The ruling Master of the Temple was Thomas Berard (1256-1273), but there is little detail about the Order in the East at this time. They had, however, considerable possessions and great influence in Cilician Armenia, and how much they were mixed up in its affairs is shown by a circumstance related by Makrizi. In 1285, when Sultan Mansúr, the successor of Bundúkdár, was besieging the Castle of Markab, there arrived in Camp the Commander of the Temple (*Kamandúr-ul Dewet*) of the Country of Armenia, charged to negotiate on the part of the King of Sis (*i.e.* of Lesser Armenia, Leon III. 1268-1289, successor of Hayton I. 1224-1268), and bringing presents from him and from the Master of the Temple, Berard's successor, William de Beaujeu (1273-1291). (III. 201.)—H. Y. and H. C.

CHAPTER XIII.

How Messer Nicolo and Messer Maffeo Polo, accompanied by Mark, travelled to the Court of the Great Kaan.

So the Two Brothers, and Mark along with them, proceeded on their way, and journeying on, summer and winter, came at length to the Great Kaan, who was then at a certain rich and great city, called KEMENFU.[1] As to what they met with on the road, whether in going or coming, we shall give no particulars at present, because we are going to tell you all those details in regular order in the after part of this Book. Their journey back to the Kaan occupied a good three years and a half, owing to the bad weather and severe cold that they encountered. And let me tell you in good sooth that when the Great Kaan heard that Messers Nicolo and Maffeo Polo were on their way back, he sent people a journey of full 40 days to meet them ; and on this journey, as on their former one, they were honourably entertained upon the road, and supplied with all that they required.

NOTE I.—The French texts read *Clemeinfu*, Ramusio *Clemenfu*. The Pucci MS. guides us to the correct reading, having *Chemensu* (*Kemensu*) for *Chemenfu*. KAIPINGFU, meaning something like "City of Peace," and called by Rashiduddin *Kaiminfu* (whereby we see that Polo as usual adopted the Persian form of the name), was a city founded in 1256, four years before Kublai's accession, some distance to the north of the Chinese wall. It became Kublai's favourite summer residence, and was styled from 1264 *Shangtu* or "Upper Court." (See *infra*, Bk. I. ch. lxi.) It was known to the Mongols, apparently by a combination of the two names, as *Shangdu Keibung*. It appears in D'Anville's map under the name of *Djao-Naiman Sumé*,

Dr. Bushell, who visited Shangtu in 1872, makes it 1103 *li* (367 miles) by road distance *via* Kalgan from Peking. The busy town of Dolonnúr lies 26 miles S.E. of it, and according to Kiepert's *Asia* that place is about 180 miles in a direct line north of Peking.

(See *Klaproth* in *J. As.* XI. 365 ; *Gaubil*, p. 115 ; *Cathay*, p. 260 ; *J. R. G. S.* vol. xliii.)

CHAPTER XIV.

How Messer Nicolo and Messer Maffeo Polo and Marco presented themselves before the Great Kaan.

And what shall I tell you? when the Two Brothers and Mark had arrived at that great city, they went to the Imperial Palace, and there they found the Sovereign attended by a great company of Barons. So they bent the knee before him, and paid their respects to him, with all possible reverence [prostrating themselves on the ground]. Then the Lord bade them stand up, and treated them with great honour, showing great pleasure at their coming, and asked many questions as to their welfare, and how they had sped. They replied that they had in verity sped well, seeing that they found the Kaan well and safe. Then they presented the credentials and letters which they had received from the Pope, which pleased him right well; and after that they produced the Oil from the Sepulchre, and at that also he was very glad, for he set great store thereby. And next, spying Mark, who was then a young gallant,[1] he asked who was that in their company? "Sire," said his father, Messer Nicolo, "'tis my son and your liegeman."[2] "Welcome is he too," quoth the Emperor. And why should I make a long story? There was great rejoicing at the Court because of their arrival ; and they met with attention and honour from everybody.

So there they abode at the Court with the other Barons.

NOTE 1.—"*Joenne Bacheler.*"

NOTE 2.—"*Sire, il est mon filz et vostre* homme." The last word in the sense which gives us the word *homage.* Thus in the miracle play of Theophilus (13th century), the Devil says to Theophilus :—

" Or joing
Tes mains, et si devien *mes hom.*
Theoph. Vez ci que je vous faz *hommage.*"

So *infra* (Bk. I. ch. xlvii.) Aung Khan is made to say of Chinghiz : " *Il est* mon homes *et mon serf.*" (See also Bk. II. ch. iv. note.) St. Lewis said of the peace he had made with Henry III. : " Il m'est mout grant honneur en la paix que je foiz au Roy d'Angleterre pour ce qu'il est *mon home,* ce que n'estoit pas devant." And Joinville says with regard to the king, " Je ne voz faire point de serement, car je n'estoie pas *son home*" (being a vassal of Champagne). A famous Saturday Reviewer quotes the term applied to a lady : " *Eddeva puella* homo *Stigandi Archiepiscopi.*" (*Théâtre Français au Moyen Age,* p. 145 ; *Joinville,* pp. 21, 37 ; *S. R.,* 6th September, 1873, p. 305.)

═══════════   ═══════════

CHAPTER XV.

How THE EMPEROR SENT MARK ON AN EMBASSY OF HIS.

Now it came to pass that Marco, the son of Messer Nicolo, sped wondrously in learning the customs of the Tartars, as well as their language, their manner of writing, and their practice of war ; in fact he came in brief space to know several languages, and four sundry written characters. And he was discreet and prudent in every way, insomuch that the Emperor held him in great esteem.[1] And so when he discerned Mark to have so much sense, and to conduct himself so well and beseemingly, he sent him on an ambassage of his, to a country which was a good six months' journey distant.[2] The young gallant executed his commission well and with discretion. Now he had taken note on several occasions that when the Prince's ambassadors returned from different parts of the world, they were able to tell him about nothing except the business on which they had gone, and that the Prince in consequence held them for no better than fools and dolts, and would say : " I had far liever hearken about the strange things, and the manners of the different countries you have seen, than merely be told of the business you went upon ; "—for he took great delight in hearing of the affairs of strange countries. Mark therefore, as he went and returned, took great pains to learn about all kinds of different matters in the countries which he visited, in order to be able to tell about them to the Great Kaan.[3]

Note 1.—The word Emperor stands here for *Seigneur*.

What the four characters acquired by Marco were is open to discussion.

The Chronicle of the Mongol Emperors rendered by Gaubil mentions, as characters used in their Empire, the Uíghúr, the Persian and Arabic, that of the Lamas (Tibetan), that of the Niuché, introduced by the Kin Dynasty, the Khitán, and the *Báshpah* character, a syllabic alphabet arranged, on the basis of the Tibetan and Sanskrit letters chiefly, by a learned chief Lama so-called, under the orders of Kublai, and established by edict in 1269 as the official character. Coins bearing this character, and dating from 1308 to 1354, are extant. The forms of the Niuché and Khitán were devised in imitation of Chinese writing, but are supposed to be syllabic. Of the Khitán but one inscription was known, and no key. "The Khitan had two national scripts, the 'small characters' (*hsiao tzŭ*) and the 'large characters' (*ta tzŭ*)." S. W. Bushell, *Insc. in the Juchen and Allied Scripts*, Cong. des Orientalistes, Paris, 1897.—*Die Sprache und Schrift der Juchen*, von Dr W. Grube, Leipzig, 1896, from a polyglot MS. dictionary, discovered by Dr F. Hirth and now kept in the Royal Library, Berlin.—H. Y. and H. C.

Chinghiz and his first successors used the Uíghúr, and sometimes the Chinese character. Of the Uíghúr character we give a specimen in Bk. IV. It is of Syriac origin, undoubtedly introduced into Eastern Turkestan by the early Nestorian missions, probably in the 8th or 9th century. The oldest known example of this character so applied, the *Kudatku Bilik*, a didactic poem in Uíghúr (a branch of Oriental Turkish), dating from A.D. 1069, was published by Prof. Vámbéry in 1870. A new edition of the *Kudatku Bilik* was published at St. Petersburg, in 1891, by Dr. W. Radloff. Vámbéry had a pleasing illustration of the origin of the Uíghúr character, when he received a visit at Pesth from certain Nestorians of Urumia on a begging tour. On being shown the original MS. of the *Kudatku Bilik*, they read the character easily, whilst much to their astonishment they could not understand a word of what was written. This Uíghúr is the basis of the modern Mongol and Manchu characters. (Cf. E. Bretschneider, *Mediæval Researches*, I. pp. 236, 263.) —H. Y. and H. C.

[At the village of Keuyung Kwan, 40 miles north of Peking, in the sub-prefecture of Ch'ang Ping, in the Chih-li province, the road from Peking to Kalgan runs beyond the pass of Nankau, under an archway, a view of which will be found at the end of this volume, on which were engraved, in 1345, two large inscriptions in six different languages : Sanskrit, Tibetan, Mongol, *Báshpah*, Uíghúr, Chinese, and a language unknown till recently. Mr Wylie's kindness enabled Sir Henry Yule to present a specimen of this. (A much better facsimile of these inscriptions than Wylie's having since been published by Prince Roland Bonaparte in his valuable *Recueil des Documents de*

Hexaglot Inscription on the East side of the Kiu-Yong Kwan.

[To face p. 29.

l'Époque Mongole, this latter is, by permission, here reproduced.) The Chinese and Mongol inscriptions have been translated by M. Ed. Chavannes; the Tibetan by M. Sylvain Lévi (*Jour. Asiat.*, Sept.-Oct. 1894, pp. 354-373); the Uíghúr, by Prof. W. Radloff (*Ibid.* Nov.-Dec. 1894, pp. 546, 550); the Mongol by Prof. G. Huth. (*Ibid.* Mars-Avril 1895, pp. 351-360.) The sixth language was supposed by A. Wylie (*J. R. A. S.* vol. xvii. p. 331, and N.S., vol. v. p. 14) to be Neuchih, Niuché, Niuchen or Juchen.   M. Devéria has shown that the inscription is written in *Si Hia*, or the language of Tangut, and gave a facsimile of a stone stèle (*pei*) in this language kept in the great Monastery of the Clouds (Ta Yun Ssŭ) at Liangchau in Kansuh, together with a translation of the Chinese text, engraved on the reverse side of the slab.   M. Devéria thinks that this writing was borrowed by the Kings of Tangut from the one derived in 920 by the Khitans from the Chinese. (*Stèle Si-Hia de Leang-tcheou.* . . . *J. As.*, 1898; *L'écriture du royaumes de Si-Hia ou Tangout*, par M. Devéria. . . . Ext. des Mém. . . . présentés à l'Ac. des. Ins. et B. Let. 1ère Sér. XI., 1898.) Dr. S. W. Bushell in two papers (*Inscriptions in the Juchen and Allied Scripts, Actes du XI. Congrés des Orientalistes*, Paris, 1897, 2nd. sect., pp. 11, 35, and the *Hsi Hsia Dynasty of Tangut, their Money and their peculiar Script, J. China Br. R. A. S.*, xxx. N.S. No. 2, pp. 142, 160) has also made a special study of the same subject.   The Si Hia writing was adopted by Yuan Ho in 1036, on which occasion he changed the title of his reign to Ta Ch'ing, *i.e.* "Great Good Fortune." Unfortunately, both the late M. Devéria and Dr. S. W. Bushell have deciphered but few of the Si Hia characters.—H. C.]

The orders of the Great Kaan are stated to have been published habitually in six languages, viz., Mongol, Uíghúr, Arabic, Persian, Tangutan (Si-Hia), and Chinese. —H. Y. and H. C.

Gházán Khan of Persia is said to have understood Mongol, Arabic, Persian, something of Kashmiri, of Tibetan, of Chinese, and a little of the *Frank* tongue (probably French).

The annals of the Ming Dynasty, which succeeded the Mongols in China, mention the establishment in the 11th moon of the 5th year Yong-lo (1407) of the *Sse yi kwan*, a linguistic office for diplomatic purposes. The languages to be studied were Niuché, Mongol, Tibetan, Sanskrit, Bokharan (Persian?) Uíghúr, Burmese, and Siamese.   To these were added by the Manchu Dynasty two languages called *Papeh* and *Pehyih*, both dialects of the S. W. frontier. (See *infra*, Bk. II. ch. lvi.-lvii., and notes.)   Since 1382, however, official interpreters had to translate Mongol texts; they were selected among the Academicians, and their service (which was independent of the *Sse yi kwan* when this was created) was under the control of the *Han-lin-yuen*. There may have been similar institutions under the Yuen, but we have no proof of it. At all events, such an office could not then be called *Sse yi kwan* (*Sse yi*, Barbarians from four sides); Niuché (Niuchen) was taught in Yong-lo's office, but not Manchu. The *Sse yi kwan* must not be confounded with the *Hui t'ong kwan*, the office for the reception of tributary envoys, to which it was annexed in 1748. (*Gaubil*, p. 148; *Gold. 'Horde*, 184; *Ilchan.* II. 147; *Lockhart* in *J. R. G. S.* XXXVI. 152; *Koeppen*, II. 99; G. Devéria, *Hist. du Collége des Interprétes de Peking* in *Mélanges* Charles de Harlez, pp. 94-102; MS. Note of Prof. A. Vissière; *The Tangut Script in the Nan-K'ou Pass*, by Dr. S. W. Bushell, *China Review*, xxiv. II. pp. 65-68.)—H. Y. and H. C.

Pauthier supposes Mark's four acquisitions to have been *Báshpah-Mongol, Arabic, Uíghúr*, and *Chinese*. I entirely reject the theory. Sir H. Yule adds: "We shall see no reason to believe that he knew either language or character" [Chinese]. The blunders Polo made in saying that the name of the city, Suju, signifies in our tongue "Earth" and Kinsay "Heaven" show he did not know the Chinese characters, but we read in Bk. II. ch. lxviii.: "And Messer Marco Polo himself, of whom this Book speaks, did govern this city (Yanju) for three full years, by the order of the Great Kaan." It seems to me [H.C.] hardly possible that Marco could have for three years been governor of so important and so Chinese a city as Yangchau, in the heart of the Empire, without acquiring a knowledge of the spoken language.—H. C. The other three languages seem highly probable. The fourth may have been Tibetan. But it is more likely that he counted separately two varieties of the same character (*e.g.* of the Arabic and Persian) as two "*lettres de leur escriptures.*"—H. Y. and H. C.

NOTE 2.—[Ramusio here adds : " Ad und città, detta Carazan," which, as we shall see, refers to the Yun-nan Province.]—H. C.

NOTE 3.—From the context no doubt Marco's employments were honourable and confidential ; but *Commissioner* would perhaps better express them than Ambassador in the modern sense.   The word *Ilchi*, which was probably in his mind, was applied to a large variety of classes employed on the commissions of Government, as we may see from a passage of Rashiduddin in D'Ohsson, which says that " there were always to be found in every city from one to two hundred *Ilchis*, who forced the citizens to furnish them with free quarters," etc., III. 404.   (See also 485.)

CHAPTER XVI.

How Mark returned from the Mission whereon he had been sent.

WHEN Mark returned from his ambassage he presented himself before the Emperor, and after making his report of the business with which he was charged, and its successful accomplishment, he went on to give an account in a pleasant and intelligent manner of all the novelties and strange things that he had seen and heard ; insomuch that the Emperor and all such as heard his story were surprised, and said: "If this young man live, he will assuredly come to be a person of great worth and ability."   And so from that time forward he was always entitled MESSER MARCO POLO, and thus we shall style him henceforth in this Book of ours, as is but right.

Thereafter Messer Marco abode in the Kaan's employment some seventeen years, continually going and coming, hither and thither, on the missions that were entrusted to him by the Lord [and sometimes, with the permission and authority of the Great Kaan, on his own private affairs.]   And, as he knew all the sovereign's ways,

Hexaglot Inscription on the West side of the Kiu-Yong Kwan.

[*To face p.* 30.

like a sensible man he always took much pains to gather knowledge of anything that would be likely to interest him, and then on his return to Court he would relate everything in regular order, and thus the Emperor came to hold him in great love and favour. And for this reason also he would employ him the oftener on the most weighty and most distant of his missions. These Messer Marco ever carried out with discretion and success, God be thanked. So the Emperor became ever more partial to him, and treated him with the greater distinction, and kept him so close to his person that some of the Barons waxed very envious thereat. And thus it came about that Messer Marco Polo had knowledge of, or had actually visited, a greater number of the different countries of the World than any other man; the more that he was always giving his mind to get knowledge, and to spy out and enquire into everything in order to have matter to relate to the Lord.

CHAPTER XVII.

How Messer Nicolo, Messer Maffeo, and Messer Marco, asked leave of the Great Kaan to go their way.

When the Two Brothers and Mark had abode with the Lord all that time that you have been told [having meanwhile acquired great wealth in jewels and gold], they began among themselves to have thoughts about returning to their own country; and indeed it was time. [For, to say nothing of the length and infinite perils of the way, when they considered the Kaan's great age, they doubted whether, in the event of his death before their departure, they would ever be able to get home.[1]]

They applied to him several times for leave to go, presenting their request with great respect, but he had such a partiality for them, and liked so much to have them about him, that nothing on earth would persuade him to let them go.

Now it came to pass in those days that the Queen BOLGANA, wife of ARGON, Lord of the Levant, departed this life. And in her Will she had desired that no Lady should take her place, or succeed her as Argon's wife, except one of her own family [which existed in Cathay]. Argon therefore despatched three of his Barons, by name respectively OULATAY, APUSCA, and COJA, as ambassadors to the Great Kaan, attended by a very gallant company, in order to bring back as his bride a lady of the family of Queen Bolgana, his late wife.[2]

When these three Barons had reached the Court of the Great Kaan, they delivered their message, explaining wherefore they were come. The Kaan received them with all honour and hospitality, and then sent for a lady whose name was COCACHIN, who was of the family of the deceased Queen Bolgana. She was a maiden of 17, a very beautiful and charming person, and on her arrival at Court she was presented to the three Barons as the Lady chosen in compliance with their demand. They declared that the Lady pleased them well.[3]

Meanwhile Messer Marco chanced to return from India, whither he had gone as the Lord's ambassador, and made his report of all the different things that he had seen in his travels, and of the sundry seas over which he had voyaged. And the three Barons, having seen that Messer Nicolo, Messer Maffeo, and Messer Marco were not only Latins, but men of marvellous good sense withal, took thought among themselves to get the three to travel with them, their intention being to return to their country by sea, on account of the great fatigue of that long land journey for a lady.   And the ambassadors were the more desirous to have their company, as being aware that those three had great knowledge and experience of the Indian Sea and the countries by which they would have to pass, and especially Messer Marco.   So they went to the Great Kaan, and begged as a favour that he would send the three Latins with them, as it was their desire to return home by sea.

The Lord, having that great regard that I have mentioned for those three Latins, was very loath to do so [and his countenance showed great dissatisfaction]. But at last he did give them permission to depart, enjoining them to accompany the three Barons and the Lady.

Note 1.—Pegolotti, in his chapters on mercantile ventures to Cathay, refers to the dangers to which foreigners were always liable on the death of the reigning sovereign. (See *Cathay*, p. 292.)

Note 2.—Several ladies of the name of BULUGHAN ("Zibellina") have a place in Mongol-Persian history.   The one here indicated, a lady of great beauty and ability, was known as the *Great Khátún* (or Lady) Bulughan, and was (according to strange Mongol custom) the wife successively of Abáka and of his son ARGHUN, the Argon of the text, Mongol sovereign of Persia.   She died on the banks of the Kur in Georgia, 7th April, 1286.   She belonged to the Mongol tribe of Bayaut, and was the daughter of Huláku's Chief Secretary Gúgah.   (*Ilchan.* I. 374 *et passim ; Erdmann's Temudschin*, p. 216.)

The names of the Envoys, ULADAI, APUSHKA, and KOJA, are all names met with in Mongol history.   And Rashiduddin speaks of an Apushka of the Mongol Tribe of Urnaut, who on some occasion was sent as Envoy to the Great Kaan from Persia,—possibly the very person.   (See *Erdmann*, 205.)

Of the Lady Cocachin we shall speak below.

Note 3.—Ramusio here has the following passage, genuine no doubt : "So everything being ready, with a great escort to do honour to the bride of King Argon, the Ambassadors took leave and set forth.   But after travelling eight months by the same way that they had come, they found the roads closed, in consequence of wars lately broken out among certain Tartar Princes ; so being unable to proceed, they were compelled to return to the Court of the Great Kaan."

CHAPTER XVIII.

How the Two Brothers and Messer Marco took leave of the
Great Kaan, and returned to their own Country.

And when the Prince saw that the Two Brothers and
Messer Marco were ready to set forth, he called them
all three to his presence, and gave them two golden
Tablets of Authority, which should secure them liberty
of passage through all his dominions, and by means of
which, whithersoever they should go, all necessaries
would be provided for them, and for all their company,
and whatever they might choose to order.[1] He charged
them also with messages to the King of France, the
King of England,[2] the King of Spain, and the other
kings of Christendom. He then caused thirteen ships
to be equipt, each of which had four masts, and often
spread twelve sails.[3] And I could easily give you all
particulars about these, but as it would be so long an
affair I will not enter upon this now, but hereafter,
when time and place are suitable. [Among the said
ships were at least four or five that carried crews of 250
or 260 men.]

And when the ships had been equipt, the Three
Barons and the Lady, and the Two Brothers and
Messer Marco, took leave of the Great Kaan, and
went on board their ships with a great company of
people, and with all necessaries provided for two years
by the Emperor. They put forth to sea, and after sailing
for some three months they arrived at a certain Island
towards the South, which is called Java,[4] and in which
there are many wonderful things which we shall tell you
all about by-and-bye. Quitting this Island they con-

tinued to navigate the Sea of India for eighteen months more before they arrived whither they were bound, meeting on their way also with many marvels, of which we shall tell hereafter.

And when they got thither they found that Argon was dead, so the Lady was delivered to CASAN, his son.

But I should have told you that it is a fact that, when they embarked, they were in number some 600 persons, without counting the mariners; but nearly all died by the way, so that only eight survived.[5]

The sovereignty when they arrived was held by KIA-CATU, so they commended the Lady to him, and executed all their commission.   And when the Two Brothers and Messer Marco had executed their charge in full, and done all that the Great Kaan had enjoined on them in regard to the Lady, they took their leave and set out upon their journey.[6]   And before their departure, Kia-catu gave them four golden tablets of authority, two of which bore gerfalcons, one bore lions, whilst the fourth was plain, and having on them inscriptions which directed that the three Ambassadors should receive honour and service all through the land as if rendered to the Prince in person, and that horses and all provisions, and every-thing necessary, should be supplied to them.   And so they found in fact; for throughout the country they received ample and excellent supplies of everything needful; and many a time indeed, as I may tell you, they were furnished with 200 horsemen, more or less, to escort them on their way in safety.   And this was all the more needful because Kiacatu was not the legitimate Lord, and therefore the people had less scruple to do mischief than if they had had a lawful prince.[7]

Another thing too must be mentioned, which does credit to those three Ambassadors, and shows for what great personages they were held. The Great Kaan regarded them with such trust and affection, that he had confided to their charge the Queen Cocachin, as well as the daughter of the King of Manzi,[8] to conduct to Argon the Lord of all the Levant. And those two great ladies who were thus entrusted to them they watched over and guarded as if they had been daughters of their own, until they had transferred them to the hands of their Lord; whilst the ladies, young and fair as they were, looked on each of those three as a father, and obeyed them accordingly. Indeed, both Casan, who is now the reigning prince, and the Queen Cocachin his wife, have such a regard for the Envoys that there is nothing they would not do for them. And when the three Ambassadors took leave of that Lady to return to their own country, she wept for sorrow at the parting.

What more shall I say? Having left Kiacatu they travelled day by day till they came to Trebizond, and thence to Constantinople, from Constantinople to Negropont, and from Negropont to Venice. And this was in the year 1295 of Christ's Incarnation.

And now that I have rehearsed all the Prologue as you have heard, we shall begin the Book of the Description of the Divers Things that Messer Marco met with in his Travels.

NOTE 1.—On these plates or tablets, which have already been spoken of, a note will be found further on. (Bk. II. ch. vii.) Plano Carpini says of the Mongol practice in reference to royal messengers : "Nuncios, quoscunque et quotcunque, et ubicunque transmittit, oportet quod dent eis sine morâ equos subductitios et expensas" (669).

NOTE 2.—The mention of the King of England appears for the first time in Pauthier's text. Probably we shall never know if the communication reached him. But we have the record of several embassies in preceding and subsequent years from the Mongol Khans of Persia to the Kings of England; all with the view of obtaining co-operation in attack on the Egyptian Sultan. Such messages came from Ábáka in 1277; from Arghún in 1289 and 1291; from Gházán in 1302; from Oljaitu in 1307. (See *Rémusat* in *Mém. de l'Acad.* VII.)

NOTE 3.—Ramusio has "*nine* sails." Marsden thinks even this lower number an error of Ramusio's, as "it is well known that Chinese vessels do not carry any kind of topsail." This is, however, a mistake, for they do sometimes carry a small topsail of cotton cloth (and formerly, it would seem from Lecomte, even a topgallant sail at times), though only in quiet weather. And the evidence as to the number of sails carried by the great Chinese junks of the Middle Ages, which evidently made a great impression on Western foreigners, is irresistible. Friar Jordanus, who saw them in Malabar, says : "With a fair wind they carry ten sails ;" Ibn Batuta : "One of these great junks carries from three sails to twelve ;" Joseph, the Indian, speaking of those

Ancient Chinese War Vessel.

that traded to India in the 15th century : "They were very great, and had sometimes twelve sails, with innumerable rowers." (*Lecomte*, I. 389 ; *Fr. Jordanus*, Hak. Soc., p. 55 ; *Ibn Batuta,* IV. 91 ; *Novus Orbis*, p. 148.) A fuller account of these vessels is given at the beginning of Bk. III.

NOTE 4.—*I.e.* in this case Sumatra, as will appear hereafter. "It is quite possible for a fleet of fourteen junks which required to keep together to take three months at the present time to accomplish a similar voyage. A Chinese trader, who has come annually to Singapore in junks for many years, tells us that he has had as long a passage as sixty days, although the average is eighteen or twenty days." (*Logan* in *J. Ind. Archip.* II. 609.)

NOTE 5.—Ramusio's version here varies widely, and looks more probable: "From the day that they embarked until their arrival there died of mariners and others on board 600 persons ; and of the three ambassadors only one survived, whose name was Goza (*Coja*) ; but of the ladies and damsels died but one."

It is worth noting that in the case of an embassy sent to Cathay a few years later by Gházán Khan, on the return by this same route to Persia, the chief of the two Persian ambassadors, and the Great Khan's envoy, who was in company, both died by the way. Their voyage, too, seems to have been nearly as long as Polo's ; for they were seven years absent from Persia, and of these only four in China. (See *Wassáf* in *Elliot*, III. 47.)

NOTE 6.—Ramusio's version states that on learning Arghún's death (which they probably did on landing at Hormuz), they sent word of their arrival to Kiacatu, who directed them to conduct the lady to Casan, who was then in the region of the *Arbre Sec* (the Province of Khorasan) guarding the frontier passes with 60,000 men, and that they did so, and then turned back to Kiacatu (probably at Tabriz), and stayed at his Court nine months. Even the Geog. Text seems to imply that they had become personally known to Casan, and I have no doubt that Ramusio's statement is an authentic expansion of the original narrative by Marco himself, or on his authority.

Arghún Khan died 10th March, 1291. He was succeeded (23rd July) by his brother Kaikhátú (*Quiacatu* of Polo), who was put to death 24th March, 1295.

We learn from Hammer's History of the Ilkhans that when Gházán, the son of Arghún (*Casan* of Polo), who had the government of the Khorasan frontier, was on his return to his post from Tabriz, where his uncle Kaikhatu had refused to see him, " he met at Abher the ambassador whom he had sent to the Great Khan to obtain in marriage a relative of the Great Lady Bulghán. This envoy brought with him the Lady KÚKÁCHIN (our author's *Cocachin*), with presents from the Emperor, and the marriage was celebrated with due festivity." Abher lies a little west of Kazvín.

Hammer is not, I find, here copying from Wassáf, and I have not been able to procure a thorough search of the work of Rashiduddin, which probably was his authority. As well as the date can be made out from the History of the Ilkhans, Gházán must have met his bride towards the end of 1293, or quite the beginning of 1294. Rashiduddin in another place mentions the fair lady from Cathay ; " The *ordu* (or establishment) of Tukiti Khatun was given to KUKACHI KHATUN, who had been brought from the Kaan's Court, and who was a kinswoman of the late chief Queen Bulghán. Kúkáchi, the wife of the Padshah of Islam, Gházán Khan, died in the month of Shaban, 695," *i.e.* in June, 1296, so that the poor girl did not long survive her promotion. (See *Hammer's Ilch.* II. 20, and 8, and I. 273 ; and *Quatremère's Rashiduddin*, p. 97.) Kukachin was the name also of the wife of Chingkim, Kublai's favourite son ; but she was of the Kungurát tribe. (*Deguignes*, IV. 179.)

NOTE 7.—Here Ramusio's text says : " During this journey Messers Nicolo, Maffeo, and Marco heard the news that the Great Khan had departed this life ; and this caused them to give up all hope of returning to those parts."

NOTE 8.—This Princess of Manzi, or Southern China, is mentioned only in the Geog. Text and in the Crusca, which is based thereon. I find no notice of her among the wives of Gházán or otherwise.

On the fall of the capital of the Sung Dynasty—the Kinsay of Polo—in 1276, the Princesses of that Imperial family were sent to Peking, and were graciously treated by Kublai's favourite Queen, the Lady Jamui. This young lady was, no doubt, one of those captive princesses who had been brought up at the Court of Khánbálik. (See *De Mailla*, IX. 376, and *infra* Bk. II. ch. lxv., note.

BOOK FIRST.

ACCOUNT OF REGIONS VISITED OR HEARD OF ON THE JOURNEY FROM THE LESSER ARMENIA TO THE COURT OF THE GREAT KAAN AT CHANDU.

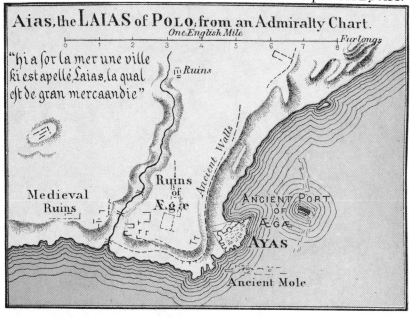

Aias, the LAIAS of POLO, from an Admiralty Chart.

One English Mile

"hi a ſor la mer une ville ki est apellé Laias, la qual est de gran mercaandie"

Ruins

Medieval Ruins

Ancient Walls

Ruins of Ægæ

Ancient Port of Ægæ

AYAS

Ancient Mole

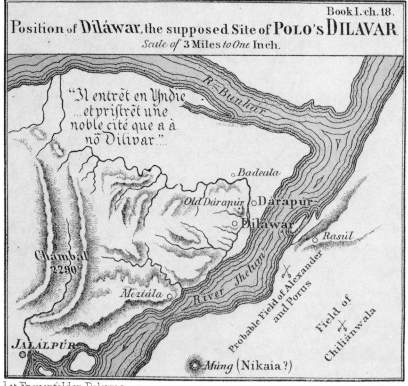

Position of Diláwar, the supposed Site of POLO's DILAVAR

Scale of 3 Miles to One Inch.

"Il entrēt en Yndie ...et priſtrēt une noble cité que a à nō Dilivar."

R. Bunhar

Badeala

Old Dárapur

Dárapur

Dilawar

Chambal 2290

Rasúl

Meziala

Probable Field of Alexander and Porus

River Jhelum

Field of Chiliánwala

JALÁLPUR

Múng (Nikaia?)

Lit. Frauenfelder, Palermo

[To face p. 41.

BOOK I.

———◆◆◆———

CHAPTER I.

HERE THE BOOK BEGINS; AND FIRST IT SPEAKS OF THE LESSER
HERMENIA.

THERE are two Hermenias, the Greater and the Less.
The Lesser Hermenia is governed by a certain King,
who maintains a just rule in his dominions, but is
himself subject to the Tartar.[1] The country contains
numerous towns and villages,[2] and has everything in
plenty; moreover, it is a great country for sport in
the chase of all manner of beasts and birds. It is,
however, by no means a healthy region, but grievously the reverse.[3] In days of old the nobles there
were valiant men, and did doughty deeds of arms; but
nowadays they are poor creatures, and good at
nought, unless it be at boozing; they are great at that.
Howbeit, they have a city upon the sea, which is called
LAYAS, at which there is a great trade. For you must
know that all the spicery, and the cloths of silk and gold,
and the other valuable wares that come from the interior,
are brought to that city. And the merchants of Venice
and Genoa, and other countries, come thither to sell their
goods, and to buy what they lack. And whatsoever persons would travel to the interior (of the East), merchants
or others, they take their way by this city of Layas.[4]

Having now told you about the Lesser Hermenia, we shall next tell you about Turcomania.

NOTE 1.—The *Petite Hermenie* of the Middle Ages was quite distinct from the Armenia Minor of the ancient geographers, which name the latter applied to the western portion of Armenia, west of the Euphrates, and immediately north of Cappadocia.

But when the old Armenian monarchy was broken up (1079-80), Rupen, a kinsman of the Bagratid Kings, with many of his countrymen, took refuge in the Taurus. His first descendants ruled as *barons*, a title adopted apparently from the Crusaders, but still preserved in Armenia. Leon, the great-great-grandson of Rupen, was consecrated King under the supremacy of the Pope and the Western Empire in 1198. The kingdom was at its zenith under Hetum or Hayton I., husband of Leon's daughter Isabel (1224-1269) ; he was, however, prudent enough to make an early submission to the Mongols, and remained ever staunch to them, which brought his territory constantly under the flail of Egypt. It included at one time all Cilicia, with many cities of Syria and the ancient Armenia Minor, of Isauria

Coin of King Hetum and his Queen Isabel.

and Cappadocia. The male line of Rupen becoming extinct in 1342, the kingdom passed to John de Lusignan, of the royal house of Cyprus, and in 1375 it was put an end to by the Sultan of Egypt. Leon VI., the ex-king, into whose mouth Froissart puts some extraordinary geography, had a pension of 1000*l*. a year granted him by our Richard II., and died at Paris in 1398.

The chief remaining vestige of this little monarchy is the continued existence of a *Catholicos* of part of the Armenian Church at Sis, which was the royal residence. Some Armenian communities still remain both in hills and plains ; and the former, the more independent and industrious, still speak a corrupt Armenian.

Polo's contemporary, Marino Sanuto, compares the kingdom of the Pope's faithful Armenians to one between the teeth of four fierce beasts, the *Lion* Tartar, the *Panther* Soldan, the Turkish *Wolf*, the Corsair *Serpent*.

(*Dulaurier*, in *J. As.* sér. V. tom. xvii. ; *St. Martin, Arm.* ; *Mar. San.* p. 32 ; *Froissart*, Bk. II. ch. xxii. *seqq.* ; *Langlois, V. en Cilicie*, 1861, p. 19.)

NOTE 2.—" *Maintes villes et maint chasteaux.*" This is a constantly recurring phrase, and I have generally translated it as here, believing *chasteaux* (*castelli*) to be used in the frequent old Italian sense of a *walled* village or small walled town, or like the Eastern *Kala'*, applied in Khorasan "to everything—town, village, or private residence—surrounded by a wall of earth." (*Ferrier*, p. 292 ; see also *A. Conolly*, I. p. 211.) Martini, in his *Atlas Sinensis*, uses " *Urbes, oppida,* castella," to indicate the three classes of Chinese administrative cities.

NOTE 3.—" *Enferme durement.*" So Marino Sanuto objects to Lesser Armenia as a place of debarkation for a crusade "*quia terra est infirma.*" Langlois, speaking of the Cilician plain : " In this region once so fair, now covered with swamps and brambles, fever decimates a population which is yearly diminishing, has nothing to oppose to the scourge but incurable apathy, and will end by disappearing altogether," etc. (*Voyage*, p. 65.) Cilician Armenia retains its reputation for sport, and is much frequented by our naval officers for that object. Ayas is noted for the extraordinary abundance of turtles.

NOTE 4.—The phrase twice used in this passage for the *Interior* is *Fra terre*, an Italianism (*Fra terra*, or, as it stands in the Geog. Latin, "*infra terram Orientis*"), which, however, Murray and Pauthier have read as an allusion to the *Euphrates*, an error based apparently on a marginal gloss in the published edition of the Soc. de Géographie. It is true that the province of Comagene under the Greek Empire got the name of *Euphratesia*, or in Arabic *Furátíyah*, but that was not in question here. The great trade of Ayas was with Tabriz, *viâ* Sivas, Erzingan, and Erzrum, as we see in Pegolotti. Elsewhere, too, in Polo we find the phrase *fra terre* used, where Euphrates could possibly have no concern, as in relation to India and Oman. (See Bk. III. chs. xxix. and xxxviii., and notes in each case.)

With regard to the phrase *spicery* here and elsewhere, it should be noted that the Italian *spezerie* included a vast deal more than ginger and other things "hot i' the mouth." In one of Pegolotti's lists of *spezerie* we find drugs, dye-stuffs, metals, wax, cotton, etc.

CHAPTER II.

CONCERNING THE PROVINCE OF TURCOMANIA.

IN Turcomania there are three classes of people. First, there are the Turcomans; these are worshippers of Mahommet, a rude people with an uncouth language of their own.[1] They dwell among mountains and downs where they find good pasture, for their occupation is cattle-keeping. Excellent horses, known as *Turquans*, are reared in their country, and also very valuable mules. The other two classes are the Armenians and the Greeks, who live mixt with the former in the towns and villages, occupying themselves with trade and handicrafts. They weave the finest and handsomest carpets in the world, and also a great quantity of fine and rich silks of cramoisy and other colours, and plenty of other stuffs. Their chief cities are CONIA, SAVAST [where the glorious Messer Saint Blaise suffered martyrdom], and CASARIA, besides many other towns and bishops' sees, of which we shall not speak at present, for it would be too long a matter. These people are subject to the

Tartar of the Levant as their Suzerain.[2]   We will now
leave this province, and speak of the Greater Armenia.

NOTE 1.—Ricold of Montecroce, a contemporary of Polo, calls the Turkmans
homines bestiales.   In our day Ainsworth notes of a Turkman village : "The dogs
were very ferocious ; . . . the people only a little better." (*J. R. G. S.* X. 292.)   The
ill report of the people of this region did not begin with the Turkmans, for the Emperor
Constantine Porphyrog. quotes a Greek proverb to the disparagement of the three
kappas, Cappadocia, Crete, and Cilicia.   (In *Banduri,* I. 6.)

NOTE 2.—In Turcomania Marco perhaps embraces a great part of Asia Minor,
but he especially means the territory of the decaying Seljukian monarchy, usually
then called by Asiatics *Rúm,* as the Ottoman Empire is now, and the capital of which
was Iconium, KUNIYAH, the Conia of the text, and Coyne of Joinville.   Ibn Batuta
calls the whole country Turkey (*Al-Turkiyah*), and the people *Turkmán;* exactly
likewise does Ricold (*Thurchia* and *Thurchimanni*).   Hayton's account of the various
classes of inhabitants is quite the same in substance as Polo's.   [The Turkmans emi-
grated from Turkestan to Asia Minor before the arrival of the Seljukid Turks.   "Their
villages," says Cuinet, *Turquie d'Asie,* II. p. 767, "are distinguished by the peculiarity
of the houses being built of sun-baked bricks, whereas it is the general habit in the country
to build them of earth or a kind of plaster, called *djès.*"—H. C.]   The migratory and
pastoral Turkmans still exist in this region, but the Kurds of like habits have taken
their place to a large extent.   The fine carpets and silk fabrics appear to be no longer
produced here, any more than the excellent horses of which Polo speaks, which must
have been the remains of the famous old breed of Cappadocia.   [It appears, however
(Vital Cuinet's *Turquie d'Asie,* I. p. 224), that fine carpets are still manufactured at
Koniah, also a kind of striped cotton cloth, called *Aladja.*—H. C.]
   A grant of privileges to the Genoese by Leon II., King of Lesser Armenia, dated
23rd December, 1288, alludes to the export of horses and mules, etc., from Ayas, and
specifies the duties upon them.   The horses now of repute in Asia as Turkman come
from the east of the Caspian.   And Asia Minor generally, once the mother of so many
breeds of high repute, is now poorer in horses than any province of the Ottoman empire.
   (*Pereg. Quat.* p. 114 ; *I. B.* II. 255 *seqq.; Hayton,* ch. xiii. ; *Liber Jurium Reip.
Januensis,* II. 184; *Tchihatcheff, As. Min.,* 2de partie, 631.)
   [The Seljukian Sultanate of Iconium or Rúm, was founded at the expense of the
Byzantines by Suleiman (1074-1081) ; the last three sovereigns of the dynasty con
temporaneous with Marco Polo are Ghiath ed-din Kaïkhosru III. (1267-1283), Ghiath
ed-din Mas'ud II. (1283-1294), Ala ed-din Kaïkobad III. (1294-1308), when this
kingdom was destroyed by the Mongols of Persia.   Privileges had been granted to
Venice by Ghiath ed-din Kaïkhosru I. (+1211), and his sons Izz ed-din Kaikaus
(1211-1220), and Ala ed-din Kaïkobad I. (1220-1237); the diploma of 1220 is unfortunately the only one of the three known to be preserved.   (Cf. Heyd, I. p. 302.)
—H. C.]
   Though the authors quoted above seem to make no distinction between Turks and
Turkmans, that which we still understand does appear to have been made in the 12th
century : ."That there may be some distinction, at least in name, between those who
made themselves a king, and thus achieved such glory, and those who still abide in
their primitive barbarism and adhere to their old way of life, the former are nowadays
termed *Turks,* the latter by their old name of *Turkomans.*" (*William of Tyre,* i. 7.)
   Casaria is KAISARÍYA, the ancient Caesareia of Cappadocia, close to the foot of
the great Mount Argaeus.   *Savast* is the Armenian form (*Sevasd*) of Sebaste, the
modern SIVAS.   The three cities, Iconium, Caesareia, and Sebaste, were metropolitan sees under the Catholicos of Sis.
   [The ruins of Sebaste are situated at about 6 miles to the east of modern Sivas,

near the village of Gavraz, on the *Kizil Irmak*. In the 11th century, the King of Armenia, Senecherim, made his capital of Sebaste. It belonged after to the Seljukid Turks, and was conquered in 1397 by Bayezid Ilderim with Tokat, Castambol and Sinope. (Cf. *Vital Cuinet.*)

One of the oldest churches in Sivas is St. George (*Sourp-Kévork*), occupied by the Greeks, but claimed by the Armenians; it is situated near the centre of the town, in what is called the "Black Earth," the spot where Timur is said to have massacred the garrison. A few steps north of St. George is the Church of St. Blasius, occupied by the Roman Catholic Armenians. The tomb of St. Blasius, however, is shown in another part of the town, near the citadel mount, and the ruins of a very beautiful Seljukian Medresseh. (From a MS. Note by Sir H. Yule. The information had been supplied by the American Missionaries to General Sir C. Wilson, and forwarded by him to Sir H. Yule.)

It must be remembered that at the time of the Seljuk Turks, there were four Medressehs at Sivas, and a university as famous as that of Amassia. Children to the number of 1000, each a bearer of a copy of the Koran, were crushed to death under the feet of the horses of Timur, and buried in the "Black Earth"; the garrison of 4000 soldiers were buried alive.

St. Blasius, Bishop of Sebaste, was martyred in 316 by order of Agricola, Governor of Cappadocia and Lesser Armenia, during the reign of Licinius. His feast is celebrated by the Latin Church on the 3rd of February, and by the Greek Church on the 11th of February. He is the patron of the Republic of Ragusa in Dalmatia, and in France of wool-carders.

At the village of Hullukluk, near Sivas, was born in 1676 Mekhitar, founder of the well-known Armenian Order, which has convents at Venice, Vienna, and Trieste.—H. C.]

CHAPTER III.

DESCRIPTION OF THE GREATER HERMENIA.

THIS is a great country. It begins at a city called ARZINGA, at which they weave the best buckrams in the world. It possesses also the best baths from natural springs that are anywhere to be found.[1] The people of the country are Armenians, and are subject to the Tartar. There are many towns and villages in the country, but the noblest of their cities is Arzinga, which is the See of an Archbishop, and then ARZIRON and ARZIZI.[2]

The country is indeed a passing great one, and in the summer it is frequented by the whole host of the Tartars of the Levant, because it then furnishes them with such excellent pasture for their cattle. But in winter the cold

is past all bounds, so in that season they quit this country
and go to a warmer region, where they find other good
pastures. [At a castle called PAIPURTH, that you pass in
going from Trebizond to Tauris, there is a very good
silver mine.[3]]

And you must know that it is in this country of
Armenia that the Ark of Noah exists on the top of a
certain great mountain [on the summit of which snow is
so constant that no one can ascend;[4] for the snow never
melts, and is constantly added to by new falls. Below,
however, the snow does melt, and runs down, producing
such rich and abundant herbage that in summer cattle
are sent to pasture from a long way round about, and it
never fails them. The melting snow also causes a great
amount of mud on the mountain].

The country is bounded on the south by a kingdom
called Mosul, the people of which are Jacobite and
Nestorian Christians, of whom I shall have more to tell
you presently. On the north it is bounded by the Land
of the Georgians, of whom also I shall speak. On the
confines towards Georgiana there is a fountain from
which oil springs in great abundance, insomuch that a
hundred shiploads might be taken from it at one time.
This oil is not good to use with food, but 'tis good to
burn, and is also used to anoint camels that have the
mange. People come from vast distances to fetch it, for
in all the countries round about they have no other oil.[5]

Now, having done with Great Armenia, we will tell
you of Georgiania.

NOTE 1.—[ERZINJAN, Erzinga, or Eriza, in the vilayet of Erzrum, was rebuilt in
1784, after having been destroyed by an earthquake. "Arzendjan," says Ibn Batuta,
II. p. 294, "is in possession of well-established markets; there are manufactured fine
stuffs, which are called after its name." It was at Erzinjan that was fought in 1244
the great battle, which placed the Seljuk Turks under the dependency of the Mongol
Khans.—H. C.] I do not find mention of its hot springs by modern travellers, but
Lazari says Armenians assured him of their existence. There are plenty of others

in Polo's route through the country, as at Ilija, close to Erzrum, and at Hássan Kalá.

The *Buckrams* of Arzinga are mentioned both by Pegolotti (*circa* 1340) and by Giov. d'Uzzano (1442). But what were they?

Buckram in the modern sense is a coarse open texture of cotton or hemp, loaded with gum, and used to stiffen certain articles of dress. But this was certainly *not* the mediæval sense. Nor is it easy to bring the mediæval uses of the term under a single explanation. Indeed Mr Marsh suggests that probably two different words have coalesced. Fr.-Michel says that *Bouqueran* was *at first* applied to a light cotton stuff of the nature of muslin, and *afterwards* to linen, but I do not see that he makes out this history of the application. Douet d'Arcq, in his *Comptes de l'Argenterie*, etc., explains the word simply in the modern sense, but there seems nothing in his text to bear this out.

A quotation in Raynouard's Romance Dictionary has " *Vestirs de polpra e de* bisso *que est* bocaran," where Raynouard renders *bisso* as *lin ;* a quotation in Ducange also makes Buckram the equivalent of Bissus ; and Michel quotes from an inventory of 1365, "*unam culcitram pinctam* (qu. punctam ?) *albam factam* de bisso *aliter* boquerant."

Mr. Marsh again produces quotations, in which the word is used as a proverbial example of *whiteness*, and inclines to think that it was a bleached cloth with a lustrous surface.

It certainly was not *necessarily* linen. Giovanni Villani, in a passage which is curious in more ways than one, tells how the citizens of Florence established races for their troops, and, among other prizes, was one which consisted of a *Bucherame di bambagine* (of cotton). Polo, near the end of the Book (Bk. III. ch. xxxiv.), speaking of Abyssinia, says, according to Pauthier's text : " *Et si y fait on moult beaux* bouquerans et autres draps de coton." The G. T. is, indeed, more ambiguous : " *Il hi se font maint biaus dras* banbacin e bocaran " (cotton *and* buckram). When, however, he uses the same expression with reference to the delicate stuffs woven on the coast of Telingana, there can be no doubt that a cotton texture is meant, and apparently a fine muslin. (See Bk. III. ch. xviii.) Buckram is *generally* named as an article of price, *chier bouquerant, rice boquerans*, etc., but not always, for Polo in one passage (Bk. II. ch. xlv.) seems to speak of it as the clothing of the poor people of Eastern Tibet.

Plano Carpini says the tunics of the Tartars were either of buckram (*bukeranum*), of *purpura* (a texture, perhaps velvet), or of *baudekin*, a cloth of gold (pp. 614-615). When the envoys of the Old Man of the Mountian tried to bully St. Lewis, one had a case of daggers to be offered in defiance, another a *bouqueran* for a winding sheet. (*Joinville*, p. 136.)

In accounts of materials for the use of Anne Boleyn in the time of her prosperity, *bokeram* frequently appears for "lyning and taynting " (?) gowns, lining sleeves, cloaks, a bed, etc., but it can scarcely have been for mere stiffening, as the colour of the buckram is generally specified as the same as that of the dress.

A number of passages seem to point to a *quilted* material. Boccaccio (Day viii. Novel 10) speaks of a quilt (*coltre*) of the whitest buckram of Cyprus, and Uzzano enters buckram quilts (*coltre di Bucherame*) in a list of *Linajuoli*, or linen-draperies. Both his handbook and Pegolotti's state repeatedly that buckrams were sold by the piece or the half-score pieces—never by measure. In one of Michel's quotations (from *Baudouin de Sebourc*) we have :

> " Gaufer li fist premiers armer d'un auqueton
> Qui fu de *bougherant* et *plaine de bon coton.*"

Mr. Hewitt would appear to take the view that Buckram meant a quilted material ; for, quoting from a roll of purchases made for the Court of Edward I., an entry for Ten Buckrams to make sleeves of, he remarks, " The sleeves appear to have been of *pourpointerie,*" *i.e.* quilting. (*Ancient Armour*, I. 240.)

This signification would embrace a large number of passages in which the term is used, though certainly not all. It would account for the mode or sale by the piece, and frequent use of the expression *a* buckram, for its habitual application to *coltre* or counterpanes, its use in the *auqueton* of Baudouin, and in the jackets of Falstaff's "men in buckram," as well as its employment in the frocks of the Mongols and Tibetans. The winter *chapkan*, or long tunic, of Upper India, a form of dress which, I believe, correctly represents that of the Mongol hosts, and is probably derived from them, is almost universally of quilted cotton.* This signification would also facilitate the transfer of meaning to the substance now called buckram, for that is used as a *kind* of quilting.

The derivation of the word is very uncertain. Reiske says it is Arabic, *Abu-Kairám*, " Pannus cum intextis figuris " ; Wedgwood, attaching the modern meaning, that it is from It., *bucherare*, to pierce full of holes, which might be if *bucherare* could be used in the sense of *puntare*, or the French *piquer ;* Marsh connects it with the *bucking* of linen ; and D'Avezac thinks it was a stuff that took its name from *Bokhara.* If the name be local, as so many names of stuffs are, the French form rather suggests *Bulgaria.* [Heyd, II. 703, says that Buckram (Bucherame) was principally manufactured at Erzinjan (Armenia), Mush, and Mardin (Kurdistan), Ispahan (Persia), and in India, etc. It was shipped to the west at Constantinople, Satalia, Acre, and Famagusta ; the name is derived from Bokhara.—H. C.]

(*Della Decima,* III. 18, 149, 65, 74, 212, etc. ; IV. 4, 5, 6, 212 ; *Reiske's* Notes to *Const. Porphyrogen.* II. ; *D'Avezac,* p. 524 ; *Vocab. Univ. Ital. ; Franc.-Michel, Recherches,* etc. II. 29 *seqq.; Philobiblon Soc. Miscell.* VI. ; *Marsh's Wedgwood's Etym. Dict.* sub voce.)

Castle of Baiburt.

NOTE 2.—Arziron is ERZRUM, which, even in Tournefort's time, the Franks called *Erzeron* (III. 126) ; [it was named *Garine*, then *Theodosiopolis*, in honour of

---

* Polo's contemporary, the Indian Poet Amír Khusrú, puts in the mouth of his king Kaikobád, a contemptuous gibe at the Mongols with their cotton-quilted dresses. (*Elliot,* III. p. 526.)

Theodosius the Great; the present name was given by the Seljukid Turks, and it means " Roman Country "; it was taken by Chinghiz Khan and Timur, but neither kept it long.   Odorico (*Cathay*, I. p. 46), speaking of this city, says it " is mighty cold." (See also on the low temperature of the place, Tournefort, *Voyage du Levant*, II. pp. 258-259.)   Arzizi, ARJISH, in the vilayet of Van, was destroyed in the middle of the 19th century; it was situated on the road from Van to Erzrum.   Arjish Kalá was one of the ancient capitals of the Kingdom of Armenia ; it was conquered by Toghrul I., who made it his residence. (Cf. VitalʹCuinet, *Turquie d'Asie*, II. p. 710). —H. C.]

Arjish is the ancient *Arsissa*, which gave the Lake Van one of its names.   It is now little more than a decayed castle, with a village inside.

Notices of Kuniyah, Kaisariya, Sivas, Arzan-ar-Rumi, Arzangan, and Arjish, will be found in Polo's contemporary Abulfeda. (See *Büsching*, IV. 303-311.)

NOTE 3.—Paipurth, or Baiburt, on the high road between Trebizond and Erzrum, was, according to Neumann, an Armenian fortress in the first century, and, according to Ritter, the castle *Baiberdon* was fortified by Justinian.   It stands on a peninsular hill, encircled by the windings of the R. Charok.   [According to Ramusio's version Baiburt was the third relay from Trebizund to Tauris, and travellers on their way from one of these cities to the other passed under this stronghold.—H. C.]   The Russians, in retiring from it in 1829, blew up the greater part of the defences.   The nearest silver mines of which we find modern notice, are those of *Gumish-Khánah* (" Silverhouse "), about 35 miles N.W. of Baiburt ; they are more correctly mines of lead rich in silver, and were once largely worked.   But the *Masálak-al-absár* (14th century), besides these, speaks of two others in the same province, one of which was near *Bajert*.   This Quatremère reasonably would read *Babert* or Baiburt. (*Not. et Extraits*, XIII. i. 337 ; *Texier, Arménie*, I. 59.)

NOTE 4.—Josephus alludes to the belief that Noah's Ark still existed, and that pieces of the pitch were used as amulets. (*Ant.* I. 3. 6.)

Ararat (16,953 feet) was ascended, first by Prof. Parrot, September 1829 ; by Spasski Aotonomoff, August 1834; by Behrens, 1835 ; by Abich, 1845 ; by Seymour in 1848 ; by Khodzko, Khanikoff, and others, for trigonometrical and other scientific purposes, in August 1850.   It is characteristic of the account from which I take these notes (*Longrimoff*, in *Bull. Soc. Géog. Paris*, sér. IV. tom. i. p. 54), that whilst the writer's countrymen, Spasski and Behrens, were " moved by a noble curiosity," the Englishman is only admitted to have " gratified a tourist's whim " !

NOTE 5.—Though Mr. Khanikoff points out that springs of naphtha are abundant in the vicinity of Tiflis, the mention of *ship-loads* (in Ramusio indeed altered, but probably by the Editor, to *camel-loads*), and the vast quantities spoken of, point to the naphtha-wells of the Baku Peninsula on the Caspian.   Ricold speaks of their supplying the whole country as far as Baghdad, and Barbaro alludes to the practice of anointing camels with the oil.   The quantity collected from the springs about Baku was in 1819 estimated at 241,000 *poods* (nearly 4000 tons), the greater part of which went to Persia. (*Pereg. Quat.* p. 122 ; *Ramusio*, II. 109 ; *El. de Laprim.* 276 ; *V. du Chev. Gamba*, I. 298.)

[The phenomenal rise in the production of the Baku oil-fields between 1890-1900, may be seen at a glance from the Official Statistics where the total output for 1900 is given as 601,000,000 poods, about 9,500,000 tons. (Cf. *Petroleum*, No. 42, vol. ii. p. 13.)]

CHAPTER IV.

Of Georgiania and the Kings thereof.

In Georgiania there is a King called David Melic, which is as much as to say " David King "; he is subject to the Tartar.[1]   In old times all the kings were born with the figure of an eagle upon the right shoulder.   The people are very handsome, capital archers, and most valiant soldiers.   They are Christians of the Greek Rite, and have a fashion of wearing their hair cropped, like Churchmen.[2]

This is the country beyond which Alexander could not pass when he wished to penetrate to the region of the Ponent, because that the defile was so narrow and perilous, the sea lying on the one hand, and on the other lofty mountains impassable to horsemen.   The strait extends like this for four leagues, and a handful of people might hold it against all the world.   Alexander caused a very strong tower to be built there, to prevent the people beyond from passing to attack him, and this got the name of the Iron Gate.   This is the place that the Book of Alexander speaks of, when it tells us how he shut up the Tartars between two mountains; not that they were really Tartars, however, for there were no Tartars in those days, but they consisted of a race of people called Comanians and many besides.[3]

[In this province all the forests are of box-wood.[4]] There are numerous towns and villages, and silk is produced in great abundance.   They also weave cloths of gold, and all kinds of very fine silk stuffs.   The country produces the best goshawks in the world [which are called *Avigi*].[5]   It has indeed no lack of anything, and

Mediæval Georgian Fortress, from a drawing dated 1634.

"La probence est toute pleine de grant montagne et d'estroit pas et de fort."

the people live by trade and handicrafts. 'Tis a very mountainous region, and full of strait defiles and of fortresses, insomuch that the Tartars have never been able to subdue it out and out.

There is in this country a certain Convent of Nuns called St. Leonard's, about which I have to tell you a very wonderful circumstance. Near the church in question there is a great lake at the foot of a mountain, and in this lake are found no fish, great or small, throughout the year till Lent come. On the first day of Lent they find in it the finest fish in the world, and great store too thereof; and these continue to be found till Easter Eve. After that they are found no more till Lent come round again; and so 'tis every year. 'Tis really a passing great miracle![6]

That sea whereof I spoke as coming so near the mountains is called the Sea of GHEL or GHELAN, and extends about 700 miles.[7] It is twelve days' journey distant from any other sea, and into it flows the great River Euphrates and many others, whilst it is surrounded by mountains. Of late the merchants of Genoa have begun to navigate this sea, carrying ships across and launching them thereon. It is from the country on this sea also that the silk called *Ghellé* is brought.[8] [The said sea produces quantities of fish, especially sturgeon, at the river-mouths salmon, and other big kinds of fish.][9]

NOTE 1.—Ramusio has: "One part of the said province is subject to the Tartar, and the other part, owing to its fortresses, remains subject to the King David." We give an illustration of one of these mediæval Georgian fortresses, from a curious collection of MS. notices and drawings of Georgian subjects in the Municipal Library at Palermo, executed by a certain P. Cristoforo di Castelli of that city, who was a Theatine missionary in Georgia, in the first half of the 17th century.

The G. T. says the King was *always* called David. The Georgian Kings of the family of Bagratidae claimed descent from King David through a prince Shampath, said to have been sent north by Nebuchadnezzar; a descent which was usually asserted in their public documents. Timur in his Institutes mentions a suit of armour given him by the King of Georgia as forged by the hand of the Psalmist King. David is a very frequent name in their royal lists. [The dynasty of the Bagratidae, which was founded in 786 by Ashod, and lasted until the annexation of Georgia by Russia on the 18th January, 1801, had nine reigning princes named David. During the second half of the 12th century the princes were : Dawith (David) IV. Narin (1247–1259), Dawith V. (1243–1272), Dimitri II. Thawdadebuli (1272–1289), Wakhtang II. (1289–1292), Dawith VI. (1292–1308).—H. C.]   There were two princes of that name, David, who shared Georgia between them under the decision of the Great Kaan in 1246, and one of them, who survived to 1269, is probably meant here.   The name of David was borne by the last titular King of Georgia, who ceded his rights to Russia in 1801.   It is probable, however, as Marsden has suggested, that the statement about the King *always* being called David arose in part out of some confusion with the title of *Dadian*, which, according to Chardin (and also to P. di Castelli), was always assumed by the Princes of Mingrelia, or Colchis as the latter calls it.   Chardin refers this title to the Persian *Dád*, "equity."   To a portrait of "Alexander, King of Iberia," or Georgia Proper, Castelli attaches the following inscription, giving apparently his official style : "With the sceptre of David, Crowned by Heaven, First King of the Orient and of the World, King of Israel," adding, "They say that he has on his shoulder a small mark of a cross, '*Factus est principatus super humerum ejus*,' and they add that he has all his ribs in one piece, and not divided."   In another place he notes that when attending the King in illness his curiosity moved him strongly to ask if these things were true, but he thought better of it! (*Khanikoff; Jour. As.* IX. 370, XI. 291, etc. ; *Tim. Instit.* p. 143 ; *Castelli* MSS.)

[A descendant of these Princes was in St. Petersburg about 1870.   He wore the Russian uniform, and bore the title of Prince Bagration—Mukransky.]

NOTE 2.—This fashion of tonsure is mentioned by Barbaro and Chardin.   The latter speaks strongly of the beauty of both sexes, as does Della Valle, and most modern travellers concur.

NOTE 3.—This refers to the Pass of Derbend, apparently the Sarmatic Gates of Ptolemy, and *Claustra Caspiorum* of Tacitus, known to the Arab geographers as the "Gate of Gates" (*Báb-ul-abwáb*), but which is still called in Turkish *Demír-Kápi*, or the Iron Gate, and to the ancient Wall that runs from the Castle of Derbend along the ridges of Caucasus, called in the East *Sadd-i-Iskandar*, the Rampart of Alexander   Bayer thinks the wall was probably built originally by one of the Antiochi, and renewed by the Sassanian Kobad or his son Naoshirwan.   It is ascribed to the latter by Abulfeda ; and according to Klaproth's extracts from the *Derbend Námah*, Naoshirwan completed the fortress of Derbend in A.D. 542, whilst he and his father together had erected 360 towers upon the Caucasian Wall which extended to the Gate of the Alans (*i.e.* the Pass of Dariel).   Mas'údi says that the wall extended for 40 parasangs over the steepest summits and deepest gorges.   The Russians must have gained some knowledge as to the actual existence and extent of the remains of this great work, but I have not been able to meet with any modern information of a very precise kind.   According to a quotation from *Reinegg's Kaukasus* (I. 120, a work which I have not been able to consult), the remains of defences can be traced for many miles, and are in some places as much as 120 feet high.   M. Moynet indeed, in the *Tour du Monde* (I. 122), states that he traced the wall to a distance of 27 versts (18 miles) from Derbend, but unfortunately, instead of describing remains of such high interest from his own observation, he cites a description written by Alex. Dumas, which he says is quite accurate.

[" To the west of Narin-Kaleh, a fortress which from the top of a promontory rises above the city, the wall, strengthened from distance to distance by large towers, follows the ridge of the mountains, descends into the ravines, and ascends the slopes to take root on some remote peak.   If the natives were to be believed, this wall, which, however, no longer has any strategetical importance, had formerly its towers bristling upon the Caucasus chain from one sea to another ; at least, this rampart did protect all the plains at the foot of the eastern Caucasus, since vestiges were found up to 30 kilometres from Derbend." (*Reclus, Asie russe*, p. 160.) It has belonged to Russia since 1813. The first European traveller who mentions it is Benjamin of Tudela.

Bretschneider (II. p. 117) observes: "Yule complains that he was not able to find any modern information regarding the famous Caucasian Wall which begins at Derbend. I may therefore observe that interesting details on the subject are found in Legkobytov's *Survey of the Russian Dominions beyond the Caucasus* (in Russian), 1836, vol. iv. pp. 158-161, and in Dubois de Montpéreux's *Voyage autour du Caucase*, 1840, vol. iv. pp. 291-298, from which I shall give here an abstract."

(He then proceeds to give an abstract, of which the following is a part :)

" The famous *Dagh bary* (mountain wall) now begins at the village of *Djelgan*, 4 versts south-west of Derbend, but we know that as late as the beginning of the last century it could be traced down to the southern gate of the city. This ancient wall then stretches westward to the high mountains of Tabasseran (it seems the Tabarestan of Mas'údi) . . . Dubois de Montpéreux enumerates the following sites of remains of the wall :—In the famous defile of *Dariel*, north-east of Kazbek. In the valley of the *Assai* river, near Wapila, about 35 versts north-east of Dariel. In the valley of the Kizil river, about 15 versts north-west of Kazbek. Farther west, in the valley of the *Fiag* or *Pog* river, between *Lacz* and *Khilak*. From this place farther west about 25 versts, in the valley of the *Arredon* river, in the district of *Valaghir*. Finally, the westernmost section of the Caucasian Wall has been preserved, which was evidently intended to shut up the maritime defile of *Gagry*, on the Black Sea."—H. C.]

There is another wall claiming the title of *Sadd-i-Iskandar* at the S.E. angle of the Caspian. This has been particularly spoken of by Vámbéry, who followed its traces from S.W. to N.E. for upwards of 40 miles. (See his *Travels in C. Asia*, 54 seqq., and *Julius Braun* in the *Ausland*, No. 22, of 1869.)

Yule (II. pp. 537-538) says, " To the same friendly correspondent [Professor Bruun] I owe the following additional particulars on this interesting subject, extracted from *Eichwald, Periplus des Kasp. M.* I. 128.

" ' At the point on the mountain, at the extremity of the fortress (of Derbend), where the double wall terminates, there begins a single wall constructed in the same style, only this no longer runs in a straight line, but accommodates itself to the contour of the hill, turning now to the north and now to the south. At first it is quite destroyed, and showed the most scanty vestiges, a few small heaps of stones or traces of towers, but all extending in a general bearing from east to west. . . . It is not till you get 4 versts from Derbend, in traversing the mountains, that you come upon a continuous wall. Thenceforward you can follow it over the successive ridges . . . and through several villages chiefly occupied by the Tartar hill-people. The wall . . . makes many windings, and every ¾ verst it exhibits substantial towers like those of the city-wall, crested with loop-holes. Some of these are still in tolerably good condition ; others have fallen, and with the wall itself have left but slight vestiges.'

" Eichwald altogether followed it up about 18 versts (12 miles) not venturing to proceed further. In later days this cannot have been difficult, but my kind correspondent had not been able to lay his hand on information.

" A letter from Mr. Eugene Schuyler communicates some notes regarding inscriptions that have been found at and near Derbend, embracing Cufic of A.D. 465, Pehlvi, and even Cuneiform. Alluding to the fact that the other *Iron-gate*, south of Shahrsabz, was called also *Kalugah*, or *Kohlugah*, he adds : ' I don't know what that means, nor do I know if the Russian Kaluga, south-west of Moscow, has anything to do with it, but I am told there is a Russian popular song, of which two lines run :

> ' " Ah Derbend, Derbend Kaluga,
> Derbend my little Treasure ! " '

View of Derbend.

"Alexandre ne poit paser quand il bost aler au Ponent . . . . ear de l'nn les est la mer, et de l'autre est gran montagne que ne se poent cabaucher. La bie est mout estroit entre la montagne et la mer."

" I may observe that I have seen it lately pointed out that *Kotuga* is a Mongol word signifying a *barrier ;* and I see that Timkowski (I. 288) gives the same explanation of *Kalgan,* the name applied by Mongols and Russians to the gate in the Great Wall, called Chang-kia-Kau by the Chinese, leading to Kiakhta."

The story alluded to by Polo is found in the mediæval romances of Alexander, and in the Pseudo-Callisthenes on which they are founded. The hero chases a number of impure cannibal nations within a mountain barrier, and prays that they may be shut up therein. The mountains draw together within a few cubits, and Alexander then builds up the gorge and closes it with gates of brass or iron. There were in all twenty-two nations with their kings, and the names of the nations were Gōth, Magōth, Anugi, Egēs, Exenach, etc. Godfrey of Viterbo speaks of them in his rhyming verses :—

> " Finibus Indorum species fuit una virorum ;
> Goth erat atque Magoth dictum cognomen eorum
>
> *   *   *   *   *   *
>
> Narrat Esias, Isidorus et Apocalypsis,
> Tangit et in titulis Magna Sibylla suis.
> Patribus ipsorum tumulus fuit venter eorum," etc.

Among the questions that the Jews are said to have put, in order to test Mahommed's prophetic character, was one series : "Who are Gog and Magog ? Where do they dwell ? What sort of rampart did Zu'lkarnain build between them and men ?" And in the Koran we find (ch. xviii. *The Cavern*) : " They will question thee, O Mahommed, regarding Zu'lkarnain. Reply : I will tell you his history"—and then follows the story of the erection of the Rampart of Yájúj and Májúj. In ch. xxi. again there is an allusion to their expected issue at the latter day. This last expectation was one of very old date. Thus the Cosmography of Aethicus, a work long believed (though erroneously) to have been abridged by St. Jerome, and therefore to be as old at least as the 4th century, says that the *Turks* of the race of Gog and Magog, a polluted nation, eating human flesh and feeding on all abominations, never washing, and never using wine, salt, nor wheat, shall come forth in the Day of Antichrist from where they lie shut up behind the Caspian Gates, and make horrid devastation. No wonder that the irruption of the Tartars into Europe, heard of at first with almost as much astonishment as such an event would produce now, was connected with this prophetic legend ! * The Emperor Frederic II., writing to Henry III. of England, says of the Tartars : " 'Tis said they are descended from the Ten Tribes who abandoned the Law of Moses, and worshipped the Golden Calf. They are the people whom Alexander Magnus shut up in the Caspian Mountains."

[See the chapter *Gog et Magog dans le roman en alexandrins*, in Paul Meyer's *Alexandre le Grand dans la Littérature française*, Paris, 1886, II. pp. 386-389.—H. C.]:

> " Gos et Margos i vienent de la tiere des Turs
> Et. cccc. m. hommes amenerent u plus,
> Il en jurent la mer dont sire est Neptunus
> Et le porte d'infier que garde Cerberus
> Que l'orguel d'Alixandre torneront a reüs
> Per çou les enclot puis es estres desus.
> Dusc' al tans Antecrist n'en istera mais nus."

According to some chroniclers, the Emperor Heraclius had already let loose the Shut-up Nations to aid him against the Persians, but it brought him no good, for he was beaten in spite of their aid, and died of grief.

---

* See Letter of Frederic to the Roman Senate, of 20th June, 1241, in *Bréholles.* Mahommedan writers, contemporary with the Mongol invasions, regarded these as a manifest sign of the approaching end of the world. (See Elliot's *Historians,* II. p. 265.)

The theory that the Tartars were Gog and Magog led to the Rampart of Alexander being confounded with the Wall of China (see *infra*, Bk. I. ch. lix.), or being relegated to the extreme N.E. of Asia, as we find it in the Carta Catalana.

These legends are referred to by Rabbi Benjamin, Hayton, Rubruquis, Ricold, Matthew Paris, and many more. Josephus indeed speaks of the Pass which Alexander fortified with gates of steel. But his saying that the King of Hyrcania was Lord of this Pass points to the Hyrcanian Gates of Northern Persia, or perhaps to the Wall of Gomushtapah, described by Vámbéry.

Ricold of Montecroce allows two arguments to connect the Tartars with the Jews who were shut up by Alexander; one that the Tartars hated the very name of Alexander, and could not bear to hear it; the other, that their manner of writing was very like the Chaldean, meaning apparently the Syriac (*ante*, p. 29). But he points out that they had no resemblance to Jews, and no knowledge of the law.

Edrisi relates how the Khalif Wathek sent one Salem the Dragoman to explore the Rampart of Gog and Magog. His route lay by Tiflis, the Alan country, and that of the Bashkirds, to the far north or north-east, and back by Samarkand. But the report of what he saw is pure fable.

In 1857, Dr. Bellew seems to have found the ancient belief in the legend still held by Afghan gentlemen at Kandahar.

At Gelath in Imeretia there still exists one valve of a large iron gate, traditionally said to be the relic of a pair brought as a trophy from Derbend by David, King of Georgia, called the Restorer (1089-1130). M. Brosset, however, has shown it to be the gate of Ganja, carried off in 1139.

(*Bayer* in *Comment. Petropol.* I. 401 *seqq.*; *Pseudo-Callisth.* by *Müller*, p. 138; *Gott. Viterb.* in *Pistorii Nidani Script. Germ.* II. 228; *Alexandriade*, pp. 310-311; *Pereg.* IV. p. 118; *Acad. des Insc. Divers Savans*, II. 483; *Edrisi*, II. 416-420, etc.)

NOTE 4.—The box-wood of the Abkhasian forests was so abundant, and formed so important an article of Genoese trade, as to give the name of *Chao de Bux* (Cavo di Bussi) to the bay of Bambor, N.W. of Sukum Kala', where the traffic was carried on. (See *Elie de Laprim.* 243.) Abulfeda also speaks of the Forest of Box (*Shard' ul-buḳs*) on the shores of the Black Sea, from which box-wood was exported to all parts of the world; but his indication of the exact locality is confused. (*Reinaud's Abulf.* I. 289.)

At the present time "Boxwood abounds on the southern coast of the Caspian, and large quantities are exported from near Resht to England and Russia. It is sent up the Volga to Tsaritzin, from thence by rail to the Don, and down that river to the Black Sea, from whence it is shipped to England." (*MS. Note*, H. Y.)

[Cf. V. Helm's *Cultivated Plants*, edited by J. S. Stallybrass, Lond., 1891, *The Box Tree*, pp. 176-179.—H. C.]

NOTE 5.—Jerome Cardan notices that "the best and biggest goshawks come from Armenia," a term often including Georgia and Caucasus. The name of the bird is perhaps the same as '*Afçi*, "Falco montanus." (See *Casiri*, I. 320.) Major St. John tells me that the *Terlán*, or goshawk, much used in Persia, is still generally brought from Caucasus. (*Cardan, de Rer. Varietate*, VII. 35.)

NOTE 6.—A letter of Warren Hastings, written shortly before his death, and after reading Marsden's Marco Polo, tells how a fish-breeder of Banbury warned him against putting pike into his fish-pond, saying, "If you should leave them where they are *till Shrove Tuesday* they will be sure to spawn, and then you will never get any other fish to breed in it." (*Romance of Travel*, I. 255.) Edward Webbe in his Travels (1590, reprinted 1868) tells us that in the "Land of Siria there is a River having great store of fish like unto Salmon-trouts, but no Jew can catch them, though either Christian and Turk shall catch them in abundance with great ease." The circumstance of fish being got only for a limited time in spring is noticed with reference to Lake Van both by Tavernier and Mr. Brant.

But the exact legend here reported is related (as M. Pauthier has already noticed) by Wilibrand of Oldenburg of a stream under the Castle of Adamodana, belonging to the Hospitallers, near Naversa (the ancient *Anazarbus*), in Cilicia under Taurus. And Khanikoff was told the same story of a lake in the district of Akhaltziké in Western Georgia, in regard to which he explains the substance of the phenomenon as a result of the rise of the lake's level by the melting of the snows, which often coincides with Lent.   I may add that Moorcroft was told respecting a sacred pond near Sir-i-Chashma, on the road from Kabul to Bamian, that the fish in the pond were not allowed to be touched, but that they were accustomed to desert it for the rivulet that ran through the valley regularly every year *on the day of the vernal equinox*, and it was then lawful to catch them.

Like circumstances would produce the same effect in a variety of lakes, and I have not been able to identify the convent of St. Leonard's.   Indeed Leonard (*Sant Lienard*, G. T.) seems no likely name for an Armenian Saint ; and the patroness of the convent (as she is of many others in that country) was perhaps Saint *Nina*, an eminent personage in the Armenian Church, whose tomb is still a place of pilgrimage ; or possibly St. *Helena*, for I see that the Russian maps show a place called *Elenovka* on the shores of Lake Sevan, N.E. of Erivan.   Ramusio's text, moreover, says that the lake was *four days in compass*, and this description will apply, I believe, to none but the lake just named.   This is, according to Monteith, 47 miles in length and 21 miles in breadth, and as far as I can make out he travelled round it in three very long marches.   Convents and churches on its shores are numerous, and a very ancient one occupies an island on the lake.   The lake is noted for its fish, especially magnificent trout.

(*Tavern.* Bk. III. ch. iii. ; *J. R. G. S.* X. 897 ; *Pereg. Quat.* p. 179 ; *Khanikoff*, 15 ; *Moorcroft*, II. 382 ; *J. R. G. S.* III. 40 *seqq.*)

Ramusio has : "In this province there is a fine city called TIFLIS, and round about it are many castles and walled villages.   It is inhabited by Christians, Armenians, Georgians, and some Saracens and Jews, but not many."

NOTE 7.—The name assigned by Marco to the Caspian, "Mer de Gheluchelan " or "Ghelachelan," has puzzled commentators.   I have no doubt that the interpretation adopted above is the correct one.   I suppose that Marco said that the sea was called "La Mer de Ghel ou (de) Ghelan," a name taken from the districts of the ancient *Gelae* on its south-western shores, called indifferently *Gil* or *Gilán*, just as many other regions of Asia have like duplicate titles (singular and plural), arising, I suppose, from the change of a *gentile* into a *local* name.   Such are Lár, Lárán, Khútl, Khutlán, etc., a class to which Badakhshán, Wakhán, Shaghnán, Mungán, Chaghánián, possibly Bámián, and many others have formerly belonged, as the adjectives in some cases surviving, *Badakhshi, Shaghni, Wákhi*, etc., show.[*]   The change exemplified in the induration of these *gentile plurals* into *local singulars* is everywhere traced in the passage from earlier to later geography.   The old Indian geographical lists, such as are preserved in the Puránas, and in Pliny's extracts from Megasthenes, are, in the main, lists of *peoples*, not of provinces, and even where the real name seems to be local a *gentile* form is often given.   So also *Tochari* and *Sogdi* are replaced by *Tokháristán* and *Sughd ;* the *Veneti* and *Taurini* by Venice and Turin ; the *Remi* and the *Parisii*, by Rheims and Paris ; *East-Saxons* and *South-Saxons* by Essex and Sussex ; not to mention the countless -*ings* that mark the tribal settlement of the Saxons in Britain.

Abulfeda, speaking of this territory, uses exactly Polo's phrase, saying that the districts in question are properly called *Kíl-o-Kílán*, but by the Arabs *Jíl-o-Jílán*. Teixeira gives the Persian name of the sea as *Darya Ghiláni*.   (See *Abulf.* in *Büsching*, v. 329.)

---

* When the first edition was published, I was not aware of remarks to like effect regarding names of this character by Sir H. Rawlinson in the *J. R. As. Soc.* vol. xi. pp. 64 and 103.

[The province of Gîl (Gílán), which is situated between the mountains and the Caspian Sea, and between the provinces of Azerbaíján and Mazandéran (H. C.)], gave name to the silk for which it was and is still famous, mentioned as *Ghelle* (*Gíli*) at the end of this chapter. This *Seta Ghella* is mentioned also by Pegolotti (pp. 212, 238, 301), and by Uzzano, with an odd transposition, as Seta *Leggi*, along with Seta *Masandroni*, *i.e.* from the adjoining province of Mazanderán (p. 192).   May not the Spanish *Geliz*, "a silk-dealer," which seems to have been a puzzle to etymologists, be connected with this ? (See *Dozy and Engelmann*, 2nd ed. p. 275.) [Prof. F. de Filippi (*Viaggo in Persia nel* 1862, . . . Milan, 1865, 8vo) speaks of the silk industry of Ghílán (pp. 295-296) as the principal product of the entire province.—H. C.]

The dimensions assigned to the Caspian in the text would be very correct if length were meant, but the Geog. Text with the same figure specifies *circuit* (*zire*).   Ramusio again has "a circuit of 2800 miles." Possibly the original reading was 2700 ; but this would be in excess.

Note 8.—The Caspian is termed by Vincent of Beauvais *Mare Seruanicum*, the Sea of Shirwan, another of its numerous Oriental names, rendered by Marino Sanuto as *Mare Salvanicum*. (III. xi. ch. ix.) But it was generally known to the Franks in the Middle Ages as the Sea of Bacu. Thus Berni :—

> " Fuor del deserto la diritta strada
> Lungo il Mar di Bacu miglior pareva."
> *(Orl. Innam.* xvii. 60.)

And in the *Sfera* of Lionardo Dati (*circa* 1390) :—

> " Da Tramontana di quest' Asia Grande
> Tartari son sotto la fredda Zona,
> Gente bestial di bestie e vivande,
> Fin dove *l'Onda di Baccù* risuona," etc.   (p. 10.)

This name is introduced in Ramusio, but probably by interpolation, as well as the correction of the statement regarding Euphrates, which is perhaps a branch of the notion alluded to in *Prologue*, ch. ii. note 5.   In a later chapter Marco calls it the *Sea of Sarai*, a title also given in the Carta Catalana.   [Odorico calls it Sea of *Bacuc* (*Cathay*) and Sea of *Bascon* (Cordier). The latter name is a corruption of Abeskun, a small town and island in the S. E. corner of the Caspian Sea, not far from Ashurada. —H. C.]

We have little information as to the Genoese navigation of the Caspian, but the great number of names exhibited along its shores in the map just named (1375) shows how familiar such navigation had become by that date.   See also *Cathay*, p. 50, where an account is given of a remarkable enterprise by Genoese buccaneers on the Caspian about that time   Mas'údi relates an earlier history of how about the beginning of the 9th century a fleet of 500 Russian vessels came out of the Volga, and ravaged all the populous southern and western shores of the Caspian.   The unhappy population was struck with astonishment and horror at this unlooked-for visitation from a sea that had hitherto been only frequented by peaceful traders or fishermen. (II. 18-24.)

Note 9.—[The enormous quantity of fish found in the Caspian Sea is ascribed to the mass of vegetable food to be found in the shallower waters of the North and the mouth of the Volga. According to Reclus, the Caspian fisheries bring in fish to the annual value of between three and four millions sterling.—H. C.]

CHAPTER V.

OF THE KINGDOM OF MAUSUL.

ON the frontier of Armenia towards the south-east is the kingdom of MAUSUL. It is a very great kingdom, and inhabited [1] by several different kinds of people whom we shall now describe.

First there is a kind of people called ARABI, and these worship Mahommet. Then there is another description of people who are NESTORIAN and JACOBITE Christians. These have a Patriarch, whom they call the JATOLIC, and this Patriarch creates Archbishops, and Abbots, and Prelates of all other degrees, and sends them into every quarter, as to India, to Baudas, or to Cathay, just as the Pope of Rome does in the Latin countries. For you must know that though there is a very great number of Christians in those countries, they are all Jacobites and Nestorians; Christians indeed, but not in the fashion enjoined by the Pope of Rome, for they come short in several points of the Faith. [2]

All the cloths of gold and silk that are called *Mosolins* are made in this country; and those great Merchants called *Mosolins*, who carry for sale such quantities of spicery and pearls and cloths of silk and gold, are also from this kingdom. [3]

There is yet another race of people who inhabit the mountains in that quarter, and are called CURDS. Some of them are Christians, and some of them are Saracens; but they are an evil generation, whose delight it is to plunder merchants. [4]

[Near this province is another called MUS and MERDIN, producing an immense quantity of cotton, from which they make a great deal of buckram [5] and other cloth. The people are craftsmen and traders, and all are subject to the Tartar King.]

NOTE I.—Polo could scarcely have been justified in calling MOSUL a very great kingdom. This is a bad habit of his, as we shall have to notice again. Badruddin Lúlú, the last Atabeg of Mosul of the race of Zenghi had at the age of 96 taken sides with Hulaku, and stood high in his favour. His son Malik Sálih, having revolted, surrendered to the Mongols in 1261 on promise of life; which promise they kept in Mongol fashion by torturing him to death. Since then the kingdom had ceased to exist as such. Coins of Badruddín remain with the name and titles of Mangku Kaan on their reverse, and some of his and of other atabegs exhibit curious imitations of Greek art. (*Quat. Rash.* p. 389; *Jour. As.* IV. VI. 141.).—H. Y. and H. C. [Mosul was pillaged by Timur at the end of

Coin of Badruddín of Mausul.

the 14th century; during the 15th it fell into the hands of the Turkomans, and during the 16th, of Ismaïl, Shah of Persia.—H. C.]

[The population of Mosul is to-day 61,000 inhabitants—(48,000 Musulmans, 10,000 Christians belonging to various churches, and 3000 Jews).—H. C.]

NOTE 2.—The Nestorian Church was at this time and in the preceding centuries diffused over Asia to an extent of which little conception is generally entertained, having a chain of Bishops and Metropolitans from Jerusalem to Peking. The Church derived its name from Nestorius, Patriarch of Constantinople, who was deposed by the Council of Ephesus in 431. The chief "point of the Faith" wherein it came short, was (at least in its most tangible form) the doctrine that in Our Lord there were two Persons, one of the Divine Word, the other of the Man Jesus; the former dwelling in the latter as in a Temple, or uniting with the latter "as fire with iron." *Nestorin*, the term used by Polo, is almost a literal transcript of the Arab form *Nastúri*. A notice of the Metropolitan sees, with a map, will be found in *Cathay*, p. ccxliv.

Játhalík, written in our text (from G. T.) *Jatolic*, by Fr. Burchard and Ricold *Jaselic*, stands for Καθολικός. No doubt it was originally *Gáthalík*, but altered in pronunciation by the Arabs. The term was applied by Nestorians to their Patriarch; among the Jacobites to the *Mafrián* or Metropolitan. The Nestorian Patriarch at this time resided at Baghdad. (*Assemani*, vol. iii. pt. 2; *Per. Quat.* 91, 127.)

The Jacobites, or Jacobins, as they are called by writers of that age (Ar. *Ya'úbkíy*), received their name from Jacob Baradaeus or James Zanzale, Bishop of Edessa (so called, Mas'údi says, because he was a maker of *barda'at* or saddle-cloths), who gave a great impulse to their doctrine in the 6th century. [At some time between the years 541 and 578, he separated from the Church and became a follower of the doctrine of Eutyches.—H. C.] The Jacobites then formed an independent Church, which at one time spread over the East as far as Sístán, where they had a see under the Sassanian Kings. Their distinguishing tenet was *Monophysitism*, viz., that Our Lord had but one Nature, the Divine. It was in fact a rebound from Nestorian doctrine, but, as might be expected in such a case, there was a vast number of shades of opinion among both bodies. The chief locality of the Jacobites was in the districts of Mosul, Tekrit, and Jazírah, and their Patriarch was at this time settled at the Monastery of St. Matthew, near Mosul, but afterwards, and to the present day, at or near Mardin. [They have at present two patriarchates: the Monastery of Zapharan near Baghdad and Etchmiadzin.—H. C.] The Armenian, Coptic, Abyssinian, and Malabar Churches all hold some shade of the Jacobite doctrine, though the first two at least have Patriarchs apart.

(*Assemani*, vol. ii. ; *Le Quien*, II. 1596; *Mas'údi*, II. 329-330; *Per. Quat.* 124-129.)

NOTE 3.—We see here that *mosolin* or *muslin* had a very different meaning from what it has now. A quotation from Ives by Marsden shows it to have been applied in the middle of last century to a strong cotton cloth made at Mosul. Dozy says the Arabs use *Mauçili* in the sense of muslin, and refers to passages in ' The Arabian Nights.' [Bretschneider (*Med. Res.* II. p. 122) observes " that in the narrative of Ch'ang Ch'un's travels to the west in 1221, it is stated that in Samarkand the men of the lower classes and the priests wrap their heads about with a piece of white *mo-sze*. There can be no doubt that mo-sze here denotes ' muslin,' and the Chinese author seems to understand by this term the same material which we are now used to call muslin."--H. C.] I have found no elucidation of Polo's application of *mosolini* to a class of merchants. But, in a letter of Pope Innocent IV. (1244) to the Dominicans in Palestine, we find classed as different bodies of Oriental Christians, "*Jacobitae, Nestoritae, Georgiani, Graeci, Armeni, Maronitae, et* Mosolini." (*Le Quien*, III. 1342.)

NOTE 4.—" The Curds," says Ricold, " exceed in malignant ferocity all the barbarous nations that I have seen. . . . . They are called *Curti*, not because they are curt in stature, but from the Persian word for *Wolves*. . . . They have three principal vices, viz., Murder, Robbery, and Treachery." Some say they have not mended since, but his etymology is doubtful. *Kúrt* is Turkish for a wolf, not Persian, which is *Gurg;* but the name (*Karduchi, Kordiaei*, etc.) is older, I imagine, than the Turkish language in that part of Asia. Quatremère refers it to the Persian *gurd*, " strong, valiant, hero." As regards the statement that some of the Kurds were Christians, Mas'údi states that the Jacobites and certain other Christians in the territory of Mosul and Mount Judi were reckoned among the Kurds. (*Not. et Ext.* XIII. i. 304.) [The Kurds of Mosul are in part nomadic and are called *Kotcheres*, but the greater number are sedentary and cultivate cereals, cotton, tobacco, and fruits. (*Cuinet.*) Old Kurdistan had Shehrizor (Kerkuk, in the sanjak of that name) as its capital.—H. C.]

NOTE 5.—Ramusio here, as in all passages where other texts have *Bucherami* and the like, puts *Boccassini*, a word which has become obsolete in its turn. I see both *Bochayrani* and *Bochasini* coupled, in a Genoese fiscal statute of 1339, quoted by Pardessus. (*Lois Maritimes*, IV. 456.)

MUSH and MARDIN are in very different regions, but as their actual interval is only about 120 miles, they *may* have been under one provincial government. Mush is essentially Armenian, and, though the seat of a Pashalik, is now a wretched place. Mardin, on the verge of the Mesopotamian Plain, rises in terraces on a lofty hill, and there, says Hammer, " Sunnis and Shias, Catholic and Schismatic Armenians, Jacobites, Nestorians, Chaldaeans, Sun-, Fire-, Calf-, and Devil-worshippers dwell one over the head of the other." (*Ilchan.* I. 191.)

CHAPTER VI.

Of the great City of Baudas, and how it was taken.

BAUDAS is a great city, which used to be the seat of the
Calif of all the Saracens in the world, just as Rome is
the seat of the Pope of all the Christians.[1] A very great
river flows through the city, and by this you can descend
to the Sea of India. There is a great traffic of merchants with their goods this way; they descend some
eighteen days from Baudas, and then come to a certain
city called KISI, where they enter the Sea of India.[2]
There is also on the river, as you go from Baudas to
Kisi, a great city called BASTRA, surrounded by woods,
in which grow the best dates in the world.[3]

In Baudas they weave many different kinds of silk
stuffs and gold brocades, such as *nasich*, and *nac*, and
cramoisy, and many another beautiful tissue richly
wrought with figures of beasts and birds. It is the
noblest and greatest city in all those regions.[4]

Now it came to pass on a day in the year of Christ
1255, that the Lord of the Tartars of the Levant, whose
name was Alaü, brother to the Great Kaan now reigning,
gathered a mighty host and came up against Baudas and
took it by storm.[5] It was a great enterprise! for in
Baudas there were more than 100,000 horse, besides
foot soldiers. And when Alaü had taken the place he
found therein a tower of the Calif's, which was full of
gold and silver and other treasure; in fact the greatest
accumulation of treasure in one spot that ever was
known.[6] When he beheld that great heap of treasure
he was astonished, and, summoning the Calif to his
presence, he said to him: "Calif, tell me now why thou

hast gathered such a huge treasure? What didst thou mean to do therewith? Knewest thou not that I was thine enemy, and that I was coming against thee with so great an host to cast thee forth of thine heritage? Wherefore didst thou not take of thy gear and employ it in paying knights and soldiers to defend thee and thy city?"

The Calif wist not what to answer, and said never a word. So the Prince continued, "Now then, Calif, since I see what a love thou hast borne thy treasure, I will e'en give it thee to eat!" So he shut the Calif up in the Treasure Tower, and bade that neither meat nor drink should be given him, saying, "Now, Calif, eat of thy treasure as much as thou wilt, since thou art so fond of it; for never shalt thou have aught else to eat!"

So the Calif lingered in the tower four days, and then died like a dog. Truly his treasure would have been of more service to him had he bestowed it upon men who would have defended his kingdom and his people, rather than let himself be taken and deposed and put to death as he was.[7] Howbeit, since that time, there has been never another Calif, either at Baudas or anywhere else.[8]

Now I will tell you of a great miracle that befell at Baudas, wrought by God on behalf of the Christians.

NOTE 1.—This form of the Mediæval Frank name of BAGHDAD, *Baudas* [the Chinese traveller, Ch'ang Te, *Si Shi Ki*, XIII. cent., says, "the kingdom of *Bao-da*," H. C.], is curiously like that used by the Chinese historians, *Paota* (*Pauthier; Gaubil*), and both are probably due to the Mongol habit of slurring gutturals. (See *Prologue*, ch. ii. note 3.) [Baghdad was taken on the 5th of February, 1258, and the Khalif surrendered to Hulaku on the 10th of February.—H. C.]

NOTE 2.—Polo is here either speaking without personal knowledge, or is so brief as to convey an erroneous impression that the Tigris flows to Kisi, whereas three-fourths of the length of the Persian Gulf intervene between the river mouth and Kisi. The latter is the island and city of KISH or KAIS, about 200 miles from the mouth of the Gulf, and for a long time one of the chief ports of trade with India and the East. The island, the *Cataea* of Arrian, now called Ghes or Kenn, is singular among the islands of the Gulf as being wooded and well supplied with fresh water. The ruins of a city [called Harira, according to Lord Curzon,] exist on the north side. According to

Wassáf, the island derived its name from one Kais, the son of a poor widow of Síráf (then a great port of Indian trade on the northern shore of the Gulf), who on a voyage to India, about the 10th century, made a fortune precisely as Dick Whittington did. The proceeds of the cat were invested in an establishment on this island. Modern attempts to nationalise Whittington may surely be given up ! It is one of the tales which, like Tell's shot, the dog Gellert, and many others, are common to many regions. (*Hammer's Ilch.* I. 239 ; *Ouseley's Travels,* I. 170 ; *Notes and Queries,* 2nd s. XI. 372.)

Mr Badger, in a postscript to his translation of the History of Omán (*Hak. Soc.* 1871), maintains that Kish or Kais was at this time a city on the mainland, and identical from Síráf. He refers to Ibn Batuta (II. 244), who certainly does speak of visiting "the city of Kais, called also Síráf." And Polo, neither here nor in Bk. III. ch. xl., speaks of Kisi as an island. I am inclined, however, to think that this was from not having visited it. Ibn Batuta says nothing of Síráf as a seat of trade ; but the historian Wassáf, who had been in the service of Jamáluddín al-Thaibi, the Lord of Kais, in speaking of the export of horses thence to India, calls it "the *Island* of Kais." (Elliot, III. 34.) Compare allusions to this horse trade in ch. xv. and in Bk. III. ch. xvii. Wassáf was precisely a contemporary of Polo.

NOTE 3.—The name is *Bascra* in the MSS., but this is almost certainly the common error of *c* for *t*. BASRA is still noted for its vast date-groves. "The whole country from the confluence of the Euphrates and Tigris to the sea, a distance of 30 leagues, is covered with these trees." (*Tav.* Bk. II. ch. iii.)

NOTE 4.—From Baudas, or Baldac, *i.e.* Baghdad, certain of these rich silk and gold brocades were called *Baldachini*, or in English *Baudekins*. From their use in the state canopies and umbrellas of Italian dignitaries, the word *Baldacchino* has come to mean a canopy, even when architectural. [*Baldekino, baldacchino,* was at first entirely made of silk, but afterwards silk was mixed (*sericum mixtum*) with cotton or thread. When Hulaku conquered Baghdad part of the tribute was to be paid with that kind of stuff. Later on, says Heyd (II. p. 697), it was also manufactured in the province of Ahwaz, at Damas and at Cyprus ; it was carried as far as France and England. Among the articles sent from Baghdad to Okkodai Khan, mentioned in the *Yüan ch'ao pi shi* (made in the 14th century), quoted by Bretschneider (*Med. Res.* II. p. 124), we note : *Nakhut* (a kind of gold brocade), *Nachidut* (a silk stuff interwoven with gold), *Dardas* (a stuff embroidered in gold). Bretschneider (p. 125) adds : "With respect to *nakhut* and *nachidut,* I may observe that these words represent the Mongol plural form of *nakh* and *nachetti. . . .* I may finally mention that in the *Yüan shi,* ch. lxxviii. (on official dresses), a stuff, *na-shi-shi,* is repeatedly named, and the term is explained there by *kin kin* (gold brocade)." —H. C.] The stuffs called *Nasich* and *Nac* are again mentioned by our traveller below (ch. lix.). We only know that they were of silk and gold, as he implies here, and as Ibn Batuta tells us, who mentions *Nakh* several times and *Nastj* once. The latter is also mentioned by Rubruquis (*Nasic*) as a present made to him at the Kaan's court. And Pegolotti speaks of both *nacchi* and *nacchetti* of silk and gold, the latter apparently answering to *Nasich*. *Nac, Nacques, Nachiz, Naciz, Nasis,* appear in accounts and inventories of the 14th century, French and English. (See *Dictionnaire des Tissus,* II. 199, and *Douet d'Arcq, Comptes de l'Argenterie des Rois de France,* etc., 334.) We find no mention of *Nakh* or *Nastj* among the stuffs detailed in the *Ain Akbari,* so they must have been obsolete in the 16th century. [Cf. Heyd, *Com. du Levant,* II. p. 698 ; *Nacco,* nachetto, comes from the Arabic *nakh* (*nekh*) ; *nassit* (*nasith*) from the Arabic *nécidj.*—H. C.] *Quermesis* or Cramoisy derived its name from the Kermes insect (Ar. *Kirmiz*) found on *Quercus coccifera,* now supplanted by cochineal. The stuff so called is believed to have been originally a crimson velvet, but apparently, like the mediæval *Purpura,* if not identical with it, it came to indicate a tissue rather than a colour. Thus Fr.-Michel quotes velvet of vermeil cramoisy, of violet, and of blue cramoisy, and *pourpres* of a variety of colours, though he says he has never met with *pourpre blanche*. I may, however, point to Plano Carpini (p. 755), who describes the courtiers at Karakorum as clad in white *purpura*.

The London prices of *Chermisi* and *Baldacchini* in the early part of the 15th century will be found in Uzzano's work, but they are hard to elucidate.

Babylon, of which Baghdad was the representative, was famous for its variegated textures in very early days. We do not know the nature of the goodly Babylonish garment which tempted Achan in Jericho, but Josephus speaks of the affluence of rich stuffs carried in the triumph of Titus, "gorgeous with life-like designs from the Babylonian loom," and he also describes the memorable Veil of the Temple as a πέπλος Βαβυλώνιος of varied colours marvellously wrought. Pliny says King Attalus invented the intertexture of cloth with gold; but the weaving of damasks of a variety of colours was perfected at Babylon, and thence they were called Babylonian.

The brocades wrought with figures of animals in gold, of which Marco speaks, are still a *spécialité* at Benares, where they are known by the name of *Shikárgáh* or hunting-grounds, which is nearly a translation of the name *Thard-wahsh* "beast-hunts," by which they were known to the mediæval Saracens. (See *Q. Makrizi*, IV. 69-70.) Plautus speaks of such patterns in carpets, the produce of Alexandria—"*Alexandrina* belluata *conchyliata tapetia*." Athenaeus speaks of Persian carpets of like description at an extravagant entertainment given by Antiochus Epiphanes; and the same author cites a banquet given in Persia by Alexander, at which there figured costly curtains embroidered with animals. In the 4th century Asterius, Bishop of Amasia in Pontus, rebukes the Christians who indulge in such attire: "You find upon them lions, panthers, bears, huntsmen, woods, and rocks; whilst the more devout display Christ and His disciples, with the stories of His miracles," etc. And Sidonius alludes to upholstery of like character:

"Peregrina det supellex
* * * *
Ubi torvus, et per artem
Resupina flexus ora,
It equo reditque telo
Simulacra bestiarum
Fugiens fugansque Parthus." (*Epist.* ix. 13.)

A modern Kashmír example of such work is shown under ch. xvii.

(*D'Avezac*, p. 524; *Pegolotti*, in *Cathay*, 295, 306; *I. B.* II. 309, 388, 422; III. 81; *Della Decima*, IV. 125-126; *Fr.-Michel*, *Recherches*, etc., II. 10-16, 204-206; *Joseph. Bell. Jud.* VII. 5, 5, and V. 5, 4; *Pliny*, VIII. 74 (or 48); *Plautus*, *Pseudolus*, I. 2; *Yonge's Athenaeus*, V. 26 and XII. 54; *Mongez* in *Mém. Acad.* IV. 275-276.)

NOTE 5.—[Bretschneider (*Med. Res.* I. p. 114) says: "Hulagu left Karakorum, the residence of his brother, on the 2nd May, 1253, and returned to his ordo, in order to organize his army. On the 19th October of the same year, all being ready, he started for the west." He arrived at Samarkand in September, 1255. For this chapter and the following of Polo, see: *Hulagu's Expedition to Western Asia, after the Mohammedan Authors*, pp. 112-122, and the *Translation of the Si Shi Ki* (Ch'ang Te), pp. 122-156, in Bretschneider's *Mediæval Researches*, I.—H.C.]

NOTE 6.—[" Hulagu proceeded to the lake of *Ormia* (Urmia), when he ordered a castle to be built on the island of *Tala*, in the middle of the lake, for the purpose of depositing here the immense treasures captured at Baghdad. A great part of the booty, however, had been sent to Mangu Khan." (*Hulagu's Exp.*, Bretschneider, *Med. Res.* I. p. 120.) Ch'ang Te says (*Si Shi Ki*, p. 139): "The palace of the Ha-li-fa was built of fragrant and precious woods. The walls of it were constructed of black and white jade. It is impossible to imagine the quantity of gold and precious stones found there."—H.C.]

NOTE 7.—

" I said to the Kalif : ' Thou art old,
Thou hast no need of so much gold.
Thou shouldst not have heaped and hidden it here,
Till the breath of Battle was hot and near,
But have sown through the land these useless hoards
To spring into shining blades of swords,
And keep thine honour sweet and clear.

*   *   *   *   *   *

Then into his dungeon I locked the drone,
And left him to feed there all alone
In the honey-cells of his golden hive :
Never a prayer, nor a cry, nor a groan
Was heard from those massive walls of stone,
Nor again was the Kalif seen alive.'
   This is the story, strange and true,
That the great Captain Alaü
Told to his brother, the Tartar Khan,
When he rode that day into Cambalu.
By the road that leadeth to Ispahan."   (*Longfellow.*) *

The story of the death of Mosta'sim Billah, the last of the Abbaside Khalifs, is told in much the same way by Hayton, Ricold, Pachymeres, and Joinville. The memory of the last glorious old man must have failed him, when he says the facts were related by some merchants who came to King Lewis, when before Saiette (or Sidon), viz. in 1253, for the capture of Baghdad occurred five years later. Mar. Sanuto says melted gold was poured down the Khalif's throat—a transfer, no doubt, from the old story of Crassus and the Parthians. Contemporary Armenian historians assert that Hulaku slew him with his own hand.

All that Rashiduddin says is : "The evening of Wednesday, the 14th of Safar, 656 (20th February, 1258), the Khalif was put to death in the village of Wakf, with his eldest son and five eunuchs who had never quitted him." Later writers say that he was wrapt in a carpet and trodden to death by horses.

[Cf. *The Story of the Death of the last Abbaside Caliph, from the Vatican MS. of Ibn-al-Furāt*, by G. le Strange (*Jour. R. As. Soc.*, April, 1900, pp. 293 - 300). This is the story of the death of the Khalif told by Ibn-al-Furāt (born in Cairo, 1335 A.D.) :

"Then Hūlagū gave command, and the Caliph was left a-hungering, until his case was that of very great hunger, so that he called asking that somewhat might be given him to eat. And the accursed Hūlagū sent for a dish with gold therein, and a dish with silver therein, and a dish with gems, and ordered these all to be set before the Caliph al Musta'sim, saying to him, ' Eat these.' But the Caliph made answer, ' These be not fit for eating.' Then said Hūlagū : ' Since thou didst so well know that these be not fit for eating, why didst thou make a store thereof? With part thereof thou mightest have sent gifts to propitiate us, and with part thou shouldst have raised an army to serve thee and defend thyself against us ! And Hūlagū commanded them to take forth the Caliph and his son to a place without the camp, and they were here bound and put into two great sacks, being afterwards trampled under foot till they both died—the mercy of Allah be upon them."—H. C.]

The foundation of the story, so widely received among the Christians, is to be found also in the narrative of Nikbi (and Mirkhond), which is cited by D'Ohsson. When the Khalif surrendered, Hulaku put before him a plateful of gold, and told him to eat it. "But one does not eat gold," said the prisoner. "Why, then,"

---

* Not that Alaü (*pace* Mr. Longfellow) ever did see Cambalu.

replied the Tartar, " did you hoard it, instead of expending it in keeping up an army?
Why did you not meet me at the Oxus?" The Khalif could only say, " Such was
God's will!" " And that which has befallen you was also God's will," said
Hulaku.

Wassáf's narrative is interesting :—" Two days after his capture the Khalif was at
his morning prayer, and began with the verse (*Koran*, III. 25), ' Say God is the
Possessor of Dominion ! It shall be given to whom He will ; it shall be taken from
whom He will : whom He will He raiseth to honour ; whom He will He casteth to
the ground.' Having finished the regular office he continued still in prayer with
tears and importunity. Bystanders reported to the Ilkhan the deep humiliation of
the Khalif's prayers, and the text which seemed to have so striking an application to
those two princes. Regarding what followed there are different stories. Some say
that the Ilkhan ordered food to be withheld from the Khalif, and that when he asked
for food the former bade a dish of gold be placed before him, etc. Eventually, after
taking counsel with his chiefs, the Padishah ordered the execution of the Khalif. It
was represented that the blood-drinking sword ought not to be stained with the gore
of Mosta'sim. He was therefore rolled in a carpet, just as carpets are usually rolled
up, insomuch that his limbs were crushed."

The avarice of the Khalif was proverbial. When the Mongol army was investing
Miafarakain, the chief, Malik Kamál, told his people that everything he had should
be at the service of those in need : " Thank God, I am not like Mosta'sim, a worshipper of silver and gold !"

(*Hayton* in *Ram.* ch. xxvi. ; *Per. Quat.* 121 ; *Pachym. Mic. Palaeol.* II. 24 ;
Joinville, p. 182 ; *Sanuto*, p. 238 ; *J. As.* sér. V. tom. xi. 490, and xvi. 291 ;
D'Ohsson, III. 243 ; *Hammer's Wassáf*, 75-76 ; *Quat. Rashid.* 305.)

Note 8.—Nevertheless Froissart brings the Khalif to life again one hundred and
twenty years later, as "*Le Galifre de Baudas.*" (Bk. III. ch. xxiv.)

CHAPTER VII.

How the Calif of Baudas took counsel to slay all the Christians in his Land.

I will tell you then this great marvel that occurred between Baudas and Mausul.

It was in the year of Christ[1] . . . that there was a
Calif at Baudas who bore a great hatred to Christians,
and was taken up day and night with the thought how
he might either bring those that were in his kingdom
over to his own faith, or might procure them all to be
slain. And he used daily to take counsel about this
with the devotees and priests of his faith,[2] for they all

bore the Christians like malice. And, indeed, it is a fact, that the whole body of Saracens throughout the world are always most malignantly disposed towards the whole body of Christians.

Now it happened that the Calif, with those shrewd priests of his, got hold of that passage in our Gospel which says, that if a Christian had faith as a grain of mustard seed, and should bid a mountain be removed, it would be removed. And such indeed is the truth. But when they had got hold of this text they were delighted, for it seemed to them the very thing whereby either to force all the Christians to change their faith, or to bring destruction upon them all. The Calif therefore called together all the Christians in his territories, who were extremely numerous. And when they had come before him, he showed them the Gospel, and made them read the text which I have mentioned. And when they had read it he asked them if that was the truth? The Christians answered that it assuredly was so. "Well," said the Calif, "since you say that it is the truth, I will give you a choice. Among such a number of you there must needs surely be this small amount of faith ; so you must either move that mountain there,"—and he pointed to a mountain in the neighbourhood—"or you shall die an ill death ; unless you choose to eschew death by all becoming Saracens and adopting our Holy Law. To this end I give you a respite of ten days ; if the thing be not done by that time, ye shall die or become Saracens." And when he had said this he dismissed them, to consider what was to be done in this strait wherein they were.

NOTE 1.—The date in the G. Text and Pauthier is 1275, which of course cannot have been intended. Ramusio has 1225.

[The Khalifs in 1225 were Abu'l Abbas Ahmed VII. en-Nassir lidini 'llah (1180-1225) and Abu Nasr Mohammed IX. ed-Dhahir bi-emri 'llah (1225-1226).—H. C.]

CHAPTER VIII.

How the Christians were in great dismay because of what the Calif had said.

THE Christians on hearing what the Calif had said were in great dismay, but they lifted all their hopes to God, their Creator, that He would help them in this their strait. All the wisest of the Christians took counsel together, and among them were a number of bishops and priests, but they had no resource except to turn to Him from whom all good things do come, beseeching Him to protect them from the cruel hands of the Calif.

So they were all gathered together in prayer, both men and women, for eight days and eight nights. And whilst they were thus engaged in prayer it was revealed in a vision by a Holy Angel of Heaven to a certain Bishop who was a very good Christian, that he should desire a certain Christian Cobler,[1] who had but one eye, to pray to God ; and that God in His goodness would grant such prayer because of the Cobler's holy life.

Now I must tell you what manner of man this Cobler was. He was one who led a life of great uprightness and chastity, and who fasted and kept from all sin, and went daily to church to hear Mass, and gave daily a portion of his gains to God. And the way how he came to have but one eye was this. It happened one day that

a certain woman came to him to have a pair of shoes
made, and she showed him her foot that he might take
her measure.   Now she had a very beautiful foot and
leg ; and the Cobler in taking her measure was conscious
of sinful thoughts.   And he had often heard it said in
the Holy Evangel, that if thine eye offend thee, pluck
it out and cast it from thee, rather than sin.   So, as
soon as the woman had departed, he took the awl that
he used in stitching, and drove it into his eye and destroyed it.   And this is the way he came to lose his eye.
So you can judge what a holy, just, and righteous man
he was.

NOTE I.—Here the G. T. uses a strange word : " *Or te vais a tel* cralantur."
It does not occur again, being replaced by *chabitier* (savetier).   It has an Oriental
look, but I can make no satisfactory suggestion as to what the word meant.

CHAPTER IX.

HOW THE ONE-EYED COBLER WAS DESIRED TO PRAY FOR THE CHRISTIANS.

Now when this vision had visited the Bishop several
times, he related the whole matter to the Christians, and
they agreed with one consent to call the Cobler before
them.   And when he had come they told him it was
their wish that he should pray, and that God had
promised to accomplish the matter by his means.   On
hearing their request he made many excuses, declaring
that he was not at all so good a man as they represented.   But they persisted in their request with so
much sweetness, that at last he said he would not tarry,
but do what they desired.

CHAPTER X.

How the Prayer of the One-eyed Cobler caused the Mountain to move.

AND when the appointed day was come, all the Christians got up early, men and women, small and great, more than 100,000 persons, and went to church, and heard the Holy Mass. And after Mass had been sung, they all went forth together in a great procession to the plain in front of the mountain, carrying the precious cross before them, loudly singing and greatly weeping as they went. And when they arrived at the spot, there they found the Calif with all his Saracen host armed to slay them if they would not change their faith ; for the Saracens believed not in the least that God would grant such favour to the Christians. These latter stood indeed in great fear and doubt, but nevertheless they rested their hope on their God Jesus Christ.

So the Cobler received the Bishop's benison, and then threw himself on his knees before the Holy Cross, and stretched out his hands towards Heaven, and made this prayer : "Blessed LORD GOD ALMIGHTY, I pray Thee by Thy goodness that Thou wilt grant this grace unto Thy people, insomuch that they perish not, nor Thy faith be cast down, nor abused nor flouted. Not that I am in the least worthy to prefer such request unto Thee ; but for Thy great power and mercy I beseech Thee to hear this prayer from me Thy servant full of sin."

And when he had ended this his prayer to God the Sovereign Father and Giver of all grace, and whilst the Calif and all the Saracens, and other people there, were looking on, the mountain rose out of its place and moved to the spot which the Calif had pointed out! And when the Calif and all his Saracens beheld, they stood amazed at the wonderful miracle that God had wrought for the Christians, insomuch that a great number of the Saracens became Christians. And even the Calif caused himself to be baptised in the name of the Father and of the Son and of the Holy Ghost, Amen, and became a Christian, but in secret. Howbeit, when he died they found a little cross hung round his neck; and therefore the Saracens would not bury him with the other Califs, but put him in a place apart. The Christians exulted greatly at this most holy miracle, and returned to their homes full of joy, giving thanks to their Creator for that which He had done.[1]

And now you have heard in what wise took place this great miracle. And marvel not that the Saracens hate the Christians; for the accursed law that Mahommet gave them commands them to do all the mischief in their power to all other descriptions of people, and especially to Christians; to strip such of their goods, and do them all manner of evil, because they belong not to their law. See then what an evil law and what naughty commandments they have! But in such fashion the Saracens act, throughout the world.

Now I have told you something of Baudas. I could easily indeed have told you first of the affairs and the customs of the people there. But it would be too long a business, looking to the great and strange things that I have got to tell you, as you will find detailed in this Book.

So now I will tell you of the noble city of Tauris.

NOTE 1.—We may remember that at a date only three years before Marco related this story (viz. in 1295), the cottage of Loreto is asserted to have changed its locality for the third and last time by moving to the site which it now occupies.

Some of the old Latin copies place the scene at Tauris. And I observe that a missionary of the 16th century does the same. The mountain, he says, is between Tauris and Nakhshiwan, and is called *Manhuc.* (*Gravina, Christianità nell' Armenia,* etc., Roma, 1605, p. 91.)

The moving of a mountain is one of the miracles ascribed to Gregory Thaumaturgus. Such stories are rife among the Mahomedans themselves. " I know," says Khanikoff, " at least half a score of mountains which the Musulmans allege to have come from the vicinity of Mecca."

Ramusio's text adds here : " All the Nestorian and Jacobite Christians from that time forward have maintained a solemn celebration of the day on which the miracle occurred, keeping a fast also on the eve thereof."

F. Göring, a writer who contributes three articles on Marco Polo to the *Neue Züricher-Zeitung,* 5th, 6th, 8th April, 1878, says : " I heard related in Egypt a report which Marco Polo had transmitted to Baghdad. I will give it here in connection with another which I also came across in Egypt.

" ' Many years ago there reigned in Babylon, on the Nile, a haughty Khalif who vexed the Christians with taxes and corvées. He was confirmed in his hate of the Christians by the Khakam Chacham Bashi or Chief Rabbi of the Jews, who one day said to him : " The Christians allege in their books that it shall not hurt them to drink or eat any deadly thing. So I have prepared a potion that one of them shall taste at my hand : if he does not die on the spot then call me no more Chacham Bashi ! " The Khalif immediately sent for His Holiness the Patriarch of Babylon, and ordered him to drink up the potion. The Patriarch just blew a little over the cup and then emptied it at a draught, and took no harm. His Holiness then on his side demanded that the Chacham Bashi should quaff a cup to the health of the Khalif, which he (the Patriarch) should first taste, and this the Khalif found only fair and right. But hardly had the Chacham Bashi put the cup to his lips than he fell down and expired.' Still the Musulmans and Jews thirsted for Christian blood. It happened at that time that a mass of the hill Mokattani became loose and threatened to come down upon Babylon. This was laid to the door of the Christians, and they were ordered to stop-it. The Patriarch in great distress has a vision that tells him summon the saintly cobbler (of whom the same story is told as here)—the cobbler bids the rock to stand still and it does so to this day. ' These two stories may still be heard in Cairo '—from whom is not said. The hill that threatened to fall on the Egyptian Babylon is called in Turkish *Dur Dagh,* ' Stay, or halt-hill.' (*L.c.* April, 1878.")—*MS. Note,* H. Y.

CHAPTER XI.

OF THE NOBLE CITY OF TAURIS.

TAURIS is a great and noble city, situated in a great province called YRAC, in which are many other towns and villages. But as Tauris is the most noble I will tell you about it.[1]

The men of Tauris get their living by trade and handicrafts, for they weave many kinds of beautiful and valuable stuffs of silk and gold. The city has such a good position that merchandize is brought thither from India, Baudas, Cremesor,[2] and many other regions; and that attracts many Latin merchants, especially Genoese, to buy goods and transact other business there; the more as it is also a great market for precious stones. It is a city in fact where merchants make large profits.[3]

The people of the place are themselves poor creatures; and are a great medley of different classes. There are Armenians, Nestorians, Jacobites, Georgians, Persians, and finally the natives of the city themselves, who are worshippers of Mahommet. These last are a very evil generation; they are known as Taurizi.[4] The city is all girt round with charming gardens, full of many varieties of large and excellent fruits.[5]

Now we will quit Tauris, and speak of the great country of Persia. [From Tauris to Persia is a journey of twelve days.]

Note 1.—Abulfeda notices that Tabríz was vulgarly pronounced *Tauriz*, and this appears to have been adopted by the Franks. In Pegolotti the name is always *Torissi*.

Tabriz is often reckoned to belong to Armenia, as by Hayton. Properly it is the chief city of *Azerbaiján*, which never was included in 'Irak. But it may be observed that Ibn Batuta generally calls the Mongol Ilkhan of Persia *Sáhib* or *Malik ul-'Irák*, and as Tabriz was the capital of that sovereign, we can account for the mistake, whilst admitting it to be one. [The destruction of Baghdad by Hulaku made Tabriz the great commercial and political city of Asia, and diverted the route of Indian products from the Mediterranean to the Euxine. It was the route to the Persian Gulf by Kashan, Yezd, and Kermán, to the Mediterranean by Lajazzo, and later on by Aleppo,—and to the Euxine by Trebizond. The destruction of the Kingdom of Armenia closed to Europeans the route of Tauris.—H. C.]

Note 2.—*Cremesor*, as Baldelli points out, is Garmsir, meaning a hot region, a term which in Persia has acquired several specific applications, and especially indicates the coast-country on the N.E. side of the Persian Gulf, including Hormuz and the ports in that quarter.

Note 3.—[Of the Italians established at Tabriz, the first whose name is mentioned is the Venetian Pietro Viglioni (Vioni); his will, dated 10th December, 1264, is still in existence. (*Archiv. Venet.* XXVI. pp. 161-165; *Heyd*, French Ed., II. p. 110.) —H. C.] At a later date (1341) the Genoese had a factory at Tabriz headed by a consul with a council of twenty-four merchants, and in 1320 there is evidence of a Venetian settlement there. (*Elie de la Prim.* 161 ; *Heyd*, II. 82.)

Rashiduddin says of Tabriz that there were gathered there under the eyes of the Padishah of Islam "philosophers, astronomers, scholars, historians, of all religions, of all sects ; people of Cathay, of Máchín, of India, of Kashmir, of Tibet, of the Uighúr and other Turkish nations, Arabs and Franks." Ibn Batuta : " I traversed the bazaar of the jewellers, and my eyes were dazzled by the varieties of precious stones which I beheld. Handsome slaves, superbly dressed, and girdled with silk, offered their gems for sale to the Tartar ladies, who bought great numbers. [Odoric (ed. Cordier) speaks also of the great trade of Tabriz.] Tabriz maintained a large population and prosperity down to the 17th century, as may be seen in Chardin. It is now greatly fallen, though still a place of importance." (*Quat. Rash.* p. 39 ; *I. B.* II. 130.)

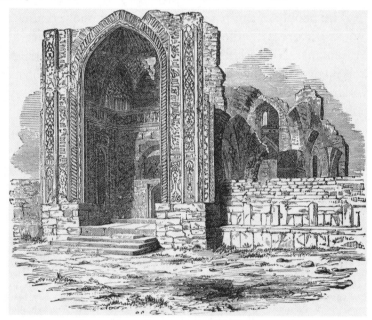

Ghazan Khan's Mosque at Tabriz.—(From Fergusson.)

NOTE 4.—In Pauthier's text this is *Touzi*, a mere clerical error, I doubt not for *Torizi*, in accordance with the G. Text (" *le peuple de la cité que sunt apelés* Tauriz"), with the Latin, and with Ramusio. All that he means to say is that the people are called *Tabrízís*. Not recondite information, but 'tis his way. Just so he tells us in ch. iii. that the people of Hermenia are called Hermins, and elsewhere that the people of Tebet are called Tebet. So Hayton thinks it not inappropriate to say that the people of Catay are called Cataini, that the people of Corasmia are called Corasmins, and that the people of the cities of Persia are called Persians.

NOTE 5.—Hamd Allah Mastaufi, the Geographer, not long after Polo's time, gives an account of Tabriz, quoted in Barbier de Meynard's *Dict. de la Perse*, p. 132. This also notices the extensive gardens round the city, the great abundance and cheapness of fruits, the vanity, insolence, and faithlessness of the Tabrízís, etc. (p. 132 *seqq.*). Our cut shows a relic of the Mongol Dynasty at Tabriz.

CHAPTER XII.

OF THE MONASTERY OF ST. BARSAMO ON THE BORDERS OF TAURIS.

ON the borders of (the territory of) Tauris there is a monastery called after Saint Barsamo, a most devout Saint. There is an Abbot, with many Monks, who wear a habit like that of the Carmelites, and these to avoid idleness are continually knitting woollen girdles. These they place upon the altar of St. Barsamo during the service, and when they go begging about the province (like the Brethren of the Holy Spirit) they present them to their friends and to the gentlefolks, for they are excellent things to remove bodily pain; wherefore every one is devoutly eager to possess them.[1]

NOTE I.—Barsauma ("The Son of Fasting") was a native of Samosata, and an Archimandrite of the Asiatic Church. He opposed the Nestorians, but became himself still more obnoxious to the orthodox as a spreader of the Monophysite Heresy. He was condemned by the Council of Chalcedon (451), and died in 458. He is a Saint of fame in the Jacobite and Armenian Churches, and several monasteries were dedicated to him ; but by far the most celebrated, and doubtless that meant here, was near Malatia. It must have been famous even among the Mahomedans, for it has an article in Bákúi's Geog. Dictionary. (*Dír-Barsúma*, see *N. et Ext.* II. 515.) This monastery possessed relics of Barsauma and of St. Peter, and was sometimes the residence of the Jacobite Patriarch and the meeting-place of the Synods.

A more marvellous story than Marco's is related of this monastery by Vincent of Beauvais: "There is in that kingdom (Armenia) a place called St. Brassamus, at which there is a monastery for 300 monks. And 'tis said that if ever an enemy attacks it, the defences of the monastery move of themselves, and shoot back the shot against the besieger."

(*Assemani* in vol. ii. *passim ; Tournefort*, III. 260 ; *Vin. Bell. Spec. Historiale.* Lib. XXX. c. cxlii. ; see also *Mar. Sanut.* III. xi. c. 16.)

CHAPTER XIII

OF THE GREAT COUNTRY OF PERSIA; WITH SOME ACCOUNT OF THE
THREE KINGS.

PERSIA is a great country, which was in old times very
illustrious and powerful; but now the Tartars have wasted
and destroyed it.

In Persia is the city of SABA, from which the Three
Magi set out when they went to worship Jesus Christ;
and in this city they are buried, in three very large and
beautiful monuments, side by side. And above them
there is a square building, carefully kept. The bodies
are still entire, with the hair and beard remaining. One
of these was called Jaspar, the second Melchior, and the
third Balthasar. Messer Marco Polo asked a great many
questions of the people of that city as to those Three
Magi, but never one could he find that knew aught of the
matter, except that these were three kings who were
buried there in days of old. However, at a place three
days' journey distant he heard of what I am going to tell
you. He found a village there which goes by the name
of CALA ATAPERISTAN,[1] which is as much as to say, " The
Castle of the Fire-worshippers." And the name is rightly
applied, for the people there do worship fire, and I will
tell you why.

They relate that in old times three kings of that
country went away to worship a Prophet that was born,
and they carried with them three manner of offerings,
Gold, and Frankincense, and Myrrh; in order to ascertain
whether that Prophet were God, or an earthly King, or a
Physician. For, said they, if he take the Gold, then he
is an earthly King; if he take the Incense he is God; if
he take the Myrrh he is a Physician.

So it came to pass when they had come to the place where the Child was born, the youngest of the Three Kings went in first, and found the Child apparently just of his own age ; so he went forth again marvelling greatly. The middle one entered next, and like the first he found the Child seemingly of his own age ; so he also went forth again and marvelled greatly. Lastly, the eldest went in, and as it had befallen the other two, so it befell him. And he went forth very pensive. And when the three had rejoined one another, each told what he had seen ; and then they all marvelled the more. So they agreed to go in all three together, and on doing so they beheld the Child with the appearance of its actual age, to wit, some thirteen days.[2] Then they adored, and presented their Gold and Incense and Myrrh. And the Child took all the three offerings, and then gave them a small closed box ; whereupon the Kings departed to return into their own land.

NOTE I.—*Kala' Atishparastán*, meaning as in the text. (*Marsden.*)

NOTE 2.—According to the Collectanea ascribed to Bede, Melchior was a hoary old man ; Balthazar in his prime, with a beard ; Gaspar young and beardless. (*Inchofer, Tres Magi Evangelici*, Romae, 1639.)

CHAPTER XIV.

WHAT BEFELL WHEN THE THREE KINGS RETURNED TO THEIR OWN COUNTRY.

AND when they had ridden many days they said they would see what the Child had given them. So they opened the little box, and inside it they found a stone.

On seeing this they began to wonder what this might be that the Child had given them, and what was the import thereof. Now the signification was this: when they presented their offerings, the Child had accepted all three, and when they saw that they had said within themselves that He was the True God, and the True King, and the True Physician.[1] And what the gift of the stone implied was that this Faith which had begun in them should abide firm as a rock. For He well knew what was in their thoughts. Howbeit, they had no understanding at all of this signification of the gift of the stone; so they cast it into a well. Then straightway a fire from Heaven descended into that well wherein the stone had been cast.

And when the Three Kings beheld this marvel they were sore amazed, and it greatly repented them that they had cast away the stone; for well they then perceived that it had a great and holy meaning. So they took of that fire, and carried it into their own country, and placed it in a rich and beautiful church. And there the people keep it continually burning, and worship it as a god, and all the sacrifices they offer are kindled with that fire. And if ever the fire becomes extinct they go to other cities round about where the same faith is held, and obtain of that fire from them, and carry it to the church. And this is the reason why the people of this country worship fire. They will often go ten days' journey to get of that fire.[2]

Such then was the story told by the people of that Castle to Messer Marco Polo; they declared to him for a truth that such was their history, and that one of the three kings was of the city called SABA, and the second of AVA, and the third of that very Castle where they still worship fire, with the people of all the country round about.[3]

Having related this story, I will now tell you of the different provinces of Persia, and their peculiarities.

NOTE 1.—"*Mire.*" This was in old French the popular word for a Leech ; the politer word was *Physicien.* (*N. et E.* V. 505.)

Chrysostom says that the Gold, Myrrh, and Frankincense were mystic gifts indicating King, Man, God ; and this interpretation was the usual one. Thus Prudentius :—

> " Regem, Deumque adnunciant
> Thesaurus et fragrans odor
> Thuris Sabaei, at myrrheus
> Pulvis sepulchrum praedocet." (*Hymnus Epiphanius.*)

And the Paris Liturgy :—

> " Offert Aurum *Caritas,*
> Et Myrrham *Austeritas,*
> Et Thus *Desiderium.*
> Auro *Rex* agnoscitur,
> *Homo* Myrrha, colitur
> Thure *Deus* gentium."

And in the " Hymns, Ancient and Modern " :—

> " Sacred gifts of mystic meaning :
> Incense doth their God disclose,
> Gold the King of Kings proclaimeth,
> Myrrh His sepulchre foreshows."

NOTE 2.—" Feruntque (Magi), si justum est credi, etiam ignem caelitus lapsum apud se sempiternis foculis custodire, cujus portionem exiguam, ut faustam praeisse quondam Asiaticis Regibus dicunt." (*Ammian. Marcell.* XXIII. 6.)

NOTE 3.—Saba or Sava still exists as SÁVAH, about 50 miles S.W. of Tehrân. It is described by Mr. Consul Abbott, who visited it in 1849, as the most ruinous town he had ever seen, and as containing about 1000 families. The people retain a tradition, mentioned by Hamd Allah Mastaufi, that the city stood on the shores of a Lake which dried up miraculously at the birth of Mahomed. Sávah is said to have possessed one of the greatest Libraries in the East, until its destruction by the Mongols on their first invasion of Persia. Both Sávah and Ávah (or Ábah) are mentioned by Abulfeda as cities of Jibal. We are told that the two cities were always at loggerheads, the former being Sunni and the latter Shiya. [We read in the *Travels* of Thévenot, a most intelligent traveller, "qu'il n'a rien écrit de l'ancienne ville de Sava qu'il trouva sur son chemin, et où il a marqué lui-même que son esprit de curiosité l'abandonna." (*Voyages,* éd. 1727, vol. v. p. 343. He died a few days after at Miana, in Armenia, 28th November, 1667). (*MS. Note.*—H. Y.)]

As regards the position of AVAH, Abbott says that a village still stands upon the site, about 16 miles S.S.E. of Sávah. He did not visit it, but took a bearing to it. He was told there was a mound there on which formerly stood a Gueber Castle. At Sávah he could find no trace of Marco Polo's legend. Chardin, in whose time Sávah was not quite so far gone to decay, heard of an alleged tomb of Samuel, at 4 leagues from the city. This is alluded to by Hamd Allah.

Keith Johnston and Kiepert put Ávah some 60 miles W.N.W. of Sávah, on the road between Kazvin and Hamadan. There seems to be some great mistake here.

Friar Odoric puts the locality of the Magi at *Kashan,* though one of the versions of Ramusio and the Palatine MS. (see Cordier's Odoric, pp. xcv. and 41 of his Itinerary), perhaps corrected in this, puts it at *Saba.*—H. Y. and H. C.

We have no means of fixing the *Kala' Atishparastán*. It is probable, however, that the story was picked up on the homeward journey, and as it seems to be implied that this castle was reached three days *after leaving* Sávah, I should look for it between Sávah and Abher. Ruins to which the name *Kila'-i-Gabr*, "Gueber Castle," attaches are common in Persia.

As regards the Legend itself, which shows such a curious mixture of Christian and Parsi elements, it is related some 350 years earlier by Mas'údi : "In the Province of Fars they tell you of a Well called the Well of Fire, near which there was a temple built. When the Messiah was born the King Koresh sent three messengers to him, the first of whom carried a bag of Incense, the second a bag of Myrrh, and the third a bag of Gold. They set out under the guidance of the Star which the king had described to them, arrived in Syria, and found the Messiah with Mary His Mother. This story of the three messengers is related by the Christians with sundry exaggerations ; it is also found in the Gospel. Thus they say that the Star appeared to Koresh at the moment of Christ's birth ; that it went on when the messengers went on, and stopped when they stopped. More ample particulars will be found in our Historical Annals, where we have given the versions of this legend as current among the Guebers and among the Christians. It will be seen that Mary gave the king's messengers a round loaf, and this, after different adventures, they hid under a rock in the province of Fars. The loaf disappeared underground, and there they dug a well, on which they beheld two columns of fire to start up flaming at the surface ; in short, all the details of the legend will be found in our Annals." The Editors say that Mas'údi had carried the story to Fars by mistaking *Shíz* in Azerbaiján (the Atropatenian Ecbatana of Sir H. Rawlinson) for *Shiraz*. A rudiment of the same legend is contained in the Arabic Gospel of the Infancy. This says that Mary gave the Magi one of the bands in which the Child was swathed. On their return they cast this into their sacred nre ; though wrapt in the flame it remained unhurt.

We may add that there was a Christian tradition that the Star descended into a well between Jerusalem and Bethlehem. Gregory of Tours also relates that in a certain well, at Bethlehem, from which Mary had drawn water, the Star was sometimes seen, by devout pilgrims who looked carefully for it, to pass from one side to the other. But only such as merited the boon could see it.

(See *Abbott* in *J. R. G. S.* XXV. 4-6; *Assemani*, III. pt. 2, 750; *Chardin*, II. 407 ; *N. et Ext.* II. 465 ; *Dict. de la Perse*, 2, 56, 298 ; *Cathay*, p. 51 ; *Mas'udi*, IV. 80 ; *Greg. Turon. Libri Miraculorum*, Paris, 1858, I. 8.)

Several of the fancies that legend has attached to the brief story of the Magi in St. Matthew, such as the royal dignity of the persons ; their location, now in Arabia, now (as here) at Saba in Persia, and again (as in Hayton and the Catalan Map) in Tarsia or Eastern Turkestan ; the notion that one of them was a Negro, and so on, probably grew out of the arbitrary application of passages in the Old Testament, such as : *Venient legati ex Aegypto :* AETHIOPIA *praevenit manus ejus Deo*" (Ps. lxviii. 31). This produced the Negro who usually is painted as one of the Three. "*Reges* THARSIS *et Insulae munera offerent : Reges* ARABUM *et* SABA *dona adducent* " (lxxii. 10). This made the Three into Kings, and fixed them in Tarsia, Arabia, and Sava. "*Mundatio Camelorum operiet te, dromedarii Madian et* EPHA : *omnes de* SABA *venient aurum et thus deferentes et laudem Domino annunciantes*" (Is. lx. 6). Here were Ava and Sava coupled, as well as the gold and frankincense.

One form of the old Church Legend was that the Three were buried at *Sessania Adrumetorum* (Hadhramaut) in Arabia, whence the Empress Helena had the bodies conveyed to Constantinople, [and later to Milan in the time of the Emperor Manuel Comnenus. After the fall of Milan (1162), Frederic Barbarossa gave them to Archbishop Rainald of Dassel (1159-1167), who carried them to Cologne (23rd July, 1164). —H. C.]

The names given by Polo, Gaspar, Melchior, and Balthasar, have been accepted from an old date by the Roman Church; but an abundant variety of other names has been assigned to them. Hyde quotes a Syriac writer who calls them Aruphon,

Hurmon, and Tachshesh, but says that some call them Gudphorbus, Artachshasht, and Labudo; whilst in Persian they were termed Amad, Zad-Amad, Drust-Amad, *i.e. Venit, Cito Venit, Sincerus Venit.* Some called them in Greek, Apellius, Amerus, and Damascus, and in Hebrew, Magaloth, Galgalath, and Saracia, but otherwise Ator, Sator, and Petatoros! The Armenian Church used the same names as the Roman, but in Chaldee they were Kaghba, Badadilma, Badada Kharida. (*Hyde, Rel. Vet. Pers.* 382-383; *Inchofer, ut supra; J. As.* sér. VI. IX. 160.)

[Just before going to press we have read Major Sykes' new book on *Persia.* Major Sykes (ch. xxiii.) does not believe that Marco visited Baghdád, and he thinks that the Venetians entered Persia near Tabriz, and travelled to Sultania, Kashán, and Yezd. Thence they proceeded to Kerman and Hormuz. We shall discuss this question in the Introduction.—H. C.]

———

CHAPTER XV.

Of the Eight Kingdoms of Persia, and how they are named.

Now you must know that Persia is a very great country, and contains eight kingdoms. I will tell you the names of them all.

The first kingdom is that at the beginning of Persia, and it is called CASVIN : the second is further to the south, and is called CURDISTAN ; the third is LOR ; the fourth [SUOLSTAN] ; the fifth ISTANIT ; the sixth SERAZY ; the seventh SONCARA ; the eighth TUNOCAIN, which is at the further extremity of Persia. All these kingdoms lie in a southerly direction except one, to wit, Tunocain ; that lies towards the east, and borders on the (country of the) Arbre Sol.[1]

In this country of Persia there is a great supply of fine horses ; and people take them to India for sale, for they are horses of great price, a single one being worth as much of their money as is equal to 20n livres Tournois ; some will be more, some less, according to the quality.[2] Here also are the finest asses in the world, one of them being worth full 30 marks of silver, for they are very large and fast, and acquire a capital amble.

Dealers carry their horses to Kisi and Curmosa, two cities on the shores of the Sea of India, and there they meet with merchants who take the horses on to India for sale.

In this country there are many cruel and murderous people, so that no day passes but there is some homicide among them. Were it not for the Government, which is that of the Tartars of the Levant, they would do great mischief to merchants; and indeed, maugre the Government, they often succeed in doing such mischief. Unless merchants be well armed they run the risk of being murdered, or at least robbed of everything; and it sometimes happens that a whole party perishes in this way when not on their guard. The people are all Saracens, *i.e.* followers of the Law of Mahommet.[3]

In the cities there are traders and artizans who live by their labour and crafts, weaving cloths of gold, and silk stuffs of sundry kinds. They have plenty of cotton produced in the country; and abundance of wheat, barley, millet, panick, and wine, with fruits of all kinds.

[Some one may say, " But the Saracens don't drink wine, which is prohibited by their law." The answer is that they gloss their text in this way, that if the wine be boiled, so that a part is dissipated and the rest becomes sweet, they may drink without breach of the commandment; for it is then no longer called wine, the name being changed with the change of flavour.[4]]

NOTE 1.—The following appear to be Polo's Eight Kingdoms :—

I. KAZVÍN; then a flourishing city, though I know not why he calls it a kingdom. Persian 'Irák, or the northern portion thereof, seems intended. Previous to Hulaku's invasion Kazvín seems to have been in the hands of the Ismailites or Assassins.

II. KURDISTAN. I do not understand the difficulties of Marsden, followed by Lazari and Pauthier, which lead them to put forth that Kurdistan is not Kurdistan but something else. The boundaries of Kurdistan according to Hamd Allah were Arabian 'Irak, Khuzistán, Persian 'Irak, Azerbaijan and Diarbekr. (*Dict. de la P.* 480.) [Cf. Curzon, *Persia pass.*—H. C.] Persian Kurdistan, in modern as in mediæval times, extends south beyond Kermanshah to the immediate border of Polo's next kingdom, viz. :

III. Lúr or Lúristán. [On Lúristán, see Curzon, *Persia*, II. pp. 273-303, with the pedigree of the Ruling Family of the Feili Lurs (Pusht-i-Kuh), p. 278.—H. C.] This was divided into two principalities, Great Lúr and Little Lúr, distinctions still existing. The former was ruled by a Dynasty called the *Faslúyah* Atabegs, which endured from about 1155 to 1424, [when it was destroyed by the Timurids ; it was a Kurd Dynasty, founded by Emad ed-din Abu Thaher (1160-1228), and the last prince of which was Ghiyas ed-din (1424). In 1258 the general Kitubuka (Hulagu's *Exp. to Persia*, Bretschneider, *Med. Res.* I. p. 121) is reported to have reduced the country of Lúr or Lúristán and its Atabeg Teghele.—H. C.]. Their territory lay in the mountainous district immediately west of Ispahan, and extended to the River of Dizfúl, which parted it from Little Lúr. The stronghold of the Atabegs was the extraordinary hill fort of Mungasht, and they had a residence also at Aidhej or Mal-Amir in the mountains south of Shushan, where Ibn Batuta visited the reigning Prince in 1327. Sir H. Rawlinson has described Mungasht, and Mr Layard and Baron de Bode have visited other parts, but the country is still very imperfectly known. Little Lúristán lay west of the R. Dizfúl, extending nearly to the Plain of Babylonia. Its Dynasty, called Kurshid, [was founded in 1184 by the Kurd Shodja ed-din Khurshid, and existed till Shah-Werdy lost his throne in 1593.—H. C.].

The Lúrs are akin to the Kurds, and speak a Kurd dialect, as do all those Ilyáts, or nomads of Persia, who are not of Turkish race. They were noted in the Middle Ages for their agility and their dexterity in thieving. The tribes of Little Lúr "do not affect the slightest veneration for Mahomed or the Koran ; their only general object of worship is their great Saint Baba Buzurg," and particular disciples regard with reverence little short of adoration holy men looked on as living representatives of the Divinity. (*Ilchan.* I. 70 *seqq.* ; *Rawlinson* in *J. R. G. S.* IX. ; *Layard* in *Do.* XVI. 75, 94; *Ld. Strangford* in *J. R. A. S.* XX. 64; *N. et E.* XIII. i. 330, *I. B.* II. 31; *D'Ohsson*, IV. 171-172.)

IV. SHÚLISTÁN, best represented by Ramusio's *Suolstan*, whilst the old French texts have *Cielstan* (*i.e.* Shelstán) ; the name applied to the country of the *Shúls*, or *Shauls*, a people who long occupied a part of Lúristán, but were expelled by the Lúrs in the 12th century, and settled in the country between Shíráz and Khuzistán (now that of the Mamaseni, whom Colonel Pelly's information identifies with the Shúls), their central points being Naobanján and the fortress called Kala' Safed or "White Castle." Ibn Batuta, going from Shiraz to Kazerun, encamped the first day in the country of the Shúls, "a Persian desert tribe which includes some pious persons." (*Q. R.* p. 385 ; *N. et E.* XIII. i. 332-333 ; *Ilch.* I. 71 ; *J. R. G. S.* XIII. Map ; *I. B.* II. 88.) ["Adjoining the Kuhgelus on the East are the tents of the Mamasenni (qy. Mohammed Huseini) Lúrs, occupying the country still known as Shúlistán, and extending as far east and south-east as Fars and the Plain of Kazerun. This tribe prides itself on its origin, claiming to have come from Seistán, and to be directly descended from Rustam, whose name is still borne by one of the Mamasenni clans." (Curzon, *Persia*, II. p. 318.)—H. C.]

V. ISPAHAN ? The name is in Ramusio *Spaan*, showing at least that he or some one before him had made this identification. The unusual combination *ff*, *i.e.* sf, in manuscript would be so like the frequent one *ft*. *i.e.* st, that the change from Isfan to Istan would be easy. But why Istan*it* ?

VI. SHÍRÁZ [(*Shir*=milk, or *Shir*=lion)—H. C.] representing the province of Fars or Persia Proper, of which it has been for ages the chief city. [It was founded after the Arab conquest in 694 A.D., by Mohammed, son of Yusuf Kekfi. (Curzon, *Persia*, II. pp. 93-110.)—H. C.] The last Dynasty that had reigned in Fars was that of the Salghur Atabegs, founded about the middle of the 12th century. Under Abu-bakr (1226-1260) this kingdom attained considerable power, embracing Fars, Kermán, the islands of the Gulf and its Arabian shores ; and Shíráz then flourished in arts and literature ; Abubakr was the patron of Saadi. From about 1262, though a Salghurian princess, married to a son of Hulaku, had the nominal title of Atabeg, the province of Fars was under Mongol administration. (*Ilch. passim.*)

VII. SHAWÁNKÁRA or Shabánkára. The G. T. has *Soucara*, but the Crusca gives the true reading *Soncara*. It is the country of the Shawánkárs, a people coupled with the Shúls and Lúrs in mediæval Persian history, and like them of Kurd affinities. Their princes, of a family Faslúyah, are spoken of as influential before the Mahomedan conquest, but the name of the people comes prominently forward only during the Mongol era of Persian history. [Shabánkára was taken in 1056 from the Buyid Dynasty, who ruled from the 10th century over a great part of Persia, by Fazl ibn Hassan (Fazluïeh-Hasunïeh). Under the last sovereign, Ardeshir, Shabánkára was taken in 1355 by the Modhafferians, who reigned in Irak, Fars, and Kermán, one of the Dynasties established at the expense of the Mongol Ilkhans after the death of Abu Saïd (1335), and were themselves subjugated by Timur in 1392.—H. C.] Their country lay to the south of the great salt lake east of Shíráz, and included Niriz and Darábjird, Fassa, Forg, and Tárum. Their capital was I/g or I/j, called also Irej, about 20 miles north-west of Daráb, with a great mountain fortress ; it was taken by Hulaku in 1259. The son of the prince was continued·in nominal authority, with Mongol administrators. In consequence of a rebellion in 1311 the Dynasty seems to have been extinguished. A descendant attempted to revive their authority about the middle of the same century. The latest historical mention of the name that I have found is in Abdurrazzák's *History of Shah Rukh*, under the year H. 807 (1404). (See *Jour. As.* 3d. s. vol. ii. 355.) But a note by Colonel Pelly informs me that the name Shabánkára is still applied (1) to the district round the towns of Runiz and Gauristan near Bandar Abbas ; (2) to a village near Maiman, in the old country of the tribe ; (3) to a *tribe* and district of Dashtistan, 38 farsakhs west of Shíráz.

With reference to the form in the text, *Soncara*, I may notice that in two passages of the *Masálak-ul-Absár*, translated by Quatremère, the name occurs as *Shankárah.* (*Q. R.* pp. 380, 440 *seqq.* ; *N. et E.* XIII. ; *Ilch.* I. 71 and *passim ; Ouseley's Travels*, II. 158 *seqq.*)

VIII. TÚN-O-KÁIN, the eastern Kuhistán or Hill country of Persia, of which Tún and Káin are chief cities. The practice of indicating a locality by combining two names in this way is common in the East. Elsewhere in this book we find *Ariora-Keshemur* and *Kes-macoran* (Kij-Makrán). Upper Sind is often called in India by the Sepoys *Rori-Bakkar,* from two adjoining places on the Indus ; whilst in former days, Lower Sind was often called *Diul-Sind.* *Karra-Mánikpúr, Uch-Multán, Kunduz-Baghlán* are other examples.

The exact expression *Tún-o-Káin* for the province here in question is used by Baber, and evidently also by some of Hammer's authorities. (*Baber*, pp. 201, 204 ; see *Ilch.* II. 190 ; I. 95, 104, and *Hist. de l'Ordre des Assassins*, p. 245.)

[We learn from (Sir) C. Macgregor's (1875) *Journey through Khorasan* (I. p. 127) that the same territory including Gháín or Kaïn is now called by the analogous name of Tabas-o-Tún. Tún and Kaïn (Gháín) are both described in their modern state, by Macgregor. (*Ibid.* pp. 147 and 161.)—H. C.]

Note that the identification of *Suolstan* is due to Quatremère (see *N. et E.* XIII. i. *circa* p. 332) ; that of *Soncara* to Defrémery (*J. As.* sér. IV. tom. xi. p. 441) ; and that of *Tunocain* to Malte-Brun. (*N. Ann. des V.* xviii. p. 261.) I may add that the *Lúrs*, the *Shúls*, and the *Shabánkáras* are the subjects of three successive sections in the *Masálak-al-Absár* of *Shihábuddín Dimishki*, a work which reflects much of Polo's geography. (See *N. et E.* XIII. i. 330-333 ; Curzon, *Persia*, II. pp. 248 and 251.)

NOTE 2.—The horses exported to India, of which we shall hear more hereafter, were probably the same class of " Gulf Arabs " that are now carried thither. But the Turkman horses of Persia are also very valuable, especially for endurance. Kinneir speaks of one accomplishing 900 miles in eleven days, and Ferrier states a still more extraordinary feat from his own knowledge. In that case one of those horses went from Tehran to Tabriz, returned, and went again to Tabriz, within twelve days, including two days' rest.   The total distance is about 1100 miles.

The *livre tournois* at this period was equivalent to a little over 18 francs of modern French silver.   But in bringing the value to our modern gold standard we must add one-third, as the ratio of silver to gold was then 1 : 12 instead of 1 : 16.   Hence the equivalent in gold of the livre tournois is very little less than 1*l.* sterling, and the price of the horse would be about 193*l.**

Mr Wright quotes an ordinance of Philip III. of France (1270-1285) fixing the maximum price that might be given for a palfrey at 60 *livres tournois*, and for a squire's *roncin* at 20 livres.   Joinville, however, speaks of a couple of horses presented to St. Lewis in 1254 by the Abbot of Cluny, which he says would at the time of his writing (1309) have been worth 500 livres (the pair, it would seem).   Hence it may be concluded in a general way that the *ordinary* price of imported horses in India approached that of the highest class of horses in Europe.   (*Hist. of Dom. Manners*, p. 317 ; *Joinville*, p. 205.)

About 1850 a very fair Arab could be purchased in Bombay for 60*l.*, or even less ; but prices are much higher now.

With regard to the donkeys, according to Tavernier, the fine ones used by merchants in Persia were imported from Arabia.   The mark of silver was equivalent to about 44*s.* of our silver money, and allowing as before for the lower relative value of gold, 30 marks would be equivalent to 88*l.* sterling.

Kisi or Kish we have already heard of.   *Curmosa* is Hormuz, of which we shall hear more.   With a Pisan, as Rusticiano was, the sound of *c* is purely and strongly aspirate.   Giovanni d'Empoli, in the beginning of the 16th century, another Tuscan, also calls it *Cormus.*   (See *Archiv. Stor. Ital.* Append. III. 81.)

Note 3.—The character of the nomad and semi-nomad tribes of Persia in those days—Kurds, Lúrs, Shúls, Karaunahs, etc.—probably deserved all that Polo says, and it is not changed now.   Take as an example Rawlinson's account of the Bakhtyáris of Luristán : " I believe them to be individually brave, but of a cruel and savage character ; they pursue their blood feuds with the most inveterate and exterminating spirit. . . . It is proverbial in Persia that the Bakhtiyaris have been compelled to forego altogether the reading of the *Fatihah* or prayer for the dead, for otherwise they would have no other occupation.   They are also most dextrous and notorious thieves." (*J. R. G. S.* IX. 105.)

Note 4.—The Persians have always been lax in regard to the abstinence from wine.

According to Athenaeus, Aristotle, in his *Treatise on Drinking* (a work lost, I imagine, to posterity), says, " If the wine be moderately boiled it is less apt to intoxicate."   In the preparation of some of the sweet wines of the Levant, such as that of Cyprus, the must is boiled, but I believe this is not the case *generally* in the East. Baber notices it as a peculiarity among the Kafirs of the Hindu Kush.   Tavernier, however, says that at Shíráz, besides the wine for which that city was so celebrated, a good deal of *boiled wine* was manufactured, and used among the poor and by travellers.   No doubt what was meant is the sweet liquor or syrup called *Dúsháb*, which Della Valle says is just the Italian *Mostocotto*, but better, clearer, and not so mawkish (I. 689).   (*Yonge's Athen.* X. 34; *Baber*, p. 145 ; *Tavernier*, Bk. V. ch. xxi.)

---

* The *Encyc. Britann.*, article " Money," gives the livre tournois of this period as 18.17 francs.   A French paper in *Notes and Queries* (4th S. IV. 485) gives it under St. Lewis and Philip III. as equivalent to 18.24 fr., and under Philip IV. to 17.95.   And lastly, experiment at the British Museum, made by the kind intervention of my friend, Mr. E. Thomas, F.R.S., gave the weights of the *sols* of St. Lewis (1226-1270) and Philip IV. (1285-1314) respectively as 63 grains and 61½ grains of remarkably pure silver.   These trials would give the *livres* (20 sols) as equivalent to 18.14 fr. and 17.70 fr. respectively.

CHAPTER XVI.

CONCERNING THE GREAT CITY OF YASDI.

YASDI also is properly in Persia; it is a good and noble city, and has a great amount of trade. They weave there quantities of a certain silk tissue known as *Yasdi*, which merchants carry into many quarters to dispose of. The people are worshippers of Mahommet.[1]

When you leave this city to travel further, you ride for seven days over great plains, finding harbour to receive you at three places only. There are many fine woods [producing dates] upon the way, such as one can easily ride through; and in them there is great sport to be had in hunting and hawking, there being partridges and quails and abundance of other game, so that the merchants who pass that way have plenty of diversion. There are also wild asses, handsome creatures. At the end of those seven marches over the plain you come to a fine kingdom which is called Kerman.[2]

NOTE I.—YEZD, an ancient city, supposed by D'Anville to be the *Isatichae* of Ptolemy, is not called by Marco a kingdom, though having a better title to the distinction than some which he classes as such. The atabegs of Yezd dated from the middle of the 11th century, and their Dynasty was permitted by the Mongols to continue till the end of the 13th, when it was extinguished by Ghazan, and the administration made over to the Mongol Diwan.

Yezd, in pre-Mahomedan times, was a great sanctuary of the Gueber worship, though now it is a seat of fanatical Mahomedanism. It is, however, one of the few places where the old religion lingers. In 1859 there were reckoned 850 families of Guebers in Yezd and fifteen adjoining villages, but they diminish rapidly.

[Heyd (*Com. du Levant*, II. p. 109) says the inhabitants of Yezd wove the finest silk of Taberistan.—H. C.] The silk manufactures still continue, and, with other weaving, employ a large part of the population. The *Yazdi*, which Polo mentions, finds a place in the Persian dictionaries, and is spoken of by D'Herbelot as *Kumásh-i-Yezdi*, "Yezd stuff." [" He [Nadir Shah] bestowed upon the ambassador [Hakeem Ataleek, the prime minister of Abulfiez Khan, King of Bokhara] a donation of a thousand mohurs of Hindostan, twenty-five pieces of *Yezdy* brocade, a rich dress, and a horse with silver harness. . . . " (*Memoirs of Khojah Abdulkurreem, a Cash-*

merian of distinction . . . transl. from the original Persian, by Francis Gladwin . . . Calcutta, 1788, 8vo, p. 36.)—H. C.]

Yezd is still a place of important trade, and carries on a thriving commerce with India by Bandar Abbási. A visitor in the end of 1865 says : " The external trade appears to be very considerable, and the merchants of Yezd are reputed to be amongst the most enterprising and respectable of their class in Persia. Some of their agents have lately gone, not only to Bombay, but to the Mauritius, Java, and China." (*Ilch.* I. 67-68 ; *Khanikoff, Mém.* p. 202 ; *Report by Major R. M. Smith,* R.E.)

Friar Odoric, who visited Yezd, calls it the third best city of the Persian Emperor, and says (*Cathay,* I. p. 52) : " There is very great store of victuals and all other good things that you can mention ; but especially is found there great plenty of figs ; and raisins also, green as grass and very small, are found there in richer profusion than in any other part of the world." [He also gives from the smaller version of Ramusio's an awful description of the Sea of Sand, one day distant from Yezd. (Cf. Tavernier, 1679, I. p. 116.)—H. C.]

NOTE 2.—I believe Della Valle correctly generalises when he says of Persian travelling that "you always travel in a plain, but you always have mountains on either hand" (I. 462). [Compare Macgregor, I. 254 : "I really cannot describe the road. Every road in Persia as yet seems to me to be exactly alike, so . . . my readers will take it for granted that the road went over a waste, with barren rugged hills in the distance, or near ; no water, no houses, no people passed."—H. C.] The distance from Yezd to Kermán is, according to Khanikoff's survey, 314 *kilomètres,* or about 195 miles. Ramusio makes the time eight days, which is probably the better reading, giving a little over 24 miles a day. Westergaard in 1844, and Khanikoff in 1859, took *ten* days ; Colonel Goldsmid and Major Smith in 1865 *twelve.* [" The distance from Yezd to Kermán by the present high road, 229 miles, is by caravans, generally made in nine stages ; persons travelling with all comforts do it in twelve stages ; travellers whose time is of some value do it easily in *seven* days." (*Houtum-Schindler, l.c.* pp. 490-491.)—H. C.]

Khanikoff observes on this chapter : " This notice of woods easy to ride through, covering the plain of Yezd, is very curious. Now you find it a plain of great extent indeed from N.W. to S.E., but narrow and arid ; indeed I saw in it only thirteen inhabited spots, counting two caravanserais. Water for the inhabitants is brought from a great distance by subterraneous conduits, a practice which may have tended to desiccate the soil, for every trace of wood has completely disappeared."

Abbott travelled from Yezd to Kermán in 1849, by a road through Báfk, *east* of the usual road, which Khanikoff followed, and parallel to it ; and it is worthy of note that he found circumstances more accordant with Marco's description. Before getting to Bafk he says of the plain that it " extends to a great distance north and south, and is probably 20 miles in breadth ;" whilst Báfk " is remarkable for its *groves of date-trees,* in the midst of which it stands, and which occupy a considerable space." Further on he speaks of "wild tufts and bushes growing abundantly," and then of " thickets of the *Ghez* tree." He heard of the wild asses, but did not see any. In his report to the Foreign Office, alluding to Marco Polo's account, he says : "It is still true that wild asses and other game are found in the *wooded spots* on the road." The ass is the *Asinus Onager,* the *Gor Khar* of Persia, or *Kulan* of the Tartars. (*Khan. Mém.* p. 200 ; *Id. sur Marco Polo,* p. 21 ; *J. R. G. S.* XXV. 20-29 ; *Mr. Abbott's MS. Report in Foreign office.*) [The difficulty has now been explained by General Houtum-Schindler in a valuable paper published in the *Jour. Roy. As. Soc.* N.S. XIII., October, 1881, p. 490. He says : "Marco Polo travelled from Yazd to Kermán *viâ* Báfk. His description of the road, seven days over great plains, harbour at three places only, is perfectly exact. The fine woods, producing dates, are at Báfk itself. (The place is generally called Báft.) Partridges and quails still abound ; wild asses I saw several on the western road, and I was told that there were a great many on the Báfk road. Travellers and caravans now always go by the eastern road *viâ*

Anár and Bahrámábád. Before the Sefavíehs (*i.e.* before A.D. 1500) the Anár road was hardly, if ever, used; travellers always took the Báfk road. The country from Yazd to Anár, 97 miles, seems to have been totally uninhabited before the Sefavíehs. Anár, as late as A.D. 1340, is mentioned as the frontier place of Kermán to the north, on the confines of the Yazd desert. When Sháh Abbás had caravanserais built at three places between Yazd and Anár (Zein ud-dín, Kermán-sháhán, and Shamsh), the eastern road began to be neglected." (Cf. Major Sykes' *Persia*, ch. xxiii.)—H. C.]

CHAPTER XVII.

CONCERNING THE KINGDOM OF KERMAN.

KERMAN is a kingdom which is also properly in Persia, and formerly it had a hereditary prince. Since the Tartars conquered the country the rule is no longer hereditary, but the Tartar sends to administer whatever lord he pleases.[1] In this kingdom are produced the stones called turquoises in great abundance; they are found in the mountains, where they are extracted from the rocks.[2] There are also plenty of veins of steel and *Ondanique*.[3] The people are very skilful in making harness of war; their saddles, bridles, spurs, swords, bows, quivers, and arms of every kind, are very well made indeed according to the fashion of those parts. The ladies of the country and their daughters also produce exquisite needlework in the embroidery of silk stuffs in different colours, with figures of beasts and birds, trees and flowers, and a variety of other patterns. They work hangings for the use of noblemen so deftly that they are marvels to see, as well as cushions, pillows, quilts, and all sorts of things.[4]

In the mountains of Kerman are found the best falcons in the world. They are inferior in size to the Peregrine, red on the breast, under the neck, and between the thighs; their flight so swift that no bird can escape them.[5]

On quitting the city you ride on for seven days,
always finding towns, villages, and handsome dwelling-
houses, so that it is very pleasant travelling; and there
is excellent sport also to be had by the way in hunting
and hawking. When you have ridden those seven days
over a plain country, you come to a great mountain;
and when you have got to the top of the pass you find
a great descent which occupies some two days to go
down. All along you find a variety and abundance of
fruits; and in former days there were plenty of inhabited
places on the road, but now there are none; and you
meet with only a few people looking after their cattle
at pasture. From the city of Kerman to this descent
the cold in winter is so great that you can scarcely abide
it, even with a great quantity of clothing.[6]

NOTE 1.—Kermán is mentioned by Ptolemy, and also by Ammianus amongst the
cities of the country so called (*Carmania*): "*inter quas nitet* Carmana *omnium
mater.*" (XXIII. 6.)

M. Pauthier's supposition that *Sirján* was in Polo's time the capital, is incorrect.
(See *N. et E.* XIV. 208, 290.) Our Author's Kermán is the city still so called; and
its proper name would seem to have been *Kuwáshír.* (See *Reinaud, Mém. sur l'Inde*,
171; also *Sprenger P. and R. R.* 77.) According to Khanikoff it is 5535 feet above
the sea.

Kermán, on the fall of the Beni Búya Dynasty, in the middle of the 11th century,
came into the hands of a branch of the Seljukian Turks, who retained it till the conquests of the Kings of Khwarizm, which just preceded the Mongol invasion. In
1226 the Amir Borák, a Kara Khitaian, who was governor on behalf of Jaláluddin of
Khwarizm, became independent under the title of Kutlugh Sultan. [He died in 1234.]
The Mongols allowed this family to retain the immediate authority, and at the time
when Polo returned from China the representative of the house was a lady known as
the *Pádishah Khátún* [who reigned from 1291], the wife successively of the Ilkhans
Abaka and Kaikhatu; an ambitious, clever, and masterful woman, who put her own
brother Siyurgutmish to death as a rival, and was herself, after the decease of Kaikhatu,
put to death by her brother's widow and daughter [1294]. The Dynasty continued,
nominally at least, to the reign of the Ilkhan Khodabanda (1304-13), when it was
extinguished. [See Major Sykes' *Persia*, chaps. v. and xxiii.]

Kermán was a Nestorian see, under the Metropolitan of Fars. (*Ilch. passim; Weil*,
III. 454; *Lequien*, II. 1256.)

["There is some confusion with regard to the names of Kermán both as a town and
as a province or kingdom. We have the names Kermán, Kuwáshír, Bardshír. I
should say the original name of the whole country was Kermán, the ancient Karamania. A province of this was called Kúreh-i-Ardeshír, which, being contracted,
became Kuwáshír, and is spoken of as the province in which Ardeshír Bábekán, the
first Sassanian monarch, resided. A part of Kúreh-i-Ardeshír was called Bardshír, or

Bard-i-Ardeshír, now occasionally Bardsír, and the present city of Kermán was situated at its north-eastern corner. This town, during the Middle Ages, was called Bardshír. On a coin of Qara Arslán Beg, King of Kermán, of A.H. 462, Mr. Stanley Lane Poole reads Yazdashír instead of Bardshír. Of Al Idrísí's Yazdashír I see no mention in histories ; Bardshír was the capital and the place where most of the coins were struck. Yazdashír, if such a place existed, can only have been a place of small importance. It is, perhaps, a clerical error for Bardshír ; without diacritical points, both words are written alike. Later, the name of the city became Kermán, the name Bardshír reverting to the district lying south-west of it, with its principal place Mashíz. In a similar manner Mashíz was often, and is so now, called Bardshír. Another old town sometimes confused with Bardshír was Sírján or Shírján, once more important than Bardshír ; it is spoken of as the capital of Kermán, of Bardshír, and of Sardsír. Its name now exists only as that of a district, with principal place S'aídábád. The history of Kermán, 'Agd-ul-'Olá, plainly says Bardshír is the capital of Kermán, and from the description of Bardshír there is no doubt of its having been the present town Kermán. It is strange that Marco Polo does not give the name of the city. In Assemanni's *Bibliotheca Orientalis* Kuwáshír and Bardashír are mentioned as separate cities, the latter being probably the old Mashíz, which as early as A.H. 582 (A.D. 1186) is spoken of in the *History of Kermán* as an important town. The Nestorian bishop of the province Kermán, who stood under the Metropolitan of Fars, resided at Hormúz." (*Houtum-Schindler, l.c.* pp. 491-492.)

There does not seem any doubt as to the identity of Bardashir with the present city of Kermán. (See *The Cities of Kirmán in the time of Hamd-Allah Mustawfi and Marco Polo*, by Guy le Strange, *Jour. R. As. Soc.* April, 1901, pp. 281, 290.) Hamd-Allah is the author of the Cosmography known as the *Nuzhat-al-Kulūb* or " Heart's Delight." (Cf. Major Sykes' *Persia*, chap. xvi., and the *Geographical Journal* for February, 1902, p. 166.)—H. C.]

Note 2.—A MS. treatise on precious stones cited by Ouseley mentions *Shebavek* in Kermán as the site of a Turquoise mine. This is probably *Shahr-i-Babek*, about 100 miles west of the city of Kermán, and not far from *Párez*, where Abbott tells us there is a mine of these stones, now abandoned. Goebel, one of Khanikoff's party, found a deposit of turquoises at Taft, near Yezd. (*Ouseley's Travels*, I. 211 ; *J. R. G. S.* XXVI. 63-65 ; *Khan. Mém.* 203.)

["The province Kermán is still rich in turquoises. The mines of Páríz or Párez are at Chemen-i-mó-aspán, 16 miles from Páríz on the road to Bahrámábád (principal place of Rafsinján), and opposite the village or garden called Gód-i-Ahmer. These mines were worked up to a few years ago ; the turquoises were of a pale blue. Other turquoises are found in the present Bardshír plain, and not far from Mashíz, on the slopes of the Chehel tan mountain, opposite a hill called the Bear Hill (tal-i-Khers). The Shehr-i-Bábek turquoise mines are at the small village Kárík, a mile from Medvár-i-Bálá, 10 miles north of Shehr-i-Bábek. They have two shafts, one of which has lately been closed by an earthquake, and were worked up to about twenty years ago. At another place, 12 miles from Shehr-i-Bábek, are seven old shafts now not worked for a long period. The stones of these mines are also of a very pale blue, and have no great value." (*Houtum-Schindler, l.c.* 1881, p. 491.)

The finest turquoises came from Khorasan ; the mines were near Maaden, about 48 miles to the north of Nishapür. (Heyd, *Com. du Levant*, II. p. 653 ; Ritter, *Erdk.* pp. 325-330.)

It is noticeable that Polo does not mention indigo at Kermán.—H. C.]

Note 3.—Edrisi says that excellent iron was produced in the "cold mountains" N.W. of Jiruft, *i.e.* somewhere south of the capital; and the *Jihán Numá*, or Great Turkish Geography, that the steel mines of Niriz, on the borders of Kermán, were famous. These are also spoken of by Teixeira. Major St. John enables me to in-

dicate their position, in the hills east of Niriz. (*Edrisi*, vol. i. p. 430; *Hammer*, *Mém.
lur la Perse*, p. 275; *Teixeira, Relaciones*, p. 378; and see Map of Itineraries,
No. II.)

[" Marco Polo's steel mines are probably the Parpa iron mines on the road from
Kermán to Shíráz, called even to-day M'aden-i-fúlád (steel mine); they are not worked
now. Old Kermán weapons, daggers, swords, old stirrups, etc., made of steel, are
really beautiful, and justify Marco Polo's praise of them." (*Houtum-Schindler*,
l.c. p. 491.)—H. C.]

Ondanique of the Geog. Text, *Andaine* of Pauthier's, *Andanicum* of the Latin,
is an expression on which no light has been thrown since Ramusio's time. The
latter often asked the Persian merchants who visited Venice, and they all agreed in
stating that it was a sort of steel of such surpassing value and excellence, that in the
days of yore a man who possessed a mirror, or sword, of *Andanic* regarded it as he
would some precious jewel. This seems to me excellent evidence, and to give the
true clue to the meaning of *Ondanique*. I have retained the latter form because it
points most distinctly to what I believe to be the real word, viz. *Hundwániy*,
" Indian Steel." * (See *Johnson's Pers. Dict.* and *De Sacy's Chrestomathie Arabe*, II.
148.) In the *Vocabulista Arabico*, of about A.D. 1200 (Florence, 1871, p. 211),
Hunduwán is explained by *Ensis*. Vüllers explains *Hundwán* as "anything peculiar
to India, especially swords," and quotes from Firdúsi, " *Khanjar-i-Hundwán*," a
hanger of Indian steel.

The like expression appears in the quotation from Edrisi below as *Hindiah*, and
found its way into Spanish in the shapes of *Alhinde, Alfinde, Alinde*, first with the
meaning of *steel*, then assuming, that of *steel mirror*, and finally that of metallic foil
of a glass mirror. (See *Dozy* and *Engelmann*, 2d ed. pp. 144-145.) *Hint* or *Al-hint*
is used in Berber also for steel. (See *J. R. A. S.* IX. 255.)

The sword-blades of India had a great fame over the East, and Indian steel,
according to esteemed authorities, continued to be imported into Persia till days quite
recent. Its fame goes back to very old times. Ctesias mentions two wonderful
swords of such material that he got from the king of Persia and his mother. It is
perhaps the *ferrum candidum* of which the Malli and Oxydracae sent a 100 talents
weight as a present to Alexander.† Indian Iron and Steel (σίδηρος 'Ινδικὸς καὶ
στόμωμα) are mentioned in the *Periplus* as imports into the Abyssinian ports.
Ferrum Indicum appears (at least according to one reading) among the Oriental
species subject to duty in the Law of Marcus Aurelius and Commodus on that matter.
Salmasius notes that among surviving Greek chemical treatises there was one περὶ
βαφῆς 'Ινδικοῦ σιδήρου, " On the Tempering of Indian Steel." Edrisi says on this
subject : "The Hindus excel in the manufacture of iron, and in the preparation of
those ingredients along with which it is fused to obtain that kind of soft Iron which
is usually styled *Indian Steel* (HINDIAH).‡ They also have workshops wherein are
forged the most famous sabres in the world. . . . It is impossible to find anything to
surpass the edge that you get from Indian Steel (*al-hadíd al-Hindí*)."

Allusions to the famous sword-blades of India would seem to be frequent in
Arabic literature. Several will be found in Hamása's collection of ancient Arabic
poems translated by Freytag. The old commentator on one of these passages says :
" *Ut optimos gladios significet* . . . *Indicos esse dixit*," and here the word used in
the original is *Hundwániyah*. In Manger's version of Arabshah's *Life of Timur*

---

* A learned friend objects to Johnson's *Hundwánty* = " Indian Steel," as too absolute ; some word
for *steel* being wanted. Even if it be so, I observe that in the three places where Polo uses *Ondanique*
(here, ch. xxi., and ch. xlii.), the phrase is always " *steel and ondanique*." This looks as if his
mental expression were *Púlád-i-Hundwáni*, rendered by an idiom like Virgil's *pocula et
aurum*.

† Kenrick suggests that the " bright iron " mentioned by Ezekiel among the wares of Tyre (ch.
xxvii. 19) can hardly have been anything else than Indian Steel, because named with cassia and
calamus.

‡ Literally rendered by Mr Redhouse : " The Indians do well the combining of the mixtures of
the chemicals with which they (smelt and) cast the soft iron, and it becomes *Indian* (steel), being referred to India (in this expression)."

are several allusions of the same kind; one, a quotation from *Antar*, recalls the *ferrum candidum* of Curtius:

> "Albi (gladii) Indici *meo in sanguine abluuntur*."

In the histories, even of the Mahomedan conquest of India, the Hindu infidels are sent to *Jihannam* with "the well-watered blade of the Hindi sword"; or the sword is personified as "a Hindu of good family." Coming down to later days, Chardin says of the steel of Persia: "They combine it with Indian steel, which is more tractable . . . . and is much more esteemed." Dupré, at the beginning of this century, tells us: "I used to believe . . . . that the steel for the famous Persian sabres came from certain mines in Khorasan. But according to all the information I have obtained, I can assert that no mine of steel exists in that province. What is used for these blades comes in the shape of disks from Lahore." Pottinger names *steel* among the imports into Kermán from India. Elphinstone the Accurate, in his *Caubul*, confirms Dupré: "Indian Steel [in Afghanistan] is most prized for the material; but the best swords are made in Persia and in Syria;" and in his *History of India*, he repeats: "The steel of India was in request with the ancients; it is celebrated in the oldest Persian poem, and is still the material of the scimitars of Khorasan and Damascus." *

Klaproth, in his *Asia Polyglotta*, gives *Andun* as the Ossetish and *Andan* as the Wotiak, for Steel. Possibly these are essentially the same with *Hundwániy* and *Alhinde*, pointing to India as the original source of supply. [In the *Sikandar Náma*, e *Bará* (or "Book of Alexander the Great," written A.D. 1200, by Abū Muhammad bin Yusuf bin Mu, Ayyid-i-Nizāmu-'d-Dīn), translated by Captain H. Wilberforce Clarke (Lond., 1881, large 8vo), steel is frequently mentioned: Canto xix. 257, p. 202; xx. 12, p. 211; xlv. 38, p. 567; lviii. 32, pp. 695, 42, pp. 697, 62, 66, pp. 699; lix. 28, p. 703.—H. C.]

Avicenna, in his fifth book *De Animâ*, according to Roger Bacon, distinguishes three very different species of iron: "1st. Iron which is good for striking or bearing heavy strokes, and for being forged by hammer and fire, but not for cutting-tools. Of this hammers and anvils are made, and this is what we commonly call *Iron* simply. 2nd. That which is purer, has more heat in it, and is better adapted to take an edge and to form cutting-tools, but is not so malleable, viz. *Steel*. And the 3rd is that which is called ANDENA. This is less known among the Latin nations. Its special character is that like silver it is malleable and ductile under a very low degree of heat. In other properties it is intermediate between iron and steel." (*Fr. R. Baconis Opera Inedita*, 1859, pp. 382-383.) The same passage, apparently, of Avicenna is quoted by Vincent of Beauvais, but with considerable differences. (See *Speculum Naturale*, VII. ch. lii. lx., and *Specul. Doctrinale*, XV. ch. lxiii.) The latter author writes *Alidena*, and I have not been able to refer to Avicenna, so that I am doubtful whether his *Andena* is the same term with the *Andaine* of Pauthier and our *Ondanique*.

The popular view, at least in the Middle Ages, seems to have regarded *Steel* as a distinct natural species, the product of a necessarily different *ore*, from iron; and some such view is, I suspect, still common in the East. An old Indian officer told me of the reply of a native friend to whom he had tried to explain the conversion of iron into steel—"What! You would have me believe that if I put an ass into the furnace it will come forth a horse." And Indian Steel again seems to have been regarded as a distinct natural species from ordinary steel. It is in fact made by a peculiar but simple process, by which the iron is converted *directly* into cast-steel, without passing through any intermediate stage analogous to that of *blister-steel*. When specimens were first examined in England, chemists concluded that the steel was made direct from the *ore*. The *Ondanique* of Marco no doubt was a fine steel resembling the

---

* In *Richardson's Pers. Dict.*, by Johnson, we have a word *Rohan, Rohina* (and other forms). "The finest Indian steel, of which the most excellent swords are made; also the swords made of that steel."

Texture, with Animals, etc., from a Cashmere Scarf in the Indian Museum.

" De deverses maineres laborés à bestes et ausiaus mout richement."

Indian article. (*Müller's Ctesias*, p. 80 ; *Curtius*, IX. 24 ; *Müller's Geog.Gr. Min.* I. 262 ; *Digest. Novum*, Lugd. 1551, Lib. XXXIX. Tit. 4 ; *Salmas. Ex. Plinian.* II. 763 ; *Edrisi*, I. 65-66 ; *J. R. S. A.* A. 387 *seqq. ; Hamasae Carmina*, I. 526 ; *Elliot*, II. 209, 394 ; *Reynolds's Utbi*, p. 216.)

NOTE 4.—Paulus Jovius in the 16th century says, I know not on what authority, that Kermán was then celebrated for the fine temper of its steel in scimitars and lance-points. These were eagerly bought at high prices by the Turks, and their quality was such that one blow of a Kermán sabre would cleave an European helmet without turning the edge. And I see that the phrase, "Kermání blade" is used in poetry by Marco's contemporary Amír Khusrú of Delhi. (*P. Jov. Hist. of his own Time*, Bk. XIV. ; *Elliot*, III. 537.)

There is, or was in Pottinger's time, still a great manufacture of *matchlocks* at Kerman ; but rose-water, shawls, and carpets are the staples of the place now. Polo says nothing that points to shawl-making, but it would seem from Edrisi that some such manufacture already existed in the adjoining district of Bamm. It is possible that the "hangings" spoken of by Polo may refer to the carpets. I have seen a genuine Kermán carpet in the house of my friend, Sir Bartle Frere. It is of very short pile, very even and dense ; the design, a combination of vases, birds, and floral tracery, closely resembling the illuminated frontispiece of some Persian MSS.

The shawls are inferior to those of Kashmir in exquisite softness, but scarcely in delicacy of texture and beauty of design. In 1850, their highest quality did not exceed 30 *tomans* (14*l.*) in price. About 2200 looms were employed on the fabric. A good deal of Kermán wool called *Kurk*, goes *viâ* Bandar Abbási and Karáchi to Amritsar, where it is mixed with the genuine Tibetan wool in the shawl manufacture. Several of the articles named in the text, including *pardahs* ("cortines") are woven in shawl-fabric. I scarcely think, however, that Marco would have confounded woven shawl with needle embroidery. And Mr. Khanikoff states that the silk embroidery, of which Marco speaks, is still performed with great skill and beauty at Kermán. Our cut illustrates the textures figured with animals, already noticed at p. 66.

The Guebers were numerous here at the end of last century, but they are rapidly disappearing now. The Musulman of Kermán is, according to Khanikoff, an epicurean gentleman, and even in regard to wine, which is strong and plentiful, his divines are liberal. "In other parts of Persia you find the scribblings on the walls of Serais to consist of philosophical axioms, texts from the Koran, or abuse of local authorities. From Kermán to Yezd you find only rhymes in praise of fair ladies or good wine."

(*Pottinger's Travels ; Khanik. Mém.* 186 *seqq.*, and *Notice*, p. 21 ; *Major Smith's Report ; Abbott's MS. Report* in F. O. ; *Notes by Major O. St. John*, R.E.)

NOTE 5.—Parez is famous for its falcons still, and so are the districts of Aktár and Sïrján. Both Mr. Abbott and Major Smith were entertained with hawking by Persian hosts in this neighbourhood. The late Sir O. St. John identifies the bird described as the *Sháhín* (Falco *Peregrinator*), one variety of which, the *Fársi*, is abundant in the higher mountains of S. Persia. It is now little used in that region, the *Terlán* or goshawk being most valued, but a few are caught and sent for sale to the Arabs of Oman. (*J. R. G. S.* XXV. 50, 63, and *Major St. John's Notes.*)

["The fine falcons, 'with red breasts and swift of flight,' come from Páríz. They are, however, very scarce, two or three only being caught every year. A well-trained Páríz falcon costs from 30 to 50 tomans (12*l.* to 20*l.*), as much as a good horse." (*Houtum-Schindler, l.c.* p. 491.) Major Sykes, *Persia*, ch. xxiii., writes : "Marco Polo was evidently a keen sportsman, and his description of the *Sháhin*, as it is termed, cannot be improved upon." Major Sykes has a list given him by a Khán of seven hawks of the province, all black and white, except the *Sháhín*, which has yellow eyes, and is the third in the order of size.—H. C.]

Note 6.—We defer geographical remarks till the traveller reaches Hormuz.

CHAPTER XVIII.

Of the City of Camadi and its Ruins; also touching the Carauna Robbers.

AFTER you have ridden down hill those two days, you find yourself in a vast plain, and at the beginning thereof there is a city called CAMADI, which formerly was a great and noble place, but now is of little consequence, for the Tartars in their incursions have several times ravaged it. The plain whereof I speak is a very hot region; and the province that we now enter is called REOBARLES.

The fruits of the country are dates, pistachioes, and apples of Paradise, with others of the like not found in our cold climate. [There are vast numbers of turtle-doves, attracted by the abundance of fruits, but the Saracens never take them, for they hold them in abomination.] And on this plain there is a kind of bird called francolin, but different from the francolin of other countries, for their colour is a mixture of black and white, and the feet and beak are vermilion colour.[1]

The beasts also are peculiar; and first I will tell you of their oxen. These are very large, and all over white as snow; the hair is very short and smooth, which is owing to the heat of the country. The horns are short and thick, not sharp in the point; and between the shoulders they have a round hump some two palms high. There are no handsomer creatures in the world. And when they have to be loaded, they kneel like the camel; once the load is adjusted, they rise. Their load is a heavy one, for they are very strong animals. Then there are sheep here as big as asses; and their tails are so large and fat, that one tail shall weigh some 30 lbs. They are fine fat beasts, and afford capital mutton.[2]

In this plain there are a number of villages and
towns which have lofty walls of mud, made as a defence
against the banditti,[3] who are very numerous, and are
called CARAONAS. This name is given them because
they are the sons of Indian mothers by Tartar fathers.
And you must know that when these Caraonas wish to
make a plundering incursion, they have certain devilish
enchantments whereby they do bring darkness over the
face of day, insomuch that you can scarcely discern your
comrade riding beside you; and this darkness they will
cause to extend over a space of seven days' journey.
They know the country thoroughly, and ride abreast,
keeping near one another, sometimes to the number of
10,000, at other times more or fewer. In this way they
extend across the whole plain that they are going to
harry, and catch every living thing that is found outside
of the towns and villages; man, woman, or beast, nothing
can escape them! The old men whom they take in this
way they butcher; the young men and the women they
sell for slaves in other countries; thus the whole land is
ruined, and has become well-nigh a desert.

The King of these scoundrels is called NOGODAR.
This Nogodar had gone to the Court of Chagatai, who
was own brother to the Great Kaan, with some 10,000
horsemen of his, and abode with him; for Chagatai was
his uncle. And whilst there this Nogodar devised a
most audacious enterprise, and I will tell you what it was.
He left his uncle who was then in Greater Armenia, and
fled with a great body of horsemen, cruel unscrupulous
fellows, first through BADASHAN, and then through
another province called PASHAI-DIR, and then through
another called ARIORA-KESHEMUR. There he lost a
great number of his people and of his horses, for the
roads were very narrow and perilous. And when he had
conquered all those provinces, he entered India at the

extremity of a province called DALIVAR.   He established himself in that city and government, which he took from the King of the country, ASEDIN SOLDAN by name, a man of great power and wealth.   And there abideth Nogodar with his army, afraid of nobody, and waging war with all the Tartars in his neighbourhood.[4]

Now that I have told you of those scoundrels and their history, I must add the fact that Messer Marco himself was all but caught by their bands in such a darkness as that I have told you of; but, as it pleased God, he got off and threw himself into a village that was hard by, called CONOSALMI.   Howbeit he lost his whole company except seven persons who escaped along with him.   The rest were caught, and some of them sold, some put to death.[5]

NOTE 1.—Ramusio has "Adam's apple" for apples of Paradise.   This was some kind of *Citrus*, though Lindley thinks it impossible to say precisely what.   According to Jacques de Vitry it was a beautiful fruit of the Citron kind, in which the bite of human teeth was plainly discernible.   (Note to *Vulgar Errors*, II. 211 ; *Bongars*, I. 1099.)   Mr. Abbott speaks of this tract as "the districts (of Kermán) lying towards the South, which are termed the Ghermseer or Hot Region, where the temperature of winter resembles that of a charming spring, and where the palm, orange, and lemon-tree flourish." (*MS. Report;* see also *J. R. G. S.* XXV. 56.)

["Marco Polo's apples of Paradise are more probably the fruits of the Konár tree. There are no plantains in that part of the country.   Turtle doves, now as then, are plentiful, and as they are seldom shot, and are said by the people to be unwholesome food, we can understand Marco Polo's saying that the people do not eat them." (*Houtum-Schindler, l.c.* pp. 492-493.)—H. C.]

The Francolin here spoken of is, as Major Smith tells me, the *Darráj* of the Persians, the *Black Partridge* of English sportsmen, sometimes called the Red-legged Francolin.   The Darráj is found in some parts of Egypt, where its peculiar call is interpreted by the peasantry into certain Arabic words, meaning "Sweet are the corn-ears !   Praised be the Lord !"   In India, Baber tells us, the call of the Black Partridge was (less piously) rendered "*Shír dáram shakrak*," "I've got milk and sugar !"   The bird seems to be the ἀτταγάς of Athenaeus, a fowl "speckled like the partridge, but larger," found in Egypt and Lydia.   The Greek version of its cry is the best of all : "τρὶς τοῖς κακούργοις κακά" ("Threefold ills to the ill-doers !").   This is really like the call of the black partridge in India as I recollect it. [*Tetrao francolinus.*—H. C.]

(*Chrestomathie Arabe*, II. 295 ; *Baber*, 320 ; *Yonge's Athen.* IX. 39.)

NOTE 2.—Abbott mentions the humped (though small) oxen in this part of Persia, and that in some of the neighbouring districts they are taught to kneel to receive the load, an accomplishment which seems to have struck Mas'udi (III. 27), who says he saw it exhibited by oxen at Rai (near modern Tehran).   The Aín Akbari also ascribes it to a very fine breed in Bengal.   The whimsical name Zebu, given to the humped or Indian ox in books of Zoology, was taken by Buffon from the exhibitors of such a beast at a French Fair, who probably invented it. That the humped breeds of oxen existed in this part of Asia in ancient times is shown by sculptures at Kouyunjik. (See cut below.)

A letter from Agassiz, printed in the Proc. As. Soc. Bengal (1865), refers to wild "zebus," and calls the species a small one. There is no wild "zebu," and some of the breeds are of enormous size.

[" White oxen, with short thick horns and a round hump between the shoulders, are now very rare between Kermán and Bender 'Abbás. They are, however, still to be found towards Belúchistán and Mekrán, and they kneel to be loaded like camels. The sheep which I saw had fine large tails; I did not, however, hear of any having so high a weight as thirty pounds." (*Houtum-Schindler, l.c.* p. 493.)—H. C.]

The fat-tailed sheep is well known in many parts of Asia and part of Africa. It is mentioned by Ctesias, and by Ælian, who says the shepherds used to extract the tallow from the live animal, sewing up the tail again; exactly the same story is told by the Chinese Pliny, Ma Twan-lin. Marco's statements as to size do not surpass those of the admirable Kämpfer: " In size they so much surpass the common sheep that it is not unusual to see them as tall as a donkey, whilst all are much more than three feet; and as to the tail I shall not exceed the truth, though I may exceed belief, if I say that it sometimes reaches 40 lbs. in weight." Captain Hutton was assured by an Afghan sheep-master that tails had occurred in his flocks weighing 12 Tabriz *mans*, upwards of 76 lbs.! The Afghans use the fat as an aperient, swallowing a dose of 4 to 6 lbs! Captain Hutton's friend testified that trucks to bear the sheep-tails were

Humped Oxen from the Assyrian Sculptures at Koyunjik.

sometimes used among the Taimúnis (north of Herat). This may help to locate that ancient and slippery story. Josafat Barbaro says he had seen the thing, but is vague as to place. (*Ælian Nat. An.* III. 3, IV. 32; *Amoen. Exoticae; Ferrier*, H. of Afghans, p. 294; *J. A. S. B.* XV. 160.)

[Rabelais says (Bk. I. ch. xvi.): "Si de ce vous efmerveillez, efmerveillez vous d'advantage de la queue des béliers de la Scythie, qui pesait plus de trente livres; et des moutons de Surie, esquels fault (si Tenaud, dict vray) affuster une charrette au cul, pour la porter tant qu'elle est longue et pesante." (See G. Capus, *A travers le roy. de Tamerlan*, pp. 21-23, on the fat sheep.)—H. C.]

NOTE 3.—The word rendered *banditti* is in Pauthier *Carans*, in G. Text *Caraunes*, in the Latin "*a scaranis et malandrinis*." The last is no doubt correct, standing for the old Italian *Scherani*, bandits. (See *Cathay*, p. 287, note.)

NOTE 4.—This is a knotty subject, and needs a long note.

The ḲARAUNAHS are mentioned often in the histories of the Mongol regime in Persia, first as a Mongol tribe forming a *Tuman*, *i.e.* a division or corps of 10,000 in the Mongol army (and I suspect it was the phrase the *Tuman of the Ḳaraunahs* in Marco's mind that suggested his repeated use of the number 10,000 in speaking of them) ; and afterwards as daring and savage freebooters, scouring the Persian provinces, and having their headquarters on the Eastern frontiers of Persia. They are described as having had their original seats on the mountains north of the Chinese wall near *Ḳaraún Jidun* or *Khidun ;* and their special accomplishment in war was the use of Naphtha Fire. Rashiduddin mentions the *Ḳaránut* as a branch of the great Mongol tribe of the Kunguráts, who certainly had their seat in the vicinity named, so these may possibly be connected with the Ḳaraunahs. The same author says that the Tuman of the Ḳaraunahs formed the *Injú* or *peculium* of Arghún Khan.

Wassáf calls them "a kind of goblins rather than human beings, the most daring of all the Mongols" ; and Mirkhond speaks in like terms.

Dr. Bird of Bombay, in discussing some of the Indo-Scythic coins which bear the word *Korano* attached to the prince's name, asserts this to stand for the name of the Ḳaraunah, "who were a Græco-Indo-Scythic tribe of robbers in the Punjab, who are mentioned by Marco Polo," a somewhat hasty conclusion which Pauthier adopts. There is, Quatremère observes, no mention of the Ḳaraunahs before the Mongol invasion, and this he regards as the great obstacle to any supposition of their having been a people previously settled in Persia. Reiske, indeed, with no reference to the present subject, quotes a passage from Hamza of Ispahan, a writer of the 10th century, in which mention is made of certain troops called *Ḳaráunahs*. But it seems certain that in this and other like cases the real reading was *Ḳazáwinah*, people of Kazvin. (See *Reiske's Constant. Porphyrog.* Bonn. ed. II. 674; *Gottwaldt's Hamza Ispahanensis*, p. 161 ; and *Quatremère* in *J. A.* sér. V. tom. xv. 173.) Ibn Batuta only once mentions the name, saying that Tughlak Sháh of Dehli was "one of those Turks called *Ḳaráunas* who dwell in the mountains between Sind and Turkestan." Hammer has suggested the derivation of the word *Carbine* from *Ḳaráwinah* (as he writes), and a link in such an etymology is perhaps furnished by the fact that in the 16th century the word *Carbine* was used for some kind of irregular horseman.

(*Gold. Horde*, 214; *Ilch.* I. 17, 344, etc. ; *Erdmann*, 168, 199, etc. ; *J. A. S.*, B. X. 96; *Q. R.* 130; *Not. et Ext.* XIV. 282; *I. B.* III. 201 ; *Ed. Webbe, his Travailes*, p. 17, 1590. Reprinted 1868.)

As regards the account given by Marco of the origin of the Caraonas, it seems almost necessarily a mistaken one. As Khanikoff remarks, he might have confounded them with the Biluchis, whose Turanian aspect (at least as regards the Brahuis) shows a strong infusion of Turki blood, and who might be rudely described as a cross between Tartars and Indians. It is indeed an odd fact that the word *Kardni* (vulgo *Cranny*) is commonly applied in India at this day to the mixed race sprung from European fathers and Native mothers, and this might be cited in corroboration of Marsden's reference to the Sanskrit *Karana*, but I suspect the coincidence arises in another way. *Karana* is the name applied to a particular class of mixt blood, whose special occupation was writing and accounts. But the prior sense of the word seems to have been "clever, skilled," and hence a writer or scribe. In this sense we find *Karáni* applied in Ibn Batuta's day to a ship's clerk, and it is used in the same sense in the *Aín Akbari*. Clerkship is also the predominant occupation of the East-Indians, and hence the term Karáni is applied to them from their business, and not from their mixt blood. We shall see hereafter that there is a Tartar term *Arghún*, applied to fair children born of a Mongol mother and *white* father ; it is possible that there may have been a correlative word like *Karáun* (from *Ḳará*, black) applied to dark children born of Mongol father and black mother, and that this led Marco to a false theory.

[Major Sykes (*Persia*) devotes a chapter (xxiv.) to *The Karwán Expedition* in which he says : " Is it not possible that the Karwánis are the Caraonas of Marco Polo ? They are distinct from the surrounding Baluchis, and pay no tribute."—H. C.]

Let us turn now to the name of Nogodar. Contempora:neously with the Karaunahs we have frequent mention of predatory bands known as *Nigúdaris*, who seem to be distinguished from the Karaunahs, but had a like character for truculence. Their headquarters were about Sijistán, and Quatremère seems disposed to look upon them as a tribe indigenous in that quarter. Hammer says they were originally the troops of

Portrait of a Hazára.

Prince Nigudar, grandson of Chaghatai, and that they were a rabble of all sorts, Mongols, Turkmans, Kurds, Shúls, and what not. We hear of their revolts and disorders down to 1319, under which date Mirkhond says that there had been one-and-twenty fights with them in four years. Again we hear of them in 1336 about Herat, whilst in Baber's time they turn up as *Nukdari*, fairly established as tribes in the mountainous tracts of Karnúd and Ghúr, west of Kabul, and coupled with the Hazáras, who still survive both in name and character. " Among both," says Baber, " there are some who speak the Mongol language." Hazáras and *Takdaris* (read *Nukdaris*) again occur coupled in the *History of Sind*. (See *Elliot*, I. 303-304.) [On the struggle against Timur of Toumen, veteran chief of the Nikoudrians (1383-84), see Major David Price's *Mahommedan History*, London, 1821, vol. iii. pp. 47-49, H. C.] In maps of the 17th century, as of Hondius and Blaeuw, we find the moun-tains north of Kabul termed *Nochdarizari*, in which we cannot miss the combination Nigudar-Hazárah, whencesoever it was got. The Hazáras are eminently Mongol in feature to this day, and it is very probable that they or some part of them are the descendants of the Karáunahs or the Nigudaris, or of both, and that the origination of the bands so called, from the scum of the Mongol inundation, is thus in degree confirmed. The Hazáras generally are said to speak an old dialect of Persian. But one tribe in Western Afghanistan retains both the name of Mongols and a language of which six-sevenths (judging from a vocabulary published by Major Leech) appear to be Mongol. Leech says, too, that the Hazáras generally are termed *Moghals* by the Ghilzais. It is worthy of notice that Abu'l Fázl, who also mentions the Nukdaris among the nomad tribes of Kabul, says the Hazáras were the remains of the Chaghataian army which Mangu Kaan sent to the aid of Hulaku, under the command of Nigudar Oghlan. (*Not. et Ext.* XIV. 284; *Ilch.* I. 284, 309, etc.; *Baber,* 134, 136, 140; *J. As.* sér. IV. tom. iv. 98; *Ayeen Akbery,* II. 192-193.)

So far, excepting as to the doubtful point of the relation between Karáunahs and Nigudaris, and as to the origin of the former, we have a general accordance with Polo's representations. But it is not very easy to identify with certainty the inroad on India to which he alludes, or the person intended by Nogodar, nephew of Chaghatai. It seems as if two persons of that name had each contributed something to Marco's history.

We find in Hammer and D'Ohsson that one of the causes which led to the war between Barka Khan and Hulaku in 1262 (see above, *Prologue,* ch. ii.) was the violent end that had befallen three princes of the House of Juji, who had accompanied Hulaku to Persia in command of the contingent of that House. When war actually broke out, the contingent made their escape from Persia. One party gained Kipchak by way of Derbend ; another, in greater force, led by NIGUDAR and Onguja, escaped to Khorasan, pursued by the troops of Hulaku, and thence eastward, where they seized upon Ghazni and other districts bordering on India.

But again : Nigudar Aghul, or Oghlan, son of (the younger) Juji, son of *Chaghatai,* was the leader of the Chaghataian contingent in Hulaku's expedition, and was still attached to the Mongol-Persian army in 1269, when Borrak Khan, of the House of Chaghatai, was meditating war against his kinsman, Abaka of Persia. Borrak sent to the latter an ambassador, who was the bearer of a secret message to Prince Nigudar, begging him not to serve against the head of his own House. Nigudar, upon this, made a pretext of retiring to his own headquarters in *Georgia,* hoping to reach Borrak's camp by way of Derbend. He was, however, intercepted, and lost many of his people. With 1000 horse he took refuge in Georgia, but was refused an asylum, and was eventually captured by Abaka's commander on that frontier. His officers were executed, his troops dispersed among Abaka's army, and his own life spared under surveillance. I find no more about him. In 1278 Hammer speaks of him as dead, and of the Nigudarian bands as having been formed out of his troops. But authority is not given.

The second Nigudar is evidently the one to whom Abu'l Fázl alludes. Khanikoff assumes that the Nigudar who went off towards India about 1260 (he puts the date earlier) was Nigudar, the grandson of Chaghatai, but he takes no notice of the second story just quoted.

In the former story we have bands under *Nigudar* going off by Ghazni, *and conquering country on the Indian frontier.* In the latter we have *Nigudar, a descendant of Chaghatai,* trying to escape from his camp *on the frontier of Great Armenia.* Supposing the Persian historians to be correct, it looks as if Marco had rolled two stories into one.

Some other passages may be cited before quitting this part of the subject. A chronicle of Herat, translated by Barbier de Meynard, says, under 1298: "The King Fakhruddin (of Herat) had the imprudence to authorise *the Amir Nigudar* to establish himself in a quarter of the city, with 300 adventurers from 'Irak. This little troop made frequent raids in Kuhistan, Sijistan, Farrah, etc., spreading terror. Khodabanda, at the request of his brother Ghazan Khan, came from Mazanderan to demand the immediate surrender of these brigands," etc. And in the account of the tremendous foray of the Chaghataian Prince Kotlogh Shah, on the east and south of Persia in 1299, we find one of his captains called *Nigudar* Bahadur. (*Gold. Horde,* 146, 157, 164; *D'Ohsson,* IV. 378 *seqq.,* 433 *seqq.,* 513 *seqq.; Ilch.* I. 216, 261, 284; II. 104; *J. A.* sér. V. tom. xvii. 455-456, 507; *Khan. Notice,* 31.)

As regards the route taken by Prince Nogodar in his incursion into India, we have no difficulty with BADAKHSHAN. PASHAI-DIR is a copulate name; the former part, as we shall see reason to believe hereafter, representing the country between the Hindu Kush and the Kabul River (see *infra,* ch. xxx.); the latter (as Pauthier already has pointed out), DIR, the chief town of Panjkora, in the hill country north of Peshawar. In *Ariora-Keshemur* the first portion only is perplexing. I will mention the most probable of the solutions that have occurred to me, and a second, due to that eminent archæologist, General A. Cunningham. (1) *Ariora* may be some corrupt or Mongol form of *Aryavartta,* a sacred name applied to the Holy Lands of Indian Buddhism, of which Kashmir was eminently one to the Northern Buddhists. *Oron,* in Mongol, is a Region or Realm, and may have taken the place of *Vartta,* giving *Aryoron* or Ariora. (2) "*Ariora,*" General Cunningham writes, "I take to be the *Harhaura* of Sanscrit—*i.e.* the Western Panjáb. Harhaura was the North-Western Division of the *Nava-Khanda,* or Nine Divisions of Ancient India. It is mentioned between *Sindhu-Sauvira* in the west (*i.e.* Sind), and *Madra* in the north (*i.e.* the Eastern Panjáb, which is still called *Madar-Des*). The name of Harhaura is, I think, preserved in the Haro River. Now, the Sind-Sagor Doab formed a portion of the kingdom of Kashmir, and the joint names, like those of Sindhu-Sauvira, describe only one State." The names of the Nine Divisions in question are given by the celebrated astronomer, Varaha Mihira, who lived in the beginning of the 6th century, and are repeated by Al Biruni. (See *Reinaud, Mém. sur l'Inde,* p. 116.) The only objection to this happy solution seems to lie in Al Biruni's remark, that the names in question were in general no longer used even in his time (A.D. 1030).

There can be no doubt that *Asidin Soldan* is, as Khanikoff has said, Ghaiassuddin Balban, Sultan of Delhi from 1266 to 1286, and for years before that a man of great power in India, and especially in the Panjáb, of which he had in the reign of Ruknuddin (1236) held independent possession.

Firishta records several inroads of Mongols in the Panjáb during the reign of Ghaiassuddin, in withstanding one of which that King's eldest son was slain; and there are constant indications of their presence in Sind till the end of the century. But we find in that historian no hint of the chief circumstances of this part of the story, viz., the conquest of Kashmir and the occupation of *Dalivar* or *Dilivar* (G. T.), evidently (whatever its identity) in the plains of India. I do find, however, in the history of Kashmir, as given by Lassen (III. 1138), that in the end of 1259, Lakshamana Deva, King of Kashmir, was killed in a campaign against the *Turushka* (Turks or Tartars), and that their leader, who is called Kajjala, got hold of the country and held it till 1287.* It is difficult not to connect this both with Polo's story and with the escapade of Nigudar about 1260, noting also that this occupation of Kashmir extended through the whole reign of Ghaiassuddin.

We seem to have a memory of Polo's story preserved in one of Elliot's extracts from Wassáf, which states that in 708 (A.D. 1308), after a great defeat of a Mongol inroad which had passed the Ganges, Sultan Ala'uddin Khilji ordered a pillar of Mongol heads to be raised before the Badáun gate, "*as was done with the* Nigudari *Moghuls*" (III. 48).

We still have to account for the occupation and locality of *Dalivar;* Marsden supposed it to be *Lahore;* Khanikoff considers it to be *Diráwal,* the ancient desert capital of the Bhattis, properly (according to Tod) *Deoráwal,* but by a transposition common in India, as it is in Italy, sometimes called *Diláwar,* in the modern State of Bháwalpúr. But General Cunningham suggests a more probable locality in DILÁWAR on the west bank of the Jelam, close to Dárápúr, and opposite to Mung. These two

---

* *Khajlak* is mentioned as a leader of the Mongol raids in India by the poet Amír Khusrú (A.D. 1289; see *Elliot,* III. 527).

sites, Diláwar-Dárápúr on the west bank, and Mung on the east, are identified by General Cunningham (I believe justly) with Alexander's Bucephala and Nicaea.   The spot, which is just opposite the battlefield of Chiliánwála, was visited (15th December, 1868) at my request, by my friend Colonel R. Maclagan, R.E.   He writes: "The present village of Diláwar stands a little above the town of Dárápúr (I mean on higher ground), looking down on Dárápúr and on the river, and on the cultivated and wooded plain along the river bank.   The remains of the Old Diláwar, in the form of quantities of large bricks, cover the low round-backed spurs and knolls of the broken rocky hills around the present village, but principally on the land side.   They cover a large area of very irregular character, and may clearly be held to represent a very considerable town.   There are no indications of the form of buildings, ... but simply large quantities of large bricks, which for a long time have been carried away and used for modern buildings. . . . . After rain coins are found on the surface. . . . . There can be no doubt of a very large extent of ground, of very irregular and uninviting character, having been covered at some time with buildings.   The position on the Jelam would answer well for the Diláwar which the Mongol invaders took and held. . . . . The strange thing is that the name should not be mentioned (I believe it is not) by any of the well-known Mahomedan historians of India.   So much for Diláwar. . . . . The people have no traditions.   But there are the remains ; and there is the name, borne by the existing village on part of the old site."   I had come to the conclusion that this was almost certainly Polo's Dalivar, and had mapped it as such, before I read certain passages in the *History of Zíyáuddín Barni*, which have been translated by Professor Dowson for the third volume of Elliot's *India*.   When the comrades of Ghaiassuddin Balban urged him to conquests, the Sultan pointed to the constant danger from the Mongols,[*] saying : "These accursed wretches have heard of the wealth and condition of Hindustan, and have set their hearts upon conquering and plundering it.   *They have taken and plundered Lahor within my territories, and no year passes that they do not come here and plunder the villages. . . . .* They even talk about the conquest and sack of Delhi."   And under a later date the historian says : "The Sultan . . . . marched to Lahor, and ordered the rebuilding of the fort which the Mughals had destroyed in the reigns of the sons of Shamsuddin.   The towns and villages of Lahor which the Mughals had devastated and laid waste he repeopled."   Considering these passages, and the fact that Polo had no personal knowledge of Upper India, I now think it probable that Marsden was right, and that *Dilivar* is really a misunderstanding of " *Città* di Livar " for *Lahàwar* or Lahore.

The *Magical darkness* which Marco ascribes to the evil arts of the Karaunas is explained by Khanikoff from the phenomenon of *Dry Fog*, which he has often experienced in Khorasan, combined with the *Dust Storm* with which we are familiar in Upper India.   In Sind these phenomena often produce a great degree of darkness. During a battle fought between the armies of Sindh and Kachh in 1762, such a fog came on, obscuring the light of day for some six hours, during which the armies were intermixed with one another and fighting desperately.   When the darkness dispersed they separated, and the consternation of both parties was so great at the events of the day that both made a precipitate retreat.   In 1844 this battle was still spoken of with wonder.   (*J. Bomb. Br. R. A. S.* I. 423.)

Major St. John has given a note on his own experience of these curious Kermán fogs (see *Ocean Highways*, 1872, p. 286) : "Not a breath of air was stirring, and the whole effect was most curious, and utterly unlike any other fog I have seen.   No deposit of dust followed, and the feeling of the air was decidedly damp.   I unfortunately could not get my hygrometer till the fog had cleared away."

[*General Houtum-Schindler, l.c.* p. 493, writes : "The magical darkness might, as Colonel Yule supposes, be explained by the curious dry fogs or dust storms, often occurring in the neighbourhood of Kermán, but it must be remarked that Marco Polo

---

[*] Professor Cowell compares the Mongol inroads in the latter part of the 13th and beginning of the 14th century, in their incessant recurrence, to the incursions of the Danes in England.   A passage in Wassáf (*Elliot*, III. 38) shows that the Mongols were, *circa* 1254-55, already in occupation of Sodra on the Chenab, and districts adjoining.

was caught in one of these storms down in Jíruft, where, according to the people I questioned, such storms now never occur. On the 29th of September, 1879, at Kermán, a high wind began to blow from S.S.W. at about 5 P.M. First there came thick heavy clouds of dust with a few drops of rain. The heavy dust then settled down, the lighter particles remained in the air, forming a dry fog of such density that large objects, like houses, trees, etc., could not even faintly be distinguished at a distance of a hundred paces. The barometers suffered no change, the three I had with me remained in *statu quo*." "The heat is over by the middle of September, and after the autumnal equinox, there are a few days of what is best described as a dense dry fog. This was undoubtedly the haze referred to by Marco Polo." (*Major Sykes*, ch. iv.)—H.C.]

"Richthofen's remarkable exposition of the phenomena of the *löss* in North China, and of the sub-aerial deposits of the steppes and of Central Asia throws some light on this. But this hardly applies to St John's experience of "no deposit of dust." (See Richthofen, *China*, pp. 96-97 s. *MS. Note*, H. Y.)

The belief that such opportune phenomena were produced by enchantment was a thoroughly Tartar one. D'Herbelot relates (art. *Giagathai*) that in an action with a rebel called Mahomed Tarabi, the Mongols were encompassed by a dust storm which they attributed to enchantment on the part of the enemy, and it so discouraged them that they took to flight.

NOTE 5.—The specification that only *seven* were saved from Marco's company is peculiar to Pauthier's Text, not appearing in the G. T.

Several names compounded of *Salm* or *Salmi* occur on the dry lands on the borders of Kermán. Edrisi, however (I. p. 428), names a place called ḲANÁT-UL-SHÁM as the first march in going from Jiruft to Walashjird. Walashjird is, I imagine, represented by *Galashkird*, Major R. Smith's third march from Jíruft (see my Map of Routes from Kermán to Hormuz); and as such an indication agrees with the view taken below of Polo's route, I am strongly disposed to identify Ḳanát-ul-Shám with his *castello* or walled village of *Canosalmi*.

["Marco Polo's Conosalmi, where he was attacked by robbers and lost the greater part of his men, is perhaps the ruined town or village Kamasal (Kahn-i-asal = the honey canal), near Kahnúj-i-pancheh and Vakílábád in Jíruft. It lies on the direct road between Shehr-i-Daqíánús (Camadi) and the Nevergún Pass. The road goes in an almost due southerly direction. The Nevergún Pass accords with Marco Polo's description of it; it is very difficult, on account of the many great blocks of sandstone scattered upon it. Its proximity to the Bashakird mountains and Mekrán easily accounts for the prevalence of robbers, who infested the place in Marco Polo's time. At the end of the Pass lies the large village Shamfl, with an old fort; the distance thence to the site of Hormúz or Bender 'Abbás (lying more to the west) is 52 miles, two days' march. The climate of Bender 'Abbás is very bad, strangers speedily fall sick, two of my men died there, all the others were seriously ill." (*Houtum-Schindler*, *l.c.* pp. 495-496.) Major Sykes (ch. xxiii.) says: "Two marches from Camadi was Kahn-i-Panchur, and a stage beyond it lay the ruins of Fariáb or Pariáb, which was once a great city, and was destroyed by a flood, according to local legend. It may have been Alexander's Salmous, as it is about the right distance from the coast, and if so, could not have been Marco's *Cono Salmi*. Continuing on, Galashkird mentioned by Edrisi, is the next stage."—H. C.]

The raids of the Mekranis and Biluchis long preceded those of the Karaunas, for they were notable even in the time of Mahmud of Ghazni, and they have continued to our own day to be prosecuted nearly on the same stage and in the same manner. About 1721, 4000 horsemen of this description plundered the town of Bander Abbasi, whilst Captain Alex. Hamilton was in the port; and Abbott, in 1850, found the dread of Bilúch robbers to extend almost to the gates of Ispahan. A striking account of the Bilúch robbers and their characteristics is given by General Ferrier. (See *Hamilton*, I. 109; *J. R.G. S.* XXV.; *Khanikoff's Mémoire*; *Macd. Kinneir*, 196; *Caravan Journeys*, p. 437 *seq.*)

CHAPTER XIX.

OF THE DESCENT TO THE CITY OF HORMOS.

THE Plain of which we have spoken extends in a
southerly direction for five days' journey, and then
you come to another descent some twenty miles in
length, where the road is very bad and full of peril,
for there are many robbers and bad characters about.
When you have got to the foot of this descent you find
another beautiful plain called the PLAIN OF FORMOSA.
This extends for two days' journey; and you find in it
fine streams of water with plenty of date-palms and other
fruit-trees. There are also many beautiful birds, franco-
lins, popinjays, and other kinds such as we have none of
in our country. When you have ridden these two days
you come to the Ocean Sea, and on the shore you find a
city with a harbour which is called HORMOS.[1] Merchants
come thither from India, with ships loaded with spicery
and precious stones, pearls, cloths of silk and gold,
elephants' teeth, and many other wares, which they sell
to the merchants of Hormos, and which these in turn
carry all over the world to dispose of again. In fact,
'tis a city of immense trade. There are plenty of towns
and villages under it, but it is the capital. The King
is called RUOMEDAM AHOMET. It is a very sickly place,
and the heat of the sun is tremendous. If any foreign
merchant dies there, the King takes all his property.

In this country they make a wine of dates mixt with
spices, which is very good. When any one not used to
it first drinks this wine, it causes repeated and violent
purging, but afterwards he is all the better for it, and
gets fat upon it. The people never eat meat and
wheaten bread except when they are ill, and if they
take such food when they are in health it makes them
ill. Their food when in health consists of dates and

salt-fish (tunny, to wit) and onions, and this kind of diet
they maintain in order to preserve their health.[2]

Their ships are wretched affairs, and many of them
get lost; for they have no iron fastenings, and are only
stitched together with twine made from the husk of the
Indian nut. They beat this husk until it becomes like
horse-hair, and from that they spin twine, and with this
stitch the planks of the ships together. It keeps well,
and is not corroded by the sea-water, but it will not stand
well in a storm. The ships are not pitched, but are
rubbed with fish-oil. They have one mast, one sail, and
one rudder, and have no deck, but only a cover spread
over the cargo when loaded. This cover consists of
hides, and on the top of these hides they put the horses
which they take to India for sale. They have no iron
to make nails of, and for this reason they use only
wooden trenails in their shipbuilding, and then stitch
the planks with twine as I have told you. Hence 'tis a
perilous business to go a voyage in one of those ships,
and many of them are lost, for in that Sea of India the
storms are often terrible.[3]

The people are black, and are worshippers of
Mahommet. The residents avoid living in the cities,
for the heat in summer is so great that it would kill
them. Hence they go out (to sleep) at their gardens in
the country, where there are streams and plenty of
water. For all that they would not escape but for one
thing that I will mention. The fact is, you see, that in
summer a wind often blows across the sands which en-
compass the plain, so intolerably hot that it would kill
everybody, were it not that when they perceive that
wind coming they plunge into water up to the neck, and
so abide until the wind have ceased.[4] [And to prove
the great heat of this wind, Messer Mark related a case
that befell when he was there. The Lord of Hormos,
not having paid his tribute to the King of Kerman the

latter resolved to claim it at the time when the people of
Hormos were residing away from the city.   So he
caused a force of 1600 horse and 5000 foot to be got
ready, and sent them by the route of Reobarles to take
the others by surprise.   Now, it happened one day that
through the fault of their guide they were not able to
reach the place appointed for their night's halt, and were
obliged to bivouac in a wilderness not far from Hormos.
In the morning as they were starting on their march
they were caught by that wind, and every man of them
was suffocated, so that not one survived to carry the
tidings to their Lord.   When the people of Hormos
heard of this they went forth to bury the bodies lest
they should breed a pestilence.   But when they laid
hold of them by the arms to drag them to the pits, the
bodies proved to be so *baked*, as it were, by that
tremendous heat, that the arms parted from the trunks,
and in the end the people had to dig graves hard by
each where it lay, and so cast them in.] [5]

The people sow their wheat and barley and other
corn in the month of November, and reap it in the
month of March.   The dates are not gathered till May,
but otherwise there is no grass nor any other green
thing, for the excessive heat dries up everything.

When any one dies they make a great business of
the mourning, for women mourn their husbands four
years.   During that time they mourn at least once a
day, gathering together their kinsfolk and friends and
neighbours for the purpose, and making a great weeping
and wailing.   [And they have women who are mourners
by trade, and do it for hire.]

Now, we will quit this country.   I shall not, however, now go on to tell you about India; but when time
and place shall suit we shall come round from the north
and tell you about it.   For the present, let us return by
another road to the aforesaid city of Kerman, for we

cannot get at those countries that I wish to tell you about except through that city.

I should tell you first, however, that King Ruomedam Ahomet of Hormos, which we are leaving, is a liegeman of the King of Kerman.[6]

On the road by which we return from Hormos to Kerman you meet with some very fine plains, and you also find many natural hot baths; you find plenty of partridges on the road; and there are towns where victual is cheap and abundant, with quantities of dates and other fruits. The wheaten bread, however, is so bitter, owing to the bitterness of the water, that no one can eat it who is not used to it. The baths that I mentioned have excellent virtues; they cure the itch and several other diseases.[7]

Now, then, I am going to tell you about the countries towards the north, of which you shall hear in regular order. Let us begin.

NOTE I.—Having now arrived at HORMUZ, it is time to see what can be made of the Geography of the route from Kermán to that port.

The port of Hormuz, [which had taken the place of Kish as the most important market of the Persian Gulf (H. C.)], stood upon the mainland. A few years later it was transferred to the island which became so famous, under circumstances which are concisely related by Abulfeda :—" Hormuz is the port of Kermán, a city rich in palms, and very hot. One who has visited it in our day tells me that the ancient Hormuz was devastated by the incursions of the Tartars, and that its people transferred their abode to an island in the sea called Zarun, near the continent, and lying west of the old city. At Hormuz itself no inhabitants remain, but some of the lowest order." (In *Büsching*, IV. 261-262.) Friar Odoric, about 1321, found Hormuz "on an island some 5 miles distant from the main." Ibn Batuta, some eight or nine years later, discriminates between Hormuz or Moghistan on the mainland, and New Hormuz on the Island of Jeraun, but describes only the latter, already a great and rich city.

The site of the Island Hormuz has often been visited and described; but I could find no published trace of any traveller having verified the site of the more ancient city, though the existence of its ruins was known to John de Barros, who says that a little fort called *Cuxstac* (*Kuhestek* of P. della Valle, II. p. 300) stood on the site. An application to Colonel Pelly, the very able British Resident at Bushire, brought me from his own personal knowledge the information that I sought, and the following particulars are compiled from the letters with which he has favoured me :—

" The ruins of Old Hormuz, well known as such, stand several miles up a creek, and in the centre of the present district of Minao. They are extensive. (though in large part obliterated by long cultivation over the site), and the traces of a long pier or Bandar were pointed out to Colonel Pelly. They are about 6 or 7 miles from the fort of Minao, and the Minao river, or its stony bed, winds down towards them. The creek is quite traceable, but is silted up, and to embark goods you have to go a farsakh towards the sea, where there is a custom-house on that part of the creek which is still navigable. Colonel Pelly collected a few bricks from the ruins. From the mouth of the Old Hormuz creek to the New Hormuz town, or town of Turumpak on the island of Hormuz, is a sail of about three farsakhs. It may be a trifle more, but any native tells you at once that it is three farsakhs from Hormuz Island to the creek where you land to go up to Minao. *Hormuzdia* was the name of the region in the days of its prosperity. Some people say that Hormuzdia was known as *Jerunia*, and Old Hormuz town as *Jerun*." (In this I suspect tradition has gone astray.) "The town and fort of Minao lie to the N.E. of the ancient city, and are built upon the lowest spur of the Bashkurd mountains, commanding a gorge through which the Rudbar river debouches on the plain of Hormuzdia." In these new and interesting particulars it is pleasing to find such precise corroboration both of Edrisi and of Ibn Batuta. The former, writing in the 12th century, says that Hormuz stood on the banks of a canal or creek from the Gulf, by which vessels came up to the city. The latter specifies the breadth of sea between Old and New Hormuz as *three farsakhs*. (*Edrisi*, I. 424; *I. B.* II. 230.)

I now proceed to recapitulate the main features of Polo's Itinerary from Kermán to Hormuz. We have :—

		Marches.
1.	From Kermán across a plain to the top of a mountain-pass, where *extreme cold was experienced*	7
2.	A descent, occupying	2
3.	A great plain, called *Reobarles*, in a much warmer climate, abounding in francolin partridge, and in dates and tropical fruit, with a ruined city of former note, called *Camadi*, near the head of the plain, which extends for	5
4.	A second very bad pass, descending for 20 miles, say . . .	1
5.	A well-watered fruitful plain, which is crossed to *Hormuz*, on the shores of the Gulf.	2
	Total	17

No European traveller, so far as I know, has described the most direct road from Kermán to Hormuz, or rather to its nearest modern representative Bander Abbási,— I mean the road by Báft. But a line to the eastward of this, and leading through the plain of Jiruft, was followed partially by Mr. Abbott in 1850, and completely by Major R. M. Smith, R.E., in 1866. The details of this route, except in one particular, correspond closely in essentials with those given by our author, and form an excellent basis of illustration for Polo's description.

Major Smith (accompanied at first by Colonel Goldsmid, who diverged to Mekran) left Kermán on the 15th of January, and reached Bander Abbási on the 3rd of February, but, as three halts have to be deducted, his total number of marches was exactly the same as Marco's, viz. 17. They divide as follows :—

		Marches.
1.	From Kermán to the caravanserai of Deh Bakri in the pass so called. "The ground as I ascended became covered with snow, and the weather bitterly cold" (*Report*)	6
2.	Two miles *over very deep snow* brought him to the top of the pass ; he then descended 14 miles to his halt. Two miles to the south of the crest he passed a second caravanserai : "The two are evidently built so near one another to afford shelter to travellers who may be unable to cross the ridge during heavy snow-storms." The next march continued the descent for 14 miles, and then carried him 10 miles along the banks of the Rudkhanah-i-Shor. The approximate height of the pass above the sea is estimated at 8000 feet. We have thus for the descent the greater part of	2
3.	"Clumps of date-palms growing near the village showed that I had now reached a totally different climate." (*Smith's Report.*) And Mr. Abbott says of the same region : "Partly wooded . . . and with thickets of reeds abounding with francolin and *Jirufti* partridge. . . . The lands yield grain, millet, pulse, French- and	

Marches.

horse-beans, rice, cotton, henna, Palma Christi, and dates, and in
part are of great fertility. . . . Rainy season from January to
March, after which a luxuriant crop of grass." Across this plain
(districts of Jiruft and Rudbar), the height of which above the sea,
is something under 2000 feet 6

4. 6½ hours, "nearly the whole way over a most difficult mountain-
pass," called the Pass of Nevergun 1

5. Two long marches over a plain, part of which is described as "continuous cultivation for some 16 miles," and the rest as a "most
uninteresting plain" 2

                                                       ——
                        Total as before   17

In the previous edition of this work I was inclined to identify Marco's route
absolutely with this Itinerary. But a communication from Major St. John, who
surveyed the section from Kermán towards Deh Bakri in 1872, shows that this first
section does not answer well to the description. The road is not all plain, for it
crosses a mountain pass, though not a formidable one. Neither is it through a
thriving, populous tract, for, with the exception of two large villages, Major St.
John found the whole road to Deh Bakri from Kermán as desert and dreary as any in
Persia. On the other hand, the more direct route to the south, which is that always
used except in seasons of extraordinary severity (such as that of Major Smith's
journey, when this route was impassable from snow), answers better, as described to
Major St. John by muleteers, to Polo's account. The first *six days* are occupied by
a gentle ascent through the districts of Bardesir and Kairat-ul-Arab, which are the
best-watered and most fertile uplands of Kermán. From the crest of the pass
reached in those six marches (which is probably more than 10,000 feet above the
sea, for it was closed by snow on 1st May, 1872), an easy descent of *two days* leads to
the Garmsir. This is traversed in four days, and then a very difficult pass is crossed
to reach the plains bordering on the sea. The cold of this route is much greater
than that of the Deh Bakri route. Hence the correspondence with Polo's description,
as far as the descent to the Garmsir, or Reobarles, seems decidedly better by this
route. It is admitted to be quite possible that on reaching this plain the two routes
coalesced. We shall assume this provisionally, till some traveller gives us a detailed
account of the Bardesir route. Meantime all the remaining particulars answer well.

[General Houtum-Schindler (*l.c.* pp. 493-495), speaking of the Itinerary from
Kermán to Hormúz and back, says : "Only two of the many routes between Kermán
and Bender 'Abbás coincide more or less with Marco Polo's description. These two
routes are the one over the Deh Bekrí Pass [see above, Colonel Smith], and the one *viâ*
Sárdú. The latter is the one, I think, taken by Marco Polo. The more direct roads
to the west are for the greater part through mountainous country, and have not
twelve stages in plains which we find enumerated in Marco Polo's Itinerary. The
road *viâ* Báft, Urzú, and the Zendán Pass, for instance, has only four stages in plains ;
the road, *viâ* Ráhbur, Rúdbár and the Nevergún Pass only six ; and the road *viâ* Sírján
also only six."

Marches.

The Sárdú route, which seems to me to be the one followed by
Marco Polo, has five stages through fertile and populous plains to
Sarvízan 5
One day's march ascends to the top of the Sarvízan Pass . . 1
Two days' descent to Ráhjird, a village close to the ruins of old
Jíruft, now called Shehr-i-Daqíánús 2
Six days' march over the "vast plain" of Jírúft and Rúdbár to
Faríáb, joining the Deh Bekrí route at Kerímábád, one stage south
of the Shehr-i-Daqíánús 6
One day's march through the Nevergún Pass to Shamíl, descending 1
Two days' march through the plain to Bender 'Abbás or Hormúz 2

                                                       ——
                        In all   17

The Sárdú road enters the Jíruft plain at the ruins of the old city, the Deh Bekrí route does so at some distance to the eastward. The first six stages performed by Marco Polo in seven days go through fertile plains and past numerous villages. Regarding the cold, "which you can scarcely abide," Marco Polo does not speak of it as existing on the mountains only; he says, "From the city of Kermán to this descent the cold in winter is very great," that is, from Kermán to near Jíruft. The winter at Kermán itself is fairly severe; from the town the ground gradually but steadily rises, the absolute altitudes of the passes crossing the mountains to the south varying from 8000 to 11,000 feet. These passes are up to the month of March always very cold; in one it froze slightly in the beginning of June. The Sárdú Pass lies lower than the others. The name is Sárdú, not Sardú from sard, "cold." Major Sykes (*Persia*, ch. xxiii.) comes to the same conclusion: "In 1895, and again in 1900, I made a tour partly with the object of solving this problem, and of giving a geographical existence to Sárdú, which appropriately means the 'Cold Country.' I found that there was a route which exactly fitted Marco's conditions, as at Sarbizan the Sárdu plateau terminates in a high pass of 9200 feet, from which there is a most abrupt descent to the plain of Jiruft, Komádin being about 35 miles, or two days' journey from the top of the pass. Starting from Kermán, the stages would be as follows:—1. Jupár (small town); 2. Bahrámjird (large village); 3. Gudar (village); 4. Ráin (small town). . . . Thence to the Sarbizan there is a distance of 45 miles, or three desert stages, thus constituting a total of 110 miles for the seven days. This is the camel route to the present day, and absolutely fits in with the description given. . . . The question to be decided by this section of the journey may then, I think, be con-sidered to be finally and most satisfactorily settled, the route proving to lie between the two selected by Colonel Yule, as being the most suitable, although he wisely left the question open."—H. C.]

In the abstract of Major Smith's Itinerary as we have given it, we do not find Polo's city of *Camadi*. Major Smith writes to me, however, that this is probably to be sought in "the ruined city, the traces of which I observed in the plain of Jiruft near Kerimabad. The name of the city is now apparently lost." It is, however, known to the natives as the *City of Dakiánús*, as Mr. Abbott, who visited the site, informs us. This is a name analogous only to the Arthur's ovens or Merlin's caves of our own country, for all over Mahomedan Asia there are old sites to which legend attaches the name of *Dakianus* or the Emperor Decius, the persecuting tyrant of the Seven Sleepers. "The spot," says Abbott, "is an elevated part of the plain on the right bank of the Hali Rúd, and is thickly strewn with kiln-baked bricks, and shreds of pottery and glass. . . . After heavy rain the peasantry search amongst the ruins for ornaments of stone, and rings and coins of gold, silver, and copper. The popular tradition concerning the city is that it was destroyed by a flood long before the birth of Mahomed."

[General Houtum-Schindler, in a paper in the *Jour. R. As. Soc.*, Jan. 1898, p. 43, gives an abstract of Dr. Houtsma's (of Utrecht) memoir, *Zur Geschichte der Saljuqen von Kerman*, and comes to the conclusion that "from these statements we can safely identify Marco Polo's Camadi with the suburb Qumádin, or, as I would read it, Qamádīn, of the city of Jíruft."—(Cf. *Major Sykes' Persia*, chap. xxiii.: "Camadi was sacked for the first time, after the death of Toghrul Shah of Kermán, when his four sons reduced the province to a condition of anarchy.")

Major P. Molesworth Sykes, *Recent Journeys in Persia* (*Geog. Journal*, X. 1897, p. 589), says: "Upon arrival in Rudbar, we turned northwards and left the Farman Farma, in order to explore the site of Marco Polo's 'Camadi.' . . . We came upon a huge area littered with yellow bricks eight inches square, while not even a broken wall is left to mark the site of what was formerly a great city, under the name of the Sher-i-Jiruft."—H. C.] The actual distance from Bamm to the City of Dakianus is, by Abbott's Journal, about 66 miles.

The name of REOBARLES, which Marco applies to the plain intermediate between the two descents, has given rise to many conjectures. Marsden pointed to *Rúdbár*, a name frequently applied in Persia to a district on a river, or intersected by streams —a suggestion all the happier that he was not aware of the fact that there *is* a district of RUDBAR exactly in the required position. The last syllable still requires explanation. I ventured formerly to suggest that it was the Arabic *Laṣṣ*, or, as Marco would certainly have written it, *Les*, a robber. Reobarles would then be RUDBAR-I-LAṢṢ, "Robber's River District." The appropriateness of the name Marco has amply illustrated ; and it appeared to me to survive in that of one of the rivers of the plain, which is mentioned by both Abbott and Smith under the title of *Rúdkhánah-i-Duzdí*, or Robbery River, a name also applied to a village and old fort on the banks of the stream. This etymology was, however, condemned as an inadmissible combination of Persian and Arabic by two very high authorities both as travellers and scholars— Sir H. Rawlinson and Mr. Khanikoff. The *Les*, therefore, has still to be explained.*

[Major Sykes (*Geog. Journal*, 1902, p. 130) heard of robbers, some five miles from Mináb, and he adds : "However, nothing happened, and after crossing the Gardan-i-Pichal, we camped at Birinti, which is situated just above the junction of Rudkhána Duzdi, or 'River of Theft,' and forms part of the district of Rudán, in Fars."

"The Jíruft and Rúdbár plains belong to the germsír (hot region), dates, pistachios, and konars (apples of Paradise) abound in them. Reobarles is Rúdbár or Rüdbáris." (*Houtum-Schindler, l.c.* 1881, p. 495.)—H. C.]

We have referred to Marco's expressions regarding the great cold experienced on the pass which formed the first descent ; and it is worthy of note that the title of "The Cold Mountains" is applied by Edrisi to these very mountains. Mr. Abbott's MS. Report also mentions in this direction, *Sardu*, said to be a cold country (as its name seems to express [see above,—H. C.]), which its population (Iliyáts) abandon in winter for the lower plains. It is but recently that the importance of this range of mountains has become known to us. Indeed the *existence* of the chain, as extending continuously from near Kashán, was first indicated by Khanikoff in 1862. More recently Major St. John has shown the magnitude of this range, which rises into summits of 15,000 feet in altitude, and after a course of 550 miles terminates in a group of volcanic hills some 50 miles S.E. of Bamm. Yet practically this chain is ignored on all our maps !

Marco's description of the "Plain of Formosa" does not apply, now at least, to the *whole* plain, for towards Bander Abbási it is barren. But to the eastward, about Minao, and therefore about Old Hormuz, it has not fallen off. Colonel Pelly writes : "The district of Minao is still for those regions singularly fertile. Pomegranates, oranges, pistachio-nuts, and various other fruits grow in profusion. The source of its fertility is of course the river, and you can walk for miles among lanes and cultivated ground, partially sheltered from the sun." And Lieutenant Kempthorne, in his notes on that coast, says of the same tract : "It is termed by the natives the Paradise of Persia. It is certainly most beautifully fertile, and abounds in orange-groves, and orchards containing apples, pears, peaches, and apricots ; with vineyards producing a delicious grape, from which was at one time made a wine called *amber-rosolli*"—a name not easy to explain. *'Ambar-i-Rasúl*, "The Prophet's Bouquet !" would be too bold a name even for Persia, though names more sacred are so profaned at Naples and on the Moselle. Sir H. Rawlinson suggests *'Ambar-'asali*, "Honey Bouquet," as possible.

When Nearchus beached his fleet on the shore of *Harmozeia* at the mouth of the *Anamis* (the River of Minao), Arrian tells us he found the country a kindly one, and

---

* It is but fair to say that scholars so eminent as Professors Sprenger and Blochmann have considered the original suggestion lawful and probable. Indeed, Mr. Blochmann says in a letter : "After studying a language for years, one acquires a natural feeling for anything un-idiomatic ; but I must confess I see nothing un-Persian in *rúdbár-i-duzd*, nor in *rúdbár-i-lass*. . . . How common *lass* is, you may see from one fact, that it occurs in children's reading-books." We must not take *Reobarles* in Marco's French as rhyming to (French) *Charles ;* every syllable sounds. It is remarkable that *Lás*, as the name of a small State near our Sind frontier, is said to mean, "in the language of the country," *a level plain.* (*J. A. S.* B. VIII. 195.) It is not clear what is meant by the language of the country. The chief is a Brahui, the people are Lumri or Numri Bilúchis, who are, according to Tod, of Jat descent.

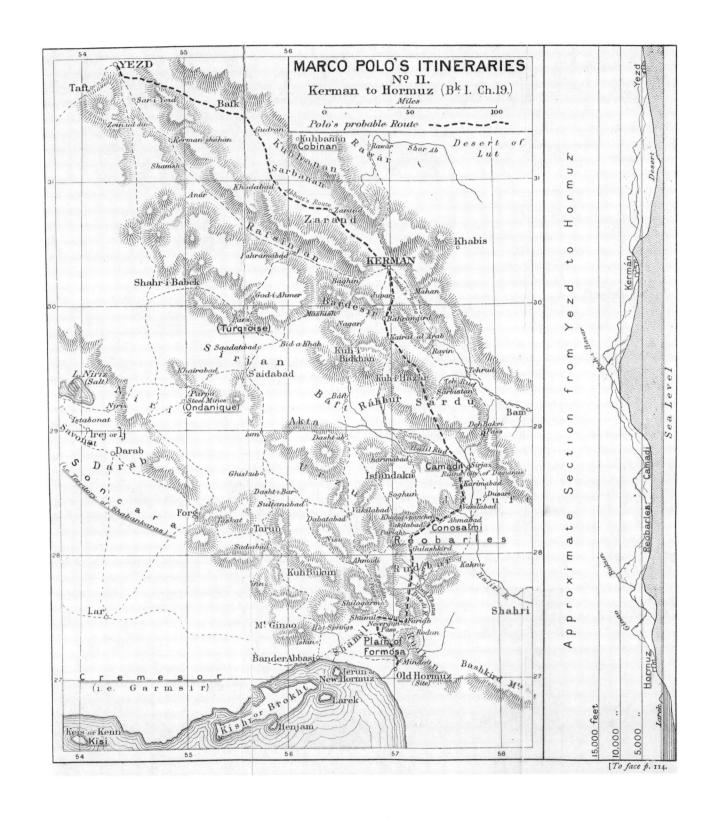

MARCO POLO'S ITINERARIES
Nº II.
Kerman to Hormuz (Bᵏ I. Ch.19.)

Miles

0   50   100

Polo's probable Route – – – – –

YEZD

Taft   *Sar-i-Yezd*   Bafk

Zein-ud-din

Shamsh   *Kermanshahan*

Gudrai   Kuhbanan
Cobinan   Ra *war*   *Rawar*   *Shor Ab*   Desert of Lut

Kuhbanan

Sarbanan

Anar   *Khudabad*   *Abbott's Route*   *Zarand*
Zarand

Raisman   Khabis

Vahramabad

KERMAN

Shahr-i-Babek   *Baghin*   *Jupar*   Mahan

God-i-Ahmer   *Bardesir*   *Candi's Fortress*
    Bahramjird
Mashish   *Nagar*   *Kairat-al-Arab*
Para   *Kuh-i-*
(Turquoise)   *Rayin*
S. Saadatabad   *Bid-a-Khah*   *Bidkhan*

Irjan   *Tehrud*
Khairabad   S'aidabad   *Kuh-i-Hazar*   *Teh-Rud*
L. Niriz   N *i r i z*   *Parpa*   *Bafu*   *Rahbur*   *Sardu*   *Sarbistan*
(Salt)   *Steel Mines*   Bam
Niriz   (Ondanique)   *Akta*
Istahonat   *Jun*   *Deh Bakri*
    Dasht-ab   *Pass*
Irej or Ij   *Halil Rud*
Savonat   Darab
D *a r a b*   *Karimabad*   Camadi   *Sirjax*
Ghisłub   *U*   Isfandaka   *Ruins City of Daganus*
Soncara   *Dasht-i-Bar*   *Karimabad*
Territory of Shabankaras   *Soghun*   *r u t*   *Dusari*
Forg   *Sultanabad*   *Vakilabad*
Taskat   *Vakilabad*
Tarun   *Dabatabad*   *Khanu-i-panchu*   *Ahmabad*
Nisu   *Vakilabad*   Conosalmi
Sadsabad   *Pariab*   Reobarles
Gulashkird
Ahmadi   Rud *bar*   *Kahnu*
KuhBukun
Jinn   *Haliri R.*
Lar   *Shilogarm*   *Dirud Rud*   Shahri
Mᵗ Ginao   *Hot Springs*   *Shamil*   *Nevergun* *Faryab*
Ishin   *Pass*   *Rodun*
Shamil   Plain of Formosa
BanderAbbasi   *Mindou*   *Bashkird Mⁿ*
Cremesor   Jerun   Old Hormuz
(i.e. Garmsir)   New Hormuz   (Site)
Kishm or Brokht   Larek
Keis or Kenn   Henjam
Kisi

Approximate Section from Yezd to Hormuz

Yezd   Desert

Kuh-i-Hazar

Kerman

Kaja Bulan   Sea Level

Camadi

Reobarles

Ginao Bulan

Hormuz I*d*

Larek

15,000 feet
10,000 ,,
5,000 ,,

[To face p. 114.

very fruitful in every way except that there were no olives. The weary mariners landed and enjoyed this pleasant rest from their toils. (*Indica*, 33 ; *J. R. G. S.* V. 274.)

The name Formosa is probably only Rusticiano's misunderstanding of *Harmuza*, aided, perhaps, by Polo's picture of the beauty of the plain. We have the same change in the old *Mafomet* for Mahomet, and the converse one in the Spanish *hermosa* for *formosa*. Teixeira's Chronicle says that the city of Hormuz was founded by Xa Mahamed Dranku, *i.e.* Shah Mahomed Dirhem-Ko, in "a plain of the same name."

The statement in Ramusio that Hormuz stood upon an island, is, I doubt not, an interpolation by himself or some earlier transcriber.

When the ships of Nearchus launched again from the mouth of the Anamis, their first day's run carried them past a certain desert and bushy island to another which was large and inhabited. The desert isle was called *Organa* ; the large one by which they anchored *Oaracta*. (*Indica*, 37.) Neither name is quite lost ; the latter greater island is Kishm or *Brakht* ; the former *Jerún*,* perhaps in old Persian *Gerún* or *Gerán*, now again desert though no longer bushy, after having been for three centuries the site of a city which became a poetic type of wealth and splendour. An Eastern saying ran, " Were the world a ring, Hormuz would be the jewel in it."

[" The *Yüan shi* mentions several seaports of the Indian Ocean as carrying on trade with China ; Hormuz is not spoken of there. I may, however, quote from the Yüan History a curious statement which perhaps refers to this port. In ch. cxxiii., biography of Arsz-lan, it is recorded that his grandson Hurdutai, by order of Kubilai Khan, accompanied *Bu-lo no-yen* on his mission to the country of *Ha-rh-ma-sz*. This latter name may be intended for Hormuz. I do not think that by the Noyen *Bulo*, M. Polo could be meant, for the title Noyen would hardly have been applied to him. But Rashid-eddin mentions a distinguished Mongol, by name *Pulad*, with whom he was acquainted in Persia, and who furnished him with much information regarding the history of the Mongols. This may be the *Bu-lo no-yen* of the Yüan History." (Bretschneider, *Med. Res.* II. p. 132.)—H. C.]

NOTE 2.—A spirit is still distilled from dates in Persia, Mekran, Sind, and some places in the west of India. It is mentioned by Strabo and Dioscorides, according to Kämpfer, who says it was in his time made under the name of a medicinal stomachic ; the rich added *Radix Chinae*, ambergris, and aromatic spices ; the poor, liquorice and Persian absinth. (*Sir B. Frere ; Amoen. Exot.* 750 ; *Macd. Kinneir*, 220.)

[" The *date* wine with spices is not now made at Bender 'Abbás. Date arrack, however, is occasionally found. At Kermán a sort of wine or arrack is made with spices and alcohol, distilled from sugar ; it is called Má-ul-Háyát (water of life), and is recommended as an aphrodisiac. Grain in the Shamíl plain is harvested in April, dates are gathered in August." (*Houtum-Schindler, l.c.* p. 496.)

See " Remarks on the Use of Wine and Distilled Liquors among the Mohammedans of Turkey and Persia," pp. 315-330 of *Narrative of a Tour through Armenia, Kurdistan, Persia, and Mesopotamia. . . .* By the Rev. Horatio Southgate, . . . London, 1840, vol. ii.—H. C.]

[Sir H. Yule quotes, in a MS. note, these lines from Moore's *Light of the Harem :*
> " Wine, too, of every clime and hue,
> Around their liquid lustre threw
> *Amber Rosolli* †—the bright dew
> From vineyards of the Green Sea gushing."] See above, p. 114.

---

* Sir Henry Rawlinson objects to this identification (which is the same that Dr. Karl Müller adopts), saying that *Organa* is more probably " Angan, formerly Argan." To this I cannot assent. Nearchus sails 300 stadia from the mouth of Anamis to Oaracta, and *on his way* passes Organa. Taking 600 stadia to the degree (Dr. Müller's value), I make it just 300 stadia from the mouth of the Hormuz creek to the eastern point of Kishm. Organa must have been either Jerún or Lárek ; Angan (*Hanjám* of Mas'udi) is out of the question. And as a straight run must have passed quite close to Jerún, not to Larek, I find the former most probable. Nearchus next day proceeds 200 stadia along Oaracta, and anchors in sight of another island (Neptune's) which was separated by 40 stadia from Oaracta. *This* was Angan ; no other island answers, and for this the distances answer with singular precision.

† Moore refers to *Persian Tales*.

The date and dry-fish diet of the Gulf people is noticed by most travellers, and P. della Valle repeats the opinion about its being the only wholesome one.   Ibn Batuta

The Double or Latin Rudder, as shown in the Navicella of Giotto.   (From Eastlake.)

says the people of Hormuz had a saying, "*Khormá wa máhí lút-i-Pádshahí,*" *i.e.* "Dates and fish make an Emperor's dish!"   A fish, exactly like the tunny of the

Mediterranean in general appearance and habits, is one of the great objects of fishery off the Sind and Mekran coasts. It comes in pursuit of shoals of anchovies, very much like the Mediterranean fish also. (*I. B.* II. 231 ; *Sir B. Frere.*)

[Friar Odoric (*Cathay*, I. pp. 55-56) says : "And there you find (before arriving at Hormuz) people who live almost entirely on dates, and you get forty-two pounds of dates for less than a groat ; and so of many other things."]

NOTE 3.—The stitched vessels of Kermán. (πλοιάρια ῥαπτὰ) are noticed in the *Periplus.* Similar accounts to those of our text are given of the ships of the Gulf and of Western India by Jordanus and John of Montecorvino. (*Jord.* p. 53 ; *Cathay*, p. 217.) "Stitched vessels," Sir B. Frere writes, "are still used. I have seen them of 200 tons burden ; but they are being driven out by iron-fastened vessels, as iron gets cheaper, except where (as on the Malabar and Coromandel coasts) the pliancy of a stitched boat is useful in a surf. Till the last few years, when steamers have begun to take all the best horses, the Arab horses bound to Bombay almost all came in the way Marco Polo describes." Some of them do still, standing over a date cargo, and the result of this combination gives rise to an extraordinary traffic in the Bombay bazaar. From what Colonel Pelly tells me, the stitched build in the Gulf is *now* confined to fishing-boats, and is disused for sea-going craft.

[Friar Odoric (*Cathay*, I. p. 57) mentioned these vessels : "In this country men make use of a kind of vessel which they call *Jase*, which is fastened only with stitching of twine. On one of these vessels I embarked, and I could find no iron at all therein." *Jase* is for the Arabic *Djehaz.*—H. C.]

The fish-oil used to rub the ships was whale-oil. The old Arab voyagers of the 9th century describe the fishermen of Siraf in the Gulf as cutting up the whale-blubber and drawing the oil from it, which was mixed with other stuff, and used to rub the joints of ships' planking. (*Reinaud*, I. 146.)

Both Montecorvino and Polo, in this passage, specify *one rudder*, as if it was a peculiarity of these ships worth noting. The fact is that, in the Mediterranean at least, the double rudders of the ancients kept their place to a great extent through the Middle Ages. A Marseilles MS. of the 13th century, quoted in Ducange, says : "A ship requires three rudders, two in place, and one to spare." Another : "Every two-ruddered bark shall pay a groat each voyage ; every one-ruddered bark shall," etc. (See Duc. under *Timonus* and *Temo.*) Numerous proofs of the use of two rudders in the 13th century will be found in "*Documenti inediti riguardanti le due Crociate di S. Ludovico IX., Re di Francia*, etc., da *L. T. Belgrano*, Genova, 1859." Thus in a specification of ships to be built at Genoa for the king (p. 7), each is to have "*Timones duo*, affaiticos, grossitudinis palmorum viiii et dimidiae, longitudinis cubitorum xxiiii." Extracts given by Capmany, regarding the equipment of galleys, show the same thing, for he is probably mistaken in saying that one of the *dos timones* specified was a spare one. Joinville (p. 205) gives incidental evidence of the same : "Those Marseilles ships have each two rudders, with each a tiller (? *tison*) attached to it in such an ingenious way that you can turn the ship right or left as fast as you would turn a horse. So on the Friday the king was sitting upon one of these tillers, when he called me and said to me," etc.[*] Francesco da Barberino, a poet of the 13th century, in the 7th part of his *Documenti d'Amore* (printed at Rome in 1640), which instructs the lover to whose lot it may fall to escort his lady on a sea-voyage (instructions carried so far as to provide even for the case of her death at sea !), alludes more than once to these plural rudders. Thus—

"—— se vedessi avenire
Che vento ti rompesse
Timoni . . .
In luogo di timoni
Fa spere † e in aqua poni." (P. 272-273.)

---

[*] This *tison* can be seen in the cuts from the tomb of St. Peter Martyr and the seal of Winchelsea.
† *Spere*, bundles of spars, etc., dragged overboard.

And again, when about to enter a port, it is needful to be on the alert and ready to run in case of a hostile reception, so the galley should enter stern foremost—a move-

12th Century Illumination. (After Pertz.)

Seal of Winchelsea.

12th Century Illumination. (After Pertz.)

From Leaning Tower. (After Jal.)

After Spinello Aretini at Siena.

From Monument of St. Peter Martyr.

ILLUSTRATIONS OF THE DOUBLE RUDDER OF THE MIDDLE AGES.

ment which he reminds his lover involves the reversal of the ordinary use of the two rudders :—

> " *L' un timon leva suso*
> *L' altro leggier tien giuso,*
> Ma convien levar mano
> Non mica com soleàno,
> Ma per contraro, e face
> Cosi 'l guidar verace." (P. 275.)

A representation of a vessel over the door of the Leaning Tower at Pisa shows this arrangement, which is also discernible in the frescoes of galley-fights by Spinello Aretini, in the Municipal Palace at Siena.

[Godinho de Eredia (1613), describing the smaller vessels of Malacca which he calls *bâlos* in ch. 13, *De Embarcações*, says : " At the poop they have two rudders, one on each side to steer with." E por poupa dos bâllos, tem 2 lêmes, hum en cada lado pera o governo. (*Malacca, l'Inde mérid. et le Cathay*, Bruxelles, 1882, 4to, f. 26.)—H. C.]

The midship rudder seems to have been the more usual in the western seas, and the double quarter-rudders in the Mediterranean. The former are sometimes styled *Navarresques* and the latter *Latins*. Yet early seals of some of the Cinque Ports show vessels with the double rudder ; one of which (that of Winchelsea) is given in the cut.

In the Mediterranean the latter was still in occasional use late in the 16th century. Captain Pantero Pantera in his book, *L'Armata Navale* (Rome, 1614, p. 44), says that the Galeasses, or great galleys, had the helm *alla Navarresca*, but also a great oar on each side of it to assist in turning the ship. And I observe that the great galeasses which precede the Christian line of battle at Lepanto, in one of the frescoes by Vasari in the Royal Hall leading to the Sistine Chapel, have the quarter-rudder very distinctly.

The Chinese appear occasionally to employ it, as seems to be indicated in a wood-cut of a vessel of war which I have traced from a Chinese book in the National Library at Paris. (See above, p. 37.) [For the Chinese words for *rudder*, see p. 126 of J. Edkins' article on *Chinese Names for Boats and Boat Gear, Jour. N. China Br. R. As. Soc.* N.S. XI. 1876.—H. C.] It is also used by certain craft of the Indian Archipelago, as appears from Mr. Wallace's description of the Prau in which he sailed from Macassar to the Aru Islands. And on the Caspian, it is stated in Smith's " Dict. of Antiquities " (art. *Gubernaculum*), the practice remained in force till late times. A modern traveller was nearly wrecked on that sea, because the two rudders were in the hands of two pilots who spoke different languages, and did not understand each other !

(Besides the works quoted see *Jal, Archéologie Navale*, II. 437-438, and *Capmany, Memorias*, III. 61.)

[Major Sykes remarks (*Persia*, ch. xxiii.) : " Some unrecorded event, probably the sight of the unseaworthy craft, which had not an ounce of iron in their composition, made our travellers decide that the risks of the sea were too great, so that we have the pleasure of accompanying them back to Kermán and thence northwards to Khorasán."—H.C.]

NOTE 4.—So also at Bander Abbási Tavernier says it was so unhealthy that foreigners could not stop there beyond March ; everybody left it in April. Not a hundredth part of the population, says Kämpfer, remained in the city. Not a beggar would stop for any reward ! The rich went to the towns of the interior or to the cool recesses of the mountains, the poor took refuge in the palm-groves at the distance of a day or two from the city. A place called 'Ishin, some 12 miles north of the city, was a favourite resort of the European and Hindu merchants. Here were fine gardens, spacious baths, and a rivulet of fresh and limpid water.

The custom of lying in water is mentioned also by Sir John Maundevile, and it was adopted by the Portuguese when they occupied Insular Hormuz, as P. della Valle and Linschoten relate. The custom is still common during great heats, in Sind and Mekran (Sir B. F.).

An anonymous ancient geography (*Liber Junioris Philosophi*) speaks of a people

in India who live in the Terrestrial Paradise, and lead the life of the Golden Age. . . .
The sun is so hot *that they remain all day in the river!*

The heat in the Straits of Hormuz drove Abdurrazzak into an anticipation of a
verse familiar to English schoolboys : " Even the bird of rapid flight was burnt up in
the heights of heaven, as well as the fish in the depths of the sea !" (*Tavern.* Bk. V.
ch. xxiii. ; *Am. Exot.* 716, 762 ; *Müller, Geog. Gr. Min.* II. 514 ; *India in XV.
Cent.* p. 49.)

NOTE 5.—A like description of the effect of the *Simúm* on the human body is
given by Ibn Batuta, Chardin, A. Hamilton, Tavernier, Thévenot, etc. ; and the first
of these travellers speaks specially of its prevalence in the desert near Hormuz, and of
the many graves of its victims ; but I have met with no reasonable account of its
poisonous action.   I will quote Chardin, already quoted at greater length by Marsden,
as the most complete parallel to the text : " The most surprising effect of the wind
is not the mere fact of its causing death, but its operation on the bodies of those who
are killed by it.   It seems as if they became decomposed without losing shape, so that
you would think them to be merely asleep, when they are not merely dead, but in
such a state that if you take hold of any part of the body it comes away in your hand.
And the finger penetrates such a body as if it were so much dust." (III. 286.)

Burton, on his journey to Medina, says : " The people assured me that this wind
never killed a man in their Allah-favoured land.   I doubt the fact.   At Bir Abbas
the body of an Arnaut was brought in swollen, and decomposed rapidly, the true
diagnosis of death by the poison-wind."   Khanikoff is very distinct as to the immediate fatality of the desert wind at Khabis, near Kermán, but does not speak of the
effect on the body after death.   This Major St. John does, describing a case that
occurred in June, 1871, when he was halting, during intense heat, at the post-house of
Pasangan, a few miles south of Kom.   The bodies were brought in of two poor men,
who had tried to start some hours before sunset, and were struck down by the poisonous blast within half-a-mile of the post-house.   " It was found impossible to wash them
before burial. . . .   Directly the limbs were touched they separated from the trunk."
(*Oc. Highways, ut. sup.*)   About 1790, when Timúr Sháh of Kabul sent an army
under the Sirdár-i-Sirdárán to put down a revolt in Meshed, this force on its return
was struck by Simúm in the Plain of Farrah, and the Sirdár perished, with a great
number of his men. (*Ferrier, H. of the Afghans,* 102 ; *J. R. G. S.* XXVI. 217 ;
Khan. Mém. 210.)

NOTE 6.—The History of Hormuz is very imperfectly known.   What I have met
with on the subject consists of—(1) An abstract by Teixeira of a chronicle of Hormuz,
written by Thurán Sháh, who was himself sovereign of Hormuz, and died in 1377 ;
(2) some contemporary notices by Wassáf, which are extracted by Hammer in his
History of the Ilkhans ; (3) some notices from Persian sources in the 2nd Decade of
De Barros (ch. ii.).   The last do not go further back than Gordun Sháh, the father
of Thurán Sháh, to whom they erroneously ascribe the first migration to the Island.

One of Teixeira's Princes is called *Ruknuddin Mahmud,* and with him Marsden
and Pauthier have identified Polo's Ruomedam Acomet, or as he is called on another
occasion in the Geog. Text, *Maimodi Acomet*.   This, however, is out of the question,
for the death of Ruknuddin is assigned to A.H. 675 (A.D. 1277), whilst there can, I
think, be no doubt that Marco's account refers to the period of his return from China,
viz. 1293 or thereabouts.

We find in Teixeira that the ruler who succeeded in 1290 was *Amir Masa'úd,*
who obtained the Government by the murder of his brother Saifuddin Nazrat.
Masa'úd was cruel and oppressive ; most of the influential people withdrew to
Baháuddin Ayaz, whom Saifuddin had made Wazir of Kalhát on the Arabian coast.
This Wazir assembled a force and drove out Masa'úd after he had reigned three years.
He fled to Kermán and died there some years afterwards.

Baháuddin, who had originally been a slave of Saifuddin Nazrat's, succeeded in

establishing his authority. But about 1300 great bodies of Turks (*i.e.* Tartars) issuing from Turkestan ravaged many provinces of Persia, including Kermán and Hormuz.
The people, unable to bear the frequency of such visitations, retired first to the
island of Kishm, and then to that of Jerún, on which last was built the city of New
Hormuz, afterwards so famous. This is Teixeira's account from Thurán Sháh, so far
as we are concerned with it. As regards the transfer of the city it agrees substantially
with Abulfeda's, which we have already quoted (*supra*, note 1).

Hammer's account from Wassáf is frightfully confused, chiefly I should suppose
from Hammer's own fault ; for among other things he assumes that Hormuz was
always on an island, and he distinguishes between the Island of Hormuz and the
Island of Jerún ! We gather, however, that Hormuz before the Mongol time formed
a government subordinate to the Salghur Atabegs of Fars (see note 1, ch. xv.), and
when the power of that Dynasty was falling, the governor Mahmúd Kalháti, established
himself as Prince of Hormuz, and became the founder of a petty dynasty, being
evidently identical with Teixeira's Ruknuddin Mahmud above-named, who is represented as reigning from 1246 to 1277. In Wassáf we find, as in Teixeira, Mahmúd's
son Masa'úd killing his brother Nazrat, and Baháuddin expelling Masa'úd. It is
true that Hammer's surprising muddle makes Nazrat kill Masa'úd ; however, as a few
lines lower we find Masa'úd alive and Nazrat dead, we may safely venture on this
correction. But we find also that Masa'úd appears as *Ruknuddin* Masa'úd, and that
Baháuddin does not assume the princely authority himself, but proclaims that of
Fakhruddin Ahmed Ben Ibrahim At-Thaibi, a personage who does not appear in
Teixeira at all. A MS. history, quoted by Ouseley, *does* mention Fakhruddin, and
ascribes to him the transfer to Jerún. Wassáf seems to allude to Baháuddin as a
sort of Sea Rover, occupying the islands of Larek and Jerún, whilst Fakhruddin
reigned at Hormuz. It is difficult to understand the relation between the two.

It is *possible* that Polo's memory made some confusion between the names of
RUKNUDDIN Masa'úd and Fakhruddin AHMED, but I incline to think the latter is
his RUOMEDAN AHMED. For Teixeira tells us that Masa'úd took refuge at the
court of Kermán, and Wassáf represents him as supported in his claims by the
Atabeg of that province, whilst we see that Polo seems to represent ʽRuomedan
Acomat as in hostility with that prince. To add to the imbroglio I find in a passage
of Wassáf Malik Fakhruddin Ahmed at-Thaibi sent by Ghazan Khan in 1297 as
ambassador to Khanbalig, staying there some years, and dying off the Coromandel
coast on his return in 1305. (Elliot, iii. pp. 45-47.)

Masa'úd's seeking help from Kermán to reinstate him is not the first case of the
same kind that occurs in Teixeira's chronicle, so there may have been some kind of
colour for Marco's representation of the Prince of Hormuz as the vassal of the Atabeg
of Kermán ("*l'homme de cest roy de Creman ;*" see *Prologue*, ch. xiv. note 2).
M. Khanikoff denies the *possibility* of the existence of any *royal dynasty* at Hormuz
at this period. That there *was* a dynasty of *Maliks* of Hormuz, however, at
this period we must believe on the concurring testimony of Marco, of Wassáf, and
of Thurán Sháh. There was also, it would seem, another *quasi* - independent
principality in the Island of Kais. (*Hammer's Ilch.* II. 50, 51 ; *Teixeira, Relacion
de los Reyes de Hormuz ; Khan. Notice*, p. 34.)

The ravages of the Tartars which drove the people of Hormuz from their city may
have begun with the incursions of the Nigudaris and Karaunahs, but they probably
came to a climax in the great raid in 1299 of the Chaghataian Prince Kotlogh Shah,
son of Dua Khan, a part of whose bands besieged the city itself, though they are said
to have been repulsed by Baháuddin Ayas.

[The Dynasty of Hormuz was founded about 1060 by a Yemen chief Mohammed
Dirhem Ko, and remained subject to Kermán till 1249, when Rokn ed-din Mahmúd
III. Kalháti (1242-1277) made himself independent. The immediate successors of
Rokn ed-din were Saif ed-din Nazraṭ (1277-1290), Masa'úd (1290-1293), Bahad ed-din
Ayaz Sayfin (1293-1311). Hormuz was captured by the Portuguese in 1510 and by
the Persians in 1622.—H. C.]

Note 7.—The indications of this alternative route to Kermán are very vague, but it may probably have been that through Finn, Tárum, and the Sírján district, passing out of the plain of Hormuz by the eastern flank of the Ginao mountain. This road would pass near the hot springs at the base of the said mountain, Sarga, Khurkhu, and Ginao, which are described by Kämpfer. Being more or less sulphureous they are likely to be useful in skin-diseases : indeed, Hamilton speaks of their efficacy in these. (I. 95.) The salt-streams are numerous on this line, and dates are abundant. The bitterness of the bread was, however, more probably due to another cause, as Major Smith has kindly pointed out to me : "Throughout the mountains in the south of Persia, which are generally covered with dwarf oak, the people are in the habit of making bread of the acorns, or of the acorns mixed with wheat or barley. It is dark in colour, and very hard, bitter, and unpalatable."

Major St. John also noticed the bitterness of the bread in Kermán, but his servants attributed it to the presence in the wheat-fields of a bitter leguminous plant, with a yellowish white flower, which the Kermánis were too lazy to separate, so that much remained in the thrashing, and imparted its bitter flavour to the grain (surely the *Tare* of our Lord's Parable !).

[General Houtum-Schindler says (*l.c.* p. 496) : "Marco Polo's return journey was, I am inclined to think, *viâ* Urzú and Báft, the shortest and most direct road. The road *viâ* Tárum and Sírján is very seldom taken by travellers intending to go to Kermán ; it is only frequented by the caravans going between Bender 'Abbás and Bahrámábád, three stages west of Kermán. Hot springs, 'curing itch,' I noticed at two places on the Urzú-Báft road. There were some near Qal'ah Asgher and others near Dashtáb ; they were frequented by people suffering from skin-diseases, and were highly sulphureous ; the water of those near Dashtáb turned a silver ring black after two hours' immersion. Another reason of my advocating the Urzú road is that the bitter bread spoken of by Marco Polo is only found on it, viz. at Báft and in Bardshír. In Sírján, to the west, and on the roads to the east, the bread is sweet. The bitter taste is from the Khúr, a bitter leguminous plant, which grows among the wheat, and whose grains the people are too lazy to pick out. There is not a single oak between Bender 'Abbás and Kermán ; none of the inhabitants seemed to know what an acorn was. A person at Báft, who had once gone to Kerbelá *viâ* Kermánsháh and Baghdad, recognised my sketch of tree and fruit immediately, having seen oak and acorn between Kermánsháh and Qasr-i-Shírín on the Baghdád road." Major Sykes writes (ch. xxiii.) : "The above description undoubtedly refers to the main winter route, which runs *viâ* Sírján. This is demonstrated by the fact that under the Kuh-i-Ginao, the summer station of Bandar Abbás, there is a magnificent sulphur spring, which, welling from an orifice 4 feet in diameter, forms a stream some 30 yards wide. Its temperature at the source is 113 degrees, and its therapeutic properties are highly appreciated. As to the bitterness of the bread, it is suggested in the notes that it was caused by being mixed with acorns, but, to-day at any rate, there are no oak forests in this part of Persia, and I would urge that it is better to accept our traveller's statement, that it was due to the bitterness of the water."—However, I prefer Gen. Houtum-Schindler's theory.—H. C.]

CHAPTER XX.

Of the Wearisome and Desert Road that has now to be Travelled.

On departing from the city of Kerman you find the road for seven days most wearisome; and I will tell you how this is.[1] The first three days you meet with no water, or next to none. And what little you do meet with is bitter green stuff, so salt that no one can drink it; and in fact if you drink a drop of it, it will set you purging ten times at least by the way. It is the same with the salt which is made from those streams; no one dares to make use of it, because of the excessive purging which it occasions. Hence it is necessary to carry water for the people to last these three days; as for the cattle, they must needs drink of the bad water I have mentioned, as there is no help for it, and their great thirst makes them do so. But it scours them to such a degree that sometimes they die of it. In all those three days you meet with no human habitation; it is all desert, and the extremity of drought. Even of wild beasts there are none, for there is nothing for them to eat.[2]

After those three days of desert [you arrive at a stream of fresh water running underground, but along which there are holes broken in here and there, perhaps undermined by the stream, at which you can get sight of it. It has an abundant supply, and travellers, worn with the hardships of the desert, here rest and refresh themselves and their beasts.][3]

You then enter another desert which extends for four days; it is very much like the former except that you do see some wild asses. And at the termination of these four days of desert the kingdom of Kerman comes to an end, and you find another city which is called Cobinan.

Note 1. ["The present road from Kermán to Kúbenán is to Zerend about 50 miles, to the Sár i Benán 15 miles, thence to Kúbenán 30 miles—total 95 miles. Marco Polo cannot have taken the direct road to Kúbenán, as it took him seven days to reach it. As he speaks of waterless deserts, he probably took a circuitous route to the east of the mountains, *viâ* Kúhpáyeh and the desert lying to the north of Khabis." (*Houtum-Schindler, l.c.* pp. 496-497.) (Cf. *Major Sykes*, ch. xxiii.)—H. C.]

Note 2.—This description of the Desert of Kermán, says Mr. Khanikoff, "is very correct. As the only place in the Desert of Lút where water is found is the dirty, salt, bitter, and green water of the rivulet called *Shor-Rúd* (the Salt River), we can have no doubt of the direction of Marco Polo's route from Kermán so far." Nevertheless I do not agree with Khanikoff that the route lay N.E. in the direction of Ambar and Kain, for a reason which will appear under the next chapter. I imagine the route to have been nearly due north from Kermán, in the direction of Tabbas or of Tún. And even such a route would, according to Khanikoff's own map, pass the Shor-Rúd, though at a higher point.

I extract a few lines from that gentleman's narrative: "In proportion as we got deeper into the desert, the soil became more and more arid; at daybreak I could still discover a few withered plants of *Caligonum* and *Salsola*, and not far from the same spot I saw a lark and another bird of a whitish colour, the last living things that we beheld in this dismal solitude. . . . The desert had now completely assumed the character of a land accursed, as the natives call it. Not the smallest blade of grass, no indication of animal life vivified the prospect; no sound but such as came from our own caravan broke the dreary silence of the void." (*Mém.* p. 176.)

[Major P. Molesworth Sykes (*Geog. Jour.* X. p. 578) writes: "At Tun, I was on the northern edge of the great Dash-i-Lut (Naked Desert), which lay between us and Kerman, and which had not been traversed, in this particular portion, since the illustrious Marco Polo crossed it, in the opposite direction, when travelling from Kerman to 'Tonocain' *viâ* Cobinan." Major Sykes (*Persia*, ch. iii.) seems to prove that geographers have, without sufficient grounds, divided the great desert of Persia into two regions, that to the north being termed Dasht-i-Kavir, and that further south the Dasht-i-Lut—and that Lut is the one name for the whole desert, Dash-i-Lut being almost a redundancy, and that *Kavir* (the arabic *Kafr*) is applied to every saline swamp. "This great desert stretches from a few miles out of Tehrán practically to the British frontier, a distance of about 700 miles."—H. C.]

Note 3.—I can have no doubt of the genuineness of this passage from Ramusio. Indeed some such passage is necessary; otherwise why distinguish between three days of desert and four days more of desert? The underground stream was probably a subterraneous canal (called *Kanát* or *Kárez*), such as is common in Persia; often conducted from a great distance. Here it may have been a relic of abandoned cultivation. Khanikoff, on the road between Kermán and Yezd, not far west of that which I suppose Marco to be travelling, says: "At the fifteen inhabited spots marked upon the map, they have water which has been brought from a great distance, and at considerable cost, by means of subterranean galleries, to which you descend by large and deep wells. Although the water flows at some depth, its course is tracked upon the surface by a line of more abundant vegetation." (*Ib.* p. 200.) Elphinstone says he has heard of such subterranean conduits 36 miles in length. (I. 398.) Polybius speaks of them: "There is no sign of water on the surface; but there are many underground channels, and these supply tanks in the desert, that are known only to the initiated. . . . . At the time when the Persians got the upper hand in Asia, they used to concede to such persons as brought spring-water to places previously destitute of irrigation, the usufruct for five generations. And Taurus being rife with springs, they incurred all the expense and trouble that was needed to form these underground channels to great distances, insomuch that in these days even the people who make use of the water don't know where the channels begin, or whence the water comes." (X. 28.)

CHAPTER XXI.

CONCERNING THE CITY OF COBINAN AND THE THINGS THAT ARE
MADE THERE.

COBINAN is a large town.[1] The people worship Mahommet. There is much Iron and Steel and *Ondanique*, and they make steel mirrors of great size and beauty. They also prepare both *Tutia* (a thing very good for the eyes) and *Spodium ;* and I will tell you the process.

They have a vein of a certain earth which has the required quality, and this they put into a great flaming furnace, whilst over the furnace there is an iron grating. The smoke and moisture, expelled from the earth of which I speak, adhere to the iron grating, and thus form *Tutia*, whilst the slag that is left after burning is the *Spodium*.[2]

NOTE 1.—KUH-BANÁN is mentioned by Moḳaddasi (A.D. 985) as one of the cities of Bardesír, the most northerly of the five circles into which he divides Kermán. (See *Sprenger, Post-und Reise-routen des Orients*, p. 77.) It is the subject of an article in the Geog. Dictionary of Yáḳút, though it has been there mistranscribed into *Kubiyán* and *Kukiyán*. (See Leipzig ed. 1869, iv. p. 316, and *Barbier de Meynard, Dict. de la Perse*, p. 498.) And it is also indicated by Mr. Abbott (*J. R. G. S.* XXV. 25) as the name of a district of Kermán, lying some distance to the east of his route when somewhat less than half-way between Yezd and Kermán. It would thus, I apprehend, be on or near the route between Kermán and Tabbas ; one which I believe has been traced by no modern traveller. We may be certain that there is now no place at Kuh-Banán deserving the title of *une cité grant*, nor is it easy to believe that there was in Polo's time ; he applies such terms so profusely. The meaning of the name is perhaps " Hill of the Terebinths, or Wild Pistachioes," " a tree which grows abundantly in the recesses of bleak, stony, and desert mountains, *e.g.* about Shamákhi, about Shiraz, and in the deserts of Luristan and Lar." (*Kämpfer*, 409, 413.)

[" It is strange that Marco Polo speaks of Kúbenán only on his return journey from Kermán ; on the down journey he must have been told that Kúbenán was in close proximity ; it is even probable that he passed there, as Persian travellers of those times, when going from Kermán to Yazd, and *vice versá*, always called at Kúbenán." (*Houtum-Schindler, l.c.* p. 490.) In all histories this name is written Kúbenán, not Kúhbenán ; the pronunciation to-day is Kóbenán and Kobenún.—H. C.]

I had thought my identification of *Cobinan* original, but a communication from Mr. Abbott, and the opportunity which this procured me of seeing his MS. Report already referred to, showed that he had anticipated me many years ago. The following is an extract : " *Districts of Kerman * * * Kooh Benan.* This is a hilly district abounding in fruits, such as grapes, peaches, pomegranates, *sinjid* (sweet-willow), walnuts, melons. A great deal of madder and some asafœtida is produced there. *This is no doubt the country alluded to by Marco Polo, under the name of Cobinam*, as producing iron, brass, and tutty, and which is still said to produce iron, copper, and tootea." There appear to be lead mines also in the district, as well as asbestos and sulphur. Mr. Abbott adds the names of nine villages, which he was not able to verify by comparison. These are Púz, Tarz, Gújard, Aspaj, Kuh-i-Gabr, Dahnah, Búghín, Bassab, Radk. The position of Kuh Banán is stated to lie between Bahabád (a place also mentioned by Yákút as producing *Tutia*) and Ráví, but this does not help us, and for approximate position we can only fall back on the note in Mr. Abbott's field-book, as published in the *J. R. G. S.*, viz. that the *District* lay in the mountains E.S.E. from a caravanserai 10 miles S.E. of Gudran. To get the seven marches of Polo's Itinerary we must carry the *Town* of Kuh Banán as far north as this indication can possibly admit, for Abbott made only five and a half marches from the spot where this observation was made to Kermán. Perhaps Polo's route deviated for the sake of the fresh water. That a district, such as Mr. Abbott's Report speaks of, should lie unnoticed, in a tract which our maps represent as part of the Great Desert, shows again how very defective our geography of Persia still is.

[" During the next stage to Darband, we passed ruins that I believe to be those of Marco Polo's 'Cobinan' as the modern Kúhbenán does not at all fit in with the great traveller's description, and it is just as well to remember that in the East the caravan routes seldom change." (Captain P. M. Sykes, *Geog. Jour.* X. p. 580.—See *Persia*, ch. xxiii.)

Kuh Banán has been visited by Mr. E. Stack, of the Indian Civil Service. (*Six Months in Persia*, London, 1882, I. 230.)—H. C.]

NOTE 2.—*Tutty* (*i.e.* Tutia) is in modern English an impure oxide of zinc, collected from the flues where brass is made ; and this appears to be precisely what Polo describes, unless it be that in his account the production of tutia from an ore of zinc is represented as the object and not an accident of the process. What he says reads almost like a condensed translation of Galen's account of *Pompholyx* and *Spodos:* " Pompholyx is produced in copper-smelting as *Cadmia* is ; and it is also produced from Cadmia (carbonate of zinc) when put in the furnace, as is done (for instance) in Cyprus. The master of the works there, having no copper ready for smelting, ordered some pompholyx to be prepared from cadmia in my presence. Small pieces of cadmia were thrown into the fire in front of the copper-blast. The furnace top was covered, with no vent at the crown, and intercepted the soot of the roasted cadmia. This, when collected, constitutes *Pompholyx*, whilst that which falls on the hearth is called *Spodos*, a great deal of which is got in copper-smelting." Pompholyx, he adds, is an ingredient in salves for eye discharges and pustules. (*Galen, De Simpl. Medic.* p. ix. in Latin ed., Venice, 1576.) Matthioli, after quoting this, says that Pompholyx was commonly known in the laboratories by the Arabic name of *Tutia*. I see that pure oxide of zinc is stated to form in modern practice a valuable eye-ointment.

Teixeira speaks of tutia as found only in Kermán, in a range of mountains twelve parasangs from the capital. The ore got here was kneaded with water, and set to bake in crucibles in a potter's kiln. When well baked, the crucibles were lifted and emptied, and the *tutia* carried in boxes to Hormuz for sale. This corresponds with a modern account in Milburne, which says that the tutia imported to India from the Gulf is made from an argillaceous ore of zinc, which is moulded into tubular cakes, and baked to a moderate hardness. The accurate Garcia da Horta is wrong for once in saying that the tutia of Kermán is no mineral, but the ash of a certain tree called *Goan*.

(*Matth. on Dioscorides*, Ven. 1565, pp. 1338-40 ; *Teixeira, Relacion de Persia*, p. 121 ; *Milburne's Or. Commerce*, I. 139 ; *Garcia*, f. 21 v. ; *Eng. Cyc.*, art. *Zinc*.)

[General A. Houtum-Schindler (*Jour. R. As. Soc.* N.S. XIII. October, 1881, p. 497) says : " The name Tútíá for collyrium is now not used in Kermán. Tútíá, when the name stands alone, is sulphate of copper, which in other parts of Persia is known as Kát-i-Kebúd ; Tútíá-i-sabz (green Tútíá) is sulphate of iron, also called Záj-i-síyah. A piece of Tútíá-i-zard (yellow Tútíá) shown to me was alum, generally called Záj-i-safíd ; and a piece of Tútíá-í-safíd (white Tútíá) seemed to be an argillaceous zinc ore. Either of these may have been the earth mentioned by Marco Polo as being put into the furnace. The lampblack used as collyrium is always called Surmah. This at Kermán itself is the soot produced by the flame of wicks, steeped in castor oil or goat's fat, upon earthenware saucers. In the high mountainous districts of the province,

Kúbenán, Páríz, and others, Surmah is the soot of the Gavan plant (Garcia's goan). This plant, a species of Astragalus, is on those mountains very fat and succulent ; from it also exudes the Tragacanth gum.   The soot is used dry as an eye-powder, or, mixed with tallow, as an eye-salve.   It is occasionally collected on iron gratings.

" Tútíá is the Arabicised word dúdhá, Persian for smokes.

" The Shems-ul-loghát calls Tútíá a medicine for eyes, and a stone used for the fabrication of Surmah.   The Tohfeh says Tútíá is of three kinds—yellow and blue mineral Tútíá, Tútíá-i-qalam (collyrium) made from roots, and Tútíá resulting from the process of smelting copper ore.   'The best Tútíá-i-qalam comes from Kermán.' It adds, 'Some authors say Surmah is sulphuret of antimony, others say it is a composition of iron' ; I should say any *black* composition used for the eyes is Surmah, be it lampblack, antimony, iron, or a mixture of all.

" Teixeira's Tútíá was an impure oxide of zinc, perhaps the above-mentioned Tútíá-i-safíd, baked into cakes ; it was probably the East India Company's Lapis Tútíá, also called Tutty.   The Company's Tutenague and Tutenage, occasionally confounded with Tutty, was the so-called 'Chinese Copper,' an alloy of copper, zinc, and iron, brought from China."

Major Sykes (ch. xxiii.) writes : " I translated Marco's description of *tutia* (which is also the modern Persian name), to a khán of Kubenán, and he assured me that the process was the same to-day ; spodium he knew nothing about, but the sulphate of zinc is found in the hills to the east of Kubenán."

Heyd (*Com.* II. p. 675) says in a note : " Il résulte de l'ensemble de ce passage que les matières désignées par Marco Polo sous le nom de 'espodie' (spodium) étaient des scories métalliques ; en général, le mot spodium désigne les résidus de la combustion des matières végétales ou des os (de l'ivoire)."—H. C.]

CHAPTER XXII.

OF A CERTAIN DESERT THAT CONTINUES FOR EIGHT DAYS' JOURNEY.

WHEN you depart from this City of Cobinan, you find yourself again in a Desert of surpassing aridity, which lasts for some eight days ; here are neither fruits nor trees to be seen, and what water there is is bitter and bad, so that you have to carry both food and water.   The cattle must needs drink the bad water, will they nill they, because of their great thirst.   At the end of those eight days you arrive at a Province which is called TONOCAIN. It has a good many towns and villages, and forms the extremity of Persia towards the North.[1]   It also contains an immense plain on which is found the ARBRE SOL, which we Christians call the *Arbre Sec ;* and I will tell you what it is like.   It is a tall and thick tree, having the bark on one side green and the other white ; and it produces a rough husk like that of a chestnut, but without anything in it.   The wood is yellow like box, and very strong, and there are no other trees near it nor within a hundred miles of it, except on one side, where you find trees within about ten miles' distance.   And there, the people of the country tell you, was fought the battle between Alexander and King Darius.[2]

The towns and villages have great abundance of everything good, for the climate is extremely temperate, being neither very hot nor very cold.   The natives all worship Mahommet, and are a very fine-looking people, especially the women, who are surpassingly beautiful.

NOTE I.—All that region has been described as "a country divided into deserts that are salt, and deserts that are not salt." (*Vigne*, I. 16.)   *Tonocain*, as we have seen (ch. xv. note I), is the Eastern Kuhistan of Persia, but extended by Polo, it would seem to include the whole of Persian Khorasan.   No city in particular is indicated as visited by the traveller, but the view I take of the position of the *Arbre Sec*, as well as his route through Kuh-Banán, would lead me to suppose that he reached the Province of TUN-O-KAIN about Tabbas.

[" Marco Polo has been said to have traversed a portion of (the Dash-i-Kavir, great Salt Desert) on his supposed route from Tabbas to Damghan, about 1272 ; although it is more probable that he marched further to the east, and crossed the northern portion of the Dash-i-Lut, Great Sand Desert, separating Khorasan in the south-east from Kermán, and occupying a sorrowful parallelogram between the towns of Neh and Tabbas on the north, and Kermán and Yezd on the south." (Curzon, *Persia*, II. pp. 248 and 251.)   Lord Curzon adds in a note (p. 248) : " The Tunogan of the text which was originally mistaken for Damghan, is correctly explained by Yule as Tun-o- (*i.e.* and) Káin."   Major Sykes writes (ch. xxiii.) : " The section of the Lut has not hitherto been rediscovered, but I know that it is desert throughout, and it is practically certain that Marco ended these unpleasant experiences at Tabas, 150 miles from Kubenán. To-day the district is known as Tun-o-Tabas, Káin being independent of it."—H. C.]

NOTE 2.—This is another subject on which a long and somewhat discursive note is inevitable.

One of the Bulletins of the Soc. de Géographie (sér. III. tom. iii. p. 187) contains a perfectly inconclusive endeavour, by M. Roux de Rochelle, to identify the *Arbre Sec* or *Arbre Sol* with a manna-bearing oak alluded to by Q. Curtius as growing in Hyrcania. There can be no doubt that the tree described is, as Marsden points out, a *Chinár* or Oriental Plane.   Mr. Ernst Meyer, in his learned *Geschichte der Botanik* (Königsberg, 1854-57, IV. 123), objects that Polo's description of the *wood* does not answer to that tree.   But, with due allowance, compare with his whole account that which Olearius gives of the Chinar, and say if the same tree be not meant.   " The trees are as tall as the pine, and have very large leaves, closely resembling those of the vine.   The fruit looks like a chestnut, but has no kernel, so it is not eatable.   The wood is of a very brown colour, and full of veins ; the Persians employ it for doors and window-shutters, and when these are rubbed with oil they are incomparably handsomer than our walnut-wood joinery." (I. 526.)   The Chinar-wood is used in Kashmir for gunstocks.

The whole tenor of the passage seems to imply that some eminent *individual*

Chinar is meant. The appellations given to it vary in the different texts. In the G. T. it is styled in this passage, "The *Arbre Seule* which the Christians call the *Arbre Sec*," whilst in ch. cci. of the same (*infra*, Bk. IV. ch. v.) it is called "*L'Arbre Sol*, which in the Book of Alexander is called *L'Arbre Seche*." Pauthier has here "*L'Arbre Solque*, que nous appelons *L'Arbre Sec*," and in the later passage "*L'Arbre Seul*, que le Livre Alexandre apelle *Arbre Sec*;" whilst Ramusio has here "*L'Albero del Sole* che si chiama per i Cristiani *L'Albor Secco*," and does not contain the later passage. So also I think all the old Latin and French printed texts, which are more or less based on Pipino's version, have "The *Tree of the Sun*, which the Latins call the *Dry Tree*."

[G. Capus says (*A travers le roy. de Tamerlan*, p. 296) that he found at Khodjakent, the remains of an enormous plane-tree or *Chinar*, which measured no less than 48 metres (52 yards) in circumference at the base, and 9 metres diameter inside the rotten trunk; a dozen tourists from Tashkent one day feasted inside, and were all at ease.—H. C.]

Pauthier, building as usual on the reading of his own text (*Solque*), endeavours to show that this odd word represents *Thoulk*, the Arabic name of a tree to which Forskal gave the title of *Ficus Vasta*, and this Ficus Vasta he will have to be the same as the Chinar. *Ficus Vasta* would be a strange name surely to give to a Plane-tree, but Forskal may be acquitted of such an eccentricity. The *Tholak* (for that seems to be the proper vocalisation) is a tree of Arabia Felix, very different from the Chinar, for it is the well-known Indian Banyan, or a closely-allied species, as may be seen in Forskal's description. The latter indeed says that the Arab botanists called it *Delb*, and that (or *Dulb*) is really a synonym for the Chinar. But De Sacy has already commented upon this supposed application of the name Delb to the *Tholak* as erroneous. (See *Flora Aegyptiaco-Arabica*, pp. cxxiv. and 179; *Abdallatif, Rel. de l'Egypte*, p. 80; *J. R. G. S.* VIII. 275; *Ritter*, VI. 662, 679.)

The fact is that the *Solque* of M. Pauthier's text is a mere copyist's error in the reduplication of the pronoun *que*. In his chief MS. which he cites as A (No. 10,260 of Bibl. Nationale, now *Fr.* 5631) we can even see how this might easily happen, for one line ends with *Solque* and the next begins with *que*. The true reading is, I doubt not, that which this MS. points to, and which the G. Text gives us in the second passage quoted above, viz. *Arbre* SOL, occurring in Ramusio as *Albero del* SOLE. To make this easier of acceptation I must premise two remarks : first, that *Sol* is "the Sun" in both Venetian and Provençal; and, secondly, that in the French of that age the prepositional sign is not *necessary* to the genitive. Thus, in Pauthier's own text we find in one of the passages quoted above, "*Le Livre Alexandre, i.e.* Liber Alexandri ;" elsewhere, "*Cazan le fils Argon*," "*à la mère sa femme*," "*Le corps Monseigneur Saint Thomas si est en ceste Province*;" in Joinville, "*le commandemant Mahommet*," "*ceux de la* Haulequa *estoient logiez entour les héberges le soudanc, et establiz pour le cors le soudanc garder*;" in Baudouin de Sebourc, "*De l'amour Bauduin esprise et enflambée*."

Moreover it is the TREE OF THE SUN that is prominent in the legendary History of Alexander, a fact sufficient in itself to rule the reading. A character in an old English play says :—

> "*Peregrine*.   Drake was a didapper to Mandevill :
> Candish and Hawkins, Frobisher, all our Voyagers
> Went short of Mandevil.   But had he reached
> To this place—here—yes, here—this wilderness,
> And seen the *Trees of the Sun and Moon*, that speak
> And told King Alexander of his death ;
> He then
> Had left a passage ope to Travellers
> That now is kept and guarded by Wild Beasts."
>
> (*Broome's Antipodes*, in *Lamb's Specimens.*)

The same trees are alluded to in an ancient Low German poem in honour of St. Anno of Cologne.   Speaking of the Four Beasts of Daniel's Vision :—

> " The third beast was a Libbard ;
> Four Eagle's Wings he had ;
> This signified the Grecian Alexander,
> Who with four Hosts went forth to conquer lands
> Even to the World's End,
> Known by its Golden Pillars.
> In India he the Wilderness broke through
> *With Trees twain he there did speak*," etc.
>
> (In *Schilteri Thesaurus Antiq. Teuton.* tom. i.*)

These oracular Trees of the Sun and Moon, somewhere on the confines of India, appear in all the fabulous histories of Alexander, from the Pseudo-Callisthenes down-wards. Thus Alexander is made to tell the story in a letter to Aristotle : " Then came some of the towns-people and said, ' We have to show thee something passing strange, O King, and worth thy visiting ; for we can show thee trees that talk with human speech.' So they led me to a certain park, in the midst of which were the Sun and Moon, and round about them a guard of priests of the Sun and Moon. And there stood the two trees of which they had spoken, like unto cypress trees ; and round about them were trees like the myrobolans of Egypt, and with similar fruit. And I addressed the two trees that were in the midst of the park, the one which was male in the Masculine gender, and the one that was female in the Feminine gender. And the name of the Male Tree was the Sun, and of the female Tree the Moon, names which were in that language *Muthu* and *Emaūsae*.† And the stems were clothed with the skins of animals ; the male tree with the skins of he-beasts, and the female tree with the skins of she-beasts. . . . And at the setting of the Sun, a voice, speaking in the Indian tongue, came forth from the (Sun) Tree ; and I ordered the Indians who were with me to interpret it. But they were afraid and would not," etc. (*Pseudo-Callisth.* ed. Müller, III. 17.)

The story as related by Firdusi keeps very near to the Greek as just quoted, but does not use the term "Tree of the Sun."   The chapter of the Sháh Námeh containing it is entitled *Dídan Sikandar dirakht-i-goydrá,* "Alexander's interview with the Speaking Tree." (*Livre des Rois,* V. 229.) In the *Chanson d'Alixandre* of Lambert le Court and Alex. de Bernay, these trees are introduced as follows :—

> " ' Signor,' fait Alixandre, 'je vus voel demander,
> Se des merveilles d'Inde me saves rien conter.'
> Cil li ont respondu : ' Se tu vius escouter
> Ja te dirons merveilles, s'es poras esprover.
> La sus en ces desers pues ii Arbres trover
> Qui c pies ont de haut, et de grossor sunt per.
> Li Solaus et La Lune les ont fait si serer
> Que sevent tous langages et entendre et parler.' "
>
> (Ed. 1861 (Dinan), p. 357.)

Maundevile informs us precisely where these trees are : "A 15 journeys in lengthe, goynge be the Deserts of the tother side of the Ryvere Beumare," if one could only

---

* "Daz dritte Dier was ein Lebarte
   Vier arin Vederich her havite ;
   Der beceichnote den Criechiskin Alexanderin,
   Der mit vier Herin vūr aftir Landin,
   Unz her die Werilt einde,
   Bi guldinin Siulin bikante.
   In India her die Wusti durchbrach,
   Mit zwein Boumin her sich da gesprach," etc.

† It is odd how near the word *Emaūsae* comes to the E. African *Mwezi;* and perhaps more odd that "the elders of U-nya-Mwezi ('the Land of the Moon') declare that their patriarchal ancestor became after death the first Tree, and afforded shade to his children and descendants. According to the Arabs the people still perform pilgrimage to a holy tree, and believe that the penalty of sacrilege in cutting off a twig would be visited by sudden and mysterious death." (*Burton* in *F. R. G. S.* XXIX. 167-168.)

tell where that is ! * A mediæval chronicler also tells us that Ogerus the Dane (*temp. Caroli Magni*) conquered all the parts beyond sea from Hierusalem to the Trees of the Sun. In the old Italian romance also of *Guerino detto il Meschino*, still a chapbook in S. Italy, the Hero (ch. lxiii.) visits the Trees of the Sun and Moon. But this is mere imitation of the Alexandrian story, and has nothing of interest. (*Maundevile*, pp. 297-298 ; *Fasciculus Temporum* in *Germ. Script. Pistorii Nidani*, II.)

It will be observed that the letter ascribed to Alexander describes the two oracular trees as resembling two cypress-trees. As such the Trees of the Sun and Moon are represented on several extant ancient medals, *e.g.* on two struck at Perga in Pamphylia in the time of Aurelian. And Eastern story tells us of two vast cypress-trees, sacred among the Magians, which grew in Khorasan, one at Kashmar near Turshiz, and the other at Farmad near Tuz, and which were said to have risen from shoots that Zoroaster brought from Paradise. The former of these was sacrilegiously cut down by the order of the Khalif Motawakkil, in the 9th century. The trunk was despatched to Baghdad on rollers at a vast expense, whilst the branches alone formed a load for 1300 camels. The night that the convoy reached within one stage of the palace, the Khalif was cut in pieces by his own guards. This tree was said to be 1450 years old, and to measure 33¾ cubits in girth. The locality of *this* "Arbor Sol" we see was in Khorasan, and possibly its fame may have been transferred to a representative of another species. The plane, as well as the cypress, was one of the distinctive trees of the Magian Paradise.

In the Peutingerian Tables we find in the N.E. of Asia the rubric " *Hic Alexander Responsum accepit,*" which looks very like an allusion to the tale of the Oracular Trees. If so, it is remarkable as a suggestion of the antiquity of the Alexandrian Legends, though the rubric may of course be an interpolation. The Trees of the Sun and Moon appear as located in India Ultima to the east of Persia, in a map which is found in MSS. (12th century) of the *Floridus* of *Lambertus ;* and they are indicated more or less precisely in several maps of the succeeding centuries. (*Ouseley's Travels*, I. 387 ; *Dabistan*, I. 307-308 ; *Santarem*, *H. de la Cosmog.* II. 189, III. 506-513, etc.)

Nothing could show better how this legend had possessed men in the Middle Ages than the fact that Vincent de Beauvais discerns an allusion to these Trees of the Sun and Moon in the blessing of Moses on Joseph (as it runs in the Vulgate), " *de pomis fructuum Solis ac Lunae.*" (Deut. xxxiii. 14.)

Marco has mixt up this legend of the Alexandrian Romance, on the authority, as we shall see reason to believe, of some of the recompilers of that Romance, with a famous subject of *Christian* Legend in that age, the ARBRE SEC or Dry Tree, one form of which is related by Maundevile and by Johan Schiltberger. " A lytille fro Ebron," says the former, " is the Mount of Mambre, of the whyche the Valeye taketh his name. And there is a Tree of Oke that the Saracens clepen *Dirpe*, that is of Abraham's Tyme, the which men clepen THE DRYE TREE." [Schiltberger adds that the heathen call it *Kurru Thereck*, *i.e.* (Turkish) *Ḳúrú Ḍirakht* = Dry Tree.] " And theye seye that it hathe ben there sithe the beginnynge of the World ; and was sumtyme grene and bare Leves, unto the Tyme that Oure Lord dyede on the Cros ; and thanne it dryede ; and so dyden alle the Trees that weren thanne in the World. And summe seyn be hire Prophecyes that a Lord, a Prynce of the West syde of the World, shalle wynnen the Lond of Promyssioun, *i.e.* the Holy Lond, withe Helpe of Cristene Men, and he schalle do synge a Masse under that Drye Tree, and than the Tree shall wexen grene and bere both Fruyt and Leves. And thorghe that Myracle manye Sarazines and Jewes schulle ben turned to Cristene Feithe. And, therefore, they dou gret Worschipe thereto, and kepen it fulle besyly. And alle be it so that it be drye, natheless yit he berethe great vertue," etc.

The tradition seems to have altered with circumstances, for a traveller of nearly two centuries later (Friar Anselmo, 1509) describes the oak of Abraham at Hebron

---

* " The River *Buemar*, in the furthest forests of India," appears to come up in one of the versions of Alexander's Letter to Aristotle, though I do not find it in Müller's edition. (See Zacher's *Pseudo-Callisthenes*, p. 160.) 'Tis perhaps Áb-i-Ámú !

as a tree of dense and verdant foliage: "The Saracens make their devotions at it, and hold it in great veneration, for it has remained thus green from the days of Abraham until now ; and they tie scraps of cloth on its branches inscribed with some of their writing, and believe that if any one were to cut a piece off that tree he would die within the year." Indeed even before Maundevile's time Friar Burchard (1283) had noticed that though the famous old tree was dry, another had sprung from its roots. And it still has a representative.

As long ago as the time of Constantine a fair was held under the Terebinth of Mamre, which was the object of many superstitious rites and excesses. The Emperor ordered these to be put a stop to, and a church to be erected at the spot. In the time of Arculph (end of 7th century) the dry trunk still existed under the roof of this church ; just as the immortal Banyan-tree of Prág exists to this day in a subterranean temple in the Fort of Allahabad.

It is evident that the story of the Dry Tree had got a great vogue in the 13th century. In the *Jus du Pelerin*, a French drama of Polo's age, the Pilgrim says :—

"S'ai puis en maint bon lieu et à maint saint esté,
S'ai esté au *Sec-Arbre* et dusc'à Duresté."

And in another play of slightly earlier date (*Le Jus de St. Nicolas*), the King of Africa, invaded by the Christians, summons all his allies and feudatories, among whom appear the Admirals of Coine (*Iconium*) and Orkenie (*Hyrcania*), and the *Amiral d'outre l'Arbre-Sec* (as it were of "the Back of Beyond") in whose country the only current coin is millstones ! Friar Odoric tells us that he heard at Tabriz that the *Arbor Secco* existed in a mosque of that city ; and Clavijo relates a confused story about it in the same locality. Of the *Dürre Baum* at Tauris there is also a somewhat pointless legend in a Cologne MS. of the 14th century, professing to give an account of the East. There are also some curious verses concerning a mystical *Dürre Bom* quoted by Fabricius from an old Low German Poem ; and we may just allude to that other mystic *Arbor Secco* of Dante—

——"una pianta dispogliata
Di fiori e d'altra fronda in ciascun ramo,"

though the dark symbolism in the latter case seems to have a different bearing.

(*Maundevile*, p. 68 ; *Schiltberger*, p. 113 ; Anselm. in *Canisii Thesaurus*, IV. 781 ; *Pereg. Quat.* p. 81 ; *Niceph. Callist.* VIII. 30 ; *Théâtre Français au Moyen Age*, pp. 97, 173 ; *Cathay*, p. 48 ; *Clavijo*, p. 90 ; *Orient und Occident*, Göttingen, 1867, vol. i. ; *Fabricii Vet. Test. Pseud.*, etc., I. 1133 ; *Dante, Purgat.* xxxii. 35.)

But why does Polo bring this *Arbre Sec* into connection with the Sun Tree of the Alexandrian Legend ? I cannot answer this to my own entire satisfaction, but I can show that such a connection had been imagined in his time.

Paulin Paris, in a notice of MS. No. 6985 (*Fonds Ancien*) of the National Library, containing a version of the *Chansons de Geste d'Alixandre*, based upon the work of L. Le Court and Alex. de Bernay, but with additions of later date, notices amongst these latter the visit of Alexander to the Valley Perilous, where he sees a variety of wonders, among others the *Arbre des Pucelles*. Another tree at a great distance from the last is called the ARBRE SEC, and reveals to Alexander the secret of the fate which attends him in Babylon. (*Les MSS. Français de la Bibl. du Roi*, III. 105.)* Again the English version of *King Alisaundre*, published in Weber's Collection, shows clearly enough that in *its* French original the term *Arbre Sec* was applied to the Oracular Trees, though the word has been miswritten, and misunderstood by

---

* It is right to notice that there may be some error in the *reference* of Paulin Paris ; at least I could not trace the *Arbre Sec* in the MS. which he cites, nor in the celebrated Bodleian Alexander, which appears to contain the same version of the story. [The fact is that Paulin Paris refers to the *Arbre*, but without the word *sec*, at the top of the first column of fol. 79 *recto* of the MS. No. *Fr.* 368 (late 6985).—H. C.]

Weber. The King, as in the Greek and French passages already quoted, meeting two old churls, asks if they know of any marvel in those parts :—

> " 'Ye, par ma fay,' quoth heo,
> ' A great merveille we wol telle the;
> That is hennes in even way
> The mountas of ten daies journey,
> Thou shalt find trowes * two :
> Seyntes and holy they buth bo ;
> Higher than in othir countray all.
> ARBESET men heom callith.'
>   *   *   *   *   *   *
> ' Sire Kyng,' quod on, ' by myn eyghe
> Either Trough is an hundrod feet hygh,
> They stondith up into the skye ;
> That on to the *Sonne*, sikirlye ;
> That othir, we tellith the nowe,
> Is sakret in the *Mone* vertue.' "
>   (*Weber*, I. 277.)

Weber's glossary gives "*Arbeset*=Strawberry Tree, *arbous, arbousier, arbutus*" ; but that is nonsense.

Further, in the French Prose Romance of Alexander, which is contained in the fine volume in the British Museum known as the Shrewsbury Book (Reg. XV. e. 6), though we do not find the Arbre Sec so named, we find it described and pictorially represented. The Romance (fol. xiiii. *v.*) describes Alexander and his chief companions as ascending a certain mountain by 2500 steps which were attached to a golden chain. At the top they find the golden Temple of the Sun and an old man asleep within. It goes on :—

" Quant le viellart les vit si leur demanda s'ils vouloient veoir les Arbres sacrez de la Lune et du Soleil que nous annuncent les choses qui sont à avenir. Quant Alexandre ouy ce si fut rempli de mult grant ioye. Si lui respondirent, ' Ouye sur, nous les voulons veoir.' Et cil lui dist, 'Se tu es nez de prince malle et de femelle il te convient entrer en celui lieu.' Et Alexandre lui respondi, ' Nous somes nez de compagne malle et de femelle.' Dont se leve le viellart du lit ou il gesoit, et leur dist, ' Hostez vos vestemens et vos chauces.' Et Tholomeus et Antigonus et Perdiacas le suivrent. Lors comencèrent à aler parmy la forest qui estoit enclose en merveilleux labour. Illec trouvèrent les arbres semblables à loriers et oliviers. Et estoient de cent pies de haults, et decouroit d'eulz incens ypobaume † à grant quantité. Après entrèrent plus avant en la forest, et trouvèrent *une arbre durement hault qui n'avoit ne fueille ne fruit.* Si seoit sur cet arbre une grant oysel qui avoit en son chief une creste qui estoit semblable au paon, et les plumes du col resplendissants come fin or. Et avoit la couleur de rose. Dont lui dist le viellart, ' Cet oysel dont vous vous merveillez est appelés Fenis, lequel n'a nul pareil en tout le monde.' Dont passèrent outre, et allèrent aux Arbres du Soleil et de la Lune. Et quant ils y furent venus, si leur dist le viellart, ' Regardez en haut, et pensez en votre coeur ce que vous vouldrez demander, et ne le dites de la bouche.' Alisandre luy demanda en quel language donnent les Arbres response aux gens. Et il lui respondit, ' L'Arbre du Soleil commence à parler Indien.' Dont baisa Alexandre les arbres, et comença en son ceur à penser s'il conquesteroit tout le monde et retourneroit en Macedonie atout son ost. Dont lui respondit l'Arbre du Soleil, ' Alexandre tu seras Roy de tout le monde, mais Macedonie tu ne verras jamais,' " etc.

The appearance of the Arbre Sec in Maps of the 15th century, such as those of Andrea Bianco (1436) and Fra Mauro (1459), may be ascribed to the influence of

---

* Trees.   † Opobalsamum.

Polo's own work; but a more genuine evidence of the prevalence of the legend is found in the celebrated Hereford Map constructed in the 13th century by Richard de Haldingham.   This, in the vicinity of India and the Terrestrial Paradise, exhibits a Tree with the rubric "*Albor Balsami est Arbor Sicca.*"

The legends of the Dry Tree were probably spun out of the words of the Vulgate in Ezekiel xvii. 24 : "*Humiliavi lignum sublime et exaltavi lignum humile ; et siccavi lignum viride et frondescere feci lignum aridum.*"   Whether the *Rue de l'Arbre Sec* in Paris derives its name from the legend I know not.   [The name of the street is taken from an old sign-board ; some say it is derived from the gibbet placed in the vicinity, but this is more than doubtful.—H. C.]

The actual tree to which Polo refers in the text was probably one of those so frequent in Persia, to which age, position, or accident has attached a character of sanctity, and which are styled *Dirakht-i-Fazl*, Trees   Excellence or Grace, and

Eomment les arbres du soleil et de la lune prophe-
tiserent la mort aliwandre.

often receive titles appropriate to Holy Persons.   Vows are made before them, and pieces torn from the clothes of the votaries are hung upon the branches or nailed to the trunks.   To a tree of such a character, imposing in decay, Lucan compares Pompey :

> " Stat magni nominis umbra.
> Qualis frugifero quercus sublimis in agro,
> *Exuvias veteres populi sacrataque gestans*
> *Dona ducum*   *   *   *   *   *
> ——Quamvis primo nutet casura sub Euro,
> Tot circum silvae firmo se robore tollant,
> Sola tamen colitur."   (*Pharsalia*, I. 135.)

The Tree of Mamre was evidently precisely one of this class; and those who have crossed the Suez Desert before railway days will remember such a *Dirakht-i-Fazl*, an aged mimosa, a veritable *Arbre Seul* (could we accept that reading), that stood just half-way across the Desert, streaming with the *exuviae veteres* of Mecca Pilgrims. The majority of such holy trees in Persia appear to be Plane-trees. Admiration for the beauty of this tree seems to have occasionally risen into superstitious veneration from a very old date. Herodotus relates that the Carians, after their defeat by the Persians on the Marsyas, rallied in the sacred grove of Plane-trees at Labranda. And the same historian tells how, some years later, Xerxes on his march to Greece decorated a beautiful Chinar with golden ornaments. Mr. Hamilton, in the same region, came on the remains of a giant of the species, which he thought might possibly be the very same. Pliny rises to enthusiasm in speaking of some noble Plane-trees in Lycia and elsewhere. Chardin describes one grand and sacred specimen, called King Hosain's Chinar, and said to be more than 1000 years old, in a suburb of Ispahan, and another hung with amulets, rags, and tapers in a garden at Shiraz.* One sacred tree mentioned by the Persian geographer Hamd Allah as distinguishing the grave of a holy man at Bostam in Khorasan (the species is not named, at least by Ouseley, from whom I borrow this) comes into striking relation with the passage in our text. The story went that it had been the staff of Mahomed; as such it had been transmitted through many generations, until it was finally deposited in the grave of Abu Abdallah Dásitáni, where it struck root and put forth branches. And it is explicitly called *Dirakht-i-Khushk, i.e.* literally *L'ARBRE SEC.*

This last legend belongs to a large class. The staff of Adam, which was created in the twilight of the approaching Sabbath, was bestowed on him in Paradise and handed down successively to Enoch and the line of Patriarchs. After the death of Joseph it was set in Jethro's garden, and there grew untouched, till Moses came and got his rod from it. In another form of the legend it is Seth who gets a branch of the Tree of Life, and from this Moses afterwards obtains his rod of power. These Rabbinical stories seem in later times to have been developed into the Christian legends of the wood destined to form the Cross, such as they are told in the Golden Legend or by Godfrey of Viterbo, and elaborated in Calderon's *Sibila del Oriente*. Indeed, as a valued friend who has consulted the latter for me suggests, probably all the Arbre Sec Legends of Christendom bore mystic reference to the Cross. In Calderon's play the Holy Rood, seen in vision, is described as a Tree :—

> ———" cuyas hojas,
> Secas mustias y marchitas,
> Desnudo el tronco dejaban
> Que, entre mil copas floridas
> De los árboles, el solo
> Sin pompa y sin bizaria
> Era cadáver del prado."

There are several Dry-Tree stories among the wonders of Buddhism; one is that of a sacred tree visited by the Chinese pilgrims to India, which had grown from the twig which Sakya, in Hindu fashion, had used as a tooth-brush; and I think there is a like story in our own country of the Glastonbury Thorn having grown from the staff of Joseph of Arimathea.

["St Francis' Church is a large pile, neere which, yet a little without the Citty, growes a tree which they report in their legend grew from the Saint's Staff, which on

---

* A recent traveller in China gives a perfectly similar description of sacred trees in Shansi. Many bore inscriptions in large letters. "If you pray, you will certainly be heard."—*Rev. A. Williamson, Journeys in N. China,* I. 163, where there is a cut of such a tree near Taiyuanfu. (See this work, I. ch. xvi.) Mr. Williamson describes such a venerated tree, an ancient acacia, known as the Acacia of the T'ang, meaning that it existed under that Dynasty (7th to 10th century). It is renowned for its healing virtues, and every available spot on its surface was crowded with votive tablets and in scriptions. (*Ib.* 303.)

going to sleepe he fixed in the ground, and at his waking found it had grown a large tree. They affirm that the wood of its decoction cures sundry diseases." (*Evelyn's Diary*, October, 1644.)—H. C.]

In the usual form of the mediæval legend, Adam, drawing near his end, sends Seth to the gate of Paradise, to seek the promised Oil of Mercy. The Angel allows Seth to put his head in at the gate. Doing so (as an old English version gives it)—

> —— "he saw a fair Well,
> Of whom all the waters on earth cometh, as the Book us doth tell ;
> Over the Well stood a Tree, with bowës broad and lere
> Ac it *ne bare leaf ne rind, but as it for-olded were ;*
> A nadder it had beclipt about, all naked withouten skin,
> That was the Tree and the Nadder that first made Adam do sin !"

The Adder or Serpent is coiled about the denuded stem : the upper branches reach to heaven, and bear at the top a new-born wailing infant, swathed in linen, whilst (here we quote a French version)—

> " Les larmes qui de lui issoient
> Contreval l'Arbre en avaloient ;
> Adonc regarda l'enfant Seth
> Tout contreval de L'ARBRE SECQ ;
> Les rachines qui le tenoient
> Jusques en Enfer s'en aloient,
> Les larmes qui de lui issirent
> Jusques dedans Enfer cheïrent."

The Angel gives Seth three kernels from the fruit of the Tree. Seth returns home and finds his father dead. He buries him in *the valley of Hebron,* and places the three grains under his tongue. A triple shoot springs up of Cedar, Cypress, and Pine, symbolising the three Persons of the Trinity. The three eventually unite into one stem, and this tree survives in various forms, and through various adventures in connection with the Scripture History, till it is found at the bottom of the Pool of Bethesda, to which it had imparted healing Virtue, and is taken thence to form the Cross on which Our Lord suffered.

The English version quoted above is from a MS. of the 14th century in the Bodleian, published by Dr. Morris in his collection of *Legends of the Holy Rood.* I have modernised the spelling of the lines quoted, without altering the words. The French citation is from a MS. in the Vienna Library, from which extracts are given by Sign. Adolfo Mussafia in his curious and learned tract (*Sulla Legenda del Legno della Croce*, Vienna, 1870), which gives a full account of the fundamental legend and its numerous variations. The examination of these two works, particularly Sign. Mussafia's, gives an astonishing impression of the copiousness with which such Christian Mythology, as it may fairly be called, was diffused and multiplied. There are in the paper referred to notices of between fifty and sixty different *works* (not MSS. or *copies* of works merely) containing this legend in various European languages.

(*Santarem*, III. 380, II. 348 ; *Ouseley*, I. 359 *seqq.* and 391 ; *Herodotus*, VII. 31 ; *Pliny*, XII. 5 ; *Chardin*, VII. 410, VIII. 44 and 426 ; *Fabricius, Vet. Test. Pseud.* I. 80 *seqq. ; Cathay*, p. 365 ; *Beal's Fah-Hian*, 72 and 78 ; *Pèlerins Bouddhistes*, II. 292 ; *Della Valle*, II. 276-277.)

He who injured the holy tree of Bostam, we are told, perished the same day : a general belief in regard to those *Trees of Grace*, of which we have already seen instances in regard to the sacred trees of Zoroaster and the Oak of Hebron. We find the same belief in Eastern Africa, where certain trees, regarded by the natives with superstitious reverence, which they express by driving in votive nails and suspending rags, are known to the European residents by the vulgar name of *Devil Trees.*

Chinar, or Oriental Plane.

Burton relates a case of the verification of the superstition in the death of an English merchant who had cut down such a tree, and of four members of his household. It is the old story which Ovid tells ; and the tree which Erisichthon felled was a *Dirakht-i-Fazl:*

> " Vittae mediam, memoresque tabellae
> Sertaque cingebant, voti argumenta potentis."
>   (*Metamorph.* VIII. 744.)

Though the coincidence with our text of Hamd Allah's Dry Tree is very striking, I am not prepared to lay stress on it as an argument for the geographical determination of Marco's *Arbre Sec.* His use of the title more than once to characterise the whole frontier of Khorasan can hardly have been a mere whim of his own: and possibly some explanation of that circumstance will yet be elicited from the Persian historians or geographers of the Mongol era.

Meanwhile it is in the vicinity of Bostam or Damghan that I should incline to place this landmark. If no one *very* cogent reason points to this, a variety of minor ones do so ; such as the direction of the traveller's journey from Kermán through Kuh Banán ; the apparent vicinity of a great Ismailite fortress, as will be noticed in the next chapter ; the connection twice indicated (see *Prologue*, ch. xviii. note 6, and Bk. IV. ch. v.) of the Arbre Sec with the headquarters of Ghazan Khan in watching the great passes, of which the principal ones debouche at Bostam, at which place also buildings erected by Ghazan still exist ; and the statement that the decisive battle between Alexander and Darius was placed there by local tradition. For though no such battle took place in that region, we know that Darius was murdered near Hecatompylos. Some place this city west of Bostam, near Damghan ; others east of it, about Jah Jerm ; Ferrier has strongly argued for the vicinity of Bostam itself. Firdusi indeed places the final battle on the confines of Kermán, and the death of Darius within that province. But this could not have been the tradition Polo met with.

I may add that the temperate climate of Bostam is noticed in words almost identical with Polo's by both Fraser and Ferrier.

The Chinar abounds in Khorasan (as far as any tree can be said to *abound* in Persia), and even in the Oases of Tun-o-Kain wherever there is water. Travellers quoted by Ritter notice Chinars of great size and age at Shahrúd, near Bostam, at Meyomid, and at Mehr, west of Sabzawar, which last are said to date from the time of Naoshirwan (7th century). There is a town to the N.W. of Meshid called *Chinárán*, "The Planes." P. Della Valle, we may note, calls Tehran "la città dei platani."

The following note by De Sacy regarding the Chinar has already been quoted by Marsden, and though it may be doubtful whether the term Arbre Sec had any relation to the idea expressed, it seems to me too interesting to be omitted : "Its sterility seems to have become proverbial among certain people of the East. For in a collection of sundry moral sentences pertaining to the Sabaeans or Christians of St. John . . . we find the following: 'The vainglorious man is like a showy Plane Tree, rich in boughs but producing nothing, and affording no fruit to its owner.' " The same reproach of sterility is cast at the Plane by Ovid's Walnut :—

> " At postquam platanis, *sterilem praebentibus umbram,*
> Uberior quâvis arbore venit honos ;
> Nos quoque fructiferae, si nux modo ponor in illis,
> Coepimus in patulas luxuriare comas."   (*Nux,* 17–20.)

I conclude with another passage from Khanikoff, though put forward in special illustration of what I believe to be a mistaken reading (*Arbre Seul*) : "Where the Chinar is of spontaneous growth, or occupies the centre of a vast and naked plain, this tree is even in our own day invested with a quite exceptional veneration, and the

locality often comes to be called 'The Place of the Solitary Tree.'" (*J. R. G. S.*
XXIX. 345; *Ferrier*, 69-76; *Fraser*, 343; *Ritter*, VIII. 332, XI. 512 *seqq.*; *Della
Valle*, I. 703; *De Sacy's Abdallatif*, p. 81; *Khanikoff, Not.* p. 38.)

[See in Fr. Zarncke, *Der Priester Johannes*, II., in the chap. *Der Baum des Seth*,
pp. 127-128, from MS. (14th century) from Cambridge, this curious passage (p. 128):
"Tandem rogaverunt eum, ut arborem siccam, de qua multum saepe loqui audierant,
liceret videre. Quibus dicebat: 'Non est appellata arbor sicca recto nomine, sed
arbor Seth, quoniam Seth, filius Adae, primi patris nostri, eam plantavit.' Et ad
arborem Seth fecit eos ducere, prohibens eos, ne arborem transmearent, sed [si?] ad
patriam suam redire desiderarent. Et cum appropinquassent, de pulcritudine arboris
mirati sunt; erat enim magnae immensitatis et miri decoris. Omnium enim colorum
varietas inerat arbori, condensitas foliorum et fructuum diversorum; diversitas avium
omnium, quae sub coelo sunt. Folia vero invicem se repercutientia dulcissimae melo-
diae modulamine resonabant, et aves amoenos cantus ultra quam credi potest promebant;
et odor suavissimus profudit eos, ita quod paradisi amoenitate fuisse. Et cum admirantes
tantam pulcritudinem aspicerent, unus sociorum aliquo eorum maior aetate, cogitans
[cogitavit?] intra se, quod senior esset et, si inde rediret, cito aliquo casu mori posset.
Et cum haec secum cogitasset, coepit arborem transire, et cum transisset, advocans
socios, iussit eos post se ad locum amoenissimum, quem ante se videbat plenum deliciis
sibi paratum [paratis?] festinare. At illi retrogressi sunt ad regem, scilicet presbiterum
Iohannem. Quos donis amplis ditavit, et qui cum eo morari voluerunt libenter et honori-
fice detinuit. Alii vero ad patriam reversi sunt."—In common with Marsden and Yule,
I have no doubt that the *Arbre Sec* is the *Chinâr*. Odoric places it at Tabriz and I
have given a very lengthy dissertation on the subject in my edition of this traveller
(pp. 21-29), to which I must refer the reader, to avoid increasing unnecessarily the size
of the present publication.—H. C.]

CHAPTER XXIII.

CONCERNING THE OLD MAN OF THE MOUNTAIN.

MULEHET is a country in which the Old Man of the
Mountain dwelt in former days; and the name means
"*Place of the Aram.*" I will tell you his whole history
as related by Messer Marco Polo, who heard it from
several natives of that region.

The Old Man was called in their language ALOADIN.
He had caused a certain valley between two mountains
to be enclosed, and had turned it into a garden, the
largest and most beautiful that ever was seen, filled with
every variety of fruit. In it were erected pavilions and
palaces the most elegant that can be imagined, all

covered with gilding and exquisite painting. And there were runnels too, flowing freely with wine and milk and honey and water; and numbers of ladies and of the most beautiful damsels in the world, who could play on all manner of instruments, and sung most sweetly, and danced in a manner that it was charming to behold. For the Old Man desired to make his people believe that this was actually Paradise. So he had fashioned it after the description that Mahommet gave of his Paradise, to wit, that it should be a beautiful garden running with conduits of wine and milk and honey and water, and full of lovely women for the delectation of all its inmates. And sure enough the Saracens of those parts believed that it *was* Paradise!

Now no man was allowed to enter the Garden save those whom he intended to be his ASHISHIN. There was a Fortress at the entrance to the Garden, strong enough to resist all the world, and there was no other way to get in. He kept at his Court a number of the youths of the country, from 12 to 20 years of age, such as had a taste for soldiering, and to these he used to tell tales about Paradise, just as Mahommet had been wont to do, and they believed in him just as the Saracens believe in Mahommet. Then he would introduce them into his garden, some four, or six, or ten at a time, having first made them drink a certain potion which cast them into a deep sleep, and then causing them to be lifted and carried in. So when they awoke, they found themselves in the Garden.[1]

NOTE I.—Says the venerable Sire de Joinville: *"Le Vieil de la Montaingne ne créoit pas en Mahommet, ainçois créoit en la Loi de Haali, qui fu Oncle Mahommet."* This is a crude statement, no doubt, but it has a germ of truth. Adherents of the family of 'Ali as the true successors of the Prophet existed from the tragical day of the death of Husain, and among these, probably owing to the secrecy with which they were compelled to hold their allegiance, there was always a tendency to all manner of strange and mystical doctrines ; as in one direction to the glorification of 'Ali as a kind of incarnation of the Divinity, a character in which his lineal representatives were held in some manner to partake ; in another direction to the development of Pantheism, and release from all positive creed and precepts.   Of these Aliites, eventually called *Shiáhs*, a chief sect, and parent of many heretical branches, were the Ismailites, who took their name, from the seventh Imam, whose return to earth they professed to expect at the end of the World.   About A.D. 1090 a branch of the Ismaili stock was established by Hassan, son of Sabah, in the mountainous districts of Northern Persia ; and, before their suppression by the Mongols, 170 years later, the power of the quasi-spiritual dynasty which Hassan founded had spread over the Eastern Kohistan, at least as far as Ḳáïn.   Their headquarters were at Alamút (" Eagle's Nest "), about 32 miles north-east of Ḳazwin, and all over the territory which they held they established fortresses of great strength.   De Sacy seems to have proved that they were called *Hashíshíya* or *Hashíshín*, from their use of the preparation of hemp called *Hashísh ;* and thence, through their system of murder and terrorism, came the modern application of the word Assassin.   The original aim of this system was perhaps that of a kind of *Vehmgericht*, to punish or terrify orthodox persecutors who were too strong to be faced with the sword.   I have adopted in the text one of the readings of the G. Text *Asciscin*, as expressing the original word with the greatest accuracy that Italian spelling admits.   In another author we find it as *Chazisii* (see *Bollandists*, May, vol. ii. p. xi.) ; Joinville calls them *Assacis ;* whilst Nangis and others corrupt the name into *Harsacidae*, and what not.

The explanation of the name MULEHET as it is in Ramusio, or *Mulcete* as it is in the G. Text (the last expressing in Rusticiano's Pisan tongue the strongly aspirated *Mulhĕtĕ*), is given by the former : " This name of Mulehet is as much as to say in the Saracen tongue ' *The Abode of Heretics*,' " the fact being that it does represent the Arabic term *Mulhid*, pl. *Muláhidah*, " Impii, heretici," which is in the Persian histories (as of Rashíduddín and Wassáf) the title most commonly used to indicate this community, and which is still applied by orthodox Mahomedans to the Nosairis, Druses, and other sects of that kind, more or less kindred to the Ismaili.   The writer of the *Tabakat-i-Násiri* calls the sectarians of Alamút *Muláhidat-ul-maut*, " Heretics of Death." *   The curious reading of the G. Text which we have preserved " *vaut à dire des* Aram," should be read as we have rendered it.   I conceive that Marco was here unconsciously using one Oriental term to explain another.   For it seems possible to explain *Aram* only as standing for *Harám*, in the sense of " wicked " or " reprobate."

In Pauthier's Text, instead of *des aram*, we find " *veult dire en françois* Diex Terrien," or Terrestrial God.   This may have been substituted, in the correction of the original rough dictation, from a perception that the first expression was unintelligible.   The new phrase does not indeed convey the meaning of *Muláhidah*, but it expresses a main characteristic of the heretical doctrine.   The correction was probably made by Polo himself ; it is certainly of very early date.   For in the romance of Bauduin de Sebourc, which I believe dates early in the 14th century, the Caliph, on witnessing the extraordinary devotion of the followers of the Old Man (see note 1, ch. xxiv.), exclaims :

> " Par Mahon .   .   .   .
> Vous estes *Diex en terre*, autre coze n'i a !" (I. p. 360.)

So also Fr. Jacopo d'Aqui in the *Imago Mundi*, says of the Assassins : " Dicitur iis quod sunt in Paradiso magno *Dei Terreni*"—expressions, no doubt, taken in both cases from Polo's book.

Khanikoff, and before him J. R. Forster, have supposed that the name *Mulehet* represents *Alamút*.   But the resemblance is much closer and more satisfactory to

---

* Elliot, II. 290.

Mulhid or *Muláhidah.* *Mulhet* is precisely the name by which the kingdom of the Ismailites is mentioned in Armenian history, and *Mulihet* is already applied in the same way by Rabbi Benjamin in the 12th century, and by Rubruquis in the 13th. The Chinese narrative of Hulaku's expedition calls it the kingdom of *Mulahi.* (*Joinville*, p. 138 ; *J. As.* sér. II., tom. xii. 285 ; *Benj. Tudela*, p. 106 ; *Rub.* p. 265 ; *Rémusat, Nouv. Mélanges*, I. 176 ; *Gaubil*, p. 128 ; *Pauthier*, pp. cxxxix.-cxli. ; *Mon. Hist. Patr. Scriptorum*, III. 1559, Turin, 1848.) [Cf. on *Mulehet, melahideh,* Heretics, plural of *molhid,* Heretic, my note, pp. 476-482 of my ed. of Friar Odoric. —H. C.]

" Old Man of the Mountain " was the title applied by the Crusaders to the chief of that branch of the sect which was settled in the mountains north of Lebanon, being a translation of his popular Arabic title *Shaikh-ul-Jibal.* But according to Hammer this title properly belonged, as Polo gives it, to the Prince of Alamût, who never called himself Sultan, Malik, or Amir ; and this seems probable, as his territory was known as the *Balad-ul-Jibal.* (See *Abulf.* in *Büsching*, V. 319.)

CHAPTER XXIV.

How the Old Man used to train his Assassins.

WHEN therefore they awoke, and found themselves in a place so charming, they deemed that it was Paradise in very truth. And the ladies and damsels dallied with them to their hearts' content, so that they had what young men would have ; and with their own good will they never would have quitted the place.

Now this Prince whom we call the Old One kept his Court in grand and noble style, and made those simple hill-folks about him believe firmly that he was a great Prophet. And when he wanted one of his *Ashishin* to send on any mission, he would cause that potion where-of I spoke to be given to one of the youths in the garden, and then had him carried into his Palace. So when the young man awoke, he found himself in the Castle, and no longer in that Paradise ; whereat he was not over well pleased. He was then conducted to the Old Man's presence, and bowed before him with great veneration as believing himself to be in the presence of a true

Prophet. The Prince would then ask whence he came, and he would reply that he came from Paradise! and that it was exactly such as Mahommet had described it in the Law. This of course gave the others who stood by, and who had not been admitted, the greatest desire to enter therein.

So when the Old Man would have any Prince slain, he would say to such a youth: "Go thou and slay So and So; and when thou returnest my Angels shall bear thee into Paradise. And shouldst thou die, natheless even so will I send my Angels to carry thee back into Paradise." So he caused them to believe; and thus there was no order of his that they would not affront any peril to execute, for the great desire they had to get back into that Paradise of his. And in this manner the Old One got his people to murder any one whom he desired to get rid of. Thus, too, the great dread that he inspired all Princes withal, made them become his tributaries in order that he might abide at peace and amity with them.[1]

I should also tell you that the Old Man had certain others under him, who copied his proceedings and acted exactly in the same manner. One of these was sent into the territory of Damascus, and the other into Curdistan.[2]

NOTE I.—Romantic as this story is, it seems to be precisely the same that was current over all the East. It is given by Odoric at length, more briefly by a Chinese author, and again from an Arabic source by Hammer in the *Mines de l'Orient*.

The following is the Chinese account as rendered by Rémusat: "The soldiers of this country (Mulahi) are veritable brigands. When they see a lusty youth, they tempt him with the hope of gain, and bring him to such a point that he will be ready to kill his father or his elder brother with his own hand. After he is enlisted, they intoxicate him, and carry him in that state into a secluded retreat, where he is charmed with delicious music and beautiful women. All his desires are satisfied for several days, and then (in sleep) he is transported back to his original position. When he awakes, they ask what he has seen. He is then informed that if he will become an Assassin, he will be rewarded with the same felicity. And with the texts and prayers that they teach him they heat him to such a pitch that whatever commission be given him he will brave death without regret in order to execute it."

The Arabic narrative is too long to extract. It is from a kind of historical romance called *The Memoirs of Hakim*, the date of which Hammer unfortunately omits to give. Its close coincidence in substance with Polo's story is quite remarkable. After a detailed description of the Paradise, and the transfer into it of the aspirant under the influence of *bang*, on his awaking and seeing his chief enter, he says, "O chief! am I awake or am I dreaming?" To which the chief: "O such an One, take heed that thou tell not the dream to any stranger. Know that Ali thy Lord hath vouchsafed to show thee the place destined for thee in Paradise. . . . Hesitate not a moment therefore in the service of the Imam who thus deigns to intimate his contentment with thee," and so on.

William de Nangis thus speaks of the Syrian Shaikh, who alone was known to the Crusaders, though one of their historians (*Jacques de Vitry*, in *Bongars*, I. 1062) shews knowledge that the headquarters of the sect was in Persia: "He was much dreaded far and near, by both Saracens and Christians, because he so often caused princes of both classes indifferently to be murdered by his emissaries. For he used to bring up in his palace youths belonging to his territory, and had them taught a variety of languages, and above all things to fear their Lord and obey him unto death, which would thus become to them an entrance into the joys of Paradise. And whosoever of them thus perished in carrying out his Lord's behests was worshipped as an angel." As an instance of the implicit obedience rendered by the *Fidāwí* or devoted disciples of the Shaikh,, Fra Pipino and Marino Sanuto relate that when Henry Count of Champagne (titular King of Jerusalem) was on a visit to the Old Man of Syria, one day as they walked together they saw some lads in white sitting on the top of a high tower. The Shaikh, turning to the Count, asked if he had any subjects as obedient as his own? and without giving time for reply made a sign to two of the boys, who immediately leapt from the tower, and were killed on the spot. The same story is told in the *Cento Novelle Antiche*, as happening when the Emperor Frederic was on a visit (imaginary) to the Veglio. And it is introduced likewise as an incident in the Romance of Bauduin de Sebourc:

"Vollés veioir merveilles? dist li Rois Seignouris"

to Bauduin and his friends, and on their assenting he makes the signal to one of his men on the battlements, and in a twinkling

"Quant le vinrent en l'air salant de tel avis,
Et aussi liément, et aussi esjois,
Qu'il deust conquester mil livres de parisis !
Ains qu'il venist a tière il fut mors et fenis,
Surles roches agues desrompis corps et pis," * etc.

(*Cathay*, 153 ; *Rémusat, Nouv. Mél.* I. 178 ; *Mines de l'Orient*, III. 201 *seqq.* ; *Nangis* in *Duchesne*, V. 332 ; *Pipino* in *Muratori*, IX. 705 : *Defrémery* in *J. As.* sér. V. tom. v. 34 *seqq.* ; *Cent. Nov. Antiche*, Firenze, 1572, p. 91 ; *Bauduin de Sebourc*, I. 359.)

The following are some of the more notable murders or attempts at murder ascribed to the Ismailite emissaries either from Syria or from Persia :—

A.D. 1092. Nizum-ul-Mulk, formerly the powerful minister of Malik Shah, Seljukian sovereign of Persia, and a little later his two sons. 1102. The Prince of Homs, in the chief Mosque of that city. 1113. Maudúd, Prince of Mosul, in the chief Mosque of Damascus. About 1114. Abul Muzafar 'Ali, Wazir of Sanjár Shah, and Chakar Beg, grand-uncle of the latter. 1116. Ahmed Yel, Prince of Maragha, at Baghdad, in the presence of Mahomed, Sultan of Persia. 1121. The Amir

---

* This story has been transferred to Peter the Great, who is alleged to have exhibited the docility of his subjects in the same way to the King of Denmark, by ordering a Cossack to jump from the Round Tower at Copenhagen, on the summit of which they were standing.

Afdhal, the powerful Wazir of Egypt, at Cairo. 1126. Kasim Aksonkor, Prince of Mosul and Aleppo, in the Great Mosque at Mosul. 1127. Moyin-uddin, Wazir of Sanjár Shah of Persia. 1129. Amír Billah, Khalif of Egypt. 1131. Taj-ul Mulúk Buri, Prince of Damascus. 1134. Shams-ul-Mulúk, son of the preceding. 1135-38. The Khalif Mostarshid, the Khalif Rashíd, and Daùd, Seljukian Prince of Azerbaijan. 1149. Raymond, Count of Tripoli. 1191. Kizil Arzlan, Prince of Azerbaijan. 1192. Conrad of Montferrat, titular King of Jerusalem ; a murder which King Richard has been accused of instigating. 1217. Oghulmish, Prince of Hamadán.

And in 1174 and 1176 attempts to murder the great Saladin. 1271. Attempt to murder Ala'uddin Juwaini, Governor of Baghdad, and historian of the Mongols. 1272. The attempt to murder Prince Edward of England at Acre.

In latter years the *Fidáwí* or Ismailite adepts appear to have let out their services simply as hired assassins. Bibars, in a letter to his court at Cairo, boasts of using them when needful. A Mahomedan author ascribes to Bibars the instigation of the attempt on Prince Edward. (*Makrizi*, II. 100 ; *J. As.* XI. 150.)

NOTE 2.—Hammer mentions as what he chooses to call "Grand Priors" under the Shaikh or "Grand Master" at Alamút, the chief, in Syria, one in the Kuhistan of E. Persia (Tun-o-Kaïn), one in Kumis (the country about Damghan and Bostam), and one in Irák ; he does not speak of any in Kurdistan. Colonel Monteith, however, says, though without stating authority or particulars, "There were several divisions of them (the Assassins) scattered throughout Syria, *Kurdistan* (near the Lake of Wan), and Asia Minor, but all acknowledging as Imaum or High Priest the Chief residing at Alamut." And it may be noted that Odoric, a generation after Polo, puts the Old Man at *Millescorte*, which looks like *Malasgird*, north of Lake Van. (*H. des Assass.* p. 104 ; *J. R. G. S.* III. 16 ; *Cathay*, p. ccxliii.)

CHAPTER XXV.

HOW THE OLD MAN CAME BY HIS END.

Now it came to pass, in the year of Christ's Incarnation, 1252, that Alaü, Lord of the Tartars of the Levant, heard tell of these great crimes of the Old Man, and resolved to make an end of him. So he took and sent one of his Barons with a great Army to that Castle, and they besieged it for three years, but they could not take it, so strong was it. And indeed if they had had food within it never would have been taken. But after being besieged those three years they ran short of victual, and were taken. The Old Man was put to death with all his men [and the Castle with its Garden of Paradise was levelled with the ground]. And since that time he has had no successor; and there was an end to all his villainies.[1]

Now let us go back to our journey.

NOTE 1.—The date in Pauthier is 1242; in the G. T. and in Ramusio 1262. Neither is right, nor certainly could Polo have meant the former.

When Mangku Kaan, after his enthronement (1251), determined at a great *Kurultai* or Diet, on perfecting the Mongol conquests, he entrusted his brother Kúblái with the completion of the subjugation of China and the adjacent countries, whilst his brother Hulaku received the command of the army destined for Persia and Syria. The complaints that came from the Mongol officers already in Persia determined him to commence with the reduction of the Ismailites, and Hulaku set out from Karakorum in February, 1254. He proceeded with great deliberation, and the Oxus was not crossed till January, 1256. But an army had been sent long in advance under "one of his Barons," Kitubuka Noyan, and in 1253 it was already actively engaged in besieging the Ismailite fortresses. In 1255, during the progress of the war, ALA'UDDIN MAHOMED, the reigning Prince of the Assassins (mentioned by Polo as Alaodin), was murdered at the instigation of his son Ruknuddin Khurshah, who succeeded to the authority. A year later (November, 1256) Ruknuddin surrendered to Hulaku. [Bretschneider (*Med. Res.* II. p. 109) says that Alamút was taken by Hulaku, 20th December, 1256.—H. C.] The fortresses given up, all well furnished with provisions and artillery engines, were 100 in number. Two of them, however, Lembeser and Girdkuh, refused to surrender. The former fell after a year; the latter is stated to have held out for *twenty years*—actually, as it would seem, about fourteen, or till December, 1270. Ruknuddin was well treated by Hulaku, and despatched to the Court of the Kaan. The accounts of his death differ, but that most commonly alleged, according to Rashiduddin, is that Mangku Kaan was irritated at hearing of his approach, asking why his post-horses should be fagged to no purpose, and sent executioners to put Ruknuddin to death on the road. Alamút had been surrendered without any substantial resistance. Some survivors of the sect got hold of it again in 1275-1276, and held out for a time. The dominion was extinguished, but the sect remained, though scattered indeed and obscure. A very strange case that came before Sir Joseph Arnould in the High Court at Bombay in 1866 threw much new light on the survival of the Ismailis.

Some centuries ago a *Dai* or Missionary of the Ismailis, named Sadruddín, made converts from the Hindu trading classes in Upper Sind. Under the name of *Khojas* the sect multiplied considerably in Sind, Kach'h, and Guzerat, whence they spread to Bombay and to Zanzibar. Their numbers in Western India are now probably not less than 50,000 to 60,000. Their doctrine, or at least the books which they revere, appear to embrace a strange jumble of Hindu notions with Mahomedan practices and Shiah mysticism, but the main characteristic endures of deep reverence, if not worship, of the person of their hereditary Imám. To his presence, when he resided in Persia, numbers of pilgrims used to betake themselves, and large remittances of what we may call *Ismail's Pence* were made to him. Abul Hassan, the last Imám but one admitted lineal descent from the later Shaikhs of Alamút, and claiming (as they did) descent from the Imám Ismail and his great ancestor 'Ali Abu Tálib, had considerable estates at Meheláti, betweeen Kúm and Hamadán, and at one time held the Government of Kermán. His son and successor, Shah Khalilullah, was killed in a brawl at Yezd in 1818. Fatteh 'Ali Sháh, fearing Ismailite vengeance, caused the homicide to be severely punished, and conferred gifts and honours on the young Imám, Agha Khan, including the hand of one of his own daughters. In 1840 Agha Khan, who had raised a revolt at Kermán, had to escape from Persia. He took refuge in Sind, and eventually rendered good service both to General Nott at Kandahár and to Sir C. Napier in Sind, for which he receives a pension from our Government.

For many years this genuine Heir and successor of the *Viex de la Montaingne* has had his headquarters at Bombay, where he devotes, or for a long time did devote, the large income that he receives from the faithful to the maintenance of a racing stable, being the chief patron and promoter of the Bombay Turf !

A schism among the Khojas, owing apparently to the desire of part of the well-to-do Bombay community to sever themselves from the peculiarities of the sect and to set up as respectable Sunnis, led in 1866 to an action in the High Court, the object of which was to exclude Agha Khan from all rights over the Khojas, and to transfer the property of the community to the charge of Orthodox Mahomedans. To the elaborate addresses of Mr. Howard and Sir Joseph Arnould, on this most singular process before an English Court, I owe the preceding particulars. The judgment was entirely in favour of the Old Man of the Mountain.

H. H. Agha Khán Meheláti, late Representative of the Old Man of the Mountain.

"𝕷𝖊 𝕾𝖊𝖎𝖌𝖓𝖊𝖚𝖗 𝖁𝖎𝖊𝖑, 𝖖𝖚𝖊 𝖏𝖊 𝖛𝖔𝖚𝖘 𝖆𝖎 𝖉𝖎𝖙 𝖘𝖎 𝖙𝖎𝖊𝖓𝖙 𝖘𝖆 𝖈𝖔𝖚𝖗𝖙 . . . . 𝖊𝖙 𝖋𝖆𝖎𝖙 à 𝖈𝖗𝖔𝖎𝖗𝖊 à 𝖈𝖊𝖑𝖊 𝖘𝖎𝖒𝖕𝖑𝖊 𝖌𝖊𝖓𝖙 𝖖𝖚𝖎 𝖑𝖎 𝖊𝖘𝖙 𝖊𝖓𝖙𝖔𝖚𝖗 𝖖𝖚𝖊 𝖎𝖑 𝖊𝖘𝖙 𝖚𝖓 𝖌𝖗𝖆𝖓𝖙 𝖕𝖗𝖔𝖕𝖍𝖊𝖙𝖊."

[Sir Bartle Frere writes of Agha Khan in 1875 : "Like his ancestor, the Old One of Marco Polo's time, he keeps his court in grand and noble style. His sons, popularly known as 'The Persian Princes,' are active sportsmen, and age has not dulled the Agha's enjoyment of horse-racing. Some of the best blood of Arabia is always to be found in his stables. He spares no expense on his racers, and no prejudice of religion or race prevents his availing himself of the science and skill of an English trainer or jockey when the races come round. If tidings of war or threatened disturbance should arise from Central Asia or Persia, the Agha is always one of the first to hear of it, and seldom fails to pay a visit to the Governor or to some old friend high in office to hear the news and offer the services of a tried sword and an experienced leader to the Government which has so long secured him a quiet refuge for his old age." Agha Khan died in April, 1881, at the age of 81. He was succeeded by his son Agha Ali Sháh, one of the members of the Legislative Council. (See *The Homeward Mail, Overland Times of India*, of 14th April, 1881.)]

The *Bohras* of Western India are identified with the Imámí-Ismáilís in some books, and were so spoken of in the first edition of this work. This is, however, an error, originally due, it would seem, to Sir John Malcolm. The nature of their doctrine, indeed, seems to be very much alike, and the Bohras, like the Ismáilís, attach a divine character to their *Mullah* or chief pontiff, and make a pilgrimage to his presence once in life. But the *persons* so reverenced are quite different ; and the Bohras recognise all the 12 Imáms of ordinary Shiahs. Their first appearance in India was early, the date which they assign being A.H. 532 (A.D. 1137-1138). Their chief seat was in Yemen, from which a large emigration to India took place on its conquest by the Turks in 1538. Ibn Batuta seems to have met with Bohras at Gandár, near Baroch, in 1342. (*Voyages*, IV. 58.)

A Chinese account of the expedition of Hulaku will be found in Rémusat's *Nouveaux Mélanges* (I.), and in Pauthier's Introduction. (*Q. R.* 115-219, esp. 213 ; *Ilch.* vol. i.; *J. A. S. B.* VI. 842 *seqq.*) [A new and complete translation has been given by Dr. E. Bretschneider, *Med. Res.* I. 112 *seqq.*—H. C.]

There is some account of the rock of Alamút and its exceedingly slender traces of occupancy, by Colonel Monteith, in *J. R. G. S.* III. 15, and again by Sir Justin Sheil in vol. viii. p. 431. There does not seem to be any specific authority for assigning the Paradise of the Shaikh to Alamút ; and it is at least worthy of note that another of the castles of the Muláhidah, destroyed by Hulaku, was called *Firdús, i.e.* Paradise. In any case, I see no reason to suppose that Polo visited Alamút, which would have been quite out of the road that he is following.

It is possible that "the Castle," to which he alludes at the beginning of next chapter, and which set him off upon this digression, was *Girdkuh.** It has not, as far as I know, been identified by modern travellers, but it stood within 10 or 12 miles of Damghan (to the west or north-west). It is probably the *Tigado* of Hayton, of which he thus speaks : " The Assassins had an impregnable castle called Tigado, which was furnished with all necessaries, and was so strong that it had no fear of attack on any side. Howbeit, Haloön commanded a certain captain of his that he should take 10,000 Tartars who had been left in garrison in Persia, and with them lay siege to the said castle, and not leave it till he had taken it. Wherefore the said Tartars continued besieging it for seven whole years, winter and summer, without being able to take it. At last the Assassins surrendered, from sheer want of clothing, but not of victuals or other necessaries." So Ramusio ; other copies read " 27 years." In any case it corroborates the fact that Girdkuh was said to have held out for an extraordinary length of time. If Rashiduddin is right in naming 1270 as the date of surrender, this would be quite a recent event when the Polo party passed, and draw special attention to the spot. (*J. As.* sér. IV. tom. xiii. 48 ; *Ilch.* I. 93, 104, 274 ; *Q. R.* p. 278 ; *Ritter*, VIII. 336.) A note which I have from *Djihan Numa* (I. 259) connects Girdkuh with a district called *Chinar*. This may be a clue to the term *Arbre Sec ;* but there are difficulties.

---

* [Ghirdkuh means "round mountain" ; it was in the district of Kumis, three parasangs west of Damghan. Under the year 1257, the *Yüan shi* mentions the taking of the fortress of *Ghi-rh-du-kie* by *K'ie-di-bu-hua*. (*Bretschneider, Med. Res.* I. p. 122 ; II. 110.)—H. C.]

CHAPTER XXVI.

CONCERNING THE CITY OF SAPURGAN.

ON leaving the Castle, you ride over fine plains and beautiful valleys, and pretty hill-sides producing excellent grass pasture, and abundance of fruits, and all other products. Armies are glad to take up their quarters here on account of the plenty that exists. This kind of country extends for six days' journey, with a goodly number of towns and villages, in which the people are worshippers of Mahommet. Sometimes also you meet with a tract of desert extending for 50 or 60 miles, or somewhat less, and in these deserts you find no water, but have to carry it along with you. The beasts do without drink until you have got across the desert tract and come to watering places.

So after travelling for six days as I have told you, you come to a city called SAPURGAN. It has great plenty of everything, but especially of the very best melons in the world. They preserve them by paring them round and round into strips, and drying them in the sun. When dry they are sweeter than honey, and are carried off for sale all over the country. There is also abundance of game here, both of birds and beasts.[1]

NOTE 1.—SAPURGAN may closely express the pronunciation of the name of the city which the old Arabic writers call *Sabúrkán* and *Shabúrkán*, now called *Shibrgán*, lying some 90 miles west of Balkh; containing now some 12,000 inhabitants, and situated in a plain still richly cultivated, though on the verge of the desert.* But I have seen no satisfactory solution of the difficulties as to the time assigned. This in the G. T. and in Ramusio is clearly six days. The point of departure is indeed uncertain, but even if we were to place that at Sharakhs on the extreme verge of

---

* The oldest form of the name is *Asapuragán,* which Rawlinson thinks traceable to its being an ancient seat of the *Asa* or *Asagartii.* (*J. R. A. S.* XI. 63.)

cultivated Khorasan, which would be quite inconsistent with other data, it would have taken the travellers something like double the time to reach Shíbrgán. Where I have followed the G. T. in its reading "*quant l'en a chevauchés six jornée tel che je vos ai contés, adunc treuve l'en une cité*," etc., Pauthier's text has "*Et quant l'en a chevauchié les vi cités, si treuve l'en une cité qui a nom Sapurgan*," and to this that editor adheres. But I suspect that *cités* is a mere lapsus for *journées*, as in the reading in one of his three MSS. What could be meant by "*chevauchier les vi cités*"? Whether the true route be, as I suppose, by Nishapúr and Meshid, or, as Khanikoff supposes, by Herat and Badghis, it is strange that no one of those famous cities is mentioned. And we feel constrained to assume that something has been misunderstood in the dictation, or has dropt out of it. As a *probable* conjecture I should apply the six days to the extent of pleasing country described in the first lines of the chapter, and identify it with the tract between Sabzawur and the cessation of fertile country beyond Meshid. The distance would agree well, and a comparison with Fraser or Ferrier will show that even now the description, allowing for the compression of an old recollection, would be well founded ; *e.g.* on the first march beyond Nishapúr : "Fine villages, with plentiful gardens full of trees, that bear fruit of the highest flavour, may be seen all along the foot of the hills, and in the little recesses formed by the ravines whence issues the water that irrigates them. It was a rich and pleasing scene, and out of question by far the most populous and cultivated tract that I had seen in Persia. . . . . Next morning we quitted Derrood . . . . by a very indifferent but interesting road, the glen being finely wooded with walnut, mulberry, poplar, and willow-trees, and fruit-tree gardens rising one above the other upon the mountain-side, watered by little rills. . . . . These gardens extended for several miles up the glen; beyond them the bank of the stream continued to be fringed with white sycamore, willow, ash, mulberry, poplar, and woods that love a moist situation," and so on, describing a style of scenery not common in Persia, and expressing diffusely (as it seems to me) the same picture as Polo's two lines. In the valley of Nishapúr, again (we quote Arthur Conolly) : "'This is Persia !' was the vain exclamation of those who were alive to the beauty of the scene ; 'this is Persia !' *Bah ! Bah !* What grass, what grain, what water ! *Bah ! Bah !*

[' If there be a Paradise on the face of the Earth,
      This is it ! This is it ! This is it ! ' "]—(I. 209.)

(See *Fraser*, 405, 432-433, 434, 436.)

With reference to the dried melons of Shibrgán, Quatremère cites a history of Herat, which speaks of them almost in Polo's words. Ibn Batuta gives a like account of the melons of Khárizm : "The surprising thing about these melons is the way the people have of slicing them, drying them in the sun, and then packing them in baskets, just as Malaga figs are treated in our part of the world. In this state they are sent to the remotest parts of India and China. There is no dried fruit so delicious, and all the while I lived at Delhi, when the travelling dealers came in, I never missed sending for these dried strips of melon." (*Q. R.* 169 ; *I. B.* III. 15.) Here, in the 14th century, we seem to recognise the Afghan dealers arriving in the cities of Hindustan with their annual camel-loads of dried fruits, just as we have seen them in our own day.

CHAPTER XXVII.

OF THE CITY OF BALC.

BALC is a noble city and a great, though it was much greater in former days. But the Tartars and other nations have greatly ravaged and destroyed it. There were formerly many fine palaces and buildings of marble, and the ruins of them still remain. The people of the city tell that it was here that Alexander took to wife the daughter of Darius.

Here, you should be told, is the end of the empire of the Tartar Lord of the Levant. And this city is also the limit of Persia in the direction between east and north-east.[1]

Now, let us quit this city, and I will tell you of another country called DOGANA.[2]

When you have quitted the city of which I have been speaking, you ride some 12 days between north-east and east, without finding any human habitation, for the people have all taken refuge in fastnesses among the mountains, on account of the Banditti and armies that harassed them. There is plenty of water on the road, and abundance of game; there are lions too. You can get no provisions on the road, and must carry with you all that you require for these 12 days.[3]

NOTE I.—BALKH, "the mother of cities," suffered mercilessly from Chinghiz. Though the city had yielded without resistance, the whole population was marched by companies into the plain, on the usual Mongol pretext of counting them, and then brutally massacred. The city and its gardens were fired, and all buildings capable of defence were levelled. The province long continued to be harried by the Chaghataian inroads. Ibn Batuta, sixty years after Marco's visit, describes the city as still in ruins, and as uninhabited: "The remains of its mosques and colleges," he says, "are still to be seen, and the painted walls traced with azure." It is no doubt the Vaeq (*Valq*) of Clavijo, "very large, and surrounded by a broad earthen wall, thirty paces across, but breached in many parts." He describes a large portion of the area within as sown with cotton. The account of its modern state in Burnes and Ferrier is much the same as Ibn Batuta's, except that they found some population ; two separate towns within the walls according to the latter. Burnes estimates the circuit of the ruins at 20 miles. The bulk of the population has been moved since 1858 to Takhtapul, 8 miles east of Balkh, where the Afghan Government is placed.

(*Erdmann,* 404-405; *I. B.* III. 59 ; *Clavijo,* p. 117 ; *Burnes,* II. 204-206; *Ferrier,* 206-207.)

According to the legendary history of Alexander, the beautiful Roxana was the daughter of Darius, and her father in a dying interview with Alexander requested the latter to make her his wife :—

> " Une fille ai mult bele ; se prendre le voles.
> Vus en seres de l'mont tout li mius maries," etc.
>
> (*Lambert Le Court,* p. 256.)

NOTE 2.—The country called *Dogana* in the G. Text is a puzzle. In the former edition I suggested *Juzgána,* a name which till our author's time was applied to a part of the adjoining territory, though not to that traversed in quitting Balkh for the east. Sir H. Rawlinson is inclined to refer the name to *Dehgán,* or " villager," a term applied in Bactria, and in Kabul, to Tajik peasantry.* I may also refer to certain passages in Baber's " Memoirs," in which he speaks of a place, and apparently a district, called *Dehánah,* which seems from the context to have lain in the vicinity of the Ghori, or Aksarai River. There is still a village in the Ghori territory, called *Dehánah.* Though this is worth mentioning, where the true solution is so uncertain, I acknowledge the difficulty of applying it. I may add also that Baber calls the River of Ghori or Aksarai, the *Dogh-*ábah. (*Sprenger, P. und R. Routen,* p. 39 and Map ; *Anderson* in *J. A. S. B.* XXII. 161 ; *Ilch.* II. 93; *Baber,* pp. 132, 134, 168, 200, also 146.)

NOTE 3.—Though Burnes speaks of the part of the road that we suppose necessarily to have been here followed from Balkh towards Taican, as barren and dreary, he adds that the ruins of *aqueducts* and houses proved that the land had at one time been peopled, though now destitute of water, and consequently of inhabitants. The country would seem to have reverted at the time of Burnes' journey, from like causes, nearly to the state in which Marco found it after the Mongol devastations.

Lions seem to mean here the real king of beasts, and not tigers, as hereafter in the book. Tigers, though found on the S. and W. shores of the Caspian, do not seem to exist in the Oxus valley. On the other hand, Rashiduddin tells us that, when Hulaku was reviewing his army after the passage of the river, several lions were started, and two were killed. The lions are also mentioned by Sidi 'Ali, the Turkish Admiral, further down the valley towards Hazárasp : " We were obliged to fight with the lions day and night, and no man dared to go alone for water." Moorcroft says of the plain between Kunduz and the Oxus : " Deer, foxes, wolves, hogs, and *lions* are numerous, the latter resembling those in the vicinity of Hariana " (in Upper India). Wood also mentions lions in Kuláb, and at Kila'chap on the Oxus. Q. Curtius tells how Alexander killed a great lion in the country north of the Oxus towards Samarkand. [A similar story is told of Timur in *The Mulfuzat Timúry,* translated by Major Charles Stewart, 1830 (p. 69) : " During the march '(near Balkh)' two lions made their appearance, one of them a male, the other a female. I (Timur) resolved to kill them myself, and having shot them both with arrows, I considered this circumstance as a lucky omen."—H. C.] (*Burnes,* II. 200 ; *Q. R.* 155 ; *Ilch.* I. 90 ; *J. As.* IX. 217 ; *Moorcroft,* II. 430 ; *Wood,* ed. 1872, pp. 259, 260 ; *Q. C.* VII. 2.)

---

* It may be observed that the careful Elphinstone distinguishes from this general application of Dehgán or Dehkán, the name *Deggán* applied to a tribe " once spread over the north-east of Afghanistan, but now as a separate people only in Kunar and Laghman."

CHAPTER XXVIII.

Of Taican, and the Mountains of Salt. Also of the Province of Casem.

AFTER those twelve days' journey you come to a fortified place called TAICAN, where there is a great corn market.[1] It is a fine place, and the mountains that you see towards the south are all composed of salt. People from all the countries round, to some thirty days' journey, come to fetch this salt, which is the best in the world, and is so hard that it can only be broken with iron picks. 'Tis in such abundance that it would supply the whole world to the end of time. [Other mountains there grow almonds and pistachioes, which are exceedingly cheap.][2]

When you leave this town and ride three days further between north-east and east, you meet with many fine tracts full of vines and other fruits, and with a goodly number of habitations, and everything to be had very cheap. The people are worshippers of Mahommet, and are an evil and a murderous generation, whose great delight is in the wine shop; for they have good wine (albeit it be boiled), and are great topers; in truth, they are constantly getting drunk. They wear nothing on the head but a cord some ten palms long twisted round it. They are excellent huntsmen, and take a great deal of game; in fact they wear nothing but the skins of the beasts they have taken in the chase, for they make of them both coats and shoes. Indeed, all of them are acquainted with the art of dressing skins for these purposes.[3]

When you have ridden those three days, you find a town called CASEM,[4] which is subject to a count. His other towns and villages are on the hills, but through this town there flows a river of some size. There are a great many porcupines hereabouts, and very large ones too. When hunted with dogs, several of them will get together and huddle close, shooting their quills at the dogs, which get many a serious wound thereby.[5]

This town of Casem is at the head of a very great province, which is also called Casem. The people have a peculiar language. The peasants who keep cattle abide in the mountains, and have their dwellings in caves, which form fine and spacious houses for them, and are made with ease, as the hills are composed of earth.[6]

After leaving the town of Casem, you ride for three days without finding a single habitation, or anything to eat or drink, so that you have to carry with you everything that you require. At the end of those three days you reach a province called Badashan, about which we shall now tell you.[7]

NOTE 1.—The *Taican* of Polo is the still existing TALIKAN in the province of Kataghan or Kunduz, but it bears the former name (*Thâîkân*) in the old Arab geographies. Both names are used by Baber, who says it lay in the *Ulugh Bágh*, or Great Garden, a name perhaps acquired by the Plains of Talikan in happier days, but illustrating what Polo says of the next three days' march. The Castle of Talikan resisted Chinghiz for seven months, and met with the usual fate (1221). [In the Travels of Sidi Ali, son of Housaïn (*Jour. Asiat.*, October, 1826, p. 203), "Talikan, in the country of Badakhschan" is mentioned.—H. C.] Wood speaks of Talikan in 1838 as a poor place of some 300 or 400 houses, mere hovels ; a recent account gives it 500 families. Market days are not usual in Upper India or Kabul, but are universal in Badakhshan and the Oxus provinces. The bazaars are only open on those days, and the people from the surrounding country then assemble to exchange goods, generally by barter. Wood chances to note : "A market was held at Talikan. . . . The thronged state of the roads leading into it soon apprised us that the day was no ordinary one." (*Abulf.* in *Büsching*, V. 352 ; *Sprenger*, p. 50 ; *P. de la Croix*, I. 63 ; *Baber*, 38, 130 ; *Burnes*, III. 8 ; *Wood*, 156 ; *Pandit Manphul's Report*.)

The distance of Talikan from Balkh is about 170 miles, which gives very short marches, if twelve days be the correct reading. Ramusio has *two* days, which is certainly wrong. XII. is easily miswritten for VII., which would be a just number.

NOTE 2.—In our day, as I learn from Pandit Manphul, the mines of rock salt are at Ak Bulák, near the Lataband Pass, and at Darúná, near the Kokcha, and these supply the whole of Badakhshan, as well as Kunduz and Chitrál. These sites are due *east* of Talikan, and are in Badakhshan. But there is a mine at *Chál*, S.E. or S.S.E. of Talikan and within the same province. There are also mines of rock-salt near the famous "stone bridge" in Kuláb, north of the Oxus, and again on the south of the Alaï steppe.   (Papers by *Manphul* and by *Faiz Baksh;* also *Notes* by *Feachenko.*)

Both pistachioes and wild almonds are mentioned by Pandit Manphul ; and see *Wood* (p. 252) on the beauty and profusion of the latter.

NOTE 3.—Wood thinks that the Tajik inhabitants of Badakhshan and the adjoining districts are substantially of the same race as the Kafir tribes of Hindu Kúsh.   At the time of Polo's visit it would seem that their conversion to Islam was imperfect.   They were probably in that transition state which obtains in our own day for some of the Hill Mahomedans adjoining the Kafirs on the south side of the mountains the reproachful title of *Nímchah Musulmán,* or Half-and-halfs.   Thus they would seem to have retained sundry Kafir characteristics ; among others that love of wine which is so strong among the Kafirs.   The boiling of the wine is noted by Baber (a connoisseur) as the custom of Nijrao, adjoining, if not then included in, Kafir-land ; and Elphinstone implies the continuance of the custom when he speaks of the Kafirs as having wine of *the consistence of jelly,* and very strong.   The wine of *Kápishí,* the Greek Kapisa, immediately south of Hindu Kúsh, was famous as early as the time of the Hindu grammarian Pánini, say three centuries B.C.   The cord twisted round the head was probably also a relic of Kafir costume : " Few of the Kafirs cover the head, and when they do, it is with a narrow band or fillet of goat's hair . . . . about a yard or a yard and a half in length, wound round the head." This style of head-dress seems to be very ancient in India, and in the Sanchi sculptures is that of the supposed Dasyas.   Something very similar, *i.e.* a scanty turban cloth twisted into a mere cord, and wound two or three times round the head, is often seen in the Panjab to this day.

The *Postín* or sheepskin coat is almost universal on both sides of the Hindu Kúsh ; and Wood notes : "The shoes in use resemble half-boots, made of goatskin, and mostly of home manufacture."   (*Baber,* 145 ; *J. A. S. B.* XXVIII. 348,   364 ; *Elphinst.* II. 384 ; *Ind. Antiquary,* I. 22 ; *Wood,* 174, 220 ; *J. R. A. S.* XIX. 2.)

NOTE 4.—Marsden was right in identifying *Scassem* or *Casem* with the *Kechem* of D'Anville's Map, but wrong in confounding the latter with the *Kishmabad* of Elphinstone—properly, I believe, *Kishnabad*—in the Anderab Valley.   Kashm, or Keshm, found its way into maps through Pétis de la Croix, from whom probably D'Anville adopted it ; but as it was ignored by Elphinstone (or by Macartney, who constructed his map), and by Burnes, it dropped out of our geography.   Indeed, Wood does not notice it except as giving name to a high hill called the Hill of Kishm, and the position even of that he omits to indicate.   The frequent mention of Kishm in the histories of Timur and Humayun (*e.g. P. de la Croix,* I. 167 ; *N. et E.* XIV. 223, 491 ; *Erskine's Baber and Humayun,* II. 330, 355, etc.) had enabled me to determine its position within tolerably narrow limits ; but desiring to fix it definitely, application was made through Colonel Maclagan to Pandit Manphul, C.S.I., a very intelligent Hindu gentleman, who resided for some time in Badakhshan as agent of the Panjab Government, and from him arrived a special note and sketch, and afterwards a MS. copy of a Report,* which set the position of Kishm at rest.

KISHM is the *Kilissemo, i.e.* Karisma or Krishma, of Hiuen Tsang ; and Sir H. Rawlinson has identified the Hill of Kishm with the Mount Kharesem of the Zend-Avesta, on which Jamshid placed the most sacred of all the fires.   It is now a small town or large village on the right bank of the Varsach river, a tributary of the Kokcha. It was in 1866 the seat of a district ruler under the Mír of Badakhshan, who was styled the Mír of Kishm, and is the modern counterpart of Marco's *Quens* or Count. The modern caravan-road between Kunduz and Badakhshan does not pass through Kishm, which is left some five miles to the right, but through the town of Mashhad, which stands on the same river.   Kishm is the warmest district of Badakhshan.   Its

---

* Since published in *J. K. G. S.* vol. xlii.

fruits are abundant, and ripen a month earlier than those at Faizabad, the capital of that country. The Varsach or Mashhad river is Marco's "*Flum auques grant.*" Wood (247) calls it "the largest stream we had yet forded in Badakhshan."

It is very notable that in Ramusio, in Pipino, and in one passage of the G. Text, the name is written *Scasem*, which has led some to suppose the *Ish-Káshm* of Wood to be meant. That place is much too far east—in fact, beyond the city which forms the subject of the next chapter. The apparent hesitation, however, between the forms *Casem* and *Scasem* suggests that the Kishm of our note may formerly have been termed S'kăshm or Ish-Kăshm, a form frequent in the Oxus Valley, *e.g. Ish-Kimish, Ish-Káshm, Ishtrakh, Ishpingao.* General Cunningham judiciously suggests (*Ladak*, 34) that this form is merely a vocal corruption of the initial *S* before a consonant, a combination which always troubles the Musulman in India, and converts every Mr. Smith or Mr. Sparks into Ismit or Ispak Sahib.

[There does not seem to me any difficulty about this note: "Shibarkhan (Afghan Turkistan), Balkh, Kunduz, Khanabad, Talikan, Kishm, Badakhshan." I am tempted to look for Dogana at Khanabad.—H. C.]

NOTE 5.—The belief that the porcupine *projected* its quills at its assailants was an ancient and persistent one—"*cum intendit cutem missiles,*" says Pliny (VIII. 35, and see also *Aelian. de Nat. An.* I. 31), and is held by the Chinese as it was held by the ancients, but is universally rejected by modern zoologists. The huddling and coiling appears to be a true characteristic, for the porcupine always tries to shield its head.

NOTE 6.—The description of Kishm as a "very great" province is an example of a bad habit of Marco's, which recurs in the next chapter. What he says of the cave-dwellings may be illustrated by Burnes's account of the excavations at Bamian, in a neighbouring district. These "still form the residence of the greater part of the population. . . . . The hills at Bamian are formed of indurated clay and pebbles, which renders this excavation a matter of little difficulty." Similar occupied excavations are noticed by Moorcroft at Heibak and other places towards Khulm.

Curiously, Pandit Manphul says of the districts about the Kokcha : "Both their hills and plains are productive, the former *being mostly composed of earth, having very little of rocky substance.*"

NOTE 7.—The capital of Badakhshan is now Faizabad, on the right bank of the Kokcha, founded, according to Manphul, by Yarbeg, the first Mír of the present dynasty. When this family was displaced for a time, by Murad Beg of Kunduz, about 1829, the place was abandoned for years, but is now re-occupied. The ancient capital of Badakhshan stood in the Dasht (or Plain) of Bahárak, one of the most extensive pieces of level in Badakhshan, in which the rivers Vardoj, Zardeo, and Sarghalan unite with the Kokcha, and was apparently termed *Jaúzgún*. This was probably the city called Badakhshan by our traveller.* As far as I can estimate, by the help of Wood and the map I have compiled, this will be from 100 to 110 miles distant from Talikan, and will therefore suit fairly with the six marches that Marco lays down.

Wood, in 1838, found the whole country between Talikan and Faizabad nearly as depopulated as Marco found that between Kishm and Badakhshan. The modern depopulation was due—in part, at least—to the recent oppressions and *razzias* of the Uzbeks of Kunduz. On their decline, between 1840 and 1850, the family of the native Mírs was reinstated, and these now rule at Faizabad, under an acknowledgment, since 1859, of Afghan supremacy.

---

* Wilford, in the end of the 18th century, speaks of Faizabad as "the new capital of Badakhshan, built near the site of the old one." The Chinese map (*vide J. R. G. S.* vol. xlii.) represents the city of *Badakhshan* to the east of Faizabad. Faiz Bakhsh, in an unpublished paper, mentions a tradition that the Lady Zobeidah, dear to English children, the daughter of Al-Mansúr and wife of Ar-Rashid, delighted to pass the spring at Jaúzgún, and built a palace there, "the ruins of which are still visible."

CHAPTER XXIX.

OF THE PROVINCE OF BADASHAN.

BADASHAN is a Province inhabited by people who worship
Mahommet, and have a peculiar language.   It forms a
very great kingdom, and the royalty is hereditary.   All
those of the royal blood are descended from King Alexander and the daughter of King Darius, who was Lord
of the vast Empire of Persia.   And all these kings call
themselves in the Saracen tongue ZULCARNIAIN, which is
as much as to say *Alexander;* and this out of regard for
Alexander the Great.[1]

It is in this province that those fine and valuable gems
the Balas Rubies are found.   They are got in certain
rocks among the mountains, and in the search for them
the people dig great caves underground, just as is done
by miners for silver.   There is but one special mountain
that produces them, and it is called SYGHINAN.   The stones
are dug on the king's account, and no one else dares dig
in that mountain on pain of forfeiture of life as well as
goods ; nor may any one carry the stones out of the
kingdom.   But the king amasses them all, and sends
them to other kings when he has tribute to render, or
when he desires to offer a friendly present ; and such only
as he pleases he causes to be sold.   Thus he acts in order
to keep the Balas at a high value ; for if he were to allow
everybody to dig, they would extract so many that the
world would be glutted with them, and they would cease
to bear any value.   Hence it is that he allows so few to
be taken out, and is so strict in the matter.[2]

There is also in the same country another mountain,
in which azure is found ; 'tis the finest in the world, and
is got in a vein like silver.   There are also other

mountains which contain a great amount of silver ore, so that the country is a very rich one; but it is also (it must be said) a very cold one.[3] It produces numbers of excellent horses, remarkable for their speed. They are not shod at all, although constantly used in mountainous country, and on very bad roads. [They go at a great pace even down steep descents, where other horses neither would nor could do the like. And Messer Marco was told that not long ago they possessed in that province a breed of horses from the strain of Alexander's horse Bucephalus, all of which had from their birth a particular mark on the forehead. This breed was entirely in the hands of an uncle of the king's; and in consequence of his refusing to let the king have any of them, the latter put him to death. The widow then, in despite, destroyed the whole breed, and it is now extinct.[4]]

The mountains of this country also supply Saker falcons of excellent flight, and plenty of Lanners likewise. Beasts and birds for the chase there are in great abundance. Good wheat is grown, and also barley without husk. They have no olive oil, but make oil from sesamé, and also from walnuts.[5]

[In the mountains there are vast numbers of sheep— 400, 500, or 600 in a single flock, and all of them wild; and though many of them are taken, they never seem to get aught the scarcer.[6]

Those mountains are so lofty that 'tis a hard day's work, from morning till evening, to get to the top of them. On getting up, you find an extensive plain, with great abundance of grass and trees, and copious springs of pure water running down through rocks and ravines. In those brooks are found trout and many other fish of dainty kinds; and the air in those regions is so pure, and residence there so healthful, that when the men who dwell below in the towns, and in the valleys and plains, find

themselves attacked by any kind of fever or other ailment
that may hap, they lose no time in going to the hills; and
after abiding there two or three days, they quite recover
their health through the excellence of that air. And
Messer Marco said he had proved this by experience: for
when in those parts he had been ill for about a year, but

Ancient Silver Patera of debased Greek art, formerly in the possession of the Princes of Badakhshan, now in the India Museum. (Four-ninths of the diameter of the Original.)

as soon as he was advised to visit that mountain, he did
so and got well at once.⁷]

In this kingdom there are many strait and perilous
passes, so difficult to force that the people have no fear of
invasion. Their towns and villages also are on lofty hills,

and in very strong positions.[8]   They are excellent archers, and much given to the chase; indeed, most of them are dependent for clothing on the skins of beasts, for stuffs are very dear among them.   The great ladies, however, are arrayed in stuffs, and I will tell you the style of their dress!   They all wear drawers made of cotton cloth, and into the making of these some will put 60, 80, or even 100 ells of stuff.   This they do to make themselves look large in the hips, for the men of those parts think that to be a great beauty in a woman.[9]

Note 1.—"The population of Badakhshan Proper is composed of Tajiks, Turks, and Arabs, who are all Sunnis, following the orthodox doctrines of the Mahomedan law, and speak Persian and Turki, whilst the people of the more mountainous tracts are Tajiks of the Shiá creed, having separate provincial dialects or languages of their own, the inhabitants of the principal places combining therewith a knowledge of Persian.   Thus, the *Shighnáni* [sometimes called *Shighni*] is spoken in Shignán and Roshán, the *Ishkáshami* in Ishkásham, the *Wakhi* in Wakhán, the *Sanglichi* in Sanglich and Zebák, and the *Minjáni* in Minján.   All these dialects materially differ from each other."   (*Pand. Manphul.*)   It may be considered almost certain that Badakhshan Proper also had a peculiar dialect in Polo's time.   Mr. Shaw speaks of the strong resemblance to *Kashmíris* of the Badakhshán people whom he had seen.

The Legend of the Alexandrian pedigree of the Kings of Badakhshan is spoken of by Baber, and by earlier Eastern authors.   This pedigree is, or was, claimed also by the chiefs of Karátegín, Darwáz, Roshán, Shighnán, Wakhán, Chitrál, Gilgít, Swát, and Khapolor in Bálti.   Some samples of those genealogies may be seen in that strange document called "Gardiner's Travels."

In Badakhshan Proper the story seems now to have died out.   Indeed, though Wood mentions one of the modern family of Mírs as vaunting this descent, these are in fact *Sáhibzádahs* of Samarkand, who were invited to the country about the middle of the 17th century, and were in no way connected with the old kings.

The traditional claims to Alexandrian descent were probably due to a genuine memory of the Graeco-Bactrian kingdom, and might have had an origin analogous to the Sultan's claim to be "Caesar of Rome"; for the real ancestry of the oldest dynasties on the Oxus was to be sought rather among the Tochari and Ephthalites than among the Greeks whom they superseded.

The cut on p. 159 presents an interesting memorial of the real relation of Bactria to Greece, as well as of the pretence of the Badakhshan princes to Grecian descent. This silver patera was sold by the family of the Mírs, when captives, to the Minister of the Uzbek chief of Kunduz, and by him to Dr. Percival Lord in 1838.   It is now in the India Museum.   On the bottom is punched a word or two in Pehlvi, and there is also a word incised in Syriac or Uighúr.   It is curious that a *pair* of paterae were acquired by Dr. Lord under the circumstances stated.   The other, similar in material and form, but apparently somewhat larger, is distinctly Sassanian, representing a king spearing a lion.

Zu-'lkarnain, "the Two-Horned," is an Arabic epithet of Alexander, with which legends have been connected, but which probably arose from the horned portraits on his coins.   [Capus, *l.c.* p. 121, says, "Iskandr Zoulcarneïn or Alexander *le Cornu,* horns being the emblem of strength."—H. C.]   The term appears in Chaucer (*Troil. and Cress.* III. 931) in the sense of *non plus*:—

> "I am, till God me better minde send,
> At *dulcarnon*, right at my wittes end."

And it is said to have still colloquial existence in that sense in some corners of England. This use is said to have arisen from the Arabic application of the term (*Bicorne*) to the 47th Proposition of Euclid. (*Baber*, 13; *N. et E.* XIV. 490; *N. An. des V.* xxvi. 296; *Burnes*, III. 186 *seqq.*; *Wood*, 155, 244; *J. A. S. B.* XXII. 300; *Ayeen Akbery*, II. 185; see *N. and Q.* 1st Series, vol. v.)

NOTE 2.—I have adopted in the text for the name of the country that one of the several forms in the G. Text which comes nearest to the correct name, viz. *Badascian*. But *Balacian* also appears both in that and in Pauthier's text. This represents *Balakhshán*, a form also sometimes used in the East. Hayton has *Balaxcen*, Clavijo *Balaxia*, the Catalan Map *Baldassia*. From the form *Balakhsh* the Balas Ruby got its name. As Ibn Batuta says: "The Mountains of Badakhshan have given their name to the Badakhshi Ruby, vulgarly called *Al Balaksh*." Albertus Magnus says the *Balagius* is the female of the Carbuncle or Ruby Proper, "and some say it is his house, and hath thereby got the name, quasi *Palatium* Carbunculi!" The Balais or Balas Ruby is, like the Spinel, a kind inferior to the real Ruby of Ava. The author of the *Masálak al Absár* says the finest Balas ever seen in the Arab countries was one presented to Malek 'Adil Ketboga, at Damascus; it was of a triangular form and weighed 50 drachms. The prices of *Balasci* in Europe in that age may be found in Pegolotti, but the needful problems are hard to solve.

> "No sapphire in Inde, no Rubie rich of price,
> There lacked than, nor Emeraud so grene,
> *Balès*, Turkès, ne thing to my device."
> >   (*Chaucer, 'Court of Love.'*)

> "L'altra letizia, che m'era già nota,
> Preclara cosa mi si fece in vista,
> Qual fin *balascio* in che lo Sol percuoto."
> >   (*Paradiso*, ix. 67.)

Some account of the Balakhsh from Oriental sources will be found in *J. As.* sér. V. tom. xi. 109.

(*I. B.* III. 59, 394; *Alb. Mag. de Mineralibus; Pegol.* p. 307; *N. et E.* XIII. i. 246.)

["The Mohammedan authors of the Mongol period mention Badakhshan several times in connection with the political and military events of that period. Guchluk, the 'gurkhan of Karakhitai,' was slain in Badakhshan in 1218 (*d'Ohsson*, I. 272). In 1221, the Mongols invaded the country (*l.c.* I. 272). On the same page, d'Ohsson translates a short account of Badakhshan by Yakut (+1229), stating that this mountainous country is famed for its precious stones, and especially rubies, called *Balakhsh*." (Bretschneider, *Med. Res.* II. p. 66.)—H. C.]

The account of the royal monopoly in working the mines, etc., has continued accurate down to our own day. When Murad Beg of Kunduz conquered Badakhshan some forty years ago, in disgust at the small produce of the mines, he abandoned working them, and sold nearly all the population of the place into slavery! They continue still unworked, unless clandestinely. In 1866 the reigning Mír had one of them opened at the request of Pandit Manphul, but without much result.

The locality of the mines is on the right bank of the Oxus, in the district of Ish Káshm and on the borders of SHIGNAN, the *Syghinan* of the text. (*P. Manph.*; *Wood*, 206; *N. Ann. des. V.* xxvi. 300.)

[The ruby mines are really in the Ghâran country, which extends along both banks of the Oxus. Barshar is one of the deserted villages ; the boundary between Gháran and Shignán is the Kuguz Parin (in Shighai dialect means " holes in the rock ") ; the Persian equivalent is " Rafak-i-Somakh." (Cf. Captain Trotter, *Forsyth's Mission*, p. 277.)—H. C.]

NOTE 3.—The mines of *Lájwurd* (whence *l'Azur* and *Lazuli*) have been, like the Ruby mines, celebrated for ages. They lie in the Upper Valley of the Kokcha, called Korán, within the Tract called *Yamgán*, of which the popular etymology is *Hamah-Kán*, or " All-Mines," and were visited by Wood in 1838. The produce now is said to be of very inferior quality, and in quantity from 30 to 60 *poods* (36 lbs each) annually. The best quality sells at Bokhara at 30 to 60 tillas, or 12*l.* to 24*l.* the pood (*Manphúl*). Surely it is ominous when a British agent writing of Badakhshan products finds it natural to express weights in Russian poods !

The Yamgán Tract also contains mines of iron, lead, alum, salammoniac, sulphur, ochre, and copper. The last are not worked. But I do not learn of any silver mines nearer than those of Paryán in the Valley of Panjshir, south of the crest of the Hindu-Kúsh, much worked in the early Middle Ages. (See *Cathay*, p. 595.)

NOTE 4.—The Kataghan breed of horses from Badakhshan and Kunduz has still a high reputation. They do not often reach India, as the breed is a favourite one among the Afghan chiefs, and the horses are likely to be appropriated in transit. (*Lumsden, Mission to Kandahar*, p. 20.)

[The Kirghiz between the Yangi Hissar River and Sirikol are the only people using the horse generally in the plough, oxen being employed in the plains, and yaks in Sirikol. (Lieutenant-Colonel Gordon, p. 222, *Forsyth's Mission*.)—H. C.]

What Polo heard of the Bucephalid strain was perhaps but another form of a story told by the Chinese, many centuries earlier, when speaking of this same region. A certain cave was frequented by a wonderful stallion of supernatural origin. Hither the people yearly brought their mares, and a famous breed was derived from the foals. (*Rém. N. Mél. As.* I. 245.)

NOTE 5.—The huskless barley of the text is thus mentioned by Burnes in the vicinity of the Hindu-Kúsh : " They rear a barley in this elevated country which has no husk, and grows like wheat ; but it is barley." It is not properly *huskless*, but when ripe it bursts the husk and remains so loosely attached as to be dislodged from it by a slight shake. It is grown abundantly in Ladak and the adjoining Hill States. Moorcroft details six varieties of it cultivated there. The kind mentioned by Marco and Burnes is probably that named by Royle *Hordeum Ægiceras*, and which has been sent to England under the name of Tartarian Wheat, though it is a genuine barley. *Naked barley* is mentioned by Galen as grown in Cappadocia ; and Matthioli speaks of it as grown in France in his day (middle of 16th century). It is also known to the Arabs, for they have a name for it—*Sult*. (*Burnes*, III. 205 ; *Moorc.* II. 148 *seqq. ; Galen, de Aliment. Facult.* Lat. ed. 13 ; *Matthioli*, Ven. 1585, p. 420 ; *Eng. Cyc.*, art. Hordeum.)

Sesamé is mentioned by P. Manphul as one of the products of Badakhshan ; linseed is another, which is also used for oil. Walnut-trees abound, but neither he nor Wood mention the oil. We know that walnut oil is largely manufactured in Kashmir. (*Moorcroft*, II. 148.)

[See on Saker and Lanner Falcons (*F. Sakar*, Briss. ; *F. lanarius*, Schlegel) the valuable paper by Edouard Blanc, *Sur l'utilisation des Oiseaux de proie en Asie centrale* in *Rev. des Sciences natur. appliquées*, 20th June, 1895.

" Hawking is the favourite sport of Central Asian Lords," says G. Capus. (*A travers le royaume de Tamerlan*, p. 132. See pp. 132-134.)

The Mirza says (*l.c.* p. 157) that the mountains of Wakhán " are only noted for producing a breed of hawks or falcons which the hardy Wâkhânis manage to catch among the cliffs. These hawks are much esteemed by the chiefs of Badakhshan,

Bokhara, etc. They are celebrated for their swiftness, and known by their white colour."—H. C.]

Note 6.—These wild sheep are probably the kind called *Kachkár*, mentioned by Baber, and described by Mr. Blyth in his Monograph of Wild Sheep, under the name of *Ovis Vignei*. It is extensively diffused over all the ramifications of Hindu-Kúsh, and westward perhaps to the Persian Elburz. "It is gregarious," says Wood, "congregating in herds of *several hundreds*." In a later chapter Polo speaks of a wild sheep apparently different and greater. (See *J. A. S. B.*, X. 858 *seqq.*)

Note 7.—This pleasant passage is only in Ramusio, but it would be heresy to doubt its genuine character. Marco's recollection of the delight of convalescence in such a climate seems to lend an unusual enthusiasm and felicity to his description of the scenery. Such a region as he speaks of is probably the cool Plateau of Shewá, of which we are told as extending about 25 miles eastward from near Faizabad, and forming one of the finest pastures in Badakhshan. It contains a large lake called by the frequent name Sar-i-Kol. No European traveller in modern times (unless Mr. Gardner) has been on those glorious table-lands. Burnes says that at Kunduz both natives and foreigners spoke rapturously of the vales of Badakhshan, its rivulets, romantic scenes and glens, its fruits, flowers, and nightingales. Wood is reticent on scenery, naturally, since nearly all his journey was made in winter. When approaching Faizabad on his return from the Upper Oxus, however, he says: "On entering the beautiful lawn at the gorge of its valley I was enchanted at the quiet loveliness of the scene. Up to this time, from the day we left Talikan, we had been moving in snow; but now it had nearly vanished from the valley, and the fine sward was enamelled with crocuses, daffodils, and snowdrops." (*P. Manphul; Burnes*, III. 176; *Wood*, 252.)

Note 8.—Yet scarcely any country in the world has suffered so terribly and repeatedly from invasion. "Enduring decay probably commenced with the wars of Chinghiz, for many an instance in Eastern history shows the permanent effect of such devastations. . . . Century after century saw only progress in decay. Even to our own time the progress of depopulation and deterioration has continued." In 1759, two of the Khojas of Kashgar, escaping from the dominant Chinese, took refuge in Badakhshan; one died of his wounds, the other was treacherously slain by Sultan Shah, who then ruled the country. The holy man is said in his dying moments to have invoked curses on Badakhshan, and prayed that it might be three times depopulated; a malediction which found ample accomplishment. The misery of the country came to a climax about 1830, when the Uzbek chief of Kunduz, Murad Beg Kataghan, swept away the bulk of the inhabitants, and set them down to die in the marshy plains of Kunduz. (*Cathay*, p. 542; *Faiz Bakhsh*, etc.)

Note 9.—This "bombasticall dissimulation of their garments," as the author of *Anthropometamorphosis* calls such a fashion, is no longer affected by the ladies of Badakhshan. But a friend in the Panjab observes that it still survives *there*. "There are ladies' trousers here which might almost justify Marco's very liberal estimate of the quantity of stuff required to make them;" and among the Afghan ladies, Dr. Bellew says, the silken trousers almost surpass crinoline in amplitude. It is curious to find the same characteristic attaching to female figures on coins of ancient kings of these regions, such as Agathocles and Pantaleon. (The last name is appropriate !)

CHAPTER XXX.

OF THE PROVINCE OF PASHAI

You must know that ten days' journey to the south of Badashan there is a Province called PASHAI, the people of which have a peculiar language, and are Idolaters, of a brown complexion. They are great adepts in sorceries and the diabolic arts. The men wear earrings and brooches of gold and silver set with stones and pearls. They are a pestilent people and a crafty; and they live upon flesh and rice. Their country is very hot.[1]

Now let us proceed and speak of another country which is seven days' journey from this one towards the south-east, and the name of which is KESHIMUR.

NOTE I.—The name of PASHAI has already occurred (see ch. xviii.) linked with DIR, as indicating a tract, apparently of very rugged and difficult character, through which the partizan leader Nigúdar passed in making an incursion from Badakhshan towards Káshmir. The difficulty here lies in the name *Pashai*, which points to the south-west, whilst *Dir* and all other indications point to the south-east. But Pashai seems to me the reading to which all texts tend, whilst it is clearly expressed in the G. T. (*Pasciai*), and it is contrary to all my experience of the interpretation of Marco Polo to attempt to torture the name in the way which has been common with commentators professed and occasional. But dropping this name for a moment, let us see to what the other indications do point.

In the meagre statements of this and the next chapter, interposed as they are among chapters of detail unusually ample for Polo, there is nothing to lead us to suppose that the Traveller ever personally visited the countries of which these two chapters treat. I believe we have here merely an amplification of the information already sketched of the country penetrated by the Nigudarian bands whose escapade is related in chapter xviii., information which was probably derived from a Mongol source. And these countries are in my belief *both* regions famous in the legends of the Northern Buddhists, viz. UDYÁNA and KÁSHMIR.

Udyána lay to the north of Peshawar on the Swát River, but from the extent assigned to it by Hiuen Tsang, the name probably covered a large part of the whole hill-region south of the Hindu-Kúsh from Chitrál to the Indus, as indeed it is represented in the Map of Vivien de St. Martin (*Pèlerins Bouddhistes*, II.). It is regarded by Fahian as the most northerly Province of India, and in his time the food and clothing of the people were similar to those of Gangetic India. It was the native country of Padma Sambhava, one of the chief apostles of Lamaism, *i.e.* of Tibetan Buddhism, and a great master of enchantments. The doctrines of Sakya, as they prevailed in Udyána in old times, were probably strongly tinged with Sivaitic magic, and the Tibetans still regard that locality as the classic ground of sorcery and witchcraft.

Hiuen Tsang says of the inhabitants : " The men are of a soft and pusillanimous character, *naturally inclined to craft and trickery.* They are fond of study, but pursue it with no ardour. *The science of magical formulae is become a regular professional business with them.* They generally wear clothes of white cotton, and rarely use any other stuff. Their spoken language, in spite of some differences, has a strong resemblance to that of India."

These particulars suit well with the slight description in our text, and the Indian atmosphere that it suggests ; and the direction and distance ascribed to Pashai suit well with Chitral, which may be taken as representing Udyána when approached from Badakhshan. For it would be quite practicable for a party to reach the town of Chitrál in ten days from the position assigned to the old capital of Badakhshan. And from Chitrál the road towards Káshmir would lie over the high Lahori pass to DIR, which from its mention in chapter xviii. we must consider an obligatory point. (*Fah-hian,* p. 26 ; *Koeppen,* I. 70 ; *Pèlerins Boud.* II. 131-132.)

[" Tao-lin (a Buddhist monk like Hiuen Tsang) afterwards left the western regions and changed his road to go to Northern India ; he made a pilgrimage to *Kia-che-mi-louo* (Káshmir), and then entered the country of *U-ch'ang-na* (Udyána). . ." (Ed. Chavannes, *I-tsing,* p. 105.)—H.C.]

We must now turn to the name *Pashai.* The Pashai Tribe are now Mahomedan, but are reckoned among the aboriginal inhabitants of the country, which the Afghans are not. Baber mentions them several times, and counts their language as one of the dozen that were spoken at Kabul in his time. Burnes says it resembles that of the Kafirs. A small vocabulary of it was published by Leech, in the seventh volume of the *J. A. S. B.,* which I have compared with vocabularies of Siah-posh Kafir, published by Raverty in vol. xxxiii. of the same journal, and by Lumsden in his *Report of the Mission to Kandahar,* in 1837. Both are Aryan, and seemingly of Professor Max Müller's class *Indic,* but not *very* close to one another.*

Ibn Batuta, after crossing the Hindu-Kúsh by one of the passes at the head of the Panjshir Valley, reaches the Mountain BASHÁI (Pashai). In the same vicinity the Pashais are mentioned by Sidi 'Ali, in 1554. And it is still in the neighbourhood of Panjshir that the tribe is most numerous, though they have other settlements in the hill-country about Nijrao, and on the left bank of the Kabul River between Kabul and Jalalabad. *Pasha* and *Pasha*-gar is also named as one of the chief divisions of the Kafirs, and it seems a fair conjecture that it represents those of the Pashais who resisted or escaped conversion to Islam. (See *Leech's Reports* in Collection pub. at Calcutta in 1839 ; *Baber,* 140 ; *Elphinstone,* I. 411 ; *J. A. S. B.* VII. 329, 731, XXVIII. 317 *seqq.,* XXXIII. 271-272 ; *I. B.* III. 86 ; *J. As.* IX. 203, and *J. R. A. S.* N.S. V. 103, 278.)

The route of which Marco had heard must almost certainly have been one of those leading by the high Valley of Zebák, and by the Doráh or the Nuksán Pass, over the watershed of Hindu-Kúsh into Chitrál, and so to Dir, as already noticed. The difficulty remains as to how he came to apply the name *Pashai* to the country south-east of Badakhshan. I cannot tell. But it is at least possible that the name of the Pashai tribe (of which the branches even now are spread over a considerable extent of country) may have once had a wide application over the southern spurs of the Hindu-Kúsh.† Our Author, moreover, is speaking here from hearsay, and hearsay geography without maps is much given to generalising. I apprehend that, along with characteristics specially referable to the Tibetan and Mongol traditions of Udyána, the term Pashai, as Polo uses it, vaguely covers the whole tract from the southern boundary of Badakhshan to the Indus and the Kabul River.

---

* The Kafir dialect of which Mr. Trumpp collected some particulars shows in the present tense of the substantive verb these remarkable forms :—*Ei sŭm, Tŭ sis, siga sĕ; Ima simis, Wi sik, Sigĕ sin.*

† In the *Tabakát-i-Násiri (Elliot,* II. 317) we find mention of the Highlands of *Pasha-Afroz,* but nothing to define their position.

But even by extending its limits to Attok, we shall not get within seven marches of Káshmir. It is 234 miles by road from Attok to Srinagar ; more than twice seven marches. And, according to Polo's usual system, the marches should be counted from Chitrál, or some point thereabouts.

Sir H. Rawlinson, in his *Monograph on the Oxus*, has indicated the probability that the name *Pashai* may have been originally connected with *Aprasin* or *Paresín*, the Zendavestian name for the Indian Caucasus, and which occurs in the Babylonian version of the Behistun Inscription as the equivalent of *Gadára* in the Persian, *i.e.* *Gandhára*, there applied to the whole country between Bactria and the Indus. (See *J. R. G. S.* XLII. 502.) Some such traditional application of the term Pashai might have survived.

CHAPTER XXXI.

OF THE PROVINCE OF KESHIMUR.

KESHIMUR also is a Province inhabited by a people who are Idolaters and have a language of their own.[1] They have an astonishing acquaintance with the devilries of enchantment ; insomuch that they make their idols to speak. They can also by their sorceries bring on changes of weather and produce darkness, and do a number of things so extraordinary that no one without seeing them would believe them.[2] Indeed, this country is the very original source from which Idolatry has spread abroad.[3]

In this direction you can proceed further till you come to the Sea of India.

The men are brown and lean, but the women, taking them as brunettes, are very beautiful. The food of the people is flesh, and milk, and rice. The clime is finely tempered, being neither very hot nor very cold. There are numbers of towns and villages in the country, but also forests and desert tracts, and strong passes, so that the people have no fear of anybody, and keep their independence, with a king of their own to rule and do justice.[4]

There are in this country Eremites (after the fashion of those parts), who dwell in seclusion and practise great abstinence in eating and drinking. They observe strict chastity, and keep from all sins forbidden in their law, so that they are regarded by their own folk as very holy persons.   They live to a very great age.[5]

There are also a number of idolatrous abbeys and monasteries.   [The people of the province do not kill animals nor spill blood ; so if they want to eat meat they get the Saracens who dwell among them to play the butcher.[6]]   The coral which is carried from our parts of

Ancient Buddhist Temple at Pandrethan in Káshmir.

the world has a better sale there than in any other country.[7]

Now we will quit this country, and not go any further in the same direction ; for if we did so we should enter India ; and that I do not wish to do at present.   For, on our return journey, I mean to tell you about India : all in regular order.   Let us go back therefore to Badashan, for we cannot otherwise proceed on our journey.

NOTE 1.—I apprehend that in this chapter Marco represents Buddhism (which is to be understood by his expression *Idolatry*, not always, but usually) as in a position of greater life and prosperity than we can believe it to have enjoyed in Káshmir at the end of the 13th century, and I suppose that his knowledge of it was derived in great part from tales of the Mongol and Tibetan Buddhists about its past glories.

I know not if the spelling *Kesciemur* represents any peculiar Mongol pronunciation of the name. Plano Carpini, probably the first modern European to mention this celebrated region, calls it *Casmir* (p. 708).

"The Cashmeerians," says Abu'l Fazl, "have a language of their own, but their books are written in the Shanskrit tongue, although the character is sometimes Cashmeerian. They write chiefly upon *Tooz* [birch-bark], which is the bark of a tree; it easily divides into leaves, and remains perfect for many years." (*Ayeen Akbery*, II. 147.) A sketch of Kashmiri Grammar by Mr. Edgeworth will be found in vol. x. of the *J. A. S. B.*, and a fuller one by Major Leech in vol. xiii. Other contributions on the language are in vol. xxxv. pt. i. p. 233 (Godwin-Austen); in vol. xxxix. pt. i. p. 95 (Dr. Elmslie); and in *Proceedings* for 1866, p. 62, *seqq.* (Sir G. Campbell and Bábú Rájendra Lál Mitra). The language, though in large measure of Sanskrit origin, has words and forms that cannot be traced in any other Indian vernacular. (*Campbell*, pp. 67, 68). The character is a modification of the Panjáb Nagari.

NOTE 2.—The Kashmirian conjurers had made a great impression on Marco, who had seen them at the Court of the Great Kaan, and he recurs in a later chapter to their weather-sorceries and other enchantments, when we shall make some remarks. Meanwhile let us cite a passage from Bernier, already quoted by M. Pauthier. When crossing the Pír Panjál (the mountain crossed on entering Káshmir from Lahore) with the camp of Aurangzíb, he met with "an old Hermit who had dwelt upon the summit of the Pass since the days of Jehangir, and whose religion nobody knew, although it was said that he could work miracles, and used at his pleasure to produce extraordinary thunderstorms, as well as hail, snow, rain, and wind. There was something wild in his countenance, and in his long, spreading, and tangled hoary beard. He asked alms fiercely, allowing the travellers to drink from earthen cups that he had set out upon a great stone, but signing to them to go quickly by without stopping. He scolded those who made a noise, 'for,' said he to me (after I had entered his cave and smoothed him down with a half rupee which I put in his hand with all humility), 'noise here raises furious storms. Aurangzíb has done well in taking my advice and prohibiting it. Shah Jehan always did the like. But Jehangir once chose to laugh at what I said, and made his drums and trumpets sound; the consequence was he nearly lost his life.'" (*Bernier*, Amst. ed. 1699, II. 290.) A successor of this hermit was found on the same spot by P. Desideri in 1713, and another by Vigne in 1837.

NOTE 3.—Though the earliest entrance of Buddhism into Tibet was from India Proper, yet Káshmir twice in the history of Tibetan Buddhism played a most important part. It was in Káshmir that was gathered, under the patronage of the great King Kanishka, soon after our era, the Fourth Buddhistic Council, which marks the point of separation between Northern and Southern Buddhism. Numerous missionaries went forth from Káshmir to spread the doctrine in Tibet and in Central Asia. Many of the Pandits who laboured at the translation of the sacred books into Tibetan were Kashmiris, and it was even in Káshmir that several of the translations were made. But these were not the only circumstances that made Káshmir a holy land to the Northern Buddhists. In the end of the 9th century the religion was extirpated in Tibet by the Julian of the Lamas, the great persecutor Langdarma, and when it was restored, a century later, it was from Káshmir in particular that fresh missionaries were procured to reinstruct the people in the forgotten Law. (See *Koeppen*, II. 12-13, 78; *J. As.* sér. VI. tom. vi. 540.)

"The spread of Buddhism to Káshmir is an event of extraordinary importance in the history of that religion.   Thenceforward that country became a mistress in the Buddhist Doctrine and the headquarters of a particular school. . . . The influence of Káshmir was very marked, especially in the spread of Buddhism beyond India. From Káshmir it penetrated to Kandahar and Kabul, . . . and thence over Bactria. Tibetan Buddhism also had its essential origin from Káshmir ; . . . so great is the importance of this region in the History of Buddhism."   (*Vassilyev, Der Budd-hismus*, I. 44.)

In the account which the Mahawanso gives of the consecration of the great Tope at Ruanwelli, by Dutthagamini, King of Ceylon (B.C. 157), 280,000 priests (!) come from Káshmir, a far greater number than is assigned to any other country except one. (*J. A. S. B.* VII. 165.)

It is thus very intelligible how Marco learned from the Mongols and the Lamas with whom he came in contact to regard Káshmir as " the very original source from which their Religion had spread abroad."   The feeling with which they looked to Káshmir must have been nearly the same as that with which the Buddhists of Burma look to Ceylon.   But this feeling towards Káshmir does not *now*, I am informed, exist in Tibet.   The reverence for the holy places has reverted to Bahar and the neighbouring " cradle-lands " of Buddhism.

It is notable that the historian Firishta, in a passage quoted by Tod, uses Marco's expression in reference to Káshmir, almost precisely, saying that the Hindoos derived their idolatry from Káshmir, " the foundry of magical superstition." (*Rajasthan*, I. 219.)

NOTE 4.—The people of Káshmir retain their beauty, but they are morally one of the most degraded races in Asia.   Long oppression, now under the Lords of Jamu as great as ever, has no doubt aggravated this.   Yet it would seem that twelve hundred years ago the evil elements were there as well as the beauty.   The Chinese traveller says: " Their manners are light and volatile, their characters effeminate and pusillanimous. . . . They are very handsome, but their natural bent is to fraud and trickery."   (*Pèl. Boud.* II. 167-168.)   Vigne's account is nearly the same.   (II. 142-143.)   " They are as mischievous as monkeys, and far more malicious," says Mr. Shaw (p. 292).

[Bernier says : " The women [of Kachemire] especially are very handsome ; and it is from this country that nearly every individual, when first admitted to the court of the Great Mogul, selects wives or concubines, that his children may be whiter than the Indians, and pass for genuine Moguls.   Unquestionably, there must be beautiful women among the higher classes, if we may judge by those of the lower orders seen in the streets and in the shops."   (*Travels in the Mogul Empire*, edited by Archibald Constable, 1891, p. 404.)]

NOTE 5.—In the time of Hiuen Tsang, who spent two years studying in Káshmir in the first half of the 7th century, though there were many Brahmans in the country, Buddhism was in a flourishing state ; there were 100 convents with about 5000 monks.   In the end of the 11th century a King (Harshadeva, 1090-1102) is mentioned *exceptionally* as a protector of Buddhism.   The supposition has been intimated above that Marco's picture refers to a traditional state of things, but I must notice that a like picture is presented in the Chinese account of Hulaku's war.   One of the thirty kingdoms subdued by the Mongols was " The kingdom of Fo (Buddha) called *Kishimi*.   It lies to the N.W. of India.   There are to be seen the men who are counted the successors of Shakia ; their ancient and venerable air recalls the countenance of Bodi-dharma as one sees it in pictures.   They abstain from wine, and content themselves with a gill of rice for their daily food, and are occupied only in reciting the prayers and litanies of Fo."   (*Rém. N. Mél. Asiat.* I. 179.)   Abu'l Fazl says that on his third visit with Akbar to Káshmir he discovered some old men of the religion of Buddha, but none of them were *literati*.   The *Rishis*, of whom he speaks with high commendation as abstaining from meat and from female society, as charitable and unfettered by traditions, were perhaps a modified remnant of the Buddhist Eremites. Colonel Newall, in a paper on the Rishis of Káshmir, traces them to a number of Shiáh Sayads, who fled to Káshmir in the time of Timur. But evidently the *genus* was of much earlier date, long preceding the introduction of Islam. (*Vie et V. de H. T.* p. 390; *Lassen*, III. 709; *Ayeen Akb.* II. 147, III. 151; *J. A. S. B.* XXXIX. pt. i. 265.)

We see from the *Dabistan* that in the 17th century Káshmir continued to be a great resort of Magian mystics and sages of various sects, professing great abstinence and credited with preternatural powers. And indeed Vámbéry tells us that even in our own day the Kashmiri Dervishes are pre-eminent among their Mahomedan brethren for cunning, secret arts, skill in exorcisms, etc. (*Dab.* I. 113 *seqq.* II. 147-148; *Vámb. Sk. of Cent. Asia*, 9.)

NOTE 6.—The first precept of the Buddhist Decalogue, or Ten Obligations of the Religious Body, is not to take life. But *animal food* is not forbidden, though restricted. Indeed it is one of the circumstances in the Legendary History of Sakya Muni, which looks as if it *must* be true, that he is related to have aggravated his fatal illness by eating a dish of pork set before him by a hospitable goldsmith. Giorgi says the butchers in Tibet are looked on as infamous; and people selling sheep or the like will make a show of exacting an assurance that these are not to be slaughtered. In Burma, when a British party wanted beef, the owner of the bullocks would decline to make one over, but would point one out that might be shot by the foreigners.

In Tibetan history it is told of the persecutor Langdarma that he compelled members of the highest orders of the clergy to become hunters and butchers. A Chinese collection of epigrams, dating from the 9th century, gives a facetious list of *Incongruous Conditions*, among which we find a poor Parsi, a sick Physician, a fat Bride, a Teacher who does not know his letters, and *a Butcher who reads the Scriptures* (of Buddhism)! (*Alph. Tib.* 445; *Koeppen*, I. 74; *N. and Q., C. and J.* III. 33.)

NOTE 7.—Coral is still a very popular adornment in the Himalayan countries. The merchant Tavernier says the people to the north of the Great Mogul's territories and in the mountains of Assam and Tibet were the greatest purchasers of coral. (*Tr. in India*, Bk. II. ch. xxiii.)

CHAPTER XXXII.

OF THE GREAT RIVER OF BADASHAN.

IN leaving Badashan you ride twelve days between east and north-east, ascending a river that runs through land belonging to a brother of the Prince of Badashan, and containing a good many towns and villages and scattered habitations. The people are Mahommetans, and valiant in war. At the end of those twelve days you come to a province of no great size, extending indeed no more than three days' journey in any direction, and this is called VOKHAN.   The people worship Mahommet, and they have a peculiar language.   They are gallant soldiers, and they have a chief whom they call NONE, which is as much as to say *Count*, and they are liegemen to the Prince of Badashan.[1]

There are numbers of wild beasts of all sorts in this region.   And when you leave this little country, and ride three days north-east, always among mountains, you get to such a height that 'tis said to be the highest place in the world!   And when you have got to this height you find [a great lake between two mountains, and out of it] a fine river running through a plain clothed with the finest pasture in the world ; insomuch that a lean beast there will fatten to your heart's content in ten days. There are great numbers of all kinds of wild beasts ; among others, wild sheep of great size, whose horns are good six palms in length.   From these horns the shepherds make great bowls to eat from, and they use the horns also to enclose folds for their cattle at night. [Messer Marco was told also that the wolves were numerous, and killed many of those wild sheep.   Hence quantities of their horns and bones were found, and these were made into great heaps by the way-side, in order to guide travellers when snow was on the ground.]

The plain is called PAMIER, and you ride across it for twelve days together, finding nothing but a desert without habitations or any green thing, so that travellers are obliged to carry with them whatever they have need of.   The region is so lofty and cold that you do not even see any birds flying.   And I must notice also that because of this great cold, fire does not burn so brightly, nor give out so much heat as usual, nor does it cook food so effectually.[2]

Now, if we go on with our journey towards the east-north-east, we travel a good forty days, continually passing over mountains and hills, or through valleys, and crossing many rivers and tracts of wilderness. And in all this way you find neither habitation of man, nor any green thing, but must carry with you whatever you require. The country is called BOLOR. The people dwell high up in the mountains, and are savage Idolaters, living only by the chase, and clothing themselves in the skins of beasts. They are in truth an evil race.[3]

NOTE 1.—[" The length of Little Pamir, according to Trotter, is 68 miles. . . . . To find the twelve days' ride in the plain of Marco Polo, it must be admitted, says Severtsof (*Bul. Soc. Géog.* XI. 1890, pp. 588-589), that he went down a considerable distance along the south-north course of the Aksu, in the Aktash Valley, and did not turn towards Tásh Kurgán, by the Neza Tash Pass, crossed by Gordon and Trotter. The descent from this pass to Tásh Kurgán finishes with a difficult and narrow defile, which may well be overflowed at the great melting of snow, from the end of May till the middle of June, even to July.

" Therefore he must have left the Aksu Valley to cross the Pass of Tagharma, about 50 or 60 kilometres to the north of the Neza Tash Pass ; thence to Kashgar, the distance, in a straight line, is about 200 kilometres, and less than 300 by the shortest route which runs from the Tagharma Pass to little Kara Kul, and from there down to Yangi Hissar, along the Ghidjik. And Marco Polo assigns *forty* days for this route, while he allows but *thirty* for the journey of 500 kilometres (at least) from Jerm to the foot of the Tagharma Pass."

Professor Paquier (*Bul. Soc. Géog.* 6e Sér. XII. pp. 121-125) remarks that the Moonshee, sent by Captain Trotter to survey the Oxus between Ishkashm and Kila Wamár, could not find at the spot marked by Yule on his map, the mouth of the Shakh-Dara, but northward 7 or 8 miles from the junction of the Murghab with the Oxus, he saw the opening of an important water-course, the Suchnan River, formed by the Shakh-Dara and the Ghund-Dara. Marco arrived at a place between Northern Wakhán and Shihgnan ; from the Central Pamir, Polo would have taken a route identical with that of the Mirza (1868-1869) by the Chichiklik Pass. Professor Paquier adds : " I have no hesitation in believing that Marco Polo was in the neighbourhood of that great commercial road, which by the *Vallis Comedarum* reached the foot of the Imaüs. He probably did not venture on a journey of fifty marches in an unknown country. At the top of the Shihgnan Valley, he doubtless found a road marked out to Little Bukharia. This was the road followed in ancient times from Bactrian to Serica ; and Ptolemy has, so to speak, given us its landmarks after Marinus of Tyre, by the *Vallis Comedarum* (Valley of actual Shihgnan) ; the *Turris Lapidea* and the *Statio Mercatorum*, neighbourhood of Tash Kurgan, capital of the present province of Sar-i-kol."

I must say that accepting, as I do, for Polo's Itinerary, the route from Wakhán to Kashgar by the Taghdum-Bash Pamir, and Tásh Kurgán, I do not agree with Professor Paquier's theory. But though I prefer Sir H. Yule's route from Badakhshan, by the River Vardoj, the Pass of Ishkashm, the Panja, to Wakhán, I do not accept his views for the Itinerary from Wakhán to Kashgar ; see p. 175.—H. C.]

The river along which Marco travels from Badakhshan is no doubt the upper stream of the Oxus, known locally as the Panja, along which Wood also travelled, followed

of late by the Mirza and Faiz Bakhsh.   It is true that the river is reached from
Badaskhshan Proper by ascending another river (the Vardoj) and crossing the Pass of
Ishkáshm, but in the brief style of our narrative we must expect such condensation.

WAKHÁN was restored to geography by Macartney, in the able map which he
compiled for Elphinstone's *Caubul*, and was made known more accurately by Wood's
journey through it.   [The district of Wakhán "comprises the valleys containing the
two heads of the Panjah branch of the Oxus, and the valley of the Panjah itself, from
the junction at Zung down to Ishkashím.   The northern branch of the Panjah has its
principal source in the Lake Victoria in the Great Pamir, which as well as the Little
Pámir, belongs to Wakhán, the Aktash River forming the well recognized boundary
between Kashgaria and Wakhán."   (Captain Trotter, *Forsyth's Mission*, p. 275.)   The
southern branch is the Sarhadd Valley.—H. C.]   The lowest part is about 8000 feet
above the sea, and the highest *Kishlak*, or village, about 11,500.   A few willows and
poplars are the only trees that can stand against the bitter blasts that blow down the
valley.   Wood estimated the total population of the province at only 1000 souls, though
it might be capable of supporting 5000.*   He saw it, however, in the depth of winter.
As to the peculiar language, see note 1, ch. xxix.   It is said to be a very old dialect
of Persian.   A scanty vocabulary was collected by Hayward.   (*J. R. G. S.* XXI. p. 29.)
The people, according to Shaw, have Aryan features, resembling those of the
Kashmiris, but harsher.

[Cf. Captain Trotter's *The Oxus below Wakhan, Forsyth's Mission*, p. 276.]

We appear to see in the indications of this paragraph precisely the same system of
government that now prevails in the Oxus valleys.   The central districts of Faizabad
and Jerm are under the immediate administration of the Mír of Badakhshan, whilst
fifteen other districts, such as *Kishm, Rusták, Zebák, Ishkáshm, Wakhán*, are
dependencies " held by the *relations of the Mír*, or by hereditary rulers, on a feudal
tenure, conditional on fidelity and military service in time of need, the holders possessing supreme authority in their respective territories, and paying little or no
tribute to the paramount power."   (*Pandit Manphul.*)   The first part of the valley
of which Marco speaks as belonging to a brother of the Prince, may correspond to
Ishkáshm, or perhaps to Vardoj ; the second, Wakhán, seems to have had a hereditary ruler ; but both were vassals of the Prince of Badakhshan, and therefore are
styled *Counts*, not kings or *Seigneurs*.

The native title which Marco gives as the equivalent of Count is remarkable.   *Non*
or *None*, as it is variously written in the texts, would in French form represent *Nono* in
Italian.   Pauthier refers this title to the " *Rao*-nana (or nano) *Rao*" which figures as
the style of Kanerkes in the Indo-Scythic coinage.   But Wilson (*Ariana Antiqua*,
p. 358) interprets *Raonano* as most probably a genitive plural of Rao, whilst the whole
inscription answers precisely to the Greek one ΒΑΣΙΛΕΤΣ ΒΑΣΙΛΕΩΝ ΚΑΝΗΡΚΟΥ,
which is found on other coins of the same prince.   General Cunningham, a very
competent authority, adheres to this view, and writes : " I do not think *None* or *Non*
can have any connection with the *Nana* of the coins."

It is remarkable, however, that NONO (said to signify "younger," or lesser) is in
Tibet the title given to a younger brother, deputy, or subordinate prince.   In
Cunningham's *Ladak* (259) we read : " *Nono* is the usual term of respect which is
used in addressing any young man of the higher ranks, and when prefixed to *Kahlon*
it means the younger or deputy minister."   And again (p. 352) : " *Nono* is the title
given to a younger brother.   Nono Sungnam was the younger brother of Chang

---

* " Yet this barren and inaccessible upland, with its scanty handful of wild people, finds a place
in Eastern history and geography from an early period, and has now become the subject of serious
correspondence between two great European Governments, and its name, for a few weeks at least, a
household word in London.   Indeed, this is a striking accident of the course of modern history.   We
see the Slav and the Englishman—representatives of two great branches of the Aryan race, but
divided by such vast intervals of space and time from the original common starting-point of their
migration—thus brought back to the lap of Pamir to which so many quivering lines point as the centre
of their earliest seats, there by common consent to lay down limits to mutual encroachment."
(*Quarterly Review*, April, 1873, p. 548.)

Raphtan, the Kahlon of Bazgo." I have recently encountered the word used independently, and precisely in Marco's application of it. An old friend, in speaking of a
journey that he had made in our Tibetan provinces, said incidentally that he had
accompanied the commissioner *to the installation of a new* NONO (I think in Spiti).
The term here corresponds so precisely with the explanation which Marco gives of
None as a Count subject to a superior sovereign, that it is difficult to regard the coincidence as accidental. The *Yuechi* or Indo-Scyths who long ruled the Oxus countries
are said to have been of Tibetan origin, and Al-Biruni repeats a report that this was
so. (*Elliot.* II. 9.)* Can this title have been a trace of their rule? Or is it Indian?

NOTE 2.—This chapter is one of the most interesting in the book, and contains one
of its most splendid anticipations of modern exploration, whilst conversely Lieutenant
John Wood's narrative presents the most brilliant confirmation in detail of Marco's
narrative.

We have very old testimony to the recognition of the great altitude of the Plateau
of PAMIR (the name which Marco gives it and which it still retains), and to the
existence of the lake (or lakes) upon its surface. The Chinese pilgrims Hwui Seng
and Sung Yun, who passed this way A.D. 518, inform us that these high lands of the
Tsung Ling were commonly said to be midway between heaven and earth. The more
celebrated Hiuen Tsang, who came this way nearly 120 years later (about 644) on his
return to China, "after crossing the mountains for 700 *li*, arrived at the valley of
Pomilo (Pamir). This valley is 1000 *li* (about 200 miles) from east to west, and 100 *li*
(20 miles) from north to south, and lies between two snowy ranges in the centre of the
Tsung Ling mountains. The traveller is annoyed by sudden gusts of wind, and the
snow-drifts never cease, spring or summer. As the soil is almost constantly frozen, you
see but a few miserable plants, and no crops can live. The whole tract is but a dreary
waste, without a trace of human kind. In the middle of the valley is a great lake
300 *li* (60 miles) from east to west, and 500 *li* from north to south. This stands in the
centre of Jambudwipa (the Buddhist οἰκουμένη) on a plateau of prodigious elevation.
An endless variety of creatures peoples its waters. When you hear the murmur and
clash of its waves you think you are listening to the noisy hum of a great market in
which vast crowds of people are mingling in excitement. . . . . The lake discharges
to the west, and a river runs out of it in that direction and joins the *Potsu* (Oxus)
. . . . . The lake likewise discharges to the east, and a great river runs out, which
flows eastward to the western frontier of *Kiesha* (Káshgar), where it joins the River Sita,
and runs eastward with it into the sea." The story of an eastern outflow from the lake
is, no doubt, legend, connected with an ancient Hindu belief (see *Cathay*, p. 347), but
Burnes in modern times heard much the same story. And the Mirza, in 1868, took up
the same impression regarding the smaller lake called Pamir Kul, in which the
southern branch of the Panja originates.

" After quitting the (frozen) surface of the river," says Wood, "we . . . . ascended
a low hill, which apparently bounded the valley to the eastward. On surmounting this,
at 3 P.M. of the 19th February, 1838, we stood, to use a native expression, upon the
Bám-i-Duniah, or ' Roof of the World,' while before us lay stretched a noble but
frozen sheet of water, from whose western end issued the infant river of the Oxus.
This fine lake (Sirikol) lies in the form of a crescent, about 14 miles long from east to
west, by an average breadth of 1 mile. On three sides it is bordered by swelling
hills about 500 feet high, while along its southern bank they rise into mountains 3500
feet above the lake, or 19,000 feet above the sea, and covered with perpetual snow,
from which never-failing source the lake is supplied. . . . . Its elevation, measured
by the temperature of boiling water, is 15,600 feet."

The absence of birds on Pamir, reported by Marco, probably shows that he passed
very late or early in the season. Hiuen Tsang, we see, gives a different account ;

---

* Ibn Haukal reckons Wakhán as an Indian country. It is a curious coincidence (it can scarcely
be more) that *Nono* in the Garo tongue of Eastern Bengal signifies "a younger brother." (*J. A. S. B.*
XXII. 153, XVIII. 208.)

Wood was there in the winter, but heard that in summer the lake swarmed with water-
fowl. [Cf. Captain Trotter, p. 263, in *Forsyth's Mission.*]

The Pamir Steppe was crossed by Benedict Goës late in the autumn of 1603, and
the narrative speaks of the great cold and desolation, and the difficulty of breathing.
We have also an abstract of the journey of Abdul Mejid, a British Agent, who passed
Pamir on his way to Kokan in 1861 :—" Fourteen weary days were occupied in crossing the steppe ; the marches were long, depending on uncertain supplies of grass and
water, which sometimes wholly failed them ; food for man and beast had to be carried
with the party, for not a trace of human habitation is to be met with in those inhospitable wilds. . . . . The steppe is interspersed with tamarisk jungle and the
wild willow, and in the summer with tracts of high grass." (*Neumann, Pilgerfahrten
Buddh. Priester*, p. 50; *V. et V. de H. T.* 271-272; *Wood*, 232; *Proc. R. G. S.* X.
150.)

There is nothing absolutely to decide whether Marco's route from Wakhán lay by
Wood's Lake "Sirikol," or Victoria, or by the more southerly source of the Oxus
in Pamir Kul.   These routes would unite in the valley of Táshkurgán, and his
road thence to Kashgar was, I apprehend, nearly the same as the Mirza's in 1868-1869,
by the lofty Chichiklik Pass and Kin Valley.   But I cannot account for the forty days
of wilderness.   The Mirza was but thirty-four days *from Faizabad to Kashgar*, and
Faiz Bakhsh only twenty-five.

[Severtsof (*Bul. Soc. Géog.* XI. 1890, p. 587), who accepts Trotter's route, by
the Pamir Khurd (Little Pamir), says there are three routes from Wakhán to Little
Pamir, going up the Sarhadd : one during the winter, by the frozen river ; the two
others available during the spring and the summer, up and down the snowy chain
along the right bank of the Sarhadd, until the valley widens out into a plain, where
a swelling is hardly to be seen, so flat is it ; this chain is the dividing ridge between
the Sarhadd and the Aksu.   From the summit, the traveller, looking towards the west,
sees *at his feet* the mountains he has crossed; to the east, the Pamir Kul and the
Aksu, the river flowing from it.   The pasture grounds around the Pamir Kul and the
sources of the Sarhadd are magnificent ; but lower down, the Aksu valley is arid,
dotted only with pasture grounds of little extent, and few and far between.   It is to
this part of Pamir that Marco Polo's description applies; more than any other part
of this *ensemble* of high valleys, this line of water parting, of the Sarhadd and the
Aksu, has the aspect of a *Roof of the World* (*Bam-i-dunya*, Persian name of Pamir).
—H. C.].

[We can trace Marco Polo's route from Wakhán, on comparing it with Captain
Younghusband's Itinerary from Kashgar, which he left on the 22nd July, 1891, for
Little Pamir : Little Pamir at Bozai-Gumbaz, joins with the Pamir-i-Wakhán at the
Wakhijrui Pass, first explored by Colonel Lockhart's mission.   Hence the route lies
by the old fort of Kurgan-i-Ujadbai at the junction of the two branches of the Tagh-
dum-bash Pamir (Supreme Head of the Mountains), the Tagh-dum-bash Pamir, Tásh
Kurgán, Bulun Kul, the Gez Defile and Kashgar. (*Proc. R. G. S.* XIV. 1892,
pp. 205-234.)—H. C.]

We may observe that Severtsof asserts *Pamir* to be a generic term, applied to all
high plateaux in the Thian Shan.*

["The Pámír plateau may be described as a great, broad, rounded ridge, extending north and south, and crossed by thick mountain chains, between which lie
elevated valleys, open and gently sloping towards the east, but narrow and confined,
with a rapid fall towards the west.   The waters which run in all, with the exception
of the eastern flow from the Tághdúngbásh, collect in the Oxus ; the Áksú from the
Little Pámír lake receiving the eastern drainage, which finds an outlet in the Áktásh
Valley, and joining the Múrgháb, which obtains that from the Alichór and Sír íz
Pámirs.   As the eastern Tághdúngbásh stream finds its way into the Yarkand river,

---

* According to Colonel Tod, the Hindu bard Chand speaks of "Pamer, chief of mountains."
(I. p. 24.) But one may like and respect Colonel Tod without feeling able to rely on such quotations
of his unconfirmed.

the watershed must be held as extending from that Pámír, down the range dividing it from the Little Pámír, and along the Neza Tásh mountains to the Kizil Art Pass, leading to the Alái." (Colonel Gordon, *Forsyth's Mission*, p. 231.)

Lieutenant-Colonel Gordon (*Forsyth's Mission*, p. 231) says also : " Regarding the name ' Pámír,' the meaning appears to be wilderness—a place depopulated, abandoned, waste, yet capable of habitation. I obtained this information on the Great Pámír from one of our intelligent guides, who said in explanation—' In former days, when this part was inhabited by Kirghiz, as is shown by the ruins of their villages and burial-grounds, the valley was not all called Pámír, as it is now. It was known by its village names, as is the country beyond Sirikol, which being now occupied by Kirghiz is not known by one name, but partly as Chárling, Bas Robát, etc. If deserted it would be Pámír." In a note Sir T. D. Forsyth adds that the same explanation of the word was given to him at Yangi-Hissar, and that it is in fact a Khokandi-Turki word.—H. C.]

It would seem, from such notices as have been received, that there is not, strictly speaking, one steppe called Pamir, but a variety of *Pamirs*, which are lofty valleys between ranges of hills, presenting luxuriant summer pasture, and with floors more or less flat, but nowhere more than 5 or 6 miles in width and often much less.

[This is quite exact ; Mr. E. Delmar Morgan writes in the *Scottish Geog. Mag.* January, 1892, p. 17 : " Following the terminology of Yule adopted by geographers, and now well established, we have (1) Pamir Alichur ; (2) Pamir Khurd (or " Little ") ;

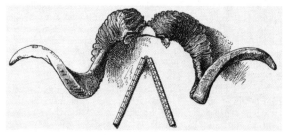

Horns of *Ovis Poli*.

(3) Pamir Kalan (or "Great") ; (4) Pamir Khargosi ("of the hare") ; (5) Pamir Sares ; (6) Pamir Rang-kul."—H. C.]

Wood speaks of the numerous wolves in this region. And the great sheep is that to which Blyth, in honour of our traveller, has given the name of *Ovis Poli*.* A pair of horns, sent by Wood to the Royal Asiatic Society, and of which a representation is given above, affords the following dimensions :—Length of one horn on the curve, 4 feet 8 inches ; round the base 14¼ inches ; distance of tips apart 3 feet 9 inches. This sheep appears to be the same as the *Rass*, of which Burnes heard that the horns were so big that a man could not lift a pair, and that foxes bred in them ; also that the carcass formed a load for two horses. Wood says that these horns supply shoes for the Kirghiz horses, and also a good substitute for stirrup-irons. "We saw numbers of horns strewed about in every direction, the spoils of the Kirghiz hunter. Some of these were of an astonishingly large size, and belonged to an animal of a species between a goat and a sheep, inhabiting the steppes of Pamir. *The ends of the horns projecting above the snow often indicated the direction of the road;* and wherever they were heaped in large quantities and disposed in a semi-circle, there our escort recognised the site of a Kirghiz summer encampment. . . . . We came in sight of a rough-looking building, decked out with the horns of the wild sheep, and all but buried amongst the snow. It was a Kirghiz burying-ground." (Pp. 223, 229, 231.)

---

* Usually written *Polii*, which is nonsense.

[With reference to Wood's remark that the horns of the *Ovis Poli* supply shoes for the Kirghiz horses, Mr. Rockhill writes to me that a Paris newspaper of 24th November, 1894, observes : " Horn shoes made of the horn of sheep are successfully used in Lyons. They are especially adapted to horses employed in towns, where the pavements are often slippery. Horses thus shod can be driven, it is said, at the most rapid pace over the worst pavement without slipping."

(Cf. Rockhill, *Rubruck*, p. 69 ; *Chasses et Explorations dans la Région des Pamirs*, par le Vte. Ed. de Poncins, Paris, 1897, 8vo.—H. C.).]

In 1867 this great sheep was shot by M. Severtsof, on the Plateau of Aksai, in the western Thian Shan. He reports these animals to go in great herds, and to be very difficult to kill. However, he brought back two specimens. The Narin River is

Ovis Poli, the Great Sheep of Pamir. (After Severtsof.)

" **El hi a grant montitude de mouton saubages qe sunt grandisme, car ont lee cornes bien six paumes** " . . . .

stated to be the northern limit of the species.* Severtsof also states that the enemies of the *Ovis Poli* are the wolves, [and Colonel Gordon says that the leopards and wolves prey almost entirely upon them. (On the *Ovis Poli*, see Captain Deasy, *In Tibet*, p. 361.)—H. C.]

Colonel Gordon, the head of the exploring party detached by Sir Douglas Forsyth, brought away a head of *Ovis Poli*, which quite bears out the account by its eponymus of horns " good 6 palms in length," say 60 inches. This head, as I learn from a letter of Colonel Gordon's to a friend, has one horn perfect which measures 65½ inches on the curves ; the other, broken at the tip, measures 64 inches ; the straight line between the tips is 55 inches.

[Captain Younghusband [1886] " before leaving the Altai Mountains, picked up several heads of the *Ovis Poli*, called Argali by the Mongols. They were somewhat

---

* [" The Tian Shan wild sheep has since been described as the *Ovis Karelini*, a species somewhat smaller than the true *Ovis Poli* which frequents the Pamirs." (Colonel Gordon, *Roof of the World*, p. 83, note.)—H. C.]

different from those which I afterwards saw at Yarkand, which had been brought in from the Pamir. Those I found in the Gobi were considerably thicker at the base, there was a less degree of curve, and a shorter length of horn. A full description of the *Ovis Poli*, with a large plate drawing of the horns, may be seen in Colonel Gordon's *Roof of the World.* (See p. 81.) (*Proc. R. G. S.* X. 1888, p. 495.) Some years later, Captain Younghusband speaks repeatedly of the great sport of shooting *Ovis Poli.* (*Proc. R. G. S.* XIV. 1892, pp. 205, 234.)—H. C.]

As to the pasture, Timkowski heard that "the pasturage of Pamir is so luxuriant and nutritious, that if horses are left on it for more than forty days they die of repletion." (I. 421.) And Wood: "The grass of Pamir, they tell you, is so rich that a sorry horse is here brought into good condition in less than twenty days; and its nourishing qualities are evidenced in the productiveness of their ewes, which almost invariably bring forth two lambs at a birth." (P. 365.)

With regard to the effect upon fire ascribed to the "great cold," Ramusio's version inserts the expression "*gli fu affermato per miracolo,*" "it was asserted to him as a wonderful circumstance." And Humboldt thinks it so strange that Marco should not have observed this personally that he doubts whether Polo himself passed the Pamir. "How is it that he does not say that he himself had seen how the flames disperse and leap about, as I myself have so often experienced at similar altitudes in the Cordilleras of the Andes, especially when investigating the boiling-point of water?" (*Cent. Asia,* Germ. Transl. I. 588.) But the words quoted from Ramusio do not exist in the old texts, and they are probably an editorial interpolation indicating disbelief in the statement.

MM. Huc and Gabet made a like observation on the high passes of north-eastern Tibet: "The *argols* gave out much smoke, but would not burn with any flame"; only they adopted the native idea that this as well as their own sufferings in respiration was caused by some pernicious exhalation.

Major Montgomerie, R.E., of the Indian Survey, who has probably passed more time nearer the heavens than any man living, sends me the following note on this passage: "What Marco Polo says as to fire at great altitudes not cooking so effectually as usual is perfectly correct as far as anything *boiled* is concerned, but I doubt if it is as to anything *roasted.* The want of brightness in a fire at great altitudes is, I think, altogether attributable to the poorness of the fuel, which consists of either small sticks or bits of roots, or of *argols* of dung, all of which give out a good deal of smoke, more especially the latter if not quite dry; but I have often seen a capital blaze made with the argols when perfectly dry. As to cooking, we found that rice, *dál,* and potatoes would never soften properly, no matter how long they were boiled. This, of course, was due to the boiling-point being only from 170° to 180°. Our tea, moreover, suffered from the same cause, and was never good when we were over 15,000 feet. This was very marked. Some of my natives made dreadful complaints about the rice and dál that they got from the village-heads in the valleys, and vowed that they only gave them what was very old and hard, as they could not soften it!"

NOTE 3.—Bolor is a subject which it would take several pages to discuss with fulness, and I must refer for such fuller discussion to a paper in the *J. R. G. S.* vol. xlii. p. 473.

The name *Bolor* is very old, occurring in Hiuen Tsang's Travels (7th century), and in still older Chinese works of like character. General Cunningham has told us that Balti is still termed *Balor* by the Dards of Gilghit; and Mr. Shaw, that *Palor* is an old name still sometimes used by the Kirghiz for the upper part of Chitrál. The indications of Hiuen Tsang are in accordance with General Cunningham's information; and the fact that Chitrál is described under the name of Bolor in Chinese works of the last century entirely justifies that of Mr. Shaw. A Pushtu poem of the 17th century, translated by Major Raverty, assigns the mountains of *Bilaur*-istán, as the northern boundary of Swát. The collation of these indications shows that the

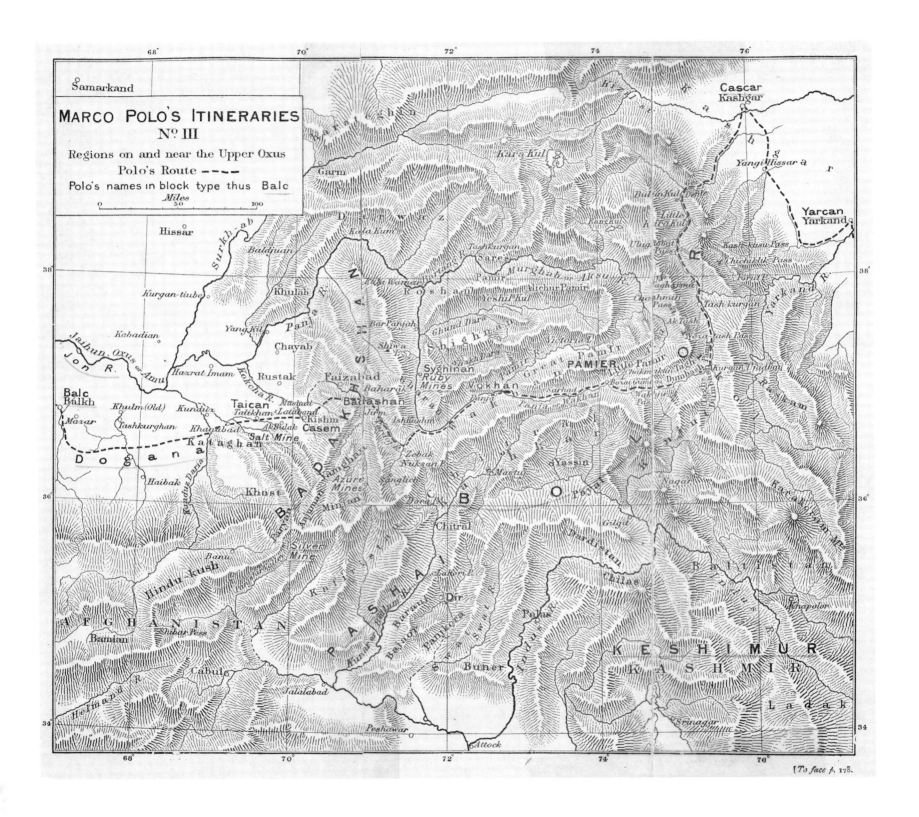

Samarkand

MARCO POLO'S ITINERARIES
Nº III

Regions on and near the Upper Oxus

Polo's Route - - - -

Polo's names in block type thus Balc

Miles
0   50   100

Cascar
Kashgar

Yangi Hissar a

Yarcan
Yarkand

Hissar

Garm

Karateghin

Kara-Kul

Darwaz

Kala Kum

Roshan

Murghab or Aksu

Alichur Pamir

Kurgan-tiube

Khulab

Surkh-ab

Baldjuan

Kabadian

Yangi-Kil

Panja R.

Bar Panjah

Shiwa T.

Tashkurgan

Sares

Pamir R.

Yeshil Kul

Ghund Dara

Chichna

Shakhara

Great Pamir

PAMIER or Pamir

Victoria L.

Tash-kurgan

Bulunkul Dehte

Little Kara-Kul

Ulug Robat Pass

Kash-kusu Pass

Chichaklik Pass

Yarkand R.

Chashnu Pass

Ak Tash

Neza-tash Pass

Jaihun, Oxus or Amu

Jon R.

Hazrat Imam

Kokcha R.

Rustak

Chayab

Faizabad

Baharak

Syghinan
Ruby Mines

Vokhan

Wakhan

Kurgan Ujudbai

Dunbash

Wakhjir Pass

Balc
Balkh

Mazar

Khulm (Old)

Kunduz

Taican
Talikhan

Mustad
Lataband

Badashan

Jirm

Ishkashm

Panja

Panja

Wakhan

Tashkurghan

Khanabad

Ak Bilak

Kishm

Casem

Khataghan

Salt Mine

Dogana

Haibak

Khost

Azure Mines

Minjan

Zebak
Nuksan P.

Singlan

Dorah P.

Mustuj

Yassin

Nagar

Karakoram Mts

Baltistan

Hindu-kush

Silver Mine

Banu

Kafir

Chitral

Gilgit

Dardistan

Chilas

Khapolor

AFGHANISTAN

Shibar Pass

Bamian

Cabul

Jalalabad

Helmand R.

Bajaog

Panjkora

Dir

Bunwal R.

Swat R.

Buner

Lakori R.

Palas

Indus R.

KESHIMUR
KASHMIR

Ladak

Peshawar

Attock

Srinagar

[To face p. 178.

term Bolor must have been applied somewhat extensively to the high regions adjoining the southern margin of Pamir. And a passage in the *Táríkh Rashídí*, written at Kashgar in the 16th century by a cousin of the great Baber, affords us a definition of the tract to which, in its larger sense, the name was thus applied : "*Malaur (i.e.* Balaur or Bolor) . . . is a country with few level spots. It has a circuit of four months' march. The eastern frontier borders on Kashgar and Yarkand ; it has Badakhshan to the north, Kabul to the west, and Kashmír to the south." The writer was thoroughly acquainted with his subject, and the region which he so defines must have embraced Sirikol and all the wild country south of Yarkand, Balti, Gilghit, Yasin, Chitrál, and perhaps Kafiristán. This enables us to understand Polo's use of the term.

The name of Bolor in later days has been in a manner a symbol of controversy. It is prominent in the apocryphal travels of George Ludwig von ————, preserved in the Military Archives at St. Petersburg. That work represents a town of Bolor as existing to the north of Badakhshan, with Wakhán still further to the north. This geography we now know to be entirely erroneous, but it is in full accordance with the maps and tables of the Jesuit missionaries and their pupils, who accompanied the Chinese troops to Kashgar in 1758-1759. The paper in the *Geographical Society's Journal*, which has been referred to, demonstrates how these erroneous data must have originated. It shows that the Jesuit geography was founded on downright accidental error, and, as a consequence, that the narratives which profess *de visu* to corroborate that geography must be downright forgeries. When the first edition was printed, I retained the belief in a *Bolor* where the Jesuits placed it.

[The Chinese traveller, translated by M. Gueluy (*Desc. de la Chine occid.* p. 53), speaks of Bolor, to the west of Yarkand, inhabited by Mahomedans who live in huts ; the country is sandy and rather poor. Severtsof says, (*Bul. Soc. Géog.* XI. 1890, p. 591) that he believes that the name of *Bolor* should be expunged from geographical nomenclature as a source of confusion and error. Humboldt, with his great authority, has too definitely attached this name to an erroneous orographical system. Lieutenant-Colonel Gordon says that he "made repeated enquiries from Kirghiz and Wakhis, and from the Mír [of Wakhán], Fatteh Ali Shah, regarding ' Bólór,' as a name for any mountain, country, or place, but all professed perfect ignorance of it." (*Forsyth's Mission.*)—H. C.]

The *J. A. S. Bengal* for 1853 (vol. xxii.) contains extracts from the diary of a Mr. Gardiner in those central regions of Asia. These read more like the memoranda of a dyspeptic dream than anything else, and the only passage I can find illustrative of our traveller is the following ; the region is described as lying twenty days south-west of Kashgar : " The Keiaz tribe live in caves on the highest peaks, subsist by hunting, keep no flocks, said to be anthropophagous, but have handsome women ; eat their flesh raw." (P. 295 ; *Pèlerins Boud.* III. 316, 421, etc. ; *Ladak*, 34, 45, 47 ; *Mag. Asiatique*, I. 92, 96-97 ; *Not. et Ext.* II. 475, XIV. 492 ; *J. A. S. B.* XXXI. 279 ; Mr. R. Shaw in *Geog. Proceedings*, XVI. 246, 400 ; *Notes regarding Bolor*, etc., *J. R. G. S.* XLII. 473.)

As this sheet goes finally to press we hear of the exploration of Pamir by officers of Mr. Forsyth's Mission. [I have made use of the information collected by them.— H. C.]

CHAPTER XXXIII.

OF THE KINGDOM OF CASCAR.

CASCAR is a region lying between north-east and east, and constituted a kingdom in former days, but now it

Head of a Native of Kashgar.

is subject to the Great Kaan. The people worship Mahommet. There are a good number of towns and

villages, but the greatest and finest is Cascar itself. The
inhabitants live by trade and handicrafts; they have

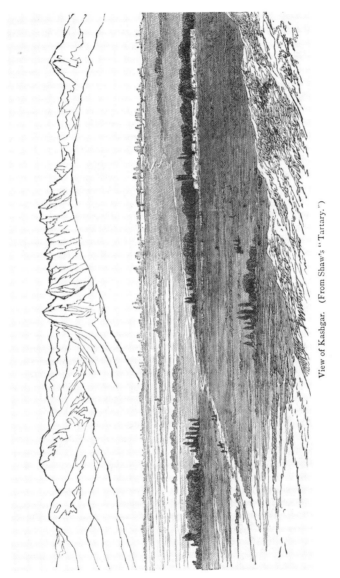

View of Kashgar. (From Shaw's "Tartary.")

beautiful gardens and vineyards, and fine estates, and
grow a great deal of cotton. From this country many

merchants go forth about the world on trading journeys. The natives are a wretched, niggardly set of people; they eat and drink in miserable fashion. There are in the country many Nestorian Christians, who have churches of their own. The people of the country have a peculiar language, and the territory extends for five days' journey.[1]

Note 1.—[There is no longer any difficulty in understanding how the travellers, after crossing Pamir, should have arrived at Kashgar if they followed the route from Táshkurgán through the Gez Defile.

The Itinerary of the Mirza from Badakhshan (Fáizabad) is the following : Zebak, Ishkashm, on the Panja, which may be considered the beginning of the Wakhán Valley, Panja Fort, in Wakhán, Raz Khan, Patur, near Lunghar (commencement of Pamir Steppe), Pamir Kul, or Barkút Yassin, 13,300 feet, Aktash, Sirikul Táshkurgán, Shukrab, Chichik Dawan, Akul, Kotul, Chahul Station (road to Yarkand) Kila Karawal, Aghiz Gah, Yangi-Hissar, Opechan, Yanga Shahr, Kashgar, where he arrived on the 3rd February, 1869. (Cf. *Report of " The Mirza's " Exploration from Caubul to Kashgar*. By Major T. G. Montgomerie, R.E. . . (*Jour. R. Geog. Soc.* XLI. 1871, pp. 132-192.)

Major Montgomerie (*l.c.* p. 144) says : "The alterations in the positions of Kashgar and Yarkund in a great measure explains why Marco Polo, in crossing from Badakhshan to Eastern Turkestan, went first to Kashgar and then to Yarkund. With the old positions of Yarkund and Kashgar it appeared that the natural route from Badakhshan would have led first to Yarkund ; with the new positions, and guided by the light of the Mirza's route, from which it is seen that the direct route to Yarkund is not a good one, it is easy to understand how a traveller might prefer going to Kashgar first, and then to Yarkund. It is satisfactory to have elicited this further proof of the general accuracy of the great traveller's account of his journey through Central Asia."

The Itinerary of Lieutenant-Colonel Gordon (*Sirikol, the Pámírs and Wakhán*, ch. vi. of *Forsyth's Mission to Yarkund in* 1873) runs thus : "Left Káshgar (21st March), Yangi-Hissar, Kaskasú Pass, descent to Chihil Gumbaz (forty Domes), where the road branches off to Yárkand (110 miles), Torut Pass, Tangi-Tár (defile), 'to the foot of a great elevated slope leading to the Chichiklik Pass, plain, and lake (14,700 feet), below the Yámbulák and Kok-Moinok Passes, which are used later in the season on the road between Yangi-Hissár and Sirikol, to avoid the Tangi-Tár and Shindi defiles. As the season advances, these passes become free from snow, while the defiles are rendered dangerous and difficult by the rush of the melting snow torrents. From the Chichiklik plain we proceeded down the Shindi ravine, over an extremely bad stony road, to the Sirikol River, up the banks of which we travelled to Táshkurgán, reaching it on the tenth day from Yangi-Hissar. The total distance is 125 miles.' Then Táshkurgán (ancient name *Várshídi*) : 'the open part of the Sirikol Valley extends from about 8 miles below Táshkurgán to apparently a very considerable distance towards the Kunjút mountain range ;' left Táshkurgán for Wákhan (2nd April, 1873); leave Sirikol Valley, enter the Shindán defile, reach the Áktásh Valley, follow the Áktásh stream (called Áksú by the Kirghiz) through the Little Pamir to the Gházkul (Little Pamir) Lake or Barkat Yássín, from which it takes its rise, four days from Táshkurgán. Little Pamir 'is bounded on the south by the continuation of the Neza Tásh range, which separates it from the Tághdúngbásh Pámir,' west of the lake, Langar, Sarhadd, 30 miles from Langar, and seven days from Sirikol, and Kila Panj, twelve days from Sirikól."—H. C.]

[I cannot admit with Professor Paquier (*l.c.* pp. 127-128) that Marco Polo did not visit Kashgar.—Grenard (II. p. 17) makes the remark that it took Marco Polo seventy days from Badakhshan to Kashgar, a distance that, in the Plain of Turkestan, he shall cross in sixteen days.—The Chinese traveller, translated by M. Gueluy (*Desc. de la Chine occidentale,* p. 45), says that the name Kashgar is made of *Kash,* fine colour, and *gar,* brick house.—H. C.]

Kashgar was the capital, from 1865 to 1877, of Ya'kúb Kúshbegi, a soldier of fortune, by descent it is said a Tajik of Shighnan, who, when the Chinese yoke was thrown off, made a throne for himself in Eastern Turkestan, and subjected the whole basin to his authority, taking the title of *Atalik Ghâzi.*

It is not easy to see how Kashgar should have been subject to the Great Kaan, except in the sense in which all territories under Mongol rule owed him homage. Yarkand, Polo acknowledges to have belonged to Kaidu, and the boundary between Kaidu's territory and the Kaan's lay between Karashahr and Komul [Bk. I. ch. xli.], much further east.

[Bretschneider, *Med. Res.* (II. p. 47), says : " Marco Polo states with respect to the kingdom of *Cascar* (I. 189) that it was subject to the Great Khan, and says the same regarding *Cotan* (I. 196), whilst *Yarcan* (I. 195), according to Marco Polo, belonged to Kaidu.   This does not agree with Rashid's statements about the boundary between Kaidu's territory and the Khan's."—H. C.]

Kashgar was at this time a Metropolitan See of the Nestorian Church. (*Cathay,* etc. 275, ccxlv.)

Many strange sayings have been unduly ascribed to our traveller, but I remember none stranger than this by Colonel Tod : " *Marco Polo calls Cashgar, where he was in the 6th century,* the birthplace of the Swedes" ! (*Rajasthan,* I. 60.) Pétis de la Croix and Tod between them are answerable for this nonsense. (See *The Hist. of Genghizcan the Great,* p. 116.)

On *cotton,* see ch. xxxvi.—On Nestorians, see Kanchau.

CHAPTER XXXIV.

OF THE GREAT CITY OF SAMARCAN.

SAMARCAN is a great and noble city towards the north-west, inhabited by both Christians and Saracens, who are subject to the Great Kaan's nephew, CAIDOU by name ; he is, however, at bitter enmity with the Kaan.[1] I will tell you of a great marvel that happened at this city.

It is not a great while ago that SIGATAY, own brother to the Great Kaan, who was Lord of this country and of many an one besides, became a Christian.[2] The

View of Samarcand. (From a sketch by Mr. Ivanoff.)
"Samarcan est une grandisme cité et noble."

Christians rejoiced greatly at this, and they built a great church in the city, in honour of John the Baptist; and by his name the church was called.   And they took a very fine stone which belonged to the Saracens, and placed it as the pedestal of a column in the middle of the church, supporting the roof.   It came to pass, however, that Sigatay died.   Now the Saracens were full of rancour about that stone that had been theirs, and which had been set up in the church of the Christians; and when they saw that the Prince was dead, they said one to another that now was the time to get back their stone, by fair means or by foul.   And that they might well do, for they were ten times as many as the Christians.   So they gat together and went to the church and said that the stone they must and would have.   The Christians acknowledged that it was theirs indeed, but offered to pay a large sum of money and so be quit.   Howbeit, the others replied that they never would give up the stone for anything in the world. And words ran so high that the Prince heard thereof, and ordered the Christians either to arrange to satisfy the Saracens, if it might be, with money, or to give up the stone.   And he allowed them three days to do either the one thing or the other.

What shall I tell you?   Well, the Saracens would on no account agree to leave the stone where it was, and this out of pure despite to the Christians, for they knew well enough that if the stone were stirred the church would come down by the run.   So the Christians were in great trouble and wist not what to do.   But they did do the best thing possible; they besought Jesus Christ that he would consider their case, so that the holy church should not come to destruction, nor the name of its Patron Saint, John the Baptist, be tarnished by its ruin.   And so when the day fixed by the Prince came round, they went to the church betimes in the morning, and lo, they found the stone removed from under the column ; the foot of the column was without support, and yet it bore the load as stoutly as before! Between the foot of the column and the ground there was a space of three palms.   So the Saracens had away their stone, and mighty little joy withal.   It was a glorious miracle, nay, it *is* so, for the column still so standeth, and will stand as long as God pleaseth.[3]

Now let us quit this and continue our journey.

NOTE I.—Of Kaidu, Kúblái Kaan's kinsman and rival, and their long wars, we shall have to speak later.   He had at this time a kind of joint occupancy of SAMARKAND and Bokhara with the Khans of Chagatai, his cousins.

[On Samarkand generally see : *Samarqand*, by W. Radloff, translated into French by L. Leger, *Rec. d'Itin. dans l'Asie Centrale*, Ecole des Langues Orient., Paris, 1878, p. 284 *et seq.* ; *A travers le royaume de Tamerlan* (*Asie Centrale*) . . . par Guillaume Capus . . . Paris, 1892, 8vo.—H. C.]

Marco evidently never was at Samarkand, though doubtless it was visited by his Father and Uncle on their first journey, when we know they were long at Bokhara. Having, therefore, little to say descriptive of a city he had not seen, he tells us a story :—

" So geographers, in Afric maps,
With savage pictures fill their gaps,
And o'er unhabitable downs
Place elephants for want of towns."

As regards the Christians of Samarkand who figure in the preceding story, we may note that the city had been one of the Metropolitan Sees of the Nestorian Church since the beginning of the 8th century, and had been a bishopric perhaps two centuries earlier.   Prince Sempad, High Constable of Armenia, in a letter written from Samarkand in 1246 or 1247, mentions several circumstances illustrative of the state of things indicated in this story : " I tell you that we have found many Christians scattered all over the East, and many fine churches, lofty, ancient, and of good architecture, which have been spoiled by the Turks.   Hence, the Christians of this country came to the presence of the reigning Kaan's grandfather (*i.e.* Chinghiz); he received them most honourably, and granted them liberty of worship, and issued orders to prevent their having any just cause of complaint by word or deed.   *And so the Saracens, who used to treat them with contempt, have now the like treatment in double measure.*"

Shortly after Marco's time, viz. in 1328, Thomas of Mancasola, a Dominican, who had come from Samarkand with a Mission to the Pope (John XXII.) from Ilchigadai, Khan of Chagatai, was appointed Latin Bishop of that city.   (*Mosheim*, p. 110, etc. ; *Cathay*, p. 192.)

NOTE 2.—CHAGATAI, here called Sigatay, was Uncle, not Brother, to the Great Kaan (Kúblái).   Nor was Kaidu either Chagatai's son or Kúblái's nephew, as Marco here and elsewhere represents him to be.   (See Bk. IV. ch. i.)   The term used to describe Chagatai's relationship is *frère charnel,* which excludes ambiguity, cousinship, or the like (such as is expressed by the Italian *fratello cugíno*), and corresponds, I believe, to the *brother german* of Scotch law documents.

NOTE 3.—One might say, These things be an allegory ! We take the fine stone that belongs to the Saracens (or Papists) to build our church on, but the day of reckoning comes at last, and our (Irish Protestant) Christians are afraid that the Church will come about their ears. May it stand, and better than that of Samarkand has done !

There is a story somewhat like this in D'Herbelot, about the Karmathian Heretics carrying off the Black Stone from Mecca, and being obliged years after to bring it back across the breadth of Arabia ; on which occasion the stone conducted itself in a miraculous manner.

There *is* a remarkable Stone at Samarkand, the *Kok-Tash* or Green Stone, on which Timur's throne was set. Tradition says that, big as it is, it was brought by him from Brusa ;—but tradition may be wrong. (See *Vámbéry's Travels,* p. 206.) [Also *H. Moser, A travers l'Asie centrale,* 114-115.—H. C.]

[The Archimandrite Palladius (*Chinese Recorder,* VI. p. 108) quotes from the *Chi shun Chin-kiang chi* (Description of Chin-Kiang), 14th century, the following passage regarding the pillar : " There is a temple (in Samarcand) supported by four enormous wooden pillars, each of them 40 feet high. One of these pillars is in a hanging position, and stands off from the floor more than a foot."—H. C.]

CHAPTER XXXV.

OF THE PROVINCE OF YARCAN.

YARCAN is a province five days' journey in extent. The people follow the Law of Mahommet, but there are also Nestorian and Jacobite Christians. They are subject to the same Prince that I mentioned, the Great Kaan's nephew. They have plenty of everything, [particularly of cotton. The inhabitants are also great craftsmen, but a large proportion of them have swoln legs, and great crops at the throat, which arises from some quality in their drinking-water.] As there is nothing else worth telling we may pass on.[1]

NOTE 1.—Yarkan or Yarken seems to be the general pronunciation of the name to this day, though we write YARKAND.

[A Chinese traveller, translated by M. Gueluy (*Desc. de la Chine occidentales,* p. 41), says that the word *Yarkand* is made of *Iar,* earth, and *Kiang (Kand?),* large, vast, but this derivation is doubtful. The more probable one is that Yarkand is made up of *Yar*, new, and *Kand, Kend*, or *Kent*, city.—H. C.]

Mir 'Izzat Ullah in modern days speaks of the prevalence of goitre at Yarkand. And Mr. Shaw informs me that during his recent visit to Yarkand (1869) he had numerous applications for iodine as a remedy for that disease. The theory which connects it with the close atmosphere of valleys will not hold at Yarkand. (*J. R. A. S.* VII. 303.)

[Dr. Sven Hedin says that three-fourths of the population of Yarkand are suffering from goitre; he ascribes the prevalence of the disease to the bad quality of the water, which is kept in large basins, used indifferently for bathing, washing, or draining. Only Hindu and "Andijdanlik" merchants, who drink well water, are free from goitre.

Lieutenant Roborovsky, the companion of Pievtsov, in 1889, says : " In the streets one meets many men and women with large goitres, a malady attributed to the bad quality of the water running in the town conduits, and drunk by the inhabitants in its natural state. It appears in men at the age of puberty, and in women when they marry." (*Proc. R. G. S.* 2 ser. XII. 1890, p. 36.)

Formerly the Mirza (*J. R. G. S.* 1871, p. 181) said : " Goitre is very common in the city [of Yarkund], and in the country round, but it is unknown in Kashgar."

General Pievtsov gives to the small oasis of Yarkand (264 square miles) a population of 150,000, that is, 567 inhabitants per square mile. He, after Prjevalsky's death, started, with V. L. Roborovsky (botanist) and P. K. Kozlov (zoologist), who were later joined by K. I. Bogdanovich (geologist), on his expedition to Tibet (1889-1890). He followed the route Yarkand, Khotan, Kiria, Nia, and Charchan.—H. C.]

CHAPTER XXXVI.

OF A PROVINCE CALLED COTAN.

COTAN is a province lying between north-east and east, and is eight days' journey in length. The people are subject to the Great Kaan,[1] and are all worshippers of Mahommet.[2] There are numerous towns and villages in the country, but Cotan, the capital, is the most noble of all, and gives its name to the kingdom. Everything is to be had there in plenty, including abundance of cotton, [with flax, hemp, wheat, wine, and the like]. The people have vineyards and gardens and estates. They live by commerce and manufactures, and are no soldiers.[3]

NOTE 1.—[The Buddhist Government of Khotan was destroyed by Boghra Khân (about 980-990); it was temporarily restored by the Buddhist Kutchluk Khân, chief of the Naïmans, who came from the banks of the Ili, destroyed the Mahomedan dynasty of Boghra Khân (1209), but was in his turn subjugated by Chinghiz Khan.

The only Christian monument discovered in Khotan is a bronze cross brought back by Grenard (III. pp. 134-135); see also Devéria, *Notes d'Epigraphie Mongole*, p. 80.—H. C.]

NOTE 2.—"*Aourent Mahommet*." Though this is Marco's usual formula to define Mahomedans, we can scarcely suppose that he meant it literally. But in other cases it was *very* literally interpreted. Thus in *Baudouin de Sebourc*, the Dame de Pontieu, a passionate lady who renounces her faith before Saladin, says :—

> " ' Et je renoïe Dieu, et le pooir qu'il a ;
> Et Marie, sa Mère, qu'on dist qui le porta ;
> *Mahom voel aourer*, aportez-le-moi chà ! '
> * * * * Li Soudans commanda
> *Qu'on aportast Mahom ; et celle l'aoura*." (I. p. 72.)

The same romance brings in the story of the Stone of Samarkand, adapted from ch. xxxiv., and accounts for its sanctity in Saracen eyes because it had long formed a pedestal for Mahound !

And this notion gave rise to the use of *Mawmet* for an idol in general ; whilst from the *Mahommerie* or place of Islamite worship the name of *mummery* came to be applied to idolatrous or unmeaning rituals ; both very unjust etymologies. Thus of mosques in *Richard Cœur de Lion* :

> "Kyrkes they made of Crystene Lawe,
> And her *Mawmettes* lete downe drawe." (*Weber*, II. 228.)

So Correa calls a golden idol, which was taken by Da Gama in a ship of Calicut, "an image of Mahomed" (372). Don Quixote too, who ought to have known better, cites with admiration the feat of Rinaldo in carrying off, in spite of forty Moors, a golden image of Mahomed.

NOTE 3.—800 *li* (160 miles) east of *Chokiuka* or Yarkand, Hiuen Tsang comes to *Kiustanna* (Kustána) or KHOTAN. "The country chiefly consists of plains covered with stones and sand. The remainder, however, is favourable to agriculture, and produces everything abundantly. From this country are got woollen carpets, fine felts, well woven taffetas, white and black jade." Chinese authors of the 10th century speak of the abundant grapes and excellent wine of Khotan.

Chinese annals of the 7th and 8th centuries tell us that the people of Khotan had chronicles of their own, a glimpse of a lost branch of history. Their writing, laws, and literature were modelled upon those of India.

Ilchi, the modern capital, was visited by Mr. Johnson, of the Indian Survey, in 1865. The country, after the revolt against the Chinese in 1863, came first under the rule of Habíb-ullah, an aged chief calling himself *Khán Bádshah* of Khotan ; and since the treacherous seizure and murder of Habíb-ullah by Ya'kub Beg of Kashgar in January 1867, it has formed a part of the kingdom of the latter.

Mr. Johnson says : "The chief grains of the country are Indian corn, wheat, barley of two kinds, *bájra, jowár* (two kinds of *holcus*), buckwheat and rice, all of which are superior to the Indian grains, and are of a very fine quality. . . . . The country is certainly superior to India, and in every respect equal to Kashmir, over which it has the advantage of being less humid, and consequently better suited to the growth of fruits. *Olives* (?), pears, apples, peaches, apricots, mulberries, grapes, currants, and melons, all exceedingly large in size and of a delicious flavour, are produced in great variety and abundance. . . . . Cotton of valuable quality, and raw silk, are produced in very large quantities."

[Khotan is the chief place of Turkestan for cotton manufactures ; its *khàm* is to be found everywhere. This name, which means raw in Persian, is given to a stuff made with cotton thread, which has not undergone any preparation ; they manufacture also two other cotton stuffs : *alatcha* with blue and red stripes, and *tchekmen*, very thick and coarse, used to make dresses and sacks ; if *khàm* is better at. Khotan, *alatcha* and *tchekmen* are superior at Kashgar. (*Grenard*, II. pp. 191-192.)

Grenard (II. pp. 175-177), among the fruits, mentions apricots (*ourouk*), ripe in June, and so plentiful that to keep them they are dried up to be used like garlic against mountain sickness ; melons (*koghoun*) ; water-melons (*tarbouz*, the best are from Hami) ; vine (*tàl*)—the best grapes (*uzum*) come from Boghâz langar, near Keria ; the best dried grapes are those from Turfan ; peaches (*shaptâlou*) ; pomegranates (*anâr*, best from Kerghalyk), etc. ; the best apples are those of Nia and Sadju ; pears are very bad ; cherries and strawberries are unknown. Grenard (II. p. 106) also says that grapes are very good, but that Khotan wine is detestable, and tastes like vinegar.

The Chinese traveller, translated by M. Gueluy (*Desc. de la Chine occidentale*, p. 45), says that all the inhabitants of Khotan are seeking for precious stones, and that melons and fruits are more plentiful than at Yarkand.—H. C.]

Mr. Johnson reports the whole country to be rich in soil and very much under-peopled. Ilchi, the capital, has a population of about 40,000, and is a great place for manufactures. The chief articles produced are silks, felts, carpets (both silk and woollen), coarse cotton cloths, and paper from the mulberry fibre. The people are strict Mahomedans, and speak a Turki dialect. Both sexes are good-looking, with a slightly Tartar cast of countenance. (*V. et V. de H. T.* 278 ; *Rémusat, H. de la V. de Khotan*, 37, 73-84 ; *Chin. Repos.* IX. 128 ; *J. R. G. S.* XXXVII. 6 *seqq.*)

[In 1891, Dutreuil de Rhins and Grenard at the small village of Yotkán, about 8 miles to the west of the present Khotan, came across what they considered the most important and probably the most ancient city of southern Chinese Turkestan. The natives say that Yotkàn is the site of the old Capital. (Cf. *Grenard*, III. p. 127 *et seq.* for a description and drawings of coins and objects found at this place.)

The remains of the ancient capital of Khotan were accidentally discovered, some thirty-five years ago, at Yotkàn, a village of the Borazân Tract. A great mass of highly interesting finds of ancient art pottery, engraved stones, and early Khotan coins with Kharoṣṭhi-Chinese legends, coming from this site, have recently been thoroughly examined in Dr. Hoernle's Report on the " British Collection of Central Asian Antiquities." *Stein.*—(See *Three further Collections of Ancient Manuscripts from Central Asia*, by Dr. A. F. R. Hoernle. . . . Calcutta, 1897, 8vo.)

" The sacred sites of Buddhist Khotan which Hiuen Tsang and Fa-hian describe, can be shown to be occupied now, almost without exception, by Mohamedan shrines forming the object of popular pilgrimages." (M. A. Stein, *Archæological Work about Khotan, Jour. R. As. Soc.*, April, 1901, p. 296.)

It may be justly said that during the last few years numerous traces of Hindu civilisation have been found in Central Asia, extending from Khotan, through the Takla-Makan, as far as Turfan, and perhaps further up.

Dr. Sven Hedin, in the year 1896, during his second journey through Takla-Makan from Khotan to Shah Yar, visited the ruins between the Khotan Daria and the Kiria Daria, where he found the remains of the city of Takla-Makan now buried in the sands. He discovered figures of Buddha, a piece of papyrus with unknown characters, vestiges of habitations. This Asiatic Pompei, says the traveller, at least ten centuries old, is anterior to the Mahomedan invasion led by Kuteïbe Ibn-Muslim, which happened at the beginning of the 8th century. Its inhabitants were Buddhist, and of Aryan race, probably originating from Hindustan.—Dutreuil de Rhins and Grenard discovered in the Kumâri grottoes, in a small hill on the right bank of the Karakash Daria, a manuscript written on birch bark in *K*harosh*t*hi characters ; these grottoes of Kumâri are mentioned in Hiuen Tsang. (II. p. 229.)

Dr. Sven Hedin followed the route Kashgar, Yangi-Hissar, Yarkand to Khotan, in 1895. He made a stay of nine days at Ilchi, the population of which he estimated at 5500 inhabitants (5000 Musulmans, 500 Chinese).

(See also Sven Hedin, *Die Geog. wissenschaft. Ergebnisse meiner Reisen in Zentralasien*, 1894-1897. *Petermann's Mitt.*, Ergänz. XXVIII. (Hft. 131), Gotha, 1900.—H. C.]

CHAPTER XXXVII.

Of the Province of Pein.

PEIN is a province five days in length, lying between east and north-east. The people are worshippers of Mahommet, and subjects of the Great Kaan. There are a good number of towns and villages, but the most noble is PEIN, the capital of the kingdom.[1] There are rivers in this country, in which quantities of Jasper and Chalcedony are found.[2] The people have plenty of all products, including cotton. They live by manufactures and trade. But they have a custom that I must relate. If the husband of any woman go away upon a journey and remain away for more than 20 days, as soon as that term is past the woman may marry another man, and the husband also may then marry whom he pleases.[3]

I should tell you that all the provinces that I have been speaking of, from Cascar forward, and those I am going to mention [as far as the city of Lop] belong to GREAT TURKEY.

NOTE I.—"In old times," says the *Haft Iklím.*, " travellers used to go from Khotan to Cathay in 14 (?) days, and found towns and villages all along the road [excepting, it may be presumed, on the terrible Gobi], so that there was no need to travel in caravans. In later days the fear of the Kalmaks caused this line to be abandoned, and the circuitous one occupied 100 days." This directer route between Khotan and China must have been followed by Fa-hian on his way to India ; by Hiuen Tsang on his way back ; and by Shah Rukh's ambassadors on their return from China in 1421. The circuitous route alluded to appears to have gone north from Khotan, crossed the Tarimgol, and fallen into the road along the base of the Thian Shan, eventually crossing the Desert southward from Komul.

Former commentators differed very widely as to the position of Pein, and as to the direction of Polo's route from Khotan. The information acquired of late years leaves the latter no longer open to doubt. It must have been nearly coincident with that of Hiuen Tsang.

The perusal of Johnson's Report of his journey to Khotan, and the Itineraries attached to it, enabled me to feel tolerable certainty as to the position of Charchan (see next chapter), and as to the fact that Marco followed a direct route from Khotan to the vicinity of Lake Lop. Pein, then, was identical with PIMA,* which was the first city reached by Hiuen Tsang on his return to China after quitting Khotan, and which lay 330 *li* east of the latter city.† Other notices of Pima appear in Rémusat's history of Khotan ; some of these agree exactly as to the distance from the capital, adding that it stood on the banks of a river flowing from the East and entering the sandy Desert ; whilst one account seems to place it at 500 *li* from Khotan. And in the Turkish map of Central Asia, printed in the *Jahán Numá*, as we learn from Sir H. Rawlinson, the town of *Pím* is placed a little way north of Khotan. Johnson found Khotan rife with stories of former cities overwhelmed by the shifting sands of the Desert, and these sands appear to have been advancing for ages ; for far to the north-east of Pima, even in the 7th century, were to be found the deserted and ruined cities of the ancient kingdoms of *Tuholo* and *Shemathona*. "Where anciently were the seats of flourishing cities and prosperous communities," says a Chinese author speaking of this region, " is nothing now to be seen but a vast desert ; all has been buried in the sands, and the wild camel is hunted on those arid plains."

Pima cannot have been very far from *Kiria*, visited by Johnson. This is a town of 7000 houses, lying east of Ilchi, and about 69 miles distant from it. The road for the most part lies through a highly cultivated and irrigated country, flanked by the sandy desert at three or four miles to the left. After passing *eastward* by Kiria it is said to make a great elbow, turning north ; and within this elbow lie the sands that have buried cities and fertile country. Here Mr. Shaw supposes Pima lay (perhaps upon the river of Kiria). At Pima itself, in A.D. 644, there was a story of the destruction of a city lying further north, a judgment on the luxury and impiety of the people and their king, who, shocked at the eccentric aspect of a holy man, had caused him to be buried in sand up to the mouth.

(*N. et E.* XIV. 477 ; *H. de la Ville de Khotan*, 63-66 ; *Klap. Tabl. Historiques*, p. 182 ; *Proc. R. G. S.* XVI. 243.)

[Dutreuil de Rhins and Grenard took the road from Khotan to Charchan ; they left Khotan on the 4th May, 1893, passed Kiria, Nia, and instead of going direct to Charchan through the desert, they passed Kara Say at the foot of the Altyn tâgh, a route three days longer than the other, but one which was less warm, and where water, meat, milk, and barley could be found. Having passed Kapa, they crossed the Karamuren, and went up from Achan due north to Charchan, where they stayed three months. Nowhere do they mention Pein, or Pima, for it appears to be *Kiria itself*, which is the only real town between Khotan and the Lobnor. Grenard says in a note (p. 54, vol. ii.): " *Pi-mo* (Keria) recalls the Tibetan *byé-ma*, which is pronounced *Péma*, or *Tchéma*, and which means *sand*. Such is perhaps also the origin of *Pialma*, a village near Khotan, and of the old name of Charchan, *Tché-mo-to-na*, of which the two last syllables would represent *grong* (pronounce *tong*=town), or *kr'om* (*t'om*=bazaar). Now, not only would this etymology be justified because these three places are indeed surrounded with sand remarkably deep, but as they were the first three important places with which the Tibetans met coming into the desert of Gobi, either by the route of Gurgutluk and of Polor, or by Karakoram and Sandju, or by Tsadam, and they had thus as good a pretext to call them 'towns of sand' as the

---

* *Pein* may easily have been miscopied for *Pem*, which is indeed the reading of some MSS. Ramusio has *Peym*.

† M. Vivien de St. Martin, in his map of Hiuen Tsang's travels, places Pima to the *west* of Khotan. Though one sees how the mistake originated, there is no real ground for this in either of the versions of the Chinese pilgrim's journey. (See *Vie et Voyages*, p. 288, and *Mémoires*, vol. ii. 242-243.)

Chinese had to give to T'un-hwang the name of *Shachau*, viz. City of Sand.   Kiria is called *Ou-mi*, under the Han, and the name of Pi-mo is found for the first time in Hiuen Tsang, that is to say, before the Tibetan invasions of the 8th century.   It is not possible to admit that the incursion of the Tu-ku-hun in the 5th century could be the cause of this change of name.   The hypothesis remains that Pi-mo was really the ancient name forced by the first Tibetan invaders spoken of by legend, that *Ou-mi* was either another name of the town, or a fancy name invented by the Chinese, like Yu-t'ien for Khotan, Su-lo for Kashgar. . . ."   Sir T. D. Forsyth (*J. R. G. S.*, XLVII., 1877, p. 3) writes : " I should say that Peim or Pima must be identical with Kiria."—H. C.]

NOTE 2.—The Jasper and Chalcedony of our author are probably only varieties of the semi-precious mineral called by us popularly *Jade*, by the Chinese *Yü*, by the Eastern Turks *Kásh*, by the Persians *Yashm*, which last is no doubt the same word with Ιασπις, and therefore with *Jasper*.   The Greek Jaspis was in reality, according to Mr. King, a green Chalcedony.

The Jade of Turkestan is largely derived from water-rolled boulders fished up by divers in the rivers of Khotan, but it is also got from mines in the valley of the Kará-kásh River.   " Some of the Jade," says Timkowski, " is as white as snow, some dark green, like the most beautiful emerald (?), others yellow, vermilion, and jet black. The rarest and most esteemed varieties are the white speckled with red and the green veined with gold."   (I. 395.)   The Jade of Khotan appears to be first mentioned by Chinese authors in the time of the Han Dynasty under Wu-ti (B.C. 140-86).   In A.D. 541 an image of Buddha sculptured in Jade was sent as an offering from Khotan ; and in 632 the process of fishing for the material in the rivers of Khotan, as practised down to modern times, is mentioned.   The importation of Jade or *Yü* from this quarter probably gave the name of *Kia-yü Kwan* or " Jade Gate " to the fortified Pass looking in this direction on the extreme N.W. of China Proper, between Shachau and Suhchau.   Since the detachment from China the Jade industry has ceased, the Musulmans having no taste for that kind of *virtù*.   (*H. de la V. de Khotan*, 2, 17, 23; also see *J. R. G. S.* XXXVI. 165, and *Cathay*, 130, 564 ; *Ritter*, II. 213 ; *Shaw's High Tartary*, pp. 98, 473.)

[On the 11th January, 1895, Dr. Sven Hedin visited one of the chief places where Jade is to be found.   It is to the north-east of Khotan, in the old bed of the Yurun Kash.   The bed of the river is divided into *claims* like gold-fields ; the workmen are Chinese for the greater part, some few are Musulmans.

Grenard (II. pp. 186-187) says that the finest Jade comes from the high Karákásh (black Jade) River and Yurungkásh (white Jade) ; the Jade River is called Su-tásh. At Khotan, Jade is polished up by sixty or seventy individuals belonging to twenty-five workshops.

" At 18 miles from Su-chau, Kia-yu-kwan, celebrated as one of the gates of China, and as the fortress guarding the extreme north-west entrance into the empire, is passed." (*Colonel M. S. Bell, Proc. R. G. S.* XII. 1890, p. 75.)

According to the Chinese characters, the name of Kia-yü Kwan does not mean " Jade Gate," and as Mr. Rockhill writes to me, it can only mean something like " barrier of the pleasant Valley."—H. C.]

NOTE 3.—Possibly this may refer to the custom of temporary marriages which seems to prevail in most towns of Central Asia which are the halting-places of cara-vans, and the morals of which are much on a par with those of seaport towns, from analogous causes.   Thus at Meshid, Khanikoff speaks of the large population of young and pretty women ready, according to the accommodating rules of Shiah Mahomedan-ism, to engage in marriages which are perfectly lawful, for a month, a week, or even twenty-four hours.   Kashgar is also noted in the East for its *chaukans*, young women with whom the traveller may readily form an alliance for the period of his stay, be it long or short.   (*Khan. Mém.* p. 98 ; *Russ. in Central Asia*, 52 ; *J. A. S. B.* XXVI. 262 ; *Burnes*, III. 195 ; *Vigne*, II. 201.)

CHAPTER XXXVIII.

OF THE PROVINCE OF CHARCHAN.

CHARCHAN is a Province of Great Turkey, lying between north-east and east. The people worship Mahommet. There are numerous towns and villages, and the chief city of the kingdom bears its name, Charchan. The Province contains rivers which bring down Jasper and Chalcedony, and these are carried for sale into Cathay, where they fetch great prices. The whole of the Province is sandy, and so is the road all the way from Pein, and much of the water that you find is bitter and bad. However, at some places you do find fresh and sweet water. When an army passes through the land, the people escape with their wives, children, and cattle a distance of two or three days' journey into the sandy waste ; and knowing the spots where water is to be had, they are able to live there, and to keep their cattle alive, whilst it is impossible to discover them ; for the wind immediately blows the sand over their track.

Quitting Charchan, you ride some five days through the sands, finding none but bad and bitter water, and then you come to a place where the water is sweet. And now I will tell you of a province called Lop, in which there is a city, also called LOP, which you come to at the end of those five days. It is at the entrance of the great Desert, and it is here that travellers repose before entering on the Desert.[1]

NOTE I.—Though the *Lake* of Lob or Lop appears on all our maps, from Chinese authority, the latter does not seem to have supplied information as to a town so called. We have, however, indications of the existence of such a place, both mediæval and recent. The History of Mirza Haidar, called the Táríkh-i-Rashídí, already referred to, in describing the Great Basin of Eastern Turkestan, says: " Formerly there were several large cities in this plain ; the names of two have survived—*Lob* and *Kank,* but of the rest there is no trace or tradition ; all is buried under the sand." [Forsyth (*J. R. G. S.* XLVII. 1877, p. 5) says that he thinks that this Kank is probably the Katak mentioned by Mirza Haidar.—H. C.] In another place the same history says that a boy heir of the house of Chaghatai, to save him from a usurper, was sent away to Sárígh Uighúr and *Lob-Kank*, far in the East. Again, in the short notices of the cities of Turkestan which Mr. Wathen collected at Bombay from pilgrims of those regions on their way to Mecca, we find the following : " *Lopp*.—Lopp is situated at a great distance from Yarkand. The inhabitants are principally Chinese ; but a few Uzbeks reside there. Lopp is remarkable for a salt-water lake in its vicinity." Johnson, speaking of a road from Tibet into Khotan, says : " This route . . . . leads not only to Ilchi and Yarkand, but also *viâ Lob* to the large and important city of Karashahr." And among the routes attached to Mr. Johnson's original Report, we have :—

" Route No. VII. *Kiria* (see note 1 to last chapter) to CHACHAN and LOB (*from native information*)."

This first revealed to me the continued existence of Marco's Charchan ; for it was impossible to doubt that in the CHACHAN and LOB of this Itinerary we had his Charchan and Lop ; and his route to the verge of the Great Desert was thus made clear.

Mr. Johnson's information made the journey from Kiria to Charchan to be 9 marches, estimated by him to amount to 154 miles, and adding 69 miles from Ilchi to Kiria (which he actually traversed) we have 13 marches or 223 miles for the distance from Ilchi to Charchan. Mr. Shaw has since obtained a route between Ilchi and Lob on very good authority. This makes the distance to Charchan, or *Charchand*, as it is called, 22 marches, which Mr. Shaw estimates at 293 miles. Both give 6 marches from Charchand to Lob, which is in fair accordance with Polo's 5, and Shaw estimates the whole distance from Ilchi to Lob at 373, or by another calculation at 384 miles, say roundly 380 miles. This higher estimate is to be preferred to Mr. Johnson's for a reason which will appear under next chapter.

Mr. Shaw's informant, Rozi of Khotan, who had lived twelve years at Charchand, described the latter as a small town with a district extending on both sides of a stream which flows to Lob, *and which affords Jade.* The people are Musulmans. They grow wheat, Indian corn, pears, and apples, etc., but no cotton or rice. It stands in a great plain, but the mountains are not far off. The nature of the products leads Mr. Shaw to think it must stand a good deal higher than Ilchi (4000), perhaps at about 6000 feet. I may observe that the Chinese hydrography of the Kashgar Basin, translated by Julien in the *N. An. des Voyages* for 1846 (vol. iii.), seems to imply that mountains from the south approach within some 20 miles of the Tarim River, between the longitude of Shayar and Lake Lop. The people of Lob are Musulman also, but very uncivilised. The Lake is salt. The hydrography calls it about 200 *li* (say 66 miles) from E. to W. and half that from N. to S., and expresses the old belief that it forms the subterranean source of the Hwang-Ho. Shaw's Itinerary shows " salt pools " at six of the stations between Kiria and Charchand, so Marco's memory in this also was exact.

Nia, a town two marches from Kiria according to Johnson, or four according to Shaw, is probably the ancient city of Ni-jang of the ancient Chinese Itineraries, which lay 30 or 40 miles on the China side of Pima, in the middle of a great marsh, and formed the eastern frontier of Khotan bordering on the Desert. (*J. R. G. S.* XXXVII. pp. 13 and 44 ; also Sir H. Rawlinson in XLII. p. 503 : *Erskine's Baber and Humayun*, I. 42 ; *Proc. R. G. S.* vol. xvi. pp. 244-249 ; *J. A. S. B.* IV. 656 ; *H. de la V. de Khotan*, u.s.)

[The Charchan of Marco Polo seems to have been built to the west of the present oasis, a little south of the road to Kiria, where ruined houses have been found. It must have been destroyed before the 16th century, since Mirza Haidar does not mention it. It was not anterior to the 7th century, as it did not exist at the time of Hiuen Tsang. (Cf. *Grenard*, III. p. 146.)

Grenard says (pp. 183-184) that he examined the remains of what is called the old town of Charchan, traces of the ancient canal, ruins of dwellings deep into the sand, of which the walls built of large and solid-baked bricks, are pretty well preserved. Save these bricks, " I found hardly anything, the inhabitants have pillaged everything long ago. I attempted some excavating, which turned out to be without result, as far as I was concerned ; but the superstitious natives declared that they were the cause of a violent storm which took place soon after. There are similar ruins in the environs, at Yantak Koudouk, at Tatrang, one day's march to the north, and at Ouadjchahari at five days to the north-east, which corresponds to the position assigned to Lop by Marco Polo." (See *Grenard's Haute Asie* on *Nia.*)

Palladius is quite mistaken (*l.c.* p. 3) in saying that the " Charchan " of Marco Polo is to be found in the present province of Karashar. (Cf. *T. W. Kingsmill's Notes on Marco Polo's Route from Khoten to China, Chinese Recorder,* VII. pp. 338-343; *Notes on Doctor Sven Hedin's Discoveries in the Valley of the Tarim, its Cities and Peoples, China Review,* XXIV. No. II. pp. 59-64.)—H. C.]

CHAPTER XXXIX.

OF THE CITY OF LOP AND THE GREAT DESERT.

LOP is a large town at the edge of the Desert, which is called the Desert of Lop, and is situated between east and north-east. It belongs to the Great Kaan, and the people worship Mahommet. Now, such persons as propose to cross the Desert take a week's rest in this town to refresh themselves and their cattle ; and then they make ready for the journey, taking with them a month's supply for man and beast. On quitting this city they enter the Desert.

The length of this Desert is so great that 'tis said it would take a year and more to ride from one end of it to the other. And here, where its breadth is least, it takes a month to cross it. 'Tis all composed of hills and valleys of sand, and not a thing to eat is to be found on it. But after riding for a day and a night you find fresh water, enough mayhap for some 50 or 100 persons with their beasts, but not for more. And all across the Desert you will find water in like manner, that is to say, in some 28 places altogether you will find good water, but in no great quantity ; and in four places also you find brackish water.[1]

Beasts there are none ; for there is nought for them to eat.   But there is a marvellous thing related of this Desert, which is that when travellers are on the move by night, and one of them chances to lag behind or to fall asleep or the like, when he tries to gain his company again he will hear spirits talking, and will suppose them to be his comrades.   Sometimes the spirits will call him by name ; and thus shall a traveller ofttimes be led astray so that he never finds his party.   And in this way many have perished.   [Sometimes the stray travellers will hear as it were the tramp and hum of a great cavalcade of people away from the real line of road, and taking this to be their own company they will follow the sound ; and when day breaks they find that a cheat has been put on them and that they are in an ill plight.[2]] Even in the day-time one hears those spirits talking. And sometimes you shall hear the sound of a variety of musical instruments, and still more commonly the sound of drums.   [Hence in making this journey 'tis customary for travellers to keep close together.   All the animals too have bells at their necks, so that they cannot easily get astray.   And at sleeping-time a signal is put up to show the direction of the next march.]

So thus it is that the Desert is crossed.[3]

NOTE I.—LOP appears to be the *Napopo, i.e. Navapa*, of Hiuen Tsang, called also the country of *Leulan*, in the Desert. (*Mém.* II. p. 247.) *Navapa* looks like Sanskrit. If so, this carries ancient Indian influence to the verge of the great Gobi. [See *supra*, p. 190.] It is difficult to reconcile with our maps the statement of a thirty days' journey across the Desert from Lop to Shachau.   Ritter's extracts, indeed, regarding this Desert, show that the constant occurrence of sandhills and deep drifts (our traveller's "hills and valleys of sand") makes the passage extremely difficult for carts and cattle.   (III. 375.)   But I suspect that there is some material error in the longitude of Lake Lop as represented in our maps, and that it should be placed *something like three degrees* more to the westward than we find it (*e.g.*) in Kiepert's Map of Asia.   By that map Khotan is not far short of 600 miles from the western extremity of Lake Lop.   By

Johnson's Itinerary (including his own journey to Kiria) it is only 338 miles from Ilchi to Lob. Mr. Shaw, as we have seen, gives us a little more, but it is only even then 380. Polo unfortunately omits his usual estimate for the extent of the "Province of Charchan," so he affords us no complete datum. But his distance between Charchan and Lob agrees fairly, as we have seen, with that both of Johnson and of Shaw, and the elbow on the road from Kiria to Charchan (*supra*, p. 192) necessitates our still further abridging the longitude between Khotan and Lop. (See Shaw's remarks in *Proc. R. G. S.* XVI. 243.)

["This desert was known in China of old by the name of *Lew-sha, i.e.* "Quicksand," or literally, "Flowing sands." (*Palladius, Jour. N. China B. R. As. Soc.* N.S. X. 1875, p. 4.)

A most interesting problem is connected with the situation of Lob-nor which led to some controversy between Baron von Richthofen and Prjevalsky. The latter placed the lake one degree more to the south than the Chinese did, and found that its water was sweet. Richthofen agreed with the Chinese Topographers and wrote in a letter to Sir Henry Yule : " I send you two tracings ; one of them is a true copy of the Chinese map, the other is made from a sketch which I constructed to-day, and on which I tried to put down the Chinese Topography together with that of Prjevalsky. It appears evident—(1) That Prjevalsky travelled by the ancient road to a point south of the true Lop-noor ; (2) that long before he reached this point he found the river courses quite different from what they had been formerly ; and (3) that following one of the new rivers which flows due south by a new road, he reached the two sweet-water lakes, one of which answers to the ancient Khas-omo. I use the word 'new' merely by way of comparison with the state of things in Kien-long's time, when the map was made. It appears that the Chinese map shows the Khas Lake too far north to cover the Kara-Koshun. The bifurcation of the roads south of the lake nearly resembles that which is marked by Prjevalsky." (Preface of E. D. Morgan's transl. of *From Kulja across the Tian Shan to Lob-nor*, by Colonel N. Prjevalsky, London, 1879, p. iv.) In this same volume Baron von Richthofen's remarks are given (pp. 135-159, with a map, p. 144), showing comparison between Chinese and Prjevalsky's Geography from tracings by Baron von Richthofen and (pp. 160-165) a translation of Prjevalsky's replies to the Baron's criticisms.

Now the Swedish traveller, Dr. Sven Hedin, claims to have settled this knotty point. Going from Korla, south-west of Kara-shahr, by a road at the foot of the Kurugh-tagh and between these mountains and the Koncheh Daria, he discovered the ruins of two fortresses, and a series of milestones (potaïs). These tall pyramids of clay and wood, indicating distances in *lis*, show the existence at an ancient period of a road with a large traffic between Korla and an unknown place to the south-east, probably on the shores of the Chinese Lob-nor. Prjevalsky, who passed between the Lower Tarim and the Koncheh Daria, could not see a lake or the remains of a lake to the east of this river. The Koncheh Daria expands into a marshy basin, the Malta Kul, from which it divides into two branches, the Kuntiekkich Tarim (East River) and the Ilek (river) to the E.S.E. Dr. Sven Hedin, after following the course of the Ilek for three days (4th April, 1896) found a large sheet of water in the valley at the very place marked by tne Chinese Topographers and Richthofen for the Lob-nor. This mass of water is divided up by the natives into Avullu Kul, Kara Kul, Tayek Kul, and Arka Kul, which are actually almost filled up with reeds. Dr. Sven Hedin afterwards visited the Lob-nor of Prjevalsky, and reached its western extremity, the Kara-buran (black storm) on the 17th April. In 1885, Prjevalsky had found the Lob-nor an immense lake ; four years later Prince Henri d'Orleans saw it greatly reduced in size, and Dr. Sven Hedin discovered but pools of water. In the meantime, since 1885, the northern (Chinese) Lob-nor has gradually filled up, so the lake is somewhat vagrant. Dr. Sven Hedin says that from his observations he can assert that Prjevalsky's ake is of recent formation.

So Marco Polo's Lob-nor should be the northern or Chinese lake.

Another proof of this given by Dr. Sven Hedin is that the Chinese give the name of

Lob to the region between Arghan and Tikkenlik, unknown in the country of the southern lake. The existence of two lakes shows what a quantity of water from the Thian Shan, the Eastern Pamir, and Northern Tibet flows into the basin of the Tarim. The Russian Lieutenant K. P. Kozlov has tried since to prove that the Chinese Lob-nor is the Kara-Koshun (Black district), which is a second lake formed by the Tarim, which discharges into and issues from the lake Kara-buran. Kozlov's arguments are published in the *Isvestia* of the Russian Geographical Society, and in a separate pamphlet. *The Geog. Jour.* (June, 1898, pp. 652-658) contains *The Lob-nor Controversy*, a full statement of the case, summarising Kozlov's pamphlet. Among the documents relating to the controversy, Kozlov "quotes passages from the Chinese work *Si-yui-shui-dao-tsi*, published in 1823, relative to the region, and gives a reduced copy of the Chinese Map published by Dr. Georg Wegener in 1863, upon which map Richthofen and Sven Hedin based their arguments." Kozlov's final conclusions (*Geog. Jour. l.c.* pp. 657-658) are the following : "The Konchehdaria, since very remote times till the present day, has moved a long way. The spot Gherelgan may be taken as a spot of relative permanence of its bed, while the basis of its delta is a line traced from the farthest northern border of the area of salt clays surrounding the Lob-nor to the Tarim. At a later period the Koncheh-daria mostly influenced the lower Tarim, and each time a change occurred in the latter's discharge, the Koncheh took a more westward course, to the detriment of its old eastern branch (Ilek). Always following the gradually receding humidity, the vegetable life changed too, while moving sands were taking its place, conquering more and more ground for the desert, and marking their conquest by remains of old shore-lines. . . .

" The facts noticed by Sven Hedin have thus another meaning—the desert to the east of the lakes, which he discovered, was formed, not by Lob-nor, which is situated 1° southwards, but by the Koncheh-daria, in its unremitted deflection to the west. The old bed Ilek, lake-shaped in places, and having a belt of salt lagoons and swamps along its eastern shores, represents remains of waters belonging, not to Lob-nor, but to the shifting river which has abandoned this old bed.

" These facts and explanations refute the second point of the arguments which were brought forward by Sven Hedin in favour of his hypothesis, asserting the existence of some other Lob-nor.

" I accept the third point of his objections, namely, that the grandfathers of the present inhabitants of the Lob-nor lived by a lake whose position was more to the north of Lob-nor ; that was mentioned already by Pievtsov, and the lake was Uchu-Kul.

" Why Marco Polo never mentioned the Lob-nor, I leave to more competent persons to decide.

" The only inference which I can make from the preceding account is that the Kara-Koshun-Kul is not only the Lob-nor of my lamented teacher, N. M. Prjevalsky, but also *the ancient, the historical, and the true Lob-nor* of the Chinese geographers. So it was during the last thousand years, and so will it remain, if ' the river of time ' in its running has not effaced it from the face of the Earth."

To Kozlov's query : " Why Marco Polo never mentioned the Lob-nor, I leave to more competent persons to decide," I have little hesitation in replying that he did not mention the Lob-nor because he did not see it. From Charchan, he followed, I believe, neither Prjevalsky's nor Pievtsov's route, but the old route from Khotan to Si-ngan fu, in the old bed of the Charchan daria, above and almost parallel to the new bed, to the Tarim,—then between Sven Hedin's and Prjevalsky's lakes, and across the desert to Shachau to join the ancient Chinese road of the Han Dynasty, partly explored by M. Bonin from Shachau.

There is no doubt as to the discovery of Prjevalsky's Lob-nor, but this does not appear to be the old Chinese Lob-nor ; in fact, there may have been several lakes co-existent ; probably there was one to the east of the mass of water described by Dr. Sven Hedin, near the old route from Korla to Shachau ; there is no fixity in these waterspreads and the soil of this part of Asia, and in the course of a few years some discrepancies will naturally arise between the observations of different travellers. But as I think that Marco Polo did not see one of the Lob-nor, but travelled between them, there is no necessity to enlarge on this question, fully treated of in this note.

See besides the works mentioned above : *Nord—Tibet und Lob-nur Gebiet.* . . . herausg. von Dr. G. Wegener. Berlin, 1893. (Sep. abd. *Zeit. Ges. f. Erdk.*)—*Die Geog. wiss. Ergebnisse meiner Reisen in Zentralasien,* 1894-1897, von Dr. Sven Hedin, Gotha, J. Perthes, 1900.

Bonvalot and Prince Henri d'Orléans (*De Paris au Tonkin, à travers le Tibet inconnu,* Paris, 1892) followed this Itinerary : Semipalatinsk, Kulja, Korla, Lob-nor, Charkalyk, Altyn Tagh, almost a straight line to Tengri Nor, then to Batang, Ta Tsien lu, Ning-yuan, Yun-nan-fu, Mong-tsŭ, and Tung-King.

Bonvalot (28th October, 1889) describes Lob in this manner : " The village of Lob is situated at some distance from [the Charchan daria]; its inhabitants come to see us ; they are miserable, hungry, *étiques ;* they offer us for sale smoked fish, duck taken with *lacet.* Some small presents soon make friends of them. They apprize us that news has spread that Pievtsov, the Russian traveller, will soon arrive " (*l.c.* p. 75). From Charkalyk, Prince Henri d'Orléans and Father Dedeken visited Lob-nor (*l.c.* p. 77 *et seq.*), but it was almost dry ; the water had receded since Prjevalsky's visit, thirteen years before. The Prince says the Lob-nor he saw was not Prjevalsky's, nor was the latter's lake the mass of water on Chinese maps ; an old sorceress gave confirmation of the fact to the travellers. According to a tradition known from one generation to another, there was at this place a large inland sea without reeds, and the elders had seen in their youth large ponds ; they say that the earth impregnated with saltpetre absorbs the water. The Prince says, according to tradition, *Lob* is a local name meaning " wild animals," and it was given to the country at the time it was crossed by Kalmuk caravans ; they added to the name *Lob* the Mongol word *Nor* (Great Lake). The travellers (p. 109) note that in fact the name Lob-nor does not apply to a Lake, but to the whole marshy part of the country watered by the Tarim, from the village of Lob to end of the river.

The Pievtsov expedition " visited the Lob-nor (2650 feet) and the Tarim, whose proper name is Yarkend-daria (*tarim* means ' a tilled field' in Kashgarian). The lake is rapidly drying up, and a very old man, 110 years old, whom Pievtsov spoke to (his son, 52 years old, was the only one who could understand the old man), said that he would not have recognized the land if he had been absent all this time. Ninety years ago there was only a narrow strip of rushes in the south-west part of the lake, and the Yarkend-daria entered it 2½ miles to the west of its present mouth, where now stands the village of Abdal. The lake was then much deeper, and several villages, now abandoned, stood on its shores. There was also much more fish, and otters, which used to live there, but have long since disappeared. As to the Yarkend-daria, tradition says that two hundred years ago it used to enter another smaller lake, Uchukul, which was connected by a channel with the Lob-nor. This old bed, named Shirga-chapkan, can still be traced by the trees which grew along it. The greater previous extension of the Lob-nor is also confirmed by the freshwater molluscs (*Limnaea uricularia,* var. *ventricosa, L. stagnalis, L. peregra,* and *Planorbis sibiricus*), which are found at a distance from its present banks. Another lake, 400 miles in circumference, Kara-boyön (*black isthmus*), lies, as is known, 27 miles to the south-west of Lob-nor. To the east of the lake, a salt desert stretches for a seven days' march, and further on begin the Kum-tagh sands, where wild camels live." (*Geog. Jour.* IX. 1897, p. 552.)

Grenard (III. pp. 194-195) discusses the Lob-nor question and the formation of four new lakes by the Koncheh-daria called by the natives beginning at the north ; Kara Kul, Tayek Kul, Sugut Kul, Tokum Kul. He does not accept Baron v. Richthofen's theory, and believes that the old Lob is the lake seen by Prjevalsky.

He says (p. 149) : "Lop must be looked for on the actual road from Charchan to Char-kalyk. Ouash Shahri, five days from Charchan, and where small ruins are to be found, corresponds well to the position of Lop according to Marco Polo, a few degrees of the compass near. But the stream which passes at this spot could never be important enough for the wants of a considerable centre of habitation and the ruins of Ouash Shahri are more of a hamlet than of a town. Moreover, Lop was certainly the meeting point of the roads of Kashgar, Urumtsi, Shachau, L'Hasa, and Khotan, and it is to this fact that this town, situated in a very poor country, owed its relative importance. Now, it is impossible that these roads crossed at Ouash Shahri. I believe that Lop was built on the site of Charkalyk itself. The Venetian traveller gives five days' journey between Charchan and Lop, whilst Charkalyk is really seven days from Charchan ; but the objection does not appear sufficient to me : Marco Polo may well have made a mistake of two days." (III. pp. 149-150.)

The Chinese Governor of Urumtsi found some years ago to the north-west of the Lob-nor, on the banks of the Tarim, and within five days of Charkalyk, a town bearing the same name, though not on the same site as the Lop of Marco Polo.—H. C.]

Note 2.—"The waste and desert places of the Earth are, so to speak, the characters which sin has visibly impressed on the outward creation ; its signs and symbols there. . . . Out of a true feeling of this, men have ever conceived of the Wilderness as the haunt of evil spirits. In the old Persian religion Ahriman and his evil Spirits inhabit the steppes and wastes of Turan, to the north of the happy Iran, which stands under the dominion of Ormuzd ; exactly as with the Egyptians, the evil Typhon is the Lord of the Libyan sand-wastes, and Osiris of the fertile Egypt." (*Archbp. Trench, Studies in the Gospels*, p. 7.) Terror, and the seeming absence of a beneficent Providence, are suggestions of the Desert which must have led men to associate it with evil spirits, rather than the figure with which this passage begins ; no spontaneous conception surely, however appropriate as a moral image.

"According to the belief of the nations of Central Asia," says I. J. Schmidt, "the earth and its interior, as well as the encompassing atmosphere, are filled with Spiritual Beings, which exercise an influence, partly beneficent, partly malignant, on the whole of organic and inorganic nature. . . . . Especially are Deserts and other wild or uninhabited tracts, or regions in which the influences of nature are displayed on a gigantic and terrible scale, regarded as the chief abode or rendezvous of evil Spirits. . . . And hence the steppes of Turan, and in particular the great sandy Desert of Gobi have been looked on as the dwelling-place of malignant beings, from days of hoar antiquity."

The Chinese historian Ma Twan-lin informs us that there were two roads from China into the Uighúr country (towards Karashahr). The longest but easiest road was by Kamul. The other was much shorter, and apparently corresponded, as far as Lop, to that described in this chapter. "By this you have to cross a plain of sand, extending for more than 100 leagues. You see nothing in any direction but the sky and the sands, without the slightest trace of a road ; and travellers find nothing to guide them but the bones of men and beasts and the droppings of camels. During the passage of this wilderness you hear sounds, sometimes of singing, sometimes of wailing ; and it has often happened that travellers going aside to see what those sounds might be have strayed from their course and been entirely lost ; for they were voices of spirits and goblins. 'Tis for these reasons that travellers and merchants often prefer the much longer route by Kamul." (*Visdelou*, p. 139.)

"In the Desert" (this same desert), says Fa-hian, "there are a great many evil demons ; there are also sirocco winds, which kill all who encounter them. There are no birds or beasts to be seen ; but so far as the eye can reach, the route is marked out by the bleached bones of men who have perished in the attempt to cross."

["The Lew-sha was the subject of various most exaggerated stories. We find more trustworthy accounts of it in the *Chow shu ;* thus it is mentioned in that history, that there sometimes arises in this desert a 'burning wind,' pernicious to men and cattle ; in such cases the old camels of the caravan, having a presentiment of its approach, flock shrieking to one place, lie down on the ground and hide their heads in the sand. On this signal, the travellers also lie down, close nose and mouth, and remain in this position until the hurricane abates. Unless these precautions are taken, men and beasts inevitably perish." (*Palladius, l.c.* p. 4.)

A friend writes to me that he thinks that the accounts of strange noises in the desert would find a remarkable corroboration in the narratives of travellers through the central desert of Australia. They conjecture that they are caused by the sudden falling of cliffs of sand as the temperature changes at night time.—H. C.]

Hiuen Tsang, in his passage of the Desert, both outward and homeward, speaks of visual illusions ; such as visions of troops marching and halting with gleaming arms and waving banners, constantly shifting, vanishing, and reappearing, "imagery created by demons." A voice behind him calls, "Fear not ! fear not !" Troubled by these fantasies on one occasion, he prays to Kwan-yin (a Buddhist divinity) ; still he could not entirely get rid of them ; but as soon as he had pronounced a few words from the *Prajna* (a holy book), they vanished in the twinkling of an eye.

These Goblins are not peculiar to the Gobi, though that appears to be their most favoured haunt. The awe of the vast and solitary Desert raises them in all similar localities. Pliny speaks of the phantoms that appear and vanish in the deserts of Africa ; Aethicus, the early Christian cosmographer, speaks, though incredulous, of the stories that were told of the voices of singers and revellers in the desert ; Mas'údi tells of the *Ghúls*, which in the deserts appear to travellers by night and in lonely hours ; the traveller, taking them for comrades, follows and is led astray. But the wise revile them and the Ghúls vanish. Thus also Apollonius of Tyana and his companions, in a desert near the Indus by moonlight, see an *Empusa* or Ghúl taking many forms. They revile it, and it goes off uttering shrill cries. Mas'údi also speaks of the mysterious voices heard by lone wayfarers in the Desert, and he gives a rational explanation of them. Ibn Batuta relates a like legend of the Western Sahara : "If the messenger be solitary, the demons sport with him and fascinate him, so that he strays from his course and perishes." The Afghan and Persian wildernesses also have their *Ghúl-i-Bedban* or Goblin of the Waste, a gigantic and fearful spectre which devours travellers ; and even the Gael of the West Highlands have the *Direach Ghlinn Eitidh*, the Desert Creature of Glen Eiti, which, one-handed, one-eyed, one-legged, seems exactly to answer to the Arabian Nesnás or *Empusa*. Nicolò Conti in the Chaldaean desert is aroused at midnight by a great noise, and sees a vast multitude pass by. The merchants tell him that these are demons who are in the habit of traversing the deserts. (*Schmidt's San. Setzen*, p. 352 ; *V. et V. de H. T.* 23, 28, 289 ; *Pliny*, VII. 2 ; *Philostratus*, Bk. II. ch. iv. ; *Prairies d'Or*, III. 315, 324 ; *Beale's Fahian ; Campbell's Popular Tales of the W. Highlands*, IV. 326 ; *I. B.* IV. 382 ; *Elphinstone*, I. 291 ; *Chodzko's Pop. Poetry of Persia*, p. 48 ; *Conti*, p. 4 ; *Forsyth, J. R. G. S.* XLVII. 1877, p. 4.)

The sound of musical instruments, chiefly of drums, is a phenomenon of another class, and is really produced in certain situations among sandhills when the sand is disturbed. [See *supra*.] A very striking account of a phenomenon of this kind regarded as supernatural is given by Friar Odoric, whose experience I fancy I have traced to the *Reg Ruwán* or "Flowing Sand" north of Kabul. Besides this celebrated example, which has been described also by the Emperor Baber, I have noted that equally well-known one of the *Jibal Nakús*, or "Hill of the Bell," in the Sinai Desert ; Wadi Hamade, in the vicinity of the same Desert ; the *Jibal-ul-Thabúl*, or "Hill of the Drums," between Medina and Mecca ; one on the Island of Eigg, in the Hebrides, discovered by Hugh Miller ; one among the Medanos or Sandhills of Arequipa, described to me by Mr. C. Markham ; the Bramador or rumbling mountain of Tarapaca ; one in hills between the Ulba and the Irtish, in the vicinity of the Altai, called the Almanac Hills, because the sounds are supposed to prognosticate weather-changes ; and a remarkable example near Kolberg on the shore of Pomerania. A Chinese narrative of the 10th century mentions the phenomenon as known near Kwachau, on the eastern border of the Lop Desert, under the name of the "Singing Sands" ; and Sir F. Goldsmid has recently made us acquainted with a second *Reg Ruwán*, on a hill near the Perso-Afghan frontier, a little to the north of Sístán.   The place is frequented in pilgrimage.   (See *Cathay*, pp. ccxliv. 156, 398; *Ritter*, II. 204; *Aus der Natur*, Leipzig, No. 47 [of 1868], p. 752; *Rémusat, H. de Khotan*, p. 74; *Proc. R. G. S.* XVII. 91.)

NOTE 3.—[We learn from Joseph Martin, quoted by Grenard, p. 170 (who met this unfortunate French traveller at Khotan, on his way from Peking to Marghelan, where he died), that from Shachau to Abdal, on the Lob-nor, there are twelve days of desert, sandy only during the first two days, stony afterwards. Occasionally a little grass is to be found for the camels; water is to be found everywhere. M. Bonin went from Shachau to the north-west towards the Kara-nor, then to the west, but lack of water compelled him to go back to Shachau. Along this road, every five *lis*, are to be found towers built with clay, and about 30 feet high, abandoned by the Chinese, who do not seem to have kept a remembrance of them in the country; this route seems to be a continuation of the Kan Suh Imperial highway. A wall now destroyed connected these towers together. " There is no doubt," writes M. Bonin, "that all these remains are those of the great route, vainly sought after till now, which, under the Han Dynasty, ran to China through Bactria, Pamir, Eastern Turkestan, the Desert of Gobi, and Kan Suh: it is in part the route followed by Marco Polo, when he went from Charchan to Shachau, by the city of Lob." The route of the Han has been also looked for, more to the south, and it was believed that it was the same as that of the Astyn Tagh, followed by Mr. Littledale in 1893, who travelled one month from Abdal (Lob-nor) to Shachau; M. Bonin, who explored also this route, and was twenty-three days from Shachau to Lob-nor, says it could not be a commercial road. Dr. Sven Hedin saw four or five towers eastward of the junction of the Tarim and the Koncheh-daria; it may possibly have been another part of the road seen by M. Bonin. (See *La Géographie*, 15th March, 1901, p. 173.)—H. C.]

CHAPTER XL.

CONCERNING THE GREAT PROVINCE OF TANGUT.

AFTER you have travelled thirty days through the Desert, as I have described, you come to a city called SACHIU, lying between north-east and east; it belongs to the Great Kaan, and is in a province called TANGUT.[1] The people are for the most part Idolaters, but there are also some Nestorian Christians and some Saracens. The Idolaters have a peculiar language, and are no traders, but live by their agriculture.[2] They have a great many abbeys and minsters full of idols of sundry fashions, to which they pay great honour and reverence, worshipping them and sacrificing to them with much ado. For example, such as have children will feed up a sheep in honour of the idol, and at the New Year, or on the day of the Idol's Feast, they will take their children and the sheep along with them into the presence of the idol with great ceremony. Then they will have the sheep slaughtered and cooked, and again present it before the idol with like reverence, and leave it there before him, whilst they are reciting the offices of their worship and their prayers for the idol's blessing on their children. And, if you will believe them, the idol feeds on the meat that is set before it! After these ceremonies they take up the flesh and carry it home, and call together all their kindred to eat it with them in great festivity [the idol-priests receiving for their portion the head, feet, entrails, and skin, with some part of the meat]. After they have eaten, they collect the bones that are left and store them carefully in a hutch.[3]

And you must know that all the Idolaters in the world burn their dead. And when they are going to carry a body to the burning, the kinsfolk build a wooden house on the way to the spot, and drape it with cloths of silk and gold. When the body is going past this building they call a halt and set before it wine and meat and other eatables; and this they do with the assurance that the defunct will be received with the like attentions in the other world. All the minstrelsy in the town goes playing before the body; and when it reaches the burning-place the kinsfolk are prepared with figures cut out of parchment and paper in the shape of men and horses and camels, and also with round pieces of paper like gold coins, and all these they burn along with the corpse. For they say that in the other world the defunct will be provided with slaves and cattle and money, just in proportion to the amount of such pieces of paper that has been burnt along with him.[4]

But they never burn their dead until they have [sent

for the astrologers, and told them the year, the day, and the hour of the deceased person's birth, and when the astrologers have ascertained under what constellation, planet, and sign he was born, they declare the day on which, by the rules of their art, he ought to be burnt]. And till that day arrive they keep the body, so that 'tis sometimes a matter of six months, more or less, before it comes to be burnt.[5]

Now the way they keep the body in the house is this : They make a coffin first of a good span in thickness, very carefully joined and daintily painted. This they fill up with camphor and spices, to keep off corruption [stopping the joints with pitch and lime], and then they cover it with a fine cloth. Every day as long as the body is kept, they set a table before the dead covered with food ; and they will have it that the soul comes and eats and drinks : wherefore they leave the food there as long as would be necessary in order that one should partake. Thus they do daily. And worse still ! Sometimes those soothsayers shall tell them that 'tis not good luck to carry out the corpse by the door, so they have to break a hole in the wall, and to draw it out that way when it is taken to the burning.[6] And these, I assure you, are the practices of all the Idolaters of those countries.

However, we will quit this subject, and I will tell you of another city which lies towards the north-west at the extremity of the desert.

NOTE I.—[The Natives of this country were called by the Chinese *T'ang-hiang*, and by the Mongols *T'angu* or *T'ang-wu*, and with the plural suffix *Tangut*. The kingdom of Tangut, or in Chinese, *Si Hia* (Western Hia), or *Ho si* (West of the Yellow River), was declared independent in 982 by Li Chi Ch'ien, who had the dynastic title or *Miao Hao* of Tai Tsu. "The rulers of Tangut," says Dr. Bushell, "were scions of the Toba race, who reigned over North China as the Wei Dynasty (A.D. 386-557), as well as in some of the minor dynasties which succeeded. Claiming descent from the ancient Chinese Hsia Dynasty of the second millennium B.C., they adopted the title of *Ta Hsia* ('Great Hsia'), and the dynasty is generally called by the Chinese Hsi Hsia, or Western Hsia." This is a list of the Tangut sovereigns, with the date of their accession to the throne : Tai Tsu (982), Tai Tsung (1002), Ching Tsung (1032), Yi Tsung (1049), Hui Tsung (1068), Ch'ung Tsung (1087), Jen Tsung (1140), Huan Tsung (1194), Hsiang Tsung (1206), Shên Tsung (1213), Hien Tsung (1223), Mo Chu (1227). In fact, the real founder of the Dynasty was Li Yuan-hao, who conquered in 1031, the cities of Kanchau and Suhchau from the Uighúr Turks, declaring himself independent in 1032, and who adopted in 1036 a special-script of which we spoke when mentioning the archway at Kiuyung Kwan. His capital was Hia chau, now Ning hia, on the Yellow River. Chinghiz invaded Tangut three times, in 1206, 1217, and at last in 1225 ; the final struggle took place the following year, when Kanchau, Liangchau, and Suhchau fell into the hands of the Mongols. After the death of Chinghiz (1227), the last ruler of Tangut, Li H'ien, who surrendered the same year to Okkodaï, son of the conqueror, was killed. The dominions of Tangut in the middle of the 11th century, according to the *Si Hia Chi Shih Pên Mo*, quoted by Dr. Bushell, "were bounded, according to the map, by the Sung Empire on the south and east, by the Liao (Khitan) on the north-east, the Tartars (Tata) on the north, the Uighúr Turks (Hui-hu) on the west, and the Tibetans on the south-west. The Alashan Mountains stretch along the northern frontier, and the western extends to the Jade Gate (Yü Mên Kwan) on the border of the Desert of Gobi." Under the Mongol Dynasty, Kan Suh was the official name of one of the twelve provinces of the Empire, and the popular name was Tangut.

(Dr. S. W. Bushell : *Inscriptions in the Juchen and Allied Scripts* and *The Hsi Hsia Dynasty of Tangut.* See above, p. 29.)

" The word Tangutan applied by the Chinese and by Colonel Prjevalsky to a Tibetan-speaking people around the Koko-nor has been explained to me in a variety of ways by native Tangutans. A very learned lama from the Gserdkog monastery, south-east of the Koko-nor, told me that Tangutan, Amdoans, and Sifan were interchange-able terms, but I fear his geographical knowledge was a little vague. The following explanation of the term Tangut is taken from the *Hsi-tsang-fu.* ' The Tangutans are descendants of the *Tang-tu-chüeh.* The origin of this name is as follows : In early days, the Tangutans lived in the Central Asian Chin-shan, where they were workers of iron. They made a model of the Chin-shan, which, in shape, resembled an iron helmet. Now, in their language, " iron helmet " is *Tang-küeh*, hence the name of the country. To the present day, the Tangutans of the Koko-nor wear a hat shaped like a pot, high crowned and narrow, rimmed with red fringe sewn on it, so that it looks like an iron helmet, and this is a proof of [the accuracy of the derivation].' Although the proof is not very satisfactory, it is as good as we are often offered by authors with greater pretension to learning.

" If I remember rightly, Prjevalsky derives the name from two words meaning ' black tents.' " (*W. W. Rockhill, China Br. R. As. Soc.*, XX. pp. 278-279.)

" Chinese authorities tell us that the name [Tangut] was originally borne by a people living in the Altaï, and that the word is Turkish. . . . The population of Tangut was a mixture of Tibetans, Turks, Uighúrs, Tukuhuns, Chinese, etc." (*Rockhill, Rubruck*, p. 150, note.—H. C.]

Sachiu is SHACHAU, " Sand-district," an outpost of China Proper, at the eastern verge of the worst part of the Sandy Desert. It is recorded to have been fortified in the 1st century as a barrier against the Hiongnu.

[The name of Shachau dates from A.D. 622, when it was founded by the first emperor of the T'ang Dynasty. Formerly, Shachau was one of the Chinese colonies established by the Han, at the expense of the Hiongnu ; it was called T'ung hoang (B.C. 111), a name still given to Shachau ; the other colonies were Kiu-kaan (Suhchau, B.C. 121) and Chang-yé (Kanchau, B.C. 111). (See *Bretschneider, Med. Res.* II. 18.)

" Sha-chow, the present *Tun-hwang-hien* (a few *li* east of the ancient town). . . . In 1820, or about that time, an attempt was made to re-establish the ancient direct way between Sha-chow and Khotan. With this object in view, an exploring party of ten men was sent from Khotan towards Sha-chow ; this party wandered in the desert over a month, and found neither dwellings nor roads, but pastures and watei everywhere. M. Polo omits to mention a remarkable place at Sha-chow, a sandy hillock (a short distance south of this town) known under the name of *Ming-sha shan* —the ' rumbling sandhill.' The sand, in rolling down the hill, produces a particular sound, similar to that of distant thunder. In M. Polo's time (1292), Khubilaï removed the inhabitants of Sha-chow to the interior of China ; fearing, probably, the aggression of the seditious princes ; and his successor, in 1303, placed there a garrison of ten thousand men." (*Palladius, l.c.* p. 5.)

"Sha-chau is one of the best oases of Central Asia. It is situated at the foot of the Nan-shan range, at a height of 3700 feet above the sea, and occupies an area of about 200 square miles, the whole of which is thickly inhabited by Chinese. Sha-chau is interesting as the meeting-place of three expeditions started independently from Russia, India, and China. Just two months before Prjevalsky reached this town, it was visited by Count Szechényi [April, 1879], and eighteen months afterwards Pundit A-k, whose report of it agrees fairly well with that of our traveller, also stayed here. Both Prejevalsky and Szechényi remark on some curious caves in a valley near Sha-chau containing Buddhistic clay idols.* These caves were in Marco Polo's time the resort of numerous worshippers, and are said to date back to the Han Dynasty." (*Prejevalsky's Journeys* . . . by E. Delmar Morgan, *Proc. R. G. S.* IX. 1887, pp. 217-218.)—H. C.]

(*Ritter*, II. 205 ; *Neumann*, p. 616 ; *Cathay*, 269, 274 ; *Erdmann*, 155 ; *Erman*, II. 267 ; *Mag. Asiat.* II. 213.)

NOTE 2.—By *Idolaters*, Polo here means Buddhists, as generally. We do not know whether the Buddhism here was a recent introduction from Tibet, or a relic of the old Buddhism of Khotan and other Central Asian kingdoms, but most probably it was the former, and the "peculiar language" ascribed to them may have been, as Neumann supposes, Tibetan. This language in modern Mongolia answers to the Latin of the Mass Book, indeed with a curious exactness, for in both cases the holy tongue is not that of the original propagators of the respective religions, but that of the hierarchy which has assumed their government. In the Lamaitic convents of China and Manchuria also the Tibetan only is used in worship, except at one privileged temple at Peking. (*Koeppen*, II. 288.) The language intended by Polo may, however, have been a Chinese dialect. (See notes 1 and 4.) The Nestorians must have been tolerably numerous in Tangut, for it formed a metropolitan province of their Church.

NOTE 3.—A practice resembling this is mentioned by Pallas as existing among the Buddhist Kalmaks, a relic of their old Shaman superstitions, which the Lamas profess to decry, but sometimes take part in. "Rich Kalmaks select from their flock a ram for dedication, which gets the name of *Tengri Tockho*, ' Heaven's Ram.' It must be a white one with a yellow head. He must never be shorn or sold, but when he gets old, and the owner chooses to dedicate a fresh one, then the old one must be sacrificed. This is usually done in autumn, when the sheep are fattest, and the neighbours are called together to eat the sacrifice. A fortunate day is selected, and the ram is slaughtered amid the cries of the sorcerer directed towards the sunrise, and the diligent sprinkling of milk for the benefit of the Spirits of the Air. The flesh is eaten, but the skeleton with a part of the fat is burnt on a turf altar erected on four pillars of an ell and a half high, and the skin, with the head and feet, is then hung up in the way practised by the Buraets." (*Sammlungen*, II. 346.)

NOTE 4.—Several of the customs of Tangut mentioned in this chapter are essentially Chinese, and are perhaps introduced here because it was on entering Tangut that the traveller first came in contact with Chinese peculiarities. This is true of the manner of forming coffins, and keeping them with the body in the house, serving food

---

* M. Bonin visited in 1899 these caves which he calls "Grottoes of Thousand Buddhas" (*Tsien Fo tung*). (*La Géographie*, 15th March, 1901, p. 171.) He found a stèle dated 1348, bearing a Buddhist prayer in six different scripts like the inscription at Kiu Yung Kwan. (*Rev. Hist. des Religions*, 1901, p. 393.)—H. C.

before the coffin whilst it is so kept, the burning of paper and papier-maché figures of slaves, horses, etc., at the tomb. Chinese settlers were very numerous at Shachau and the neighbouring Kwachau, even in the 10th century. (*Ritter*, II. 213.) ["Keeping a body unburied for a considerable time is called *khñg koan*, 'to conceal or store away a coffin,' or *thîng koan*, 'to detain a coffin.' It is, of course, a matter of necessity in such cases to have the cracks and fissures, and especially the seam where the case and the lid join, hermetically caulked. This is done by means of a mixture of chunam and oil. The seams, sometimes even the whole coffin, are pasted over with linen, and finally everything is varnished black, or, in case of a mandarin of rank, red. In process of time, the varnishing is repeated as many times as the family think desirable or necessary. And in order to protect the coffin still better against dust and moisture, it is generally covered with sheets of oiled paper, over which comes a white pall." (*De Groot*, I. 106.)—H. C.] Even as regards the South of China many of the circumstances mentioned here are strictly applicable, as may be seen in *Doolittle's Social Life of the Chinese*. (See, for example, p. 135 ; also *Astley*, IV. 93-95, or Marsden's quotations from *Duhalde*.) The custom of burning the dead has been for several centuries disused in China, but we shall see hereafter that Polo represents it as general in his time. On the custom of burning gilt paper in the form of gold coin, as well as of paper clothing, paper houses, furniture, slaves, etc., see also *Medhurst*, p. 213, and *Kidd*, 177-178. No one who has read Père Huc will forget his ludicrous account of the Lama's charitable distribution of paper horses for the good of disabled travellers. The manufacture of mock money is a large business in Chinese cities. In Fuchau there are more than thirty large establishments where it is kept for sale. (*Doolittle*, 541.) [The Chinese believe that sheets of paper, partly tinned over on one side, are, "according to the prevailing conviction, turned by the process of fire into real silver currency available in the world of darkness, and sent there through the smoke to the soul ; they are called *gûn-tsoá*, 'silver paper.' Most families prefer to previously fold every sheet in the shape of a hollow ingot, a 'silver ingot,' *gûn-khò*, as they call it. This requires a great amount of labour and time, but increases the value of the treasure immensely." (*De Groot*, I. 25.) "Presenting paper money when paying a visit of condolence is a custom firmly established, and accordingly compʹied with by everybody with great strictness . . . . The paper is designed for the equipment of the coffin, and, accordingly, always denoted by the term *koan-thaô-tsoá*, 'coffin paper.' But as the receptacle of the dead is, of course, not spacious enough to hold the whole mass offered by so many friends, it is regularly burned by lots by the side of the corpse, the ashes being carefully collected to be afterwards wrapped in paper and placed in the coffin, or at the side of the coffin, in the tomb." (*De Groot*, I. 31-32.)—H. C.] There can be little doubt that these latter customs are symbols of the ancient sacrifices of human beings and valuable property on such occasions ; so Manetho states that the Egyptians in days of yore used human sacrifices, but a certain King Amosis abolished them and substituted images of wax. Even when the present Manchu Dynasty first occupied the throne of China, they still retained the practice of human sacrifice. At the death of Kanghi's mother, however, in 1718, when four young girls offered themselves for sacrifice on the tomb of their mistress, the emperor would not allow it, and prohibited for the future the sacrifice of life or the destruction of valuables on such occasions. (*Deguignes, Voy.* I. 304.)

NOTE 5.—Even among the Tibetans and Mongols burning is only one of the modes of disposing of the dead. "They sometimes bury their dead : often they leave them exposed in their coffins, or cover them with stones, paying regard to the sign under which the deceased was born, his age, the day and hour of his death, which determine the mode in which he is to be interred (or otherwise disposed of). For this purpose they consult some books which are explained to them by the Lamas." (*Timk.* II. 312.) The extraordinary and complex absurdities of the books in question are given in detail by Pallas, and curiously illustrate the paragraph in the text. (See *Sammlungen*, II.

254 *seqq.*) ["The first seven days, including that on which the demise has taken place, are generally deemed to be lucky for the burial, especially the odd ones. But when they have elapsed, it becomes requisite to apply to a day-professor. . . . . The popular almanac which chiefly wields sway in Amoy and the surrounding country, regularly stigmatises a certain number of days as *ting-sng jít:* ' days of reduplication of death,' because encoffining or burying a dead person on such a day will entail another loss in the family shortly afterwards." (*De Groot,* I. 103, 99-100.)—H. C.]

NOTE 6.—The Chinese have also, according to Duhalde, a custom of making a new opening in the wall of a house by which to carry out the dead; and in their prisons a special hole in the wall is provided for this office. This same custom exists among the Esquimaux, as well as, according to Sonnerat, in Southern India, and it used to exist in certain parts both of Holland and of Central Italy. In the " clean village of Broek," near Amsterdam, those special doors may still be seen. And in certain towns of Umbria, such as Perugia, Assisi, and Gubbio, this opening was common, elevated some feet above the ground, and known as the " Door of the Dead."

I find in a list, printed by Liebrecht, of popular French superstitions, amounting to 479 in number, condemned by Maupas du Tour, Bishop of Evreux in 1664, the following : " When a woman lies in of a dead child, it must not be taken out by the door of the chamber but by the window, for if it were taken out by the door the woman would never lie in of any but dead children." The Samoyedes have the superstition mentioned in the text, and act exactly as Polo describes.

["The body [of the Queen of Bali, 17th century] was drawn out of a large aperture made in the wall to the right hand side of the door, in the absurd opinion of *cheating the devil,* whom these islanders believe to lie in wait in the ordinary passage." (*John Crawfurd, Hist. of the Indian Archipelago,* II. p. 245.)—H. C.]

And the Rev. Mr. Jaeschke writes to me from Lahaul, in British Tibet : "Our Lama (from Central Tibet) tells us that the owner of a house and the members of his family when they die are carried through the house-door ; but if another person dies in the house his body is removed by some other aperture, such as a window, or the smokehole in the roof, or a hole in the wall dug expressly for the purpose. Or a wooden frame is made, fitting into the doorway, and the body is then carried through ; it being considered that by this contrivance the evil consequences are escaped that might ensue, were it carried through the ordinary, and, so to say, *undisguised* house-door ! Here, in Lahaul and the neighbouring countries, we have not heard of such a custom."

(*Duhalde,* quoted by Marsden ; *Semedo,* p. 175 ; *Mr. Sala* in *N. and Q.,* 2nd S. XI. 322 ; *Lubbock,* p. 500; *Sonnerat,* I. 86; *Liebrecht's Gervasius of Tilbury,* Hanover, 1856, p. 224; *Mag. Asiat.* II. 93.)

CHAPTER XLI.

OF THE PROVINCE OF CAMUL.

CAMUL is a province which in former days was a kingdom. It contains numerous towns and villages, but the chief city bears the name of CAMUL. The province lies between the two deserts ; for on the one side is the

Great Desert of Lop, and on the other side is a small desert of three days' journey in extent.[1]   The people are all Idolaters, and have a peculiar language.   They live by the fruits of the earth, which they have in plenty, and dispose of to travellers.   They are a people who take things very easily, for they mind nothing but playing and singing, and dancing and enjoying themselves.[2]

And it is the truth that if a foreigner comes to the house of one of these people to lodge, the host is delighted, and desires his wife to put herself entirely at the guest's disposal, whilst he himself gets out of the way, and comes back no more until the stranger shall have taken his departure.   The guest may stay and enjoy the wife's society as long as he lists, whilst the husband has no shame in the matter, but indeed considers it an honour.   And all the men of this province are made wittols of by their wives in this way.[3]   The women themselves are fair and wanton.

Now it came to pass during the reign of MANGU KAAN, that as lord of this province he came to hear of this custom, and he sent forth an order commanding them under grievous penalties to do so no more [but to provide public hostelries for travellers].   And when they heard this order they were much vexed thereat.   [For about three years' space they carried it out.   But then they found that their lands were no longer fruitful, and that many mishaps befell them.]   So they collected together and prepared a grand present which they sent to their Lord, praying him graciously to let them retain the custom which they had inherited from their ancestors; for it was by reason of this usage that their gods bestowed upon them all the good things that they possessed, and without it they saw not how they could continue to exist.[4] When the Prince had heard their petition his reply was " Since ye must needs keep your shame, keep it then," and so he left them at liberty to maintain their naughty custom. And they always have kept it up, and do so still.

Now let us quit Camul, and I will tell you of another province which lies between north-west and north, and belongs to the Great Kaan.

NOTE 1.—Kamul (or Komul) does not fall into the great line of travel towards Cathay which Marco is following. His notice of it, and of the next province, forms a digression like that which he has already made to Samarkand. It appears very doubtful if Marco himself had visited it ; his father and uncle may have done so on their first journey, as one of the chief routes to Northern China from Western Asia lies through this city, and has done so for many centuries. This was the route described by Pegolotti as that of the Italian traders in the century following Polo ; it was that followed by Marignolli, by the envoys of Shah Rukh at a later date, and at a much later by Benedict Goës. The people were in Polo's time apparently Buddhist, as the Uighúrs inhabiting this region had been from an old date : in Shah Rukh's time (1420) we find a mosque and a great Buddhist Temple cheek by jowl ; whilst Ramusio's friend Hajji Mahomed (*circa* 1550) speaks of Kamul as the first Mahomedan city met with in travelling from China.

Kamul stands on an oasis carefully cultivated by aid of reservoirs for irrigation, and is noted in China for its rice and for some of its fruits, especially melons and grapes. It is still a place of some consequence, standing near the bifurcation of two great roads from China, one passing north and the other south of the Thian Shan, and it was the site of the Chinese Commissariat depôts for the garrisons to the westward. It was lost to the Chinese in 1867.

Kamul appears to have been the see of a Nestorian bishop. A Bishop of Kamul is mentioned as present at the inauguration of the Catholicos Denha in 1266. (*Russians in Cent. Asia*, 129 ; *Ritter*, II. 357 *seqq.* ; *Cathay, passim* ; *Assemani*, II. 455-456.)

[*Kamul* is the Turkish name of the province called by the Mongols *Khamil*, by the Chinese *Hami ;* the latter name is found for the first time in the *Yuen Shi*, but it is first mentioned in Chinese history in the 1st century of our Era under the name of *I-wu-lu* or *I-wu* (*Bretschneider, Med. Res.* II. p. 20) ; after the death of Chinghiz, it belonged to his son Chagataï. From the Great Wall, at the Pass of Kia Yü, to Hami there is a distance of 1470 *li*. (*C. Imbault-Huart. Le Pays de Hami ou Khamil . . . d'après les auteurs chinois, Bul. de Géog. hist. et desc.*, Paris, 1892, pp. 121-195.) The Chinese general Chang Yao was in 1877 at Hami, which had submitted in 1867 to the Athalik Ghazi, and made it the basis of his operations against the small towns of Chightam and Pidjam, and Yakúb Khan himself stationed at Turfan. The Imperial Chinese Agent in this region bears the title of *K'u lun Pan She Ta Ch'en* and resides at K'urun (Urga) ; of lesser rank are the agents (*Pan She Ta Ch'en*) of Kashgar, Kharashar, Kuché, Aksu, Khotan, and Hami. (See a description of Hami by Colonel M. S. Bell, *Proc. R. G. S.* XII. 1890, p. 213.)—H. C.]

NOTE 2.—Expressed almost in the same words is the character attributed by a Chinese writer to the people of Kuché in the same region. (*Chin. Repos.* IX. 126.) In fact, the character seems to be generally applicable to the people of East Turkestan, but sorely kept down by the rigid Islam that is now enforced. (See *Shaw, passim*, and especially the Mahrambáshi's lamentations over the jolly days that were no more, pp. 319, 376.)

Note 3.—Pauthier's text has "*sont si* honni *de leur moliers comme vous avez ouy.*" Here the Crusca has "*sono* bozzi *delle loro moglie,*" and the Lat. Geog. "*sunt* bezzi *de suis uxoribus.*" The Crusca Vocab. has inserted *bozzo* with the meaning we have given, on the strength of this passage. It occurs also in Dante (*Paradiso,* XIX. 137), in the general sense of *disgraced.*

The shameful custom here spoken of is ascribed by Polo also to a province of Eastern Tibet, and by popular report in modern times to the Hazaras of the Hindu-Kush, a people of Mongolian blood, as well as to certain nomad tribes of Persia, to say nothing of the like accusation against our own ancestors which has been drawn from Laonicus Chalcondylas. The old Arab traveller Ibn Muhalhal (10th century) also relates the same of the Hazlakh (probably *Kharlikh*) Turks : "Ducis alicujus uxor vel filia vel soror, quum mercatorum agmen in terram venit, eos adit, eorumque lustrat faciem. Quorum siquis earum afficit admiratione hunc domum suam ducit, eumque apud se hospitio excipit, eique benigne facit. Atque marito suo et filio fratrique rerum necessariarum curam demandat ; neque dum hospes apud eam habitat, nisi necessarium est, maritus eam adit." A like custom prevails among the Chukchis and Koryaks in the vicinity of Kamtchatka. (*Elphinstone's Caubul ; Wood,* p. 201 ; *Burnes,* who discredits, II. 153, III. 195 ; *Laon. Chalcond.* 1650, pp. 48-49 ; *Kurd de Schloezer,* p. 13 ; *Erman,* II. 530.)

[" It is remarkable that the Chinese author, *Hung Hao,* who lived a century before M. Polo, makes mention in his memoirs nearly in the same words of this custom of the Uighúrs, with whom he became acquainted during his captivity in the kingdom of the *Kin.* According to the chronicle of the Tangut kingdom of Si-hia, Hami was the nursery of Buddhism in Si-hia, and provided this kingdom with Buddhist books and monks." (*Palladius, l.c.* p. 6.)—H. C.]

Note 4.—So the Jewish rabble to Jeremiah : "Since we left off to burn incense to the Queen of Heaven, and to pour out drink-offerings to her, we have wanted all things, and have been consumed by the sword and by famine." (*Jerem.* xliv. 18.)

CHAPTER XLII.

Of the Province of Chingintalas.

Chingintalas is also a province at the verge of the Desert, and lying between north-west and north. It has an extent of sixteen days' journey, and belongs to the Great Kaan, and contains numerous towns and villages. There are three different races of people in it—Idolaters, Saracens, and some Nestorian Christians.[1] At the northern extremity of this province there is a mountain in which are excellent veins of steel and ondanique.[2] And you must know that in the same mountain there is a vein of the substance from which Salamander is made.[3]

For the real truth is that the Salamander is no beast, as they allege in our part of the world, but is a substance found in the earth; and I will tell you about it.

Everybody must be aware that it can be no animal's nature to live in fire, seeing that every animal is composed of all the four elements.[4] Now I, Marco Polo, had a Turkish acquaintance of the name of Zurficar, and he was a very clever fellow. And this Turk related to Messer Marco Polo how he had lived three years in that region on behalf of the Great Kaan, in order to procure those Salamanders for him.[5] He said that the way they got them was by digging in that mountain till they found a certain vein. The substance of this vein was then taken and crushed, and when so treated it divides as it were into fibres of wool, which they set forth to dry. When dry, these fibres were pounded in a great copper mortar, and then washed, so as to remove all the earth and to leave only the fibres like fibres of wool. These were then spun, and made into napkins. When first made these napkins are not very white, but by putting them into the fire for a while they come out as white as snow. And so again whenever they become dirty they are bleached by being put in the fire.

Now this, and nought else, is the truth about the Salamander, and the people of the country all say the same. Any other account of the matter is fabulous nonsense. And I may add that they have at Rome a napkin of this stuff, which the Grand Kaan sent to the Pope to make a wrapper for the Holy Sudarium of Jesus Christ.[6]

We will now quit this subject, and I will proceed with my account of the countries lying in the direction between north-east and east.

Note 1.—The identification of this province is a difficulty, because the geographical definition is vague, and the name assigned to it has not been traced in other authors. It is said to lie *between north-west and north,* whilst Kamul was said to lie *towards the north-west.* The account of both provinces forms a digression, as is clear from the last words of the present chapter, where the traveller returns to take up his regular route " in the direction between north-east and east." The point from which he digresses, and to which he reverts, is Shachau, and 'tis presumably from Shachau that he assigns bearings to the two provinces forming the subject of the digression. Hence, as Kamul lies *vers maistre, i.e.* north-west, and Chingintalas *entre maistre et tramontaine, i.e.* nor'-nor'-west, Chingintalas can scarcely lie due west of Kamul, as M. Pauthier would place it, in identifying it with an obscure place called *Saiyintala,* in the territory of Urumtsi. Moreover, the province is said to belong to the Great Kaan. Now, *Urumtsi* or Bishbalik seems to have belonged, not to the Great Kaan, but to the empire of Chagatai, or possibly at this time to Kaidu. Rashiduddin, speaking of the frontier between the Kaan and Kaidu, says :—" From point to point are posted bodies of troops under the orders of princes of the blood or other generals, and they often come to blows with the troops of Kaidu. Five of these are cantoned on the verge of the Desert ; a sixth in Tangut, near Chagan-Nor (White Lake) ; a seventh in the vicinity of Karakhoja, a city of the Uighúrs, which lies between the two States, and maintains neutrality."

Karakhoja, this neutral town, is near Turfan, to the south-east of Urumtsi, which thus would lie *without* the Kaan's boundary ; Kamul and the country north-east of it would lie within it. This country, to the north and north-east of Kamul, has remained till quite recently unexplored by any modern traveller, unless we put faith in Mr. Atkinson's somewhat hazy narrative. But it is here that I would seek for Chingintalas.

Several possible explanations of this name have suggested themselves or been suggested to me. I will mention two.

1. Klaproth states that the Mongols applied to Tibet the name of *Baron-tala,* signifying the " Right Side," *i.e.* the south-west or south quarter, whilst Mongolia was called *Dzöhn* (or *Dzegun*) *Tala, i.e.* the " Left," or north-east side. It is possible that *Chigin-talas* might represent *Dzegun Tala* in some like application. The etymology of *Dzungaria,* a name which in modern times covers the territory of which we are speaking, is similar.

2. Professor Vámbéry thinks that it is probably *Chingin Tala,* " The Vast Plain." But nothing can be absolutely satisfactory in such a case except historical evidence of the application of the name.

I have left the identity of this name undecided, though pointing to the general position of the region so-called by Marco, as indicated by the vicinity of the Tangnu-Ola Mountains (p. 215). A passage in the Journey of the Taouist Doctor, Changchun, as translated by Dr. Bretschneider (*Chinese Recorder and Miss. Journ.,* Shanghai, Sept.-Oct., 1874, p. 258), suggests to me the strong probability that it may be the *Kem-kém-jút* of Rashiduddin, called by the Chinese teacher *Kien-kien*-chau.

Rashiduddin couples the territory of the Kirghiz with Kemkemjút, but defines the country embracing both with some exactness: " On one side (south-east ?), it bordered on the Mongol country ; on a second (north-east ?), it was bounded by the Selenga; on a third (north), by the ' great river called Angara, which flows on the confines of Ibir-Sibir ' (*i.e.* of Siberia) ; on a fourth side by the territory of the Naimans. This great country contained *many towns and villages,* as well as many nomad inhabitants." Dr. Bretschneider's Chinese Traveller speaks of it as a country where *good iron was found,* where (grey) squirrels abounded, and wheat was cultivated. Other notices quoted by him show that it lay to the south-east of the Kirghiz country, and had its name from the *Kien* or *Ken* R. (*i.e.* the Upper Yenisei).

The name (*Kienkien*), the general direction, the existence of good iron ("steel and ondanique "), the many towns and villages in a position where we should little look for such an indication, all point to the identity of this region with the Chingintalas of our text.   The only alteration called for in the Itinerary Map (No. IV.) would be to spell the name *Hinkin*, or *Ghinghin* (as it *is* in the Geographic Text), and to shift it a very little further to the north.

(See *Chingin* in *Kovalevski's Mongol Dict.*, No. 2134 ; and for *Baron-tala*, etc., see *Della Penna, Breve Notizia del Regno del Thibet*, with Klaproth's notes, p. 6; *D'Avezac*, p. 568 ; *Relation* prefixed to D'Anville's Atlas, p. 11 ; *Alphabetum Tibetanum*, 454 ; and *Kircher, China Illustrata*, p. 65.)

Since the first edition was published, Mr. Ney Elias has traversed the region in question from east to west ; and I learn from him that at Kobdo he found the most usual name for that town among Mongols, Kalmaks, and Russians to be SANKIN-hoto.   He had not then thought of connecting this name with Chinghin-talas, and has therefore no information as to its origin or the extent of its application.   But he remarks that Polo's bearing of between north and north-west, if understood to be *from Kamul*, would point exactly to Kobdo.   He also calls attention to the Lake *Sankin-dalai*, to the north-east of Uliasut'ai, of which Atkinson gives a sketch.   The recurrence of this name over so wide a tract may have something to do with the Chinghin-talas of Polo.   But we must still wait for further light. *

["Supposing that M. Polo mentions this place on his way from Sha-chow to Su-chow, it is natural to think that it is *Chi-kin-talas*, *i.e.* 'Chi-kin plain' or valley ; Chi-kin was the name of a lake, called so even now, and of a defile, which received its name from the lake.   The latter is on the way from Kia-yü kwan to Ansi chow." (*Palladius, l.c.* p. 7.)   "*Chikin*, or more correctly *Chigin*, is a Mongol word meaning 'ear.'" (*Ibid.*)   Palladius (p. 8) adds : "The Chinese accounts of Chi-kin are not in contradiction to the statements given by M. Polo regarding the same subject ; but when the distances are taken into consideration, a serious difficulty arises ; Chi-kin is two hundred and fifty or sixty *li* distant from Su-chow, whilst, according to M. Polo's statement, ten days are necessary to cross this distance.   One of the three following explanations of this discrepancy must be admitted : either Chingintalas is not Chi-kin, or the traveller's memory failed, or, lastly, an error crept into the number of days' journey.   The two last suppositions I consider the most probable ; the more so that similar difficulties occur several times in Marco Polo's narrative." (*L.c.* p. 8.) —H. C.]

NOTE 2.—[*Ondanique*.—We have already referred to this word, *Kermán*, p. 90. *Cobinan*, p. 124.   La Curne de Sainte-Palaye (*Dict.*), F. Godefroy (*Dict.*), Du Cange (*Gloss.*), all give to *andain* the meaning of *enjambée*, from the Latin *andare*.   Godefroy, *s.v. andaine*, calls it *sorte d'acier ou de fer*, and quotes besides Marco Polo :

> "I. espiel, ou ot fer d'andaine,
>   Dont la lamele n'iert pas trouble."

(Huon de Mery, *Le Tornoiement de l'Antechrist*, p. 3, Tarbé.)

There is a forest in the department of Orne, arrondissement of Domfront, which belonged to the Crown before 1669, and is now State property, called Forêt d'Andaine ; it is situated near some bed of iron.   Is this the origin of the name ?—H. C.]

NOTE 3.—The Altai, or one of its ramifications, is probably the mountain of the text, but so little is known of this part of the Chinese territory that we can learn scarcely anything of its mineral products.   Still Martini does mention that asbestos is found "in the Tartar country of *Tangu*," which probably is the *Tangnu Oola* branch of the Altai to the south of the Upper Yenisei, and in the very region we have indicated as Chingintalas.   Mr. Elias tells me he inquired for asbestos by its Chinese name at Uliasut'ai, but without success.

---

* The late Mr. Atkinson has been twice alluded to in this note.   I take the opportunity of saying that Mr. Ney Elias, a most competent judge, who has travelled across the region in question whilst admitting, as every one must, Atkinson's vagueness and sometimes very careless statements, is not at all disposed to discredit the truth of his narrative.

NOTE 4.—

> "Degli elementi quattro principali,
> Che son la Terra, e l'Acqua, e l'Aria, e'l Foco,
> Composti sono gli universi Animali,
> Pigliando di ciascuno assai o poco."
>
> (*Dati, La Sfera*, p. 9.)

Zurficar in the next sentence is a Mahomedan name, *Zu'lfikár*, the title of [the edge of] Ali's sword.

NOTE 5.—Here the G. Text adds : "*Et je meisme le vi*," intimating, I conceive, his having himself seen specimens of the asbestos—not to his having been at the place.

NOTE 6.—The story of the Salamander passing unhurt through fire is at least as old as Aristotle. But I cannot tell when the fable arose that asbestos was a substance derived from the animal. This belief, however, was general in the Middle Ages, both in Asia and Europe. " The fable of the Salamander," says Sir Thomas Browne, " hath been much promoted by stories of incombustible napkins and textures which endure the fire, whose materials are called by the name of Salamander's wool, which many, too literally apprehending, conceive some investing part or integument of the Salamander. . . . Nor is this Salamander's wool desumed from any animal, but a mineral substance, metaphorically so called for this received opinion."

Those who knew that the Salamander was a lizard-like animal were indeed perplexed as to its woolly coat. Thus the Cardinal de Vitry is fain to say the creature "*profert ex cute* quasi quamdam lanam *de quá zonae contextae comburi non possunt igne.*" A Bestiary, published by Cahier and Martin, says of it : " *De lui naist une cose qui n'est ne soie ne lin ne laine.*" Jerome Cardan looked in vain, he says, for hair on the Salamander ! Albertus Magnus calls the incombustible fibre *pluma Salamandri ;* and accordingly Bold Bauduin de Sebourc finds the Salamander in the Terrestrial Paradise *a kind of bird covered with the whitest plumage ;* of this he takes some, which he gets woven into a cloth ; this he presents to the Pope, and the Pontiff applies it to the purpose mentioned in the text, viz. to cover the holy napkin of St. Veronica.

Gervase of Tilbury writes : " I saw, when lately at Rome, a broad strap of Salamander skin, like a girdle for the loins, which had been brought thither by Cardinal Peter of Capua. When it had become somewhat soiled by use, I myself saw it cleaned perfectly, and without receiving harm, by being put in the fire."

In Persian the creature is called *Samandar, Samandal,* etc., and some derive the word from *Sam,* "fire," and *Andar,* "within." Doubtless it is a corruption of the Greek Σαλαμάνδρα, whatever be the origin of that. Bakui says the animal is found at Ghur, near Herat, and is *like a mouse.* Another author, quoted by D'Herbelot, says it is *like a marten.*

[Sir T. Douglas Forsyth, in his *Introductory Remarks* to Prjevalsky's *Travels to Lob-nor* (p. 20), at Aksu says : " The asbestos mentioned by Marco Polo as a utilized product of this region is not even so known in this country."—H. C.]

+ Interesting details regarding the fabrication of cloth and paper from amianth or asbestos are contained in a report presented to the French Institute by M. Sage (*Mém. Ac. Sciences,* 2e Sem., 1806, p. 102), of which large extracts are given in the *Diction. général des Tissus,* par M. Bezon, 2e éd. .vol. ii. Lyon, 1859, p. 5. He mentions that a *Sudarium* of this material is still shown at the Vatican ; we hope it is the cover which Kúblái sent.

[This hope is not to be realized. Mgr. Duchesne, of the Institut de France, writes to me from Rome, from information derived from the keepers of the Vatican Museum, that there is no sudarium from the Great Khan, that indeed part of a sudarium made of asbestos is shown (under glass) in this Museum, about 20 inches long, but it is ancient, and was found in a Pagan tomb of the Appian Way.—H. C.]

M. Sage exhibited incombustible paper made from this material, and had himself seen a small furnace of Chinese origin made from it. Madame Perpenté, an Italian lady, who experimented much with asbestos, found that from a crude mass of that substance threads could be elicited which were ten times the length of the mass itself, and were indeed sometimes several metres in length, the fibres seeming to be involved, like silk in a cocoon. Her process of preparation was much like that described by Marco. She succeeded in carding and reeling the material, made gloves and the like, as well as paper, from it, and sent to the Institute a work printed on such paper.

The Rev. A. Williamson mentions asbestos as found in Shantung. The natives use it for making stoves, crucibles, and so forth.

(*Sir T. Browne*, I. 293; *Bongars*, I. 1104; *Cahier et Martin*, III. 271; *Cardan, de Rer. Varietate*, VII. 33; *Alb. Mag. Opera*, 1551, II. 227, 233; *Fr. Michel, Recherches*, etc., II. 91; *Gerv. of Tilbury*, p. 13; *N. et E.* II. 493; *D. des Tissus*, II. 1-12; *J. N. China Branch R.A.S.*, December, 1867, p. 70.) [*Berger de Xivrey, Traditions tératologiques*, 457-458, 460-463.—H. C.]

CHAPTER XLIII.

OF THE PROVINCE OF SUKCHUR.

On leaving the province of which I spoke before,[1] you ride ten days between north-east and east, and in all that way you find no human dwelling, or next to none, so that there is nothing for our book to speak of.

At the end of those ten days you come to another province called SUKCHUR, in which there are numerous towns and villages. The chief city is called SUKCHU.[2] The people are partly Christians and partly Idolaters, and all are subject to the Great Kaan.

The great General Province to which all these three provinces belong is called TANGUT.

Over all the mountains of this province rhubarb is found in great abundance, and thither merchants come to buy it, and carry it thence all over the world.[3] [Travellers, however, dare not visit those mountains with any cattle but those of the country, for a certain plant grows there which is so poisonous that cattle which eat it lose their hoofs. The cattle of the country

know it and eschew it.[4]]   The people live by agriculture,
and have not much trade.   [They are of a brown complexion.   The whole of the province is healthy.]

NOTE 1.—Referring apparently to Shachau; see Note 1 and the closing words of
last chapter.

NOTE 2.—There is no doubt that the province and city are those of SUHCHAU,
but there is a great variety in the readings, and several texts have a marked difference
between the name of the province and that of the city, whilst others give them as the
same.   I have adopted those to which the resultants of the readings of the best texts
seem to point, viz. *Succiur* and *Succiu*, though with considerable doubt whether they
should not be identical.   Pauthier declares that *Suctur*, which is the reading of his
favourite MS., is the exact pronunciation, after the vulgar Mongol manner, of *Suh-
chau-lu*, the *Lu* or circuit of Suhchau; whilst Neumann says that the Northern
Chinese constantly add an euphonic particle *or* to the end of words.   I confess to
little faith in such refinements, when no evidence is produced.

[Suhchau had been devastated and its inhabitants massacred by Chinghiz Khan in
1226.—H. C.]

Suhchau is called by Rashiduddin, and by Shah Rukh's ambassadors, *Sukchú*, in
exact correspondence with the reading we have adopted for the name of the city,
whilst the Russian Envoy Boikoff, in the 17th century, calls it "*Suktsey*, where the
rhubarb grows"; and Anthony Jenkinson, in Hakluyt, by a slight metathesis,
Sowchick.   Suhchau lies just within the extreme north-west angle of the Great Wall.   It
was at Suhchau that Benedict Goës was detained, waiting for leave to go on to
Peking, eighteen weary months, and there he died just as aid reached him.

NOTE 3.—The real rhubarb [*Rheum palmatum*] grows wild, on very high mountains.
The central line of its distribution appears to be the high range dividing the head waters
of the Hwang-Ho, Yalung, and Min-Kiang.   The chief markets are Siningfu (see ch.
lvii.), and Kwan-Kian in Szechwan.   In the latter province an inferior kind is grown
in fields, but the genuine rhubarb defies cultivation.   (See *Richthofen*, Letters, No.
VII. p. 69.)   Till recently it was almost all exported by Kiakhta and Russia, but
some now comes *via* Hankau and Shanghai.

["See, on the preparation of the root in China, Gemelli-Careri. (*Churchill's Collect.*,
Bk. III. ch. **v**. 365.)   It is said that when Chinghiz Khan was pillaging Tangut, the
only things his minister, Yeh-lü Ch'u-ts'ai, would take as his share of the booty were
a few Chinese books and a supply of rhubarb, with which he saved the lives of a
great number of Mongols, when, a short time after, an epidemic broke out in the
army." (*D'Ohsson*, I. 372.—*Rockhill, Rubruck*, p. 193, note.)

"With respect to rhubarb . . . the *Suchowchi* also makes the remark, that the
best rhubarb, with golden flowers in the breaking, is gathered in this province
(district of *Shan-tan*), and that it is equally beneficial to men and beasts, preserving
them from the pernicious effects of the heat." (*Palladius, l.c.* p. 9.)—H. C.]

NOTE 4.—*Erba* is the title applied to the poisonous growth, which may be either
"plant" or "grass."   It is not unlikely that it was a plant akin to the *Andromeda
ovalifolia*, the tradition of the poisonous character of which prevails everywhere along
the Himalaya from Nepal to the Indus.

It is notorious for poisoning sheep and goats at Simla and other hill sanitaria;
and Dr. Cleghorn notes the same circumstance regarding it that Polo heard of the
plant in Tangut, viz. that its effects on flocks imported from the plains are highly
injurious, whilst those of the hills do not appear to suffer, probably because they shun
the young leaves, which alone are deleterious.   Mr. Marsh attests the like fact regard-

ing the *Kalmia angustifolia* of New England, a plant of the same order (*Ericaceae*). Sheep bred where it abounds almost always avoid browsing on its leaves, whilst those brought from districts where it is unknown feed upon it and are poisoned.

Firishta, quoting from the *Zafar-Námah*, says: "On the road from Kashmir towards Tibet there is a plain on which no other vegetable grows but a poisonous grass that destroys all the cattle that taste of it, and therefore no horsemen venture to travel that route." And Abbé Desgodins, writing from E. Tibet, mentions that sheep and goats are poisoned by rhododendron leaves. (*Dr. Hugh Cleghorn* in *J. Agricultural and Hortic. Society of India*, XIV. part 4 ; *Marsh's Man and Nature*, p. 40 ; *Brigg's Firishta*, IV. 449 ; *Bul. de la Soc. de Géog.* 1873, I. 333.)

[" This poisonous plant seems to be the *Stipa inebrians* described by the late Dr. Hance in the *Journal of Bot.* 1876, p. 211, from specimens sent to me by Belgian Missionaries from the Ala Shan Mountains, west of the Yellow River." (*Bretschneider*, *Hist. of Bot. Disc.* I. p. 5.)

" M. Polo notices that the cattle not indigenous to the province lose their hoofs in the Suh-chau Mountains ; but that is probably not on account of some poisonous grass, but in consequence of the stony ground." (*Palladius, l.c.* p. 9.)—H. C.]

CHAPTER XLIV.

OF THE CITY OF CAMPICHU.

CAMPICHU is also a city of Tangut, and a very great and noble one. Indeed it is the capital and place of government of the whole province of Tangut.[1] The people are Idolaters, Saracens, and Christians, and the latter have three very fine churches in the city, whilst the Idolaters have many minsters and abbeys after their fashion. In these they have an enormous number of idols, both small and great, certain of the latter being a good ten paces in stature ; some of them being of wood, others of clay, and others yet of stone. They are all highly polished, and then covered with gold. The great idols of which I speak lie at length.[2] And round about them there are other figures of considerable size, as if adoring and paying homage before them.

Now, as I have not yet given you particulars about the customs of these Idolaters, I will proceed to tell you about them.

You must know that there are among them certain
religious recluses who lead a more virtuous life than the
rest. These abstain from all lechery, though they do
not indeed regard it as a deadly sin; howbeit if any
one sin against nature they condemn him to death.
They have an Ecclesiastical Calendar as we have; and
there are five days in the month that they observe
particularly; and on these five days they would on no
account either slaughter any animal or eat flesh meat.
On those days, moreover, they observe much greater
abstinence altogether than on other days.[3]

Among these people a man may take thirty wives,
more or less, if he can but afford to do so, each having
wives in proportion to his wealth and means; but the
first wife is always held in highest consideration. The
men endow their wives with cattle, slaves, and money,
according to their ability. And if a man dislikes any
one of his wives, he just turns her off and takes another.
They take to wife their cousins and their fathers' widows
(always excepting the man's own mother), holding to be
no sin many things that we think grievous sins, and, in
short, they live like beasts.[4]

Messer Maffeo and Messer Marco Polo dwelt a
whole year in this city when on a mission.[5]

Now we will leave this and tell you about other provinces towards the north, for we are going to take you
a sixty days' journey in that direction.

NOTE 1.—Campichiu is undoubtedly Kanchau, which was at this time, as Pauthier
tells us, the chief city of the administration of *Kansuh*, corresponding to Polo's
Tangut. *Kansuh* itself is a name compounded of the names of the two cities *Kan*-
chau and *Suh*-chau.

[Kanchau fell under the Tangut dominion in 1208. (*Palladius*, p. 10.) The Musulmans mentioned by Polo at Shachau and Kanchau probably came from Khotan.—
H. C.]

The difficulties that have been made about the form of the name *Campiciou*, etc., in
Polo, and the attempts to explain these, are probably alike futile. Quatremère writes
the Persian form of the name after Abdurrazzak as *Kamtcheou*, but I see that Erdmann

writes it after Rashid, I presume on good grounds, as *Ckamidschu, i.e. Ḳamiju* or *Ḳamichu.* And that this *was* the Western pronunciation of the name is shown by the form which Pegolotti uses, *Camexu, i.e.* Camechu. The *p* in Polo's spelling is probably only a superfluous letter, as in the occasional old spelling of *dampnum, contempnere, hympnus, tirampnus, sompnour, Dampne Deu.* In fact, Marignolli writes Polo's *Quinsai* as *Campsay.*

It is worthy of notice that though Ramusio's text prints the names of these two cities as *Succuir* and *Campion,* his own pronunciation of them appears to have been quite well understood by the Persian traveller Hajji Mahomed, for it is perfectly clear that the latter recognized in these names Suhchau and Kanchau. (See *Ram.* II. f. 14v.) The second volume of the *Navigationi,* containing Polo, was published after Ramusio's death, and it is possible that the names as he himself read them were more correct (*e.g. Succiur, Campjou*).

NOTE 2.—This is the meaning of the phrase in the G. T. : " *Ceste grande ydie gigent,*" as may be seen from Ramusio's *giaciono distesi.* Lazari renders the former expression, " giganteggia un idolo," etc., a phrase very unlike Polo. The circumstance is interesting, because this recumbent Colossus at Kanchau is mentioned both by Hajji Mahomed and by Shah Rukh's people. The latter say : " In this city of Kanchú there is an Idol-Temple 500 cubits square. In the middle is an idol lying at length which measures 50 paces. The sole of the foot is nine paces long, and the instep is 21 cubits in girth. Behind this image and overhead are other idols of a cubit (?) in height, besides figures of *Bakshis* as large as life. The action of all is hit off so admirably that you would think they were alive." These great recumbent figures are favourites in Buddhist countries still, *e.g.* in Siam, Burma, and Ceylon. They symbolise Sakya Buddha entering *Nirvána.* Such a recumbent figure, perhaps the prototype of these, was seen by Hiuen Tsang in a Vihara close to the Sál Grove at Kusinágara, where Sakya entered that state, *i.e.* died. The stature of Buddha was, we are told, 12 cubits; but Brahma, Indra, and the other gods vainly tried to compute his dimensions. Some such rude metaphor is probably embodied in these large images. I have described one 69 feet long in Burma (represented in the cut), but others exist of much greater size, though probably none equal to that which Hiuen Tsang, in the 7th century, saw near Bamian, which was 1000 feet in length ! I have heard of but one such image remaining in India, viz. in one of the caves at Dhamnár in Málwa. This is 15 feet long, and is popularly known as " Bhim's Baby." (*Cathay,* etc.,

Colossal Figure, Buddha entering Nirvana.

" Et si voz di qu'il ont de pdres que sunt grant dix pas. . . . Ceste grant pdres gigent." . . .

pp. cciii., ccxviii. ; *Mission to Ava,* p. 52 ; *V. et V. de H. T.,* p. 374 : *Cunningham's Archæl. Reports,* ii. 274 ; *Tod,* ii. 273.)

[" The temple, in which M. Polo saw an idol of Buddha, represented in a lying position, is evidently *Wo-fo-sze, i.e.* 'Monastery of the lying Buddha.' It was built in 1103 by a Tangut queen, to place there three idols representing Buddha in this posture, which have since been found in the ground on this very spot." (*Palladius, l.c.* p. 10.)

Rubruck (p. 144) says : "A Nestorian, who had come from Cathay told me that in that country there is an idol so big that it can be seen from two days off." Mr. Rockhill (*Rubruck*, p. 144, *note*) writes : " The largest stone image I have seen is in a cave temple at Yung-kán, about 10 miles north-west of Ta-t'ung Fu in Shan-si. Père Gerbillon says the Emperor K'ang-hsi measured it himself and found it to be 57 *chih* high (61 feet). (*Duhalde, Description*, IV. 352.) I have seen another colossal statue in a cave near Pinchou in north-west Shan-si ; and there is another about 45 miles south of Ning-hsia Fu, near the left bank of the Yellow River. (*Rockhill, Land of the Lamas*, 26, and *Diary*, 47.) The great recumbent figure of the ' Sleeping Buddha' in the Wo Fo ssŭ, near Peking, is of clay."

King Haython (Brosset's ed. p. 181) mentions the statue in clay, of an extra-ordinary height, of a God (Buddha) aged 3040 years, who is to live 370,000 years more, when he will be superseded by another god called *Madri* (Maitreya).—H. C.]

Great Lama Monastery.

NOTE 3.—Marco is now speaking of the Lamas, or clergy of Tibetan Buddhism. The customs mentioned have varied in details, both locally and with the changes that the system has passed through in the course of time.

The institutes of ancient Buddhism set apart the days of new and full moon to be observed by the *Sramanas* or monks, by fasting, confession, and listening to the reading of the law. It became usual for the laity to take part in the observance, and the number of days was increased to three and then to four, whilst Hiuen Tsang himself speaks of " the six fasts of every month," and a Chinese authority quoted by Julien gives the days as the 8th, 14th, 15th, 23rd, 29th, and 30th. Fahian says that in Ceylon preaching took place on the 8th, 14th, and 15th days of the month. **Four** is the number now most general amongst Buddhist nations, and the days may be re-garded as a kind of Buddhist Sabbath.   In the southern countries and in Nepal they occur at the moon's changes.   In Tibet and among the Mongol Buddhists they are not at equal intervals, though I find the actual days differently stated by different authorities.   Pallas says the Mongols observed the 13th, 14th, and 15th, the three days being brought together, he thought, on account of the distance many Lamas had to travel to the temple—just as in some Scotch country parishes they used to give two sermons in one service for like reason !   Koeppen, to whose work this note is much indebted, says the Tibetan days are the 14th, 15th, 29th, 30th, and adds as to the manner of observance : "On these days, by rule, among the Lamas, nothing should be tasted but farinaceous food and tea; the very devout refrain from all food from sunrise to sunset.   The Temples are decorated, and the altar tables set out with the holy symbols, with tapers, and with dishes containing offerings in corn, meal, tea, butter, etc., and especially with small pyramids of dough, or of rice or clay, and ac-companied by much burning of incense-sticks.   The service performed by the priests is more solemn, the music louder and more exciting, than usual.   The laity make their offerings, tell their beads, and repeat *Om mani padma hom*," etc.   In the *con-cordat* that took place between the Dalai-Lama and the Altun Khaghan, on the re-conversion of the Mongols to Buddhism in the 16th century, one of the articles was the entire prohibition of hunting and the slaughter of animals on the monthly fast days.   The practice varies much, however, even in Tibet, with different provinces and sects—a variation which the Ramusian text of Polo implies in these words : "For five days, or *four days*, or *three* in each month, they shed no blood," etc.

In Burma the Worship Day, as it is usually called by Europeans, is a very gay scene, the women flocking to the pagodas in their brightest attire.   (*H. T. Mémoires*, I. 6, 208 ; *Koeppen*, I. 563-564, II. 139, 307-308; *Pallas, Samml.* II. 168-169).

NOTE 4.—These matrimonial customs are the same that are afterwards ascribed to the Tartars, so we defer remark.

NOTE 5.—So Pauthier's text, "*en legation.*"   The G. Text includes Nicolo Polo, and says, "on business of theirs that is not worth mentioning," and with this Ramusio agrees.

CHAPTER XLV.

Of the City of Etzina.

WHEN you leave the city of Campichu you ride for twelve days, and then reach a city called ETZINA, which is towards the north on the verge of the Sandy Desert; it belongs to the Province of Tangut.[1]   The people are Idolaters, and possess plenty of camels and cattle, and the country produces a number of good falcons, both Sakers and Lanners.   The inhabitants live by their cultivation and their cattle, for they have no trade.

At this city you must needs lay in victuals for forty
days, because when you quit Etzina, you enter on a
desert which extends forty days' journey to the north,
and on which you meet with no habitation nor baiting-
place.[2] In the summer-time, indeed, you will fall in
with people, but in the winter the cold is too great.
You also meet with wild beasts (for there are some small
pine-woods here and there), and with numbers of wild
asses.[3] When you have travelled these forty days across
the Desert you come to a certain province lying to the
north.   Its name you shall hear presently.

Note i.—Deguignes says that YETSINA is found in a Chinese Map of Tartary of
the Mongol era, and this is confirmed by Pauthier, who reads it *Itsinai*, and adds

Wild Ass of Mongolia.

that the text of the Map names it as one of the seven *Lu* or Circuits of the Province
of Kansuh (or Tangut).   Indeed, in D'Anville's Atlas we find a river called *Etsina
Pira,* running northward from Kanchau, and a little below the 41st parallel joining

another from Suhchau. Beyond the junction is a town called *Hoa-tsiang*, which probably represents Etzina. Yetsina is also mentioned in Gaubil's History of Chinghiz as taken by that conqueror in 1226, on his last campaign againt Tangut. This capture would also seem from Pétis de la Croix to be mentioned by Rashiduddin. Gaubil says the Chinese Geography places Yetsina north of Kanchau and north-east of Suhchau, at a distance of 120 leagues from Kanchau, but observes that this is certainly too great. (*Gaubil*, p. 49.)

[I believe there can be no doubt that Etzina must be looked for on the river *Hei-shui*, called *Etsina* by the Mongols, east of Suhchau. This river empties its waters into the two lakes Soho-omo and Sopo-omo. Etzina would have been therefore situated on the river on the border of the Desert, at the top of a triangle whose bases would be Suhchau and Kanchau. This river was once part of the frontier of the kingdom of Tangut. (Cf. *Devéria, Notes d'épigraphie mongolo-chinoise*, p. 4.) Reclus (*Géog. Univ., Asie Orientale*, p. 159) says: "To the east [of Hami], beyond the Chukur Gobi, are to be found also some permanent villages and the remains of cities. One of them is perhaps the 'cité d'Etzina' of which Marco Polo speaks, and the name is to be found in that of the river Az-sind."

"Through Kanchau was the shortest, and most direct and convenient road to *I-tsi-nay*. . . . I-tsi-nay, or *Echiné*, is properly the name of a lake. Khubilaï, disquieted by his factious relatives on the north, established a military post near lake I-tsi-nay, and built a town, or a fort on the south-western shore of this lake. The name of I-tsi-nay appears from that time; it does not occur in the chronicle of the Tangut kingdom; the lake had then another name. Vestiges of the town are seen to this day; the buildings were of large dimensions, and some of them were very fine. In Marco Polo's time there existed a direct route from I-tsi-nay to Karakorum; traces cf this road are still noticeable, but it is no more used. This circumstance, *i.e.* the existence of a road from I-tsi-nay to Karakorum, probably led Marco Polo to make an excursion (a mental one, I suppose) to the residence of the Khans in Northern Mongolia." (*Palladius, l.c.* pp. 10-11.)—H. C.]

NOTE 2.—"*Erberge*" (G. T.). Pauthier has *Herbage.*

NOTE 3.—The Wild Ass of Mongolia is the *Dshiggetai* of Pallas (*Asinus hemionus* of Gray), and identical with the Tibetan *Kyang* of Moorcroft and Trans-Himalayan sportsmen. It differs, according to Blyth, only in shades of colour and unimportant markings from the *Ghor Khar* of Western India and the Persian Deserts, the *Kulan* of Turkestan, which Marco has spoken of in a previous passage (*suprà*, ch. xvi.; *J. A. S. B.* XXVIII. 229 *seqq.*). There is a fine Kyang in the Zoological Gardens, whose portrait, after Wolf, is given here. But Mr. Ney Elias says of this animal that he has little of the aspect of his nomadic brethren. [The wild ass (Tibetan *Kyang*, Mongol *Holu* or *Hulan*) is called by the Chinese *yeh ma*, "wild horse," though "every one admits that it is an ass, and should be called *yeh lo-tzŭ*." (*Rockhill, Land of the Lamas*, 151, note.)—H. C.]

[Captain Younghusband (1886) saw in the Altaï Mountains "considerable numbers of wild asses, which appeared to be perfectly similar to the Kyang of Ladak and Tibet, and wild horses too—the *Equus Prejevalskii*—roaming about these great open plains." (*Proc. R. G. S.* X. 1888, p. 495.) Dr. Sven Hedin says the *habitat* of the *Kulan* is the heights of Tibet as well as the valley of the Tarim; it looks like a mule with the mane and tail of an ass, but shorter ears, longer than those of a horse: he gives a picture of it.—H. C.]

CHAPTER XLVI.

OF THE CITY OF CARACORON.

CARACORON is a city of some three miles in compass.
[It is surrounded by a strong earthen rampart, for stone
is scarce there. And beside it there is a great citadel
wherein is a fine palace in which the Governor resides.]
'Tis the first city that the Tartars possessed after they
issued from their own country. And now I will tell
you all about how they first acquired dominion and
spread over the world.[1]

Originally the Tartars[2] dwelt in the north on the
borders of CHORCHA.[3] Their country was one of great
plains; and there were no towns or villages in it, but
excellent pasture-lands, with great rivers and many
sheets of water; in fact it was a very fine and extensive
region. But there was no sovereign in the land. They
did, however, pay tax and tribute to a great prince
who was called in their tongue UNC CAN, the same
that we call Prester John, him in fact about whose
great dominion all the world talks.[4] The tribute he had
of them was one beast out of every ten, and also a tithe
of all their other gear.

Now it came to pass that the Tartars multiplied
exceedingly. And when Prester John saw how great
a people they had become, he began to fear that he
should have trouble from them. So he made a scheme
to distribute them over sundry countries, and sent one
of his Barons to carry this out. When the Tartars
became aware of this they took it much amiss, and
with one consent they left their country and went off
across a desert to a distant region towards the north,
where Prester John could not get at them to annoy

them. Thus they revolted from his authority and paid him tribute no longer. And so things continued for a time.

NOTE I.—KARÁKORUM, near the upper course of the River Orkhon, is said by Chinese authors to have been founded by Búkú Khan of the Hoei-Hu or Uigúrs, in the 8th century. In the days of Chinghiz, we are told that it was the headquarters of his ally, and afterwards enemy, Togrul Wang Khan, the Prester John of Polo. ["The name of this famous city is Mongol, *Kara*, 'black,' and *Kuren*, 'a camp,' or properly 'pailing.'"] It was founded in 1235 by Okkodai, who called it Ordu Balik, or "the City of the Ordu," otherwise "The Royal City." Mohammedan authors say it took its name of Karákorum from the mountains to the south of it, in which the Orkhon had its source. (*D'Ohsson*, ii. 64.) The Chinese mention a range of mountains from which the Orkhon flows, called *Wu-tê kien shan*. (*T'ang shu*, bk. 43*b*.) Probably these are the same. Rashiduddin speaks of a tribe of Utikien Uigúrs living in this country. (*Bretschneider, Med. Geog.* 191 ; *D'Ohsson*, i. 437. *Rockhill, Rubruck*, 220, note.)—Karákorum was called by the Chinese *Ho-lin* and was chosen by Chinghiz, in 1206, as his capital; the full name of it, *Ha-la Ho-lin*, was derived from a river to the west. (*Yuen shi*, ch. lviii.) Gaubil (*Holin*, p. 10) says that the river, called in his days in Tartar *Karoha*, was, at the time of the Mongol Emperors, named by the Chinese *Ha-la Ho-lin*, in Tartar language *Ka la Ko lin*, or *Cara korin*, or *Kara Koran*. In the spring of 1235, Okkodai had a wall raised round Ho-lin and a palace called *Wang an*, built inside the city. (*Gaubil, Gentchiscan*, 89.) After the death of Kúblái, *Ho-lin* was altered into *Ho-Ning*, and, in 1320, the name of the province was changed into *Ling-pé* (mountainous north, *i.e.* the *Yin-shan* chain, separating China Proper from Mongolia). In 1256, Mangu Kaan decided to transfer the seat of government to Kaiping-fu, or Shangtu, near the present Dolonnor, north of Peking. (*Suprà* in Prologue, ch. xiii. note 1.) In 1260, Kúblái transferred his capital to *Ta-Tu* (Peking).

Plano Carpini (1246) is the first Western traveller to mention it by name which he writes *Caracoron* ; he visited the Sira Orda, at half a day's journey from Karákorum, where Okkodai used to pass the summer ; it was situated at a place Ormektua. (*Rockhill, Rubruck*, 21, 111.) Rubruquis (1253) visited the city itself ; the following is his account of it : "As regards the city of Caracoron, you must understand that if you set aside the Kaan's own Palace, it is not as good as the Borough of St. Denis ; and as for the Palace, the Abbey of St. Denis is worth ten of it ! There are two streets in the town ; one of which is occupied by the Saracens, and in that is the market-place. The other street is occupied by the Cathayans, who are all craftsmen. Besides these two streets there are some great palaces occupied by the court secretaries. There are also twelve idol temples belonging to different nations, two Mahummeries in which the Law of Mahomet is preached, and one church of the Christians at the extremity of the town. The town is enclosed by a mud-wall and has four gates. At the east gate they sell millet and other corn, but the supply is scanty ; at the west gate they sell rams and goats ; at the south gate oxen and waggons ; at the north gate horses. . . . Mangu Kaan has a great Court beside the Town Rampart, which is enclosed by a brick wall, just like our priories. Inside there is a big palace, within which he holds a drinking-bout twice a year ; . . . there are also a number of long buildings like granges, in which are kept his treasures and his stores of victual" (345-6 ; 334).

Where was Karákorum situated ?

The Archimandrite Palladius is very prudent (*l.c.* p. 11) : "Everything that the studious Chinese authors could gather and say of the situation of Karakhorum is collected in two Chinese works, *Lo fung low wen kao* (1849), and *Mungku yew mu ki*

(1859). However, no positive conclusion can be derived from these researches, chiefly
in consequence of the absence of a tolerably correct map of Northern Mongolia."

Abel Rémusat (*Mém. sur Géog. Asie Centrale*, p. 20) made a confusion between
Karábalgasun and Karákorum which has misled most writers after him.

Sir Henry Yule says: "The evidence adduced in Abel Rémusat's paper on
Karákorum (*Mém. de l'Acad. R. des Insc.* VII. 288) establishes the site on the north
bank of the Orkhon, and about five days' journey above the confluence of the Orkhon
and Tula. But as we have only a very loose knowledge of these rivers, it is impossible
to assign the geographical position with accuracy. Nor is it likely that ruins exist
beyond an outline perhaps of the Kaan's Palace walls."

In the *Geographical Magazine* for July, 1874 (p. 137), Sir Henry Yule has been
enabled, by the kind aid of Madame Fedtchenko in supplying a translation from the
Russian, to give some account of Mr. Paderin's visit to the place, in the summer of
1873, along with a sketch-map.

" The site visited by Mr. Paderin is shown, by the particulars stated in that paper,
to be sufficiently identified with Karákorum. It is precisely that which Rémusat indicated, and which bears in the Jesuit maps, as published by D'Anville, the name of
Talarho Hara Palhassoun (*i.e.* Kará Balghásun), standing 4 or 5 miles from the left
bank of the Orkhon, in lat. (by the Jesuit Tables) 47° 32″ 24″. It is now known as Kara-
Khărăm (Rampart) or Kara Balghasun (city). The remains consist of a quadrangular
rampart of mud and sun-dried brick, of about 500 paces to the side, and now about
9 feet high, with traces of a higher tower, and of an inner rampart parallel to the
other. But these remains probably appertain to the city as re-occupied by the
descendants of the Yuen in the end of the 14th century, after their expulsion from
China."

Dr. Bretschneider (*Med. Res.* I. p. 123) rightly observes: " It seems, however,
that Paderin is mistaken in his supposition. At least it does not agree with the
position assigned to the ancient Mongol residence in the Mongol annals *Erdenin
erikhe*, translated into Russian, in 1883, by Professor Pozdneiev. It is there positively
stated (p. 110, note 2) that the monastery of *Erdenidsu*, founded in 1585, was
erected on the ruins of that city, which once had been built by order of Ogotai Khan,
and where he had established his residence; and where, after the expulsion of the
Mongols from China, Togontemur again had fixed the Mongol court. This vast
monastery still exists, one English mile, or more, east of the Orkhon. It has even
been astronomically determined by the Jesuit missionaries, and is marked on our maps
of Mongolia. Pozdneiev, who visited the place in 1877, obligingly informs me that
the square earthen wall surrounding the monastery of Erdenidsu, and measuring
about an English mile in circumference, may well be the very wall of ancient
Karákorum."

Recent researches have fully confirmed the belief that the Erdeni Tso, or Erdeni
Chao, Monastery occupies the site of Karákorum, near the bank of the Orkhon,
between this river and the Kokchin (old) Orkhon. (See map in *Inscriptions de
l'Orkhon*, Helsingfors, 1892; a plan of the vicinity and of the Erdeni Tso is given
(plate 36) in *W. Radloff's Atlas der Alterthümer der Mongolei*, St. Pet., 1892.)

According to a work of the 13th century quoted by the late Professor G. Devéria,
the distance between the old capital of the Uighúr, Kara Balgasún, on the left bank
of the Orkhon, north of Erdeni Tso, and the Ho-lin or Karákorum of the Mongols,
would be 70 *li* (about 30 miles), and such is the space between Erdeni Tso and Kara
Balgasún. M. Marcel Monnier (*Itinéraires*, p. 107) estimates the bird's-eye distance
from Erdeni Tso to Kara Balgasún at 33 kilom. (about 20½ miles). "When the
brilliant epoch of the power of the Chinghizkhanides," says Professor Axel Heikel,
"was at an end, the city of Karákorum fell into oblivion, and towards the year 1590
was founded, in the centre of this historically celebrated region of the Orkhon, the
most ancient of Buddhist monasteries of Mongolia, this of Erdeni Tso [Erdeni Chao].
It was built, according to a Mongol chronicle, on the ruins of the town built by
Okkodaï, son of Chinghiz Khan, that is to say, on the ancient Karákorum."

(*Inscriptions de l'Orkhon.*)   So Professor Heikel, like Professor Pozdneiev, concludes that Erdeni Tso was built on the site of Karákorum and cannot be
mistaken for Karabalgásun.   Indeed it is highly probable that one of the walls
of the actual convent belonged to the old Mongol capital.   The travels and
researches by expeditions from Finland and Russia have made these questions pretty
clear.   Some most interesting inscriptions have been brought home and have been
studied by a number of Orientalists : G. Schlegel, O. Donner, G. Devéria, Vasiliev,

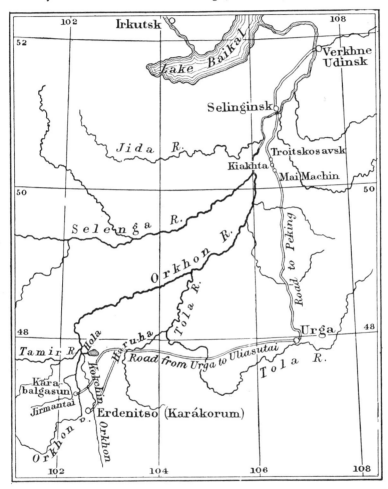

G. von der Gabelentz, Dr. Hirth, G. Huth, E. H. Parker, W. Bang, etc., and
especially Professor Vilh. Thomsen, of Copenhagen, who deciphered them (*Déchiffre-
ment des Inscriptions de l'Orkhon et de l'Iénissei*, Copenhague, 1894, 8vo ;
Inscriptions de l'Orkhon déchiffrées, *par* V. Thomsen, Helsingfors, 1894, 8vo), and
Professor W. Radloff of St. Petersburg (*Atlas der Alterthümer der Mongolei*, 1892-6,
fol. ; *Die alttürkischen Inschriften der Mongolei*, 1894-7, etc.).   There is an immense
literature on these inscriptions, and for the bibliography, I must refer the reader to

H. Cordier, Etudes Chinoises (1891-1894), Leide, 1895, 8vo; *Id.* (1895-1898), Leide, 1898, 8vo. The initiator of these discoveries was N. Iarindsev, of Irkutsk, who died at Barnaoul in 1894, and the first great expedition was started from Finland in 1890, under the guidance of Professor Axel Heikel. (*Inscriptions de l' Orkhon recueillies par l'expédition finnoise*, 1890, *et publiées par la Société Finno-Ougrienne*, Helsingfors, 1892, fol.) The Russian expedition left the following year, 1891, under the direction of the Academician W. Radloff.

M. Chaffanjon (*Nouv. Archiv. des Missions Scient.* IX., 1899, p. 81), in 1895, does not appear to know that there is a difference between Kará Korum and Kará Balgásun, as he writes: "Forty kilometres south of Kara Korum *or* Kara Balgásun, the convent of Erdin Zoun."

A plan of Kara Balgásun is given (plate 27) in *Radloff's Atlas*. See also *Henri Cordier et Gaubil, Situation de Holin en Tartarie*, Leide, 1893.

In Rubruquis's account of Karákorum there is one passage of great interest: "Then master William [Guillaume L'Orfèvre] had made for us an iron to make wafers . . . he made also a silver box to put the body of Christ in, with relics in little cavities made in the sides of the box." Now M. Marcel Monnier, who is one of the last, if not the last traveller who visited the region, tells me that he found in the large temple of Erdeni Tso an iron (the cast bore a Latin cross; had the wafer been Nestorian, the cross should have been Greek) and a silver box, which are very likely the objects mentioned by Rubruquis. It is a new proof of the identity of the sites of Erdeni Tso and Karákorum.—H. C.]

Entrance to the Erdeni Tso Great Temple.

NOTE 2.—[Mr. Rockhill (*Rubruck*, 113, note) says: "The earliest date to which I have been able to trace back the name Tartar is A.D. 732. We find mention made in a Turkish inscription found on the river Orkhon and bearing that date, of the *Tokuz Tatar*, or 'Nine (tribes of) Tatars,' and of the *Otuz Tatar*, or 'Thirty (tribes of) Tatars.' It is probable that these tribes were then living between the Oguz or Uigúr Turks on the west, and the Kitan on the east. (*Thomsen, Inscriptions de l'Orkhon*, 98, 126, 140.) Mr. Thos. Watters tells me that the Tartars are first mentioned by the Chinese in the period extending from A.D. 860 to 874; the earliest mention I have discovered, however, is under date of A.D. 880. (*Wu tai shih*, Bk. 4.)

We also read in the same work (Bk. 74, 2) that 'The Ta-ta were a branch of the Mo-ho (the name the Nû-chēn Tartars bore during the Sui and T'ang periods : *Ma Tuan-lin*, Bk. 327, 5). They first lived to the north of the Kitan. Later on they were conquered by this people, when they scattered, a part becoming tributaries of the Kitan, another to the P'o-hai (a branch of the Mo-ho), while some bands took up their abode in the Yin Shan in Southern Mongolia, north of the provinces of Chih-li and Shan-si, and took the name of *Ta-ta.*' In 981 the Chinese ambassador to the Prince of Kao-chang (Karakhodjo, some 20 miles south-east of Turfan) traversed the Ta-ta country. They then seem to have occupied the northern bend of the Yellow River. He gives the names of some nine tribes of Ta-ta living on either side of the river. He notes that their neighbours to the east were Kitan, and that for a long time they had been fighting them after the occupation of Kan-chou by the Uigúrs. (*Ma Tuan-lin*, Bk. 336, 12-14.) We may gather from this that these Tartars were already settled along the Yellow River and the Yin Shan (the valley in which is now the important frontier mart of Kwei-hua Ch'eng) at the beginning of the ninth century, for the Uigúrs, driven southward by the Kirghiz, first occupied Kan-chou in north-western Kan-suh, somewhere about A.D. 842."]

Note 3.—Chorcha (*Ciorcia*) is the Manchu country, whose people were at that time called by the Chinese *Yuché* or *Niuché*, and by the Mongols *Churché*, or as it is in Sanang Setzen, *Jurchid.* The country in question is several times mentioned by Rashiduddin as Churché. The founders of the *Kin* Dynasty, which the Mongols superseded in Northern China, were of Churché race. [It was part of Nayan's appanage. (See Bk. II. ch. v.)—H. C.]

Note 4.—The idea that a Christian potentate of enormous wealth and power, and bearing this title, ruled over vast tracts in the far East, was universal in Europe from the middle of the 12th to the end of the 13th century, after which time the Asiatic story seems gradually to have died away, whilst the Royal Presbyter was assigned to a locus in Abyssinia; the equivocal application of the term *India* to the East of Asia and the East of Africa facilitating this transfer. Indeed I have a suspicion, contrary to the view now generally taken, that the term may from the first have belonged to the Abyssinian Prince, though circumstances led to its being applied in another quarter for a time. It appears to me almost certain that the letter of Pope Alexander III., preserved by R. Hoveden, and written in 1177 to the *Magnificus Rex Indorum, Sacerdotum sanctissimus*, was meant for the King of Abyssinia.

Be that as it may, the inordinate report of Prester John's magnificence became especially diffused from about the year 1165, when a letter full of the most extravagant details was circulated, which purported to have been addressed by this potentate to the Greek Emperor Manuel, the Roman Emperor Frederick, the Pope, and other Christian sovereigns. By the circulation of this letter, glaring fiction as it is, the idea of this Christian Conqueror was planted deep in the mind of Europe, and twined itself round every rumour of revolution in further Asia. Even when the din of the conquests of Chinghiz began to be audible in the West, he was invested with the character of a Christian King, and more or less confounded with the mysterious Prester John.

The first notice of a conquering Asiatic potentate so styled had been brought to Europe by the Syrian Bishop of Gabala (*Jibal*, south of Laodicea in Northern Syria), who came, in 1145, to lay various grievances before Pope Eugene III. He reported that not long before a certain John, inhabiting the extreme East, king and Nestorian priest, and claiming descent from the Three Wise Kings, had made war on the *Samiard* Kings of the Medes and Persians, and had taken Ecbatana their capital. He was then proceeding to the deliverance of Jerusalem, but was stopped by the Tigris, which he could not cross, and compelled by disease in his host to retire.

M. d'Avezac first showed to whom this account must apply, and the subject has more recently been set forth with great completeness and learning by Dr. Gustavus Oppert. The conqueror in question was the founder of Kara Khitai, which existed as a great Empire in Asia during the last two-thirds of the 12th century. This chief was a prince of the Khitan dynasty of Liao, who escaped with a body of followers from Northern China on the overthrow of that dynasty by the *Kin* or Niuchen about 1125. He is called by the Chinese historians Yeliu Tashi; by Abulghazi, Nuzi Taigri Ili; and by Rashiduddin, Nushi (or Fushi) Taifu. Being well received by the Uighúrs and other tribes west of the Desert who had been subject to the Khitan Empire, he gathered an army and commenced a course of conquest which eventually extended over Eastern and Western Turkestan, including Khwarizm, which became tributary to him. He took the title of *Gurkhan*, said to mean Universal or Suzerain Khan, and fixed at Bala Sagun, north of the Thian Shan, the capital of his Empire, which became known as *Kará* (Black) *Khitai.* * [The dynasty being named by the Chinese *Si-Liao* (Western Liao) lasted till it was destroyed in 1218.—H. C.] In 1141 he came to the aid of the King of Khwarizm against *Sanjar* the Seljukian sovereign of Persia (whence the *Samiard* of the Syrian Bishop), who had just taken Samarkand, and defeated that prince with great slaughter. Though the Gurkhan himself is not described to have extended his conquests into Persia, the King of Khwarizm followed up the victory by an invasion of that country, in which he plundered the treasury and cities of Sanjar.

Admitting this Karacathayan prince to be the first conqueror (in Asia, at all events) to whom the name of Prester John was applied, it still remains obscure how that name arose. Oppert supposes that *Gurkhan* or *Kurkhan*, softened in West Turkish pronunciation into *Yurkan*, was confounded with *Yochanan* or *Johannes;* but he finds no evidence of the conqueror's profession of Christianity except the fact, notable certainly, that the daughter of the last of his brief dynasty is recorded to have been a Christian. Indeed, D'Ohsson says that the first Gurkhan was a Buddhist, though on what authority is not clear. There seems a probability at least that it was an error in the original ascription of Christianity to the Karacathayan prince, which caused the confusions as to the identity of Prester John which appear in the next century, of which we shall presently speak. Leaving this doubtful point, it has been plausibly suggested that the title of Presbyter Johannes was connected with the legends of the immortality of John the Apostle (ὁ πρεσβύτερος, as he calls himself in the 2nd and 3rd epistles), and the belief referred to by some of the Fathers that he would be the Forerunner of our Lord's second coming, as John the Baptist had been of His first.

A new theory regarding the original Prester John has been propounded by Professor Bruun of Odessa, in a Russian work entitled *The Migrations of Prester John*. The author has been good enough to send me large extracts of this essay in (French) translation; and I will endeavour to set forth the main points as well as the small space that can be given to the matter will admit. Some remarks and notes shall be added, but I am not in a position to do justice to Professor Bruun's views, from the

---

* A passage in Mirkhond extracted by Erdmann (*Temudschín*, p. 532) seems to make Bálá Sághún the same as Bishbálik, now Urumtsi, but this is inconsistent with other passages abstracted by Oppert (*Presbyter Johan.* 131-32); and Vámbéry indicates a reason for its being sought very much further west (*H. of Bokhara*, 116). [Dr. Bretschneider (*Med. Res.*) has a chapter on Kara-Khitaí (I. 208 *seqq.*), and in a long note on Bala Sagun, which he calls Belasagun, he says (p. 226) that "according to the Tarikh Djihan Kúshai (*d'Ohsson*, i. 433), the city of Belasagun had been founded by Buku Khan, sovereign of the Uigurs, in a well-watered plain of Turkestan with rich pastures. The Arabian geographers first mention Belasagun, in the ninth or tenth century, as a city beyond the Sihun or Yaxartes, depending on *Isfidjab* (Sairam, according to Lerch), and situated east of Taras. They state that the people of Turkestan considered Belasagun to represent 'the navel of the earth,' on account of its being situated in the middle between east and west, and likewise between north and south." (*Sprenger's Poststr. d. Or., Mavarannahar*). Dr. Bretschneider adds (p. 227): "It is not improbable that ancient Belasagun was situated at the same place where, according to the T'ang history, the Khan of one branch of the Western T'u Kuě (Turks) had his residence in the seventh century. It is stated in the T'ang shu that *Ibi Shabolo Shehu Khan*, who reigned in the first half of the seventh century, placed his ordo on the northern border of the river *Sui ye*. This river, and a city of the same name, are frequently mentioned in the T'ang annals of the seventh and eighth centuries, in connection with the warlike expeditions of the Chinese in Central Asia. *Sui ye* was situated on the way from the river *Ili* to the city of Ta-lo-sz' (Talas). In 679 the Chinese had built on the Sui ye River a fortress; but in 748 they were constrained to destroy it." (Comp. *Visdelou* in *Suppl. Bibl. Orient.* pp. 110-114; *Gaubil's Hist. de la Dyn. des Thang*, in *Mém. conc. Chin.* xv. p. 403 *seqq.*).—H. C.]

want of access to some of his most important authorities, such as Brosset's *History of Georgia*, and its appendices.

It will be well, before going further, to give the essential parts of the passage in the History of Bishop Otto of Freisingen (referred to in vol i. p. 229), which contains the first allusion to a personage styled Prester John :

"We saw also there [at Rome in 1145] the afore-mentioned Bishop of Gabala, from Syria. . . . We heard him bewailing with tears the peril of the Church beyond-sea since the capture of Edessa, and uttering his intention on that account to cross the Alps and seek aid from the King of the Romans and the King of the Franks. He was also telling us how, not many years before, one JOHN, KING and PRIEST, who dwells in the extreme Orient beyond Persia and Armenia, and is (with his people) a Christian, but a Nestorian, had waged war against the brother Kings of the Persians and Medes who are called the Samiards, and had captured Ecbatana, of which we have spoken above, the seat of their dominion. The said Kings having met him with their forces made up of Persians, Medes, and Assyrians, the battle had been maintained for 3 days, either side preferring death to flight. But at last PRESBYTER JOHN (for so they are wont to style him), having routed the Persians, came forth the victor from a most sanguinary battle. After this victory (he went on to say) the aforesaid John was advancing to fight in aid of the Church at Jerusalem ; but when he arrived at the Tigris, and found there no possible means of transport for his army, he turned northward, as he had heard that the river in that quarter was frozen over in winter-time. Halting there for some years* in expectation of a frost, which never came, owing to the mildness of the season, he lost many of his people through the unaccustomed climate, and was obliged to return homewards. This personage is said to be of the ancient race of those Magi who are mentioned in the Gospel, and to rule the same nations that they did, and to have such glory and wealth that he uses (they say) only an emerald sceptre. It was (they say) from his being fired by the example of his fathers, who came to adore Christ in the cradle, that he was proposing to go to Jerusalem, when he was prevented by the cause already alleged."

Professor Bruun will not accept Oppert's explanation, which identifies this King and Priest with the Gur-Khan of Karacathay, for whose profession of Christianity there is indeed (as has been indicated—*supra*) no real evidence ; who could not be said to have made an attack upon any pair of brother Kings of the Persians and the Medes, nor to have captured Ecbatana (a city, whatever its identity, of Media) ; who could never have had any intention of coming to Jerusalem ; and whose geographical position in no way suggested the mention of Armenia.

Professor Bruun thinks he finds a warrior much better answering to the indications in the Georgian prince John Orbelian, the general-in-chief under several successive Kings of Georgia in that age.

At the time when the Gur-Khan defeated Sanjar the real brothers of the latter had been long dead ; Sanjar had withdrawn from interference with the affairs of Western Persia ; and Hamadán (if this is to be regarded as Ecbatana) was no residence of his. But it was the residence of Sanjar's nephew Mas'úd, in whose hands was now the dominion of Western Persia ; whilst Mas'úd's nephew, Dáúd, held Media, *i.e.* Azerbeiján, Arrán, and Armenia. It is in these two princes that Professor Bruun sees the *Samiardi fratres* of the German chronicler.

Again the expression "extreme Orient" is to be interpreted by local usage. And with the people o Little Armenia, through whom probably such intelligence reached the Bishop of Gabala, the expression the *East* signified specifically Great Armenia (which was then a part of the kingdom of Georgia and Abkhasia), as Dulaurier has stated.†

It is true that the Georgians were not really Nestorians, but followers of the Greek Church. It was the fact, however, that in general, the Armenians, whom the

---

* Sic : *per aliquot annos*, but an evident error.
† *I. As.* sér. V. tom. xi. 449.

Greeks accused of following the Jacobite errors, retorted upon members of the Greek Church with the reproach of the opposite heresy of Nestorianism. And the attribution of Nestorianism to a Georgian Prince is, like the expression "*extreme East*," an indication of the Armenian channel through which the story came.

The intention to march to the aid of the Christians in Palestine is more like the act of a Georgian General than that of a Karacathayan Khan; and there are in the history of the Kingdom of Jerusalem several indications of the proposal at least of Georgian assistance.

The personage in question is said to have come from the country of the Magi, from whom he was descended. But these have frequently been supposed to come from Great Armenia. *E.g.* Friar Jordanus says they came from Moghán.*

The name *Ecbatana* has been so variously applied that it was likely to lead to ambiguities. But it so happens that, in a previous passage of his History, Bishop Otto of Freisingen, in rehearsing some Oriental information gathered apparently from the same Bishop of Gabala, has shown what was the place that he had been taught to identify with Ecbatana, viz. the old Armenian city of ANI.† Now this city was captured from the Turks, on behalf of the King of Georgia, David the Restorer, by his great *sbasalar*,‡ John Orbelian, in 1123-24.

Professor Bruun also lays stress upon a passage in a German chronicle of date some years later than Otho's work:

"1141. Liupoldus dux Bawariorum obiit, Henrico fratre ejus succedente in ducatu. Iohannes Presbyter Rex Armeniæ et Indiæ cum duobus regibus fratribus Persarum et Medorum pugnavit et vicit."§

He asks how the Gur-Khan of Karakhitai could be styled King of *Armenia* and of India? It may be asked, *per contra*, how either the King of Georgia or his *Peshwa* (to use the Mahratta analogy of John Orbelian's position) could be styled King of Armenia and of *India*? In reply to this, Professor Bruun adduces a variety of quotations which he considers as showing that the term *India* was applied to some Caucasian region.

My own conviction is that the report of Otto of Freisingen is not merely the *first mention* of a great Asiatic potentate called Prester John, but that his statement is the whole and sole basis of good faith on which the story of such a potentate rested; and I am quite as willing to believe, on due evidence, that the nucleus of fact to which his statement referred, and on which such a pile of long-enduring fiction was erected, occurred in Armenia as that it occurred in Turan. Indeed in many respects the story would thus be more comprehensible. One cannot attach any value to the quotation from the Annalist in Pertz, because there seems no reason to doubt that the passage is a mere adaptation of the report by Bishop Otto, of whose work the Annalist makes other use, as is indeed admitted by

---

* The Great Plain on the Lower Araxes and Cyrus. The word Moghán=*Magi*: and Abulfeda quotes this as the etymology of the name. (*Reinaud's Abulf.* I. 300.)—Y. [*Cordier, Odoric*, 36.]

† Here is the passage, which is worth giving for more reasons than one:

"That portion of ancient Babylon which is still occupied is (as we have heard from persons of character from beyond sea) styled BALDACH, whilst the part that lies, according to the prophecy, deserted and pathless extends some ten miles to the Tower of Babel. The inhabited portion called Baldach is very large and populous; and though it should belong to the Persian monarchy it has been conceded by the Kings of the Persians to their High Priest, whom they call the *Caliph*; in order that in this also a certain analogy [*quaedam habitudo*], such as has been often remarked before, should be exhibited between Babylon and Rome. For the same (privilege) that here in the city of Rome has been made over to our chief Pontiff by the Christian Emperor, has there been conceded to their High Priest by the Pagan Kings of Persia, to whom Babylonia has for a long time been subject. But the Kings of the Persians (just as our Kings have their royal city, like Aachen) have themselves established the seat of their kingdom at Egbatana, which, in the Book of Judith, Arphaxat is said to have founded, and which in their tongue is called HANI, containing as they allege 100,000 or more fighting men, and have reserved to themselves nothing of Babylon except the nominal dominion. Finally, the place which is now vulgarly called Babylonia, as I have mentioned, is not upon the Euphrates (at all) as people suppose, but on the Nile, about 6 days' journey from Alexandria; and is the same as Memphis, to which Cambyses, the son of Cyrus, anciently gave the name of Babylon."—Ottonis Frising. Lib. VII. cap. 3, in *Germanic Hist. Illust. etc. Christiani Urstisii Basiliensis,*. Francof. 1585.—Y.

‡ Sbasalar, or "General-in-chief," =Pers. *Sipáhsálár.*—Y.

§ *Continuatio Ann. Admutensium*, in Pertz, Scriptores, IX. 580

Professor Bruun, who (be it said) is a pattern of candour in controversy. But much else that the Professor alleges is interesting and striking. The fact that Azerbeijan and the adjoining regions were known as "the East" is patent to the readers of this book in many a page, where the Khan and his Mongols in occupation of that region are styled by Polo *Lord of the* LEVANT, *Tartars of the* LEVANT (*i.e.* of the East), even when the speaker's standpoint is in far Cathay.* The mention of *Ani* as identical with the Ecbatana of which Otto had heard is a remarkable circumstance which I think even Oppert has overlooked. That this Georgian hero *was* a Christian and that his name *was* John are considerable facts. Oppert's conversion of Korkhan into Yokhanan or John is anything but satisfactory. The identification proposed again makes it quite intelligible how the so-called Prester John should have talked about coming to the aid of the Crusaders; a point so difficult to explain on Oppert's theory, that he has been obliged to introduce a duplicate John in the person of a Greek Emperor to solve that knot; another of the weaker links in his argument. In fact, Professor Bruun's thesis seems to me more than fairly successful in *paving the way* for the introduction of a Caucasian Prester John; the barriers are removed, the carpets are spread, the trumpets sound royally—but the conquering hero comes not!

He does very nearly come. The almost royal power and splendour of the Orbelians at this time is on record: "They held the office of *Sbasalar* or Generalissimo of all Georgia. All the officers of the King's Palace were under their authority. Besides that they had 12 standards of their own, and under each standard 1000 warriors mustered. As the custom was for the King's flag to be white and the pennon over it red, it was ruled that the Orpelian flag should be red and the pennon white. . . . At banquets they alone had the right to couches whilst other princes had cushions only. Their food was served on silver; and to them it belonged to crown the kings."† Orpel Ivané, *i.e.* John Orbelian, Grand *Sbasalar*, was for years the pride of Georgia and the hammer of the Turks. In 1123-1124 he wrested from them Tiflis and the whole country up to the Araxes, including *Ani*, as we have said. His King David, the Restorer, bestowed on him large additional domains from the new conquests; and the like brilliant service and career of conquest was continued under David's sons and successors, Demetrius and George; his later achievements, however, and some of the most brilliant, occurring after the date of the Bishop of Gabala's visit to Rome. But still we hear of no actual conflict with the chief princes of the Seljukian house, and of no event in his history so important as to account for his being made to play the part of Presbyter Johannes in the story of the Bishop of Gabala. Professor Bruun's most forcible observation in reference to this rather serious difficulty is that the historians have transmitted to us extremely little detail concerning the reign of Demetrius II., and do not even agree as to its duration. Carebat vate sacro : "It was," says Brosset, "long and glorious, but it lacked a commemorator." If new facts can be alleged, the identity may still be proved. But meantime the conquests of the Gur-Khan and his defeat of Sanjar, just at a time which suits the story, are indubitable, and this great advantage Oppert's thesis retains. As regards the claim to the title of *Presbyter* nothing worth mentioning is alleged on either side.

When the Mongol Conquests threw Asia open to Frank travellers in the middle of the 13th century, their minds were full of Prester John; they sought in vain for an adequate representative, but it was not in the nature of things but they should find *some* representative. In fact they found *several*. Apparently no real tradition existed among the Eastern Christians of any such personage, but the persistent demand produced a supply, and the honour of identification with Prester John, after hovering over one head and another, settled finally upon that of the King of the Keraits, whom we find to play the part in our text.

Thus in Plano Carpini's single mention of Prester John as the King of the

Christians of India the Greater, who defeats the Tartars by an elaborate stratagem, Oppert recognizes Sultan Jaláluddín of Khwarizm and his temporary success over the Mongols in Afghanistan.   In the Armenian Prince Sempad's account, on the other hand, this Christian King of India is *aided* by the Tartars to defeat and harass the neighbouring Saracens, his enemies, and becomes the Mongol's vassal.   In the statement of Rubruquis, though distinct reference is made to the conquering Gurkhan (under the name of Coir Cham of Caracatay), the title of *King John* is assigned to the Naiman Prince (*Kushluk*), who had married the daughter of the last lineal sovereign of Karakhitai, and usurped his power, whilst, with a strange complication of confusion, UNC, Prince of the Crit and Merkit (Kerait and Merkit, two great tribes of Mongolia) * and Lord of Karákorum, is made the brother and successor of this Naiman Prince.   His version of the story, as it proceeds, has so much resemblance to Polo's, that we shall quote the words.   The Crit and Merkit, he says, were Nestorian Christians.   " But their Lord had abandoned the worship of Christ to follow idols, and kept by him those priests of the idols who are all devil-raisers and sorcerers.   Beyond his pastures, at the distance of ten or fifteen days' journey, were the pastures of the MOAL (Mongol), who were a very poor people, without a leader and without any religion except sorceries and divinations, such as all the people of those parts put so much faith in.   Next to Moal was another poor tribe called TARTAR.   King John having died without an heir, his brother Unc got his wealth, and caused himself to be proclaimed Cham, and sent out his flocks and herds even to the borders of Moal.   At that time there was a certain blacksmith called Chinghis among the tribe of Moal, and he used to lift the cattle of Unc Chan as often as he had a chance, insomuch that the herdsmen of Unc Chan made complaint to their master.   The latter assembled an army, and invaded the land of the Moal in search of Chinghis, but he fled and hid himself among the Tartars.   So Unc, having plundered the Moal and Tartars, returned home.   And Chinghis addressed the Tartars and Moal, saying : ' It is because we have no leader that we are thus oppressed by our neighbours.'   So both Tartars and Moal made Chinghis himself their leader and captain.   And having got a host quietly together, he made a sudden onslaught upon Unc and conquered him, and compelled him to flee into Cathay.   On that occasion his daughter was taken, and given by Chinghis to one of his sons, to whom she bore Mangu, who now reigneth. . . .   The land in which they (the Mongols) first were, and where the residence of Chinghis still exists, is called *Onan Kerule.*†   But because Caracoran is in the country which was their first conquest, they regard it as a royal city, and there hold the elections of their Chan."

Here we see plainly that the Unc Chan of Rubruquis is the Unc Can or Unecan of Polo.   In the narrative of the former, Unc is only *connected* with King or Prester John ; in that of the latter, rehearsing the story as heard some 20 or 25 years later, the two are *identified*.   The shadowy *rôle* of Prester John has passed from the Ruler of Kara Khitai to the Chief of the Keraits.   This transfer brings us to another history.

---

* ["The Keraits," says Mr. Rockhill (*Rubruck,*111, note), "lived on the Orkhon and the Tula, southeast of Lake Baikal ; Abulfaraj relates their conversion to Christianity in 1007 by the Nestorian Bishop of Merv.   Rashideddin, however, says their conversion took place in the time of Chingis Khan. (*D'Ohsson*, I. 48 ; *Chabot, Mar Jabalaha, III.* 14.) D'Avezac (536) identifies, with some plausibility, I think, the Keraits with the *Kĭ-lê* (or *T'teh-lê*) of the early Chinese annals.   The name K'ĭ-lê was applied in the 3rd century A.D. to *all* the Turkish tribes, such as the *Hui-hu* (Uigúrs), *Kieh-Ku* (Kirghiz) Alans, etc., and they are said to be the same as the *Kao-ch'ê*, from whom descended the *Cangle* of Rubruck. (*T'ang shu*, Bk. 217, i.; *Ma Tuan-lin*, Bk. 344, 9, Bk. 347, 4.) As to the Merkits, or Merkites, they were a nomadic people of Turkish stock, with a possible infusion of Mongol blood. They are called by Mohammedan writers Uduyut, and were divided into four tribes.   They lived on the Lower Selinga and its feeders. (*D'Ohsson*, i. 54 ; *Howorth, History*, I., pt. i. 22, 698.)"—H. C.]

† [*Onan Kerule* is "the country watered by the Orkhon and Kerulun Rivers, *i.e.* the country to the south and south-east of Lake Baikal.   The headquarters (*ya-chang*) of the principal chief of the Uigurs in the eighth century was 500 *li* (about 165 miles) south-west of the confluence of the Wen-Kun ho (Orkhon) and the Tu-lo ho (Tura).   Its ruins, sometimes, but wrongly, confounded with those of the Mongol city of Karakorum, some 20 miles from it, built in 1235 by Ogodai, are now known by the name of Kara Balgasun, ' Black City.' " [See p. 228.] The name *Onankerule* seems to be taken from the form *Onan-ou-Keloran*, which occurs in Mohammedan writers. (*Quatremère*, 115 *et seq.* ; see also *T'ang shu*, Bk. 43b ; *Rockhill, Rubruck*, 116, note.)—H. C.]

We have already spoken of the extensive diffusion of Nestorian Christianity in Asia during the early and Middle Ages. The Christian historian Gregory Abulfaraj relates a curious history of the conversion, in the beginning of the 11th century, of the King of *Kerith* with his people, dwelling in the remote north-east of the land of the Turks. And that the Keraits continued to profess Christianity down to the time of Chinghiz is attested by Rashiduddin's direct statement, as well as by the numerous Christian princesses from that tribe of whom we hear in Mongol history. It is the chief of this tribe of whom Rubruquis and Polo speak under the name of Unc Khan, and whom the latter identifies with Prester John. His proper name is called Tuli by the Chinese, and Togrul by the Persian historians, but the Kin sovereign of Northern China had conferred on him the title of *Wang* or King, from which his people gave him the slightly corrupted cognomen of اونک خان , which some scholars read *Awang*, and *Avenk* Khan, but which the spelling of Rubruquis and Polo shows probably to have been pronounced as *Aung* or *Ung* Khan.* The circumstance stated by Rubruquis of his having abandoned the profession of Christianity, is not alluded to by Eastern writers; but in any case his career is not a credit to the Faith. I cannot find any satisfactory corroboration of the claims of supremacy over the Mongols which Polo ascribes to Aung Khan. But that his power and dignity were considerable, appears from the term *Pádsháh* which Rashiduddin applies to him. He had at first obtained the sovereignty of the Keraits by the murder of two of his brothers and several nephews. Yessugai, the father of Chinghiz, had been his staunch friend, and had aided him effectually to recover his dominion from which he had been expelled. After a reign of many years he was again ejected, and in the greatest necessity sought the help of Temujin (afterwards called Chinghiz Khan), by whom he was treated with the greatest consideration. This was in 1196. For some years the two chiefs conducted their forays in alliance, but differences sprang up between them; the son of Aung Khan entered into a plot to kill Temujin, and in 1202-1203 they were in open war. The result will be related in connection with the next chapters.

We may observe that the idea which Joinville picked up in the East about Prester John corresponds pretty closely with that set forth by Marco. Joinville represents him as one of the princes to whom the Tartars were tributary in the days of their oppression, and as "their ancient enemy"; one of their first acts, on being organized under a king of their own, was to attack him and conquer him, slaying all that bore arms, but sparing all monks and priests. The expression used by Joinville in speaking of the original land of the Tartars, "*une grande* berrie *de sablon,*" has not been elucidated in any edition that I have seen. It is the Arabic بَرِّية , *Băríya,* "a Desert." No doubt Joinville learned the word in Palestine. (See *Joinville,* p. 143 *seqq.*; see also *Oppert, Der Presb. Johannes in Sage und Geschichte,* and *Cathay,* etc., pp. 173-182.) [*Fried. Zarncke, Der Priester Johannes; Cordier, Odoric.* —H. C.]

---

* Vámbéry makes *Ong* an Uighúr word, signifying "right." [Palladius (*l.c.* 23) says: "The consonance of the names of Wang-Khan and Wang-Ku (Ung-Khan and Ongu—Ongot of Rashiduddin, a Turkish Tribe) led to the confusion regarding the tribes and persons, which at M. Polo's time seems to have been general among the Europeans in China; M. Polo and Johannes de Monte Corvino transfer the title of Prester John from Wang-Khan, already perished at that time, to the distinguished family of Wang-Ku."—H. C.]

CHAPTER XLVII.

Of Chinghis, and how he became the First Kaan of the Tartars.

Now it came to pass in the year of Christ's Incarnation 1187 that the Tartars made them a King whose name was Chinghis Kaan.[1] He was a man of great worth, and of great ability (eloquence), and valour. And as soon as the news that he had been chosen King was spread abroad through those countries, all the Tartars in the world came to him and owned him for their Lord. And right well did he maintain the Sovereignty they had given him. What shall I say? The Tartars gathered to him in astonishing multitude, and when he saw such numbers he made a great furniture of spears and arrows and such other arms as they used, and set about the conquest of all those regions till he had conquered eight provinces. When he conquered a province he did no harm to the people or their property, but merely established some of his own men in the country along with a proportion of theirs, whilst he led the remainder to the conquest of other provinces. And when those whom he had conquered became aware how well and safely he protected them against all others, and how they suffered no ill at his hands, and saw what a noble prince he was, then they joined him heart and soul and became his devoted followers. And when he had thus gathered such a multitude that they seemed to cover the earth, he began to think of conquering a great part of the world. Now in the year of Christ 1200 he sent an embassy to Prester John, and desired to have his daughter to wife. But when Prester John heard that Chinghis Kaan demanded his daughter in marriage he waxed very wroth, and said to the Envoys, "What impudence is this, to ask my daughter to wife! Wist he not well that he was my liegeman and serf? Get ye back to him and tell him that I had liever set my daughter in the fire than give her in marriage to him, and that he deserves death at my hand, rebel and traitor that he is!" So he bade the Envoys begone at once, and never come into his presence again. The Envoys, on receiving this reply, departed straightway, and made haste to their master, and related all that Prester John had ordered them to say, keeping nothing back.[2]

NOTE 1.—Temujin was born in the year 1155, according to all the Persian historians, who are probably to be relied on; the Chinese put the event in 1162. 1187 does not appear to be a date of special importance in his history. His inauguration as sovereign under the name of Chinghiz Kaan was in 1202 according to the Persian authorities, in 1206 according to the Chinese.

In a preceding note (p. 236) we have quoted a passage in which Rubruquis calls Chinghiz "a certain blacksmith." This mistaken notion seems to have originated in the resemblance of his name *Temújin* to the Turki *Temúrjí*, a blacksmith; but it was common throughout Asia in the Middle Ages, and the story is to be found not only in Rubruquis, but in the books of Hayton, the Armenian prince, and of Ibn Batuta, the Moor. That cranky Orientalist, Dr. Isaac Jacob Schmidt, positively reviles William Rubruquis, one of the most truthful and delightful of travellers, and certainly not inferior to his critic in mother-wit, for adopting this story, and rebukes Timkowski—not for adopting it, but for merely telling us the very interesting fact that the story was still, in 1820, current in Mongolia. (*Schmidt's San. Setz.* 376, and *Timkowski*, I. 147.)

NOTE 2.—Several historians, among others Abulfaraj, represent Chinghiz as having married a daughter of Aung Khan; and this is current among some of the mediæval European writers, such as Vincent of Beauvais. It is also adopted by Pétis de la Croix in his history of Chinghiz, apparently from a comparatively late Turkish historian; and both D'Herbelot and St. Martin state the same; but there seems to be no foundation for it in the best authorities: either Persian or Chinese. (See *Abulfaragius*, p. 285; *Speculum Historiale*, Bk. XXIX. ch. lxix.; *Hist. of Genghiz Can*, p. 29; and *Golden Horde*, pp. 61-62.) But there is a real story at the basis of Polo's, which seems to be this: About 1202, when Aung Khan and Chinghiz were still acting in professed alliance, a double union was proposed between Aung Khan's daughter Jaur Bigi and Chinghiz's son Juji, and between Chinghiz's daughter Kijin Bigi and Togrul's grandson Kush Buka. From certain circumstances this union fell through, and this was one of the circumstances which opened the breach between the two chiefs. There were, however, several marriages between the families. (*Erdmann*, 283; others are quoted under ch. lix., note 2.)

CHAPTER XLVIII.

How Chinghis mustered his People to march against Prester John.

WHEN Chinghis Kaan heard the brutal message that Prester John had sent him, such rage seized him that his heart came nigh to bursting within him, for he was a man of a very lofty spirit. At last he spoke, and that so loud that all who were present could hear him : "Never more might he be prince if he took not revenge for the brutal message of Prester John, and such revenge that insult never in this world was so dearly paid for. And before long Prester John should know whether he were his serf or no!"

So then he mustered all his forces, and levied such a host as never before was seen or heard of, sending word to Prester John to be on his defence. And when Prester John had sure tidings that Chinghis was really coming against him with such a multitude, he still professed to treat it as a jest and a trifle, for, quoth he, "these be no soldiers." Natheless he marshalled his forces and mustered his people, and made great preparations, in order that if Chinghis did come, he might take him and put him to death. In fact he marshalled such an host of many different nations that it was a world's wonder.

And so both sides gat them ready to battle. And why should I make a long story of it? Chinghis Kaan with all his host arrived at a vast and beautiful plain which was called TANDUC, belonging to Prester. John, and there he pitched his camp; and so great was the multitude of his people that it was impossible to number them. And when he got tidings that Prester

John was coming, he rejoiced greatly, for the place afforded a fine and ample battle-ground, so he was right glad to tarry for him there, and greatly longed for his arrival.

But now leave we Chinghis and his host, and let us return to Prester John and his people.

CHAPTER XLIX.

How Prester John marched to meet Chinghis.

Now the story goes that when Prester John became aware that Chinghis with his host was marching against him, he went forth to meet him with all his forces, and advanced until he reached the same plain of Tanduc, and pitched his camp over against that of Chinghis Kaan at a distance of 20 miles. And then both armies remained at rest for two days that they might be fresher and heartier for battle.[1]

So when the two great hosts were pitched on the plains of Tanduc as you have heard, Chinghis Kaan one day summoned before him his astrologers, both Christians and Saracens, and desired them to let him know which of the two hosts would gain the battle, his own or Prester John's. The Saracens tried to ascertain, but were unable to give a true answer; the Christians, however, did give a true answer, and showed manifestly beforehand how the event should be. For they got a cane and split it lengthwise, and laid one half on this side and one half on that, allowing no one to touch the pieces. And one piece of cane they called *Chinghis Kaan*, and the other piece they called *Prester John*. And then they said to Chinghis: "Now mark! and you will see the event of the battle, and who shall have the best of it; for whose cane soever shall get above the other, to him shall victory be." He replied that he would fain see it, and bade them begin. Then the Christian astrologers read a Psalm out of the Psalter, and went through other incantations. And lo! whilst all were beholding, the cane that bore the name of Chinghis Kaan, without being touched by anybody, advanced to the other that bore the name of Prester John, and got on the top of it. When the Prince saw that he was greatly delighted, and seeing how in this matter he found the Christians to tell the truth, he always treated them with great respect, and held them for men of truth for ever after.[2]

Note 1.—Polo in the preceding chapter has stated that this plain of Tanduc was in Prester John's country. He plainly regards it as identical with the Tanduc of which he speaks more particularly in ch. lix. as belonging to Prester John's descendants, and which must be located near the Chinese Wall. He is no doubt wrong in placing the battle there. Sanang Setzen puts the battle between the two, the only one which he mentions, "at the outflow of the Onon near Kulen Buira." The same action is placed by De Mailla's authorities at Calantschan, by P. Hyacinth at Kharakchin Schatu, by Erdmann after Rashid in the vicinity of Hulun Barkat and Kalanchinalt, which latter was on the borders of the Churché or Manchus. All this points to the vicinity of Buir Nor and Hulan or Kalon Nor (though the Onon is far from these). But this was *not* the final defeat of Aung Khan or Prester John, which took place some time later (in 1203) at a place called the Chacher Ondur (or Heights), which Gaubil places between the Tula and the Kerulun, therefore near the modern Urga. Aung Khan was wounded, and fled over the frontier of the Naiman; the officers of that tribe seized and killed him. (*Schmidt*, 87, 383; *Erdmann*, 297; *Gaubil*, p. 10.)

Note 2.—A Tartar divination by twigs, but different from that here employed, is older than Herodotus, who ascribes it to the Scythians. We hear of one something like the last among the Alans, and (from Tacitus) among the Germans. The words of Hosea (iv. 12), "My people ask counsel at their stocks, and their staff declareth unto them," are thus explained by Theophylactus: "They stuck up a couple of sticks, whilst murmuring certain charms and incantations; the sticks then, by the operation of devils, direct or indirect, would fall over, and the direction of their fall was noted," etc. The Chinese method of divination comes still nearer to that in the text. It is conducted by tossing in the air two symmetrical pieces of wood or bamboo of a peculiar form. It is described by Mendoza, and more particularly, with illustrations, by Doolittle.*

But Rubruquis would seem to have witnessed nearly the same process that Polo describes. He reprehends the conjuring practices of the Nestorian priests among the Mongols, who seem to have tried to rival the indigenous *Kâms* or Medicine-men.

---

* [On the Chinese divining-twig, see *Dennys, Folk-lore of China*, 57.—H. C.]

Visiting the Lady Kuktai, a Christian Queen of Mangu Kaan, who was ill, he says: " The Nestorians were repeating certain verses, I know not what (they said it was part of a Psalm), over two twigs which were brought into contact in the hands of two men.   The monk stood by during the operation" (p. 326).* Pétis de la Croix quotes from Thévenot's travels, a similar mode of divination as much used, before a fight, among the Barbary corsairs.   Two men sit on the deck facing one another and each holding two arrows by the points, and hitching the notches of each pair of arrows into the other pair.   Then the ship's writer reads a certain Arabic formula, and it is pretended that whilst this goes on, the two sets of arrows, *of which one represents the Turks and the other the Christians*, struggle together in spite of the resistance of the holders, and finally one rises over the other.   This is perhaps the divination by arrows which is prohibited in the Koran. (*Sura*, V. v. 92.)   It is related by Abulfeda that Mahomed found in the Kaaba an image of Abraham with such arrows in his hand.

P. della Valle describes the same process, conducted by a Mahomedan conjuror of Aleppo : " By his incantations he made the four points of the arrows come together without any movement of the holders, and by the way the points spontaneously placed themselves, obtained answers to interrogatories."

And Mr. Jaeschke writes from Lahaul: "There are many different ways of divination practised among the Buddhists ; and that also mentioned by Marco Polo is known to our Lama, but in a slightly different way, making use of *two arrows* instead of a cane split up, wherefore this kind is called *da-mo*, 'Arrow-divination.'" Indeed the practice is not extinct in India, for in 1833 Mr. Vigne witnessed its application to detect the robber of a government chest at Lodiana.

As regards Chinghiz's respect for the Christians there are other stories.   Abulfaragius has one about Chinghiz seeing in a dream a religious person who promised him success.   He told the dream to his wife, Aung Khan's daughter, who said the description answered to that of the bishop who used to visit her father.   Chinghiz then inquired for a bishop among the Uighúr Christians in his camp, and they indicated Mar Denha.   Chinghiz thenceforward was milder towards the Christians, and showed them many distinctions (p. 285).   Vincent of Beauvais also speaks of Rabbanta, a Nestorian monk, who lived in the confidence of Chinghiz's wife, daughter of "the Christian King David or Prester John," and who used by divination to make many revelations to the Tartars.   We have already said that there seems no ground for assigning a daughter of Aung Khan as wife to Chinghiz.   But there was a *niece* of the former, named Abika, among the wives of Chinghiz.   And Rashiduddin *does* relate a dream of the Kaan's in relation to her.   But it was to the effect that he was divinely commanded to give her away ; and this he did next morning !

(*Rawlins. Herod.* IV. 67 ; *Amm. Marcell.* XXXI. 2 ; *Delvio, Disq. Magic.* 558 ; *Mendoza*, Hak. Soc. I. 47; *Doolittle*, 435-436 ; *Hist. of Genghizcan*, pp. 52-53; *Preston's al-Hariri*, p. 183 ; *P. della V.* II. 865-866; *Vigne*, I. 46 ; *D'Ohsson*, I. 418-419).

---

* [With reference to this passage from *Rubruck*, Mr. Rockhill says (195, note): "The mode of divining here referred to is apparently the same as that described by Polo.   It must not however be confounded with rabdomancy, in which bundles of wands or arrows were used." Ammianus Marcellinus (XXXI. 2. 350) says this mode of divination was practised by the Alans. "They have a singular way of divining : they take straight willow wands and make bundles of them, and on examining them at a certain time, with certain secret incantations, they know what is going to happen."—H. C.]

CHAPTER L.

The Battle between Chinghis Kaan and Prester John.

AND after both sides had rested well those two days, they armed for the fight and engaged in desperate combat; and it was the greatest battle that ever was seen. The numbers that were slain on both sides were very great, but in the end Chinghis Kaan obtained the victory. And in the battle Prester John was slain.

Death of Chinghiz Khan. (From a miniature in the *Livre des Merveilles.*)

And from that time forward, day by day, his kingdom passed into the hands of Chinghis Kaan till the whole was conquered.

I may tell you that Chinghis Kaan reigned six years after this battle, engaged continually in conquest, and taking many a province and city and stronghold. But at the end of those six years he went against a certain castle that was called CAAJU, and there he was shot with an arrow in the knee, so that he died of his wound.   A great pity it was, for he was a valiant man and a wise.[1]

I will now tell you who reigned after Chinghis, and then about the manners and customs of the Tartars.

Note 1.—Chinghiz in fact survived Aung Khan some 24 years, dying during his fifth expedition against Tangut, 18th August 1227, aged 65 according to the Chinese accounts, 72 according to the Persian.   Sanang Setzen says that Kurbeljin Goa Khatún, the beautiful Queen of Tangut, who had passed into the tents of the conqueror, did him some bodily mischief (it is not said what), and then went and drowned herself in the Karamuren (or Hwang-ho), which thenceforth was called by the Mongols the *Khátún-gol*, or Lady's River, a name which it in fact still bears. Carpini relates that Chinghiz was killed by lightning.   The Persian and Chinese historians, however, agree in speaking of his death as natural.   Gaubil calls the place of his death Lou-pan, which he says was in lat. 38°.   Rashiduddin calls it Leung-Shan, which appears to be the mountain range still so called in the heart of Shensi.

The name of the place before which Polo represents him as mortally wounded is very variously given.   According to Gaubil, Chinghiz was in reality dangerously wounded by an arrow-shot at the siege of Taitongfu in 1212.   And it is possible, as Oppert suggests, that Polo's account of his death before *Caagiu* (as I prefer the reading), arose out of a confusion between this circumstance and those of the death of *Mangku Kaan*, which is said to have occurred at the assault of Hochau in Sze-ch'uan, a name which Polo would write *Caagiu*, or nearly so.   Abulfaragius specifically says that Mangku Kaan died *by an arrow*; though it is true that other authors say he died of disease, and Haiton that he was drowned ; all which shows how excusable were Polo's errors as to events occurring 50 to 100 years before his time.   (See. *Oppert's Presbyter Johannes*, p. 76 ; *De Mailla*, IX. 275, and note ; *Gaubil*, 18, 50, 52, 121 ; *Erdmann*, 443 ; *Ss. Setzen*, 103.)

It is only by referring back to ch. xlvii., where we are told that Chinghiz "began to think of conquering a great part of the world," that we see Polo to have been really aware of the vast extent and aim of the conquests of Chinghiz ; the *aim* being literally the conquest of the world as he conceived it ; the *extent* of the empire which he initiated actually covering (probably) one half of the whole number of the human race.   (See remarks in *Koeppen, Die Relig. des Buddha*, II. 86.)

CHAPTER LI.

Of those who did Reign after Chinghis Kaan, and of the Customs of the Tartars.

Now the next that reigned after Chinghis Kaan, their first Lord,[1] was Cuy Kaan, and the third Prince was Batuy Kaan, and the fourth was Alacou Kaan, the fifth Mongou Kaan, the sixth Cublay Kaan, who is the sovereign now reigning, and is more potent than any of the five who went before him ; in fact, if you were to take all those five together, they would not be so powerful as he is[2] Nay, I will say yet more ; for if you were to put together all the Christians in the world, with their Emperors and their Kings, the whole of these Christians,—aye, and throw in the Saracens to boot,—would not have such power, or be able to do so much as this Cublay, who is the Lord of all the Tartars in the world, those of the Levant and of the Ponent included ; for these are all his liegemen and subjects. I mean to show you all about this great power of his in this book of ours.

You should be told also that all the Grand Kaans, and all the descendants of Chinghis their first Lord, are carried to a mountain that is called Altay to be interred. Wheresoever the Sovereign may die, he is carried to his burial in that mountain with his predecessors ; no matter an the place of his death were 100 days' journey distant, thither must he be carried to his burial.[3]

Let me tell you a strange thing too. When they are carrying the body of any Emperor to be buried with the others, the convoy that goes with the body doth put to the sword all whom they fall in with on the road, saying : "Go and wait upon your Lord in the other world!" For they do in sooth believe that all such as they slay in this manner do go to serve their Lord in the other world. They do the same too with horses ; for when the Emperor dies, they kill all his best horses, in order that he may have the use of them in the other world, as they believe. And I tell you as a certain truth, that when Mongou Kaan died, more than 20,000 persons, who chanced to meet the body on its way, were slain in the manner I have told.[4]

NOTE 1.—Before parting with Chinghiz let me point out what has not to my knowledge been suggested before, that the name of " *Cambuscan* bold " in Chaucer's tale is only a corruption of the name of Chinghiz. The name of the conqueror appears in Fr. Ricold as *Camiuscan*, from which the transition to Cambuscan presents no difficulty. *Camius* was, I suppose, a clerical corruption out of *Canjus* or *Cianjus*. In the chronicle of St. Antonino, however, we have him called " *Chinghiscan rectius* Tamgius *Cam* " (XIX. c. 8). If this is not merely the usual blunder of *t* for *c*, it presents a curious analogy to the form *Tankiz Khán* always used by Ibn Batuta. I do not know the origin of the latter, unless it was suggested by *tankis* (Ar.) " Turning upside down." (See *Pereg. Quat.*, p. 119 ; *I. B.* III. 22, etc.)

NOTE 2.—Polo's history here is inadmissible. He introduces into the list of the supreme Kaans *Batu*, who was only Khan of Kipchak (the Golden Horde), and *Hulaku*, who was Khan of Persia, whilst he omits *Okkodai*, the immediate successor of Chinghiz. It is also remarkable that he uses the form *Alacou* here instead of *Alaü* as elsewhere ; nor does he seem to mean the same person, for he was quite well aware that *Alaü* was Lord of the Levant, who sent ambassadors to the Great Khan Cúbláy, and could not therefore be one of his predecessors. The real succession ran : 1. Chinghiz ; 2. Okkodai ; 3. Kuyuk ; 4. Mangku ; 5. Kúblái.

There are quite as great errors in the history of Haiton, who had probably greater advantages in this respect than Marco. And I may note that in Teixeira's abridgment of Mirkhond, Hulaku is made to succeed Mangku Kaan on the throne of Chinghiz. (*Relaciones*, p. 338.)

NOTE 3.—The ALTAI here certainly does not mean the Great South Siberian Range to which the name is now applied. Both *Altai* and *Altun-Khan* appear sometimes to be applied by Sanang Setzen to the Khingan of the Chinese, or range running immediately north of the Great Wall near Kalgan. (See ch. lxi. note 1.) But in reference to this matter of the burial of Chinghiz, he describes the place as " the district of Yekeh Utek, between the shady side of the Altai-Khan and the sunny side of the Kentei-Khan." Now the Kentei-Khan (*khan* here meaning " mountain ") is near the sources of the Onon, immediately to the north-east of Urga ; and Altai-Khan in this connection cannot mean the hills near the Great Wall, 500 miles distant.

According to Rashiduddin, Chinghiz was buried at a place called *Búrkán Káldún* (" God's Hill "), or *Yekeh Kúrúk* (" The Great Sacred or Tabooed Place ") ; in another passage he calls the spot *Búdah Undúr* (which means, I fancy, the same as Búrkán Káldún), near the River Selenga. Búrkán Kaldún is often mentioned by Sanang Setzen, and Quatremère seems to demonstrate the identity of this place with the mountain called by Pallas (and Timkowski) *Khanoolla*. This is a lofty mountain near Urga, covered with dense forest, and is indeed the first woody mountain reached in travelling from Peking. It is still held sacred by the Mongols and guarded from access, though the tradition of Chinghiz's grave seems to be extinct. Now, as this Khanoolla (" Mount Royal," for *khan* here means " sovereign," and *oolla* " mountain ") stands immediately to the south of the *Kentei* mentioned in the quotation from S. Setzen, this identification agrees with his statement, on the supposition that the Khanoolla is the Altai of the same quotation. The Khanoolla must also be the *Han* mountain which Mongol chiefs claiming descent from Chinghiz named to Gaubil as the burial-place of that conqueror. Note that the Khanoolla, which we suppose to be the Altai of Polo, and here of Sanang Setzen, belongs to a range known as *Khingan*, whilst we see that Setzen elsewhere applies Altai and Altan-Khan to the other Khingan near the Great Wall.

Erdmann relates, apparently after Rashiduddin, that Chinghiz was buried at the foot of a tree which had taken his fancy on a hunting expedition, and which he had then pointed out as the place where he desired to be interred. It was then conspicuous, but afterwards the adjoining trees shot up so rapidly, that a dense wood covered the whole locality, and it became impossible to identify the spot. (*Q. R.* 117 *seqq.; Timk.* I. 115 *seqq.*, II. 475-476; *San. Setz.* 103, 114-115, 108-109; *Gaubil,* 54; *Erd.* 444.)

["There are no accurate indications," says Palladius (*l.c.* pp. 11-13), "in the documents of the Mongol period on the burial-places of Chingiz Khan and of the Khans who succeeded him. The *Yuan-shi* or 'History of the Mongol Dynasty in China,' in speaking of the burial of the Khans, mentions only that they used to be conveyed from Peking to the north, to their common burial-ground in the *K'i-lien* Valley. This name cannot have anything in common with the ancient *K'i-lien* of the Hiungnu, a hill situated to the west of the Mongol desert ; the *K'i-lien* of the Mongols is to be sought more to the east. When Khubilai marched out against Prince Nayan, and reached the modern Talnor, news was received of the occupation of the Khan's burial-ground by the rebels. They held out there very long, which exceedingly afflicted Khubilai [*Yuan shi lui pien*]; and this goes to prove that the tombs could not be situated much to the west. Some more positive information on this subject is found in the diary of the campaign in Mongolia in 1410, of the Ming Emperor Yunglo [*Pe ching lu*]. He reached the Kerulen at the place where this river, after running south, takes an easterly direction. The author of the diary notes, that from a place one march and a half before reaching the Kerulen, a very large mountain was visible to the north-east, and at its foot a solitary high and pointed hillock, covered with stones. The author says, that the sovereigns of the house of Yuan used to be buried near this hill. It may therefore be plausibly supposed that the tombs of the Mongol Khans were near the Kerulen, and that the 'K'i-lien' of the *Yuan shi* is to be applied to this locality ; it seems to me even, that K'i-lien is an abbreviation, customary to Chinese authors, of Kerulen. The way of burying the Mongol Khans is described in the *Yuan shi* (ch. 'On the national religious rites of the Mongols'), as well as in the *Ch'ue keng lu,* 'Memoirs of the time of the Yuan Dynasty.' When burying, the greatest care was taken to conceal from outside people the knowledge of the locality of the tomb. With this object in view, after the tomb was closed, a drove of horses was driven over it, and by this means the ground was, for a considerable distance, trampled down and levelled. It is added to this (probably from hearsay) in the *Ts'ao mu tze Memoirs* (also of the time of the Yuan Dynasty), that a young camel used to be killed (in the presence of its mother) on the tomb of the deceased Khan; afterwards, when the time of the usual offerings of the tomb approached, the mother of this immolated camel was set at liberty, and she came crying to the place where it was killed; the locality of the tomb was ascertained in this way."

The Archimandrite Palladius adds in a footnote : "Our well-known Mongolist N. Golovkin has told us, that according to a story actually current among the Mongols, the tombs of the former Mongol Khans are situated near Tas-ola Hill, equally in the vicinity of the Kerulen. He states also that even now the Mongols are accustomed to assemble on that hill on the seventh day of the seventh moon (according to an ancient custom), in order to adore Chingiz Khan's tomb. Altan tobchi (translated into Russian by Galsan Gomboeff), in relating the history of the Mongols after their expulsion from China, and speaking of the Khans' tombs, calls them *Naiman tzagan gher, i.e.* 'Eight White Tents' (according to the number of chambers for the souls of the chief deceased Khans in Peking), and sometimes simply *Tzagan gher,* 'the White Tent,' which, according to the translator's explanation, denotes only Chingiz Khan's tomb."

"According to the Chinese Annals (*T'ung kien kang mu*), quoted by Dr. E. Bretschneider (*Med. Res.* I. p. 157), Chinghiz died near the *Liu p'an shan* in 1227, after having subdued the Tangut empire. On modern Chinese maps *Liu p'an shan* is marked south of the city of *Ku yüan chou,* department of *P'ing liang,* in *Kan suh.* The *Yüan shí,* however, implies that he died in Northern Mongolia. We read there, in the annals, *s.a.* 1227, that in the fifth intercalary month the Emperor moved to the mountain *Liu p'an shan* in order to avoid the heat of the summer. In the sixth month the empire of the *Hia* (Tangut) submitted. Chinghiz rested on the river *Si Kiang* in the district of *Ts'ing shui* (in Kansuh ; it has still the same name). In autumn, in the seventh month (August), on the day *jen wu*, the Emperor fell ill, and eight days later died in his palace *Ha-lao-t'u* on the River *Sa-li*. This river Sali is repeatedly mentioned in the *Yüan shi*, viz. in the first chapter, in connection with the first military doings of Chinghiz. Rashid reports (*D'Ohsson*, I. 58) that Chinghiz in 1199 retired to his residence *Sari Kihar*. The *Yüan chao pi shi* (Palladius' transl., 81) writes the same name *Saari Keher* (*Keher* in modern Mongol means 'a plain'). On the ancient map of Mongolia found in the *Yüan shi lei pien*, *Sa-li K'ie-rh* is marked south of the river *Wa-nan* (the *Onon* of our maps), and close to *Sa-li K'ie-rh* we read: ' Here was the original abode of the Yüan ' (Mongols). Thus it seems the passage in the Yüan history translated above intimates that Chinghiz died in Mongolia, and not near the *Liu p'an shan*, as is generally believed. The *Yüan ch'ao pi shi* (Palladius' transl., 152) and the '*Ts'in cheng lu* (Palladius' transl., 195) both agree in stating that, after subduing the Tangut empire, Chinghiz returned home, and then died. Colonel Yule, in his *Marco Polo* (I. 245), states ' that Rashid calls the place of Chinghiz' death *Leung shan*, which appears to be the mountain range still so-called in the heart of Shensi.' I am not aware from what translation of Rashid, Yule's statement is derived, but d'Ohsson (I. 375, note) seems to quote the same passage in translating from Rashid : ' *Liu-p'an-shan* was situated on the frontiers of the *Churche* (empire of the *Kin*), *Nangias* (empire of the *Sung*) and *Tangut* ;' which statement is quite correct."

We now come to the Mongol tradition, which places the tomb of Chinghiz in the country of the Ordos, in the great bend of the Yellow River.

Two Belgian missionaries, MM. de Vos and Verlinden, who visited the tomb of Chinghiz Khan, say that before the Mahomedan invasion, on a hill a few feet high, there were two courtyards, one in front of the other, surrounded by palisades. In the second courtyard, there were a building like a Chinese dwelling-house and six tents. In a double tent are kept the remains of the *bokta* (the Holy). The neighbouring tents contained various precious objects, such as a gold saddle, dishes, drinking-cups, a tripod, a kettle, and many other utensils, all in solid silver. (*Missions Catholiques*, No. 315, 18th June, 1875.)—This periodical gives (p. 293) a sketch of the tomb of the Conqueror, according to the account of the two missionaries.

Prjevalsky (*Mongolia and Tangut*) relates the story of the *Khatún Gol* (see *supra*, p. 245), and says that her tomb is situated at 11 versts north-east of lake of Dzaïdemin Nor, and is called by the Mongols Tumir-Alku, and by the Chinese Djiou-Djin Fu ; one of the legends mentioned by the Russian traveller gives the Ordo country as the burial-place of Chinghiz, 200 versts south of lake Dabasun Nor ; the remains are kept in two coffins, one of wood, the other of silver ; and the Khan prophesied that after eight or ten centuries he would come to life again and fight the Emperor of China, and being victorious, would take the Mongols from the Ordos back to their country of Khalka ; Prjevalsky did not see the tomb, nor did Potanin.

" Their holiest place [of the Mongols of Ordos] is a collection of felt tents called ' Edjen-joro,' reputed to contain the bones of Jenghiz Khan. These sacred relics are entrusted to the care of a caste of Darhats, numbering some fifty families. Every summer, on the twenty-first day of the sixth moon, sacrifices are offered up in his honour, when numbers of people congregate to join in the celebration, such gatherings being called *táilgan*." On the southern border of the Ordos are the ruins of Boro-balgasun [Grey town], said to date from Jenghiz Khan's time. (*Potanin, Proc. R. G. S.* IX. 1887, p. 233.)

The last traveller who visited the tomb of Chinghiz is M. C. E. Bonin, in July 1896 ; he was then on the banks of the Yellow River in the northern part of the Ordo country, which is exclusively inhabited by nomadic and pastoral Mongols, forming seven tribes or hords, Djungar, Talat, Wan, Ottok, Djassak, Wushun and Hangkin, among which are eastward the Djungar and in the centre the Wan ; according to their own tradition, these tribes descend from the seven armies encamped in the country at the time of Chinghiz's death; the King of Djungar was 67 years of age, and was the chief of all the tribes, being considered the 37th descendant of the conqueror in a direct line. His predecessor was the Wushun Wang. M. Bonin gives (*Revue de Paris*, 15th February 1898) the following description of the tomb and of the country surrounding it. Between the *yamen* (palace) of the King (Wang) of Djungar and the tomb of Chinghiz-Khan, there are five or six marches made difficult by the sands of the Gobi, but horses and camels may be used for the journey. The road, southward through the desert, passes near the great lama-monastery called *Barong-tsao* or *Si-tsao* (Monastery of the West), and in Chinese *San-t'ang sse* (Three Temples). This celebrated monastery was built by the King of Djungar to hold the tablets of his ancestors—on the ruins of an old temple, said to have been erected by Chinghiz himself. More than a thousand lamas are registered there, forty of them live at the expense of the Emperor of China. Crossing afterwards the two upper branches of the Ulan Múren (Red River) on the banks of which Chinghiz was murdered, according to local tradition, close to the lake of Chahan Nor (White Lake), near which are the tents of the Prince of Wan, one arrives at last at the spot called *Yeke-Etjen-Koro*, in Mongol: the abode of the Great Lord, where the tomb is to be found. It is erected to the south-east of the village, comprising some twenty tents or tent-like huts built of earth. Two large white felt tents, placed side by side, similar to the tents of the modern Mongols, but much larger, cover the tomb; a red curtain, when drawn, discloses the large and low silver coffin, which contains the ashes of the Emperor, placed on the ground of the second tent; it is shaped like a big trunk, with great rosaces engraved upon it. The Emperor, according to local tradition, was cremated on the bank of the Ulan Muren, where he is supposed to have been slain. On the twenty-first day of the third moon the anniversary fête of Mongolia takes place; on this day of the year only are the two mortuary tents opened, and the coffin is exhibited to be venerated by people coming from all parts of Mongolia. Many other relics, dispersed all over the Ordo land, are brought thither on this occasion; these relics called in Mongol *Chinghiz Bogdo* (Sacred remains of Chinghiz) number ten; they are in the order adopted by the Mongols: the saddle of Chinghiz, hidden in the Wan territory; the bow, kept at a place named Hu-ki-ta-lao Hei, near Yeke Etjen-Koro; the remains of his war-horse, called Antegan-tsegun (more), preserved at Kebere in the Djungar territory; a fire-arm kept in the palace of the King of Djungar; a wooden and leather vase called Pao-lao-antri, kept at the place Shien-ni-chente; a wax figure containing the ashes of the Khan's equerry, called Altaqua-tosu, kept at Ottok (one of the seven tribes); the remains of the second wife, who lay at Kiasa, on the banks of the Yellow River, at a place called on Prjevalsky's map in Chinese Djiou-Djin-fu, and in Mongol Tumir-Alku; the tomb of the third wife of Chinghiz, who killed him, and lay to-day at Bagha-Ejen-Koro, "the abode of the little Sovereign," at a day's march to the south of the Djungar King's palace; the very tomb of Yeke-Etjen-Koro, which is supposed to contain also the ashes of the first wife of the Khan; and last, his great standard, a black wood spear planted in the desert, more than 150 miles to the south of the tomb; the iron of it never gets rusty; no one dares touch it, and therefore it is not carried to Yeke-Etjen-Koro with the other relics for the yearly festival. (See also *Rockhill, Diary*, p. 29.)—H. C.]

NOTE 4.—Rashiduddin relates that the escort, in carrying Chinghiz to his burial, slew all whom they met, and that forty noble and beautiful girls were despatched to serve him in the other world, as well as superb horses. As Mangku Kaan died in the heart of China, any attempt to carry out the barbarous rule in his case would involve great slaughter. (*Erd.* 443; *D'Ohsson*, I. 381, II. 13; and see *Cathay*, 507-508.)

Sanang Setzen ignores these barbarities. He describes the body of Chinghiz as removed to his native land on a two-wheeled waggon, the whole host escorting it, and wailing as they went: " And Kiluken Bahadur of the Sunid Tribe (one of the Khan's old comrades) lifted up his voice and sang—

Whilom Thou didst swoop like a Falcon : A rumbling waggon now trundles thee off:
                              O My King !
Hast thou in truth then forsaken thy wife and thy children and the Diet of thy People?
                              O My King !
Circling in pride like an Eagle whilom Thou didst lead us,
                              O My King !
But now Thou hast stumbled and fallen, like an unbroken Colt,
                              O My King !' " (p. 108.)

[" The burying of living men with the dead was a general custom with the tribes of Eastern Asia. Favourite servants and wives were usually buried in this way. In China, the chief wives and those concubines who had already borne children, were exempted from this lot. The Tunguz and other tribes were accustomed to kill the selected victims by strangulation. In China they used to be buried alive ; but the custom of burying living men ceased in A.D. 1464. [*Hwang ming ts'ung sin lu.*] In the time of the present Manchu Dynasty, the burying of living men was prohibited by the Emperor Kang-hi, at the close of the 17th century, *i.e.* the forced burying ; but voluntary sepulture remained in force [*Yu chi wen*]. Notwithstanding this prohibition, cases of forced burying occurred again in remote parts of Manchuria ; when a concubine refused to follow her deceased master, she was forcibly strangled with a bow-string [*Ninguta chi*]. I must observe, however, that there is no mention made in historical documents of the existence of this custom with the Mongols; it is only an hypothesis based on the analogy between the religious ideas and customs of the Mongols and those of other tribes." (*Palladius*, p. 13.)

In his *Religious System of China*, II., Dr. J. J. M. de Groot devotes a whole chapter (ix. 721 *seqq.*), *Concerning the Sacrifice of Human Beings at Burials, and Usages connected therewith.* The oldest case on record in China dates as far back as B.C. 677, when sixty-six men were killed after the ruler Wu of the state of Ts'in died.

The Official Annals of the Tartar Dynasty of Liao, quoted by Professor J. J. M. de Groot (*Religious System of China*, vol. ii. 698), state that "in the tenth year of the T'ung hwo period (A.D. 692) the killing of horses for funeral and burial rites was interdicted, as also the putting into the tombs of coats of mail, helmets, and articles and trinkets of gold and silver." Professor de Groot writes (*l.c.* 709) : "But, just as the placing of victuals in the graves was at an early date changed into sacrifices of food outside the graves, so burying horses with the dead was also modified under the Han Dynasty into presenting them to the dead without interring them, and valueless counterfeits were on such occasions substituted for the real animals."—H. C.]

CHAPTER LII.

CONCERNING THE CUSTOMS OF THE TARTARS.

Now that we have begun to speak of the Tartars, I have plenty to tell you on that subject. The Tartar custom is to spend the winter in warm plains, where they find good pasture for their cattle, whilst in summer they betake themselves to a cool climate among the mountains and valleys, where water is to be found as well as woods and pastures.

Their houses are circular, and are made of wands covered with felts.[1] These are carried along with them whithersoever they go; for the wands are so strongly bound together, and likewise so well combined, that the frame can be made very light. Whenever they erect these huts the door is always to the south. They also have waggons covered with black felt so efficaciously that no rain can get in. These are drawn by oxen and camels, and the women and children travel in them.[2] The women do the buying and selling, and whatever is necessary to provide for the husband and household; for the men all lead the life of gentlemen, troubling themselves about nothing but hunting and hawking, and looking after their goshawks and falcons, unless it be the practice of warlike exercises.

They live on the milk and meat which their herds supply, and on the produce of the chase; and they eat all kinds of flesh, including that of horses and dogs, and Pharaoh's rats, of which last there are great numbers in burrows on those plains.[3] Their drink is mare's milk.

They are very careful not to meddle with each other's wives, and will not do so on any account, holding that to be an evil and abominable thing. The women too are very good and loyal to their husbands, and notable housewives withal.[4] [Ten or twenty of them will dwell together in charming peace and unity, nor shall you ever hear an ill word among them.]

The marriage customs of Tartars are as follows. Any man may take a hundred wives an he so please, and if he be able to keep them. But the first wife is ever held most in honour, and as the most legitimate [and the same applies to the sons whom she may bear].

The husband gives a marriage payment to his wife's mother, and the wife brings nothing to her husband. They have more children than other people, because they have so many wives. They may marry their cousins, and if a father dies, his son may take any of the wives, his own mother always excepted; that is to say the eldest son may do this, but no other. A man may also take the wife of his own brother after the latter's death. Their weddings are celebrated with great ado.[5]

NOTE 1.—The word here in the G. T. is "*fennes*," which seems usually to mean *ropes*, and in fact Pauthier's text reads: "*Il ont mesons de verges et les cueuvrent de cordes.*" Ramusio's text has *feltroni*, and both Müller and the Latin of the S. G. have *filtro*. This is certainly the right reading. But whether *fennes* was ever used as a form of *feltres* (as *pennes* means *peltry*) I cannot discover. Perhaps some words have dropped out. A good description of a Kirghiz hut (35 feet in diameter), and exactly corresponding to Polo's account, will be found in *Atkinson's Siberia*, and another in *Vámbéry's Travels*. How comfortable and civilised the aspect of such a hut may be, can be seen also in Burnes's account of a Turkoman dwelling of this kind. This description of hut or tent is common to nearly all the nomade tribes of Central Asia. The trellis-work forming the skeleton of the tent-walls is (at least among the Turkomans) loosely pivoted, so as to draw out and compress like "lazy-tongs."

Dressing up a tent.

Rubruquis, Pallas, Timkowski, and others, notice the custom of turning the door to the south; the reason is obvious. (*Atkinson*, 285; *Vámb.* 316; *Burnes*, III. 51; *Conolly*, I. 96.) But throughout the Altai, Mr. Ney Elias informs me, K'alkas,

Kirghiz, and Kalmaks all pitch their tents facing *east*.   The prevailing winter wind is there *westerly*.

[Mr. Rockhill (*Rubruck*, p. 56, note) says that he has often seen Mongol tents facing east and south-east.   He adds : "It is interesting to find it noted in the *Chou Shu* (Bk. 50, 3) that the Khan of the Turks, who lived always on the Tu-kin mountains, had his tent invariably facing south, so as to show reverence to the sun's rising place." —H. C.]

Note 2.—Æschylus already knows the

> "wandering Scyths who dwell
>   In latticed huts high-poised on easy wheels."
>
> (*Prom. Vinct.* 709-710.)

And long before him Hesiod says Phineus was carried by the Harpies—

> "To the Land of the Milk-fed nations, whose houses are waggons."
>
> (*Strabo*, vii. 3-9.)

Ibn Batuta describes the Tartar waggon in which he travelled to Sarai as mounted on four great wheels, and drawn by two or more horses :—

"On the waggon is put a sort of pavilion of wands laced together with narrow thongs.   It is very light, and is covered with felt or cloth, and has latticed windows, so that the person inside can look out without being seen.   He can change his position at pleasure, sleeping or eating, reading or writing, during the journey."   These waggons were sometimes of enormous size.   Rubruquis declares that he measured between the wheel-tracks of one and found the interval to be 20 feet.   The axle was like a ship's mast, and twenty-two oxen were yoked to the waggon, eleven abreast. (See opposite cut.)   He describes the huts as not usually taken to pieces, but carried all standing.   The waggon just mentioned carried a hut of 30 feet diameter, for it projected beyond the wheels at least 5 feet on either side.   In fact, Carpini says explicitly, "Some of the huts are speedily taken to pieces and put up again ; such are packed on the beasts. Others cannot be taken to pieces, but are carried bodily on the waggons.   To carry the smaller tents on a waggon one ox may serve ; for the larger ones three oxen or four, or even more, according to the size."   The carts that were used to transport the Tartar valuables were covered with felt soaked in tallow or ewe's milk, to make them waterproof.   The tilts of these were rectangular, in the form of a large trunk.   The carts used in Kashgar, as described by Mr. Shaw, seem to resemble these latter. (*I. B.* II. 381-382 ; *Rub.* 221 ; *Carp.* 6, 16.)

The words of Herodotus, speaking generally of the Scyths, apply perfectly to the Mongol hordes under Chinghiz : "Having neither cities nor forts, and carrying their dwellings with them wherever they go ; accustomed, moreover, one and all, to shoot from horseback ; and living not by husbandry but on their cattle, their waggons the only houses that they possess, how can they fail of being unconquerable?" (Bk. IV. ch. 46, p. 41, *Rawlins.*)   Scythian prisoners in their waggons are represented on the Column of Theodosius at Constantinople ; but it is difficult to believe that these waggons, at least as figured in Banduri, have any really Scythian character.

It is a curious fact that the practice of carrying these *yurts* or felt tents upon waggons appears to be entirely obsolete in Mongolia.   Mr. Ney Elias writes : "I frequently showed your picture [that opposite] to Mongols, Chinese, and Russian border-traders, but none had ever seen anything of the kind.   The only cart I have ever seen used by Mongols is a little low, light, roughly-made bullock-dray, *certainly* of Chinese importation."   The old system would, however, appear to have been kept up to our own times by the Nogai Tartars, near the Sea of Azof.   (See note from Heber, in *Clark's Travels*, 8vo ed. I. 440, and Dr. Clark's vignette at p. 394 in the same volume.)

Note 3.—*Pharaoh's Rat* was properly the Gerboa of Arabia and North Africa,

Mediæval Tartar Huts and Waggons.

which the Arabs also regard as a dainty. There is a kindred animal in Siberia, called *Alactaga*, and a kind of Kangaroo-rat (probably the same) is mentioned as very abundant on the Mongolian Steppe. There is also the *Zieselmaus* of Pallas, a Dormouse, I believe, which he says the Kalmaks, even of distinction, count a delicacy, especially cooked in sour milk. "They eat not only the flesh of all their different kinds of cattle, including horses and camels, but also that of many wild animals which other nations eschew, *e.g.* marmots and *zieselmice*, beavers, badgers, otters, and lynxes, leaving none untouched except the dog and weasel kind, and also (unless *very* hard pressed) the flesh of the fox and the wolf." (*Pallas, Samml.* I. 128; also *Rubr.* 229-230.)

["In the Mongol biography of Chinghiz Khan (Mongol text of the *Yuan ch'ao pi shi*), mention is made of two kinds of animals (mice) used for food; the tarbagat (*Aritomys Bobac*) and *kuchugur.*" (*Palladius, l.c.* p. 14.) Regarding the marmots called *Sogur* by Rubruquis, Mr. Rockhill writes (p. 69): "Probably the *Mus citillus*, the *Suslik* of the Russians. . . . M. Grenard tells me that *Soghur*, more usually written *sour* in Turki, is the ordinary name of the marmot."—H. C.]

NOTE 4.—"Their wives are chaste; nor does one ever hear any talk of their immodesty," says Carpini;—no Boccaccian and Chaucerian stories.

NOTE 5.—"The Mongols are not prohibited from having a plurality of wives; the first manages the domestic concerns, and is the most respected." (*Timk.* II. 310.) Naturally Polygamy is not so general among the Mongols as when Asia lay at their feet. The Buraets, who seem to retain the old Mongol customs in great completeness, are polygamists, and have as many wives as they choose. Polygamy is also very prevalent among the Yakuts, whose lineage seems to be Eastern Turk. (*Ritter,* III. 125; *Erman,* II. 346.)

Of the custom that entitled the son on succeeding to take such as he pleased of his deceased father's wives, we have had some illustration (see *Prologue,* ch. xvii. note 2), and many instances will be found in Hammer's or other Mongol Histories. The same custom seems to be ascribed by Herodotus to the Scyths (IV. 78). A number of citations regarding the practice are given by Quatremère. (*Q. R.* p. 92.) A modern Mongol writer in the *Mélanges Asiatiques* of the Petersburg Academy, states that the custom of taking a deceased brother's wives is now obsolete, but that a proverb preserves its memory (II. 656). It is the custom of some Mahomedan nations, notably of the Afghans, and is one of those points that have been cited as a supposed proof of their Hebrew lineage.

"The Kalin is a present which the Bridegroom or his parents make to the parents of the Bride. All the Pagan nations of Siberia have this custom; they differ only in what constitutes the present, whether money or cattle." (*Gmelin,* I. 29; see also *Erman,* II. 348.)

CHAPTER LIII.

CONCERNING THE GOD OF THE TARTARS.

THIS is the fashion of their religion. [They say there is a Most High God of Heaven, whom they worship daily with thurible and incense, but they pray to Him only for health of mind and body.   But] they have [also] a certain [other] god of theirs called NATIGAY, and they say he is the god of the Earth, who watches over their children, cattle, and crops.   They show him great worship and honour, and every man hath a figure of him in his house, made of felt and cloth ; and they also make in the same manner images of his wife and children.   The wife they put on the left hand, and the children in front. And when they eat, they take the fat of the meat and grease the god's mouth withal, as well as the mouths of his wife and children.   Then they take of the broth and sprinkle it before the door of the house ; and that done, they deem that their god and his family have had their share of the dinner.[1]

Their drink is mare's milk, prepared in such a way that you would take it for white wine ; and a right good drink it is, called by them *Kemiz*.[2]

The clothes of the wealthy Tartars are for the most part of gold and silk stuffs, lined with costly furs, such as sable and ermine, vair and fox-skin, in the richest fashion.

NOTE 1.—There is no reference here to Buddhism, which was then of recent introduction among the Mongols ; indeed, at the end of the chapter, Polo speaks of their new adoption of the Chinese idolatry, *i.e.* Buddhism.   We may add here that the Buddhism of the Mongols decayed and became practically extinct after their expulsion from China (1368-1369).   The old Shamanism then apparently revived ; nor was it till 1577 that the great reconversion of Mongolia to Lamaism began.   This reconversion is the most prominent event in the Mongol history of Sanang Setzen, whose great-grandfather Khutuktai Setzen, Prince of the Ordos, was a chief agent in the movement.

The Supreme Good Spirit appears to have been called by the Mongols *Tengri* (Heaven), and *Khormuzda*, and is identified by Schmidt with the Persian Hormuzd. In Buddhist times he became identified with Indra.

Plano Carpini's account of this matter is very like Marco's : " They believe in one God, the Maker of all things, visible and invisible, and the Distributor of good and evil in this world ; but they worship Him not with prayers or praises or any kind of service.   Natheless, they have certain idols of felt, imitating the human face, and having underneath the face something resembling teats ; these they place on either side of the door.   These they believe to be the guardians of the flocks, from whom they have the boons of milk and increase.   Others they fabricate of bits of silk, and these are highly honoured ; . . . . and whenever they begin to eat or drink, they first offer these idols a portion of their food or drink."

The account agrees generally with what we are told of the original Shamanism of the Tunguses, which recognizes a Supreme Power over all, and a small number of potent spirits called *Ongot.*   These spirits among the Buraets are called, according to one author, *Nougait* or *Nogat*, and according to Erman *Ongotui.*   In some form of this same word, *Nogait, Ongot, Onggod, Ongotui*, we are, I imagine, to trace the *Natigay* of Polo.   The modern representative of this Shamanist *Lar* is still found among the Buraets, and is thus described by Pallas under the name of *Immegiljin:* "He is honoured as the tutelary god of the sheep and other cattle.   Properly, the divinity consists of *two* figures, hanging side by side, one of whom represents the god's wife.   These two figures are merely a pair of lanky flat bolsters with the upper part shaped into a round disk, and the body hung with a long woolly fleece ; eyes, nose, breasts, and navel, being indicated by leather knobs stitched on.   The male figure commonly has at his girdle the foot-rope with which horses at pasture are fettered, whilst the female, which is sometimes accompanied by smaller figures representing her children, has all sorts of little nicknacks and sewing implements."   Galsang

Tartar Idols and Kumis Churn.

Czomboyef, a recent Russo-Mongol writer already quoted, says also : "Among the Buryats, in the middle of the hut and place of honour, is the *Dsaiagaçhi* or 'Chief Creator of Fortune.'   At the door is the *Emelgelji*, the Tutelary of the Herds and Young Cattle, made of sheepskins.   Outside the hut is the *Chandaghatu*, a name implying that the idol was formed of a white hare-skin, the Tutelary of the Chase and perhaps of War.   All these have been expelled by Buddhism except Dsaiagachi, who is called *Tengri*, and introduced among the Buddhist divinities."

[Dorji Banzaroff, in his dissertation *On the Black Religion, i.e.* Shamanism, 1846, "is disposed to see in Natigay of M. Polo, the Ytoga of other travellers, *i.e.* the Mongol *Etugen*—'earth,' as the object of veneration of the Mongol Shamans. They look upon it as a divinity, for its power as *Delegei in echen, i.e.* 'the Lord of Earth,' and on account of its productiveness, *Altan delegei, i.e.* 'Golden Earth.'" Palladius (*l.c.* pp. 14-16) adds one new variant to what the learned Colonel Yule has collected and set forth with such precision, on the Shaman household gods.   "The Dahurs and Barhus have in their dwellings, according to the number of the male members of the family, puppets made of straw, on which eyes, eyebrows, and mouth are drawn ; these puppets are dressed up to the waist.   When some one of the family dies, his puppet is taken out of the house, and a new puppet is made for every newly-born member of the family.   On New Year's Day offerings are made to the puppets, and care is taken not to disturb them (by moving them, etc.), in order to avoid bringing sickness upon the family."   (*He lung kiang wai ki.*)

(Cf. *Rubruck*, 58-59, and Mr. Rockhill's note, 59-60.)—H. C.]

NOTE 2.—KIMIZ or KUMIZ, the habitual drink of the Mongols, as it still is of most of the nomads of Asia.   It is thus made.   Fresh mare's milk is put in a well-seasoned bottle-necked vessel of horse-skin ; a little *kurút* (see note 5, ch. liv.) or some sour cow's milk is added ; and when acetous fermentation is commencing it is violently churned with a peculiar staff which constantly stands in the vessel.   This interrupts fermentation and introduces a quantity of air into the liquid.   It is customary for visitors who may drop in to give a turn or two at the churn-stick.   After three or four days the drink is ready.

Kumiz keeps long ; it is wonderfully tonic and nutritious, and it is said that it has cured many persons threatened with consumption.   The tribes using it are said to be remarkably free from pulmonary disease ; and indeed I understand there is a regular *Galactopathic* establishment somewhere in the province of Orenburg for treating pulmonary patients with Kumiz diet.

It has a peculiar fore- and after-taste which, it is said, everybody does not like. Yet I have found no confession of a dislike to Kumiz.   Rubruquis tells us it is pungent on the tongue, like *vinum raspei* (*vin rapé* of the French), whilst you are drinking it, but leaves behind a pleasant flavour like milk of almonds.   It makes a man's inside feel very cosy, he adds, even turning a weak head, and is strongly diuretic.   To this last statement, however, modern report is in direct contradiction. The Greeks and other Oriental Christians considered it a sort of denial of the faith to drink Kumiz.   On the other hand, the Mahomedan converts from the nomad tribes seem to have adhered to the use of Kumiz even when strict in abstinence from wine ; and it was indulged in by the early Mamelukes as a public solemnity.   Excess on such an occasion killed Bibars Bundukdari, who was passionately fond of this liquor.

The intoxicating power of Kumiz varies according to the *brew*.   The more advanced is the vinous fermentation the less acid is the taste and the more it sparkles. The effect, however, is always slight and transitory, and leaves no unpleasant sensation, whilst it produces a strong tendency to refreshing sleep.   If its good qualities amount to half what are ascribed to it by Dr. W. F. Dahl, from whom we derive some of these particulars, it must be the pearl of all beverages.   "With the nomads it is the drink of all from the suckling upwards, it is the solace of age and illness, and the greatest of treats to all !"

There was a special kind called *Ḳard Ḳumiz*, which is mentioned both by Rubruquis and in the history of Wassáf.   It seems to have been strained and clarified.   The modern Tartars distil a spirit from Kumiz of which Pallas gives a detailed account.   (*Dahl, Ueber den Kumyss* in *Baer's Beiträge*, VII. ; *Lettres sur le Caucase et la Crimée*, Paris, 1859, p. 81 ; *Makrizi*, II. 147 ; *J. As.* XI. 160 ; *Levchine*, 322-323 ; *Rubr.* 227-228, 335 ; *Gold. Horde*, p. 46 ; *Erman*, I. 296 ; *Pallas, Samml.* I. 132 *seqq.*)

[In the *Si yu ki*, Travels to the West of Ch'ang ch'un, we find a drink called *tung lo*.   "The Chinese characters, *tung lo*," says Bretschneider (*Med. Res.* I. 94), " denote according to the dictionaries preparations from mare's or cow's milk, as Kumis, sour milk, etc.   In the *Yuan shi* (ch. cxxviii.) biography of the Kipchak prince *Tú-tú-ha*, it is stated that ' black mare's milk ' (evidently the cara cosmos of Rubruck), very pleasant to the taste, used to be sent from Kipchak to the Mongol court in China."   (On the drinks of the Mongols, see Mr. Rockhill's note, *Rubruck*, p. 62.)— The Mongols indulge in sour milk (*tarak*) and distilled mare's milk (*arreki*), but

Mr. Rockhill (*Land of the Lamas*, 130) says he never saw them drink *kumiz*.—H. C.]

The mare's-milk drink of Scythian nomads is alluded to by many ancient authors. But the manufacture of Kumiz is particularly spoken of by Herodotus. "The (mare's) milk is poured into deep wooden casks, about which the blind slaves are placed, and then the milk is stirred round. That which rises to the top is drawn off, and considered the best part; the under portion is of less account." Strabo also speaks of the nomads beyond the Cimmerian Chersonesus, who feed on horse-flesh and other flesh, mare's-milk cheese, mare's milk, and sour milk (ὀξυγάλακτα) "*which they have a particular way of preparing.*" Perhaps Herodotus was mistaken about the wooden tubs. At least all modern attempts to use anything but the orthodox skins have failed. Priscus, in his narrative of the mission of himself and Maximin to Attila, says the Huns brought them a drink made from *barley* which they called Κάμος. The barley was, no doubt, a misapprehension of his. (*Herod.* Bk. iv. p. 2, in *Rawl.*; *Strabo*, VII. 4, 6: *Excerpta de Legationibus*, in *Corp. Hist. Byzant.* I. 55.)

CHAPTER LIV.

CONCERNING THE TARTAR CUSTOMS OF WAR.

ALL their harness of war is excellent and costly. Their arms are bows and arrows, sword and mace; but above all the bow, for they are capital archers, indeed the best that are known. On their backs they wear armour of cuirbouly, prepared from buffalo and other hides, which is very strong.[1] They are excellent soldiers, and passing valiant in battle. They are also more capable of hardships than other nations; for many a time, if need be, they will go for a month without any supply of food, living only on the milk of their mares and on such game as their bows may win them. Their horses also will subsist entirely on the grass of the plains, so that there is no need to carry store of barley or straw or oats; and they are very docile to their riders. These, in case of need, will abide on horseback the livelong night, armed at all points, while the horse will be continually grazing.

Of all troops in the world these are they which en-dure the greatest hardship and fatigue, and which cost the least; and they are the best of all for making wide conquests of country. And this you will perceive from what you have heard and shall hear in this book; and (as a fact) there can be no manner of doubt that now they are the masters of the biggest half of the world. Their troops are admirably ordered in the manner that I shall now relate.

You see, when a Tartar prince goes forth to war, he takes with him, say, 100,000 horse. Well, he appoints an officer to every ten men, one to every hundred, one to every thousand, and one to every ten thousand, so that his own orders have to be given to ten persons only, and each of these ten persons has to pass the orders only to other ten, and so on; no one having to give orders to more than ten. And every one in turn is responsible only to the officer immediately over him; and the discipline and order that comes of this method is marvellous, for they are a people very obedient to their chiefs. Further, they call the corps of 100,000 men a *Tuc;* that of 10,000 they call a *Toman;* the thousand they call . . . . ; the hundred *Guz;* the ten . . . .[2] And when the army is on the march they have always 200 horsemen, very well mounted, who are sent a distance of two marches in advance to reconnoitre, and these always keep ahead. They have a similar party detached in the rear, and on either flank, so that there is a good look-out kept on all sides against a surprise. When they are going on a distant expedition they take no gear with them except two leather bottles for milk; a little earthenware pot to cook their meat in, and a little tent to shelter them from rain.[3] And in case of great urgency they will ride ten days on end without lighting a fire or taking a meal. On such an occasion they will sustain themselves on the blood of their horses, opening a vein and letting the blood jet into their mouths, drinking till they have had enough, and then staunching it.[4]

They also have milk dried into a kind of paste to carry with them ; and when they need food they put this in water, and beat it up till it dissolves, and then drink it. [It is prepared in this way ; they boil the milk, and when the rich part floats on the top they skim it into another vessel, and of that they make butter ; for the milk will not become solid till this is removed. Then they put the milk in the sun to dry. And when they go on an expedition, every man takes some ten pounds of this dried milk with him. And of a morning he will take a half pound of it and put it in his leather bottle, with as much water as he pleases. So, as he rides along, the milk-paste and the water in the bottle get well churned together into a kind of pap, and that makes his dinner.[5]]

When they come to an engagement with the enemy, they will gain the victory in this fashion. [They never let themselves get into a regular medley, but keep perpetually riding round and shooting into the enemy. And] as they do not count it any shame to run away in battle, they will [sometimes pretend to] do so, and in running away they turn in the saddle and shoot hard and strong at the foe, and in this way make great havoc. Their horses are trained so perfectly that they will double hither and thither, just like a dog, in a way that is quite astonishing. Thus they fight to as good purpose in running away as if they stood and faced the enemy, because of the vast volleys of arrows that they shoot in this way, turning round upon their pursuers, who are fancying that they have won the battle. But when the Tartars see that they have killed and wounded a good many horses and men, they wheel round bodily, and return to the charge in perfect order and with loud cries ;

and in a very short time the enemy are routed. In truth they are stout and valiant soldiers, and inured to war. And you perceive that it is just when the enemy sees them run, and imagines that he has gained the battle, that he has in reality lost it ; for the Tartars wheel round in a moment when they judge the right time has come. And after this fashion they have won many a fight.[6]

All this that I have been telling you is true of the manners and customs of the genuine Tartars. But I must add also that in these days they are greatly degenerated ; for those who are settled in Cathay have taken up the practices of the Idolaters of the country, and have abandoned their own institutions ; whilst those who have settled in the Levant have adopted the customs of the Saracens.[7]

NOTE I.—The bow was the characteristic weapon of the Tartars, insomuch that the Armenian historians often call them "The Archers." (*St. Martin.* II. 133.) "CUIRBOULY, leather softened by boiling, in which it took any form or impression required, and then hardened." (*Wright's Dict.*) The English adventurer among the Tartars, whose account of them is given by Archbishop Ivo of Narbonne, in Matthew Paris (*sub.* 1243), says : "De coriis bullitis sibi arma levia quidem, sed tamen impenetrabilia coaptarunt." This armour is particularly described by Plano Carpini (p. 685). See the tail-piece to Book IV.

[Mr. E. H. Parker (*China Review*, XXIV. iv. p. 205) remarks that "the first coats of mail were made in China in 1288 : perhaps the idea was obtained from the Malays or Arabs."—H. C.]

NOTE 2.—M. Pauthier has judiciously pointed out the omissions that have occurred here, perhaps owing to Rusticiano's not properly catching the foreign terms applied to the various grades. In the G. Text the passage runs : "*Et sachiés que les cent mille est apellé un* Tut (read *tuc*) *et les dix mille un* Toman, *et les por milier et por centenier et por desme.*" In Pauthier's (uncorrected) text one of the missing words is supplied : "*Et appellent les C.M. un* Tuc ; *et les X.M. un* Toman ; *et un millier* Guz *por centenier et por disenier.*" The blanks he supplies thus from Abulghazi : "*.Et un millier* : [un Miny] ; Guz, *por centenier et* [Un] *por disenier.*" The words supplied are Turki, but so is the *Guz*, which appears already in Pauthier's text, whilst *Toman* and *Tuc* are common to Turki and Mongol. The latter word, *Túk* or *Túgh*, is the horse-tail or yak-tail standard which among so many Asiatic nations has marked the supreme military command. It occurs as *Taka* in ancient Persian, and Cosmas Indicopleustes speaks of it as *Tupha*. The Nine Orloks or Marshals under Chinghiz were entitled to the *Tuk*, and theirs is probably the class of command here indicated as of 100,000, though the figure must not be strictly taken. Timur ordains that every Amir who should conquer a kingdom or command in a victory should receive a title of honour, the *Tugh* and the *Nakkárá*. (*Infra*, Bk. II. ch. iv. note 3.) Baber on several occasions speaks of conferring the *Tugh* upon his generals for dis-

tinguished service.   One of the military titles at Bokhara is still *Tokhsabai*, a corruption of *Túgh-Sáhibi* (Master of the Tugh).

We find the whole gradation except the *Tuc* in a rescript of Janibeg, Khan of
Sarai, in favour of Venetian merchants dated February 1347.   It begins in the
Venetian version :   " *La parola de Zanibeck allo puovolo di Mogoli, alli* Baroni di
Thomeni,* delli miera, delli centenera, delle dexiene." (*Erdmann*, 576 ; *D'Avezac*,
577-578 ; *Rémusat, Langues Tartares*, 303 ; *Pallas, Samml.* I. 283 ; *Schmidt*, 379, 381 ;
Baber, 260, etc. ; *Vámbéry*, 374 ; *Timour Inst.* pp. 283 and 292-293 ; *Bibl. de l'Ec.
des Chartes*, tom. lv. p. 585.)

The decimal division of the army was already made by Chinghiz at an early period
of his career, and was probably much older than his time.   In fact we find the
Myriarch and Chiliarch already in the Persian armies of Darius Hystaspes.   From the
Tartars the system passed into nearly all the Musulman States of Asia, and the titles
Min-bashi or *Bimbashi, Yuzbashi, Onbashi*, still subsist not only in Turkestan, but
also in Turkey and Persia.   The term *Tman* or *Tma* was, according to Herberstein,
still used in Russia in his day for 10,000.   (*Ramus.* II. 159.)

[The King of An-nam, Dinh Tiên-hòang (A.D. 968) had an army of 1,000,000
men forming 10 corps of 10 legions ; each legion forming 10 cohorts of 10 centuries ;
each century forming 10 squads of 10 men.—H. C.]

NOTE 3.—Ramusio's edition says that what with horses and mares there will be
an average of eighteen beasts (?) to every man.

NOTE 4.—See the Oriental account quoted below in Note 6.

So Dionysius, combining this practice with that next described, relates of the
Massagetæ that they have no delicious bread nor native wine :

> " But with horse's blood
> And white milk mingled set their banquets forth."
> (*Orbis Desc.* 743-744.)

And Sidonius :

> " Solitosque cruentum
> Lac potare Getas, et pocula tingere venis."
> (*Parag. ad Avitum.*)

["The Scythian soldier drinks the blood of the first man he overthrows in
battle." (*Herodotus, Rawlinson*, Bk. IV. ch. 64, p. 54.)—H. C.]   " When in
lack of food, they bleed a horse and suck the vein.   If they need something
more solid, they put a sheep's pudding full of blood under the saddle ; this in time
gets coagulated and cooked by the heat, and then they devour it." (*Georg. Pachymeres*,
V. 4.)   The last is a well-known story, but is strenuously denied and ridiculed by
Bergmann. (*Streifereien, etc.* I. 15.)   Joinville tells the same story.   Hans
Schiltberger asserts it very distinctly : " Ich hon och gesehen wann sie in reiss
ylten, das sie ein fleisch nemen, und es dunn schinden und legents unter den
sattel, und riten doruff; und essents wann sie hungert" (ch. 35).   Botero had
" heard from a trustworthy source that a Tartar of Perekop, travelling on the
steppes, lived for some days on the blood of his horse, and then, not daring to bleed it
more, cut off and ate its *ears !*" (*Relazione Univers.* p. 93.)   The Turkmans speak of
such practices, but Conolly says he came to regard them as hyperbolical talk (I. 45).

[Abul-Ghazi Khan, in his History of Mongols, describing a raid of Russian
(*Ourous*) Cossacks, who were hemmed in by the Uzbeks, says : " The Russians had
in continued fighting exhausted all their water.   They began to drink blood ; the
fifth day they had not even blood remaining to drink." (*Transl. by Baron Des
Maisons*, St. Petersburg, II. 295.)]

NOTE 5.—Rubruquis thus describes this preparation, which is called *Kurút:*

---

* This is *Chomeni* in the original, but I have ventured to correct it.

" The milk that remains after the butter has been made, they allow to get as sour as sour can be, and then boil it.   In boiling, it curdles, and that curd they dry in the sun ; and in this way it becomes as hard as iron-slag.   And so it is stored in bags against the winter.   In the winter time, when they have no milk, they put that sour curd, which they call *Griut*, into a skin, and pour warm water on it, and they shake it violently till the curd dissolves in the water, to which it gives an acid flavour ; that water they drink in place of milk.   But above all things they eschew drinking plain water."   From Pallas's account of the modern practice, which is substantially the same, these cakes are also made from the leavings of distillation in making milk-arrack. The Kurút is frequently made of ewe-milk.   Wood speaks of it as an indispensable article in the food of the people of Badakhshan, and under the same name it is a staple food of the Afghans.   (*Rubr.* 229 ; *Samml.* I. 136 ; *Dahl*, u.s. ; *Wood*, 311.)

[It is the *ch'ura* of the Tibetans.   " In the Kokonor country and Tibet, this *krut* or *chura* is put in tea to soften, and then eaten either alone or mixed with parched barley meal (*tsamba*)."   (*Rockhill, Rubruck*, p. 68, note.)—H. C.]

NOTE 6.—Compare with Marco's account the report of the Mongols, which was brought by the spies of Mahomed, Sultan of Khwarizm, when invasion was first menaced by Chinghiz : " The army of Chinghiz is countless, as a swarm of ants or locusts.   Their warriors are matchless in lion-like valour, in obedience, and endurance. They take no rest, and flight or retreat is unknown to them.   On their expeditions they are accompanied by oxen, sheep, camels, and horses, and sweet or sour milk suffices them for food.   Their horses scratch the earth with their hoofs and feed on the roots and grasses they dig up, so that they need neither straw nor oats.   They themselves reck nothing of the clean or the unclean in food, and eat the flesh of all animals, even of dogs, swine, and bears.   They will open a horse's vein, draw blood, and drink it. . . . . In victory they leave neither small nor great alive ; they cut up women great with child and cleave the fruit of the womb.   If they come to a great river, as they know nothing of boats, they sew skins together, stitch up all their goods therein, tie the bundle to their horses' tails, mount with a hard grip of the mane, and so swim over."   This passage is an absolute abridgment of many chapters of Carpini. Still more terse was the sketch of Mongol proceedings drawn by a fugitive from Bokhara after Chinghiz's devastations there.   It was set forth in one unconscious hexameter :

" *Ámdand u khandand u sokhtand u kushtand u burdand u raftand !* "

"They came and they sapped, they fired and they slew, trussed up their loot and were gone ! "

Juwaini, the historian, after telling the story, adds : " The cream and essence of whatever is written in this volume might be represented in these few words."

A Musulman author quoted by Hammer, Najmuddin of Rei, gives an awful picture of the Tartar devastations, " Such as had never been heard of, whether in the lands of unbelief or of Islam, and can only be likened to those which the Prophet announced as signs of the Last Day, when he said : ' The Hour of Judgment shall not come until ye shall have fought with the Turks, men small of eye and ruddy of countenance, whose noses are flat, and their faces like hide-covered shields.   Those shall be Days of Horror ! ' ' And what meanest thou by horror ? ' said the Companions ; and he replied, ' SLAUGHTER ! SLAUGHTER ! '   This beheld the Prophet in vision 600 years ago.   And could there well be worse slaughter than there was in Rei, where I, wretch that I am, was born and bred, and where the whole population of five hundred thousand souls was either butchered or dragged into slavery ? "

Marco habitually suppresses or ignores the frightful brutalities of the Tartars, but these were somewhat less, no doubt, in Kúblái's time.

The Hindustani poet Amir Khosru gives a picture of the Mongols more forcible than elegant, which Elliot has translated (III. 528).

This is Hayton's account of the Parthian tactics of the Tartars : " They will run away, but always keeping their companies together ; and it is very dangerous to give them chase, for as they flee they shoot back over their heads, and do great execution among their pursuers. They keep very close rank, so that you would not guess them for half their real strength." Carpini speaks to the same effect. Baber, himself of Mongol descent, but heartily hating his kindred, gives this account of their military usage in his day: "Such is the uniform practice of these wretches the Moghuls; if they defeat the enemy they instantly seize the booty; if they are defeated, they plunder and dismount their own allies, and, betide what may, carry off the spoil." (*Erdmann*, 364, 383, 620; *Gold. Horde*, 77, 80; *Elliot*, II. 388; *Hayton* in *Ram.* ch. xlviii.; *Baber*, 93; *Carpini*, p. 694.)

NOTE 7.—"The Scythians" (*i.e.* in the absurd Byzantine pedantry, *Tartars*), says Nicephorus Gregoras, "from converse with the Assyrians, Persians, and Chaldæans, in time acquired their manners and adopted their religion, casting off their ancestral atheism. . . . . And to such a degree were they changed, that though in former days they had been wont to cover the head with nothing better than a loose felt cap, and for other clothing had thought themselves well off with the skins of wild beasts or ill-dressed leather, and had for weapons only clubs and slings, or spears, arrows, and bows extemporised from the oaks and other trees of their mountains and forests, now, forsooth, they will have no meaner clothing than brocades of silk and gold! And their luxury and delicate living came to such a pitch that they stood far as the poles asunder from their original habits" (II. v. 6).

CHAPTER LV.

CONCERNING THE ADMINISTERING OF JUSTICE AMONG THE TARTARS.

THE way they administer justice is this. When any one has committed a petty theft, they give him, under the orders of authority, seven blows of a stick, or seventeen, or twenty-seven, or thirty-seven, or forty-seven, and so forth, always increasing by tens in proportion to the injury done, and running up to one hundred and seven. Of these beatings sometimes they die.[1] But if the offence be horse-stealing, or some other great matter, they cut the thief in two with a sword. Howbeit, if he be able to ransom himself by paying nine times the value of the thing stolen, he is let off. Every Lord or other person who possesses beasts has them marked with his peculiar brand, be they horses, mares, camels, oxen, cows, or other great cattle, and then they are sent abroad to graze over the plains

without any keeper. They get all mixt together, but
eventually every beast is recovered by means of its
owner's brand, which is known. For their sheep and
goats they have shepherds. All their cattle are remarkably fine, big, and in good condition.[2]

They have another notable custom, which is this.
If any man have a daughter who dies before marriage,
and another man have had a son also die before
marriage, the parents of the two arrange a grand
wedding between the dead lad and lass. And marry
them they do, making a regular contract! And when
the contract papers are made out they put them in
the fire, in order (as they will have it) that the parties
in the other world may know the fact, and so look on
each other as man and wife. And the parents thenceforward consider themselves sib to each other, just as
if their children had lived and married. Whatever
may be agreed on between the parties as dowry, those
who have to pay it cause to be painted on pieces of
paper and then put these in the fire, saying that in
that way the dead person will get all the real articles
in the other world.[3]

Now I have told you all about the manners and
customs of the Tartars ; but you have heard nothing
yet of the great state of the Grand Kaan, who is the
Lord of all the Tartars and of the Supreme Imperial
Court. All that I will tell you in this book in proper
time and place, but meanwhile I must return to my
story which I left off in that great plain when we
began to speak of the Tartars.[4]

NOTE 1.—The cudgel among the Mongols was not confined to thieves and such
like. It was the punishment also of military and state offences, and even princes
were liable to it without fatal disgrace. "If they give any offence," says Carpini, "or
omit to obey the slightest beck, the Tartars themselves are beaten like donkeys."
The number of blows administered was, according to Wassáf, always odd, 3, 5, and
so forth, up to 77. (*Carp.* 712 ; *Ilchan.* I. 37.)

["They also punish with death grand larceny, but as for petty thefts, such as that of a sheep, so long as one has not repeatedly been taken in the act, they beat him cruelly, and if they administer an hundred blows they must use an hundred sticks." (*Rockhill, Rubruck*, p. 80.)—H. C.]

NOTE 2.—"They have no herdsmen or others to watch their cattle, because the laws of the Turks (*i.e.* Tartars) against theft are so severe. . . . . A man in whose possession a stolen horse is found is obliged to restore it to its owner, *and to give nine of the same value;* if he cannot, his children are seized in compensation; if he have no children, he is slaughtered like a mutton." (*Ibn Batuta*, II. 364.)

NOTE 3.—This is a Chinese custom, though no doubt we may trust Marco for its being a Tartar one also. "In the province of Shansi they have a ridiculous custom, which is to marry dead folks to each other. F. Michael Trigault, a Jesuit, who lived several years in that province, told it us whilst we were in confinement. It falls out that one man's son and another man's daughter die. Whilst the coffins are in the house (and they used to keep them two or three years, or longer) the parents agree to marry them; they send the usual presents, as if the pair were alive, with much ceremony and music. After this they put the two coffins together, hold the wedding dinner in their presence, and, lastly, lay them together in one tomb. The parents, from this time forth, are looked on not merely as friends but as relatives—just as they would have been had their children been married when in life." (*Navarrete*, quoted by *Marsden.*) Kidd likewise, speaking of the Chinese custom of worshipping at the tombs of progenitors, says: "So strongly does veneration for this tribute after death prevail that parents, in order to secure the memorial of the sepulchre for a daughter who has died during her betrothal, give her in marriage after her decease to her intended husband, who receives with nuptial ceremonies at his own house a paper effigy made by her parents, and after he has burnt it, erects a tablet to her memory—an honour which usage forbids to be rendered to the memory of unmarried persons. The law seeks without effect to abolish this absurd custom." (*China*, etc., pp. 179-180.)

[Professor J. J. M. de Groot (*Religious System of China*) gives several instances of marriages after death; the following example (II. 804-805) will illustrate the custom: "An interesting account of the manner in which such *post-mortem* marriages were concluded at the period when the Sung Dynasty governed the Empire, is given by a contemporary work in the following words: 'In the northern parts of the Realm it is customary, when an unmarried youth and an unmarried girl breathe their last, that the two families each charge a match-maker to demand the other party in marriage. Such go-betweens are called match-makers for disembodied souls. They acquaint the two families with each other's circumstances, and then cast lots for the marriage by order of the parents on both sides. If they augur that the union will be a happy one, (wedding) garments for the next world are cut out, and the match-makers repair to the grave of the lad, there to set out wine and fruit for the consummation of the marriage. Two seats are placed side by side, and a small streamer is set up near each seat. If these streamers move a little after the libation has been performed, the souls are believed to approach each other; but if one of them does not move, the party represented thereby is considered to disapprove of the marriage. Each family has to reward its match-maker with a present of woven stuffs. Such go-betweens make a regular livelihood out of these proceedings.'"—H. C.]

The Ingushes of the Caucasus, according to Klaproth, have the same custom: "If a man's son dies, another who has lost his daughter goes to the father and says, 'Thy son will want a wife in the other world; I will give him my daughter; pay me the price of the bride.' Such a demand is never refused, even though the purchase of the bride amount to thirty cows." (*Travels, Eng. Trans.* 345.)

NOTE 4.—There is a little doubt about the reading of this last paragraph. The G. T. has—"*Mès desormès volun retorner à nostre conte en* la grant plaingne *où nos estion quant nos comechames des fais des Tartars*," whilst Pauthier's text has "*Mais desormais vueil retourner à mon conte que Je lessai* d'or plain *quant nous commençames des faiz des Tatars.*" The former reading looks very like a misunderstanding of one similar to the latter, where *d'or plain* seems to be an adverbial expression, with some such meaning as "just now," "a while ago." I have not, however, been able to trace the expression elsewhere. Cotgrave has *or primes*, "but even now," etc. ; and has also *de plain*, "presently, immediately, out of hand." It seems quite possible that *d'or plain* should have had the meaning suggested.

CHAPTER LVI.

Sundry Particulars of the Plain beyond Caracoron.

AND when you leave Caracoron and the Altay, in which they bury the bodies of the Tartar Sovereigns, as I told you, you go north for forty days till you reach a country called the PLAIN OF BARGU.[1] The people there are called MESCRIPT ; they are a very wild race, and live by their cattle, the most of which are stags, and these stags, I assure you, they used to ride upon. Their customs are like those of the Tartars, and they are subject to the Great Kaan. They have neither corn nor wine. [They get birds for food, for the country is full of lakes and pools and marshes, which are much frequented by the birds when they are moulting, and when they have quite cast their feathers and can't fly, those people catch them. They also live partly on fish.[2]]

And when you have travelled forty days over this great plain you come to the ocean, at the place where the mountains are in which the Peregrine falcons have their nests. And in those mountains it is so cold that you find neither man or woman, nor beast nor bird, except one kind of bird called *Barguerlac*, on which the falcons feed. They are as big as partridges, and have feet like those of parrots and a tail like a swallow's,

and are very strong in flight. And when the Grand
Kaan wants Peregrines from the nest, he sends thither
to procure them.[3] It is also on islands in that sea that
the Gerfalcons are bred. You must know that the
place is so far to the north that you leave the North
Star somewhat behind you towards the south! The
gerfalcons are so abundant there that the Emperor
can have as many as he likes to send for. And you
must not suppose that those gerfalcons which the
Christians carry into the Tartar dominions go to the
Great Kaan; they are carried only to the Prince of
the Levant.[4]

Now I have told you all about the provinces northward as far as the Ocean Sea, beyond which there is
no more land at all; so I shall proceed to tell you
of the other provinces on the way to the Great Kaan.
Let us, then, return to that province of which I
spoke before, called Campichu.

NOTE 1.—The readings differ as to the length of the journey. In Pauthier's text
we seem to have first a journey of forty days from near Karakorúm to the Plain of Bargu,
and then a journey of forty days more across the plain to the Northern Ocean. The
G. T. seems to present only *one* journey of forty days (Ramusio, of sixty days), but
leaves the interval from Karakorúm undefined. I have followed the former, though
with some doubt.

NOTE 2.—This paragraph from Ramusio replaces the following in Pauthier's text:
"In the summer they got abundance of game, both beasts and birds, but in winter,
there is none to be had because of the great cold."

Marco is here dealing, I apprehend, with hearsay geography, and, as is common
in like cases, there is great compression of circumstances and characteristics, analogous
to the like compression of little-known regions in mediæval maps.

The name *Bargu* appears to be the same with that often mentioned in Mongol
history as BARGUCHIN TUGRUM or BARGUTI, and which Rashiduddin calls the
northern limit of the inhabited earth. This commenced about Lake Baikal, where
the name still survives in that of a river (*Barguzin*) falling into the Lake on the east
side, and of a town on its banks (*Barguzinsk*). Indeed, according to Rashid himself,
BARGU was the name of one of the tribes occupying the plain; and a quotation from
Father Hyacinth would seem to show that the country is still called *Barakhu*.

[The Archimandrite Palladius (*Elucidations*, 16-17) writes :—" In the Mongol text
of Chingis Khan's biography, this country is called Barhu and Barhuchin; it is to be
supposed, according to Colonel Yule's identification of this name with the modern
Barguzin, that this country was near Lake Baikal. The fact that Merkits were in
Bargu is confirmed by the following statement in Chingis Khan's biography:
'When Chingis Khan defeated his enemies, the Merkits, they fled to Barhuchin

tokum.' *Tokum* signifies 'a hollow, a low place,' according to the Chinese translation of the above-mentioned biography, made in 1381 ; thus Barhuchin tokum undoubtedly corresponds to M. Polo's Plain of Bargu. As to M. Polo's statement that the inhabitants of Bargu were Merkits, it cannot be accepted unconditionally. The Merkits were not indigenous to the country near Baikal, but belonged originally, —according to a division set forth in the Mongol text of the *Yuan ch'ao pi shi*,— to the category of tribes *living in yurts, i.e.* nomad tribes, or tribes of the desert. Meanwhile we find in the same biography of Chingis Khan, mention of a people called Barhun, which belonged to the category of tribes *living in the forests;* and we have therefore reason to suppose that the Barhuns were the aborigines of Barhu. After the time of Chingis Khan, this ethnographic name disappears from Chinese history ; it appears again in the middle of the 16th century. The author of the *Yyu* (1543-1544), in enumerating the tribes inhabiting Mongolia and the adjacent countries, mentions the Barhu, as a strong tribe, able to supply up to several tens of thousands (?) of warriors, armed with steel swords ; but the country inhabited by them is not indicated. The Mongols, it is added, call them Black Ta-tze (Khara Mongols, *i.e.* 'Lower Mongols').

" At the close of the 17th century, the Barhus are found inhabiting the western slopes of the interior Hing'an, as well as between Lake Kulon and River Khalkha, and dependent on a prince of eastern Khalkhas, Doro beile. (Manchu title.)

" At the time of Galdan Khan's invasion, a part of them fled to Siberia with the eastern Khalkhas, but afterwards they returned. [*Mung ku yew mu ki* and *Lung sha ki lio.*] After their rebellion in 1696, quelled by a Manchu General, they were included with other petty tribes (regarding which few researches have been made) in the category *butkha*, or hunters, and received a military organisation. They are divided into Old and New Barhu, according to the time when they were brought under Manchu rule. The Barhus belong to the Mongolian, not to the Tungusian race ; they are sometimes considered even to have been in relationship with the Khalkhas. (*He lung kiang wai ki* and *Lung sha ki lio.*)

" This is all the substantial information we possess on the Barhu. Is there an affinity to be found between the modern Barhus and the Barhuns of Chingis Khan's biography?—and is it to be supposed, that in the course of time, they spread from Lake Baikal to the Hing'an range? or is it more correct to consider them a branch of the Mongol race indigenous to the Hing'an Mountains, and which received the general archaic name of Bargu, which might have pointed out the physical character of the country they inhabited [*Kin Shi*], just as we find in history the Urianhai of Altai and the Urianhai of Western Manchuria? It is difficult to solve this question for want of historical data."—H. C.]

Mescript, or *Mecri*, as in G. T. The *Merkit*, a great tribe to the south-east of the Baikal, were also called *Mekrit*, and sometimes *Megrin*. The Mekrit are spoken of also by Carpini and Rubruquis. D'Avezac thinks that the *Kerait*, and not the *Merkit*, are intended by all three travellers. As regards Polo, I see no reason for this view. The name he uses is *Mekrit*, and the position which he assigns to them agrees fairly with that assigned on good authority to the Merkit or Mekrit. Only, as in other cases, where he is rehearsing hearsay information, it does not follow that the identification of the name involves the correctness of all the circumstances which he connects with that name. We saw in ch. xxx. that under *Pashai* he seemed to lump circumstances belonging to various parts of the region from Badakhshan to the Indus; so here under *Mekrit* he embraces characteristics belonging to tribes extending far beyond the Mekrit, and which in fact are appropriate to the Tunguses. Rashiduddin seems to describe the latter under the name of *Uriangkut* of the Woods, a people dwelling beyond the frontier of Barguchin, and in connection with whom he speaks of their Reindeer obscurely, as well as of their tents of birch bark, and their hunting on snow-shoes.

The mention of the Reindeer by Polo in this passage is one of the interesting points which Pauthier's text omits. Marsden objects to the statement that the stags are ridden upon, and from this motive mis-renders "*li qual' anche* cavalcano," as, "which they make use of for the purpose of travelling." Yet he might have found in Witsen that the Reindeer are *ridden* by various Siberian Tribes, but especially by the Tunguses. Erman is very full on the reindeer-riding of the latter people, having himself travelled far in that way in going to Okhotsk, and gives a very detailed description of the saddle, etc., employed. The reindeer of the Tunguses are stated by the same traveller to be much larger and finer animals than those of Lapland. They are also used for pack-carriage and draught. Old Richard Eden says that the "olde wryters" relate that "certayne Scythians doe ryde on Hartes." I have not traced to what he refers, but if the statement be in any ancient author it is very remarkable. Some old editions of Olaus Magnus have curious cuts of Laplanders and others riding on reindeer, but I find nothing in the text appropriate. We hear from travellers of the Lapland deer being occasionally mounted, but only it would seem in sport, not as a practice. (*Erdmann*, 189, 191 ; *D'Ohsson*, I. 103 ; *D'Avezac*, 534 *seqq.* ; *J. As.* sér. II. tom. xi. ; sér. IV. tom. xvii. 107 ; *N. et E.* XIII. i. 274-276 ; *Witsen*, II. 670, 671, 680 ; *Erman*, II. 321, 374, 429, 449 *seqq.*, and original German, II. 347 *seqq.* ; *Notes on Russia*, Hac. Soc. II. 224 ; *J. A. S. B.* XXIX. 379.)

The numerous lakes and marshes swarming with water-fowl are very characteristic of the country between Yakutsk and the Kolyma. It is evident that Marco had his information from an eye-witness, though the whole picture is compressed. Wrangell, speaking of Nijni Kolyma, says : " It is at the moulting season that the great bird-hunts take place. The sportsmen surround the nests, and slip their dogs, which drive the birds to the water, on which they are easily knocked over with a gun or arrow, or even with a stick. . . . This chase is divided into several periods. They begin with the ducks, which moult first ; then come the geese ; then the swans. . . . In each case the people take care to choose the time when the birds have lost their feathers." The whole calendar with the Yakuts and Russian settlers on the Kolyma is a succession of fishing and hunting seasons which the same author details. (I. 149, 150 ; 119-121.)

NOTE 3.—What little is said of the *Barguerlac* points to some bird of the genus, *Pterocles*, or Sand Grouse (to which belong the so-called Rock Pigeons of India), or to the allied *Tetrao paradoxus* of Pallas, now known as *Syrrhaptes Pallasii*. Indeed, we find in Zenker's Dictionary that *Boghurtlák* (or *Baghírtlák*, as it is in Pavet de Courteille's) in Oriental Turkish is the *Kata*, *i.e.* I presume, the *Pterocles alchata* of Linnæus, or Large Pin-tailed Sand Grouse. Mr. Gould, to whom I referred the point, is clear that the *Syrrhaptes* is Marco's bird, and I believe there can be no question of it.

[Passing through Ch'ang-k'ou, Mr. Rockhill found the people praying for rain. " The people told me," he says, in his *Journey* (p. 9), " that they knew long ago the year would be disastrous, for the sand grouse had been more numerous of late than for years, and the saying goes *Sha-ch'i kuo, mai lao-po*, ' when the sand grouse fly by, wives will be for sale.' "—II. C.]

The chief difficulty in identification with the Syrrhaptes or any known bird, would be "the feet like a parrot's." The feet of the Syrrhaptes are not indeed like a parrot's, though its awkward, slow, and waddling gait on the ground, may have suggested the comparison ; and though it has very odd and anomalous feet, a circumstance which the Chinese indicate in another way by calling the bird (according to Huc) *Lung Kio*, or " Dragon-foot." [Mr. Rockhill (*Journey*) writes in a note (p. 9) : " I, for my part, never heard any other name than *sha-ch'i*, 'sand-fowl,' given them. This name is used, however, for a variety of birds, among others the partridge."—H. C.] The hind-toe is absent, the toes are unseparated, recognisable only by the broad flat nails, and fitted below with a callous couch, whilst the whole foot is covered with short dense feathers like hair, and is more like a quadruped's paw than a bird's foot.

The home of the Syrrhaptes is in the Altai, the Kirghiz Steppes, and the country round Lake Baikal, though it also visits the North of China in great flights. "On plains of grass and sandy deserts," says Gould (*Birds of Great Britain*, Part IV.), "at one season covered with snow, and at another sun-burnt and parched by drought, it finds a congenial home ; in these inhospitable and little-known regions it breeds, and when necessity compels it to do so, wings its way . . . . over incredible distances to obtain water or food." Huc says, speaking of the bird on the northern frontier of China : " They generally arrive in great flights from the north, especially when much snow has fallen, flying with astonishing rapidity, so that the movement of their wings produces a noise like hail." It is said to be very delicate eating. The bird owes its place in Gould's *Birds of Great Britain* to the fact—strongly illustrative of its being *moult volant*, as Polo says it is—that it appeared in England in 1859, and since then, at least up to 1863, continued to arrive annually in pairs or companies in nearly all parts of our island, from Penzance to Caithness. And Gould states that it was breeding in the Danish islands. A full account by Mr. A. Newton of this remarkable immigration is contained in the *Ibis* for April, 1864, and many details in *Stevenson's Birds of Norfolk*, I. 376 *seqq.*

Syrrhaptes Pallasii.

There are plates of *Syrrhaptes* in *Radde's Reisen im Süden von Ost-Sibirien*, Bd. II. ; in vol. v. of *Temminck*, Planches Coloriées, Pl. 95 ; in *Gould*, as above ; in *Gray*, *Genera of Birds*, vol. iii. p. 517 (life size) ; and in the *Ibis* for April, 1860. From the last our cut is taken.

[See *A. David et Oustalet, Oiseaux de la Chine*, 389, on *Syrrhaptes Pallasii* or *Syrrhaptes Paradoxus.*—H. C.]

NOTE 4.—Gerfalcons (*Shonḳár*) were objects of high estimation in the Middle Ages, and were frequent presents to and from royal personages. Thus among the presents sent with an embassy from King James II. of Aragon to the Sultan of Egypt, in 1314, we find three white gerfalcons. They were sent in homage to Chinghiz and to Kúblái, by the Kirghiz, but I cannot identify the mountains where they or the Peregrines were found. The Peregrine falcon was in Europe sometimes termed *Faucon Tartare*. (See *Ménage* s. v. *Sahin.*) The Peregrine of Northern Japan, and probably therefore that of Siberia, is identical with that of Europe. Witsen speaks of an island in the Sea of Tartary, from which falcons were got, apparently referring to a Chinese map as his authority ; but I know nothing more of it. (*Capmany*, IV. 64-65 ; *Ibis*, 1862, p. 314 ; *Witsen*, II. 656.)

[On the *Falco peregrinus*, Lin., and other Falcons, see Ed. Blanc's paper mentioned on p. 162. The *Falco Saker* is to be found all over Central Asia ; it is called by the Pekingese *Hwang-yng* (yellow falcon). (*David et Oustalet, Oiseaux de la Chine*, 31-32.)—H. C.]

CHAPTER LVII.

OF THE KINGDOM OF ERGUIUL, AND PROVINCE OF SINJU.

ON leaving Campichu, then, you travel five days across a tract in which many spirits are heard speaking in the night season; and at the end of those five marches, towards the east, you come to a kingdom called ERGUIUL, belonging to the Great Kaan. It is one of the several kingdoms which make up the great Province of Tangut. The people consist of Nestorian Christians, Idolaters, and worshippers of Mahommet.[1]

There are plenty of cities in this kingdom, but the capital is ERGUIUL. You can travel in a south-easterly direction from this place into the province of Cathay. Should you follow that road to the south-east, you come to a city called SINJU, belonging also to Tangut, and subject to the Great Kaan, which has under it many towns and villages.[2] The population is composed of Idolaters, and worshippers of Mahommet, but there are some Christians also. There are wild cattle in that country [almost] as big as elephants, splendid creatures, covered everywhere but on the back with shaggy hair a good four palms long. They are partly black, partly white, and really wonderfully fine creatures [and the hair or wool is extremely fine and white, finer and whiter than silk. Messer Marco brought some to Venice as a great curiosity, and so it was reckoned by those who saw it]. There are also plenty of them tame, which have been caught young. [They also cross these with the common cow, and the cattle from this cross are wonderful beasts, and better for work than other animals.] These the people use commonly for burden and general work, and in the plough as well; and at the latter they will do full twice as much work as any other cattle, being such very strong beasts.[3]

In this country too is found the best musk in the world; and I will tell you how 'tis produced. There exists in that region a kind of wild animal like a gazelle. It has feet and tail like the gazelle's, and stag's hair of a very coarse kind, but no horns. It has four tusks, two below and two above, about three inches long, and slender in form, one pair growing upwards, and the other downwards. It is a very pretty creature. The musk is found in this way. When the creature has been taken, they find at the navel between the flesh and the skin something like an impostume full of blood, which they cut out and remove with all the skin attached to it. And the blood inside this impostume is the musk that produces that powerful perfume. There is an immense number of these beasts in the country we are speaking of. [The flesh is very good to eat. Messer Marco brought the dried head and feet of one of these animals to Venice with him.[4]]

The people are traders and artizans, and also grow abundance of corn. The province has an extent of 26 days' journey. Pheasants are found there twice as big as ours, indeed nearly as big as a peacock, and having tails of 7 to 10 palms in length; and besides them other pheasants in aspect like our own, and birds of many other kinds, and of beautiful variegated plumage.[5] The people, who are Idolaters, are fat folks with little noses and black hair, and no beard, except a few hairs on the upper lip. The women too have very smooth

and white skins, and in every respect are pretty
creatures. The men are very sensual, and marry
many wives, which is not forbidden by their religion.
No matter how base a woman's descent may be, if
she have beauty she may find a husband among the
greatest men in the land, the man paying the girl's
father and mother a great sum of money, according to
the bargain that may be made.

NOTE 1.—No approximation to the name of Erguiul in an appropriate position
has yet been elicited from Chinese or other Oriental sources. We cannot go widely
astray as to its position, five days east of Kanchau. Klaproth identifies it with
Liangchau-fu ; Pauthier with the neighbouring city of Yungchang, on the ground
that the latter was, in the time of Kúblái, the head of one of the *Lús*, or Circles, of
Kansuh or Tangut, which he has shown some reason for believing to be the
"kingdoms" of Marco.

It is probable, however, that the *town* called by Polo Erguiul lay north of both
the cities named, and more in line with the position assigned below to *Egrigaya*. (See
note 1, ch. lviii.)

I may notice that the structure of the name Ergui-ul or Ergiu-ul, has a look of
analogy to that of *Tang-keu-ul*, named in the next note.

[" Erguiul is Erichew of the Mongol text of the *Yuen ch'ao pi shi*, Si-liang in the
Chinese history, the modern *Liang chow fu*. Klaproth, on the authority of Rashid-
eddin, has already identified this name with that of Si-liang." (*Palladius*, p. 18.)
M. Bonin left Ning-h'ia at the end of July, 1899, and he crossed the desert to
Liangchau in fifteen days from east to west ; he is the first traveller who took this
route : Prjevalsky went westward, passing by the residence of the Prince of Alashan,
and Obrutchev followed the route south of Bonin's.—H. C.]

NOTE 2.—No doubt Marsden is right in identifying this with SINING-CHAU, now
Sining-fu, the Chinese city nearest to Tibet and the Kokonor frontier. Grueber and
Dorville, who passed it on their way to Lhasa, in 1661, call it *urbs ingens*. Sining
was visited also by Huc and Gabet, who are unsatisfactory, as usually on geographical matters. They also call it "an immense town," but thinly peopled, its
commerce having been in part transferred to Tang-keu-ul, a small town closer to the
frontier.

[Sining belonged to the country called Hwang chung ; in 1198, under the Sung
Dynasty, it was subjugated by the Chinese, and was named Si-ning chau ; at the beginning of the Ming Dynasty (from 1368), it was named Si-ning wei, and since 1726
Si-ning fu. (Cf. Gueluy, *Chine*, p. 62.) From Liangchau, M. Bonin went to Sining
through the Lao kou kau pass and the Ta-Tung ho. Obrutchev and Grum Grijmaïlo
took the usual route from Kanchau to Sining. After the murder of Dutreuil de Rhins
at Tung bu *m*do, his companion, Grenard, arrived at Sining, and left it on the 29th
July, 1894. Dr. Sven Hedin gives in his book his own drawing of a gate of Sining-
fu, where he arrived on the 25th November, 1896.—H. C.]

Sining is called by the Tibetans *Ziling* or Jiling, by the Mongols *Seling Khoto*.
A shawl wool texture, apparently made in this quarter, is imported into Kashmir and
Ladak, under the name of *S'ling*. I have supposed Sining to be also the *Zilm* of
which Mr. Shaw heard at Yarkand, and am answerable for a note to that effect on
p. 38 of his *High Tartary*. But Mr. Shaw, on his return to Europe, gave some rather
strong reasons against this. (See *Proc. R. G. S.* XVI. 245 ; *Kircher*, pp. 64, 66 ;

Della Penna, 27 ; *Davies's Report*, App. p. ccxxix. ; *Vigne*, II. 110, 129.) [At present Sining is called by the Tibetans Seling K'ar or Kuar, and by the Mongols, Seling K'utun, *K'ar* and *K'utun* meaning "fortified city." (*Rockhill, Land of the Lamas*, 49, note.)—H. C.]

[Mr. Rockhill (*Diary of a Journey*, 65) writes : "There must be some Scotch blood in the Hsi-ningites, for I find they are very fond of oatmeal and of cracked wheat. The first is called *yen-mei ch'en*, and is eaten boiled with the water in which mutton has been cooked, or with neat's-foot oil (*yang-t'i yu*). The cracked wheat (*mei-tzŭ fan*) is eaten prepared in the same way, and is a very good dish."—H. C.]

Note 3.—The *Dong*, or Wild Yak, has till late years only been known by vague rumour. It has always been famed in native reports for its great fierceness. The *Haft Iklím* says that "it kills with its horns, by its kicks, by treading under foot, and by tearing with its teeth," whilst the Emperor Humáyún himself told Sidi 'Ali, the Turkish admiral, that when it had knocked a man down it skinned him from head to heels by licking him with its tongue ! Dr. Campbell states, in the *Journal of the As. Soc. of Bengal*, that it was said to be four times the size of the domestic Yak. The horns are alleged to be sometimes three feet long, and of immense girth ; they are handed round full of strong drink at the festivals of Tibetan grandees, as the Urus horns were in Germany, according to Cæsar.

A note, with which I have been favoured by Dr. Campbell (long the respected Superintendent of British Sikkim) says : "Captain Smith, of the Bengal Army, who had travelled in Western Tibet, told me that he had shot many wild Yaks in the neighbourhood of the Mansarawar Lake, and that he measured a bull which was 18 hands high, *i.e.* 6 feet. All that he saw were *black* all over. He also spoke to the fierceness of the animal. He was once charged by a bull that he had wounded, and narrowly escaped being killed. Perhaps my statement (above referred to) in regard to the relative size of the Wild and Tame Yak, may require modification if applied to all the countries in which the Yak is found. At all events, the finest specimen of the tame Yak I ever saw, was not in Nepal, Sikkim, Tibet, or Bootan, but in the *Jardin des Plantes* at Paris ; and that one, a male, was brought from Shanghai. The best drawing of a Yak I know is that in Turner's *Tibet*."

[Lieutenant Samuel Turner gave a very good description of the Yak of Tartary, which he calls *Soora-Goy*, or the Bushy-tailed Bull of Tibet. (*Asiat. Researches*, No. XXIII, pp. 351-353, with a plate.) He says with regard to the colour : "There is a great variety of colours amongst them, but black or white are the most prevalent. It is not uncommon to see the long hair upon the ridge of the back, the tail, tuft upon the chest, and the legs below the knee white, when all the rest of the animal is jet black." A good drawing of "an enormous" Yak is to be found on p. 183 of Captain Wellby's *Unknown Tibet*. (See also Captain Deasy's work on *Tibet*, p. 363.) Prince Henri d'Orléans brought home a fine specimen, which he shot during his journey with Bonvalot ; it is now exhibited in the galleries of the Muséum d'Histoire Naturelle. Some Yaks were brought to Paris on the 1st April, 1854, and the celebrated artist, Mme. Rosa Bonheur, made sketches after them. (See *Jour. Soc. Acclimatation*, June, 1900, 39-40.)—H. C.]

Captain Prjevalsky, in his recent journey (1872-1873), shot twenty wild Yaks south of the Koko Nor. He specifies one as 11 feet in length exclusive of the tail, which was 3 feet more ; the height 6 feet. He speaks of the Yak as less formidable than it looks, from apathy and stupidity, but very hard to kill ; one having taken eighteen bullets before it succumbed.

[Mr. Rockhill (*Rubruck*, 151, note) writes : "The average load carried by a Yak is about 250 lbs. The wild Yak bull is an enormous animal, and the people of Turkestan and North Tibet credit him with extraordinary strength. Mirza Haidar, in the *Tarikhi Rashidi*, says of the wild Yak or *kutás:* 'This is a very wild and ferocious beast. In whatever manner it attacks one it proves fatal. Whether it strikes with its horns, or kicks, or overthrows its victim. If it has no opportunity of doing any of these things, it tosses its enemy with its tongue twenty *gaz* into the air, and he is dead before reaching the ground. One male *kutás* is a load for twelve horses. One man cannot possibly raise a shoulder of the animal.'"—Captain Deasy (*In Tibet*, 363) says : ' In a few places on lofty ground in Tibet we found Yaks in herds numbering from ten to thirty, and sometimes more. Most of the animals are black, brown specimens being very rare. Their roving herds move with great agility over the steep and stony ground, apparently enjoying the snow and frost and wind, which seldom fail. . . . Yaks are capable of offering formidable resistance to the sportsman. . . .'"—H. C.]

The tame Yaks are never, I imagine, "caught young," as Marco says; it is a domesticated *breed*, though possibly, as with buffaloes in Bengal, the breed may occasionally be refreshed by a cross of wild blood. They are employed for riding, as beasts of burden, and in the plough. [Lieutenant S. Turner, *l.c.*, says, on the other hand : " They are never employed in agriculture, but are extremely useful as beasts of burthen."—H. C.] In the higher parts of our Himalayan provinces, and in Tibet, the Yak itself is most in use ; but in the less elevated tracts several breeds crossed with the common Indian cattle are more used. They have a variety of names according to their precise origin. The inferior Yaks used in the plough are ugly enough, and "have more the appearance of large shaggy bears than of oxen," but the Yak used for riding, says Hoffmeister, "is an infinitely handsomer animal. It has a stately hump, a rich silky hanging tail nearly reaching the ground, twisted horns, a noble bearing, and an erect head." Cunningham, too, says that the *Dso*, one of the mixed breeds, is "a very handsome animal, with long shaggy hair, generally black and white." Many of the various tame breeds appear to have the tail and back white, and also the fringe under the body, but black and red are the prevailing colours. Some of the crossbred cows are excellent milkers, better than either parent stock.

Notice in this passage the additional and interesting particulars given by Ramusio, *e.g.* the use of the mixed breeds. "Finer than silk," is an exaggeration, or say an *hyberbole*, as is the following expression, " As big as elephants," even with Ramusio's apologetic *quasi*. Cæsar says the Hercynian Urus was *magnitudine paullo infra elephantos*.

The tame Yak is used across the breadth of Mongolia. Rubruquis saw them at Karakorum, and describes them well. Mr. Ney Elias tells me he found Yaks common everywhere along his route in Mongolia, between the Tui river (long. *circa* 101°) and the upper valleys of the Kobdo near the Siberian frontier. At Uliasut'ai they were used occasionally by Chinese settlers for drawing carts, but he never saw them used for loads or for riding, as in Tibet. He has also seen Yaks in the neighbourhood of Kwei-hwa-ch'eng. (*Tenduc*, see ch. lix. note 1.) This may be taken as the eastern limit of the employment of the Yak ; the western limit is in the highlands of Khokand.

These animals had been noticed by Cosmas [who calls them *agriobous*] in the 6th century, and by Ælian in the 3rd. The latter speaks of them as black cattle with white tails, from which fly-flappers were made for Indian kings. And the great Kalidása thus sang of the Yak, according to a learned (if somewhat rugged) version ascribed to Dr. Mill. The poet personifies the Himálaya :—

> " For Him the large Yaks in his cold plains that bide
> Whisk here and there, playful, their tails' bushy pride,
> And evermore flapping those fans of long hair
> Which borrowed moonbeams have made splendid and fair,
> Proclaim at each stroke (what our flapping men sing)
> His title of Honour, ' The Dread Mountain King.'"

Who can forget Père Huc's inimitable picture of the hairy Yaks of their caravan, after passing a river in the depth of winter, "walking with their legs wide apart, and bearing an enormous load of stalactites, which hung beneath their bellies quite to the ground. The monstrous beasts *looked exactly as if they were preserved in sugar-candy.*" Or that other, even more striking, of a great troop of wild Yaks, caught in the upper waters of the Kin-sha Kiang, as they swam, in the moment of congelation, and thus preserved throughout the winter, gigantic "flies in amber."

(*N. et E.* XIV. 478; *J. As.* IX. 199; *J. A. S. B.* IX. 566, XXIV. 235; *Shaw*, p. 91; *Ladak*, p. 210; *Geog. Magazine*, April, 1874; *Hoffmeister's Travels*, p. 441; *Rubr.* 288; *Æl. de Nat. An.* XV. 14; *J. A. S. B.* I. 342; *Mrs. Sinnett's Huc*, pp. 228, 235.)

NOTE 4.—Ramusio adds that the hunters seek the animal at New Moon, at which time the musk is secreted.

The description is good except as to the *four* tusks, for the musk deer has canine teeth only in the upper jaw, slender and prominent as he describes them. The flesh of the animal is eaten by the Chinese, and in Siberia by both Tartars and Russians, but that of the males has a strong musk flavour.

The "immense number" of these animals that existed in the Himalayan countries may be conceived from Tavernier's statement, that on one visit to Patna, then the great Indian mart for this article, he purchased 7673 pods of musk. These presumably came by way of Nepal; but musk pods of the highest class were also imported from Khotan *viâ* Yarkand and Leh, and the lowest price such a pod fetched at Yarkand was 250 tankas, or upwards of 4*l.* This import has long been extinct, and indeed the trade in the article, except towards China, has altogether greatly declined, probably (says Mr. Hodgson) because its repute as a medicine is becoming fast exploded. In Sicily it is still so used, but apparently only as a sort of decent medical *viaticum*, for when it is said "the Doctors have given him musk," it is as much as to say that they have given up the patient.

["Here Marco Polo speaks of musk; musk and rhubarb (which he mentions before, Sukchur, ch. xliii.) are the most renowned and valuable of the products of the province of Kansu, which comparatively produces very little; the industry in both these articles is at present in the hands of the Tanguts of that province [*Su chow chi*]." (*Palladius*, p. 18.)

Writing under date 15th February, 1892, from Lusar (coming from Sining), Mr. Rockhill says: "The musk trade here is increasing, Cantonese and Ssŭ-ch'uanese traders now come here to buy it, paying for good musk four times its weight in silver (*ssŭ huan*, as they say). The best test of its purity is an examination of the colour. The Tibetans adulterate it by mixing tsamba and blood with it. The best time to buy it is from the seventh to the ninth moon (latter part of August to middle of November)." Mr. Rockhill adds in a note: "Mongols call musk *owo;* Tibetans call it *latsé.* The best musk they say is 'white musk,' *tsahan owo* in Mongol, in Tibetan *latsé karpo.* I do not know whether white refers to the colour of the musk itself or to that of the hair on the skin covering the musk pouch." (*Diary of a Journey*, p. 71.)—H. C.]

Three species of the *Moschus* are found in the Mountains of Tibet, and *M. Chrysogaster*, which Mr. Hodgson calls "the loveliest," and which chiefly supplies the highly-prized pod called *Kághazi*, or "Thin-as-paper," is almost exclusively confined to the Chinese frontier. Like the Yak, the *Moschus* is mentioned by Cosmas (*circa* A.D. 545), and *musk* appears in a Greek prescription by Aëtius of Amida, a physician practising at Constantinople about the same date.

(*Martini*, p. 39; *Tav., Des Indes*, Bk. II. ch. xxiv.; *J. A. S. B.* XI. 285; *Davies's Rep.* App. p. ccxxxvii.; *Dr. Flückiger in Schweiz. Wochenschr. für Pharmacie*, 1867; *Heyd, Commerce du Levant*, II. 636-640.)

NOTE 5.—The China pheasant answering best to the indications in the text, appears to be *Reeves's Pheasant*.   Mr. Gould has identified this bird with Marco's in his magnificent *Birds of Asia*, and has been kind enough to show me a specimen which, with the body, measured 6 feet 8 inches. The tail feathers alone, however, are said to reach to 6 and 7 feet, so that Marco's ten palms was scarcely an exaggeration. These tail-feathers are often seen on the Chinese stage in the cap of the hero of the drama, and also decorate the hats of certain civil functionaries.

Size is the point in which the bird fails to meet Marco's description. In that respect the latter would rather apply to the *Crossoptilon auritum*, which is nearly as big as a turkey, or to the glorious *Múnál* (*Lopophorus impeyanus*), but then that has no length of tail. The latter seems to be the bird described by Ælian : "Magnificent cocks which have the crest variegated and ornate like a crown of flowers, and the tail feathers not curved like a cock's, but broad and carried in a train like a peacock's ; the feathers are partly golden, and partly azure or emerald-coloured." (*Wood's Birds*, 610, from which I have copied the illustration; *Williams, M. K.* I. 261 ; *Æl. De Nat. An.* XVI. 2.) A species of *Crossoptilon* has recently been found by Captain Prjevalsky in Alashan, the Egrigaia (as I believe) of next chapter, and one also by Abbé Armand David at the Koko Nor.

[See on the Phasianidæ family in Central and Western Asia, *David et Oustalet, Oiseaux de la Chine*, 401-421 ; the *Phasianus Reevesii* or *veneratus* is called by the Chinese of Tung-lin, near Peking, *Djeu-ky* (hen-arrow) ; the *Crossoptilon auritum* is named *Ma-ky.*—H. C.]

Reeves's Pheasant.

CHAPTER LVIII.

OF THE KINGDOM OF EGRIGAIA.

STARTING again from Erguiul you ride eastward for eight days, and then come to a province called EGRIGAIA, containing numerous cities and villages, and belonging to Tangut.[1] The capital city is called CALACHAN.[2] The people are chiefly Idolaters, but there are fine churches belonging to the Nestorian Christians. They are all subjects of the Great Kaan. They make in this city great quantities of camlets of camel's wool, the finest in the world; and some of the camlets that they make are white, for they have white camels, and these are the best of all. Merchants purchase these stuffs here, and carry them over the world for sale.[3]

We shall now proceed eastward from this place and enter the territory that was formerly Prester John's.

NOTE I.—Chinghiz invaded Tangut in all five times, viz. in 1205, 1207, 1209 (or according to Erdmann, 1210-1211), 1218, and 1226-1227, on which last expedition he died.

A. In the third invasion, according to D'Ohsson's Chinese guide (Father Hyacinth), he took the town of *Uiraca*, and the fortress of Imen, and laid siege to the capital, then called Chung-sing or Chung-hing, now Ning-hsia.

Rashid, in a short notice of this campaign, calls the first city *Erica*, *Erlaca*, or, as Erdmann has it, *Artacki*. In De Mailla it is *Ulahai*.

B. On the last invasion (1226), D'Ohsson's Chinese authority says that Chinghiz took Kanchau and Suhchau, Cholo and Khola in the province of Liangcheu, and then proceeded to the Yellow River, and invested Lingchau, south of Ning-hsia.

Erdmann, following his reading of Rashiduddin, says Chinghiz took the cities of Tangut, called *Arucki*, *Kachu*, *Sichu*, and *Kamichu*, and besieged Deresgai (D'Ohsson, *Derssekai*), whilst Shidergu, the King of Tangut, betook himself to his capital *Artackin.*

D'Ohsson, also professing to follow Rashid, calls this "his capital *Irghai*, which the Mongols call *Ircaya.*" Klaproth, illustrating Polo, reads "Eyircai, which the Mongols call *Eyircayá.*"

Pétis de la Croix, relating the same campaign and professing to follow Fadlallah, *i.e.* Rashiduddin, says the king "retired to his fortress of *Arbaca.*"

C. Sanang Setzen several times mentions a city called *Irghai*, *apparently* in Tangut; but all we can gather as to his position is that it seems to have lain east of Kanchau.

We perceive that the *Arbaca* of P. de la Croix, the *Eyircai* of Klaproth, the *Uiraca* of D'Ohsson, the *Artacki* or *Artackin* of Erdmann, are all various readings or forms of the same name, and are the same with the Chinese form *Ulahai* of De Mailla, and most probably the place is the *Egrigaia* of Polo.

We see also that Erdmann mentions another place *Aruki* (اروکی؟) in connection with Kanchau and Suhchau. This is, I suspect, the *Erguiul* of Polo, and perhaps the Irghai of Sanang Setzen.

Rashiduddin seems wrong in calling Ircayá the capital of the king, a circumstance which leads Klaproth to identify it with Ning-hsia. Pauthier, identifying Ulahai with Egrigaya, shows that the former was one of the circles of Tangut, but *not* that of Ning-hsia. Its position, he says, is uncertain. Klaproth, however, inserts it in his map of Asia, in the era of Kúblái (*Tabl. Hist.* pl. 22), as *Ulakhai* to the north of Ning-hsia, near the great bend eastward of the Hwang-Ho. Though it may have extended in this direction, it is probable, from the name referred to in next note, that Egrigaia or Ulahai is represented by the modern principality of ALASHAN, visited by Prjevalsky in 1871 and 1872.

[New travels and researches enable me to say that there can be no doubt that *Egrigaia = Ning-hsia.* Palladius (*l.c.* 18) says: "*Egrigaia* is Erigaia of the Mongol text. Klaproth was correct in his supposition that it is modern Ning-h'ia. Even now the Eleuths of Alashan call Ning-h'ia, *Yargai*. In M. Polo's time this department was famous for the cultivation of the Safflower (*carthamus tinctorius*). [*Siu t'ung kien*, A.D. 1292.]" Mr. Rockhill (cf. his *Diary of a Journey*) writes to me that Ninghsia is still called *Irge Khotun* by Mongols at the present day. M. Bonin (*J. As.*, 1900, I. 585) mentions the same fact.

Palladius (19) adds: "*Erigaia* is not to be confounded with *Urahai*, often mentioned in the history of Chingis Khan's wars with the Tangut kingdom. Urahai was a fortress in a pass of the same name in the Alashan Mountains. Chingis Khan spent five months there (an. 1208), during which he invaded and plundered the country in the neighbourhood. [*Si hia shu shi.*] The Alashan Mountains form r semicircle 500 *li* in extent, and have over forty narrow passes leading to the department of Ning-hia; the broadest and most practicable of these is now called Ch'i-mu-K'ów; it is not more than 80 feet broad. [*Ning hia fu chi.*] It may be that the Urahai fortress existed near this pass."

"From Liang-chow fu, M. Polo follows a special route, leaving the modern postal route on his right; the road he took has, since the time of the Emperor K'ang-hi, been called the courier's route." (*Palladius*, 18.)—H. C.]

NOTE 2.—*Calachan*, the chief town of Egrigaia, is mentioned, according to Klaproth, by Rashiduddin, among the cities of Tangut, as KALAJÁN. The name and approximate position suggest, as just noticed, identity with Alashan, the modern capital of which, called by Prjevalsky Dyn-yuan-yin, stands some distance west of the Hwang-Ho, in about lat. 39°. Polo gives no data for the interval between this and his next stage.

[The *Dyn-yuan-yin* of Prjevalsky is the camp of *Ting-yuan-yng* or Fu-ma-fu of M. Bonin, the residence of the Si-wang (western prince), of Alashan, an abbreviation of Alade-shan (*shan*, mountain in Chinese), Alade=Eleuth or Œlöt; the sister of this prince married a son of Prince Tuan, the chief of the *Boxers*. (*La Géographie*, 1901, I. 118.) Palladius (*l.c.* 19) says: "Under the name of Calachan, Polo probably means the summer residence of the Tangut kings, which was 60 *li* from Ning-hia, at the foot of the Alashan Mountains. It was built by the famous Tangut king Yuen-hao, on a large scale, in the shape of a castle, in which were high terraces and magnificent buildings. Traces of these buildings are visible to this day. There are often found coloured tiles and iron nails 1 foot, and even 2 feet long. The last Tangut kings made this place their permanent residence, and led there an indolent and sensual life. The Chinese name of this residence was Ho-lan shan *Li-Kung*.

There is sufficient reason to suppose that this very residence is named (under the year 1226) in the Mongol text *Alashai nuntuh;* and in the chronicles of the Tangut Kingdom, *Halahachar,* otherwise *Halachar,* apparently in the Tangut language. Thus M. Polo's Calachan can be identified with the Halachar of the *Si hia shu shi,* and can be taken to designate the Alashan residence of the Tangut kings."—H. C.]

NOTE 3.—Among the Buraets and Chinese at Kiakhta snow-white camels, without albino character, are often seen, and probably in other parts of Mongolia. (See *Erdmann,* II. 261.) Philostratus tells us that the King of Taxila furnished white camels to Apollonius. I doubt if the present King of Taxila, whom Anglo-Indians call the Commissioner of Ráwal Pindi, could do the like.

Cammellotti appear to have been fine woollen textures, by no means what are now called camlets, nor were they necessarily of camel's wool, for those of Angora goat's wool were much valued. M. Douet d'Arcq calls it "a fine stuff of wool approaching to our Cashmere, and sometimes of silk." Indeed, as Mr. Marsh points out, the word is Arabic, and has nothing to do with *Camel* in its origin ; though it evidently came to be associated therewith. *Khamlat* is defined in F. Johnson's Dict. : "Camelot, silk and camel's hair ; also all silk or velvet, especially pily and plushy," and *Khaml* is "pile or plush." *Camelin* was a different and inferior material. There was till recently a considerable import of different kinds of woollen goods from this part of China into Ladakh, Kashmir, and the northern Panjáb. [Leaving Ning-hsia, Mr. Rockhill writes (*Diary,* 1892, 44): "We passed on the road a cart with Jardine and Matheson's flag, coming probably from Chung-Wei Hsien, where camel's wool is sold in considerable quantities to foreigners. This trade has fallen off very much in the last three or four years on account of the Chinese middlemen rolling the wool in the dirt so as to add to its weight, and practising other tricks on buyers."— H. C.] Among the names of these were *Sling, Shirum, Gurun,* and *Khoza,* said to be the names of the towns in China where the goods were made. We have supposed *Sling* to be Sining (note 2, ch. lvii.), but I can make nothing of the others. Cunningham also mentions "camlets of camel's hair," under the name of *Suklát,* among imports from the same quarter. The term *Suklát* is, however, applied in the Panjáb trade returns to *broadcloth.* Does not this point to the real nature of the *siclatoun* of the Middle Ages ? It is, indeed, often spoken of as used for banners, which implies that it was not a *heavy* woollen :

> " There was mony gonfanoun
> Of gold, sendel, and siclatoun."
>   (*King Alisaundre,* in Weber, I. 85.)

But it was also a material for ladies' robes, for quilts, leggings, housings, pavilions. Franc. Michel does not decide what it was, only that it was generally *red* and wrought with gold. Dozy renders it "silk stuff brocaded with gold " ; but this seems conjectural. Dr. Rock says it was a thin glossy silken stuff, often with a woof of gold thread, and seems to derive it from the Arabic ṣaḳl, "polishing" (a sword), which is improbable. Perhaps the name is connected with *Ṣiḳiliyat,* "Sicily."

(*Marsh on Wedgwood,* and *on Webster* in *N. Y. Nation,* 1867 ; *Douet D'Arcq,* p. 355 ; *Punjab Trade Rep.,* App. ccxix.-xx. ; *Ladak,* 242 ; *Fr.-Michel Rech.* I. 221 *seqq. ; Dozy, Dict. des Vêtements, etc. ; Dr. Rock's Kens. Catal.* xxxix.-xl.)

CHAPTER LIX.

CONCERNING THE PROVINCE OF TENDUC, AND THE DESCENDANTS OF PRESTER JOHN.

TENDUC is a province which lies towards the east, and contains numerous towns and villages; among which is the chief city, also called TENDUC. The king of the province is of the lineage of Prester John, George by name, and he holds the land under the Great Kaan; not that he holds anything like the whole of what Prester John possessed.[1] It is a custom, I may tell you, that these kings of the lineage of Prester John always obtain to wife either daughters of the Great Kaan or other princesses of his family.[2]

In this province is found the stone from which Azure is made. It is obtained from a kind of vein in the earth, and is of very fine quality.[3] There is also a great manufacture of fine camlets of different colours from camel's hair. The people get their living by their cattle and tillage, as well as by trade and handicraft.

The rule of the province is in the hands of the Christians, as I have told you; but there are also plenty of Idolaters and worshippers of Mahommet. And there is also here a class of people called *Argons*, which is as much as to say in French *Guasmul*, or, in other words, sprung from two different races: to wit, of the race of the Idolaters of Tenduc and of that of the worshippers of Mahommet. They are handsomer men than the other natives of the country, and having more ability, they come to have authority; and they are also capital merchants.[4]

You must know that it was in this same capital city

of Tenduc that Prester John had the seat of his government when he ruled over the Tartars, and his heirs still abide there; for, as I have told you, this King George is of his line, in fact, he is the sixth in descent from Prester John.

Here also is what *we* call the country of Gog and Magog; *they*, however, call it Ung and Mungul, after the names of two races of people that existed in that Province before the migration of the Tartars. *Ung* was the title of the people of the country, and *Mungul* a name sometimes applied to the Tartars.[5]

And when you have ridden seven days eastward through this province you get near the provinces of Cathay. You find throughout those seven days' journey plenty of towns and villages, the inhabitants of which are Mahommetans, but with a mixture also of Idolaters and Nestorian Christians. They get their living by trade and manufactures; weaving those fine cloths of gold which are called *Nasich* and *Naques*, besides silk stuffs of many other kinds. For just as we have cloths of wool in our country, manufactured in a great variety of kinds, so in those regions they have stuffs of silk and gold in like variety.[6]

All this region is subject to the Great Kaan. There is a city you come to called Sindachu, where they carry on a great many crafts such as provide for the equipment of the Emperor's troops. In a mountain of the province there is a very good silver mine, from which much silver is got: the place is called Ydifu. The country is well stocked with game, both beast and bird.[7]

Now we will quit that province and go three days' journey forward.

Note 1.—Marco's own errors led commentators much astray about Tanduc or Tenduc, till Klaproth put the matter in its true light.

Our traveller says that Tenduc had been the seat of Aung Khan's sovereignty ; he has already said that it had been the scene of his final defeat, and he tells us that it was still the residence of his descendants in their reduced state. To the last piece of information he can speak as a witness, and he is corroborated by other evidence ; but the second statement we have seen to be almost certainly erroneous ; about the first we cannot speak positively.

Klaproth pointed out the true position of Tenduc in the vicinity of the great northern bend of the Hwang-Ho, quoting Chinese authorities to show that *Thianté* or *Thianté-Kiun* was the name of a district or group of towns to the north of that bend, a name which he supposes to be the original of Polo's *Tenduc*. The general position entirely agrees with Marco's indications ; it lies on his way eastward from Tangut towards Chagannor, and Shangtu (see ch. lx., lxi.), whilst in a later passage (Bk. II. ch. lxiv.), he speaks of the Caramoran or Hwang-Ho in its lower course, as " coming from the lands of Prester John."

M. Pauthier finds severe fault with Klaproth's identification of the *name* Tenduc with the Thianté of the Chinese, belonging to a city which had been destroyed 300 years before, whilst he himself will have that name to be a corruption of *Tathung*. The latter is still the name of a city and Fu of northern Shansi, but in Mongol time its circle of administration extended beyond the Chinese wall, and embraced territory on the left of the Hwang-Ho, being in fact the first *Lu*, or circle, entered on leaving Tangut, and therefore, Pauthier urges, the " Kingdom of Tanduc " of our text.

I find it hard to believe that Marco could get no nearer TATHUNG than in the form of *Tanduc* or *Tenduc*. The origin of the last may have been some Mongol name, not recovered. But it is at least conceivable that a name based on the old *Thianté-Kiun* might have been retained among the Tartars, from whom, and not from the Chinese, Polo took his nomenclature. Thianté had been, according to Pauthier's own quotations, the *military post of Tathung ;* Klaproth cites a Chinese author of the Mongol era, who describes the Hwang-Ho as passing through *the territory of the ancient Chinese city of Thianté ;* and Pauthier's own quotation from the Modern Imperial Geography seems to imply that a place in that territory was recently known as Fung-chau-*Thianté-Kiun.*

In the absence of preciser indications, it is reasonable to suppose that the Plain of Tenduc, with its numerous towns and villages, was the extensive and well-cultivated plain which stretches from the Hwang-Ho, past the city of Kuku-Khotan, or " Blue Town." This tract abounds in the remains of cities attributed to the Mongol era. And it is not improbable that the city of Tenduc was Kuku-Khotan itself, now called by the Chinese Kwei-hwa Ch'eng, but which was known to them in the Middle Ages as *Tsing-chau*, and to which we find the Kin Emperor of Northern China sending an envoy in 1210 to demand tribute from Chinghiz. The city is still an important mart and a centre of Lamaitic Buddhism, being the residence of a *Khutukhtu*, or personage combining the characters of cardinal and voluntarily re-incarnate saint, as well as the site of five great convents and fifteen smaller ones. Gerbillon notes that Kuku Khotan had been a place of great trade and population during the Mongol Dynasty.

[The following evidence shows, I think, that we must look for the city of Tenduc to *Tou Ch'eng* or *Toto Ch'eng*, called *Togto* or *Tokto* by the Mongols. Mr. Rockhill (*Diary*, 18) passed through this place, and 5 *li* south of it, reached on the Yellow River, Ho-k'ou (in Chinese) or Dugus or Dugei (in Mongol). Gerbillon speaks of Toto in his sixth voyage in Tartary. (*Du Halde*, IV. 345.) Mr. Rockhill adds that he cannot but think that Yule overlooked the existence of Togto when he identified Kwei-hwa Ch'eng with Tenduc. Tou Ch'eng is two days' march west of Kwei-hwa Ch'eng. " On the loess hill behind this place are the ruins of a large camp, Orch'eng, in all likelihood the site of the old town " (*l.c.* 18). M. Bonin (*J. As.* XV. 1900, 589) shares Mr. Rockhill's opinion. From Kwei-hwa Ch'eng, M. Bonin went by the valley of the Hei Shui River to the Hwang Ho ; at the junction of the two rivers stands the village of Ho-k'au (Ho-k'ou), south of the small town To Ch'eng, surmounted by the ruins of the old square Mongol stronghold of Tokto, the walls of which are still in a good state of preservation.—(*La Géographie*, I. 1901, p. 116.)

On the other hand, it is but fair to state that Palladius (21) says : " The name of Tenduc obviously corresponds to T'ien-te Kiun, a military post, the position of which Chinese geographers identify correctly with that of the modern Kuku-hoton (*Ta tsing y t'ung chi*, ch. on the Tumots of Kuku-hoton). The T'ien-te Kiun post existed under this name during the K'itan (Liao) and Kin Dynasties up to Khubilaï's time (1267) ; when under the name of Fung-chow it was left only a district town in the department of Ta-t'ung fu. The Kin kept in T'ien-te Kiun a military chief, *Chao-t'ao-shi*, whose duty it was to keep an eye on the neighbouring tribes, and to use, if needed, military force against them. The T'ien-te Kiun district was hardly greater in extent than the modern aïmak of Tumot, into which Kuku-hoton was included since the 16th century, *i.e.* 370 *li* from north to south, and 400 *li* from east to west ; during the Kin it had a settled population, numbering 22,600 families."

In a footnote, Palladius refers to the geographical parts of the *Liao shi*, *Kin shi*, and *Yuen shi*, and adds : " M. Polo's commentators are wrong in suspecting an anachronism in his statement, or trying to find Tenduc elsewhere."

We find in the *North-China Herald* (29th April, 1887, p. 474) the following note from the *Chinese Times :* " There are records that the position of this city [Kwei-hwa Ch'eng] was known to the builder of the Great Wall. From very remote times, it appears to have been a settlement of nomadic tribes. During the last 1000 years it has been alternately possessed by the Mongols and Chinese. About A.D. 1573, Emperor Wan-Li reclaimed it, enclosed a space within walls, and called it Kwei-hwa Ch'êng."

Potanin left Peking on the 13th May, 1884, for Kuku-khoto (or Kwei-hwa-Ch'eng), passing over the triple chain of mountains dividing the Plain of Peking from that on which Kuku-khoto is situate. The southernmost of these three ridges bears the Chinese name of Wu-tai-shan, " the mountain of five sacrificial altars," after the group of five peaks, the highest of which is 10,000 feet above the sea, a height not exceeded by any mountain in Northern China. · At its southern foot lies a valley remarkable for its Buddhist monasteries and shrines, one of which, " Shing-tung-tze," is entirely made of brass, whence its name.

" Kuku-Khoto is the depôt for the Mongolian trade with China. It contains two hundred tea-shops, five theatres, fifteen temples, and six Mongol monasteries. Among its sights are the Buddhist convent of Utassa, with its five pinnacles and bas-reliefs, the convent of Fing-sung-si, and a temple containing a statue erected in honour of the Chinese general, Pai-jin-jung, who avenged an insult offered to the Emperor of China." (*Proc. R. G. S.* IX. 1887, p. 233.)—H. C.]

A passage in Rashiduddin does seem to intimate that the Kerait, the tribe of Aung Khan, *alias* Prester John, did occupy territory close to the borders of Cathay or Northern China ; but neither from Chinese nor from other Oriental sources has any illustration yet been produced of the existence of Aung Khan's descendants as rulers in this territory under the Mongol emperors. There is, however, very positive evidence to that effect supplied by other European travellers, to whom the fables prevalent in the West had made the supposed traces of Prester John a subject of strong interest.

Thus John of Monte Corvino, afterwards Archbishop of Cambaluc or Peking, in his letter of January, 1305, from that city, speaks of Polo's King George in these terms : " A certain king of this part of the world, by name George, belonging to the sect of the Nestorian Christians, and of the illustrious lineage of that great king who was called Prester John of India, in the first year of my arrival here [*circa* 1295-1296] attached himself to me, and, after he had been converted by me to the verity of the Catholic faith, took the Lesser Orders, and when I celebrated mass used to attend me wearing his royal robes. Certain others of the Nestorians on this account accused him of apostasy, but he brought over a great part of his people with him to· the true Catholic faith, and built a church of royal magnificence in honour of our God, of the

Holy Trinity, and of our Lord, the Pope, giving it the name of *the Roman Church*. This King George, six years ago, departed to the Lord, a true Christian, leaving as his heir a son scarcely out of the cradle, and who is now nine years old. And after King George's death, his brothers, perfidious followers of the errors of Nestorius, perverted again all those whom he had brought over to the Church, and carried them back to their original schismatical creed. And being all alone, and not able to leave His Majesty the Cham, I could not go to visit the church above-mentioned, which is twenty days' journey distant. . . . I had been in treaty with the late King George, if he had lived, to translate the whole Latin ritual, that it might be sung throughout the extent of his territory; and whilst he was alive I used to celebrate mass in his church according to the Latin rite." The distance mentioned, twenty days' journey from Peking, suits quite well with the position assigned to Tenduc, and no doubt the Roman Church was in the city to which Polo gives that name.

Friar Odoric, travelling from Peking towards Shensi, about 1326-1327, also visits the country of Prester John, and gives to its chief city the name of *Tozan*, in which perhaps we may trace *Tathung*. He speaks as if the family still existed in authority.

King George appears again in Marco's own book (Bk. IV. ch. ii.) as one of Kúblái's generals against Kaidu, in a battle fought near Karakorúm. (*Journ. As.* IX. 299 *seqq.*; *D'Ohsson*, I. 123; *Huc's Tartary, etc.* I. 55 *seqq.*; *Koeppen*, II. 381; *Erdmann's Temudschin; Gerbillon* in *Astley*, IV. 670; *Cathay*, pp. 146 and 199 *seqq.*)

Note 2.—Such a compact is related to have existed reciprocally between the family of Chinghiz and that of the chief of the Ḳunguráts; but I have not found it alleged of the Kerait family except by Friar Odoric. We find, however, many *princesses* of this family married into that of Chinghiz. Thus three nieces of Aung Khan became wives respectively of Chinghiz himself and of his sons Juji and Tului; she who was the wife of the latter, Serḳuḳteni Bigi, being the mother of Mangú, Hulaku, and Kúblái. Duḳuz Khatun, the Christian wife of Hulaku, was a granddaughter of Aung Khan.

The name *George*, of Prester John's representative, may have been actually Jirjis, Yurji, or some such Oriental form of Georgius. But it is possible that the title was really *Gurgán*, "Son-in-Law," a title of honour conferred on those who married into the imperial blood, and that this title may have led to the statements of Marco and Odoric about the nuptial privileges of the family. Gurgán in this sense was one of the titles borne by Timur.*

[The following note by the Archimandrite Palladius (*Eluc.* 21-23) throws a great light on the relations between the families of Chinghiz Khan and of Prester John.

"T'ien-te Kiun was bounded on the north by the *Yn-shan* Mountains, in and beyond which was settled the Sha-t'o Tu-K'iu tribe, *i.e.* Tu-K'iu of the sandy desert. The K'itans, when they conquered the northern borders of China, brought also under their rule the dispersed family of these Tu-K'iu. With the accession of the Kin, a Wang Ku [Ongot] family made its appearance as the ruling family of those

---

* Mr. Ney Elias favours me with a curious but tantalising communication on this subject: "An old man called on me at Kwei-hwa Ch'eng (Tenduc), who said he was neither Chinaman, Mongol, nor Mahomedan, and lived on ground a short distance to the north of the city, especially allotted to his ancestors by the Emperor, and where there now exist several families of the same origin. He then mentioned the connection of his family with that of the Emperor, but in what way I am not clear, and said that he ought to be, or had been, a prince. Other people coming in, he was interrupted and went away. . . . . He was not with me more than ten minutes, and the incident is a specimen of the difficulty in obtaining interesting information, except by mere chance. . . . The idea that struck me was, that he was perhaps a descendant of King George of Tenduc; for I had your M. P. before me, and had been inquiring as much as I dared about subjects it suggested. . . . . At Kwei-hwa Ch'eng I was very closely spied, and my servant was frequently told to warn me against asking too many questions."

I should mention that Oppert, in his very interesting monograph, *Der Presbyter Johannes*, refuses to recognise the Kerait chief at all in that character, and supposes Polo's King George to be the representative of a prince of the Liao (*supra*, p. 205), who, as we learn from De Mailla's History, after the defeat of the Kin, in which he had assisted Chinghiz, settled in Liaotung, and received from the conqueror the title of King of the Liao. This seems to me geographically and otherwise quite inadmissible.

tribes ; it issued from those Sha-t'o Tu-K'iu, who once reigned in the north of China as the How T'ang Dynasty (923-936 A.D.). It split into two branches, the Wang-Ku of the Yn-shan, and the Wang-Ku of the Lin-t'ao (west of Kan-su). The Kin removed the latter branch to Liao-tung (in Manchuria). The Yn-shan Wang-Ku guarded the northern borders of China belonging to the Kin, and watched their herds. When the Kin, as a protection against the inroads of the tribes of the desert, erected a rampart, or new wall, from the boundary of the Tángut Kingdom down to Manchuria, they intrusted the defence of the principal places of the Yn-shan portion of the wall to the Wang-Ku, and transferred there also the Liao-tung Wang-Ku. At the time Chingiz Khan became powerful, the chief of the Wang-Ku of the Yn-shan was Alahush ; and at the head of the Liao-tung Wang-Ku stood *Pa-sao-ma-ie-li*. Alahush proved a traitor to the Kin, and passed over to Chinghiz Khan ; for this he was murdered by the malcontents of his family, perhaps by Pa-sao-ma-ie-li, who remained true to the Kin. Later on, Chingiz Khan married one of his daughters to the son of Alahush, by name Po-yao-ho, who, however, had no children by her. He had three sons by a concubine, the eldest of whom, Kiun-pu-hwa, was married to Kuyuk Khan's daughter. Kiun-pu-hwa's son, Ko-li-ki-sze, had two wives, both of imperial blood. During a campaign against Haidu, he was made prisoner in 1298, and murdered. His title and dignities passed over in A.D. 1310 to his son *Chuan*. Nothing is known of Alahush's later descendants ; they probably became entirely Chinese, like their relatives of the Liao-tung branch.

"The Wang-Ku princes were thus *de jure* the sons-in-law of the Mongol Khans, and they had, moreover, the hereditary title of Kao-t'ang princes (Kao-t'ang wang) ; it is very possible that they had their residence in ancient T'ien-te Kiun (although no mention is made of it in history), just as at present the Tumot princes reside in Kuku-hoton.

"The consonance of the names of Wang-Khan and Wang-Ku (Ung-Khan and Ongu) led to the confusion regarding the tribes and persons, which at Marco Polo's time seems to have been general among the Europeans in China ; Marco Polo and Johannes de Monte Corvino transfer the title of Prester John from Wang-Khan, already perished at that time, to the distinguished family of Wang-Ku. Their Georgius is undoubtedly Ko-li-ki-sze, Alahush's great-grandson. That his name is a Christian one is confirmed by other testimonies ; thus in the Asu (Azes) regiment of the Khan's guards was Ko-li-ki-sze, *aliàs* Kow-r-ki (†1311), and his son Ti-mi-ti-r. There is no doubt that one of them was Georgius, and the other Demetrius. Further, in the description of *Chin-Kiang* in the time of the Yuen, mention is made of Ko-li-ki-sze Ye-li-ko-wen, *i.e* Ko-li-ki-sze, the Christian, and of his son Lu-ho (Luke).

" Ko-li-ki-sze of Wang-ku is much praised in history for his valour and his love for Confucian doctrine ; he had in consequence of a special favour of the Khan two Mongol princesses for wives at the same time (which is rather difficult to conciliate with his being a Christian). The time of his death is correctly indicated in a letter of Joannes de M. Corvino of the year 1305 : *ante sex annos migravit ad Dominum.* He left a young son *Chu-an*, who probably is the Joannes of the letter of Ioannes (Giovani) de M Corvino, so called *propter nomen meum*, says the missionary. In another Wang-ku branch, Si-li-ki-sze reminds one also of the Christian name *Sergius*."—H. C.]

NOTE 3.—" The *Lapis Armenus*, or Azure, . . . is produced in the district of Tayton-fu (*i.e. Tathung*), belonging to Shansi." (*Du Halde* in *Astley*, IV. 309 ; see also *Martini*, p. 36.)

NOTE 4.—This is a highly interesting passage, but difficult, from being corrupt in the G. Text, and over-curt in Pauthier's MSS. In the former it runs as follows : " *Hil hi a une jeneration de jens que sunt appellés* Argon, *qe vaut à dire en françois* Guasmul, *ce est à dire qu'il sunt né del deus generasions de la lengnée des celz* Argon Tenduc et des celz reduc et des celz que aorent Maomet. *Il sunt biaus homes plus que le autre dou païs et plus sajes et plus mercaant.*" Pauthier's text runs thus : "*Il ont une generation de gens, ces Crestiens qui ont la Seigneurie, qui s'appellent* Argon, *qui vaut a dire* Gasmul ; *et sont plus beaux hommes que les autres mescreans et plus sages. Et pour ce ont il la seigneurie et sont bons marchans.*" And Ramusio : "*Vi è anche una sorte di gente che si chiamano* Argon, *per che sono nati di due generazioni,* cioè da quella di Tenduc che adorano gl' idoli, e da quella che osservano la legge di Macometto. *E questi sono i piu belli uomini che si trovino in quel paese e più savi, e più accorti nella mercanzia.*"

In the first quotation the definition of the *Argon* as sprung *de la lengnée, etc.*, is not intelligible as it stands, but seems to be a corruption of the same definition that has been rendered by Ramusio, viz. that the Argon were half-castes between the race of the Tenduc Buddhists and that of the Mahomedan settlers. These two texts do not assert that the Argon were Christians. Pauthier's text at first sight seems to assert this, and to identify them with the Christian rulers of the province. But I doubt if it means more than that the Christian rulers *have under them* a people called Argon, etc. The passage has been read with a bias, owing to an erroneous interpretation of the word *Argon* in the teeth of Polo's explanation of it.

Klaproth, I believe, first suggested that *Argon* represents the term *Arkhaiún*, which is found repeatedly applied to Oriental Christians, or their clergy, in the histories of the Mongol era.[*] No quite satisfactory explanation has been given of the origin of that term. It is barely possible that it may be connected with that which Polo uses here ; but he tells us as plainly as possible that he means by the term, not a Christian, but a *half-breed.*

And in this sense the word is still extant in Tibet, probably also in Eastern Turkestan, precisely in Marco's form, ARGON. It is applied in Ladak, as General Cunningham tells us, specifically to the mixt race produced by the marriages of Kashmirian immigrants with Bŏt (Tibetan) women. And it was apparently to an analogous cross between Caucasians and Turanians that the term was applied in Tenduc. Moorcroft also speaks of this class in Ladak, calling them *Argands.* Mr. Shaw styles them "a set of ruffians called *Argoons,* half-bred between Toorkistan fathers and Ladak mothers. . . . They possess all the evil qualities of both races, without any of their virtues." And the author of the Dabistan, speaking of the Tibetan Lamas, says : "Their king, if his mother be not of royal blood, is by them called *Arghún,* and not considered their true king." [See p. 291, my reference to *Wellby's Tibet.*—H. C.] Cunningham says the word is probably Turki, اُرغُون, *Arghún,* "Fair," "not *white,*" as he writes to me, "but *ruddy* or *pink,* and therefore 'fair.' *Arghún* is both Turki and Mogholi, and is applied to all fair children, both male and female, as *Arghun Beg, Arghuna Khatun,* etc.[†] We find an *Arghún* tribe named in Timur's Institutes, which probably derived its descent from such half-breeds. And though the Arghún Dynasty of Kandahar and Sind claimed their descent and name from Arghún Khan of Persia, this may have had no other foundation.

---

[*] The term *Arkaiun*, or *Arkaun*, in this sense, occurs in the Armenian History of Stephen Orpelian, quoted by St. Martin. The author of the *Tárikh Jahán Kushai,* cited by D'Ohsson, says that Christians were called by the Mongols *Arkáún.* When Hulaku invested Baghdad we are told that he sent a letter to the Judges, Shaikhs, Doctors and *Arkauns,* promising to spare such as should act peaceably. And in the subsequent sack we hear that no houses were spared except those of a few *Arkauns* and foreigners. In Rashiduddin's account of the Council of State at Peking, we are told that the four *Fanchan,* or Ministers of the Second Class, were taken from the four nations of Tájiks, Cathayans, Uighúrs, and *Arkaun.* Sabadin *Arkaun* was the name of one of the Envoys sent by Arghun Khan of Persia to the Pope in 1288. Traces of the name appear also in Chinese documents of the Mongol era, as denoting *some* religious body. Some of these have been quoted by Mr. Wylie ; but I have seen no notice taken of a very curious extract given by Visdelou. This states that Kúblái in 1289 established a Board of nineteen chief officers to have surveillance of the affairs of the Religion of the Cross, of the *Marha,* the *Siliepan,* and the *Yelikhawen.* This Board was raised to a higher rank in 1315 : and at that time 72 minor courts presiding over the religion of the *Yelikhawen* existed under its supervision. Here we evidently have the word *Arkhaiun* in a Chinese form ; and we may hazard the suggestion that *Marha, Siliepan* and *Yelikhawen* meant respectively the Armenian, Syrian, or Jacobite, and Nestorian Churches. (*St. Martin, Mém.* II. 133, 143; 279; *D'Ohsson,* II. 264 ; *Ilchan,* I. 150, 152 ; *Cathay,* 264 ; *Acad.* VII. 359 ; Wylie in *J. As.* V. xix. 466. Suppt. to *D'Herbelot,* 142.)

[†] The word is not in Zenker or Pavet de Courteille.

There are some curious analogies between these Argons of whom Marco speaks and those Mahomedans of Northern China and Chinese Turkestan lately revolted against Chinese authority, who are called *Tungăni*, or as the Russians write it *Dungen*, a word signifying, according to Professor Vámbéry, in Turki, "a convert." *
These Tungani are said by one account to trace their origin to a large body of Uighúrs, who were transferred *to the vicinity of the Great Wall* during the rule of the Thang Dynasty (7th to 10th century). Another tradition derives their origin from Samarkand. And it is remarkable that Rashiduddin speaks of a town to the west or north-west of Peking, "most of the inhabitants of which are natives of Samarkand, and have planted a number of gardens in the Samarkand style."† The former tradition goes on to say that marriages were encouraged between the Western settlers and the Chinese women. In after days these people followed the example of their kindred in becoming Mahomedans, but they still retained the practice of marrying Chinese wives, though bringing up their children in Islam. The Tungani are stated to be known in Central Asia for their commercial integrity; and they were generally selected by the Chinese for police functionaries. They are passionate and ready to use the knife; but are distinguished from both Manchus and Chinese by their strength of body and intelligent countenances. Their special feature is their predilection for mercantile speculations.

Looking to the many common features of the two accounts—the origin as a half-breed between Mahomedans of Western extraction and Northern Chinese, the position in the vicinity of the Great Wall, the superior physique, intelligence, and special capacity for trade, it seems highly probable that the Tungani of our day are the descendants of Marco's Argons. Otherwise we may at least point to these analogies as a notable instance of like results produced by like circumstances on the same scene; in fact, of history repeating itself. (See *The Dungens*, by *Mr. H. K. Heins*, in the *Russian Military Journal* for August, 1866, and *Western China*, in the *Ed. Review* for April, 1868; ‡ *Cathay*, p. 261.)

[Palladius (pp. 23-24) says that "it is impossible to admit that Polo had meant to designate by this name the Christians, who were called by the Mongols *Erkeun* [*Ye li ke un*]. He was well acquainted with the Christians in China, and of course could not ignore the name under which they were generally known to such a degree as to see in it a designation of a cross-race of Mahommetans and heathens." From the *Yuen ch'ao pi shi* and the *Yuen shi*, Palladius gives some examples which refer to Mahommedans.

Professor Devéria (*Notes d'Épig.* 49) says that the word "Αρχων was used by the Mongol Government as a designation for the members of the Christian clergy at large; the word is used between 1252 and 1315 to speak of *Christian* priests by the historians of the Yuen Dynasty; it is not used before nor is it to be found in the Singan-fu inscription (*l.c.* 82). Mr. E. H. Parker (*China Review*, xxiv. p. 157) supplies a few omissions in Devéria's paper; we note among others: "Ninth moon of 1329. Buddhist services ordered to be held by the Uighúr priests, and by the Christians [*Ye li ke un*]."

Captain Wellby writes (*Unknown Tibet*, p. 32): "We impressed into our service six other muleteers, four of them being Argoons, who are really half-castes, arising

---

* Mr. Shaw writes *Toongánee*. The first mention of this name that I know of is in Izzat Ullah's Journal. (Vide *J. R. A. S.* VII. 310.) The people are there said to have got the name from having first settled in *Tungan*. Tung-gan is in the same page the name given to the strong city of T'ung Kwan on the Hwang-ho. (See Bk. II. ch. xli. note 1.) A variety of etymologies have been given, but Vámbéry's seems the most probable.

† Probably no man could now say what this means. But the following note from Mr. Ney Elias is very interesting in its suggestion of analogy: "In my report to the Geographical Society I have noticed the peculiar Western appearance of Kwei-hwa-ch'eng, and the little gardens of creepers and flowers in pots which are displayed round the porches in the court-yards of the better class of houses, and which I have seen in no other part of China. My attention was especially drawn to these by your quotation from Rashiduddin."

‡ A translation of *Heins'* was kindly lent me by the author of this article, the lamented Mr. J. W. S. Wyllie.

VOL. I.   T 2

from the merchants of Turkestan making short marriages with the Ladakhi women."
—H. C.]

Our author gives the odd word *Guasmul* as the French equivalent of Argon. M. Pauthier has first, of Polo's editors, given the true explanation from Ducange. The word appears to have been in use in the Levant among the Franks as a name for the half-breeds sprung from their own unions with Greek women. It occurs three times in the history of George Pachymeres. Thus he says (*Mich. Pal.* III. 9), that the Emperor Michael "depended upon the *Gasmuls*, or mixt breeds (συμμίκτοι), which is the sense of this word of the Italian tongue, for these were born of Greeks and Italians, and sent them to man his ships; for the race in question inherited at once the military wariness and quick wit of the Greeks, and the dash and pertinacity of the Latins." Again (IV. 26) he speaks of these "Gasmuls, whom a Greek would call διγενεῖς, men sprung from Greek mothers and Italian fathers." Nicephorus Gregoras also relates how Michael Palaeologus, to oppose the projects of Baldwin for the recovery of his fortunes, manned 60 galleys, chiefly with the tribe of Gasmuls (γένος τοῦ Γασμουλικοῦ), to whom he assigns the same characteristics as Pachymeres. (**IV.** v. 5, also VI. iii. 3, and XIV. x. 11.) One MS. of Nicetas Choniates also, in his annals of Manuel Comnenus (see Paris ed. p. 425), speaks of "the light troops whom we call *Basmuls.*" Thus it would seem that, as in the analogous case of the *Turcopuli*, sprung from Turk fathers and Greek mothers, their name had come to be applied technically to a class of troops. According to Buchon, the laws of the Venetians in Candia mention, as different races in that island, the *Vasmulo*, Latino, Blaco, and Griego.

Ducange, in one of his notes on Joinville, says: "During the time that the French possessed Constantinople, they gave the name of *Gas-moules* to those who were born of French fathers and Greek mothers; or more probably *Gaste-moules*, by way of derision, as if such children by those irregular marriages . . . . had in some sort debased the wombs of their mothers!" I have little doubt (*pace tanti viri*) that the word is in a Gallicized form the same with the surviving Italian *Guazzabúglio*, a hotch-potch, or mish-mash. In Davanzati's *Tacitus*, the words "Colluviem *illam nationum*" (*Annal.* II. 55) are rendered "*quello* guazzabuglio *di nazioni*," in which case we come very close to the meaning assigned to *Guasmul*. The Italians are somewhat behind in matters of etymology, and I can get no light from them on the history of this word. (See *Buchon, Chroniques Etrangères*, p. xv. ; *Ducange, Gloss. Graecitatis*, and his note on *Joinville*, in *Bohn's Chron. of the Crusades*, 466.)

NOTE 5.—It has often been cast in Marco's teeth that he makes no mention of the Great Wall of China, and that is true; whilst the apologies made for the omission have always seemed to me unsatisfactory. [I find in Sir G. Staunton's account of Macartney's Embassy (II. p. 185) this most amusing explanation of the reason why Marco Polo did not mention the wall: "A copy of Marco Polo's route to China, taken from the Doge's Library at Venice, is sufficient to decide this question. By this route it appears that, in fact, that traveller did not pass through Tartary to Pekin, but that after having followed the usual track of the caravans, as far to the eastward from Europe as Samarcand and Cashgar, he bent his course to the south-east across the River Ganges to Bengal (!), and, keeping to the southward of the Thibet mountains, reached the Chinese province of Shensee, and through the adjoining province of Shansee to the capital, without interfering with the line of the Great Wall."—H. C.] We shall see presently that the Great Wall is spoken of by Marco's contemporaries Rashiduddin and Abulfeda. Yet I think, if we read "between the lines," we shall see reason to believe that the Wall *was* in Polo's mind at this point of the dictation, whatever may have been his motive for withholding distincter notice of it.* I cannot conceive why he should say : "Here is what we call the country of Gog and Magog," except as intimating "Here we are

---

* I owe the suggestion of this to a remark in *Oppert's Presbyter Johannes*, p. 77.

The Rampart of Gog and Magog.

beside the GREAT WALL known as the Rampart of Gog and Magog," and being there he tries to find a reason why those names should have been applied to it. Why they were really applied to it we have already seen. (*Supra*, ch. iv. note 3.) Abulfeda says : " The Ocean turns northward along the east of China, and then expands in the same direction till it passes China, and comes opposite to the Rampart of Yájúj and Májúj ; " whilst the same geographer's definition of the boundaries of China exhibits that country as bounded on the west by the Indo-Chinese wildernesses ; on the south, by the seas ; on the east, by the Eastern Ocean ; on the north, by the *land of Yájúj and Májúj*, and other countries unknown. Ibn Batuta, with less accurate geography in his head than Abulfeda, maugre his travels, asks about the Rampart of Gog and Magog (*Sadd Yájúj wa Majúj*) when he is at Sin Kalán, *i.e.* Canton, and, as might be expected, gets little satisfaction.

Apart from this interesting point Marsden seems to be right in the general bearing of his explanation of the passage, and I conceive that the two classes of people whom Marco tries to identify with Gog and Magog do substantially represent the two genera or species, TURKS and MONGOLS, or, according to another nomenclature used by Rashiduddin, the *White* and *Black* Tartars. To the latter class belonged Chinghiz and his MONGOLS proper, with a number of other tribes detailed by Rashiduddin, and these I take to be in a general way the MUNGUL of our text. The *Ung*, on the other hand, are the UNG-*ḳut*, the latter form being presumably only the Mongol plural of UNG. The Ung-ḳút were a Turk tribe who were vassals of the Kin Emperors of Cathay, and were intrusted with the defence of the Wall of China, or an important portion of it, which was called by the Mongols *Ungu*, a name which some connect with that of the tribe. [See note pp. 288-9.] Erdmann indeed asserts that the wall by which the Ung-ḳut dwelt was not the Great Wall, but some other. There are traces of other great ramparts in the steppes north of the present wall. But Erdmann's arguments seem to me weak in the extreme.

[Mr. Rockhill (*Rubruck*, p. 112) writes : " The earliest mention I have found of the name *Mongol* in Oriental works occurs in the Chinese annals of the After T'ang period (A.D. 923-934), where it occurs in the form *Meng-ku*. In the annals of the Liao Dynasty (A.D. 916-1125) it is found under the form *Meng-ku-li*. The first occurrence of the name in the *Tung chien kang mu* is, however, in the 6th year Shao-hsing of Kao-tsung of the Sung (A.D. 1136). It is just possible that we may trace the word back a little earlier than the After T'ang period; and that the *Meng-wa* (or *ngo*, as this character may have been pronounced at the time), a branch of the Shih-wei, a Tungusic or Kitan people living around Lake Keule, to the east of the Baikal, and along the Kerulun, which empties into it, during the 7th and subsequent centuries, and referred to in the *T'ang shu* (Bk. 219), is the same as the later Meng-ku. Though I have been unable to find, as stated by Howorth (*History*, i. pt. I. 28), that the name *Meng-ku* occurs in the T'ang shu, his conclusion that the northern Shih-wei of that time constituted the Mongol nation proper is very likely correct. . . . . I. J. Schmidt (*Ssanang Setzen*, 380) derives the name *Mongol* from *mong*, meaning ' brave, daring, bold,' while Rashideddin says it means ' simple, weak ' (*d'Ohsson*, i. 22). The Chinese characters used to transcribe the name mean ' dull, stupid,' and ' old, ancient,' but they are used purely phonetically. . . . . The Mongols of the present day are commonly called by the Chinese *Ta-tzŭ*, but this name is resented by the Mongols as opprobrious, though it is but an abbreviated form of the name *Ta-ta-tzŭ*, in which, according to Rubruck, they once gloried."—H. C.]

Vincent of Beauvais has got from some of his authorities a conception of the distinction of the Tartars into two races, to which, however, he assigns no names : " *Sunt autem duo genera Tartarorum, diversa quidem habentia idiomata, sed unicam legem ac ritum, sicut Franci et Theutonici.*" But the result of *his* effort to find a realisation of Gog and Magog is that he makes *Guyuk Kaan* into Gog, and *Mangu Kaan* into Magog. Even the intelligent Friar Ricold says of the Tartars : " They say themselves that they are descended from Gog and Magog : and on this account they are called *Mogoli*, as if from a corruption of *Magogoli*." (*Abulfeda* in *Büsching*, IV.

140, 274-275; *I. B.* IV. 274; *Golden Horde,* 34, 68; *Erdmann,* 241-242, 257-258; *Timk.* I. 259, 263, 268; *Vinc. Bellov. Spec. Hist.* XXIX. 73, XXXI. 32-34; *Pereg. Quat.* 118; *Not. et Ext.* II. 536.)

NOTE 6.—The towns and villages were probably those immediately north of the Great Wall, between 112° and 115° East longitude, of which many remains exist, ascribed to the time of the Yuen or Mongol Dynasty. This tract, between the Great Wall and the volcanic plateau of Mongolia, is extensively colonised by Chinese, and has resumed the flourishing aspect that Polo describes. It is known now as the *Ku-wei,* or extramural region.

[After Kalgan, Captain Younghusband, on the 12th April, 1886, "passed through the [outer] Great Wall . . . . entering what Marco Polo calls the land of Gog and Magog. For the next two days I passed through a hilly country inhabited by Chinese, though it really belongs to Mongolia; but on the 14th I emerged on to the real steppes, which are the characteristic features of Mongolia Proper." (*Proc. R. G. S.* X., 1888, p. 490.)—H. C.]

Of the cloths called *nakh* and *nasij* we have spoken before (*supra* ch. vi. note 4). These stuffs, or some such as these, were, I believe, what the mediæval writers called *Tartary cloth,* not because they were made in Tartary, but because they were brought from China and its borders through the Tartar dominions; as we find that for like reason they were sometimes called stuffs of *Russia.* Dante alludes to the supposed skill of Turks and Tartars in weaving gorgeous stuffs, and Boccaccio, commenting thereon, says that Tartarian cloths are so skilfully woven that no painter with his brush could equal them: Maundevile often speaks of cloths of Tartary (*e.g.* pp. **175, 247**). So also Chaucer:

> " On every trumpe hanging a broad banere
> Of fine *Tartarium.*"

Again, in the French inventory of the *Garde-Meuble* of 1353 we find two pieces of *Tartary,* one green and the other red, priced at 15 crowns each. (*Flower and Leaf,* 211; *Dante, Inf.* XVII. 17, and *Longfellow,* p. 159; *Douet d'Arcq,* p. 328; *Fr.-Michel, Rech.* I. 315, II. 166 *seqq.*)

NOTE 7.—SINDACHU (Sindacui, Suidatui, etc., of the MSS.) is SIUEN-HWA-FU, called under the Kin Dynasty *Siuen-te-chau,* more than once besieged and taken by Chinghiz. It is said to have been a summer residence of the later Mongol Emperors, and fine parks full of grand trees remain on the western side. It is still a large town and the capital of a *Fu,* about 25 miles south of the Gate on the Great Wall at Chang Kia Kau, which the Mongols and Russians call Kalgan. There is still a manufacture of felt and woollen articles here.

[Mr. Rockhill writes to me that this place is noted for the manufacture of buck-skins.—H. C.]

Ydifu has not been identified. But Baron Richthofen saw old mines north-east of Kalgan, which used to yield argentiferous galena; and Pumpelly heard of silver-mines near Yuchau, in the same department.

[In the *Yuen-shi* it is " stated that there were gold and silver mines in the districts of Siuen-te-chow and Yuchow, as well as in the Kiming shan Mountains. These mines were worked by the Government itself up to 1323, when they were transferred to private enterprise. Marco Polo's *Ydifu* is probably a copyist's error, and stands instead of Yuchow." (*Palladius,* 24, 25.)—H. C.]

CHAPTER LX.

CONCERNING THE KAAN'S PALACE OF CHAGANNOR.

AT the end of those three days you find a city called
CHAGAN NOR [which is as much as to say White Pool],
at which there is a great Palace of the Grand Kaan's;[1]
and he likes much to reside there on account of the
Lakes and Rivers in the neighbourhood, which are the
haunt of swans[2] and of a great variety of other birds.
The adjoining plains too abound with cranes, partridges,
pheasants, and other game birds, so that the Emperor
takes all the more delight in staying there, in order to
go a-hawking with his gerfalcons and other falcons, a
sport of which he is very fond.[3]

There are five different kinds of cranes found in
those tracts, as I shall tell you. First, there is one
which is very big, and all over as black as a crow; the
second kind again is all white, and is the biggest of all;
its wings are really beautiful, for they are adorned with
round eyes like those of a peacock, but of a resplendent
golden colour, whilst the head is red and black on a
white ground. The third kind is the same as ours.
The fourth is a small kind, having at the ears beautiful
long pendent feathers of red and black. The fifth
kind is grey all over and of great size, with a handsome
head, red and black.[4]

Near this city there is a valley in which the Emperor
has had several little houses erected in which he keeps
in mew a huge number of *cators*, which are what we
call the Great Partridge. You would be astonished
to see what a quantity there are, with men to take
charge of them. So whenever the Kaan visits the
place he is furnished with as many as he wants.[5]

NOTE 1.—[According to the *Siu t'ung kien*, quoted by Palladius, the palace in Chagannor was built in 1280.—H.C.]

NOTE 2.—"*Ou demeurent* sesnes." *Sesnes, Cesnes, Cecini, Cesanae,* is a mediæval form of *cygnes, cigni,* which seems to have escaped the dictionary-makers. It occurs in the old Italian version of *Brunetto Latini's Tresor,* Bk. V. ch. xxv., as *cecino;* and for other examples, see *Cathay,* p. 125.

NOTE 3.—The city called by Polo CHAGAN-NOR (meaning in Mongol, as he says, "White Lake") is the *Chaghan Balghasun* mentioned by Timkowski as an old city of the Mongol era, the ruined rampart of which he passed about 30 miles north of the Great Wall at Kalgan, and some 55 miles from Siuen-hwa, adjoining the Imperial pastures. It stands near a lake still called Chaghan-Nor, and is called by the Chinese Pe-ching-tzu, or White City, a translation of Chaghan Balghasun. Dr. Bushell says of one of the lakes (Ichi-Nor), a few miles east of Chaghan-Nor : "We . . . . found the water black with waterfowl, which rose in dense flocks, and filled the air with discordant noises. *Swans,* geese, and ducks predominated, and *three different species of cranes* were distinguished."
The town appears as *Tchahan Toloho* in D'Anville. It is also, I imagine, the *Arulun Tsaghan Balghasun* which S. Setzen says Kúblái built about the same time with Shangtu and another city "on the shady side of the Altai," by which here he seems to mean the Khingan range adjoining the Great Wall. (*Timk.* II. 374, 378-379; *J. R. G. S.* vol. xliii. ; *S. Setz.* 115.) I see Ritter has made the same identification of Chaghan-Nor (II. 141).

NOTE 4.—The following are the best results I can arrive at in the identification of these five cranes.
1. Radde mentions as a rare crane in South Siberia *Grus monachus,* called by the Buraits *Kará Togorü,* or "Black Crane." Atkinson also speaks of "a beautiful black variety of crane," probably the same. The *Grus monachus* is not, however, jet black, but brownish rather. (*Radde, Reisen,* Bd. II. p. 318 ; *Atkinson. Or. and W. Sib.* 548.)
2. *Grus leucogeranus* (?) whose chief habitat is Siberia, but which sometimes comes as far south as the Punjab. It is the largest of the genus, snowy white, with red face and beak ; the ten largest quills are black, but this barely shows as a narrow black line when the wings are closed. The resplendent golden eyes on the wings remain unaccounted for ; no naturalist whom I have consulted has any knowledge of a crane or crane-like bird with such decorations. When 'tis discovered, let it be the *Grus Poli!*
3. *Grus cinerea.*
4. The colour of the pendants varies in the texts. Pauthier's and the G. Text have *red and black;* the Lat. S. G. *black* only, the Crusca *black and white,* Ramusio *feathers red and blue* (not pendants). The *red and black* may have slipt in from the preceding description. I incline to believe it to be the Demoiselle, *Anthropoïdes Virgo,* which is frequently seen as far north as Lake Baikal. It has a tuft of pure *white* from the eye, and a beautiful black pendent ruff or collar ; the general plumage purplish-grey.
5. Certainly the Indian *Sáras* (vulgo Cyrus), or *Grus antigone,* which answers in colours and grows to 52 inches high.

NOTE 5.—*Cator* occurs only in the G. Text and the Crusca, in the latter with the interpolated explanation "*cioè contornici*" (*i.e.* quails), whilst the S. G. Latin has *coturnices* only. I suspect this impression has assisted to corrupt the text, and that it was originally written or dictated *ciacor* or *çacor,* viz. *chakór,* a term applied in the East to more than one kind of "Great Partridge." Its most common application in India is to the Himalayan red-legged partridge, much resembling on a somewhat larger scale the bird so called in Europe. It is the "Francolin" of Moorcroft's

Travels, and the *Caccabis Chukor* of Gray. According to Cunningham the name is applied in Ladak to the bird sometimes called the Snow-pheasant, Jerdan's Snow-cock, *Tetraogallus himalayensis* of Gray. And it must be the latter which Moorcroft speaks of as "the gigantic Chukor, much larger than the common partridge, found in large coveys on the edge of the snow ; . . . . one plucked and drawn weighed 5 lbs." ; described by Vigne as "a partridge as large as a hen-turkey " ; the original perhaps of that partridge "larger than a vulture" which formed one of the presents from an Indian King to Augustus Caesar. [With reference to the large Tibetan partridge found in the Nan-shan Mountains in the meridian of Sha-chau by Prjevalsky, M. E. D. Morgan in a note (*P. R. Geog. S.* ix. 1887, p. 219), writes : " *Megaloperdrix thibetanus.* Its general name in Asia is *ullar*, a word of Kirghiz or Turkish origin ; the Mongols call it *hailik*, and the Tibetans *kung-mo*. There are two other varieties of this bird found in the Himalaya and Altai Mountains, but the habits of life and call-note of all three are the same."] From the extensive diffusion of the term, which seems to be common to India, Tibet, and Persia (for the latter, see *Abbott* in *J. R. G. S.* XXV. 41), it is likely enough to be of Mongol origin, not improbably *Tsokhor*, "dappled or pied." (*Kovalevsky*, No. 2196, and *Strahlenberg's* Vocabulary ; see also *Ladak*, 205; *Moorcr.* I. 313, 432 ; *Jerdan's Birds of India*, III. 549, 572 ; *Dunlop*, *Hunting in Himalaya*, 178 ; *J. A. S. B.* VI. 774.)

The chakór is mentioned by Baber (p. 282) ; and also by the Hindi poet Chand (*Rás Mála*, I. 230, and *Ind. Antiquary*, I. 273). If the latter passage is genuine, it is adverse to my Mongol etymology, as Chand lived before the Mongol era.

The keeping of partridges for the table is alluded to by Chaucer in his portrait of the Franklin, *Prologue, Cant. Tales:*

> " It snewed in his hous of mete and drinke,
> Of alle deyntees that men coud of thinke,
> After the sondry sesons of the yere,
> So changed he his mete and his soupere.
> *Full many a fat partrich hadde he in mewe,*
> · And many a breme and many a luce in stewe."

CHAPTER LXI.

Of the City of Chandu, and the Kaan's Palace there.

AND when you have ridden three days from the city last mentioned, between north-east and north, you come to a city called CHANDU,[1] which was built by the Kaan now reigning. There is at this place a very fine marble Palace, the rooms of which are all gilt and painted with figures of men and beasts and birds, and with a variety of trees and flowers, all executed with such exquisite art that you regard them with delight and astonishment.[2]

Round this Palace a wall is built, inclosing a compass

of 16 miles, and inside the Park there are fountains and
rivers and brooks, and beautiful meadows, with all kinds
of wild animals (excluding such as are of ferocious
nature), which the Emperor has procured and placed
there to supply food for his gerfalcons and hawks,
which he keeps there in mew.   Of these there are
more than 200 gerfalcons alone, without reckoning the
other hawks.   The Kaan himself goes every week to
see his birds sitting in mew, and sometimes he rides
through the park with a leopard behind him on his
horse's croup ; and then if he sees any animal that
takes his fancy, he slips his leopard at it,[3] and the
game when taken is made over to feed the hawks in
mew.   This he does for diversion.

Moreover [at a spot in the Park where there is a
charming wood] he has another Palace built of cane, of
which I must give you a description.   It is gilt all over,
and most elaborately finished inside.   [It is stayed on
gilt and lackered columns, on each of which is a dragon
all gilt, the tail of which is attached to the column
whilst the head supports the architrave, and the claws
likewise are stretched out right and left to support the
architrave.]   The roof, like the rest, is formed of canes,
covered with a varnish so strong and excellent that no
amount of rain will rot them.   These canes are a good
3 palms in girth, and from 10 to 15 paces in length.
[They are cut across at each knot, and then the pieces
are split so as to form from each two hollow tiles, and
with these the house is roofed ; only every such tile of
cane has to be nailed down to prevent the wind from
lifting it.]   In short, the whole Palace is built of these
canes, which (I may mention) serve also for a great
variety of other useful purposes.   The construction of
the Palace is so devised that it can be taken down and
put up again with great celerity ; and it can all be taken

to pieces and removed whithersoever the Emperor may command. When erected, it is braced [against mishaps from the wind] by more than 200 cords of silk.[4]

The Lord abides at this Park of his, dwelling some-times in the Marble Palace and sometimes in the Cane Palace for three months of the year, to wit, June, July, and August; preferring this residence because it is by no means hot; in fact it is a very cool place. When the 28th day of [the Moon of] August arrives he takes his departure, and the Cane Palace is taken to pieces.[5] But I must tell you what happens when he goes away from this Palace every year on the 28th of the August [Moon].

You must know that the Kaan keeps an immense stud of white horses and mares; in fact more than 10,000 of them, and all pure white without a speck. The milk of these mares is drunk by himself and his family, and by none else, except by those of one great tribe that have also the privilege of drinking it. This privilege was granted them by Chinghis Kaan, on account of a certain victory that they helped him to win long ago. The name of the tribe is HORIAD.[6]

Now when these mares are passing across the country, and any one falls in with them, be he the greatest lord in the land, he must not presume to pass until the mares have gone by; he must either tarry where he is, or go a half-day's journey round if need so be, so as not to come nigh them; for they are to be treated with the greatest respect. Well, when the Lord sets out from the Park on the 28th of August, as I told you, the milk of all those mares is taken and sprinkled on the ground. And this is done on the injunction of the Idolaters and Idol-priests, who say that it is an excellent thing to sprinkle that milk on the ground every 28th of August, so that the Earth and the Air

and the False Gods shall have their share of it, and
the Spirits likewise that inhabit the Air and the Earth.
And thus those beings will protect and bless the Kaan
and his children and his wives and his folk and his gear,
and his cattle and his horses, his corn and all that is his.
After this is done, the Emperor is off and away.[7]

But I must now tell you a strange thing that hitherto
I have forgotten to mention.   During the three months
of every year that the Lord resides at that place, if it
should happen to be bad weather, there are certain
crafty enchanters and astrologers in his train, who are
such adepts in necromancy and the diabolic arts, that
they are able to prevent any cloud or storm from
passing over the spot on which the Emperor's Palace
stands.   The sorcerers who do this are called TEBET
and KESIMUR, which are the names of two nations of
Idolaters.   Whatever they do in this way is by the
help of the Devil, but they make those people believe
that it is compassed by dint of their own sanctity and
the help of God.[8]   [They always go in a state of dirt
and uncleanness, devoid of respect for themselves, or
for those who see them, unwashed, unkempt, and
sordidly attired.]

These people also have a custom which I must tell
you.   If a man is condemned to death and executed by
the lawful authority, they take his body and cook and
eat it.   But if any one die a natural death then they
will not eat the body.[9]

There is another marvel performed by those BACSI,
of whom I have been speaking as knowing so many
enchantments.[10]   For when the Great Kaan is at his
capital and in his great Palace, seated at his table,
which stands on a platform some eight cubits above
the ground, his cups are set before him [on a great
buffet] in the middle of the hall pavement, at a distance

of some ten paces from his table, and filled with wine,
or other good spiced liquor such as they use. Now
when the Lord desires to drink, these enchanters by
the power of their enchantments cause the cups to
move from their place without being touched by anybody, and to present themselves to the Emperor! This
every one present may witness, and there are ofttimes
more than 10,000 persons thus present. 'Tis a truth
and no lie! and so will tell you the sages of our own
country who understand necromancy, for they also can
perform it.[11]

And when the Idol Festivals come round, these
Bacsi go to the Prince and say: "Sire, the Feast of
such a god is come" (naming him). "My Lord, you
know," the enchanter will say, "that this god, when
he gets no offerings, always sends bad weather and
spoils our seasons. So we pray you to give us such
and such a number of black-faced sheep," naming
whatever number they please. "And we beg also,
good my lord, that we may have such a quantity of
incense, and such a quantity of lignaloes, and"—so much
of this, so much of that, and so much of t'other, according to their fancy—"that we may perform a solemn
service and a great sacrifice to our Idols, and that so
they may be induced to protect us and all that is
ours."

The *Bacsi* say these things to the Barons entrusted
with the Stewardship, who stand round the Great Kaan,
and these repeat them to the Kaan, and he then orders
the Barons to give everything that the Bacsi have asked
for. And when they have got the articles they go and
make a great feast in honour of their god, and hold
great ceremonies of worship with grand illuminations
and quantities of incense of a variety of odours, which
they make up from different aromatic spices. And

then they cook the meat, and set it before the idols, and sprinkle the broth hither and thither, saying that in this way the idols get their bellyful. Thus it is that they keep their festivals. You must know that each of the idols has a name of his own, and a feast-day, just as our Saints have their anniversaries.[12]

They have also immense Minsters and Abbeys, some of them as big as a small town, with more than two thousand monks (*i.e.* after their fashion) in a single abbey.[13] These monks dress more decently than the rest of the people, and have the head and beard shaven. There are some among these *Bacsi* who are allowed by their rule to take wives, and who have plenty of children.[14]

Then there is another kind of devotees called SENSIN, who are men of extraordinary abstinence after their fashion, and lead a life of such hardship as I will describe. All their life long they eat nothing but bran,[15] which they take mixt with hot water. That is their food: bran, and nothing but bran; and water for their drink. 'Tis a lifelong fast! so that I may well say their life is one of extraordinary asceticism. They have great idols, and plenty of them; but they sometimes also worship fire. The other Idolaters who are not of this sect call these people heretics—*Patarins* as we should say[16]—because they do not worship their idols in their own fashion. Those of whom I am speaking would not take a wife on any consideration.[17] They wear dresses of hempen stuff, black and blue,[18] and sleep upon mats; in fact their asceticism is something astonishing. Their idols are all feminine, that is to say, they have women's names.[19]

Now let us have done with this subject, and let me tell you of the great state and wonderful magnificence of the Great Lord of Lords; I mean that great Prince

who is the Sovereign of the Tartars, CUBLAY by name,
that most noble and puissant Lord.

NOTE 1.—[There were two roads to go from Peking to Shangtu : the eastern road
through Tu-shi-k'ow, and the western (used for the return journey) road by Ye-hu ling.
Polo took this last road, which ran from Peking to Siuen-te chau through the same
places as now ; but from the latter town it led, not to Kalgan as it does now, but
more to the west, to a place called now Shan-fang pú where the pass across the Ye-hu
ling range begins. "On both these roads *nabo*, or temporary palaces, were built, as
resting-places for the Khans ; eighteen on the eastern road, and twenty-four on the
western." (*Palladius*, p. 25.) The same author makes (p. 26) the following remarks :
"M. Polo's statement that he travelled three days from Siuen-te chau to Chagannor,
and three days also from the latter place to Shang-tu, agrees with the information
contained in the ' Researches on the Routes to Shangtu.' The Chinese authors have
not given the precise position of Lake Chagannor ; there are several lakes in the desert
on the road to Shangtu, and their names have changed with time. The palace in
Chagannor was built in 1280" (according to the *Siu t'ung kien*).—H. C.]

NOTE 2.—*Chandu*, called more correctly in Ramusio *Xandu*, *i.e.* SHANDU, and
by Fr. Odorico *Sandu*, viz. SHANG-TU or "Upper Court," the Chinese title of Kúblái's
summer residence at Kaipingfu, *Mongolicè* Keibung (see ch. xiii. of Prologue) [is
called also *Loan king*, *i.e.* "the capital on the Loan River," according to Palladius,
p. 26.—H. C.]. The ruins still exist, in about lat. 40° 22', and a little west of the
longitude of Peking. The site is 118 miles in direct line from Chaghan-nor, making
Polo's three marches into rides of unusual length.* The ruins bear the Mongol name
of *Chao Naiman Sumé Khotan*, meaning "city of the 108 temples," and are about
26 miles to the north-west of Dolon-nor, a bustling, dirty town of modern origin,
famous for the manufactory of idols, bells, and other ecclesiastical paraphernalia of
Buddhism. The site was visited (though not described) by Père Gerbillon in 1691,
and since then by no European traveller till 1872, when Dr. Bushell of the British
Legation at Peking, and the Hon. T. G. Grosvenor, made a journey thither from
the capital, by way of the Nan-kau Pass (*supra* p. 26), Kalgan, and the vicinity of
Chaghan-nor, the route that would seem to have been habitually followed, in their
annual migration, by Kúblái and his successors.

The deserted site, overgrown with rank weeds and grass, stands but little above
the marshy bed of the river, which here preserves the name of Shang-tu, and about a
mile from its north or left bank. The walls, of earth faced with brick and unhewn
stone, still stand, forming, as in the Tartar city of Peking, a double *enceinte*, of which
the inner line no doubt represents the area of the "Marble Palace" of which Polo
speaks. This forms a square of about 2 *li* (⅔ of a mile) to the side, and has three
gates—south, east, and west, of which the southern one still stands intact, a perfect
arch, 20 ft. high and 12 ft. wide. The outer wall forms a square of 4 *li* (1⅓ mile) to
the side, and has six gates. The foundations of temples and palace-buildings can be
traced, and both enclosures are abundantly strewn with blocks of marble and fragments of lions, dragons, and other sculptures, testifying to the former existence of a
flourishing city, but exhibiting now scarcely one stone upon another. A broken
memorial tablet was found, half buried in the ground, within the north-east angle of
the outer rampart, bearing an inscription in an antique form of the Chinese character,
which proves it to have been erected by Kúblái, in honour of a Buddhist ecclesiastic
called Yun-Hien. Yun - Hien was the abbot of one of those great minsters and

* This distance is taken from a tracing of the map prepared for Dr. Bushell's paper quoted below.
But there is a serious discrepancy between this tracing and the observed position of Dolon-nor, which
determines that of Shang-tu, as stated to me in a letter from Dr. Bushell. [See Note 1.]

Heading
In the Old Chinese Seal-Character, of an INSCRIPTION on a Memorial raised by KÚBLÁI-KAAN
to a Buddhist Ecclesiastic in the vicinity of his SUMMER-PALACE
at SHANG-TU in Mongolia.
Reduced from a facsimile obtained on the spot by Dr. *S. W. Bushell*,
1872.
(About one-Fourth the Length and Breadth of Original.)

[*To face p.* 305.

abbeys of *Bacsis*, of which Marco speaks, and the exact date (no longer visible) of the monument was equivalent to A.D. 1288.*

This city occupies the south-east angle of a more extensive enclosure, bounded by what is now a grassy mound, and embracing, on Dr. Bushell's estimate, about 5 square miles. Further knowledge may explain the discrepancy from Marco's dimension, but this must be the park of which he speaks.† The woods and fountains have disappeared, like the temples and palaces; all is dreary and desolate, though still abounding in the game which was one of Kúblái's attractions to the spot. A small monastery, occupied by six or seven wretched Lamas, is the only building that remains in the vicinity. The river Shangtu, which lower down becomes the Lan [or Loan] -Ho, was formerly navigated from the sea up to this place by flat grain-boats.

[Mgr. de Harlez gave in the *T'oung Pao* (x. p. 73) an inscription in *Chuen* character on a *stele* found in the ruins of Shangtu, and built by an officer with the permission of the Emperor; it is probably a token of imperial favour; the inscription means: *Great Longevity*.—H. C.]

In the wail which Sanang Setzen, the poetical historian of the Mongols, puts, perhaps with some traditional basis, into the mouth of Toghon Temur, the last of the Chinghizide Dynasty in China, when driven from his throne, the changes are rung on the lost glories of his capital *Daïtu* (see *infra*, Book II. ch. xi.) and his summer palace *Shangtu;* thus (I translate from Schott's amended German rendering of the Mongol):

" My vast and noble Capital, My Daïtu, My splendidly adorned !
And Thou my cool and delicious Summer-seat, my Shangtu-Keibung !
Ye, also, yellow plains of Shangtu, Delight of my godlike Sires !
I suffered myself to drop into dreams,—and lo ! my Empire was gone !
Ah Thou my Daïtu, built of the nine precious substances !
Ah my Shangtu-Keibung, Union of all perfections !
Ah my Fame ! Ah my Glory, as Khagan and Lord of the Earth !
When I used to awake betimes and look forth, how the breezes blew loaded with
    fragrance !
And turn which way I would all was glorious perfection of beauty !

　.　　　.　　　.　　.　　.　　.　　.　　.

Alas for my illustrious name as the Sovereign of the World !
Alas for my Daïtu, seat of Sanctity, Glorious work of the Immortal KÚBLÁI !
    All, all is rent from me ! "

It was, in 1797, whilst reading this passage of Marco's narrative in old Purchas that Coleridge fell asleep, and dreamt the dream of Kúblái's Paradise, beginning:

　　　" In Xanadu did Kubla Khan
　　　　A stately pleasure-dome decree :
　　　　Where Alph, the sacred River, ran
　　　　Through caverns measureless to man
　　　　　Down to a sunless sea.
　　　　So twice five miles of fertile ground
　　　　With walls and towers were girdled round :
　　　　And there were gardens bright with sinuous rills
　　　　　Where blossomed many an incense-bearing tree ;
　　　　And here were forests ancient as the hills,
　　　　　Enfolding sunny spots of greenery."

---

* These particulars were obtained by Dr. Bushell through the Archimandrite Palladius, from the MS. account of a Chinese traveller who visited Shangtu about two hundred years ago, when probably the whole inscription was above ground. The inscription is also mentioned in the Imp. Geography of the present Dynasty, quoted by Klaproth. This work gives the interior wall 5 *li* to the side, instead of 2 *li*, and the outer wall 10 *li*, instead of 4 *li*. By Dr. Bushell's kindness, I give a reduction of his sketch plan (see *Itinerary Map*, No. IV. at end of this volume), and also a plate of the heading of the inscription. The translation of this is: "Monument conferred by the Emperor of the August Yuen (Dynasty) in memory of His High Eminence Yun Hien (styled) Chang-Lao (canonised as) Shou-Kung (Prince of Longevity)." [See *Missions de Chine et du Congo*, No. 28, Mars, 1891, Bruxelles.]

† Ramusio's version runs thus: "The palace presents one side to the centre of the city and the other to the city wall. And from either extremity of the palace where it touches the city wall, there

It would be a singular coincidence in relation to this poem were Klaproth's reading correct of a passage in Rashiduddin which he renders as saying that the palace at Kaiminfu was "called Langtin, and was built after a plan that Kúblái had seen in a dream, and had retained in his memory." But I suspect D'Ohsson's reading is more accurate, which runs : "Kúblái caused a Palace to be built for him east of Kaipingfu, called Lengten ; *but he abandoned it in consequence of a dream.*" For we see from Sanang Setzen that the Palaces of Lengten and Kaiming or Shangtu were distinct : "Between the year of the Rat (1264), when Kúblái was fifty years old, and the year of the Sheep (1271), in the space of eight years, he built four great cities, viz. for Summer Residence SHANGTU KEIBUNG Kürdu Balgasun, for Winter Residence Yeke DAÏTU Khotan, and on the shady side of the Altai (see ch. li. note 3, *supra*) Arulun TSAGHAN BALGASUN, and Erchügin LANGTING Balgasun." A valuable letter from Dr. Bushell enables me now to indicate the position of Langtin : "The district through which the river flows eastward from Shangtu is known to the Mongolians of the present day by the name of *Lang-tírh* (*Lang-ting'rh*). . . . The ruins of the city are marked on a Chinese map in my possession Pai-dseng-tzu, *i.e.* 'White City,' implying that it was formerly an Imperial residence. The remains of the wall are 7 or 8 *li* in diameter, of stone, and situated about 40 *li* north-north-west from Dolon-nor."

(*Gerbillon* in *Astley*, IV. 701-716 ; Klaproth, in *J. As.* sèr. II. tom. xi. 345-350; *Schott, Die letzten Jahre der Mongolenherrschaft in China* (Berl. Acad. d. Wissensch. 1850, pp. 502-503) ; *Huc's Tartary*, etc., p. 14 *seqq.* ; *Cathay*, 134, 261 ; *S. Setzen*, p. 115 ; *Dr. S. W. Bushell, Journey outside the Great Wall*, in *J. R. G. S.* for 1874, and MS. notes.)

One of the pavilions of the celebrated Yuen-ming-Yuen may give some idea of the probable style, though not of the scale, of Kúblái's Summer Palace.

Hiuen Tsang's account of the elaborate and fantastic ornamentation of the famous Indian monasteries at Nalanda in Bahár, where Mr. Broadley has lately made such remarkable discoveries, seems to indicate that these fantasies of Burmese and Chinese architecture may have had a direct origin in India, at a time when timber was still a principal material of construction there : "The pavilions had pillars adorned with dragons, and posts that glowed with all the colours of the rainbow, sculptured frets, columns set with jade, richly chiselled and lackered, with balustrades of vermilion, and carved open work. The lintels of the doors were tastefully ornamented, and the roofs covered with shining tiles, the splendours of which were multiplied by mutual reflection and from moment to moment took a thousand forms." (*Vie et Voyages*, 157.)

NOTE 3.—[Rubruck says, (*Rockhill*, p. 248) : "I saw also the envoy of a certain Soldan of India, who had brought eight leopards and ten *greyhounds*, taught to sit on horses' backs, as leopards sit."—H. C.]

NOTE 4.—Ramusio's is here so much more lucid than the other texts, that I have adhered mainly to his account of the building. The roof described is of a kind in use in the Indian Archipelago, and in some other parts of Transgangetic India, in which the semi-cylinders of bamboo are laid just like Roman tiles.

Rashiduddin gives a curious account of the way in which the foundations of the terrace on which this palace stood were erected in a lake. He says, too, in accord with Polo : "Inside the city itself a second palace was built, about a bowshot from the first : but the Kaan generally takes up his residence in the palace outside the town," *i.e.*, as I imagine, in Marco's Cane Palace. (*Cathay*, pp. 261-262.)

["The *Palace of canes* is probably the Palm Hall, *Tsung tien*, alias *Tsung mao tien*, of the Chinese authors, which was situated in the western palace garden of Shangtu. Mention is made also in the *Altan Tobchi* of a cane tent in Shangtu." (*Palladius*, p. 27.)—H. C.]

---

runs another wall, which fetches a compass and encloses a good 16 miles of plain, and so that no one can enter this enclosure except by passing through the palace."

Marco might well say of the bamboo that "it serves also a great variety of other purposes." An intelligent native of Arakan who accompanied me in wanderings on duty in the forests of the Burmese frontier in the beginning of 1853, and who used to ask many questions about Europe, seemed able to apprehend almost everything except the possibility of existence in a country without bamboos! "When I speak of bamboo huts, I mean to say that posts and walls, wall-plates and rafters, floor and thatch, and

Pavilion at Yuen-ming-Yuen.

the withes that bind them, are all of bamboo. In fact, it might almost be said that among the Indo-Chinese nations the staff of life is *a bamboo!* Scaffolding and ladders, landing-jetties, fishing apparatus, irrigation wheels and scoops, oars, masts, and yards [and in China, sails, cables, and caulking, asparagus, medicine, and works of fantastic art], spears and arrows, hats and helmets, bow, bowstring and quiver, oil-cans, waterstoups and cooking-pots, pipe-sticks [tinder and means of producing fire], conduits, clothes-boxes, pawn-boxes, dinner-trays, pickles, preserves, and melodious musical instruments, torches, footballs, cordage, bellows, mats, paper ; these are but a few of the articles that are made from the bamboo ;" and in China, to sum up the whole, as Barrow observes, it maintains order throughout the Empire ! (*Ava Mission*, p. 153 ; and see also *Wallace, Ind. Arch.* I. 120 *seqq.*)

NOTE 5.—" The Emperor . . . . began this year (1264) to depart from Yenking (Peking) in the second or third month for Shangtu, not returning until the eighth month. Every year he made this passage, and all the Mongol emperors who succeeded him followed his example." (*Gaubil*, p. 144.)

[" The Khans usually resorted to Shangtu in the 4th moon and returned to Peking in the 9th. On the 7th day of the 7th moon there were libations performed in honour of the ancestors ; a shaman, his face to the north, uttered in a loud voice the names of Chinghiz Khan and of other deceased Khans, and poured mare's milk on the ground. The propitious day for the return journey to Peking was also appointed then." (*Palladius*, p. 26.)—H. C.]

NOTE 6.—White horses were presented in homage to the Kaan on New Year's Day (*the White Feast*), as we shall see below. (Bk. II. ch. xv.) Odoric also mentions this practice ; and, according to Huc, the Mongol chiefs continued it at least to the time of the Emperor K'ang-hi. Indeed Timkowski speaks of annual tributes of white camels and white horses from the Khans of the Kalkas and other Mongol dignitaries, in the present century. (*Huc's Tartary*, etc. ; *Tim.* II. 33.)

By the HORIAD are no doubt intended the UIRAD or OIRAD, a name usually interpreted as signifying the "Closely Allied," or Confederates ; but Vámbéry explains it as (Turki) *Oyurat*, "Grey horse," to which the statement in our text appears to lend colour. They were not of the tribes properly called Mongol, but after their submission to Chinghiz they remained closely attached to him. In Chinghiz's victory over Aung-Khan, as related by S. Setzen, we find Turulji Taishi, the son of the chief of the Oirad, one of Chinghiz's three chief captains ; perhaps that is the victory alluded to. The seats of the Oirad appear to have been about the head waters of the Kem, or Upper Yenisei.

In A.D. 1295 there took place a curious desertion from the service of Gházán Khan of Persia of a vast corps of the Oirad, said to amount to 18,000 *tents*. They made their way to Damascus, where they were well received by the Mameluke Sultan. But their heathenish practices gave dire offence to the Faithful. They were settled in the *Sáhil*, or coast districts of Palestine. Many died speedily ; the rest embraced Islam, spread over the country, and gradually became absorbed in the general population. Their sons and daughters were greatly admired for their beauty. (*S. Setz.* p. 87 ; *Erdmann,* 187 ; *Pallas, Samml.* I. 5 *seqq. ; Makrizi,* III. 29 ; *Bretschneider, Med. Res.* II. p. 159 *seqq.*)

[With reference to Yule's conjecture, I may quote Palladius (*l.c.* p. 27): " It is, however, strange that the Oirats alone enjoyed the privilege described by Marco Polo ; for the highest position at the Mongol Khan's court belonged to the Kunkrat tribe, out of which the Khans used to choose their first wives, who were called Empresses of the first *ordo*."—H. C.]

NOTE 7.—Rubruquis assigns such a festival to the month of May : " On the 9th day of the May Moon they collect all the white mares of their herds and consecrate them. The Christian priests also must then assemble with their thuribles. They then sprinkle new cosmos (*kumíz*) on the ground, and make a great feast that day, for according to their calendar, it is their time of first drinking new cosmos, just as we reckon of our new wine at the feast of St. Bartholomew (24th August), or that of St. Sixtus (6th August), or of our fruit on the feast of St. James and St. Christopher " (25th July). [With reference to this feast, Mr. Rockhill gives (*Rubruck*, p. 241, note) extracts from *Pallas, Voyages,* IV. 579, and *Professor Radloff, Aus Siberien,* I. 378.

—H. C.] The Yakuts also hold such a festival in June or July, when the mares foal, and immense wooden goblets of kumíz are emptied on that occasion. They also pour out kumíz for the Spirits to the four quarters of heaven.

The following passage occurs in the narrative of the Journey of Chang Te-hui, a Chinese teacher, who was summoned to visit the camp of Kúblái in Mongolia, some twelve years before that Prince ascended the throne of the Kaans : *

" On the 9th day of the 9th Moon (October), the Prince, having called his subjects before his chief tent, performed the libation of the milk of a white mare. This was the customary sacrifice at that time. The vessels used were made of birch-bark, not ornamented with either silver or gold. Such here is the respect for simplicity. . . . .

" At the last day of the year the Mongols suddenly changed their camping-ground to another place, for the mutual congratulation on the 1st Moon. Then there was every day feasting before the tents for the lower ranks. Beginning with the Prince, all dressed themselves in white fur clothing. . . . . †

" On the 9th day of the 4th Moon (May) the Prince again collected his vassals before the chief tent for the libation of the milk of a white mare. This sacrifice is performed twice a year."

It has been seen (p. 308) that Rubruquis also names the 9th day of the May moon as that of the consecration of the white mares. The autumn libation is described by Polo as performed on the 28th day of the August moon, probably because it was un‐ suited to the circumstances of the Court at Cambaluc, where the Kaan was during October, and the day named was the last of his annual stay in the Mongolian uplands.

Baber tells that among the ceremonies of a Mongol Review the Khan and his staff took kumiz and sprinkled it towards the standards. An Armenian author of the Mongol era says that it was the custom of the Tartars, before drinking, to sprinkle drink towards heaven, and towards the four quarters. Mr. Atkinson notices the same practice among the Kirghiz : and I found the like in old days among the Kasias of the eastern frontier of Bengal.

The time of year assigned by Polo for the ceremony implies some change. Perhaps it had been made to coincide with the Festival of Water Consecration of the Lamas, with which the time named in the text seems to correspond. On that occasion the Lamas go in procession to the rivers and lakes and consecrate them by benediction and by casting in offerings, attended by much popular festivity.

Rubruquis seems to intimate that the Nestorian priests were employed to con‐ secrate the white mares by incensing them. In the rear of Lord Canning's camp in India I once came upon the party of his *Shutr Suwárs*, or dromedary-express-riders, busily engaged in incensing with frankincense the whole of the dromedaries, which were kneeling in a circle. I could get no light on the practice, but it was very probably a relic of the old Mongol custom. (*Rubr.* 363 ; *Erman*, II. 397 ; *Billings' Journey*, Fr. Tr. I. 217 ; *Baber*, 103 ; *J. As.* sèr. V. tom. xi. p. 249 ; *Atk. Amoor*, p. 47 ; *J. A. S. B.* XIII. 628 ; *Koeppen*, II. 313.)

NOTE 8.—The practice of weather-conjuring was in great vogue among the Mongols, and is often alluded to in their history.

The operation was performed by means of a stone of magical virtues, called *Yadah* or *Jadah-Tásh*, which was placed in or hung over a basin of water with sundry ceremonies. The possession of such a stone is ascribed by the early Arab traveller Ibn Mohalhal to the *Kímák*, a great tribe of the Turks. In the war raised against Chinghiz and Aung Khan, when still allies, by a great confederation of the Naiman and other tribes in 1202, we are told that Sengun, the son of Aung Khan, when sent to meet the enemy, caused them to be enchanted, so that all their attempted move-

---

* This narrative, translated from Chinese into Russian by Father Palladius, and from the Russian into English by Mr. Eugene Schuyler, Secretary of the U.S. Legation at St. Petersburg, was oblig‐ ingly sent to me by the latter gentleman, and appeared in the *Geographical Magazine* for January, 1875, p. 7.

† See Bk. II. chap. xiv. note 3.

ments against him were defeated by snow and mist. The fog and darkness were indeed so dense that many men and horses fell over precipices, and many also perished with cold. In another account of (apparently) the same matter, given by Mir-Khond, the conjuring is set on foot by the *Yadachi* of Buyruk Khan, Prince of the Naiman, but the mischief all rebounds on the conjurer's own side.

In Tului's invasion of Honan in 1231-1232, Rashiduddin describes him, when in difficulty, as using the *Jadah* stone with success.

Timur, in his Memoirs, speaks of the Jets using incantations to produce heavy rains which hindered his cavalry from acting against them. A *Yadachi* was captured, and when his head had been taken off the storm ceased.

Baber speaks of one of his early friends, Khwaja Ka Mulai, as excelling in falconry and acquainted with *Yadagari* or the art of bringing on rain and snow by means of enchantment. When the Russians besieged Kazan in 1552 they suffered much from the constant heavy rains, and this annoyance was universally ascribed to the arts of the Tartar Queen, who was celebrated as an enchantress. Shah Abbas believed he had learned the Tartar secret, and put much confidence in it. (*P. Della V.* I. 869.)

[Grenard says (II. p. 256) the most powerful and most feared of sorcerers [in Chinese Turkestan] is the *djâduger*, who, to produce rain or fine weather, uses a jade stone, given by Noah to Japhet. Grenard adds (II. 406-407) there are sorcerers (Ngag-pa-snags-pa) whose specialty is to make rain fall; they are similar to the Turkish *Yadachi* and like them use a stone called " water cristal," *chu shel;* probably jade stone.

Mr. Rockhill (*Rubruck*, p. 245, note) writes : " Rashideddin states that when the Urianghit wanted to bring a storm to an end, they said injuries to the sky, the lightning and thunder. I have seen this done myself by Mongol storm-dispellers. (See *Diary*, 201, 203.) 'The other Mongol people,' he adds, 'do the contrary. When the storm rumbles, they remain shut up in their huts, full of fear.' The subject of storm-making, and the use of stones for that purpose, is fully discussed by Quatremère, *Histoire*, 438-440." (Cf. also *Rockhill, l.c.* p. 254.)—H. C.]

An edict of the Emperor Shi-tsung, of the reigning dynasty, addressed in 1724-1725 to the Eight Banners of Mongolia, warns them against this rain-conjuring : " If I," indignantly observes the Emperor, "offering prayer in sincerity have yet room to fear that it may please Heaven to leave MY prayer unanswered, it is truly intolerable that mere common people wishing for rain should at their own caprice set up altars of earth, and bring together a rabble of Hoshang (Buddhist Bonzes) and Taossé to conjure the spirits to gratify their wishes."

[" Lamas were of various extraction ; at the time of the great assemblies, and of the Khan's festivities in Shangtu, they erected an altar near the Khan's tent and prayed for fine weather ; the whistling of shells rose up to heaven." These are the words in which Marco Polo's narrative is corroborated by an eye-witness who has celebrated the remarkable objects of Shangtu (*Loan king tsa yung*). These Lamas, in spite of the prohibition by the Buddhist creed of bloody sacrifices, used to sacrifice sheep's hearts to Mahakala. It happened, as it seems, that the heart of an executed criminal was also considered an agreeable offering ; and as the offerings could be, after the ceremony, eaten by the sacrificing priests, Marco Polo had some reason to accuse the Lamas of cannibalism." (*Palladius*, 28.)—H. C.]

The practice of weather-conjuring is not yet obsolete in Tartary, Tibet, and the adjoining countries.*

Weather-conjuring stories were also rife in Europe during the Middle Ages. One

---

* In the first edition I had supposed a derivation of the Persian words *Jádú* and *Jádúgari*, used commonly in India for conjuring, from the Tartar use of *Yadah*. And Pallas says the Kirghiz call their witches *Jádugar*. (*Voy.* II. 298.) But I am assured by Sir H. Rawlinson that this etymology is more than doubtful, and that at any rate the Persian (*Jádú*) is probably older than the Turkish term. I see that M. Pavet de Courteille derives *Yadah* from a Mongol word signifying "change of weather," etc.

such is conspicuously introduced in connection with a magical fountain in the romance of the *Chevalier au Lyon :*

> " Et s'i pant uns bacins d'or fin
> A une si longue chaainne
> Qui dure jusqu'a la fontainne.
> Lez la fontainne troveras
> Un perron tel con tu verras
>   *   *   *   *
> S'au bacin viaus de l'iaue prandre
> Et dessor le perron espandre,
> La verras une tel tanpeste
> Qu'an cest bois ne remandra beste,"
>   etc.   etc.*

The effect foretold in these lines is the subject of a woodcut illustrating a Welsh version of the same tale in the first volume of the *Mabinogion.* And the existence of such a fountain is alluded to by Alexander Neckam. (*De Naturis Rerum*, Bk. II. ch. vii.)

In the *Cento Novelle Antiche* also certain necromancers exhibit their craft before the Emperor Frederic (Barbarossa apparently) : " The weather began to be overcast, and lo ! of a sudden rain began to fall with continued thunders and lightnings, as if the world were come to an end, and hailstones that looked like steel-caps," etc. Various other European legends of like character will be found in *Liebrecht's Gervasius von Tilbury*, pp. 147-148.

Rain-makers there are in many parts of the world ; but it is remarkable that those also of Samoa in the Pacific operate by means of a *rain-stone.*

Such weather conjurings as we have spoken of are ascribed by Ovid to Circe :

> " Concipit illa preces, et verba venefica dicit ;
> Ignotosque Deos ignoto carmine adorat,
>   *   *   *   *
> *Tunc quoque cantato densetur carmine caelum,*
> *Et nebulas exhalat humus."—Metam.* XIV. 365.

And to Medea :—

> ——" Quum volui, ripis mirantibus, amnes
> In fontes rediere suos . . . . (another feat of the Lamas)
>   . . . . *Nubila pello,*
> *Nubilaque induco ; ventos abigoque, vocoque."—Ibid.* VII. 199.

And by Tibullus to the *Saga (Eleg.* I. 2, 45) ; whilst Empedocles, in verses ascribed to him by Diogenes Laertius, claims power to communicate like secrets of potency :—

> " By my spells thou may'st
> To timely sunshine turn the purple rains,
> And parching droughts to fertilising floods."

(See *Cathay*, p. clxxxvii. ; *Erdm.* 282 ; *Oppert*, 182 *seqq. ; Erman*, I. 153 ; *Pallas, Samml.* II. 348 *seqq. ; Timk.* I. 402 ; *J. R. A. S.* VII. 305-306 ; *D'Ohsson*, II. 614 ; and for many interesting particulars, *Q. R.* p. 428 *seqq.*, and *Hammer's Golden Horde*, 207 and 435 *seqq.*)

NOTE 9.—It is not clear whether Marco attributes this cannibalism to the Tibetans and Kashmirians, or brings it in as a particular of Tartar custom which he had forgotten to mention before.

---

* [See W. Foerster's ed., *Halle*, 1887, p. 15, 386.—H. C.]

The accusations of cannibalism indeed against the Tibetans in old accounts are frequent, and I have elsewhere (see *Cathay*, p. 151) remarked on some singular Tibetan practices which go far to account for such charges.   Della Penna, too, makes a statement which bears curiously on the present passage.   Remarking on the great use made by certain classes of the Lamas of human skulls for magical cups, and of human thigh bones for flutes and whistles, he says that to supply them with these *the bodies of executed criminals were stored up at the disposal of the Lamas;* and a Hindu account of Tibet in the *Asiatic Researches* asserts that when one is killed in a fight both parties rush forward and struggle for the liver, which they eat (vol. xv).

[Carpini says of the people of Tibet: "They are pagans; they have a most astonishing, or rather horrible, custom, for, when any one's father is about to give up the ghost, all the relatives meet together, and they eat him, as was told to me for certain." Mr. Rockhill (*Rubruck*, p. 152, note) writes: "So far as I am aware, this charge [of cannibalism] is not made by any Oriental writer against the Tibetans, though both Arab travellers to China in the ninth century and Armenian historians of the thirteenth century say the Chinese practised cannibalism. The Armenians designate China by the name *Nankas*, which I take to be Chinese *Nan-kuo*, 'southern country,' the *Manzi* country of Marco Polo."—H. C.]

But like charges of cannibalism are brought against both Chinese and Tartars very positively.   Thus, without going back to the Anthropophagous Scythians of Ptolemy and Mela, we read in the *Relations* of the Arab travellers of the ninth century: "In China it occurs sometimes that the governor of a province revolts from his duty to the emperor.   In such a case he is slaughtered and eaten.   *In fact, the Chinese eat the flesh of all men who are executed by the sword.*"   Dr. Rennie mentions a superstitious practice, the continued existence of which in our own day he has himself witnessed, and which might perhaps have given rise to some such statement as that of the Arab travellers, if it be not indeed a relic, in a mitigated form, of the very practice they assert to have prevailed.   After an execution at Peking certain large pith balls are steeped in the blood, and under the name of *blood-bread* are sold as a medicine for consumption.   *It is only to the blood of decapitated criminals that any such healing power is attributed.*   It has been asserted in the annals of the *Propagation de la Foi* that the Chinese executioners of M. Chapdelaine, a missionary who was martyred in Kwang-si in 1856 (28th February), were seen to eat the heart of their victim; and M. Huot, a missionary in the Yun-nan province, recounts a case of cannibalism which he witnessed.   Bishop Chauveau, at Ta Ts'ien-lu, told Mr. Cooper that he had seen men in one of the cities of Yun-nan eating the heart and brains of a celebrated robber who had been executed.   Dr. Carstairs Douglas of Amoy also tells me that the like practices have occurred at Amoy and Swatau.

[With reference to cannibalism in China see *Medical Superstitions an Incentive to Anti-Foreign Riots in China*, by D. J. Macgowan, *North China Herald*, 8th July, 1892, pp. 60-62.   Mr. E. H. Parker (*China Review*, February-March, 1901; 136) relates that the inhabitants of a part of Kwang-si boiled and ate a Chinese officer who had been sent to pacify them.   "The idea underlying this horrible act [cannibalism] is, that by eating a portion of the victim, especially the heart, one acquires the valour with which he was endowed." (*Dennys' Folk-lore of China*, 67.)—H. C.]

Hayton, the Armenian, after relating the treason of a Saracen, called Parwana (he was an Iconian Turk), against Abaka Khan, says: "He was taken and cut in two, and orders were issued that in all the food eaten by Abaka there should be put a portion of the traitor's flesh.   Of this Abaka himself ate, and caused all his barons to partake.   *And this was in accordance with the custom of the Tartars.*"   The same story is related independently and differently by Friar Ricold, thus: "When the army of Abaga ran away from the Saracens in Syria, a certain great Tartar baron was arrested who had been guilty of treason.   And when the Emperor Khan was giving the order for his execution the Tartar ladies and women interposed, and begged that he might be made over to them.   Having got hold of the prisoner they boiled him alive, and cutting his body up into mince-meat gave it to eat to the whole army, as an example to others." Vincent of Beauvais makes a like statement : "When they capture any one who is at bitter enmity with them, they gather together and eat him in vengeance of his revolt, and like infernal leeches suck his blood," a custom of which a modern Mongol writer thinks that he finds a trace in a surviving proverb. Among more remote and ignorant Franks the cannibalism of the Tartars was a general belief. Ivo of Narbonne, in his letter written during the great Tartar invasion of Europe (1242), declares that the Tartar chiefs, with their dog's head followers and other *Lotophagi* (!), ate the bodies of their victims like so much bread ; whilst a Venetian chronicler, speaking of the council of Lyons in 1274, says there was a discussion about making a general move against the Tartars, "*porce qu'il manjuent la char humaine.*" These latter writers no doubt rehearsed mere popular beliefs, but Hayton and Ricold were both intelligent persons well acquainted with the Tartars, and Hayton at least not prejudiced against them.

The old belief was revived in Prussia during the Seven Years' War, in regard to the Kalmaks of the Russian army ; and Bergmann says the old Kalmak warriors confessed to him that they had done what they could to encourage it by cutting up the bodies of the slain in presence of their prisoners, and roasting them ! But Levchine relates an act on the part of the Kirghiz Kazaks which was no jest. They drank the blood of their victim if they did not eat his flesh.

There is some reason to believe that cannibalism was in the Middle Ages generally a less strange and unwonted horror than we should at first blush imagine, and especially that it was an idea tolerably familiar in China. M. Bazin, in the second part of *Chine Moderne*, p. 461, after sketching a Chinese drama of the Mongol era ("The Devotion of Chao-li"), the plot of which turns on the acts of a body of cannibals, quotes several other passages from Chinese authors which indicate this. Nor is this wonderful in the age that had experienced the horrors of the Mongol wars.

That was no doubt a fable which Carpini heard in the camp of the Great Kaan, that in one of the Mongol sieges in Cathay, when the army was without food, one man in ten of their own force was sacrificed to feed the remainder.[*] But we are told in sober history that the force of Tului in Honan, in 1231-1232, was reduced to such straits as to eat grass and human flesh. At the siege of the Kin capital Kaifongfu, in 1233, the besieged were reduced to the like extremity ; and the same occurred the same year at the siege of Tsaichau ; and in 1262, when the rebel general Litan was besieged in Tsinanfu. The Taiping wars the other day revived the same horrors in all their magnitude. And savage acts of the same kind by the Chinese and their Turk partisans in the defence of Kashgar were related to Mr. Shaw.

Probably, however, nothing of the kind in history equals what Abdallatif, a sober and scientific physician, describes as having occurred before his own eyes in the great Egyptian famine of A.H. 597 (1200). The horrid details fill a chapter of some length, and we need not quote from them.

Nor was Christendom without the rumour of such barbarities. The story of King Richard's banquet in presence of Saladin's ambassadors on the head of a Saracen curried (for so it surely was),—

> "soden full hastily
> With powder and with spysory,
> And with saffron of good colour "—

fable as it is, is told with a zest that makes one shudder ; but the tale in the *Chanson d'Antioche*, of how the licentious bands of ragamuffins, who hung on the army of the First Crusade, and were known as the *Tafurs*,[†] ate the Turks whom they killed at

---

[*] A young Afghan related in the presence of Arthur Conolly at Herat that on a certain occasion when provisions ran short the Russian General gave orders that 50,000 men should be killed and served out as rations ! (I. 346.)

[†] Ar. *Táfir*, a sordid, squalid fellow.

the siege, looks very like an abominable truth, corroborated as it is by the prose chronicle of worse deeds at the ensuing siege of Marrha :—

> "A lor cotiaus qu'il ont trenchans et afilés
> Escorchoient les Turs, aval parmi les près.
> Voiant Paiens, les ont par pièces découpés.
> En l'iave et el carbon les ont bien quisinés,
> Volontiers les menjuent sans pain et dessalés." *

(*Della Penna*, p. 76; *Reinaud, Rel.* I. 52; *Rennie's Peking*, II. 244; *Ann. de la Pr. de la F.* XXIX. 353, XXI. 298; *Hayton* in *Ram.* ch. xvii.; *Per. Quat.* p. 116; *M. Paris*, sub. 1243; *Mél. Asiat. Acad. St. Pétersb.* II. 659; *Canale* in *Arch. Stor. Ital.* VIII.; *Bergm. Nomad. Streifereien*, I. 14; *Carpini*, 638; *D'Ohsson*, II. 30, 43, 52; *Wilson's Ever Victorious Army*, 74; *Shaw*, p. 48; *Abdallatif*, p. 363 seqq.; *Weber*, II. 135; *Littré, H. de la Langue Franç.* I. 191; *Gesta Tancredi* in *Thes. Nov. Anecd.* III. 172.)

NOTE 10.—*Bakhshi* is generally believed to be a corruption of *Bhikshu*, the proper Sanscrit term for a religious mendicant, and in particular for the Buddhist devotees of that character. *Bakhshi* was probably applied to a class only of the Lamas, but among the Turks and Persians it became a generic name for them all. In this sense it is habitually used by Rachiduddin, and thus also in the Ain Akbari : "The learned among the Persians and Arabians call the priests of this (Buddhist) religion *Bukshee*, and in Tibbet they are styled Lamas."

According to Pallas the word among the modern Mongols is used in the sense of *Teacher*, and is applied to the oldest and most learned priest of a community, who is the local ecclesiastical chief. Among the Kirghiz Kazzaks again, who profess Mahomedanism, the word also survives, but conveys among them just the idea that Polo seems to have associated with it, that of a mere conjuror or "medicine-man"; whilst in Western Turkestan it has come to mean a Bard.

The word Bakhshi has, however, wandered much further from its original meaning. From its association with persons who could read and write, and who therefore occasionally acted as clerks, it came in Persia to mean a clerk or secretary. In the Petrarchian Vocabulary, published by Klaproth, we find *scriba* rendered in *Comanian*, *i.e.* Turkish of the Crimea, by *Bacsi*. The transfer of meaning is precisely parallel to that in regard to our *Clerk*. Under the Mahomedan sovereigns of India, *Bakhshi* was applied to an officer performing something like the duties of a quartermaster-general; and finally, in our Indian army, it has come to mean a paymaster. In the latter sense, I imagine it has got associated in the popular mind with the Persian *bakhshídan*, to bestow, and *bakhshísh*. (See a note in *Q. R.* p. 184 seqq.; *Cathay*, p. 474; *Ayeen Akbery*, III. 150; *Pallas, Samml.* II. 126; *Levchine*, p. 355; *Klap. Mém.* III.; *Vámbéry, Sketches*, p. 81.)

The sketch from the life, on p. 326, of a wandering Tibetan devotee, whom I met once at Hardwár, may give an idea of the sordid *Bacsis* spoken of by Polo.

NOTE 11.—This feat is related more briefly by Odoric: "And jugglers cause cups of gold full of good wine to fly through the air, and to offer themselves to all who list to drink." (*Cathay*, p. 143.) In the note on that passage I have referred to a somewhat similar story in the *Life of Apollonius*. "Such feats," says Mr. Jaeschke, "are often mentioned in ancient as well as modern legends of Buddha and other saints; and our Lamas have heard of things very similar performed by conjuring *Bonpos.*" (See p. 323.) The moving of cups and the like is one of the sorceries ascribed in old legends to Simon Magus : "He made statues to walk; leapt into the fire without being burnt; flew in the air; made bread of stones; changed his shape; assumed two faces at once; converted himself into a pillar; caused closed doors to fly open spontaneously; made the vessels in a house seem to move of themselves," etc. The

---

* [Cf. Paulin Paris's ed., 1848, II. p. 5.—H. C.]

Jesuit Delrio laments that credulous princes, otherwise of pious repute, should have allowed diabolic tricks to be played before them, "as, for example, things of iron, and silver goblets, or other heavy articles, to be moved by bounds from one end of a table to the other, without the use of a magnet or of any attachment." The pious prince appears to have been Charles IX., and the conjuror a certain Cesare Maltesio. Another Jesuit author describes the veritable mango-trick, speaking of persons who " within three hours' space did cause a genuine shrub of a span in length to grow out of the table, besides other trees that produced both leaves and fruit."

In a letter dated 1st December, 1875, written by Mr. R. B. Shaw, after his last return from Kashgar and Lahore, this distinguished traveller says : " I have heard stories related regarding a Buddhist high priest whose temple is said to be not far to the east of Lanchau, which reminds me of Marco Polo and Kúblái Khan.   This high priest is said to have the magic power of attracting cups and plates to him from a distance, so that things fly through the air into his hands."   (*MS. Note.*—H. Y.)

The profession and practice of exorcism and magic in general is greatly more prominent in Lamaism or Tibetan Buddhism than in any other known form of that religion.   Indeed, the old form of Lamaism as it existed in our traveller's day, and till the reforms of Tsongkhapa (1357-1419), and as it is still professed by the *Red* sect in Tibet, seems to be a kind of compromise between Indian Buddhism and the old indigenous Shamanism.   Even the reformed doctrine of the Yellow sect recognises an orthodox kind of magic, which is due in great measure to the combination of Sivaism with the Buddhist doctrines, and of which the institutes are contained in the vast collection of the *Jud* or Tantras, recognised among the holy books.   The magic arts of this code open even a short road to the Buddhahood itself.   To attain that perfection of power and wisdom, culminating in the cessation of sensible existence, requires, according to the ordinary paths, a period of three *asankhyas* (or say Uncountable Time × 3), whereas by means of the magic arts of the *Tantras* it may be reached in the course of three *rebirths* only, nay, of one !   But from the Tantras also can be learned how to acquire miraculous powers for objects entirely selfish and secular, and how to exercise these by means of *Dhárani* or mystic Indian charms.

Still the orthodox Yellow Lamas professedly repudiate and despise the grosser exhibitions of common magic and charlatanism which the Reds still practise, such as knife-swallowing, blowing fire, cutting off their own heads, etc.   But as the vulgar will not dispense with these marvels, every great orthodox monastery in Tibet *keeps a conjuror*, who is a member of the unreformed, and does not belong to the brotherhood of the convent, but lives in a particular part of it, bearing the name of *Choichong*, or protector of religion, and is allowed to marry.   The magic of these Choichong is in theory and practice different from the orthodox Tantrist magic.   The practitioners possess no literature, and hand down their mysteries only by tradition.   Their fantastic equipments, their frantic bearing, and their cries and howls, seem to identify them with the grossest Shamanist devil dancers.

Sanang Setzen enumerates a variety of the wonderful acts which could be performed through the *Dhárani*.   Such were, sticking a peg into solid rock ; restoring the dead to life ; turning a dead body into gold ; penetrating everywhere as air does ; flying ; catching wild beasts with the hand ; reading thoughts ; making water flow backwards ; eating tiles ; sitting in the air with the legs doubled under, etc.   Some of these are precisely the powers ascribed to Medea, Empedocles, and Simon Magus, in passages already cited.   Friar Ricold says on this subject : " There are certain men whom the Tartars honour above all in the world, viz. the *Baxitae* (*i.e. Bakhshis*), who are a kind of idol-priests.   These are men from India, persons of deep wisdom, well-conducted, and of the gravest morals.   They are usually acquainted with magic arts, and depend on the counsel and aid of demons ; they exhibit many illusions, and predict some future events.   For instance, one of eminence among them was said to fly ; the truth, however, was (as it proved), that he did not fly, but did walk close to the surface of the ground without touching it ; and *would seem to sit down without having any substance to support him.*"   This last performance was witnessed by Ibn

Batuta at Delhi, in the presence of Sultan Mahomed Tughlak ; and it was professedly exhibited by a Brahmin at Madras in the present century, a descendant doubtless of those Brahmans whom Apollonius saw walking two cubits from the ground. It is also described by the worthy Francis Valentyn as a performance known and practised in his own day in India. It is related, he says, that "a man will first go and sit on three sticks put together so as to form a tripod ; after which, first one stick, then a second, then the third shall be removed from under him, and the man shall not fall but shall still remain sitting in the air ! Yet I have spoken with two friends who had seen this at one and the same time ; and one of them, I may add, mistrusting his own eyes, had taken the trouble to feel about with a long stick if there were nothing on which the body rested ; yet, as the gentleman told me, he could neither feel nor see any such thing. Still, I could only say that I could not believe it, as a thing too manifestly contrary to reason."

Akin to these performances, though exhibited by professed jugglers without claim to religious character, is a class of feats which might be regarded as simply inventions if told by one author only, but which seem to deserve prominent notice from their being recounted by a series of authors, certainly independent of one another, and writing at long intervals of time and place. Our first witness is Ibn Batuta, and it will be necessary to quote him as well as the others in full, in order to show how closely their evidence tallies. The Arab Traveller was present at a great entertainment at the Court of the Viceroy of Khansa (*Kinsay* of Polo, or Hang-chau fu) : "That same night a juggler, who was one of the Kán's slaves, made his appearance, and the Amír said to him, 'Come and show us some of your marvels.' Upon this he took a wooden ball, with several holes in it, through which long thongs were passed, and, laying hold of one of these, slung it into the air. It went so high that we lost sight of it altogether. (It was the hottest season of the year, and we were outside in the middle of the palace court.) There now remained only a little of the end of a thong in the conjuror's hand, and he desired one of the boys who assisted him to lay hold of it and mount. He did so, climbing by the thong, and we lost sight of him also ! The conjuror then called to him three times, but getting no answer, he snatched up a knife as if in a great rage, laid hold of the thong, and disappeared also ! By and bye he threw down one of the boy's hands, then a foot, then the other hand, and then the other foot, then the trunk, and last of all the head ! Then he came down himself, all puffing and panting, and with his clothes all bloody, kissed the ground before the Amír, and said something to him in Chinese. The Amír gave some order in reply, and our friend then took the lad's limbs, laid them together in their places, and gave a kick, when, presto ! there was the boy, who got up and stood before us ! All this astonished me beyond measure, and I had an attack of palpitation like that which overcame me once before in the presence of the Sultan of India, when he showed me something of the same kind. They gave me a cordial, however, which cured the attack. The Kazi Afkharuddin was next to me, and quoth he, '*Wallah !* 'tis my opinion there has been neither going up nor coming down, neither marring nor mending ; 'tis all hocus pocus !'"

Now let us compare with this, which Ibn Batuta the Moor says he saw in China about the year 1348, the account which is given us by Edward Melton, an Anglo-Dutch traveller, of the performances of a Chinese gang of conjurors, which he witnessed at Batavia about the year 1670 (I have forgotten to note the year). After describing very vividly the *basket-murder* trick, which is well known in India, and now also in Europe, and some feats of bamboo balancing similar to those which were recently shown by Japanese performers in England, only more wonderful, he proceeds : " But now I am going to relate a thing which surpasses all belief, and which I should scarcely venture to insert here had it not been witnessed by thousands before my own eyes. One of the same gang took a ball of cord, and grasping one end of the cord in his hand slung the other up into the air with such force that its extremity was beyond reach of our sight. He then immediately climbed up the cord with indescribable swiftness, and got so high that we could no longer see him. I stood full of astonishment, not conceiving what was to come of this; when lo! a leg came tumbling down out of the air.   One of the conjuring company instantly snatched it up and threw it into the basket whereof I have formerly spoken.   A moment later a hand came down, and immediately on that another leg.   And in short all the members of the body came thus successively tumbling from the air and were cast together into the basket.   The last fragment of all that we saw tumble down was the head, and no sooner had that touched the ground than he who had snatched up all the limbs and put them in the basket turned them all out again topsy-turvy.   Then straightway we saw with these eyes all those limbs creep together again, and in short, form a whole

Chinese Conjuring Extraordinary.

man, who at once could stand and go just as before, without showing the least damage!   Never in my life was I so astonished as when I beheld this wonderful performance, and I doubted now no longer that these misguided men did it by the help of the Devil.   For it seems to me totally impossible that such things should be accomplished by natural means."   The same performance is spoken of by Valentyn, in a passage also containing curious notices of the basket-murder trick, the mango trick, the sitting in the air (quoted above), and others; but he refers to Melton, and I am not sure whether he had any other authority for it.   The cut on this page is taken from Melton's plate.

Again we have in the Memoirs of the Emperor Jahángir a detail of the wonderful performances of seven jugglers from Bengal who exhibited before him.   Two of their feats are thus described : "*Ninth.* They produced a man whom they divided limb from limb, actually severing his head from the body. They scattered these mutilated members along the ground, and in this state they lay for some time. They then extended a sheet or curtain over the spot, and one of the men putting himself under the sheet, in a few minutes came from below, followed by the individual supposed to have been cut into joints, in perfect health and condition, and one might have safely sworn that he had never received wound or injury whatever. . . . *Twenty-third.* They produced a chain of 50 cubits in length, and in my presence threw one end of it towards the sky, *where it remained as if fastened to something in the air.* A dog was then brought forward, and being placed at the lower end of the chain, immediately ran up, and reaching the other end, *immediately disappeared in the air.* In the same manner a hog, a panther, a lion, and a tiger were successively sent up the chain, and all equally disappeared at the upper end of the chain. At last they took down the chain and put it into a bag, no one ever discovering in what way the different animals were made to vanish into the air in the mysterious manner above described."

[There would appear (says the *Times of India*, quoted by the *Weekly Dispatch*, 15th September, 1889) to be a fine field of unworked romance in the annals of Indian jugglery. One Siddeshur Mitter, writing to the Calcutta paper, gives a thrilling account of a conjurer's feat which he witnessed recently in one of the villages of the Hooghly district. He saw the whole thing himself, he tells us, so there need be no question about the facts. On the particular afternoon when he visited the village the place was occupied by a company of male and female jugglers, armed with bags and boxes and musical instruments, and all the mysterious paraphernalia of the peripatetic *Jadugar.* While Siddeshur was looking on, and in the broad, clear light of the afternoon, a man was shut up in a box, which was then carefully nailed up and bound with cords. Weird spells and incantations of the style we are all familiar with were followed by the breaking open of the box, which, "to the unqualified amazement of everybody, was found to be perfectly empty." All this is much in the usual style ; but what followed was so much superior to the ordinary run of modern Indian jugglery that we must give it in the simple Siddeshur's own words. When every one was satisfied that the man had really disappeared, the principal performer, who did not seem to be at all astonished, told his audience that the vanished man had gone up to the heavens to fight Indra. "In a few moments," says Siddeshur, "he expressed anxiety at the man's continued absence in the aerial regions, and said that he would go up to see what was the matter. A boy was called, who held upright a long bamboo, up which the man climbed to the top, whereupon we suddenly lost sight of him, and the boy laid the bamboo on the ground. Then there fell on the ground before us the different members of a human body, all bloody,—first one hand, then another, a foot, and so on, until complete. The boy then elevated the bamboo, and the principal performer, appearing on the top as suddenly as he had disappeared, came down, and seeming quite disconsolate, said that Indra had killed his friend before he could get there to save him. He then placed the mangled remains in the same box, closed it, and tied it as before. Our wonder and astonishment reached their climax when, a few minutes later, on the box being again opened, the man jumped out perfectly hearty and unhurt." Is not this rather a severe strain on one's credulity, even for an Indian jugglery story?]

In Philostratus, again, we may learn the antiquity of some juggling tricks that have come up as novelties in our own day. Thus at Taxila a man set his son against a board, and then threw darts tracing the outline of the boy's figure on the board. This feat was shown in London some fifteen or twenty years ago, and humorously commemorated in *Punch* by John Leech.

(*Philostratus,* Fr. Transl. Bk. III. ch. xv. and xxvii. ; *Mich. Glycas,* Ann. II. 156, Paris ed. ; *Delrio, Disquis. Magic.* pp. 34, 100 ; *Koeppen,* I. 31, II. 82, 114-115, 260, 262, 280; *Vassilyev,* 156; *Della Penna,* 36; *S. Setzen,* 43, 353; *Pereg. Quat.* 117 ; *I. B.* IV. 39 and 290 *seqq.;* *Asiat. Researches,* XVII. 186; *Valentyn,* V. 52-54; *Edward Melton, Engelsch Edelmans, Zeldzaame en Gedenkwaardige Zee en Land*

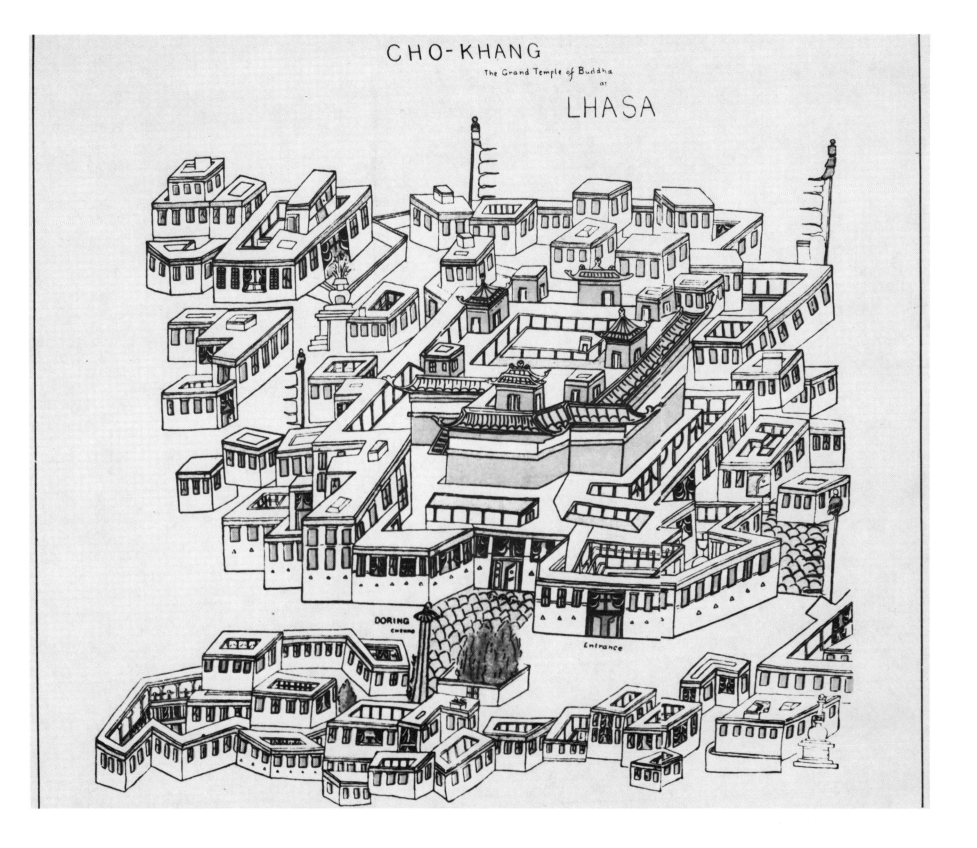

CHO-KHANG

The Grand Temple of Buddha

at

LHASA

Reizen, etc., aangevangen in den Jaare 1660 *en geendigd in den Jaare* 1677, Amsterdam, 1702, p. 468; *Mem. of the Emp. Jahangueir*, pp. 99, 102.)

NOTE 12.—["The maintenance of the Lamas, of their monasteries, the expenses for the sacrifices and for transcription of sacred books, required enormous sums. The Lamas enjoyed a preponderating influence, and stood much higher than the priests of other creeds, living in the palace as if in their own house. The perfumes, which M. Polo mentions, were used by the Lamas for two purposes; they used them for joss-sticks, and for making small turrets, known under the name of *ts'a-ts'a;* the joss-sticks used to be burned in the same way as they are now; the *ts'a-ts'a* were inserted in *suburgas* or buried in the ground. At the time when the *suburga* was built in the garden of the Peking palace in 1271, there were used, according to the Empress' wish, 1008 turrets made of the most expensive perfumes, mixed with pounded gold, silver, pearls, and corals, and 130,000 *ts'a-ts'a* made of ordinary perfumes." (*Palladius*, 29.—H. C.]

NOTE 13.—There is no exaggeration in this number. Turner speaks of 2500 monks in one Tibetan convent. Huc mentions Chorchi, north of the Great Wall, as containing 2000; and Kúnbúm, where he and Gabet spent several months, on the borders of Shensi and Tibet, had nearly 4000. The missionary itinerary from Nepal to L'hasa given by Giorgi, speaks of a group of convents at a place called Brephung, which formerly contained 10,000 inmates, and at the time of the journey (about 1700) still contained 5000, including attendants. Dr. Campbell gives a list of twelve chief convents in L'hasa and its vicinity (not including the Potala or Residence of the Grand Lama), of which one is said to have 7500 members, resident and itinerary. Major Montgomerie's Pandit gives the same convent 7700 Lamas. In the great monastery at L'hasa called *Labrang*, they show a copper kettle holding more than 100 buckets, which was used to make tea for the Lamas who performed the daily temple service. The monasteries are usually, as the text says, like small towns, clustered round the great temples. That represented at p. 224 is at Jehol, and is an imitation of the Potala at L'hasa. (*Huc's Tartary, etc.*, pp. 45, 208, etc.; *Alph. Tibetan*, 453; *J. A. S. B.* XXIV. 219; *J. R. G. S.* XXXVIII. 168; *Koeppen*, II. 338.) [*La Géographie*, II. 1901, pp. 242-247, has an article by Mr. J. Deniker, *La Première Photographie de Lhassa*, with a view of *Potala*, in 1901, from a photograph by M. O. Norzunov; it is interesting to compare it with the view given by Kircher in 1670.—H. C.]

["The monasteries with numbers of monks, who, as M. Polo asserts, behaved decently, evidently belonged to Chinese Buddhists, *ho-shang;* in Kúblái's time they had two monasteries in Shangtu, in the north-east and north-west parts of the town." (*Palladius*, 29.) Rubruck (*Rockhill's* ed. p. 145) says: "All the priests (of the idolaters) shave their heads, and are dressed in saffron colour, and they observe chastity from the time they shave their heads, and they live in congregations of one or two hundred."—H. C.]

NOTE 14.—There were many anomalies in the older Lamaism, and it permitted, at least in some sects of it which still subsist, the marriage of the clergy under certain limitations and conditions. One of Giorgi's missionaries speaks of a Lama of high *hereditary* rank as a spiritual prince who marries, but separates from his wife as soon as he has a son, who after certain trials is deemed worthy to be his successor. ["A good number of Lamas were married, as M. Polo correctly remarks; their wives were known amongst the Chinese, under the name of *Fan-sao*." (*Ch'ue keng lu*, quoted by *Palladius*, 28.)—H. C.] One of the "*reforms*" of Tsongkhapa was the absolute prohibition of marriage to the clergy, and in this he followed the institutes of the oldest Buddhism. Even the *Red Lamas*, or unreformed, cannot now marry without a dispensation.

But even the oldest orthodox Buddhism had its Lay brethren and Lay sisters (*Upásaka* and *Upásiká*), and these are to be found in Tibet and Mongolia (*Voués au blanc*, as it were). They are called by the Mongols, by a corruption of the Sanskrit,

Monastery of Lamas.

Ubashi and *Ubashanza*. Their vows extend to the strict keeping of the five great commandments of the Buddhist Law, and they diligently ply the rosary and the prayer-wheel, but they are not pledged to celibacy, nor do they adopt the tonsure. As a sign of their amphibious position, they commonly wear a red or yellow girdle. These are what some travellers speak of as the lowest order of Lamas, permitted to marry ; and Polo may have regarded them in the same light.

(*Koeppen*, II. 82, 113, 276, 291 ; *Timk.* II. 354 ; *Erman*, II. 304 ; *Alph. Tibet.* 449.)

Note 15.—[Mr. Rockhill writes to me that "bran" is certainly Tibetan *tsamba* (parched barley).—H. C.]

Note 16.—Marco's contempt for *Patarins* slips out in a later passage (Bk. III. ch. xx.). The name originated in the eleventh century in Lombardy, where it came to be applied to the "heretics," otherwise called "Cathari." Muratori has much on the origin of the name Patarini, and mentions a monument, which still exists, in the Piazza de' Mercanti at Milan, in honour of Oldrado Podestà of that city in 1233, and which thus, with more pith than grammar, celebrates his meritorious acts :—

" Qui solium struxit Catharos *ut debuit* UXIT."

Other cities were as piously Catholic. A Mantuan chronicler records under 1276 : "Captum fuit Sermionum seu redditum fuit Ecclesiæ, et capti fuerunt cercha CL Patarini contra fidem, inter masculos et feminas ; qui omnes ducti fuerunt Veronam, et ibi incarcerati, *et pro magna parte* COMBUSTI." (*Murat. Dissert.* III. 238 ; *Archiv. Stor. Ital.* N.S. I. 49.)

Note 17.—Marsden, followed by Pauthier, supposes these unorthodox ascetics to be Hindu Sanyasis, and the latter editor supposes even the name *Sensi* or *Sensin* to represent that denomination. Such wanderers do occasionally find their way to Tartary ; Gerbillon mentions having encountered five of them at Kuku Khotan (*supra*, p. 286), and I think John Bell speaks of meeting one still further north. But what is said of the great and numerous idols of the *Sensin* is inconsistent with such a notion, as is indeed, it seems to me, the whole scope of the passage. Evidently no occasional vagabonds from a far country, but some indigenous sectaries, are in question. Nor would bran and hot water be a Hindu regimen. The staple diet of the Tibetans is *Chamba*, the meal of toasted barley, mixed sometimes with warm water, but more frequently with hot tea, and I think it is probable that these were the elements of the ascetic diet rather than the mere *bran* which Polo speaks of. Semedo indeed says that some of the Buddhist devotees professed never to take any food but tea ; knowing people said they mixed with it pellets of sun-dried beef. The determination of the sect intended in the text is, I conceive, to be sought in the history of Chinese or Tibetan Buddhism and their rivals.

Both Baldelli and Neumann have indicated a general opinion that the *Taossé* or some branch of that sect is meant, but they have entered into no particulars except in a reference by the former to *Shien-sien*, a title of perfection affected by that sect, as the origin of Polo's term *Sensin*. In the substance of this I think they are right. But I believe that in the text this Chinese sect are, rightly or wrongly, identified with the ancient Tibetan sect of *Boṅ-po*, and that part of the characters assigned belong to each.

First with regard to the *Taossé*. These were evidently the *Patarini* of the Buddhists in China at this time, and Polo was probably aware of the persecution which the latter had stirred up Kúblái to direct against them in 1281—persecution at least it is called, though it was but a mild proceeding in comparison with the thing contemporaneously practised in Christian Lombardy, for in heathen Cathay, books, and not human creatures, were the subjects doomed to burn, and even that doom was not carried out.

["The Tao-sze," says M. Polo, "were looked upon as heretics by the other sects; that is, of course, by the Lamas and Ho-shangs; in fact in his time a passionate struggle was going on between Buddhists and Tao-sze, or rather a persecution of the latter by the former; the Buddhists attributed to the doctrine of the Tao-sze a pernicious tendency, and accused them of deceit; and in support of these assertions they pointed to some of their sacred books. Taking advantage of their influence at Court, they persuaded Kúblái to decree the burning of these books, and it was carried out in Peking." (*Palladius*, 30.)—H. C.]

The term which Polo writes as *Sensin* appears to have been that popularly applied to the Taossé sect at the Mongol Court. Thus we are told by Rashíduddín in his History of Cathay: "In the reign of Din-Wang, the 20th king of this (the 11th) Dynasty, TAI SHANG LÁI KÚN, was born. This person is stated to have been accounted a prophet by the people of Khitá; his father's name was Hán; like Shákmúni he is said to have been conceived by light, and it is related that his mother bore him in her womb no less a period than 80 years. The people who embraced his doctrine were called شِنْشِين شِنْشِين (*Shăn-shăn* or *Shinshin*)." This is a correct epitome of the Chinese story of *Laokiun* or *Lao-tsé*, born in the reign of *Ting Wang* of the Cheu Dynasty. The whole title used by Rashíduddín, *Tai Shang Lao Kiun*, "The Great Supreme Venerable Ruler," is that formerly applied by the Chinese to this philosopher.

Further, in a Mongol [and Chinese] inscription of the year 1314 from the department of Si-ngan fu, which has been interpreted and published by Mr. Wylie, the Taossé priests are termed *Senshing*. [See *Devéria, Notes d'Épigraphie*, pp. 39-43, and Prince *R. Bonaparte's Recueil*, Pl. xii. No. 3.—H. C.]

Seeing then that the very term used by Polo is that applied by both Mongol and Persian authorities of the period to the Taossé, we can have no doubt that the latter are indicated, whether the facts stated about them be correct or not.

The word Senshing-ud (the Mongol plural) is represented in the Chinese version of Mr. Wylie's inscription by *Sín-săng*, a conventional title applied to literary men, and this perhaps is sufficient to determine the Chinese word which *Sensin* represents. I should otherwise have supposed it to be the *Shin-sian* alluded to by Baldelli, and mentioned in the quotations which follow; and indeed it seems highly probable that two terms so much alike should have been confounded by foreigners. Semedo says of the Taossé: "They pretend that by means of certain exercises and meditations one shall regain his youth, and others shall attain to be *Shien-sien, i.e.* 'Terrestrial Beati,' in whose state every desire is gratified, whilst they have the power to transport themselves from one place to another, however distant, with speed and facility." Schott, on the same subject, says: "By *Sian* or *Shin-sian* are understood in the old Chinese conception, and particularly in that of the Tao-Kiao [or Taossé] sect, persons who withdraw to the hills to lead the life of anchorites, and who have attained, either through their ascetic observances or by the power of charms and elixirs, to the possession of miraculous gifts and of terrestrial immortality." And M. Pauthier himself, in his translation of the Journey of Khieu, an eminent doctor of this sect, to the camp of the Great Chinghiz in Turkestan, has related how Chinghiz bestowed upon this personage "a seal with a tiger's head and a diploma" (surely a lion's head, *P'aizah* and *Yarligh*; see *infra*, Bk. II. ch. vii. note 2), "wherein he was styled *Shin Sien* or Divine Anchorite." *Sian-jin* again is the word used by Hiuen Tsang as the equivalent to the name of the Indian *Rishis*, who attain to supernatural powers.

["*Sensin* is a sufficiently faithful transcription of *Sien-seng* (Sien-shing in Pekingese); the name given by the Mongols in conversation as well as in official documents, to the Tao-sze, in the sense of preceptors, just as Lamas were called by them *Bacshi*, which corresponds to the Chinese *Sien-seng*. M. Polo calls them fasters and ascetics. It was one of the sects of Taouism. There was another one which practised cabalistic and other mysteries. The Tao-sze had two monasteries in

Shangtu, one in the eastern, the other in the western part of the town." (*Palladius*, 30.)—H. C.]

One class of the Tao priests or devotees does marry, but another class never does. Many of them lead a wandering life, and derive a precarious subsistence from the sale of charms and medical nostrums. They shave the sides of the head, and coil the remaining hair in a tuft on the crown, in the ancient Chinese manner; moreover, says Williams, they "*are recognised by their slate-coloured robes.*" On the feast of one of their divinities whose title Williams translates as "High Emperor of the Sombre Heavens," they assemble before his temple, "and having made a great fire, about 15 or 20 feet in diameter, go over it barefoot, preceded by the priests and bearing the gods in their arms. They firmly assert that if they possess a sincere mind they will not be injured by the fire; but both priests and people get miserably burr* on these occasions." Escayrac de Lauture says that on those days they leap, dance, and whirl round the fire, striking at the devils with a straight Roman-like sword, and sometimes wounding themselves as the priests of Baal and Moloch used to do.

(*Astley*, IV. 671; *Morley* in *J. R. A. S.* VI. 24; *Semedo*, 111, 114; *De Mailla*, IX. 410; *J. As.* sér. V. tom. viii. 138; *Schott über den Buddhismus*, etc. 71; *Voyage de Khieou* in *J. As.* sér. VI. tom. ix. 41; *Middle Kingdom*, II. 247; *Doolittle*, 192; *Esc. de Lauture, Mém. sur la Chine, Religion*, 87, 102; *Pèler. Boudd.* II. 370, and III. 468.)

Let us now turn to the *Bon-po*. Of this form of religion and its sectaries not much is known, for it is now confined to the eastern and least known part of Tibet. It is, however, believed to be a remnant of the old pre-Buddhistic worship of the powers of nature, though much modified by the Buddhistic worship with which it has so long been in contact. Mr. Hodgson also pronounces a collection of drawings of Bonpo divinities, which were made for him by a mendicant friar of the sect from the neighbourhood of Tachindu, or Ta-t'sien-lu, to be saturated with *Sakta* attributes, *i.e.* with the spirit of the Tantrika worship, a worship which he tersely defines as "a mixture of lust, ferocity, and mummery," and which he believes to have originated in an incorporation with the Indian religions of the rude superstitions of the primitive Turanians. Mr. Hodgson was told that the Bonpo sect still possessed numerous and wealthy Vihars (or abbeys) in Tibet. But from the information of the Catholic missionaries in Eastern Tibet, who have come into closest contact with the sect, it appears to be now in a state of great decadence, "oppressed by the Lamas of other sects, the *Peunbo* (Bonpo) think only of shaking off the yoke, and getting deliverance from the vexations which the smallness of their number forces them to endure." In June, 1863, apparently from such despairing motives, the Lamas of Tsodam, a Bonpo convent in the vicinity of the mission settlement of Bonga in E. Tibet, invited the Rev. Gabriel Durand to come and instruct them. "In this temple," he writes, "are the *monstrous idols* of the sect of Peunbo; horrid figures, whose features only Satan could have inspired. They are disposed about the enclosure according to their power and their seniority. Above the pagoda is a loft, the nooks of which are crammed with all kinds of diabolical trumpery; little idols of wood or copper, hideous masques of men and animals, superstitious Lama vestments, drums, trumpets of human bones, sacrificial vessels, in short, all the utensils with which the devil's servants in Tibet honour their master. And what will become of it all? The Great River, whose waves roll to Martaban (the Lu-kiang or Salwen), is not more than 200 or 300 paces distant. . . . Besides the infernal paintings on the walls, eight or nine monstrous idols, seated at the inner end of the pagoda, were calculated by their size and aspect to inspire awe. In the middle was *Tamba-Shi-Rob*, the great doctor of the sect of the Peunbo, squatted with his right arm outside his red scarf, and holding in his left the vase of knowledge. . . . On his right hand sat *Keumta-Zon-bo*, 'the All-Good,' . . . with ten hands and three heads, one over the other. . . . At his right is *Dreuma*, the most celebrated goddess of the sect. On the left of Tamba-Shi-Rob was another goddess, whose name they never could tell me. On the left again of this anonymous goddess appeared *Tam-pla-mi-ber*, ... a monstrous dwarf environed by flames and his head garnished with a diadem of skulls. *He trod with one foot on the head of Shakia-tupa* [*Shakya Thubba, i.e.* 'the Mighty Shakya,' the usual Tibetan appellation of Sakya Buddha himself]. . . . The idols are made of a coarse composition of mud and stalks kneaded together, on which they put first a coat of plaster and then various colours, or even silver or gold. . . . *Four oxen would scarcely have been able to draw one of the idols."* Mr. Emilius Schlagintweit, in a paper on the subject of this sect, has explained some of the names used by the missionary. *Tamba-Shi-Rob* is "*bs*tanpa *g*Shen-rabs," *i.e.* the doctrine of Shen-rabs, who is regarded as the founder of the Bon religion. [Cf. *Grenard,* II. 407.—H. C.] *Keun-tu-zon-bo* is "Kun-tu-*b*zang-po," "*the All Best."*

[*Bon-po* seems to be (according to Grenard, II. 410) a "coarse naturism combined with ancestral worship" resembling Taoism. It has, however, borrowed a good deal from Buddhism. "I noticed," says Mr. Rockhill (*Journey,* 86), "a couple of grimy volumes of Bönbo sacred literature. One of them I examined ; it was a funeral service, and was in the usual Bönbo jargon, three-fourths Buddhistic in its nomenclature." The Bon-po Lamas are above all sorcerers and necromancers, and are very similar to the *kam* of the Northern Turks, the *bö* of the Mongols, and lastly to the *Shamans.* During their operations, they wear a tall pointed black hat, surmounted by the feather of a peacock, or of a cock, and a human skull. Their principal divinities are the White God of Heaven, the Black Goddess of Earth, the Red Tiger and the Dragon ; they worship an idol called *Kye'-p'ang* formed of a mere block of wood covered with garments. Their sacred symbol is the *svastika* turned from right to left ⌐⌐ . The most important of their monasteries is Zo-chen gum-pa, in the north-east of Tibet, where they print most of their books. The Bonpos Lamas "are very popular with the agricultural Tibetans, but not so much so with the pastoral tribes. who nearly all belong to the Gélupa sect of the orthodox Buddhist Church." A. K. says, "Buddhism is the religion of the country ; there are two sects, one named Mangba and the other Chiba or Baimbu." *Explorations made by A——K——,* 34. *Mangba* means "Esoteric," *Chiba* (*p'yi-ba*), "Exoteric," and *Baimbu* is Bönbo. *Rockhill, Journey,* 289, *et passim. ; Land of the Lamas,* 217-218 ; *Grenard, Mission Scientifique,* II. 407 *seqq.*—H. C.]

There is an indication in Koeppen's references that the followers of the *Bon* doctrine are sometimes called in Tibet *Nag-choi,* or "Black Sect," as the old and the reformed Lamas are called respectively the "Red" and the "Yellow." If so, it is reasonable to conclude that the first appellation, like the two last, has a reference to the colour of clothing affected by the priesthood.

The Rev. Mr. Jaeschke writes from Lahaul : "There are no Bonpos in our part of the country, and as far as we know there cannot be many of them in the whole of Western Tibet, *i.e.* in Ladak, Spiti, and all the non-Chinese provinces together ; we know, therefore, not much more of them than has been made known to the European public by different writers on Buddhism in Tibet, and lately collected by Emil de Schlagintweit. . . . Whether they can be with certainty identified with the Chinese *Taossé* I cannot decide, as I don't know if anything like historical evidence about their Chinese origin has been detected anywhere, or if it is merely a conclusion from the similarity of their doctrines and practices. . . . But the Chinese author of the *Wei-tsang-tu-Shi,* translated by Klaproth, under the title of *Description du Tubet* (Paris, 1831), renders *Bonpo* by *Taossé.* So much seems to be certain that it was the ancient religion of Tibet, before Buddhism penetrated into the country, and that even at later periods it several times gained the ascendancy when the secular power was of a disposition averse to the Lamaitic hierarchy. Another opinion is that the Bon religion was originally a mere fetishism, and related to or identical with Shamanism ; this appears to me very probable and easy to reconcile with the former supposition, for it may afterwards, on becoming acquainted with the Chinese doctrine of the 'Taossé,' have adorned itself with many of its tenets. . . . With regard to the following particulars, I have got most of my information from our Lama, a native of the neighbourhood of Tashi Lhunpo, whom we consulted about all your questions. The extraordinary asceticism which struck Marco Polo so much is of course not to be understood as being practised by all members of the sect, but exclusively, or more especially, by the *priests*. That these *never* marry, and are consequently more strictly celibatary than many sects of the Lamaitic priesthood, was confirmed by our Lama." (Mr. Jaeschke then remarks upon the *bran* to much the same effect as I have done above.) "The Bonpos are by all Buddhists regarded as heretics. Though they worship idols partly the same, at least in name, with those of the Buddhists, . . . their rites seem to be very different. The most conspicuous and most generally known of their customs, futile in itself, but in the eyes of the common people the greatest sign of their sinful heresy, is that they perform the religious ceremony of making a turn round a sacred object *in the opposite direction* to that prescribed by Buddhism. As to their dress, our Lama said that they had no particular colour of garments, but their priests frequently wore red clothes, as some sects of the Buddhist priesthood do. Mr Heyde, however, once on a journey in our neighbouring county of Langskar, saw a man *clothed in black with blue borders*, who the people said was a *Bonpo*."

[Mr. Rockhill (*Journey*, 63) saw at Kao miao-tzŭ "a *red*-gowned, long-haired Bönbo Lama," and at Kumbum (p. 68), "was surprised to see quite a large number of Bönbo Lamas, recognisable by their huge mops of hair and their *red* gowns, and also from their being dirtier than the ordinary run of people."—H. C.]

The identity of the Bonpo and Taossé seems to have been accepted by Csoma de Körös, who identifies the Chinese founder of the latter, Lao-tseu, with the Shen-rabs of the Tibetan Bonpos. Klaproth also says, " Bhonbp'o, Bhanpo, and *Shen*, are the names by which are commonly designated (in Tibetan) the Taoszu, or follower of the Chinese philosopher Laotseu." * Schlagintweit refers to Schmidt's Tibetan Grammar (p. 209) and to the Calcutta edition of the *Fo-kouè-ki* (p. 218) for the like identification, but I do not know how far any two of these are independent testimonies. General Cunningham, however, fully accepts the identity, and writes to me : " Fahian (ch. xxiii.) calls the heretics who assembled at Râmagrâma *Taossé*,† thus identifying them with the Chinese Finitimists. The Taossé are, therefore, the same as the *Swâstikas*, or worshippers of the mystic cross *Swasti*, who are also *Tirthakaras*, or 'Pure-doers.' The synonymous word *Punya* is probably the origin of *Pon* or *Bon*, the Tibetan Finitimists. From the same word comes the Burmese *P'ungyi* or *Pungi*." I may add that the Chinese envoy to Cambodia in 1296, whose narrative Rémusat has translated, describes a sect which he encountered there, apparently Brahminical, as *Taossé*. And even if the Bonpo and the Taossé were not fundamentally identical, it is extremely probable that the Tibetan and Mongol Buddhists should have applied to them one name and character. Each played towards them the same part in Tibet and in China respectively ; both were heretic sects and hated rivals ; both made high pretensions to asceticism and supernatural powers ; both, I think we see reason to believe, affected the dark clothing which Polo assigns to the *Sensin ;* both, we may add, had "great idols and plenty of them." We have seen in the account of the Taossé the ground that certain of their ceremonies afford for the allegation that they "sometimes also worship fire," whilst the whole account of that rite and of others mentioned by Duhalde,‡ shows what a powerful element of the old devil-dancing Shamanism there is in their practice. The French Jesuit, on the other hand, shows us what a prominent place female

---

* *Shen*, or coupled with *jin* "people," *Shenjin*, in this sense affords another possible origin of the word *Sensin ;* but it may in fact be at bottom, as regards the first syllable, the same with the etymology we have preferred.

† I do not find this allusion in Mr. Beal's new version of Fahian. [See Rémusat's éd. p. 227 ; Klaproth says (*Ibid.* p. 230) that the *Tao-szu* are called in Tibetan *Bonbò* and *Youngdhroungpa.*—H. C.]

‡ Apparently they had at their command the whole encyclopædia of modern "Spiritualists." Duhalde mentions among their sorceries the art of producing by their invocations the figures of Lao-tseu and their divinities in the air, and of *making a pencil to write answers to questions without anybody touching it.*

divinities occupied in the Bon-po Pantheon,* though we cannot say of either sect that "their idols are all feminine." A strong symptom of relation between the two religions, by the way, occurs in M. Durand's account of the Bon Temple. We see there that *Shen-rabs*, the great doctor of the sect, occupies a chief and central place among the idols. Now in the Chinese temples of the Taossé the figure of *their* Doctor *Lao-tseu* is one member of the triad called the "Three Pure Ones," which constitute the chief objects of worship. This very title recalls General Cunningham's etymology of Bonpo.

Tibetan Bacsi.

[At the quarterly fair (*yueh kai*) of Ta-li (Yun-Nan), Mr. E. C. Baber (*Travels*, 158-159) says : "A Fakir with a praying machine, which he twirled for the salvation of the pious at the price of a few cash, was at once recognised by us ; he was our old acquaintance, the Bakhsi, whose portrait is given in *Colonel Yule's Marco Polo*." —H. C.]

(*Hodgson*, in *J. R. A. S.* XVIII. 396 *seqq.* ; *Ann. de la Prop. de la Foi*, XXXVI.

---

* It is possible that this may point to some report of the mystic impurities of the Tantrists. The *Saktián*, or Tantrists, according to the Dabistan, hold that the worship of a female divinity affords a greater recompense. (II. 155.)

301-302, 424-427 ; *E. Schlagintweit, Ueber die Bon-pa Sekte in Tibet*, in the *Sitzens-berichte* of the Munich Acad. for 1866, Heft I. pp. 1-12 ; *Koeppen*, II. 260 ; *Ladak*, p. 358 ; *J. As.* sér. II. tom. i. 411-412 ; *Rémusat. Nouv. Mél. Asiat.* I. 112 ; *Astley*, IV. 205 ; *Doolittle*, 191.)

NOTE 18.—Pauthier's text has *blons*, no doubt an error for *blous*.   In the G. Text it is *bloies*.   Pauthier interprets the latter term as "blond ardent," whilst the glossary to the G. Text explains it as both *blue* and *white*.   *Raynouard's Romance Dict.* explains *Bloi* as "Blond."   Ramusio has *biave*, and I have no doubt that *blue* is the meaning.   The same word (*bloie*) is used in the G. Text, where Polo speaks of the bright colours of the Palace tiles at Cambaluc, and where Pauthier's text has "*vermeil et jaune et vert* et blou," and again (*infra*, Bk. II. ch. xix.), where the two corps of huntsmen are said to be clad respectively in *vermeil* and in *bloie*.   Here, again, Pauthier's text has *bleu*.   The Crusca in the description of the *Sensin* omits the colours altogether ; in the two other passages referred to it has *bioda, biodo.*

["The Tao-sze, says Marco Polo, wear dresses of black and blue linen ; *i.e.* they wear dresses made of tatters of black and blue linen, as can be seen also at the present day." (*Palladius*, 30.)—H. C.]

NOTE 19.—["The idols of the Tao-sze, according to Marco Polo's statement, have female names ; in fact, there are in the pantheon of Taoism a great many female divinities, still enjoying popular veneration in China ; such are *Tow Mu* (the 'Ursa major,' constellation), *Pi-hia-yuen Kiun* (the celestial queen), female divinities for lying-in women, for children, for diseases of the eyes ; and others, which are to be seen everywhere.   The Tao-sze have, besides these, a good number of male divinities, bearing the title of *Kiun* in common with female divinities ; both these circumstances might have led Marco Polo to make the above statement." (*Palladius*, p. 30.)—H. C.]

BOOK SECOND.

----•◦•----

(1.) ACCOUNT OF THE GREAT KAAN CUBLAÝ; OF HIS PALACES AND CAPITAL; HIS COURT, GOVERNMENT, AND SPORTS.

(2.) CITIES AND PROVINCES VISITED BY THE TRAVELLER ON ONE JOURNEY WESTWARD FROM THE CAPITAL TO THE FRONTIERS OF MIEN IN THE DIRECTION OF INDIA.

(3.) AND ON ANOTHER SOUTHWARD FROM THE CAPITAL TO FUCHU AND ZAYTON.

BOOK II.

---·◇·---

PART I.—THE KAAN, HIS COURT AND CAPITAL.

CHAPTER I.

OF CUBLAY KAAN, THE GREAT KAAN NOW REIGNING, AND OF HIS GREAT PUISSANCE.

Now am I come to that part of our Book in which I shall tell you of the great and wonderful magnificence of the Great Kaan now reigning, by name CUBLAY KAAN; *Kaan* being a title which signifyeth "The Great Lord of Lords," or Emperor. And of a surety he hath good right to such a title, for·all men know for a certain truth that he is the most potent man, as regards forces and lands and treasure, that existeth in the world, or ever hath existed from the time of our First Father Adam until this day. All this I will make clear to you for truth, in this book of ours, so that every one shall be fain to acknowledge that he is the greatest Lord that is now in the world, or ever hath been. And now ye shall hear how and wherefore.[1]

NOTE I.—According to Sanang Setzen, Chinghiz himself discerned young Kúblái's superiority. On his deathbed he said: "The words of the lad Kúblái are well worth attention; see, all of you, that ye heed what he says! One day he will sit in my seat and bring you good fortune such as you have had in my day!" (p. 105).

The Persian history of Wassáf thus exalts Kúblái : "Although from the frontiers of this country ('Irák) to the Centre of Empire, the Focus of the Universe, the genial abode of the ever-Fortunate Emperor and Just Kaan, is a whole year's journey, yet the stories that have been spread abroad, even in these parts, of his glorious deeds, his institutes, his decisions, his justice, the largeness and acuteness of his intellect, his correctness of judgment, his great powers of administration, from the mouths of credible witnesses, of well-known merchants and eminent travellers, are so surpassing, that one beam of his glories, one fraction of his great qualities, suffices to eclipse all that history tells of the Cæsars of Rome, of the Chosroes of Persia, of the Khagans of China, of the (Himyarite) Kails of Arabia, of the Tobbas of Yemen, and the Rajas of India, of the monarchs of the houses of Sassan and Búya, and of the Seljukian Sultans." (*Hammer's Wassaf*, orig. p. 37.)

Some remarks on Kúblái and his government by a Chinese author, in a more rational and discriminative tone, will be found below under ch. xxiii., note 2.

A curious Low-German MS. at Cologne, giving an account of the East, says of the "Keyser von Kathagien—syn recht Name is der groisse *Hunt !*" (Magnus Canis, the Big Bow-wow as it were. See *Orient und Occident*, vol. i. p. 640.)

CHAPTER II.

CONCERNING THE REVOLT OF NAYAN, WHO WAS UNCLE TO THE GREAT KAAN CUBLAY.

Now this Cublay Kaan is of the right Imperial lineage, being descended from Chinghis Kaan, the first sovereign of all the Tartars. And he is the sixth Lord in that succession, as I have already told you in this book. He came to the throne in the year of Christ, 1256, and the Empire fell to him because of his ability and valour and great worth, as was right and reason.[1] His brothers, indeed, and other kinsmen disputed his claim, but his it remained, both because maintained by his great valour, and because it was in law and right his, as being directly sprung of the Imperial line.

Up to the year of Christ now running, to wit 1298, he hath reigned two-and-forty years, and his age is about eighty-five, so that he must have been about forty-three years of age when he first came to the throne.[2] Before that time he had often been to the wars, and had shown himself a gallant soldier and an excellent captain. But after coming to the throne he never went to the wars in person save once.[3] This befel in the year of Christ, 1286, and I will tell you why he went.

There was a great Tartar Chief, whose name was NAYAN,[4] a young man [of thirty], Lord over many lands and many provinces; and he was Uncle to the Emperor Cublay Kaan of whom we are speaking. And when he found himself in authority this Nayan waxed proud in the insolence of his youth and his great power; for indeed he could bring into the field 300,000 horsemen, though all the time he was liegeman to his nephew, the Great Kaan Cublay, as was right and reason. Seeing then what great power he had, he took it into his head that he would be the Great Kaan's vassal no longer; nay more, he would fain wrest his empire from him if he could. So this Nayan sent envoys to another Tartar Prince called CAIDU, also a great and potent Lord, who was a kinsman of his, and who was a nephew of the Great Kaan and his lawful liegeman also, though he was in rebellion and at bitter enmity with his sovereign Lord and Uncle. Now the message that Nayan sent was this: That he himself was making ready to march against the Great Kaan with all his forces (which were great), and he begged Caidu to do likewise from his side, so that by attacking Cublay on two sides at once with such great forces they would be able to wrest his dominion from him.

And when Caidu heard the message of Nayan, he was right glad thereat, and thought the time was come at last to gain his object. So he sent back answer that he would do as requested; and got ready his host, which mustered a good hundred thousand horsemen.

Now let us go back to the Great Kaan, who had news of all this plot.

NOTE 1.—There is no doubt that Kúblái was proclaimed Kaan in 1260 (4th month), his brother Mangku Kaan having perished during the seige of Hochau in Ssechwan in August of the preceding year. But Kúblái had come into Cathay some years before as his brother's Lieutenant.

He was the *fifth*, not sixth, Supreme Kaan, as we have already noticed. (Bk. I. ch. li. note 2.)

NOTE 2.—Kúblái was born in the eighth month of the year corresponding to 1216, and had he lived to 1298 would have been eighty-two years old. [According to Dr. E. Bretschneider (*Peking*, 30), quoting the *Yuen-Shi*, Kúblái died at Khanbaligh, in the Tze-t'an tien in February, 1294.—H. C.] But by Mahomedan reckoning he would have been close upon eighty-five. He was the fourth son of Tuli, who was the youngest of Chinghiz's four sons by his favourite wife Burté Fujin. (See *De Mailla*, IX. 255, etc.)

NOTE 3.—This is not literally true ; for soon after his accession (in 1261) Kúblái led an army against his brother and rival Arikbuga, and defeated him. And again in his old age, if we credit the Chinese annalist, in 1289, when his grandson Kanmala (or Kambala) was beaten on the northern frontier by Kaidu, Kúblái took the field himself, though on his approach the rebels disappeared.

Kúblái and his brother Hulaku, young as they were, commenced their military career on Chinghiz's last expedition (1226-1227). His most notable campaign was the conquest of Yunnan in 1253-1254. (*De Mailla*, IX. 298, 441.)

NOTE 4.—NAYAN was no "uncle" of Kúblái's, but a cousin in a junior generation. For Kúblái was the grandson of Chinghiz, and Nayan was the great-great-grandson of Chinghiz's brother Uchegin, called in the Chinese annals Pilgutai. [Belgutai was Chinghiz's step-brother. (*Palladius*.)—H. C.] On this brother, the great-uncle of Kúblái, and the commander of the latter's forces against Arikbuga in the beginning of the reign, both Chinghiz and Kúblái had bestowed large territories in Eastern Tartary towards the frontier of Corea, and north of Liaotong towards the Manchu country. ["The situation and limits of his appanage are not clearly defined in history. According to Belgutai's biography, it was between the Onon and Kerulen (*Yuen shi*), and according to Shin Yao's researches (*Lo fung low wen kao*), at the confluence of the Argun and Shilka. Finally, according to Harabadur's biography, it was situated in Abalahu, which geographically and etymologically corresponds to modern Butkha (*Yuen shi*); Abalahu, as Kúblái himself said, was rich in fish ; indeed, after the suppression of Nayan's rebellion, the governor of that country used to send to the Peking Court fishes weighing up to a thousand Chinese pounds (*kin.*). It was evidently a country near the Amur River." (*Palladius, l.c.* 31.)—H. C.] Nayan had added to his inherited territory, and become very powerful. ["History has apparently connected Nayan's appanage with that of Hatan (a grandson of Hachiun, brother of Chinghiz Khan), whose *ordo* was contiguous to Nayan's, on the left bank of the Amur, opposite east of Blagovietschensk, on the spot, where still the traces of an ancient city can be seen. Nayan's possessions stretched south to Kwangning, which belonged to his appanage, and it was from this town that he had the title of prince of Kwang-ning (*Yuen shi*)." (*Palladius, l.c.* 31.)—H. C.] Kaidu had gained influence over Nayan, and persuaded him to rise against Kúblái. A number of the other Mongol princes took part with him. Kúblái was much disquieted at the rumours, and sent his great lieutenant BAYAN to reconnoitre. Bayan was nearly captured, but escaped to court and reported to his master the great armament that Nayan was preparing. Kúblái succeeded by diplomacy in detaching some of the princes from the enterprise, and resolved to march in person to the scene of action, whilst despatching Bayan to the Karakorum frontier to intercept Kaidu. This was in the summer of 1287. What followed will be found in a subsequent note (ch. iv. note 6). (For Nayan's descent, see the Genealogical Table in the Appendix (A).)

CHAPTER III.

How the Great Kaan marched against Nayan.

When the Great Kaan heard what was afoot, he made his preparations in right good heart, like one who feared not the issue of an attempt so contrary to justice. Confident in his own conduct and prowess, he was in no degree disturbed, but vowed that he would never wear crown again if he brought not those two traitorous and disloyal Tartar chiefs to an ill end. So swiftly and secretly were his preparations made, that no one knew of them but his Privy Council, and all were completed within ten or twelve days. In that time he had assembled good 360,000 horsemen, and 100,000 footmen, —but a small force indeed for him, and consisting only of those that were in the vicinity. For the rest of his vast and innumerable forces were too far off to answer so hasty a summons, being engaged under orders from him on distant expeditions to conquer divers countries and provinces. If he had waited to summon all his troops, the multitude assembled would have been beyond all belief, a multitude such as never was heard of or told of, past all counting. In fact, those 360,000 horsemen that he got together consisted merely of the falconers and whippers-in that were about the court![1]

And when he had got ready this handful (as it were) of his troops, he ordered his astrologers to declare whether he should gain the battle and get the better of his enemies. After they had made their observations, they told him to go on boldly, for he would conquer and gain a glorious victory : whereat he greatly rejoiced.

So he marched with his army, and after advancing for 20 days they arrived at a great plain where Nayan lay with all his host, amounting to some 400,000 horse.

Now the Great Kaan's forces arrived so fast and so
suddenly that the others knew nothing of the matter.
For the Kaan had caused such strict watch to be made
in every direction for scouts that every one that appeared
was instantly captured. Thus Nayan had no warning
of his coming and was completely taken by surprise;
insomuch that when the Great Kaan's army came up,
he was asleep in the arms of a wife of his of whom
he was extravagantly fond. So thus you see why it was
that the Emperor equipped his force with such speed
and secrecy.

NOTE 1.—I am afraid Marco, in his desire to impress on his readers the great
power of the Kaan, is here giving the reins to exaggeration on a great scale.

Ramusio has here the following explanatory addition :—" You must know that in
all the Provinces of Cathay and Mangi, and throughout the Great Kaan's dominions,
there are too many disloyal folk ready to break into rebellion against their Lord, and
hence it is needful in every province containing large cities and much population, to
maintain garrisons. These are stationed four or five miles from the cities, and the
latter are not allowed to have walls or gates by which they might obstruct the
entrance of the troops at their pleasure. These garrisons as well as their commanders the Great Khan causes to be relieved every two years ; and bridled in this
way the people are kept quiet, and can make no disturbance. The troops are
maintained not only by the pay which the Kaan regularly assigns from the revenues
of each province, but also by the vast quantities of cattle which they keep, and by the
sale of milk in the cities, which furnishes the means of buying what they require.
They are scattered among their different stations, at distances of 30, 40, or 60 days
(from the capital) ; and had Cublay decided to summon but the half of them, the
number would have been incredible," etc.

[Palladius says (p. 37) that in the Mongol-Chinese documents, the Mongol
garrisons cantoned near the Chinese towns are mentioned under the name of *Aolu,*
but no explanation of the term is given.—H. C.]

The system of controlling garrisons, quartered at a few miles from the great cities,
is that which the Chinese followed at Kashgar, Yarkand, etc. It is, in fact, our own
system in India, as at Barrackpúr, Dinapúr, Sikandarábád, Mián Mír.

CHAPTER IV.

OF THE BATTLE THAT THE GREAT KAAN FOUGHT WITH NAYAN.

WHAT shall I say about it? When day had well
broken, there was the Kaan with all his host upon a
hill overlooking the plain where Nayan lay in his tent,

in all security, without the slightest thought of any one
coming thither to do him hurt. In fact, this confidence
of his was such that he kept no vedettes whether in front
or in rear; for he knew nothing of the coming of the
Great Kaan, owing to all the approaches having been
completely occupied as I told you. Moreover, the place
was in a remote wilderness, more than thirty marches
from the Court, though the Kaan had made the distance
in twenty, so eager was he to come to battle with
Nayan.

And what shall I tell you next? The Kaan was
there on the hill, mounted on a great wooden bartizan,[1]
which was borne by four well-trained elephants, and
over him was hoisted his standard, so high aloft that
it could be seen from all sides. His troops were ordered
in battles of 30,000 men apiece; and a great part of the
horsemen had each a foot-soldier armed with a lance set
on the crupper behind him (for it was thus that the footmen were disposed of);[2] and the whole plain seemed
to be covered with his forces. So it was thus that the
Great Kaan's army was arrayed for battle.

When Nayan and his people saw what had happened,
they were sorely confounded, and rushed in haste to
arms. Nevertheless they made them ready in good
style and formed their troops in an orderly manner.
And when all were in battle array on both sides as I
have told you, and nothing remained but to fall to
blows, then might you have heard a sound arise of
many instruments of various music, and of the voices
of the whole of the two hosts loudly singing. For this
is a custom of the Tartars, that before they join battle
they all unite in singing and playing on a certain two-
stringed instrument of theirs, a thing right pleasant to
hear. And so they continue in their array of battle,
singing and playing in this pleasing manner, until the

great Naccara of the Prince is heard to sound. As
soon as that begins to sound the fight also begins on
both sides; and in no case before the Prince's Naccara
sounds dare any commence fighting.[3]

So then, as they were thus singing and playing,
though ordered and ready for battle, the great Naccara
of the Great Khan began to sound. And that of Nayan
also began to sound. And thenceforward the din of
battle began to be heard loudly from this side and from
that. And they rushed to work so doughtily with their
bows and their maces, with their lances and swords,
and with the arblasts of the footmen, that it was a
wondrous sight to see. Now might you behold such
flights of arrows from this side and from that, that
the whole heaven was canopied with them and they
fell like rain. Now might you see on this side and
on that full many a cavalier and man-at-arms fall
slain, insomuch that the whole field seemed covered
with them. From this side and from that such cries
arose from the crowds of the wounded and dying that
had God thundered, you would not have heard Him!
For fierce and furious was the battle, and quarter there
was none given.[4]

But why should I make a long story of it? You
must know that it was the most parlous and fierce and
fearful battle that ever has been fought in our day.
Nor have there ever been such forces in the field in
actual fight, especially of horsemen, as were then engaged—for, taking both sides, there were not fewer
than 760,000 horsemen, a mighty force! and that
without reckoning the footmen, who were also very
numerous. The battle endured with various fortune
on this side and on that from morning till noon. But
at the last, by God's pleasure and the right that was
on his side, the Great Khan had the victory, and Nayan

lost the battle and was utterly routed. For the army
of the Great Kaan performed such feats of arms that
Nayan and his host could stand against them no longer,
so they turned and fled.   But this availed nothing for
Nayan ; for he and all the barons with him were taken
prisoners, and had to surrender to the Kaan with all
their arms.

Now you must know that Nayan was a baptized
Christian, and bore the cross on his banner ; but this
nought availed him, seeing how grievously he had done
amiss in rebelling against his Lord.   For he was the
Great Kaan's liegeman,[5] and was bound to hold his
lands of him like all his ancestors before him.[6]

NOTE 1.—"*Une grande* bretesche." *Bretesche, Bertisca* (whence old English
Brattice, and *Bartizan*), was a term applied to any boarded structure of defence or
attack, but especially to the timber parapets and roofs often placed on the top of the
flanking-towers in mediæval fortifications ; and this use quite explains the sort of
structure here intended.   The term and its derivative *Bartizan* came later to be
applied to projecting *guérites* or watch-towers of masonry.   *Brattice* in English is
now applied to a fence round a pit or dangerous machinery.   (See *Muratori, Dissert.*
I. 334 ; *Wedgwood's Dict. of Etym.* sub. v. *Brattice; Viollet le Duc,* by *Macdermott,*
p. 40 ; *La Curne de Sainte—Palaye, Dict.; F. Godefroy, Dict.*)

[John Ranking (*Hist. Res. on the Wars and Sports of the Mongols and Romans*)
in a note regarding this battle writes (p. 60) : "It appears that it is an old custom in
Persia, to use four elephants a-breast."   The Senate decreed Gordian III. to represent him triumphing after the Persian mode, with chariots drawn with four elephants.
Augustan Hist. vol. ii. p. 65.   See plate, p. 52.—H. C.]

NOTE 2.—This circumstance is mentioned in the extract below from Gaubil.   He
may have taken it from Polo, as it is not in Pauthier's Chinese extracts ; but Gaubil
has other facts not noticed in these.

[Elephants came from the Indo-Chinese Kingdoms, Burma, Siam, Ciampa.—H. C.]

NOTE 3.—The specification of the Tartar instrument of two strings is peculiar to
Pauthier's texts.   It was no doubt what Dr. Clarke calls "the *balalaika* or two-
stringed lyre," the most common instrument among the Kalmaks.

The sounding of the Nakkára as the signal of action is an old Pan-Asiatic custom,
but I cannot find that this very striking circumstance of the whole host of Tartars
playing and singing in chorus, when ordered for battle and waiting the signal from
the boom of the Big Drum, is mentioned by any other author.

The *Nakkárah* or *Nagárah* was a great kettledrum, formed like a brazen caldron,
tapering to the bottom and covered with buffalo-hide—at least 3½ or 4 feet in
diameter.   Bernier, indeed, tells of *Nakkáras* in use at the Court of Delhi that were
not less than a fathom across ; and Tod speaks of them in Rájpútána as "about 8 or
10 feet in diameter."   The Tartar Nakkárahs were usually, I presume, carried on a
camel ; but as Kúblái had begun to use elephants, his may have been carried on an

elephant, as is sometimes the case in India. Thus, too, P. della Valle describes those of an Indian Embassy at Ispahan : "The Indian Ambassador was also accompanied by a variety of warlike instruments of music of strange kinds, and particularly by certain Naccheras of such immense size that each pair had an elephant to carry them, whilst an Indian astride upon the elephant between the two Naccheras played upon them with both hands, dealing strong blows on this one and on that ; what a din was made by these vast drums, and what a spectacle it was, I leave you to imagine."

Joinville also speaks of the Nakkara as the signal for action : " So he was setting his host in array till noon, and then he made those drums of theirs to sound that they call *Nacaires*, and then they set upon us horse and foot." The Great Nakkara of the Tartars appears from several Oriental histories to have been called *Kúrkah*. I cannot find this word in any dictionary accessible to me, but it is in the *Ain Akbari* (*Kawargah*) as distinct from the *Nakkárah*. Abulfazl tells us that Akbar not only had a rare knowledge of the science of music, but was likewise an excellent performer —especially on the *Nakkárah*!

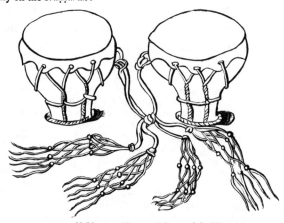

Nakkaras. (From a Chinese original.)

The privilege of employing the Nakkara in personal state was one granted by the sovereign as a high honour and reward.

The crusades naturalised the word in some form or other in most European languages, but in our own apparently with a transfer of meaning. For Wright defines *Naker* as "a cornet or horn of brass." And Chaucer's use seems to countenance this :—

" Pipes, Trompes, Nakeres, and Clariounes,
That in the Bataille blowen blody sounes."
—*The Knight's Tale.*

On the other hand, Nacchera, in Italian, seems always to have retained the meaning of *kettle-drum*, with the slight exception of a local application at Siena to a metal circle or triangle struck with a rod. The fact seems to be that there is a double origin, for the Arabic dictionaries not only have *Nakkarah*, but *Nakír*, and *Nákúr*, "cornu, tuba." The orchestra of Bibars Bundukdári, we are told, consisted of 40 pairs of kettle-drums, 4 drums, 4 hautbois, and 20 trumpets (*Nakír*). (*Sir B. Frere; Della Valle*, II. 21 ; *Tod's Rájasthán*, I. 328 ; *Joinville*, p. 83 ; *N. et E.* XIV. 129, and following note ; Blochmann's *Ain-i-Akbari*, pp. 50-51 ; *Ducange*, by Haenschel, s.v.; *Makrizi*, I. 173.)

[Dozy (*Supp. aux Dict. Arabes*) has نَقَّارَة [*naqqārè*] " petit tambour ou timbale, bassin de cuivre ou de terre recouvert d'une peau tendue," and " grosses timbales en cuivre portées sur un chameau ou un mulet."—Devic (*Dict. Étym.*) writes : " Bas Latin, *nacara ;* bas grec, ἀνάχαρα. Ce n'est point comme on l'a dit, l'Arabe نَقِلِير *naqïr* ou نَاقُر *nâqör*, qui signifient *trompette, clairon*, mais le persan نَاقَة, en arabe, نَقَّارَة *naqāra, timbale*." It is to be found also in Abyssinia and south of Gondokoro ; it is mentioned in the *Sedjarat Malayu.*

In French, it gives *nacaire* and *gnacare* from the Italian *gnacare.* " Quatre jouent de la guitare, quatre des castagnettes, quatre des gnacares." (MOLIÈRE, *Pastorale Comique.*)—H. C.]

Nakkaras. (From an Indian original.)

NOTE 4.—This description of a fight will recur again and again till we are very tired of it. It is difficult to say whether the style is borrowed from the historians of the East or the romancers of the West. Compare the two following parallels. First from an Oriental history :—

" The Ear of Heaven was deafened with the din of the great *Kurkahs* and Drums, and the Earth shook at the clangour of the Trumpets and Clarions. The shafts began to fall like the rain-drops of spring, and blood flowed till the field looked like the Oxus." (*J. A S.* sér. IV. tom. xix. 256)

Next from an Occidental Romance :—

" Now rist grete tabour betyng,
Blaweyng of pypes, and ek trumpyng,
Stedes lepyng, and ek arnyng,
Of sharp speres, and avalyng
Of stronge knighttes, and wyghth meetyng ;
Launces breche and increpyng ;
Knighttes fallyng, stedes lesyng ;
Herte and hevedes thorough kervyng ;
Swerdes draweyng, lymes lesyng
Hard assaylyng, strong defendyng,
Stiff withstondyng and wighth fleigheyng.
Sharp of takyng armes spoylyng ;
So gret bray, so gret crieyng,

Ifor the folk there was dyeyng;
So muche dent, noise of sweord,
The thondur blast no myghte beo hirde,
No the sunne hadde beo seye,
For the dust of the poudré!
No the weolkyn seon be myght,
So was arewes and quarels flyght."
—*King Alisaunder, in Weber,* I. 93-94.

And again :—

" The eorthe quaked heom undur,
No scholde mon have herd the thondur."
—*Ibid.* 142.

Also in a contemporary account of the fall of Acre (1291): " Renovatur ergo bellum terribile inter alterutros . . . . clamoribus interjectis hinc et inde ad terrorem; *ita ut nec Deus tonans in sublime coaudiri potuisset."* (*De Excidio Acconis,* in *Martene et Durand,* V. 780.)

NOTE 5.—" *Car il estoit* homme *au Grant Kaan."* (See note **2**, ch. xiv., in Prologue.)

NOTE 6.—In continuation of note 4, chap. ii., we give Gaubil's conclusion of the story of Nayan : " The Emperor had gone ahead with a small force, when Nayan's General came forward with 100,000 men to make a reconnaissance. The Sovereign, however, put on a bold front, and though in great danger of being carried off, showed no trepidation. It was night, and an urgent summons went to call troops to the Emperor's aid. They marched at once, the horsemen taking the foot soldiers on the crupper behind them. Nayan all this while was taking it quietly in his camp, and his generals did not venture to attack the Emperor, suspecting an ambuscade. Liting then took ten resolute men, and on approaching the General's camp, caused a Fire-*Pao* to be discharged ; the report caused a great panic among Nayan's troops, who were very ill disciplined at the best. Meanwhile the Chinese and Tartar troops had all come up, and Nayan was attacked on all sides : by Liting at the head of the Chinese, by Yusitemur at the head of the Mongols, by Tutuha and the Emperor in person at the head of his guards and the troops of *Kincha* (Kipchak). The presence of the Emperor rendered the army invincible, and Nayan's forces were completely defeated. That prince himself was taken, and afterwards put to death. The battle took place in the vicinity of the river Liao, and the Emperor returned in triumph to Shangtu " (207). The Chinese record given in detail by Pauthier is to the like effect, except as to the Kaan's narrow escape, of which it says nothing.

As regards the Fire-*Pao* (the latter word seems to have been applied to military machines formerly, and now to artillery), I must refer to Favé and Reinaud's very curious and interesting treatise on the Greek fire (*du Feu Grégeois*). They do not seem to assent to the view that the arms of this description which are mentioned in the Mongol wars were cannon, but rather of the nature of rockets.

[Dr. G. Schlegel (*Toung Pao,* No. 1, 1902), in a paper entitled, *On the Invention and Use of Fire-Arms and Gunpowder in China, prior to the Arrival of Europeans,* says that " now, notwithstanding all what has been alleged by different European authors against the use of gunpowder and fire-arms in China, I maintain that not only the Mongols in 1293 had cannon, but that they were already acquainted with them in 1232." Among his many examples, we quote the following from the Books of the Ming Dynasty : " What were anciently called *P'ao* were all machines for hurling stones. In the beginning of the Mongol Dynasty (A.D. 1260), *p'ao* (catapults) of the Western regions were procured. In the siege [in 1233] of the city of *Ts'ai chow* of the *Kin* (Tatars), fire was for the first time employed (in these *p'ao*), but the art of making them was not handed down, and they were afterwards seldom used."—H. C.]

CHAPTER V.

How the Great Kaan caused Nayan to be put to death.

And when the Great Kaan learned that Nayan was taken right glad was he, and commanded that he should be put to death straightway and in secret, lest endeavours should be made to obtain pity and pardon for him, because he was of the Kaan's own flesh and blood. And this was the way in which he was put to death: he was wrapt in a carpet, and tossed to and fro so mercilessly that he died. And the Kaan caused him to be put to death in this way because he would not have the blood of his Line Imperial spilt upon the ground or exposed in the eye of Heaven and before the Sun.[1]

And when the Great Kaan had gained this battle, as you have heard, all the Barons and people of Nayan's provinces renewed their fealty to the Kaan. Now these provinces that had been under the Lordship of Nayan were four in number; to wit, the first called CHORCHA; the second CAULY; the third BARSCOL; the fourth SIKINTINJU. Of all these four great provinces had Nayan been Lord; it was a very great dominion.[2]

And after the Great Kaan had conquered Nayan, as you have heard, it came to pass that the different kinds of people who were present, Saracens and Idolaters and Jews,[3] and many others that believed not in God, did gibe those that were Christians because of the cross that Nayan had borne on his standard, and that so grievously that there was no bearing it. Thus they would say to the Christians: "See now what precious help this God's Cross of yours hath rendered Nayan, who was a Christian and a worshipper thereof." And such a din arose about the matter that it reached the Great Kaan's own ears. When it did so, he sharply rebuked those who cast these gibes at the Christians; and he also bade the Christians be of good heart, "for if the Cross had rendered no help to Nayan, in that It had done right well; nor could that which was good, as It was, have done otherwise; for Nayan was a disloyal and traitorous Rebel against his Lord, and well deserved that which had befallen him. Wherefore the Cross of your God did well in that It gave him no help against the right." And this he said so loud that everybody heard him. The Christians then replied to the Great Kaan: "Great King, you say the truth indeed, for our Cross can render no one help in wrong-doing; and therefore it was that It aided not Nayan, who was guilty of crime and disloyalty, for It would take no part in his evil deeds."

And so thenceforward no more was heard of the floutings of the unbelievers against the Christians; for they heard very well what the Sovereign said to the latter about the Cross on Nayan's banner, and its giving him no help.

NOTE 1.—Friar Ricold mentions this Tartar maxim : "One Khan will put another to death, to get possession of the throne, but he takes great care that the blood be not spilt. For they say that it is highly improper that the blood of the Great Khan should be spilt upon the ground ; so they cause the victim to be smothered somehow or other." The like feeling prevails at the Court of Burma, where a peculiar mode of execution without bloodshed is reserved for Princes of the Blood. And Kaempfer, relating the conspiracy of Faulcon at the Court of Siam, says that two of the king's brothers, accused of participation, were beaten to death with clubs of sandal-wood, "for the respect entertained for the blood-royal forbids its being shed." See also note 6, ch. vi. Bk. I., on the death of the Khalif Mosta'sim Billah. (*Pereg. Quat.* p. 115; *Mission to Ava*, p. 229; *Kaempfer*, I. 19.)

NOTE 2.—CHORCHA is the Manchu country, Niuché of the Chinese. (*Supra*, note 2, ch. xlvi. Bk. I.) ["Chorcha is Churchin.—Nayan, as vassal of the Mongol khans, had the commission to keep in obedience the people of Manchuria (subdued in 1233), and to care for the security of the country (*Yuen shi*); there is no doubt that he shared these obligations with his relative Hatan, who stood nearer to the native tribes of Manchuria." (*Palladius*, 32.)—H. C.]

KAULI is properly Corea, probably here a district on the frontier thereof, as it is improbable that Nayan had any rule over Corea. ["The Corean kingdom proper could not be a part of the prince's appanage. Marco Polo might mean the northern part of Corea, which submitted to the Mongols in A.D. 1269, with sixty towns, and which was subordinated entirely to the central administration in Liao-yang. As to the southern part of Corea, it was left to the king of Corea, who, however, was a vassal of the Mongols." (*Palladius*, 32.) The king of Corea (*Ko rye, Kao-li*) was in 1288 Chyoung ryel wang (1274-1298); the capital was Syong-to, now Kăi syeng (K'ai-ch'eng). —H. C.]

BARSKUL, "Leopard-Lake," is named in Sanang Setsen (p. 217), but seems there to indicate some place in the west of Mongolia, perhaps the *Barkul* of our maps. This Barskul must have been on the Manchu frontier. [There are in the *Yuen-shi* the names of the department of *P'u-yü-lu*, and of the place *Pu-lo-ho*, which, according to the system of Chinese transcription, approach to Barscol; but it is difficult to prove this identification, since our knowledge of these places is very scanty; it only remains to identify Barscol with Abalahu, which is already known; a conjecture all the more probable as the two names of P'u-yü-lu and Pu-lo-ho have also some resemblance to Abalahu. (*Palladius*, 32.) Mr. E. H. Parker says (*China Review*, xviii. p. 261) that Barscol may be Pa-la ssŭ or Bars Koto [in Tsetsen]. "This seems the more probable in that Cauly and Chorcha are clearly proved to be Corea and Niuché or Manchuria, so that Bars Koto would naturally fall within Nayan's appanage." —H. C.]

The reading of the fourth name is doubtful, *Sichuigiu, Sichingiu* (G. T.), *Sichin-tingiu*, etc. The Chinese name of Mukden is *Shing-king*, but I know not if it be so old as our author's time. I think it very possible that the real reading is *Sinchin-tingin*, and that it represents SHANGKING-TUNGKING, expressing the two capitals of the Khitan Dynasty in this region, the position of which will be found indicated in No. IV. map of Polo's itineraries. (See *Schott, Aelteste Nachrichten von Mongolen und Tartaren*, Berlin Acad. 1845, pp. 11-12.)

[Sikintinju is Kien chau "belonging to a town which was in Nayan's appanage, and is mentioned in the history of his rebellion. There were two Kien-chow, one in the time of the Kin in the modern aimak of Khorchin; the other during the Mongol Dynasty, on the upper part of the river Ta-ling ho, in the limits of the modern aimak of Kharachin (*Man chow yuen lew k'ao*); the latter depended on Kuang-ning (*Yuen-shi*). Mention is made of Kien-chow, in connection with the following circumstance. When Nayan's rebellion broke out, the Court of Peking sent orders to the King of Corea, requiring from him auxiliary troops; this circumstance is mentioned in the Corean Annals, under the year 1288 (*Kao li shi*, ch. xxx. f. 11) in the following words:—'In the present year, in the fourth month, orders were received from Peking to send five thousand men with provisions to Kien-chow, which is 3000 *li* distant from the King's residence.' This number of *li* cannot of course be taken literally; judging by the distances estimated at the present day, it was about 2000 *li* from the Corean K'ai-ch'eng fu (then the Corean capital) to the Mongol Kien-chow; and as much to the Kien-chow of the Kin (through Mukden and the pass of Fa-k'u mun in the willow palisade). It is difficult to decide to which of these two cities of the same name the troops were ordered to go, but at any rate, there are sufficient reasons to identify Sikintinju of Marco Polo with Kien-chow." (*Palladius*, 33.)—H. C.]

We learn from Gaubil that the rebellion did not end with the capture of Nayan. In the summer of 1288 several of the princes of Nayan's league, under Hatan (apparently the *Abkan* of Erdmann's genealogies), the grandson of Chinghiz's brother Kajyun [Hachiun], threatened the provinces north-east of the wall. Kúbláí sent his grandson and designated heir, Teimur, against them, accompanied by some of his best generals. After a two days' fight on the banks of the River Kweilei, the rebels were completely beaten. The territories on the said River *Kweilei*, the *Tiro*, or *Torro*, and the *Liao*, are mentioned both by Gaubil and De Mailla as among those which

had belonged to Nayan. As the Kweilei and Toro appear on our maps and also
the better-known Liao, we are thus enabled to determine with tolerable precision
Nayan's country. (See *Gaubil*, p. 209, and *De Mailla*, 431 *seqq.*)

[" The rebellion of Nayan and Hatan is incompletely and contradictorily related in
Chinese history. The suppression of both these rebellions lasted four years. In
1287 Nayan marched from his *ordo* with sixty thousand men through Eastern
Mongolia. In the 5th moon (*var.* 6th) of the same year Khubilai marched against
him from Shangtu. The battle was fought in South-Eastern Mongolia, and gained by
Khubilai, who returned to Shangtu in the 8th month. Nayan fled to the south-east,
across the mountain range, along which a willow palisade now stands ; but forces
had been sent beforehand from Shin-chow (modern Mukden) and Kuang-ning
(probably to watch the pass), and Nayan was made prisoner.

" Two months had not passed, when Hatan's rebellion broke out (so that it took
place in the same year 1287). It is mentioned under the year 1288, that Hatan was
beaten, and that the whole of Manchuria was pacified ; but in 1290, it is again
recorded that Hatan disturbed Southern Manchuria, and that he was again defeated. It is to this time that the narratives in the biographies of Liting, Yuesi
Femur, and Mangwu ought to be referred. According to the first of these biographies,
Hatan, after his defeat by Liting on the river Kui lui (Kuilar ?), fled, and perished.
According to the second biography, Hatan's dwelling (on the Amur River) was
destroyed, and he disappeared. According to the third, Mangwu and Naimatai
pursued Hatan to the extreme north, up to the eastern sea-coast (the mouth of the
Amur). Hatan fled, but two of his wives and his son Lao-ti were taken ; the latter
was executed, and this was the concluding act of the suppression of the rebellion in
Manchuria. We find, however, an important *variante* in the history of Corea ; it
is stated there that in 1290, Hatan and his son Lao-ti were carrying fire and
slaughter to Corea, and devastated that country ; they slew the inhabitants and fed on
human flesh. The King of Corea fled to the Kiang-hwa island. The Coreans were
not able to withstand the invasion. The Mongols sent to their aid in 1291, troops
under the command of two generals, Seshekan (who was at that time governor of
Liao-tung) and Namantai (evidently the above-mentioned Naimatai). The Mongols
conjointly with the Coreans defeated the insurgents, who had penetrated into the
very heart of the country ; their corpses covered a space 30 *li* in extent ; Hatan
and his son made their way through the victorious army and fled, finding a refuge in
the Niuchi (Djurdji) country, from which Laotai made a later incursion into Corea.
Such is the discrepancy between historians in relating the same fact. The statement
found in the Corean history seems to me more reliable than the facts given by
Chinese history." (*Palladius*, 35-37.)—H. C.]

NOTE 3.—This passage, and the extract from Ramusio's version attached to the
following chapter, contain the only allusions by Marco to Jews in China. John of
Monte Corvino alludes to them, and so does Marignolli, who speaks of having held
disputations with them at Cambaluc ; Ibn Batuta also speaks of them at Khansa or
Hangchau. Much has been written about the ancient settlement of Jews at Kaifungfu,
in Honan. One of the most interesting papers on the subject is in the *Chinese
Repository*, vol. xx. It gives the translation of a Chinese-Jewish Inscription, which
in some respects forms a singular parallel to the celebrated Christian Inscription of
Si-ngan fu, though it is of far more modern date (1511). It exhibits, as that inscription does, the effect of Chinese temperament or language, in modifying or diluting
doctrinal statements. Here is a passage : " With respect to the Israelitish religion,
we find on inquiry that its first ancestor, Adam, came originally from India, and
that during the (period of the) Chau State the Sacred Writings were already in existence. The Sacred Writings, embodying Eternal Reason, consist of 53 sections.
The principles therein contained are very abstruse, and the Eternal Reason therein
revealed is very mysterious, being treated with the same veneration as Heaven.
The founder of the religion is Abraham, who is considered the first teacher of it.

Then came Moses, who established the Law, and handed down the Sacred Writings. After his time, during the Han Dynasty (B.C. 206 to A.D. 221), this religion entered China. In (A.D.) 1164, a synagogue was built at P'ien. In (A.D.) 1296, the old Temple was rebuilt, as a place in which the Sacred Writings might be deposited with veneration."

[According to their oral tradition, the Jews came to China from *Si Yih* (Western Regions), probably Persia, by Khorasan and Samarkand, during the first century of our era, in the reign of the Emperor Ming-ti (A.D. 58-75) of the Han Dynasty. They were at times confounded with the followers of religions of India, *T'ien Chu kiao*, and very often with the Mohammedans *Hwui-Hwui* or *Hwui-tsü*; the common name of their religion was *T'iao kin kiao*, "Extract Sinew Religion." However, three lapidary inscriptions, kept at Kaï-fung, give different dates for the arrival of the Jews in China : one dated 1489 (2nd year Hung Che, Ming Dynasty) says that seventy Jewish families arrived at P'ien liang (Kaï-fung) at the time of the Sung (A.D. 960-1278); one dated 1512 (7th year Chêng Têh) says that the Jewish religion was introduced into China under the Han Dynasty (B.C. 206-A.D. 221), and the last one dated 1663 (2nd year K'ang-hi) says that this religion was first preached in China under the Chau Dynasty (B.C. 1122-255); this will not bear discussion.

The synagogue, according to these inscriptions, was built in 1163, under the Sung Emperor Hiao ; under the Yuen, in 1279, the rabbi rebuilt the ancient temple known as *Ts'ing Chen sse*, probably on the site of a ruined mosque ; the synagogue was re-built in 1421 during the reign of Yung-lo ; it was destroyed by an inundation of the Hwang-ho in 1642, and the Jews began to rebuild it once more in 1653.

The first knowledge Europeans had of a colony of Jews at K'aï-fung fu, in the Ho-nan province, was obtained through the Jesuit missionaries at Peking, at the beginning of the 17th century ; the celebrated Matteo Ricci having received the visit of a young Jew, the Jesuits Aleni (1613), Gozani (1704), Gaubil and Domenge who made in 1721 two plans of the synagogue, visited Kaï-fung and brought back some documents. In 1850, a mission of enquiry was sent to that place by the *London Society for promoting Christianity among the Jews ;* the results of this mission were published at Shang-haï, in 1851, by Bishop G. Smith of Hongkong ; fac-similes of the Hebrew manuscripts obtained at the synagogue of Kaï-fung were also printed at Shang-haï at the London Missionary Society's Press, in the same year. The Jewish merchants of London sent in 1760 to their brethren of Kaï-fung a letter written in Hebrew ; a Jewish merchant of Vienna, J. L. Liebermann, visited the Kaï-fung colony in 1867. At the time of the T'aï-P'ing rising, the rebels marched against Kaï-fung in 1857, and with the rest of the population, the Jews were dispersed. (*J. Tobar, Insc. juives de Kaï-fong-fou,* 1900 ; *Henri Cordier, Les Juifs en Chine,* and *Fung and Wagnall's Jewish Encyclopedia.*) Palladius writes (p. 38), "The Jews are mentioned for the first time in the *Yuen shi* (ch. xxxiii. p. 7), under the year 1329, on the occasion of the re-establishment of the law for the collection of taxes from dissidents. Mention of them is made again under the year 1354, ch. xliii. fol. 10, when on account of several insurrections in China, rich Mahommetans and Jews were invited to the capital in order to join the army. In both cases they are named *Chu hu* (Djuhud)."—H. C.]

The synagogue at Kaifungfu has recently been demolished for the sake of its materials, by the survivors of the Jewish community themselves, who were too poor to repair it. The tablet that once adorned its entrance, bearing in gilt characters the name ESZLOYIH (Israel), has been appropriated by a mosque. The 300 or 400 survivors seem in danger of absorption into the Mahomedan or heathen population. The last Rabbi and possessor of the sacred tongue died some thirty or forty years ago, the worship has ceased, and their traditions have almost died away.

(*Cathay*, 225, 341, 497 ; *Ch. Rep.* XX. 436 ; *Dr. Martin,* in *J. N. China Br. R.A.S.* 1866, pp. 32-33.)

CHAPTER VI.

How the Great Kaan went back to the City of Cambaluc.

And after the Great Kaan had defeated Nayan in the way you have heard, he went back to his capital city of Cambaluc and abode there, taking his ease and making festivity. And the other Tartar Lord called Caydu was greatly troubled when he heard of the defeat and death of Nayan, and held himself in readiness for war; but he stood greatly in fear of being handled as Nayan had been.[1]

I told you that the Great Kaan never went on a campaign but once, and it was on this occasion; in all other cases of need he sent his sons or his barons into the field. But this time he would have none go in command but himself, for he regarded the presumptuous rebellion of Nayan as far too serious and perilous an affair to be otherwise dealt with.

Note 1.—Here Ramusio has a long and curious addition. Kúblái, it says, remained at Cambaluc till March, "in which our Easter occurs; and learning that this was one of our chief festivals, he summoned all the Christians, and bade them bring with them the Book of the Four Gospels. This he caused to be incensed many times with great ceremony, kissing it himself most devoutly, and desiring all the barons and lords who were present to do the same. And he always acts in this fashion at the chief Christian festivals, such as Easter and Christmas. And he does the like at the chief feasts of the Saracens, Jews, and Idolaters. On being asked why, he said: 'There are Four Prophets worshipped and revered by all the world. The Christians say their God is Jesus Christ; the Saracens, Mahommet; the Jews, Moses; the Idolaters, Sogomon Borcan [*Sakya-Muni Burkhan* or Buddha], who was the first god among the idols; and I worship and pay respect to all four, and pray that he among them who is greatest in heaven in very truth may aid me.' But the Great Khan let it be seen well enough that he held the Christian Faith to be the truest and best—for, as he says, it commands nothing that is not perfectly good and holy. But he will not allow the Christians to carry the Cross before them, because on it was scourged and put to death a person so great and exalted as Christ.

"Some one may say: 'Since he holds the Christian faith to be best, why does he not attach himself to it, and become a Christian?' Well, this is the reason that he gave to Messer Nicolo and Messer Maffeo, when he sent them as his envoys to the Pope, and when they sometimes took occasion to speak to him about the faith of Christ. He said: 'How would you have me to become a Christian? You see that the Christians of these parts are so ignorant that they achieve nothing and can achieve nothing, whilst you see the Idolaters can do anything they please, insomuch that when I sit at table the cups from the middle of the hall come to me full of wine or other liquor without being touched by anybody, and I drink from them. They control storms, causing them to pass in whatever direction they please, and do many other marvels ; whilst, as you know, their idols speak, and give them predictions on whatever subjects they choose. But if I were to turn to the faith of Christ and become a Christian, then my barons and others who are not converted would say : "What has moved you to be baptised and to take up the faith of Christ? What powers or miracles have you witnessed on His part?" (You know the Idolaters here say that their wonders are performed by the sanctity and power of their idols.) Well, I should not know what answer to make ; so they would only be confirmed in their errors, and the Idolaters, who are adepts in such surprising arts, would easily compass my death. But now you shall go to your Pope, and pray him on my part to send hither an hundred men skilled in your law, who shall be capable of rebuking the practices of the Idolaters to their faces, and of telling them that they too know how to do such things but will not, because they are done by the help of the devil and other evil spirits, and shall so control the Idolaters that these shall have no power to perform such things in their presence. When we shall witness this we will denounce the Idolaters and their religion, and then I will receive baptism ; and when I shall have been baptised, then all my barons and chiefs shall be baptised also, and their followers shall do the like, and thus in the end there will be more Christians here than exist in your part of the world !'

"And if the Pope, as was said in the beginning of this book, had sent men fit to preach our religion, the Grand Kaan would have turned Christian ; for it is an undoubted fact that he greatly desired to do so."

In the simultaneous patronage of different religions, Kúblái followed the practice of his house. Thus Rubruquis writes of his predecessor Mangku Kaan : "It is his custom, on such days as his diviners tell him to be festivals, or any of the Nestorian priests declare to be holydays, to hold a court. On these occasions the Christian priests enter first with their paraphernalia, and pray for him, and bless his cup. They retire, and then come the Saracen priests and do likewise ; the priests of the Idolaters follow. He all the while believes in none of them, though they all follow his court as flies follow honey. He bestows his gifts on all of them, each party believes itself to be his favourite, and all prophesy smooth things to him." Abulfaragius calls Kúblái "a just prince and a wise, who loved Christians and honoured physicians of learning, whatsoever their nation."

There is a good deal in Kúblái that reminds us of the greatest prince of that other great Mongol house, Akbar. And if we trusted the first impression of the passage just quoted from Ramusio, we might suppose that the grandson of Chinghiz too had some of that real wistful regard towards the Lord Jesus Christ, of which we seem to see traces in the grandson of Baber. But with Kúblái, as with his predecessors, religion seems to have been only a political matter ; and this aspect of the thing will easily be recognised in a re-perusal of his conversation with Messer Nicolas and Messer Maffeo. The Kaan must be obeyed ; how man shall worship God is indifferent ; this was the constant policy of his house in the days of its greatness. Kúblái, as Koeppen observes, the first of his line to raise himself above the natural and systematic barbarism of the Mongols, probably saw in the promotion of Tibetan Buddhism, already spread to some extent among them, the readiest means of civilising his countrymen. But he may have been quite sincere in saying what is here ascribed to him in *this* sense, viz. : that if the Latin Church, with its superiority of character and acquirement, had come to his aid as he had once requested, he would gladly have used *its* missionaries as his civilising instruments instead of the Lamas and their trumpery. (*Rubr.* 313 ; *Assemani*, III. pt. ii. 107 ; *Koeppen*, II. 89, 96.)

CHAPTER VII.

HOW THE KAAN REWARDED THE VALOUR OF HIS CAPTAINS.

So we will have done with this matter of Nayan, and go on with our account of the great state of the Great Kaan.

We have already told you of his lineage and of his age; but now I must tell you what he did after his return, in regard to those barons who had behaved well in the battle. Him who was before captain of 100 he made captain of 1000; and him who was captain of 1000 men he made to be captain of 10,000, advancing every man according to his deserts and to his previous rank. Besides that, he also made them presents of fine silver plate and other rich appointments; gave them Tablets of Authority of a higher degree than they held before; and bestowed upon them fine jewels of gold and silver, and pearls and precious stones; insomuch that the amount that fell to each of them was something astonishing. And yet 'twas not so much as they had deserved; for never were men seen who did such feats of arms for the love and honour of their Lord, as these had done on that day of the battle.[1]

Now those Tablets of Authority, of which I have spoken, are ordered in this way. The officer who is a captain of 100 hath a tablet of silver; the captain of 1000 hath a tablet of gold or silver-gilt; the commander of 10,000 hath a tablet of gold, with a lion's head on it. And I will tell you the weight of the different tablets, and what they denote. The tablets of the captains of 100 and 1000 weigh each of them 120 *saggi;* and the tablet with the lion's head engraven on it, which is that of the commander of 10,000, weighs 220 *saggi.* And on

each of the tablets is inscribed a device, which runs:
"*By the strength of the great God, and of the great grace
which He hath accorded to our Emperor, may the name
of the Kaan be blessed; and let all such as will not obey
him be slain and be destroyed.*" And I will tell you
besides that all who hold these tablets likewise receive
warrants in writing, declaring all their powers and
privileges.

I should mention too that an officer who holds the
chief command of 100,000 men, or who is general-in-
chief of a great host, is entitled to a tablet that weighs
300 *saggi*. It has an inscription thereon to the same
purport that I have told you already, and below the
inscription there is the figure of a lion, and below the
lion the sun and moon. They have warrants also of
their high rank, command, and power.[2] Every one,
moreover, who holds a tablet of this exalted degree is
entitled, whenever he goes abroad, to have a little
golden canopy, such as is called an umbrella, carried
on a spear over his head in token of his high command.
And whenever he sits, he sits in a silver chair.[3]

To certain very great lords also there is given a
tablet with gerfalcons on it; this is only to the very
greatest of the Kaan's barons, and it confers on them
his own full power and authority; so that if one of those
chiefs wishes to send a messenger any whither, he can
seize the horses of any man, be he even a king, and any
other chattels at his pleasure.[4]

NOTE 1.—So Sanang Setzen relates that Chinghiz, on returning from one of his
great campaigns, busied himself in reorganising his forces and bestowing rank and
title, according to the deserts of each, on his nine *Orlok*, or marshals, and all who
had done good service. "He named commandants over hundreds, over thousands,
over ten thousands, over hundred thousands, and opened his treasury to the multitude
of the people" (p. 91).

NOTE 2.—We have several times already had mention of these tablets. (See
Prologue, ch. viii. and xviii.) The earliest European allusion to them is in
Rubruquis : "And Mangu gave to the Moghul (whom he was going to send to the

King of France) a bull of his, that is to say, a golden plate of a palm in breadth and half a cubit in length, on which his orders were inscribed. Whosoever is the bearer of that may order what he pleases, and his order shall be executed straightway."

These golden bulls of the Mongol Kaans appear to have been originally tokens of high favour and honour, though afterwards they became more frequent and conventional. They are often spoken of by the Persian historians of the Mongols under the name of *Páïzah*, and sometimes *Páïzah Sir-i-Sher*, or "Lion's Head Paizah." Thus, in a firmán of Ghazan Khan, naming a viceroy to his conquests in Syria, the Khan confers on the latter "the sword, the august standard, the drum, and *the Lion's Head Paizah*." Most frequently the grant of this honour is coupled with *Yarlígh;* "to such an one were granted *Yarlígh* and *Páïzah*," the former word (which is still applied in Turkey to the Sultan's rescripts) denoting the written patent which accompanies the grant of the tablet, just as the sovereign's warrant accompanies the badge of a modern Order. Of such written patents also Marco speaks in this passage, and as he uttered it, no doubt the familiar words *Yarlígh u Páïzah* were in his mind. The Armenian history of the Orpelians, relating the visit of Prince Sempad, brother of King Hayton, to the court of Mangku Kaan, says: "They gave him also a *P'haiza* of gold, *i.e.* a tablet whereon the name of God is written by the Great Kaan himself; and this constitutes the greatest honour known among the Mongols. Farther, they drew up for him a sort of patent, which the Mongols call *Iarlekh*," etc. The Latin version of a grant by Uzbek Khan of Kipchak to the Venetian Andrea Zeno, in 1333,* ends with the words: "*Dedimus* baisa *et* privilegium *cum bullis rubeis*," where the latter words no doubt represent the *Yarlígh altamghá*, the warrant with the red seal or stamp,† as it may be seen upon the letter of Arghun Khan. (See plate at ch. xvii. of Bk. IV.). So also Janibek, the son of Uzbek, in 1344, confers privileges on the Venetians, "*eisdem dando* baissinum *de auro*"; and again Bardibeg, son, murderer, and successor of Janibeg, in 1358, writes: "Avemo dado comandamento [*i.e.* Yarlíg] cum le bolle rosse, et lo *paysam*."

Under the Persian branch, at least, of the house the degree of honour was indicated by the *number* of lions' heads upon the plate, which varied from 1 to 5. The Lion and Sun, a symbol which survives, or has been revived, in the modern

Seljukian Coin with the Lion and Sun.

Persian decoration so called, formed the emblem of the Sun in Leo, *i.e.* in highest power. It had already been used on the coins of the Seljukian sovereigns of Persia and Iconium; it appears on coins of the Mongol Ilkhans Ghazan, Oljaïtu, and Abusaid, and it is also found on some of those of Mahomed Uzbek Khan of Kipchak.

Hammer gives regulations of Ghazan Khan's on the subject of the Paizah, from which it is seen that the latter were of different *kinds* as well as degrees. Some were held by great governors and officers of state, and these were cautioned against letting the Paizah out of their own keeping; others were for officers of inferior order; and, again, "for persons travelling on state commissions with post-horses, particular paizah (which Hammer says were of brass) are appointed, on which their names are inscribed." These last would seem therefore to be merely such permissions to travel by the Government post-horses as are still required in Russia, perhaps in lineal derivation from Mongol practice. The terms of Ghazan's decree and other contemporary notices show that great abuses were practised with the Paizah, as an authority for living at free quarters and making other arbitrary exactions.

The word *Paizah* is said to be Chinese, *Pai-tseu*, "a tablet." A trace of the name and the thing still survives in Mongolia. The horse-*Bai* is the name applied to

---

* "In anno Simiae, octavâ lunâ, die quarto exeunte, juxta fluvium Cobam (*the Kuban*), apud Ripam Rubeam existentes scripsimus." The original was in *linguâ Persaycâ*.
† See *Golden Horde*, p. 218.

HALF THE LENGTH AND BREADTH OF ORIGINAL

"TABLE D'OR DE COMMANDEMENT,"

THE PAIZA OF THE MONGOLS

FROM A SPECIMEN FOUND IN
E. SIBERIA

a certain ornament on the horse caparison, which gives the rider a title to be furnished with horses and provisions on a journey.

Where I have used the Venetian erm *saggio*, the French texts have here and elsewhere *saics* and *saies*, and sometimes *pois*. *Saic* points to *saiga*, which, according to Dupré de St. Maur, is in the Salic laws the equivalent of a denier or the twelfth part of a sol. *Saggio* is possibly the same word, or rather may have been confounded with it, but the saggio was a recognised Venetian weight equal to ⅙ of an ounce. We shall see hereafter that Polo appears to use it to indicate the *miskál*, a weight which may be taken at 74 grains Troy. On that supposition the smallest tablet specified in the text would weigh 18½ ozs. Troy.

I do not know if any gold Paizah has been discovered, but several of silver have been found in the Russian dominions; one near the Dnieper, and two in Eastern Siberia. The first of our plates represents one of these, which was found in the Minusinsk circle of the Government of Yenisei in 1846, and is now in the Asiatic Museum of the Academy of St. Petersburg. For the sake of better illustration of our text, I have taken the liberty to represent the tablet as of gold, instead of silver with only the inscription gilt. The moulded ring inserted in the orifice, to suspend the plate by, is of iron. On the reverse side the ring bears some Chinese characters engraved, which are interpreted as meaning "Publication No. 42." The inscription on the plate itself is in the Mongol language and Baspa character (*supra*, Prologue, note 1, ch. xv.), and its purport is a remarkable testimony to the exactness of Marco's account, and almost a proof of his knowledge of the language and character in which the inscriptions were engraved. It runs, according to Schmidt's version: "*By the strength of the eternal heaven! May the name of the Khagan be holy! Who pays him not reverence is to be slain, and must die!*" The inscriptions on the other plates discovered were essentially similar in meaning. Our second plate shows one of them with the inscription in the Uighúr character.

The superficial dimensions of the Yenisei tablet, as taken from Schmidt's full-size drawing, are 12·2 in. by 3·65 in. The weight is not given.

In the French texts nothing is said of the size of the tablets. But Ramusio's copy in the Prologue, where the tables given by Kiacatu are mentioned (*supra*, p. 35), says that they were a cubit in length and 5 fingers in breadth, and weighed 3 to 4 marks each, *i.e.* 24 to 32 ounces.

(*Dupré de St. Maur, Essai sur les Monnoies*, etc., 1746, p. viii.; also (on *saiga*) see *Pertz, Script.* XVII. 357; *Rubruq.* 312; *Golden Horde*, 219-220, 521; *Ilch.* II. 166 *seqq.*, 355-356; *D'Ohsson*, III. 412-413; *Q. R.* 177-180; *Ham. Wassáf*, 154, 176; *Makrizi*, IV. 158; *St. Martin, Mém. sur l'Arménie*, II. 137, 169; *M. Mas Latrie* in *Bibl. de l'Éc. des Chartes*, IV. 585 *seqq.*; *J. As.* sér. V. tom. xvii. 536 *seqq*; *Schmidt, über eine Mongol. Quadratinschrift*, etc., Acad. St. P., 1847; Russian paper by *Grigorieff* on same subject, 1846.)

["The History tells us (*Liao Shih*, Bk. LVII. f. 2) that the official silver tablets *p'ai tzŭ* of the period were 600 in number, about a foot in length, and that they were engraved with an inscription like the above ['Our imperial order for post horses. Urgent.'] in national characters (*kuo tzŭ*), and that when there was important state business the Emperor personally handed the tablet to the envoy, which entitled him to demand horses at the post stations, and to be treated as if he were the Emperor himself travelling. When the tablet was marked 'Urgent,' he had the right to take private horses, and was required to ride, night and day, 700 *li* in twenty-four hours. On his return he had to give back the tablet to the Emperor, who handed it to the prince who had the custody of the state tablets and seals." (*Dr. S. W. Bushell, Actes XI. Cong. Int. Orient.*, Paris, p. 17.)

"The Kin, in the thirteenth century, used badges of office made of silver. They were rectangular, bore the imperial seal, and an inscription indicative of the duty of the bearer. (*Chavannes, Voyageurs chez les Khitans*, 102.) The Nü-chên at an earlier date used wooden *pai-tzŭ* tied to each horseman and horse, to distinguish them by. (*Ma Tuan-lin*, Bk. 327, 11.)" (*Rockhill, Rubruck*, p. 181, note.)

"Tiger's tablets—*Sinice Hu fu*, and *p'ai tsze* in the common language. The Mongols had them of several kinds, which differed by the metal, of which they were made, as well as by the number of pearls (one, two, or three in number), which were incrusted in the upper part of the tablet. Falcon's tablets with the figure of a falcon were round, and used to be given only to special couriers and envoys of the Khan. [*Yuen shi lui pien* and *Yuen ch'ao tien chang.*] The use of the *Hu-fu* was adopted by the Mongols probably from the Kin." (*Palladius, l.c.* p. 39.)

Rubruquis (Rockhill's ed. pp. 153-154) says:—"And whenever the principal envoy [of Longa] came to court he carried a highly-polished tablet of ivory about a cubit long and half a palm wide. Every time he spoke to the chan or some great personage, he always looked at that tablet as if he found there what he had to say, nor did he look to the right or the left, nor in the face of him with whom he was talking. Likewise, when coming into the presence of the Lord, and when leaving it, he never looked at anything but his tablet." Mr. Rockhill observes : "These tablets are called *hu* in Chinese, and were used in China and Korea ; in the latter country down to quite recent times. They were made of jade, ivory, bamboo, etc., according to the rank of the owner, and were about three feet long. The *hu* was originally used to make memoranda on of the business to be submitted by the bearer to the Emperor or to write the answers to questions he had had submitted to them. Odoric also refers to 'the tablets of white ivory which the Emperor's barons held in their hands as they stood silent before him.'"

(Cf. the golden tablets which were of various classes with a tiger for image and pearls for ornaments, *Devéria, Epigraphie*, p. 15 *et seq.*)—H. C.]

NOTE 3.—*Umbrella*. The phrase in Pauthier's text is " *Palieque que on dit ombrel.*" The Latin text of the Soc. de Géographie has "*unum pallium* de auro," which I have adopted as probably correct, looking to Burma, where the old etiquettes as to umbrellas are in full force. These etiquettes were probably in both countries of old Hindu origin. *Pallium*, according to Muratori, was applied in the Middle Ages to a kind of square umbrella, by which is probably meant rather a canopy on four staves, which was sometimes assigned by authority as an honourable privilege.

But the genuine umbrella would seem to have been used also, for Polo's contemporary, Martino da Canale, says that, when the Doge goes forth of his palace, " *si vait apres lui un damoiseau qui porte une umbrele de dras à or sur son chief,*" which umbrella had been given by "*Monseigneur l'Apostoille.*" There is a picture by Girolamo Gambarota, in the Sala del Gran Consiglio, at Venice, which represents the investiture of the Doge with the umbrella by Pope Alexander III., and Frederick Barbarossa (concerning which see *Sanuto* Junior, in *Muratori*, XXII. 512).

The word *Parasol* also occurs in the Petrarchian vocabulary (14th century) as the equivalent of *saioual* (Pers. *sáyában* or *sáiwán*, an umbrella). Carpini notices that umbrellas (*solinum vel tentoriolum in hastâ*) were carried over the Tartar nobles and their wives, even on horseback ; and a splendid one, covered with jewels, was one of the presents made to Kuyuk Kaan on his enthronement.

With respect to the honorary character attaching to umbrellas in China, I may notice that recently an English resident of Ningpo, on his departure for Europe, was presented by the Chinese citizens, as a token of honour, with a pair of *Wan min sàn*, umbrellas of enormous size.

The umbrella must have gone through some curious vicissitudes ; for at one time we find it familiar, at a later date apparently unknown, and then reintroduced as some strange novelty. Arrian speaks of the σκιάδια, or umbrellas, as used by all Indians of any consideration ; but the thing of which he spoke was familiar to the use of Greek and Roman ladies, and many examples of it, borne by slaves behind their mistresses, are found on ancient vase-paintings. Athenaeus quotes from Anacreon the description of a "beggar on horseback" who

"like a woman bears
An ivory parasol over his delicate head."

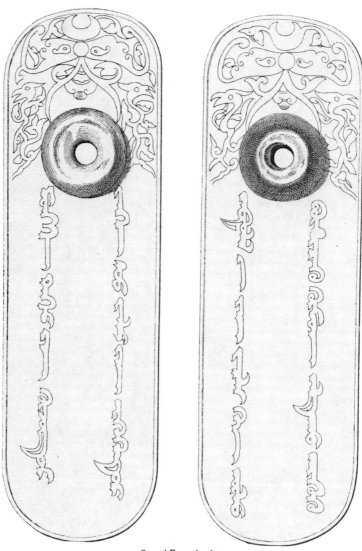

Second Example of a
MONGOL PAÏZA,
with Superscription in the *Uighúr* Character,
found near the River Dnieper,
1845.
(Half the Length and Breadth of the Original.)

[*To face p.* 355.

An Indian prince, in a Sanskrit inscription of the 9th century, boasts of having wrested from the King of Márwár the two umbrellas pleasing to Parvati, and white as the summer moonbeams. Prithi Ráj, the last Hindu king of Delhi, is depicted by the poet Chand as shaded by a white umbrella on a golden staff. An unmistakable umbrella, copied from a Saxon MS. in the Harleian collection, is engraved in *Wright's History of Domestic Manners*, p. 75. The fact that the gold umbrella is one of the paraphernalia of high church dignitaries in Italy seems to presume acquaintance with the thing from a remote period. A decorated umbrella also accompanies the host when sent out to the sick, at least where I write, in Palermo. Ibn Batuta says that in his time all the people of Constantinople, civil and military, great and small, carried great umbrellas over their heads, summer and winter. Ducange quotes, from a MS. of the Paris Library, the Byzantine court regulations about umbrellas, which are of the genuine Pan-Asiatic spirit;—σκιάδια χρυσοκόκκινα extend from the Hypersebastus to the grand Stratopedarchus, and so on; exactly as used to be the case, with different titles, in Java. And yet it is curious that John Marignolli, Ibn Batuta's contemporary in the middle of the 14th century, and Barbosa in the 16th century, are alike at pains to describe the umbrella as some strange object. And in our own country it is commonly stated that the umbrella was first used in the last century, and that Jonas Hanway (died 1786) was one of the first persons who made a practice of carrying one. The word *umbrello* is, however, in Minsheu's dictionary. [See *Hobson-Jobson*, s.v. *Umbrella.*—H. C.]

(*Murat. Dissert.* II. 229; *Archiv. Storic. Ital.* VIII. 274, 560; *Klapr. Mém.* III.; *Carp.* 759; *N. and Q.*, *C. and J.* II. 180; *Arrian, Indica*, XVI.; *Smith's Dict.*, *G. and R. Ant.*, s. v. *umbraculum*; *J. R. A. S.* v. 351; *Rás Mála*, I. 221; *I. B.* II. 440; *Cathay*, 381; *Ramus.* I. f. 301.)

Alexander, according to Athenaeus, feasted his captains to the number of 6000, and made them all sit upon silver chairs. The same author relates that the King of Persia, among other rich presents, bestowed upon Entimus the Gortynian, who went up to the king in imitation of Themistocles, *a silver chair and a gilt umbrella*. (Bk. I. Epit. ch. 31, and II. 31.)

The silver chair has come down to our own day in India, and is much affected by native princes.

NOTE 4.—I have not been able to find any allusion, except in our author, to tablets, with gerfalcons (*shonkár*). The *shonkár* appears, however, according to Erdmann, on certain coins of the Golden Horde, struck at Sarai.

There is a passage from Wassáf used by Hammer, in whose words it runs that the Sayad Imámuddín, appointed (A.D. 683) governor of Shiraz by Arghun Khan, "was

Sculptured Gerfalcon. (From the Gate of Iconium.)

invested with *both* the Mongol symbols of delegated sovereignty, the Golden Lion's
Head, and the golden *Cat's Head*." It would certainly have been more satisfactory
to find "Gerfalcon's Head" in lieu of the latter; but it is probable that the same
object is meant. The cut below exhibits the conventional effigy of a gerfalcon as
sculptured over one of the gates of Iconium, Polo's Conia. The head might easily
pass for a conventional representation of a cat's head, and is indeed strikingly like the
grotesque representation that bears that name in mediæval architecture. (*Erdmann,
Numi Asiatici*, I. 339; *Ilch.* I. 370.)

CHAPTER VIII.

CONCERNING THE PERSON OF THE GREAT KAAN.

THE personal appearance of the Great Kaan, Lord of
Lords, whose name is Cublay, is such as I shall now tell
you. He is of a good stature, neither tall nor short, but
of a middle height. He has a becoming amount of
flesh, and is very shapely in all his limbs. His complexion is white and red, the eyes black and fine,[1] the
nose well formed and well set on. He has four wives,
whom he retains permanently as his legitimate consorts;
and the eldest of his sons by those four wives ought by
rights to be emperor;—I mean when his father dies.
Those four ladies are called empresses, but each is distinguished also by her proper name. And each of them
has a special court of her own, very grand and ample;
no one of them having fewer than 300 fair and charming
damsels. They have also many pages and eunuchs,
and a number of other attendants of both sexes; so that
each of these ladies has not less than 10,000 persons
attached to her court.[2]

When the Emperor desires the society of one of
these four consorts, he will sometimes send for the lady
to his apartment and sometimes visit her at her own.
He has also a great number of concubines, and I will
tell you how he obtains them.

You must know that there is a tribe of Tartars called UNGRAT, who are noted for their beauty. Now every year an hundred of the most beautiful maidens of this tribe are sent to the Great Kaan, who commits them to the charge of certain elderly ladies dwelling in his palace. And these old ladies make the girls sleep

Portrait of Kúblái Kaan. (From a Chinese Engraving.)

with them, in order to ascertain if they have sweet breath [and do not snore], and are sound in all their limbs. Then such of them as are of approved beauty, and are good and sound in all respects, are appointed to attend on the Emperor by turns. Thus six of these damsels take their turn for three days and nights, and wait on him when he is in his chamber and when he is in his bed, to serve him in any way, and to be entirely at his orders. At the end of the three days and nights they are relieved by other six. And so throughout the year, there are reliefs of maidens by six and six, changing every three days and nights.[3]

———

NOTE 1.—We are left in some doubt as to the colour of Kúblái's eyes, for some of the MSS. read *vairs* and *voirs*, and others *noirs*. The former is a very common epithet for eyes in the mediæval romances. And in the ballad on the death of St. Lewis, we are told of his son Tristram :—

> " Droiz fu comme un rosel, *iex vairs comme faucon*,
> Dès le tens Moysel ne nasqui sa façon."

The word has generally been interpreted *bluish-grey*, but in the passage just quoted, Fr.-Michel explains it by *brillans*. However, the evidence for *noirs* here seems strongest. Rashiduddin says that when Kúblái was born Chinghiz expressed surprise at the child's being so *brown*, as its father and all his other sons were fair. Indeed, we are told that the descendants of Yesugai (the father of Chinghiz) were in general distinguished by blue eyes and reddish hair. (*Michel's Joinville*, p. 324 ; *D'Ohsson*, II. 475 ; *Erdmann*, 252.)

NOTE 2.—According to Hammer's authority (Rashid ?) Kúblái had *seven* wives ; Gaubil's Chinese sources assign him *five*, with the title of empress (*Hwang-heu*). Of these the best beloved was the beautiful Jamúi Khátún (Lady or Empress Jamúi, illustrating what the text says of the manner of styling these ladies), who bore him four sons and five daughters. Rashiduddin adds that she was called *Kún Kú*, or the great consort, evidently the term *Hwang-heu*. (Gen. Tables in *Hammer's Ilkhans ; Gaubil*, 223 ; *Erdmann*, 200.)

["Kúblái's four wives, *i.e.* the empresses of the first, second, third, and fourth *ordos*. *Ordo* is, properly speaking, a separate palace of the Khan, under the management of one of his wives. Chinese authors translate therefore the word *ordo* by 'harem.' The four *Ordo* established by Chingis Khan were destined for the empresses, who were chosen out of four different nomad tribes. During the reign of the first four Khans, who lived in Mongolia, the four *ordo* were considerably distant one from another, and the Khans visited them in different seasons of the year ; they existed nominally as long as China remained under Mongol domination. The custom of choosing the empress out of certain tribes, was in the course of time set aside by the Khans. The empress, wife of the last Mongol Khan in China, was a Corean princess by birth ; and she contributed in a great measure to the downfall of the Mongol Dynasty." (*Palladius*, 40.)

I do not believe that Rashiduddin's *Kún Kú* is the term *Hwang-heu ;* it is the term *Kiün Chu*, King or Queen, a sovereign.—H. C.]

NOTE 3.—*Ungrat*, the reading of the Crusca, seems to be that to which the others point, and I doubt not that it represents the great Mongol tribe of KUNGURAT, which gave more wives than any other to the princes of the house of Chinghiz ; a conclusion in which I find I have been anticipated by De Mailla or his editor (IX. 426). To this tribe (which, according to Vámbéry, took its name from (Turki) *Kongur-At*, " Chestnut Horse ") belonged Burteh Fujin, the favourite wife of Chinghiz himself, and mother of his four heirs ; to the same tribe belonged the two wives of Chagatai, two of Hulaku's seven wives, one of Mangku Kaan's, two at least of Kúblái's including the beloved Jamúi Khátún, one at least of Abaka's, two of Ahmed Tigudar's, two of Arghun's, and two of Ghazan's.

The seat of the Kungurats was near the Great Wall. Their name is still applied to one of the tribes of the Uzbeks of Western Turkestan, whose body appears to have been made up of fractions of many of the Turk and Mongol tribes. Kungurat is also the name of a town of Khiva, near the Sea of Aral, perhaps borrowed from the Uzbek clan.

The conversion of *Kungurat* into *Ungrat* is due, I suppose, to that Mongol tendency to soften gutturals which has been before noticed. (*Erdm.* 199-200; *Hammer, passim ; Burnes,* III. 143, 225.)

The Ramusian version adds here these curious and apparently genuine particulars :—

"The Great Kaan sends his commissioners to the Province to select four or five hundred, or whatever number may be ordered, of the most beautiful young women, according to the scale of beauty enjoined upon them. And they set a value upon the comparative beauty of the damsels in this way. The commissioners on arriving assemble all the girls of the province, in presence of appraisers appointed for the purpose. These carefully survey the points of each girl in succession, as (for example) her hair, her complexion, eyebrows, mouth, lips, and the proportion of all her limbs. They will then set down some as estimated at 16 carats, some at 17, 18, 20, or more or less, according to the sum of the beauties or defects of each. And whatever standard the Great Kaan may have fixed for those that are to be brought to him, whether it be 20 carats or 21, the commissioners select the required number from those who have attained that standard, and bring them to him. And when they reach his presence he has them appraised anew by other parties, and has a selection made of 30 or 40 of those, who then get the highest valuation."

Marsden and Murray miss the meaning of this curious statement in a surprising manner, supposing the carat to represent some absolute value, 4 grains of gold according to the former, whence the damsel of 20 carats was estimated at 13*s.* 4*d.* ! This is sad nonsense ; but Marsden would not have made the mistake had he not been fortunate enough to live before the introduction of Competitive Examinations. This Kungurat business was in fact a competitive examination in beauty ; total marks attainable 24 ; no candidate to pass who did not get 20 or 21. *Carat* expresses *n* ÷ 24, not any absolute value.

Apart from the mode of valuation, it appears that a like system of selection was continued by the Ming, and that some such selection from the daughters of the Manchu nobles has been maintained till recent times. Herodotus tells that the like custom prevailed among the Adyrmachidae, the Libyan tribe next Egypt. Old Eden too relates it of the "Princes of Moscovia." (*Middle Km.* I. 318; *Herod.* IV. 168, Rawl. ; *Notes on Russia,* Hak. Soc. II. 253.)

CHAPTER IX.

CONCERNING THE GREAT KAAN'S SONS.

THE Emperor hath, by those four wives of his, twenty-two male children ; the eldest of whom was called CHINKIN for the love of the good Chinghis Kaan, the

first Lord of the Tartars. And this Chinkin, as the
Eldest Son of the Kaan, was to have reigned after his
father's death; but, as it came to pass, he died. He
left a son behind him, however, whose name is TEMUR,
and he is to be the Great Kaan and Emperor after the
death of his Grandfather, as is but right; he being the
child of the Great Kaan's eldest son. And this Temur
is an able and brave man, as he hath already proven on
many occasions.[1]

The Great Kaan hath also twenty-five other sons
by his concubines; and these are good and valiant
soldiers, and each of them is a great chief. I tell you
moreover that of his children by his four lawful wives
there are seven who are kings of vast realms or
provinces, and govern them well; being all able and
gallant men, as might be expected. For the Great
Kaan their sire is, I tell you, the wisest and most
accomplished man, the greatest Captain, the best to
govern men and rule an Empire, as well as the most
valiant, that ever has existed among all the Tribes of
Tartars.[2]

NOTE 1.—Kúblái had a son older than CHIMKIN or CHINGKIM, to whom
Hammer's Genealogical Table gives the name of *Jurji*, and attributes a son called
Ananda. The Chinese authorities of Gaubil and Pauthier call him *Turchi* or *Torchi*,
i.e. Dorjé, "Noble Stone," the Tibetan name of a sacred Buddhist emblem in the
form of a dumb-bell, representing the *Vajra* or Thunderbolt. Probably Dorjé died
early, as in the passage we shall quote from Wassáf also Chingkim is styled the
Eldest Son: Marco is probably wrong in connecting the name of the latter with that
of Chinghiz. Schmidt says that he does not know what *Chingkim* means.

[Mr. Parker says that Chen kim was the *third* son of Kúblái (*China Review*, xxiv.
p. 94.) Teimur, son of Chen kim, wore the temple name (*miao-hao*) of *Ch'êng Tsung*
and the title of reign (*nien-hao*) of *Yuen Chêng* and *Ta Têh.*—H. C.]

Chingkim died in the 12th moon of 1284-1285, aged 43. He had received a
Chinese education, and the Chinese Annals ascribe to him all the virtues which so
often pertain in history to heirs apparent who have not reigned.

"When Kúblái approached his 70th year," says Wassáf, "he desired to raise his
eldest son Chimkin to the position of his representative and declared successor,
during his own lifetime; so he took counsel with the chiefs, in view to giving the
Prince a share of his authority and a place on the Imperial Throne. The chiefs, who
are the Pillars of Majesty and Props of the Empire, represented that His Majesty's
proposal to invest his Son, during his own lifetime, with Imperial authority, was not
in accordance with the precedents and Institutes (*Yasa*) of the World-conquering

Padshah Chinghiz Khan; but still they would consent to execute a solemn document, securing the Kaanship to Chimkin, and pledging themselves to lifelong obedience and allegiance to him.   It was, however, the Divine Fiat that the intended successor should predecease him who bestowed the nomination. . . . . . The dignitaries of the Empire then united their voices in favour of TEIMUR, the son of Chimkin."

Teimur, according to the same authority, was the third son of Chimkin; but the eldest, Kambala, *squinted;* the second, Tarmah (properly *Tarmabala* for *Dharma-phala*, a Buddhist Sanskrit name) was rickety in constitution; and on the death of the old Kaan (1294) Teimur was unanimously named to the Throne, after some opposition from Kambala, which was put down by the decided bearing of the great soldier Bayan. (*Schmidt*, p. 399; *De Mailla*, IX. 424; *Gaubil*, 203; *Wassáf*, 46.)

[The Rev. W. S. Ament (*Marco Polo in Cambaluc*, p. 106), makes the following remarks regarding this young prince (Chimkin): "The historians give good reasons for their regard for Chen Chin.   He had from early years exhibited great promise and had shown great proficiency in the military art, in government, history, mathematics, and the Chinese classics.   He was well acquainted with the condition and numbers of the inhabitants of Mongolia and China, and with the topography and commerce of the Empire (Howorth).   He was much beloved by all, except by some of his father's own ministers, whose lives were anything but exemplary.   That Kúblái had full confidence in his son is shown by the fact that he put the collecting of taxes in his hands.   The native historians represent him as economical in the use of money and wise in the choice of companions.   He carefully watched the officers in his charge, and would tolerate no extortion of the people.   After droughts, famines or floods, he would enquire into the condition of the people and liberally supply their needs, thus starting them in life again.   Polo ascribes all these virtues to the Khan himself.   Doubtless he possessed them in greater or less degree, but father and son were one in all these benevolent enterprises."—H. C.]

NOTE 2.—The Chinese Annals, according to Pauthier and Gaubil, give only *ten* sons to Kúblái, at least by his legitimate wives; Hammer's Table gives *twelve*.   It is very probable that xxii. was an early clerical error in the texts of Polo for xii. *Dodeci* indeed occurs in one MS. (No. 37 of our Appendix F), though not one of much weight.

Of these legitimate sons Polo mentions, in different parts of his work, five by name.   The following is the list from Hammer and D'Ohsson, with the Chinese forms from Pauthier in parentheses.   The seven whose names are in capitals had the title of *Wang* or "King" of particular territories, as M. Pauthier has shown from the Chinese Annals, thus confirming Marco's accuracy on that point.

I. Jurji or Dorjé (Torchi).   II. CHIMKIN or CHINGKIM (Yu Tsung, King of Yen, *i.e.* Old Peking).   III. MANGALAI (Mankola, "King of the Pacified West"), mentioned by Polo (*infra*, ch. xli.) as King of Kenjanfu or Shensi.   IV. NUMUGAN (Numukan, "Pacifying King of the North"), mentioned by Polo (Bk. IV. ch. ii.) as with King George joint leader of the Kaan's army against Kaidu.   V. Kuridai (not in Chinese List).   VI. HUKAJI (Hukochi, "King of Yunnan"), mentioned by Polo (*infra*, ch. xlix.) as King of Carajan.   VII. AGHRUKJI or UKURUJI (Gaoluchi, "King of Siping" or Tibet).   VIII. Abaji (Gaiyachi?).   IX. KUKJU or GEUKJU (Khokhochu, "King of Ning" or Tangut).   X. Kutuktemur (Hutulu Temurh).   XI. TUKAN (Thohoan, "King of Chinnan").   His command lay on the Tungking frontier, where he came to great grief in 1288, in consequence of which he was disgraced. (See *Cathay*, p. 272.)   XII. Temkan (not in Chinese List).   Gaubil's Chinese List omits *Hutulu Temurh*, and introduces a prince called *Gantanpouhoa* as 4th son.

M. Pauthier lays great stress on Polo's intimate knowledge of the Imperial affairs (p. 263) because he knew the name of the Hereditary Prince to be Teimur; this being, he says, the private name which could not be known until after the owner's death, except by those in the most confidential intimacy.   The public only then discovered that, like the Irishman's dog, his real name was Turk, though he had always been called Toby! But M. Pauthier's learning has misled him. At least the secret must have been very badly kept, for it was known in Teimur's lifetime not only to Marco, but to Rashiduddin in Persia, and to Hayton in Armenia; to say nothing of the circumstance that the name *Temur Khaghan* is also used during that Emperor's life by Oljaitu Khan of Persia in writing to the King of France a letter which M. Pauthier himself republished and commented upon. (See his book, p. 780.)

CHAPTER X.

CONCERNING THE PALACE OF THE GREAT KAAN.

YOU must know that for three months of the year, to wit December, January, and February, the Great Kaan resides in the capital city of Cathay, which is called CAMBALUC, [and which is at the north-eastern extremity of the country]. In that city stands his great Palace, and now I will tell you what it is like.

It is enclosed all round by a great wall forming a square, each side of which is a mile in length; that is to say, the whole compass thereof is four miles. This you may depend on; it is also very thick, and a good ten paces in height, whitewashed and loop-holed all round.[1] At each angle of the wall there is a very fine and rich palace in which the war-harness of the Emperor is kept, such as bows and quivers,[2] saddles and bridles, and bowstrings, and everything needful for an army. Also midway between every two of these Corner Palaces there is another of the like; so that taking the whole compass of the enclosure you find eight vast Palaces stored with the Great Lord's harness of war.[3] And you must understand that each Palace is assigned to only one kind of article; thus one is stored with bows, a second with saddles, a third with bridles, and so on in succession right round.[4]

The great wall has five gates on its southern face, the middle one being the great gate which is never opened on any occasion except when the Great Kaan himself goes forth or enters.   Close on either side of this great gate is a smaller one by which all other people pass ; and then towards each angle is another great gate, also open to people in general ; so that on that side there are five gates in all.[5]

Inside of this wall there is a second, enclosing a space that is somewhat greater in length than in breadth. This enclosure also has eight palaces corresponding to those of the outer wall, and stored like them with the Lord's harness of war.   This wall also hath five gates on the southern face, corresponding to those in the outer wall, and hath one gate on each of the other faces, as the outer wall hath also.   In the middle of the second enclosure is the Lord's Great Palace, and I will tell you what it is like.[6]

You must know that it is the greatest Palace that ever was.   [Towards the north it is in contact with the outer wall, whilst towards the south there is a vacant space which the Barons and the soldiers are constantly traversing.[7]   The Palace itself] hath no upper story, but is all on the ground floor, only the basement is raised some ten palms above the surrounding soil [and this elevation is retained by a wall of marble raised to the level of the pavement, two paces in width and projecting beyond the base of the Palace so as to form a kind of terrace-walk, by which people can pass round the building, and which is exposed to view, whilst on the outer edge of the wall there is a very fine pillared balustrade ; and up to this the people are allowed to come].   The roof is very lofty, and the walls of the Palace are all covered with gold and silver.   They are also adorned with representations of dragons [sculptured and gilt],

beasts and birds, knights and idols, and sundry other
subjects. And on the ceiling too you see nothing but
gold and silver and painting. [On each of the four
sides there is a great marble staircase leading to the top
of the marble wall, and forming the approach to the
Palace.][8]

The Hall of the Palace is so large that it could
easily dine 6000 people; and it is quite a marvel to see
how many rooms there are besides. The building is
altogether so vast, so rich, and so beautiful, that no
man on earth could design anything superior to it.
The outside of the roof also is all coloured with vermilion
and yellow and green and blue and other hues, which
are fixed with a varnish so fine and exquisite that they
shine like crystal, and lend a resplendent lustre to the
Palace as seen for a great way round.[9] This roof is
made too with such strength and solidity that it is fit to
last for ever.

[On the interior side of the Palace are large buildings with halls and chambers, where the Emperor's
private property is placed, such as his treasures of gold,
silver, gems, pearls, and gold plate, and in which reside
the ladies and concubines. There he occupies himself
at his own convenience, and no one else has access.]

Between the two walls of the enclosure which I have
described, there are fine parks and beautiful trees bearing a variety of fruits. There are beasts also of sundry
kinds, such as white stags and fallow deer, gazelles and
roebucks, and fine squirrels of various sorts, with
numbers also of the animal that gives the musk, and
all manner of other beautiful creatures,[10] insomuch that
the whole place is full of them, and no spot remains
void except where there is traffic of people going and
coming. [The parks are covered with abundant grass;
and the roads through them being all paved and raised

two cubits above the surface, they never become muddy,
nor does the rain lodge on them, but flows off into the
meadows, quickening the soil and producing that abundance of herbage.]

From that corner of the enclosure which is towards
the north-west there extends a fine Lake, containing
foison of fish of different kinds which the Emperor hath
caused to be put in there, so that whenever he desires
any he can have them at his pleasure. A river enters
this lake and issues from it, but there is a grating of
iron or brass put up so that the fish cannot escape in
that way.[11]

Moreover on the north side of the Palace, about a
bow-shot off, there is a hill which has been made by art
[from the earth dug out of the lake]; it is a good
hundred paces in height and a mile in compass. This
hill is entirely covered with trees that never lose their
leaves, but remain ever green. And I assure you that
wherever a beautiful tree may exist, and the Emperor
gets news of it, he sends for it and has it transported
bodily with all its roots and the earth attached to them,
and planted on that hill of his. No matter how big the
tree may be, he gets it carried by his elephants; and in
this way he has got together the most beautiful collection
of trees in all the world. And he has also caused the
whole hill to be covered with the ore of azure,[12] which
is very green. And thus not only are the trees all
green, but the hill itself is all green likewise; and there
is nothing to be seen on it that is not green; and hence
it is called the GREEN MOUNT; and in good sooth 'tis
named well.[13]

On the top of the hill again there is a fine big palace
which is all green inside and out; and thus the hill, and
the trees, and the palace form together a charming
spectacle; and it is marvellous to see their uniformity

of colour! Everybody who sees them is delighted. And the Great Kaan had caused this beautiful prospect to be formed for the comfort and solace and delectation of his heart.

You must know that beside the Palace (that we have been describing), *i.e.* the Great Palace, the Emperor has caused another to be built just like his own in every respect, and this he hath done for his son when he shall reign and be Emperor after him.[14] Hence it is made just in the same fashion and of the same size, so that everything can be carried on in the same manner after his own death. [It stands on the other side of the lake from the Great Kaan's Palace, and there is a bridge crossing the water from one to the other.][15] The Prince in question holds now a Seal of Empire, but not with such complete authority as the Great Kaan, who remains supreme as long as he lives.

Now I am going to tell you of the chief city of Cathay, in which these Palaces stand; and why it was built, and how.

NOTE 1.—[According to the *Ch'ue keng lu*, translated by Bretschneider, 25, "the wall surrounding the palace . . . is constructed of bricks, and is 35 *ch'i* in height. The construction was begun in A.D. 1271, on the 17th of the 8th month, between three and five o'clock in the afternoon, and finished next year on the 15th of the 3rd month."—H. C.]

NOTE 2.—*Tarcasci* (G. T.) This word is worthy of note as the proper form of what has become in modern French *carquois*. The former is a transcript of the Persian *Tärkäsh;* the latter appears to be merely a corruption of it, arising perhaps clerically from the constant confusion of *c* and *t* in MSS. (See *Defrémery*, quoted by Pauthier, *in loco*.) [Old French *tarquais* (13th century), Hatzfeldt and Darmesteter's *Dict.* gives: "Coivres orent ceinz et tarchais." (WACE, *Rou*, III., 7698; 12th century).]

NOTE 3.—["It seems to me [Dr. Bretschneider] that Polo took the towers, mentioned by the Chinese author, in the angles of the galleries and of the Kung-ch'eng for palaces; for further on he states, that 'over each gate [of Cambaluc] there is a great and handsome palace.' I have little doubt that over the gates of Cambaluc, stood lofty buildings similar to those over the gates of modern Peking. These tower-like buildings are called *lou* by the Chinese. It may be very likely, that at the time of Marco Polo, the war harness of the Khan was stored in these towers of the palace wall. The author of the *Ch'ue keng lu*, who wrote more than fifty years later, assigns to it another place." (*Bretschneider, Peking*, 32.)—H. C.]

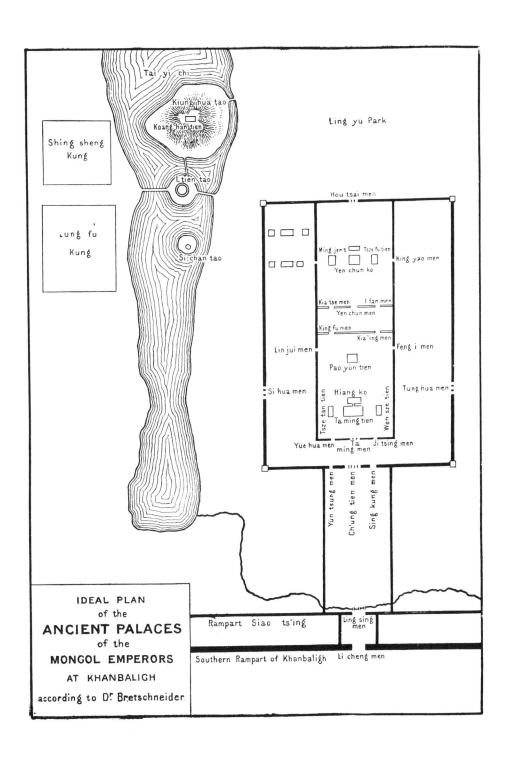

Tai yi chi

Kiung hua tao

Koang han tien

Ling yu Park

Shing sheng Kung

I tien tao

Lung fu Kung

Si chan tao

Hou tsai men

Ming jen t   Tsze fu tien

Yen chun ko

King yao men

Kia tse men   I fan men

Yen chun men

King fu men

Kia ing men

Feng i men

Lin jui men

Pao yun tien

Si hua men

Tsze tan tien

Hiang ko

Tung hua men

Ta ming tien

Wen sze tien

Yüe hua men   Ta ming men   Ji tsing men

Yün tsung men

Ch'ung tien men

Sing kung men

IDEAL PLAN
of the
ANCIENT PALACES
of the
MONGOL EMPERORS
AT KHANBALIGH
according to Dr Bretschneider

Rampart   Siao   ts'ing

Ling sing men

Southern Rampart of Khanbaligh   Li cheng men

NOTE 4.—The stores are now outside the walls of the "Prohibited City," corresponding to Polo's Palace-Wall, but within the walls of the "Imperial City." (*Middle Kingdom*, I. 61.) See the cut at p. 376.

NOTE 5.—The two gates near the corners apparently do not exist in the Palace now. "On the south side there are three gates to the Palace, both in the inner and the outer walls. The middle one is absolutely reserved for the entrance or exit of the Emperor ; all other people pass in and out by the gate to the right or left of it." (*Trigautius*, Bk. I. ch. vii.) This custom is not in China peculiar to Royalty. In private houses it is usual to have three doors leading from the court to the guest-rooms, and there is a great exercise of politeness in reference to these ; the guest after much pressing is prevailed on to enter the middle door, whilst the host enters by the side. (See *Deguignes, Voyages*, I. 262.) [See also *H. Cordier's Hist. des Relat. de la Chine*, III. ch. x. *Audience Impériale*.]

["It seems Polo took the three gateways in the middle gate (*Ta-ming men*) for three gates, and thus speaks of five gates instead of three in the southern wall." (*Bretschneider, Peking*, 27, note.)—H. C.]

NOTE 6.—Ramusio's version here diverges from the old MSS. It makes the inner enclosure a mile square ; and the second (the city of Taidu) six miles square, as here, but adds, at a mile interval, a third of eight miles square. Now it is remarkable that Mr. A. Wylie, in a letter dated 4th December 1873, speaking of a recent visit to Peking, says : "I found from various inquiries that there are several remains of a very much larger city wall, inclosing the present city ; but time would not allow me to follow up the traces."

Pauthier's text (which I have corrected by the G. T.), after describing the *outer inclosure* to be a *mile every way*, says that the inner inclosure lay at *an interval of a mile within it !*

[Dr. Bretschneider observes "that in the ancient Chinese works, three concentric inclosures are mentioned in connection with the palace. The innermost inclosed the *Ta-nei*, the middle inclosure, called *Kung-ch'eng* or *Huang-ch'eng*, answering to the wall surrounding the present prohibited city, and was about 6 *li* in circuit. Besides this there was an outer wall (a rampart apparently) 20 *li* in circuit, answering to the wall of the present imperial city (which now has 18 *li* in circuit." The *Huang-ch'eng* of the Yuen was measured by imperial order, and found to be 7 *li* in circuit ; the wall of the Mongol palace was 6 *li* in circuit, according to the *Ch'ue keng lu*. (*Bretschneider, Peking*, 24.)—Marco Polo's mile could be approximately estimated = 2·77 Chinese *li*. (*Ibid*. 24, note.) The common Chinese *li* = 360 *pu*, or 180 chang, or 1800 *ch'i* (feet) ; 1 *li* = 1894 English feet or 575 mètres ; at least according to the old Venice measures quoted in *Yule's Marco Polo*, II., one pace = 5 feet. Besides the common *li*, the Chinese have another *li*, used for measuring fields, which has only 240 *pu* or 1200 *ch'i*. This is the *li* spoken of in the *Ch'ue keng lu*. (*Ibid*. 13, note.) —H. C.]

NOTE 7.—["Near the southern face of the wall are barracks for the Life Guards." (*Ch'ue keng lu*, translated by Bretschneider, 25.)—H. C.]

NOTE 8.—This description of palace (see opposite cut), an elevated basement of masonry with a superstructure of timber (in general carved and gilded), is still found in Burma, Siam, and Java, as well as in China. If we had any trace of the palaces of the ancient Asokas and Vikramadityas of India, we should probably find that they were of the same character. It seems to be one of those things that belonged to some ancient Panasiatic fashion, as the palaces of Nineveh were of a somewhat similar construction. In the Audience Halls of the Moguls at Delhi and Agra we can trace the ancient form, though the superstructure has there become an arcade of marble instead of a pavilion on timber columns.

[" The *Ta-ming tien* (Hall of great brightness) is without doubt what Marco Polo calls ' the Lord's Great Palace.' . . . He states, that it ' hath no upper story' ; and indeed, the palace buildings which the Chinese call *tien* are always of one story. Polo speaks also of a ' very fine pillared balustrade' (the *chu lang*, pillared verandah,

Palace at Khan-baligh.   (From the *Livre des Merveilles*.)

of the Chinese author).   Marco Polo states that the basement of the great palace ' is raised some ten palms above the surrounding soil.' We find in the *Ku kung i lu :* ' The basement of the Ta-ming tien is raised about 10 *ch'i* above the soil.' There can also be no doubt that the Ta-ming tien stood at about the same place

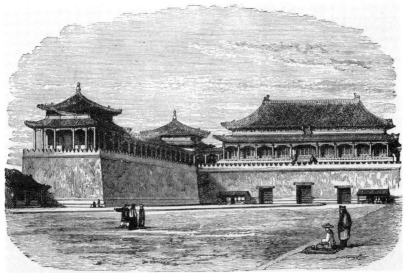

Winter Palace at Peking.

where now the *T'ai-ho tien*, the principal hall of the palace, is situated." (*Bret-schneider, Peking*, 28, note.)

The *Ch'ue keng lu*, translated by Bretschneider, 25, contains long articles devoted to the description of the palace of the Mongols and the adjacent palace grounds. They are too long to be reproduced here.—II. C.]

NOTE 9.—"As all that one sees of these palaces is varnished in those colours, when you catch a distant view of them at sunrise, as I have done many a time, you would think them all made of, or at least covered with, pure gold enamelled in azure and green, so that the spectacle is at once majestic and charming." (*Magaillans*, p. 353.)

NOTE 10.—[This is the *Ling yu* or "Divine Park," to the east of the *Wan-sui shan*, "in which rare birds and beasts are kept. Before the Emperor goes to Shang-tu, the officers are accustomed to be entertained at this place." (*Ch'ue keng lu*, quoted by Bretschneider, 36.)—H. C.]

NOTE 11.—"On the west side, where the space is amplest, there is a lake very full of fish. It is in the form of a fiddle, and is an Italian mile and a quarter in length. It is crossed at the narrowest part, which corresponds to gates in the walls, by a handsome bridge, the extremities of which are adorned by two triumphal arches of three openings each. . . . The lake is surrounded by palaces and pleasure houses, built partly in the water and partly on shore, and charming boats are provided on it for the use of the Emperor when he chooses to go a-fishing or to take an airing." (*Ibid.* 282-283.) The marble bridge, as it now exists, consists of nine arches, and is 600 feet long. (*Rennie's Peking*, II. 57.)

Ramusio specifies another lake in the *city*, fed by the same stream before it enters the palace, and used by the public for watering cattle.

["The lake which Marco Polo saw is the same as the *T'ai-yi ch'i* of our days. It has, however, changed a little in its form. This lake and also its name *T'ai-yi ch'i* date from the twelfth century, at which time an Emperor of the Kin first gave orders to collect together the water of some springs in the hills, where now the summer palaces stand, and to conduct it to a place north of his capital, where pleasure gardens were laid out. The river which enters the lake and issues from it exists still, under its ancient name *Kin-shui*." (*Bretschneider, Peking*, 34.)—H. C.]

NOTE 12.—The expression here is in the Geog. Text, "*Roze de l'açur*," and in Pauthier's "*de rose et de l'asur*." *Rose Minerale*, in the terminology of the alchemists, was a red powder produced in the sublimation of gold and mercury, but I can find no elucidation of the term Rose of Azure. The Crusca Italian has in the same place *Terra dello Azzurro*. Having ventured to refer the question to the high authority of Mr. C. W. King, he expresses the opinion that *Roze* here stands for *Roche*, and that probably the term *Roche de l'azur* may have been used loosely for *blue-stone, i.e.* carbonate of copper, which would assume a green colour through moisture. He adds: "Nero, according to Pliny, actually used *chrysocolla*, the siliceous carbonate of copper, in powder, for strewing the circus, to give the course the colour of his favourite faction, the *prasine* (or green). There may be some analogy between this device and that of Kúblái Khan." This parallel is a very happy one.

NOTE 13.—Friar Odoric gives a description, short, but closely agreeing in sub-stance with that in the Text, of the Palace, the Park, the Lake, and the Green Mount.

A green mount, answering to the description, and about 160 feet in height, stands immediately in rear of the palace buildings. It is called by the Chinese *King-Shan*, "Court Mountain," *Wan-su-Shan*, "Ten Thousand Year Mount," and *Mei-Shan*, "Coal Mount," the last from the material of which it is traditionally said to be com-posed (as a provision of fuel in case of siege).* Whether this is Kúblái's Green Mount

---

* Some years ago, in Calcutta, I learned that a large store of charcoal existed under the soil of Fort William, deposited there, I believe, in the early days of that fortress.

["The *Jihia* says that the name of *Mei shan* (Coal hill) was given to it from the stock of coal buried at its foot, as a provision in case of siege." (*Bretschneider, Peking*, 38.)—H. C.]

Mei Shan.

does not seem to be quite certain. Dr. Lockhart tells me that, according to the information he collected when living at Peking, it is not so, but was formed by the Ming Emperors from the excavation of the existing lake on the site which the Mongol Palace had occupied. There is another mount, he adds, adjoining the east shore of the lake, which must be of older date even than Kúblái, for a Dagoba standing on it is ascribed to the *Kin.*

[The "Green Mount" was an island called *K'iung-hua* at the time of the Kin ; in 1271 it received the name of *Wan-sui shan ;* it is about 100 feet in height, and is the only hill mentioned by Chinese writers of the Mongol time who refer to the palace grounds. It is not the present *King-shan,* north of the palace, called also *Wan-sui-shan* under the Ming, and now the *Mei-shan,* of more recent formation. "I have no doubt," says Bretschneider (*Peking, l.c.* 35), "that Marco Polo's handsome palace on the top of the Green Mount is the same as the *Kuang-han tien*" of the *Ch'ue keng lu.* It was a hall in which there was a jar of black jade, big enough to hold more than 30 piculs of wine ; this jade had white veins, and in accordance with these veins, fish and animals have been carved on the jar. (*Ibid.* 35.) "The *Ku kung i lu,* in describing the *Wan-sui-shan,* praises the beautiful shady green of the vegetation there." (*Ibid.* 37.)—H. C.]

["Near the eastern end of the bridge (*Kin-ao yü-tung* which crosses the lake) the visitor sees a circular wall, which is called *yüan ch'eng* (round wall). It is about 350 paces in circuit. Within it is an imperial building *Ch'eng-kuang tien,* dating from the Mongol time. From this circular enclosure, another long and beautifully executed marble bridge leads northwards, to a charming hill, covered with shady trees, and capped by a magnificent white *suburga.*" (*Bretschneider,* p. 22.)—H. C.]

In a plate attached to next chapter, I have drawn, on a small scale, the existing cities of Peking, as compared with the Mongol and Chinese cities in the time of Kúblái. The plan of the latter has been constructed (1) from existing traces, as exhibited in the Russian Survey republished by our War Office ; (2) from information kindly afforded by Dr. Lockhart ; and (3) from Polo's description and a few slight notices by Gaubil and others. It will be seen, even on the small scale of these plans, that the general arrangement of the palace, the park, the lakes (including that in the city, which appears in Ramusio's version), the bridge, the mount, etc., in the existing Peking, very closely correspond with Polo's indications ; and I think the strong probability is that the Ming really built on the old traces, and that the lake, mount, etc., as they now stand, are substantially those of the Great Mongol, though Chinese policy or patriotism may have spread the belief that the foreign traces were obliterated. Indeed, if that belief were true, the Mongol Palace must have been very much out of the axis of the City of Kúblái, which is in the highest degree improbable. The *Bulletin de la Soc. de Géographie* for September 1873, contains a paper on Peking by the physician to the French Embassy there. Whatever may be the worth of the meteorological and hygienic details in that paper, I am bound to say that the historical and topographical part is so inaccurate as to be of no value.

NOTE 14.—For son, read grandson. But the G. T. actually names the Emperor's son Chingkim, whose death our traveller has himself already mentioned.

NOTE 15.—["Marco Polo's bridge, crossing the lake from one side to the other, must be identified with the wooden bridge mentioned in the *Ch'ue keng lu.* The present marble bridge spanning the lake was only built in 1392." "A marble bridge connects this island (an islet with the hall *I-t'ien tien*) with the *Wan-sui shan.* Another bridge, made of wood, 120 *ch'i* long and 22 broad, leads eastward to the wall of the Imperial Palace. A third bridge, a wooden draw-bridge 470 *ch'i* long, stretches to the west over the lake to its western border, where the palace *Hing-sheng kung* [built in 1308] stan is." (*Bretschneider, Peking,* 36.)—H. C.]

Yüan ch'eng.

CHAPTER XI.

Concerning the City of Cambaluc.

Now there was on that spot in old times a great and noble city called CAMBALUC, which is as much as to say in our tongue "The city of the Emperor."[1]   But the Great Kaan was informed by his Astrologers that this city would prove rebellious, and raise great disorders against his imperial authority.   So he caused the present city to be built close beside the old one, with only a river between them.[2]   And he caused the people of the old city to be removed to the new town that he had founded ; and this is called TAIDU. [However, he allowed a portion of the people which he did not suspect to remain in the old city, because the new one could not hold the whole of them, big as it is.]

As regards the size of this (new) city you must know that it has a compass of 24 miles, for each side of it hath a length of 6 miles, and it is four-square.   And it is all walled round with walls of earth which have a thickness of full ten paces at bottom, and a height of more than 10 paces ;[3] but they are not so thick at top, for they diminish in thickness as they rise, so that at top they are only about 3 paces thick.   And they are provided throughout with loop-holed battlements, which are all whitewashed.

There are 12 gates, and over each gate there is a great and handsome palace, so that there are on each side of the square three gates and five palaces ; for (I ought to mention) there is at each angle also a great and handsome palace.   In those palaces are vast halls in which are kept the arms of the city garrison.[4]

The streets are so straight and wide that you can

see right along them from end to end and from one gate to the other. And up and down the city there are beautiful palaces, and many great and fine hostelries, and fine houses in great numbers. [All the plots of ground on which the houses of the city are built are four-square, and laid out with straight lines; all the plots being occupied by great and spacious palaces, with courts and gardens of proportionate size. All these plots were assigned to different heads of families. Each square plot is encompassed by handsome streets for traffic; and thus the whole city is arranged in squares just like a chess-board, and disposed in a manner so perfect and masterly that it is impossible to give a description that should do it justice.][5]

Moreover, in the middle of the city there is a great clock—that is to say, a bell—which is struck at night. And after it has struck three times no one must go out in the city, unless it be for the needs of a woman in labour, or of the sick.[6] And those who go about on such errands are bound to carry lanterns with them. Moreover, the established guard at each gate of the city is 1000 armed men; not that you are to imagine this guard is kept up for fear of any attack, but only as a guard of honour for the Sovereign, who resides there, and to prevent thieves from doing mischief in the town.[7]

NOTE 1.— -!- The history of the city on the site of Peking goes back to very old times, for it had been [under the name of *Kí*] the capital of the kingdom of Yen, previous to B.C. 222, when it was captured by the Prince of the T'sin Dynasty. [Under the T'ang dynasty (618-907) it was known under the name of Yu-chau.] It became one of the capitals of the Khitans in A.D. 936, and of the Kin sovereigns, who took it in 1125, in 1151 under the name of Chung-tu. Under the name of Yenking, [given to this city in 1013] it has a conspicuous place in the wars of Chinghiz against the latter dynasty. He captured it in 1215. In 1264, Kúblái adopted it as his chief residence, and founded in 1267, the new city of TATU ("Great Court"), called by the Mongols TAIDU or DAITU since 1271 (see Bk. I. ch. lxi. note 1), at a little distance —Odoric says half a mile—to the north-east of the old Yenking. Tatu was completed in the summer of 1267.

Old Yenking had, when occupied by the Kin, a circuit of 27 *li* (commonly estimated at 9 miles, but in early works the *li* is not more than ⅛ of a mile), afterwards increased to 30 *li*. But there was some kind of outer wall about the city and its suburbs, the circuit of which is called 75 *li*. ["At the time of the Yuen the walls still existed, and the ancient city of the Kin was commonly called Nan-ch'eng (Southern city), whilst the Mongol capital was termed the northern city." *Bretschneider, Peking*, 10.—H. C.] (*Lockhart;* and see *Amyot*, II. 553, and note 6 to last chapter.)

Polo correctly explains the name *Cambaluc*, i.e. *Kaan-baligh*, "The City of the Kaan."

NOTE 2.—The river that ran between the old and new city must have been the little river *Yu*, which still runs through the modern Tartar city, and fills the city ditches.

[Dr. Bretschneider (*Peking*, 49) thinks that there is a strong probability that Polo speaks of the *Wen-ming ho*, a river which, according to the ancient descriptions, ran near the southern wall of the Mongol capital.—H. C.]

NOTE 3.—This height is from Pauthier's Text; the G. Text has, "*twenty* paces,"

South Gate of Imperial City at Peking.

"Elle a douze portes, et sor chascune porte a une grandisme palais et biaus."

i.e. 100 feet. A recent French paper states the dimensions of the existing walls as 14 mètres (45½ feet) high, and 14·50 (47¼ feet) thick, "the top forming a paved promenade, unique of its kind, and recalling the legendary walls of Thebes and Babylon." (*Ann. d'Hygiène Publique*, 2nd s. tom. xxxii. for 1869, p. 21.)

[According to the French astronomers (Fleuriais and Lapied) sent to Peking for the Transit of Venus in December, 1875, the present Tartar city is 23 kil. 55 in circuit, viz. if 1 *li* = 575 m., 41 *li;* from the north to the south 5400 mètres; from east to west 6700 mètres; the wall is 13 mètres in height and 12 mètres in width.—H. C.]

NOTE 4.—Our attempted plan of Cambaluc, as in 1290, differs somewhat from this description, but there is no getting over certain existing facts.

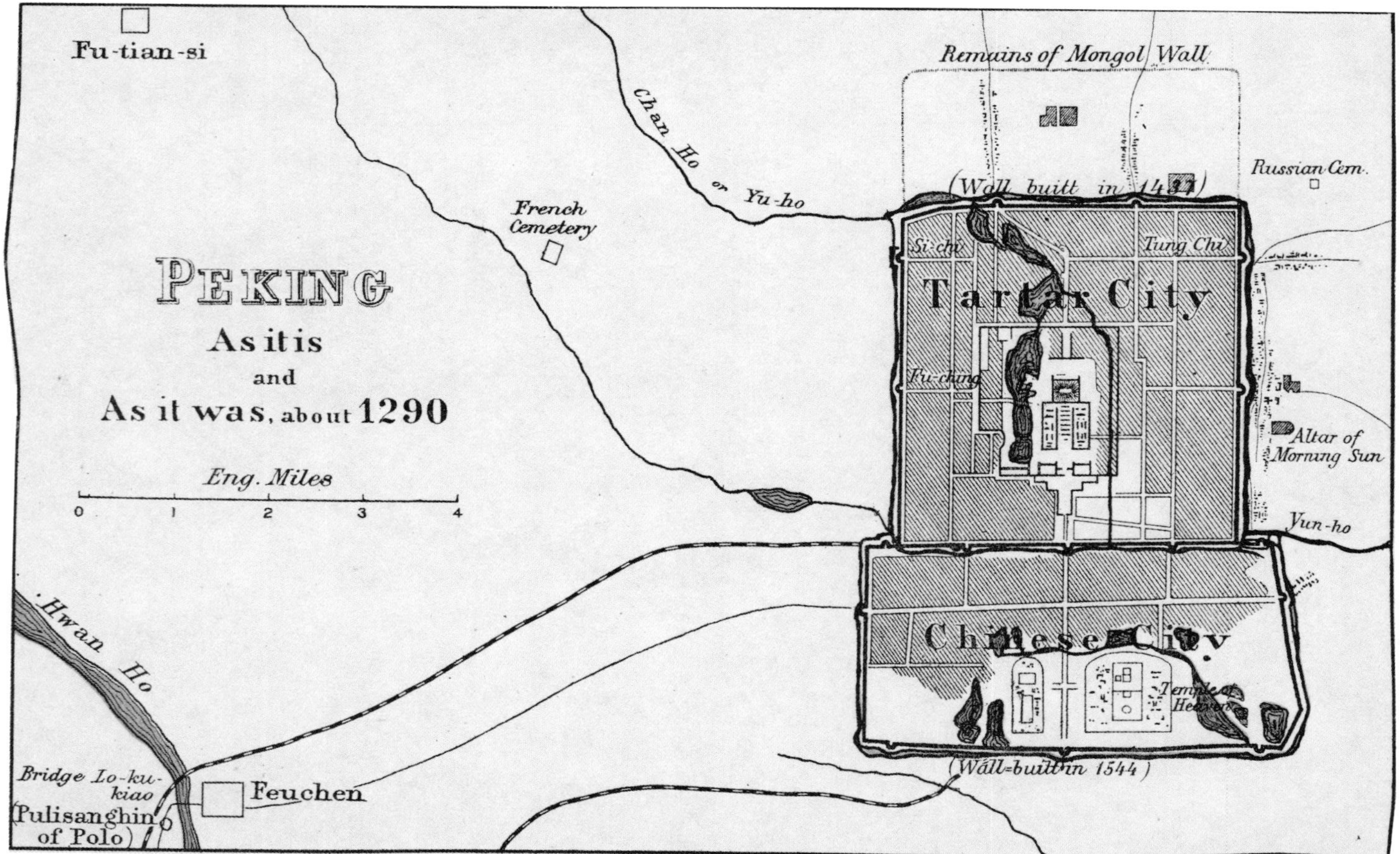

PEKING
As it is
and
As it was, about 1290

Eng. Miles

0   1   2   3   4

Fu-tian-si

Remains of Mongol Wall.

Russian Cem.

Altar of Morning Sun

Yun-ho

(Wall built in 1437)

Tung Che

Tartar City

Sz-chi

Hwang

Chinese City

Temple of Heaven

(Wall built in 1544)

Chan Ho or Yu-ho

French Cemetery

Hwan Ho

Bridge Lo-kou-kiao
Pulisanghin of Polo
Feuchen

[To face p. 376]

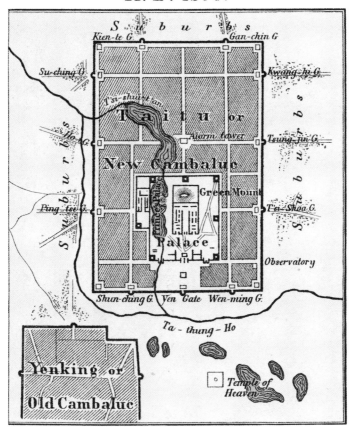

A. D. 1290.

The existing Tartar city of Peking (technically *Neï-ch'ing,* " The Interior City," or *King-ch'ing,* "City of the Court") stands on the site of Taidu, and represents it. After the expulsion of the Mongols (1368) the new native Dynasty of Ming established their capital at Nanking. But this was found so inconvenient that the third sovereign of the Dynasty re-occupied Taidu or Cambaluc, the repairs of which began in 1409. He reduced it in size by cutting off nearly a third part of the city at the north end. The remains of this abandoned portion of wall are, however, still in existence, approaching 30 feet in height all round. This old wall is called by the Chinese *The Wall of the Yuen (i.e.* the Mongol Dynasty), and it is laid down in the Russian Survey. [The capital of the Ming was 40 *li* in circuit, according to the *Ch'ang an k'o hua.*] The existing walls were built, or restored rather (the north wall being in any case, of course, entirely new), in 1437. There seems to be no doubt that the present south front of the Tartar city was the south front of Taidu. The whole outline of Taidu is therefore still extant, and easily measurable. If the scale on the War Office edition of the Russian Survey be correct, the long sides measure close upon 5 miles and 500 yards; the short sides, 3 miles and 1200 yards. Hence the whole perimeter was just about 18 English miles, or less than 16 Italian miles. If, however, a pair of compasses be run round Taidu and Yenking (as we have laid the latter down from such data as could be had) *together,* the circuit will be something like 24 Italian miles, and this may have to do with Polo's error.

[" The *Yuen shi* states that *Ta-tu* was 60 *li* in circumference. The *Ch'ue keng lu,* a work published at the close of the Yuen Dynasty, gives the same number of *li* for the circuit of the capital, but explains that *li* of 240 *pu* each are meant. If this statement be correct, it would give only 40 common or geographical *li* for the circuit of the Mongol town." (*Bretschneider, Peking,* 13.) Dr. Bretschneider writes (p. 20): " The outlines of Khanbaligh, partly in contradiction with the ancient Chinese records, if my view be correct, would have measured about 50 common *li* in circuit (13 *li* and more from north to south, 11·64 from east to west.")—H. C.]

Polo [and Odoric] again says that there were 12 gates—3 to every side. Both Gaubil and Martini also say that there were 12 gates. But I believe that both are trusting to Marco. There are 9 gates in the present Tartar city—viz. 3 on the south side and 2 on each of the other sides. The old Chinese accounts say there were 11 gates in Taidu. (See *Amyot, Mém.* II. 553.) I have in my plan, therefore, assumed that one gate on the east and one on the west were obliterated in the reduction of the *enceinte* by the Ming. But I must observe that Mr. Lockhart tells me he did not find the traces of gates in those positions, whilst the 2 gates on the *north* side of the old Mongol rampart are quite distinct, with the barbicans in front, and the old Mongol bridge over the ditch still serving for the public thoroughfare.*

[" The *Yuen shi* as well as the *Ch'ue keng lu,* and other works of the Yuen, agree in stating that the capital had eleven gates. They are enumerated in the following order : Southern wall—(1) The gate direct south (mid.) was called *Li-cheng men ;* (2) the gate to the left (east), *Wen-ming men ;* (3) the gate to the right (west), *Shun-ch'eng men.* Eastern wall—(4) The gate direct east (mid.), *Ch'ung-jen men ;* (5) the gate to the south-east, *Ts'i-hua men ;* (6) the gate to the north-east, *Kuang-hi men.* Western wall—(7) The gate direct west (mid.), *Ho-i men ;* (8) the gate to the south-west, *P'ing-tse men ;* (9) the gate to the north-west, *Su-ts'ing men.* Northern Wall—(10) The gate to the north-west, *K'ien-te men ;* (11) the gate to the north-east, *An-chen men.*" (*Bretschneider, Peking,* 13-14.)—H. C.]

When the Ming established themselves on the old Mongol site, population seems to have gathered close about the southern wall, probably using material from the remains of Yenking. This excrescence was inclosed by a new wall in 1554, and was

---

* Mr. Wylie confirms my assumption : " Whilst in Peking I traced the old mud wall, ... and found it quite in accordance with the outline in your map. Mr. Gilmour (a missionary to the Mongols) and I rode round it, he taking the outside and I the inside. . . . . Neither of us observed the arch that Dr. Lockhart speaks of. . . . . *There* are *gate-openings about the middle of the east and west sides,* but no barbicans." (4th December 1873.)

called the "Outer Town." It is what is called by Europeans the *Chinese City*. Its western wall exhibits in the base sculptured stones, which seem to have belonged to the old palace of Yenking. Some traces of Yenking still existed in Gaubil's time : the only relic of it now pointed out is a pagoda outside of the Kwang-An-Măn, or western gate of the Outer City, marked in the War Office edition of the Russian Map as "Tower." (Information from *Dr. Lockhart.*)

The "Great Palaces" over the gates and at the corner bastions are no doubt well illustrated by the buildings which still occupy those positions. There are two such lofty buildings at each of the gates of the modern city, the outer one (shown on p. 376) forming an elevated redoubt.

NOTE 5.—The French writer cited under note 3 says of the city as it stands : " La ville est de la sorte coupée en échiquier à peu près régulier dont les quadres circonscrits par des larges avenues sont percés eux-mêmes d'une multitude de rues et ruelles . . . qui toutes à peu prés sont orientées N. et S., E. et O. Une seule volonté a évidemment présidé à ce plan, et jamais édilité n'a eu à exécuter d'un seul coup aussi vaste entreprise."

NOTE 6.—Martini speaks of the public clock-towers in the Chinese cities, which in his time were furnished with water-clocks. A watchman struck the hour on a great gong, at the same time exhibiting the hour in large characters. The same person watched for fires, and summoned the public with his gong to aid in extinguishing them.

[The Rev. G. B. Farthing mentions (*North-China Herald*, 7th September, 1884) at T'ai-yuen fu the remains of an object in the bell-tower, which was, and is still known, as one of the eight wonders of this city ; it is a vessel of brass, a part of a water-clock from which water formerly used to flow down upon a drum beneath and mark off time into equal divisions.—H. C.]

The tower indicated by Marco appears still to exist. It occupies the place which I have marked as Alarm Tower in the plan of Taidu. It was erected in 1272, but probably rebuilt on the Ming occupation of the city. ["The *Yuen yi t'ung chi*, or 'Geography of the Mongol Empire' records : 'In the year 1272, the bell-tower and the drum-tower were built in the *middle* of the capital.' A bell-tower (*chung-lou*) and a drum-tower (*ku-lou*) exist still in Peking, in the northern part of the Tartar City. The *ku-lou* is the same as that built in the thirteenth century, but the bell-tower dates only from the last century. The bell-tower of the Yuen was a little to the east of the drum-tower, where now the temple *Wan-ning sse* stands. This temple is nearly in the middle of the position I (Bretschneider) assign to Khanbaligh." (*Bretschneider, Peking*, 20.)—H. C.] In the Court of the Old Observatory at Peking there is pre-served, with a few other ancient instruments, which date from the Mongol era, a very elaborate water-clock, provided with four copper basins embedded in brickwork, and rising in steps one above the other. A cut of this courtyard, with its instruments and aged trees, also ascribed to the Mongol time, will be found in ch. xxxiii. (*Atlas Sinensis*, p. 10 ; *Magaillans*, 149-151 ; *Chine Moderne*, p. 26 ; *Tour du Monde* for 1864, vol. ii. p. 34.)

NOTE 7.—"Nevertheless," adds the Ramusian, "there does exist I know not what uneasiness about the people of Cathay."

CHAPTER XII.

HOW THE GREAT KAAN MAINTAINS A GUARD OF TWELVE THOUSAND HORSE, WHICH ARE CALLED KESHICAN.

YOU must know that the Great Kaan, to maintain his state, hath a guard of twelve thousand horsemen, who are styled KESHICAN, which is as much as to say " Knights devoted to their Lord." Not that he keeps these for fear of any man whatever, but merely because of his own exalted dignity. These 12,000 men have four captains, each of whom is in command of 3000 ; and each body of 3000 takes a turn of three days and nights to guard the palace, where they also take their meals. After the expiration of three days and nights they are relieved by another 3000, who mount guard for the same space of time, and then another body takes its turn, so that there are always 3000 on guard. Thus it goes until the whole 12,000, who are styled (as I said) Keshican, have been on duty ; and then the tour begins again, and so runs on from year's end to year's end." [1]

NOTE 1.—I have *deduced* a reading for the word *Quescican* (Keshican), which is not found precisely in any text. Pauthier reads *Questiau* and *Quesitau ;* the G. Text has *Quesitam* and *Quecitain ;* the Crusca *Questi Tan ;* Ramusio, *Casitan ;* the Riccardiana, *Quescitam.* Recollecting the constant clerical confusion between *c* and *t*, what follows will leave no doubt I think that the true reading to which all these variations point is *Quescican.* *

In the Institutes of Ghazan Khan, we find established among other formalities for the authentication of the royal orders, that they should be stamped on the back, in black ink, with the seals of the *Four Commanders* of the *Four Kiziks*, or *Corps of the Life Guard.*

Wassáf also, in detailing the different classes of the great dignitaries of the Mongol monarchy, names (1) the *Noyáns* of the Ulus, or princes of the blood ; (2) the great chiefs of the tribes ; (3) the *Amírs of the four Keshik*, or *Corps of the Body Guard;* (4) the officers of the army, commanding ten thousands, thousands, and so on.

Moreover, in Rashiduddin, we find the identical plural form used by our author. He says that, after the sack of Baghdad, Hulaku, who had escaped from the polluted atmosphere of the city, sent " Ilká Noyán and Ḳarábúgá, with 3000 Moghul horse

---

* One of the nearest readings is that of the Brandenburg Latin collated by Müller, which has *Quaesicam.*

into Baghdad, in order to have the buildings repaired, and to put things generally in order. These chiefs posted sentries from the KISHÍKÁN (ڪشــگان), and from their own followings in the different quarters of the town, had the carcases of beasts removed from the streets, and caused the bazaars to be rebuilt."

We find *Kishik* still used at the court of Hindustan, under the great kings of Timur's House, for the corps on tour of duty at the palace ; and even for the sets of matchlocks and sabres, which were changed weekly from Akbar's armoury for the royal use. The royal guards in Persia, who watch the king's person at night, are termed *Keshikchi*, and their captain *Keshikchi Bashi.* [" On the night of the 11th of Jemady ul Sany, A.H. 1160 (or 8th June, 1747), near the city of Khojoon, three days' journey from Meshed, Mohammed Kuly Khan Ardemee, who was of the same tribe with Nadir Shah, his relation, and Kushukchee Bashee, with seventy of the *Kukshek* or guard, . . . bound themselves by an oath to assassinate Nadir Shah." (*Memoirs of Khojeh Abdulkurreem . . . transl. by F. Gladwin*, Calcutta, 1788, pp. 166-167).]

Friar Odoric speaks of the four barons who kept watch by the Great Kaan's side as the *Cuthé*, which probably represents the Chinese form *Kiesie* (as in De Mailla), or *Kuesie* (as in Gaubil). The latter applies the term to four devoted champions of Chinghiz, and their descendants, who were always attached to the Kaan's body-guard, and he identifies them with the *Quesitan* of Polo, or rather with the captains of the latter ; adding expressly that the word *Kuesie* is Mongol.

I see *Kishik* is a proper name among the Kalmak chiefs ; and *Keshikten* also is the name of a Mongol tribe, whose territory lies due north of Peking, near the old site of Shangtu. (Bk. I. ch. lxi.) [*Keshikhteng*, a tribe (*pu ;* mong. *aimak*) of the Chao Uda League (*mêng ;* mong. *chogolgân*) among the twenty-four tribes of the *Nei Mung-ku* (Inner Mongols). (See *Mayers' Chinese Government*, p. 81.)— H. C.] In Kovalevsky, I find the following :—

(No. 2459) " *Keshik*, grace, favour, bounty, benefit, good fortune, charity."
(No. 2461) " *Keshikten*, fortunate, happy, blessed."
(No. 2541) " *Kichyeku*, to be zealous, assiduous, devoted."
(No. 2588) " *Kushiku*, to hinder, to bar the way to," etc.

The third of these corresponds closely with Polo's etymology of " knights devoted to their lord," but perhaps either the first or the last may afford the real derivation.

In spite of the different initials (ق instead of ك), it can scarcely be doubted that the *Ḳalchi* and *Ḳalaḳchi* of Timur's Institutes are mere mistranscriptions of the same word, *e.g.:* " I ordered that 12,000 *Ḳalchi*, men of the sword completely armed, should be cantoned in the Palace ; to the right and to the left, to the front, and in the rear of the imperial diwán ; thus, that 1000 of those 12,000 should be every night upon guard," etc. The translator's note says of *Ḳalchi*, " A Mogul word supposed to mean *guards.*" We see that even the traditional number of 12,000, and its division into four brigades, are maintained. (See *Timour's Inst.*, pp. 299 and 235, 237.)

I must add that Professor Vámbéry does not assent to the form *Keshikán*, on the ground that this Persian plural is impossible in an old Tartar dialect, and he supposes the true word to be *Kechilan* or *Kechiklen*, " the night-watchers," from *Kiche* or *Kichek* (Chag. and Uighúr), = " night."

I believe, however, that Persian was the colloquial language of foreigners at the Kaan's court, who would not scruple to make a Persian plural when wanted ; whilst Rashid has exemplified the actual use of this one.

(*D'Ohsson*, IV. 410 ; *Gold. Horde*, 228, 238 ; *Ilch.* II. 184 ; *Q. R.* pp. 308-309 ; *Ayeen Akb.* I. 270, and *Blochmann's*, p. 115 ; *J. As.* sèr. IV. tom. xix. 276 ; *Olearius*, ed. 1659, I. 656 ; *Cathay*, 135 ; *De Mailla*, ix. 106 ; *Gaubil*, p. 6 ; *Pallas, Samml.* I. 35.)

["By *Keshican* in *Colonel Yule's Marco Polo, Keshikten* is evidently meant. This is a general Mongol term to designate the Khan's lifeguard. It is derived from the word *Keshik,* meaning a guard by turns ; a corps on tour of duty. *Keshik* is one of the archaisms of the Mongol language, for now this word has another meaning in Mongol. Colonel Yule has brought together several explanations of the term. It seems to me that among his suppositions the following is the most consistent with the ancient meaning of the word : —

" We find *Kishik* still used at the court of Hindustan, under the great kings of Timur's House, for the corps on tour of duty at the palace. . . . . The royal guards in Persia, who watch the King's person at night, are termed *Keshikchi.*"

" The Keshikten was divided into a day-watch called *Turgaut* and a night-watch *Kebteul.* The Kebte-ul consisted of pure Mongols, whilst the Turgaut was composed of the sons of the vassal princes and governors of the provinces, and of hostages. The watch of the Khan was changed every three days, and contained 400 men. In 1330 it was reduced to 100 men." (*Palladius,* 42-43.) Mr. E. H. Parker writes in the *China Review,* XVIII. p. 262, that they "are evidently the 'body guards' of the modern viceroys, now pronounced Kashiha, but, evidently, originally *Keshigha.*" —H. C.]

CHAPTER XIII.

THE FASHION OF THE GREAT KAAN'S TABLE AT HIS HIGH FEASTS.

AND when the Great Kaan sits at table on any great court occasion, it is in this fashion. His table is elevated a good deal above the others, and he sits at the north end of the hall, looking towards the south, with his chief wife beside him on the left. On his right sit his sons and his nephews, and other kinsmen of the Blood Imperial, but lower, so that their heads are on a level with the Emperor's feet. And then the other Barons sit at other tables lower still. So also with the women ; for all the wives of the Lord's sons, and of his nephews and other kinsmen, sit at the lower table to his right ; and below them again the ladies of the other Barons and Knights, each in the place assigned by the Lord's orders. The tables are so disposed that the Emperor can see the whole of them from end to end, many as they are.[1] [Further, you are not to suppose that everybody sits at table ; on the contrary, the greater part of the soldiers and their officers sit at their meal in the hall on the carpets.] Outside the hall will be found more than 40,000 people; for there is a great concourse of folk bringing presents to the Lord, or come from foreign countries with curiosities.

In a certain part of the hall near where the Great Kaan holds his table, there [is set a large and very beautiful piece of workmanship in the form of a square coffer, or buffet, about three paces each way, exquisitely wrought with figures of animals, finely carved and gilt. The middle is hollow, and in it] stands a great vessel of pure gold, holding as much as an ordinary butt; and at each corner of the great vessel is one of smaller size [of the capacity of a firkin], and from the former the wine or beverage flavoured with fine and costly spices is drawn off into the latter. [And on the buffet aforesaid are set all the Lord's drinking vessels, among which are certain pitchers of the finest gold,] which are called *verniques*,[2] and are big enough to hold drink for eight or ten persons. And one of these is put between every two persons, besides a couple of golden cups with handles, so that every man helps himself from the pitcher that stands between him and his neighbour. And the ladies are supplied in the same way. The value of these pitchers and cups is something immense; in fact, the Great Kaan has such a quantity of this kind of plate, and of gold and silver in other shapes, as no one ever before saw or heard tell of, or could believe.[3]

[There are certain Barons specially deputed to see that foreigners, who do not know the customs of the Court, are provided with places suited to their rank; and these Barons are continually moving to and fro in the hall, looking to the wants of the guests at table, and causing the servants to supply them promptly with wine, milk, meat, or whatever they lack. At every door of the hall (or, indeed, wherever the Emperor may be) there stand a couple of big men like giants, one on each side, armed with staves. Their business is to see that no one steps upon the threshold in entering, and if this does happen, they strip the offender of his clothes, and he must pay a forfeit to have them back again ; or in lieu of taking his clothes, they give him a certain number of blows. If they are foreigners ignorant of the order, then there are Barons appointed to introduce them, and explain it to them. They think, in fact, that it brings bad luck if any one touches the threshold. Howbeit, they are not expected to stick at this in going forth again, for at that time some are like to be the worse for liquor, and incapable of looking to their steps.[4]]

And you must know that those who wait upon the Great Kaan with his dishes and his drink are some of the great Barons. They have the mouth and nose muffled with fine napkins of silk and gold, so that no breath nor odour from their persons should taint the dish or the goblet presented to the Lord. And when the Emperor is going to drink, all the musical instruments, of which he has vast store of every kind, begin to play. And when he takes the cup all the Barons and the rest of the company drop on their knees and make the deepest obeisance before him, and then the Emperor doth drink. But each time that he does so the whole ceremony is repeated.[5]

I will say nought about the disnes, as you may easily conceive that there is a great plenty of every possible kind. But you should know that in every case where a Baron or Knight dines at those tables, their wives also dine there with the other ladies. And when all have dined and the tables have been removed, then come in a great number of players and jugglers, adepts at all sorts of wonderful feats,[6] and perform before the Emperor and the rest of the company, creating great diversion and mirth, so that everybody is full of laughter and enjoyment.   And when the performance is over, the company breaks up and every one goes to his quarters.

Note 1.—We are to conceive of rows of small tables, at each of which were set probably but two guests.   This seems to be the modern Chinese practice, and to go back to some very old accounts of the Tartar nations.   Such tables we find in use in the tenth century, at the court of the King of Bolghar (see *Prologue*, note 2, ch. ii.), and at the Chinese entertainments to Shah Rukh's embassy in the fifteenth century.   Megasthenes described the guests at an Indian banquet as having a table set before each individual.   (*Athenaeus*, IV. 39, *Yonge's Transl.*)

[Compare Rubruck's account, Rockhill's ed., p. 210: "The Chan sits in a high place to the north, so that he can be seen by all . . . ." (See also Friar Odoric, *Cathay*, p. 141.)—H. C.]

Note 2.—This word (G. T. and Ram.) is in the Crusca Italian transformed into an adjective, "*vaselle* vernicate *d'oro*," and both Marsden and Pauthier have substantially adopted the same interpretation, which seems to me in contradiction with the text.   In Pauthier's text the word is *vernigal*, pl. *vernigaux*, which he explains, I know not on what authority, as "*coupes sans anses vernies ou laquées d'or.*"   There is, indeed, a Venetian sea-term, *Vernegal*, applied to a wooden bowl in which the food of a mess is put, and it seems possible that this word may have been substituted for the unknown *Vernique.*   I suspect the latter was some Oriental term, but I can find nothing nearer than the Persian *Bărni*, Ar. *Al-Bărníya*, "vas fictile in quo quid recondunt," whence the Spanish word *Albornia*, "a great glazed vessel in the shape of a bowl, with handles."   So far as regards the form, the change of *Barniya* into *Vernique* would be quite analogous to that change of *Hundwáníy* into *Ondanique*, which we have already met with. (See *Dozy et Engelmann, Glos. des Mots Espagnols*, etc., 2nd ed., 1867, p. 73; and *Boerio, Diz. del. Dial. Venez.*)

[*F. Godefroy, Dict., s.v. Vernigal*, writes: "Coupe sans anse, vernie ou laquée d'or," and quotes, besides Marco Polo, the *Regle du Temple*, p. 214, éd. Soc. Hist. de France:

"Les *vernigaus* et les escuelles."

About *vernegal*, cf. *Rockhill, Rubruck*, p. 86, note.   Rubruck says (*Soc. de Géog.* p. 241): "Implevimus unum *veringal* de biscocto et platellum unum de pomis et aliis fructibus."   Mr. Rockhill translates *veringal* by basket.

Dr. Bretschneider (*Peking*, 28) mentions "a large jar made of wood and *varnished*, the inside lined with silver," and he adds in a note "perhaps this statement may serve to explain Marco Polo's *verniques* or *vaselle* vernicate *d'oro*, big enough to hold drink for eight or ten persons."—H. C.]

A few lines above we have "of the capacity of a *firkin*."   The word is *bigoncio*, which is explained in the *Vocab. Univ. Ital.* as a kind of tub used in the vintage, and containing 3 *mine*, each of half a *stajo*.   This seems to point to the *Tuscan* mina, or half stajo, which is = ⅓ of a bushel.   Hence the *bigoncio* would = a bushel, or, in old liquid measure, about a firkin.

Note 3.—A buffet, with flagons of liquor and goblets, was an essential feature in the public halls or tents of the Mongols and other Asiatic races of kindred manners. The ambassadors of the Emperor Justin relate that in the middle of the pavilion of Dizabulus, the Khan of the Turks, there were set out drinking-vessels, and flagons and great jars, all of gold; corresponding to the *coupes* (or *hanas à mances*), the *verniques*, and the *grant peitere* and *petietes peiteres* of Polo's account.   Rubruquis describes in Batu Khan's tent a buffet near the entrance, where *Kumiz* was set forth, with great goblets of gold and silver, etc., and the like at the tent of the Great Kaan. At a festival at the court of Oljaitu, we are told, "Before the throne stood golden buffets . . . set out with full flagons and goblets." Even in the private huts of the Mongols there was a buffet of a humbler kind exhibiting a skin of *Kumiz*, with other kinds of drink, and cups standing ready ; and in a later age at the banquets of Shah Abbas we find the great buffet in a slightly different form, and the golden flagon still set to every two persons, though it no longer contained the liquor, which was handed round. (*Cathay*, clxiv., cci. ; *Rubr.* 224, 268, 305 ; *Ilch.* II. 183 ; *Della Valle*, I. 654 and 750-751.)

[Referring to the "large and very beautiful piece of workmanship," Mr. Rockhill, *Rubruck*, 208-209, writes : "Similar works of art and mechanical contrivances were often seen in Eastern courts. The earliest I know of is the golden plane-tree and grape vine with bunches of grapes in precious stones, which was given to Darius by Pythius the Lydian, and which shaded the king's couch. (Herodotus, IV. 24.) The most celebrated, however, and that which may have inspired Mangu with the desire to have something like it at his court, was the famous Throne of Solomon (Σολομῶντεος Θρόνος) of the Emperor of Constantinople, Theophilus (A.D. 829-842). . . . Abulfeda states that in A.D. 917 the envoys of Constantine Porphyrogenitus to the Caliph el Moktader saw in the palace of Bagdad a tree with eighteen branches, some of gold, some of silver, and on them were gold and silver birds, and the leaves of the tree were of gold and silver. By means of machinery, the leaves were made to rustle and the birds to sing. Mirkhond speaks also of a tree of gold and precious stones in the city of Sultanieh, in the interior of which were conduits through which flowed drinks of different kinds. Clavijo describes a somewhat similar tree at the court of Timur."

Dr. Bretschneider (*Peking*, 28, 29) mentions a clepsydra with a lantern. By means of machinery put in motion by water, at fixed times a little man comes forward exhibit-ing a tablet, which announces the hours. He speaks also of a musical instrument which is connected, by means of a tube, with two peacocks sitting on a cross-bar, and when it plays, the mechanism causes the peacocks to dance.—H. C.]

Odoric describes the great jar of liquor in the middle of the palace hall, but in his time it was made of a great mass of jade (p. 130).

NOTE 4.—This etiquette is specially noticed also by Odoric, as well as by Makrizi, by Rubruquis, and by Plano Carpini. According to the latter the breach of it was liable to be punished with death. The prohibition to tread on the threshold is also specially mentioned in a Mahomedan account of an embassy to the court of Barka Khan. And in regard to the tents, Rubruquis says he was warned not to touch the ropes, for these were regarded as representing the threshold. A Russo-Mongol author of our day says that the memory of this etiquette or superstition is still preserved by a Mongol pro-verb : "Step not on the threshold ; it is a sin !" But among some of the Mongols more than this survives, as is evident from a passage in Mr. Michie's narrative : "There is a right and a wrong way of approaching a *yourt* also. Outside the door there are generally ropes lying on the ground, held down by stakes, for the purpose of tying up the animals when they want to keep them together. There is a way of getting over or round these ropes that I never learned, but on one occasion the ignorant breach of the rule on our part excluded us from the hospitality of the family." The feeling or superstition was in full force in Persia in the 17th century, at least in regard to the threshold of the king's palace. It was held a sin to tread upon it in entering. (*Cathay*, 132 ; *Rubr.* 255, 268, 319 ; *Plan. Carp.* 625, 741 ; *Makrizi*, I. 214 ; *Mél. Asiat. Ac. St. Petersb.* II. 660 ; *The Siberian Overland Route*, p. 97 ; *P. Della Valle*, II. 171.)

[Mr. Rockhill writes (*Rubruck*, p. 104) : "The same custom existed among the Fijians, I believe. I may note that it also prevailed in ancient China. It is said of Confucius 'when he was standing he did not occupy the middle of the gate-way ; when he passed in or out, he did not tread on the threshold.' (*Lun-yü*, Bk. X. ch.

iv. 2.) In China, the bride's feet must not touch the threshold of the bridegroom's house. (Cf. *Dennys' Folk-lore in China*, p. 18.)

"The author of the *Ch'ue keng lu* mentions also the athletes with clubs standing at the door, at the time of the khan's presence in the hall. He adds, that next to the Khan, two other life-guards used to stand, who held in their hands 'natural' axes of jade (axes found fortuitously in the ground, probably primitive weapons)." (*Palladius*, p. 43.)—H. C.]

NOTE 5.—Some of these etiquettes were probably rather Chinese than Mongol, for the regulations of the court of Kúblái apparently combined the two. In the visit of Shah Rukh's ambassadors to the court of the Emperor Ch'êng Tsu of the Ming Dynasty in 1421, we are told that by the side of the throne, at an imperial banquet, "there stood two eunuchs, each having a band of thick paper over his mouth, and extending to the tips of his ears. . . . . Every time that a dish, or a cup of *darassun* (rice-wine) was brought to the emperor, all the music sounded." (*N. et Ext.* XIV. 408, 409.) In one of the Persepolitan sculptures, there stands behind the King an eunuch bearing a fan, and with his mouth covered ; at least so says Heeren. (*Asia*, I. 178.)

NOTE 6.—"*Jougleours et entregetours de maintes plusieurs manieres de granz experimenz*" (P.); "*de Giuculer et de Tregiteor*" (G. T.). Ital. *Tragettatore*, a juggler ; Romance, *Trasjitar, Tragitar*, to juggle. Thus Chaucer :—

> "There saw I playing Jogelours,
> Magiciens, and *Tragetours*,
> And Phetonisses, Charmeresses,
> Old Witches, Sorceresses," etc.
>   —*House of Fame*, III. 169.

And again:—

> "For oft at festes have I wel herd say,
> That *Tregetoures*, within an halle large,
> Have made come in a water and a barge,
> And in the halle rowen up and doun.
> Somtime hath semed come a grim leoun ;
> *   *   *   *   *
> Somtime a Castel al of lime and ston,
> And whan hem liketh, voideth it anon."
>   —*The Franklin's Tale*, II. 454.

Performances of this kind at Chinese festivities have already been spoken of in note 9 to ch. lxi. of Book I. Shah Rukh's people, Odoric, Ysbrandt Ides, etc., describe them also. The practice of introducing such *artistes* into the dining-hall after dinner seems in that age to have been usual also in Europe. See, for example, *Wright's Domestic Manners*, pp. 165-166, and the Court of the Emperor Frederic II., in *Kington's Life* of that prince, I. 470. (See also *N. et E.* XIV. 410; *Cathay*, 143; *Ysb. Ides*, p. 95.)

CHAPTER XIV.

CONCERNING THE GREAT FEAST HELD BY THE GRAND KAAN EVERY YEAR ON HIS BIRTHDAY.

YOU must know that the Tartars keep high festival yearly on their birthdays. And the Great Kaan was born on the 28th day of the September moon, so on that day is held the greatest feast of the year at the Kaan's Court, always excepting that which he holds on New Year's Day, of which I shall tell you afterwards.[1]

Now, on his birthday, the Great Kaan dresses in the best of his robes, all wrought with beaten gold;[2] and full 12,000 Barons and Knights on that day come forth dressed in robes of the same colour, and precisely like those of the Great Kaan, except that they are not so costly; but still they are all of the same colour as his, and are also of silk and gold. Every man so clothed has also a girdle of gold; and this as well as the dress is given him by the Sovereign. And I will aver that there are some of these suits decked with so many pearls and precious stones that a single suit shall be worth full 10,000 golden bezants.

And of such raiment there are several sets. For you must know that the Great Kaan, thirteen times in the year, presents to his Barons and Knights such suits of raiment as I am speaking of.[3] And on each occasion they wear the same colour that he does, a different colour being assigned to each festival. Hence you may see what a huge business it is, and that there is no prince in the world but he alone who could keep up such customs as these.

On his birthday also, all the Tartars in the world, and all the countries and governments that owe allegiance to the Kaan, offer him great presents according to their several ability, and as prescription or orders have fixed the amount. And many other persons also come with great presents to the Kaan, in order to beg for some employment from him. And the Great Kaan has chosen twelve Barons on whom is laid the charge of assigning to each of these supplicants a suitable answer.

On this day likewise all the Idolaters, all the

Saracens, and all the Christians and other descriptions of people make great and solemn devotions, with much chaunting and lighting of lamps and burning of incense, each to the God whom he doth worship, praying that He would save the Emperor, and grant him long life and health and happiness.

And thus, as I have related, is celebrated the joyous feast of the Kaan's birthday.[4]

Now I will tell you of another festival which the Kaan holds at the New Year, and which is called the White Feast.

NOTE 1.—The Chinese Year commences, according to Duhalde, with the New Moon nearest to the Sun's Passage of the middle point of Aquarius; according to Pauthier, with the New Moon immediately preceding the Sun's entry into Pisces. (These would almost always be identical, but not always.) Generally speaking, the first month will include part of February and part of March. The eighth month will then be September-October (*v. ante*, ch. ii. note 2).

[According to Dr. S. W. Williams (*Middle Kingdom*, II. p. 70): "The year is lunar, but its commencement is regulated by the sun. New Year falls on the first new moon after the sun enters Aquarius, which makes it come not before January 21st nor after February 19th." "The beginning of the civil year, writes Peter Hoang (*Chinese Calendar*, p. 13), depends upon the good pleasure of the Emperors. Under the Emperor Hwang-ti (2697 B.C.) and under the Hsia Dynasty (2205 B.C.), it was made to commence with the 3rd month *yin-yüeh* [Pisces]; under the Shang Dynasty (1766 B.C.) with the 2nd month *ch'ou-yüeh* [Aquarius], and under the Chou Dynasty (1122 B.C.) with the 1st month *tzu-yüeh* [Capricorn]."—H. C.]

NOTE 2.—The expression "*à or batuz*" as here applied to robes, is common among the mediæval poets and romance-writers, *e.g.* Chaucer :—

> " Full yong he was and merry of thought,
> And in samette with birdes wrought
> And with gold beaten full fetously,
> His bodie was clad full richely."
> —*Rom. of the Rose*, 836-839.

M. Michel thinks that in a stuff so termed the gold wire was *beaten out* after the execution of the embroidery, a process which widened the metallic surface and gave great richness of appearance. The fact was rather, however, according to Dr. Rock, that the gold used in weaving such tissues was *not* wire but beaten sheets of gold cut into narrow strips. This would seem sufficient to explain the term "beaten gold," though Dr. Rock in another passage refers it to a custom which he alleges of sewing goldsmith's work upon robes. (*Fr. Michel, Recherches*, II. 389, also I. 371; *Rock's Catalogue*, pp. xxv. xxix. xxxviii. cvi.)

NOTE 3.—The number of these festivals and distributions of dresses is *thirteen* in all the old texts, except the Latin of the Geog. Soc., which has *twelve*. Thirteen would seem therefore to have been in the original copy. And the Ramusian version expands this by saying, "Thirteen great feasts that the Tartars keep with much solemnity to each of the thirteen moons of the year." * It is possible, however, that this latter sentence is an interpolated gloss ; for, besides the improbability of munificence so frequent, Pauthier has shown some good reasons why *thirteen* should be regarded as an error for *three*. The official History of the Mongol Dynasty, which he quotes, gives a detail of raiment distributed in presents on great state occasions *three* times a year. Such a mistake might easily have originated in the first dictation, *treize* substituted for *trois*, or rather for the old form *tres ;* but we must note that the number 13 is repeated and corroborated in ch. xvi. Odoric speaks of *four* great yearly festivals, but there are obvious errors in what he says on this subject. Hammer says the great Mongol Feasts were three, viz. New Year's Day, the Kaan's Birthday, and the Feast of the Herds.

Something like the changes of costume here spoken of is mentioned by Rubruquis at a great festival of four days' duration at the court of Mangku Kaan : " Each day of the four they appeared in different raiment, suits of which were given them for each day of a different colour, but everything on the same day of one colour, from the boots to the turban." So also Carpini says regarding the assemblies of the Mongol nobles at the inauguration of Kuyuk Kaan : " The first day they were all clad in white pourpre (? *albis purpuris*, see Bk. I. ch. vi. note 4), the second day in ruby pourpre, the third day in blue pourpre, the fourth day in the finest baudekins." (*Cathay*, 141 ; *Rubr.* 368 ; *Pl. Car.* 755.)

[Mr. Rockhill (*Rubruck*, p. 247, note) makes the following remarks : " Odoric, however, says that the colours differed according to the rank. The custom of presenting *khilats* is still observed in Central Asia and Persia. I cannot learn from any other authority that the Mongols ever wore turbans. Odoric says the Mongols of the imperial feasts wore ' coronets' (*in capite coronati*)."—H. C.]

NOTE 4.—[" The accounts given by Marco Polo regarding the feasts of the Khan and the festival dresses at his Court, agree perfectly with the statements on the same subject of contemporary Chinese writers. Banquets were called in the common Mongol language *chama*, and festival dresses *chisun*. General festivals used to be held at the New Year and at the Birthday of the Khan. In the *Mongol-Chinese Code*, the ceremonies performed in the provinces on the Khan's Birthday are described. One month before that day the civil and military officers repaired to a temple, where a service was performed for the Khan's health. On the morning of the Birthday a sumptuously adorned table was placed in the open air, and the representatives of all classes and all confessions were obliged to approach the table, to prostrate themselves and exclaim three times : *Wan-sui* (*i.e.* ' Ten thousand years' life to the Khan). After that the banquet took place. In the same code (in the article on the *Ye li ke un* [Christians, *Erke-un*]) it is stated, that in the year 1304,—owing to a dispute, which had arisen in the province of Kiang-nan between the *ho-shang* (Buddhist priests) and the Christian missionaries, as to precedence in the above-mentioned ceremony,—a special edict was published, in which it was decided that in the rite of supplication, Christians should follow the Buddhist and Taouist priests." (*Palladius*, pp. 44-45.) —H. C.]

---

* There are thirteen months to the Chinese year in seven out of every nineteen.
[" This interval of 10 years comprises 235 lunar months generally 125 *long* months of 30 days 110 *short* months of 29 days, (but sometimes 124 *long* and 111 *short* months), and 7 *intercalary* months. The year of twelve months is called a common year, that of thirteen months, an *intercalary* year." (*P. Hoang, Chinese Calendar,* p. 12.—H. C.)]

CHAPTER XV.

OF THE GREAT FESTIVAL WHICH THE KAAN HOLDS ON NEW YEAR'S DAY.

THE beginning of their New Year is the month of February, and on that occasion the Great Kaan and all his subjects made such a Feast as I now shall describe.

It is the custom that on this occasion the Kaan and all his subjects should be clothed entirely in white; so, that day, everybody is in white, men and women, great and small. And this is done in order that they may thrive all through the year, for they deem that white clothing is lucky.[1] On that day also all the people of all the provinces and governments and kingdoms and countries that own allegiance to the Kaan bring him great presents of gold and silver, and pearls and gems, and rich textures of divers kinds. And this they do that the Emperor throughout the year may have abundance of treasure and enjoyment without care. And the people also make presents to each other of white things, and embrace and kiss and make merry, and wish each other happiness and good luck for the coming year. On that day, I can assure you, among the customary presents there shall be offered to the Kaan from various quarters more than 100,000 white horses, beautiful animals, and richly caparisoned. [And you must know 'tis their custom in offering presents to the Great Kaan (at least when the province making the present is able to do so), to present nine times nine articles. For instance, if a province sends horses, it sends nine times nine or 81 horses; of gold, nine times nine pieces of gold, and so with stuffs or whatever else the present may consist of.][2]

On that day also, the whole of the Kaan's elephants, amounting fully to 5000 in number, are exhibited, all covered with rich and gay housings of inlaid cloth repre-. senting beasts and birds, whilst each of them carries on his back two splendid coffers; all of these being filled. with the Emperor's plate and other costly furniture required for the Court on the occasion of the White Feast.[3] And these are followed by a vast number of camels which are likewise covered with rich housings and laden with things needful for the Feast. All these are paraded before the Emperor, and it makes the finest sight in the world.

Moreover, on the morning of the Feast, before the tables are set, all the Kings, and all the Dukes, Marquesses, Counts, Barons, Knights, and Astrologers, and Philosophers, and Leeches, and Falconers, and other officials of sundry kinds from all the places round about, present themselves in the Great Hall before the Emperor; whilst those who can find no room to enter stand outside in such a position that the Emperor can see them all well. And the whole company is marshalled in this wise. First are the Kaan's sons, and his nephews, and the other Princes of the Blood Imperial; next to them all Kings; then Dukes, and then all others in succession according to the degree of each. And when they are all seated, each in his proper place, then a great prelate rises and says with a loud voice: " Bow and adore!" And as soon as he has said this, the company bow down until their foreheads touch the earth in adoration towards the Emperor as if he were a god. And this adoration they repeat four times, and then go to a highly decorated altar, on which is a vermilion tablet with the name of the Grand Kaan inscribed thereon, and a beautiful censer of gold. So they incense the tablet and the altar with great reverence, and then return each man to his seat.[4]

When all have performed this, then the presents are offered, of which I have spoken as being so rich and costly. And after all have been offered and been seen by the Emperor, the tables are set, and all take their places at them with perfect order as I have already told you. And after dinner the jugglers come in and amuse the Court as you have heard before; and when that is over, every man goes to his quarters.

NOTE 1.—The first month of the year is still called by the Mongols *Chaghan* or *Chaghan Sara*, "the White" or the "White Month"; and the wearing of white clothing on this festive occasion must have been purely a Mongol custom. For when Shah Rukh's ambassadors were present at the New Year's Feast at the Court of the succeeding *Chinese* Dynasty (2nd February, 1421) they were warned that *no one* must wear white, as that among the Chinese was the colour of mourning. (*Koeppen*, I. 574, II. 309; *Cathay*, p. ccvii.)

NOTE 2.—On the mystic importance attached to the number 9 on all such occasions among the Mongols, see *Hammer's Golden Horde*, p. 208; *Hayton*, ch. iii. in Ramusio II.; *Not. et Ext.* XIV. Pt. I. 32; and *Strahlenberg* (II. 210 of Amsterd. ed. 1757). Vámbéry, speaking of the *Kálín* or marriage price among the Uzbegs, says: "The question is always how many times *nine* sheep, cows, camels, or horses, or how many times nine ducats (as is the custom in a town), the father is to receive for giving up his daughter." (*Sketches of Cent. Asia*, p. 103.) Sheikh Ibrahim of Darband, making offerings to Timur, presented *nines* of everything else, but of slaves *eight* only. "Where is the ninth?" enquired the court official. "Who but I myself?" said the Sheikh, and so won the heart of Timur. (*A. Arabsiadis . . . . Timuri Hist.* p. 357.)

NOTE 3.—The elephant stud of the Son of Heaven had dwindled till in 1862 Dr. Rennie found but one animal; now none remain. [Dr. S. W. Williams writes (*Middle Kingdom*, I. pp. 323-324): "Elephants are kept at Peking for show, and are used to draw the state chariot when the Emperor goes to worship at the Altars of Heaven and Earth, but the sixty animals seen in the days of Kienlung, by Bell, have since dwindled to one or two. Van Braam met six going into Peking, sent thither from Yun-Nan." These were no doubt carrying tribute from Burmah.—H. C.] It is worth noticing that the housings of cut cloth or *appliqué* work ("*draps entaillez*") are still in fashion in India for the caparison of elephants.

NOTE 4.—In 1263 Kúblái adopted the Chinese fashion of worshipping the tablets of his own ancestors, and probably at the same time the adoration of his own tablet by his subjects was introduced. Van Braam ingenuously relates how he and the rest of the Dutch Legation of 1794 performed the adoration of the Emperor's Tablet on first entering China, much in the way described in the text.

There is a remarkable amplification in the last paragraph of the chapter as given by Ramusio: "When all are in their proper places, a certain great personage, or high prelate as it were, gets up and says with a loud voice : ' Bow yourselves and adore ! ' On this immediately all bend and bow the forehead to the ground. Then the prelate says again : ' God save and keep our Lord the Emperor, with length of years and with mirth and happiness.' And all answer : ' So may it be ! ' And then again the prelate says : ' May God increase and augment his Empire and its prosperity more and more, and keep all his subjects in peace and goodwill, and may all things go well throughout his Dominion ! ' And all again respond : ' So may it be ! ' And this adoration is repeated four times."

One of Pauthier's most interesting notes is a long extract from the official Directory of Ceremonial under the Mongol Dynasty, which admirably illustrates the chapters we have last read. I borrow a passage regarding this adoration : " The Musician's Song having ceased, the Ministers shall recite with a loud voice the following Prayer : ' Great Heaven, that extendest over all ! Earth which art under the guidance of Heaven ! We invoke You and beseech You to heap blessings upon the Emperor and the Empress ! Grant that they may live ten thousand, a hundred thousand years ! '

" Then the first Chamberlain shall respond : ' May it be as the prayer hath said ! ' The Ministers shall then prostrate themselves, and when they rise return to their places, and take a cup or two of wine."

The K'o-tow (*Khêu-thêu*) which appears repeatedly in this ceremonial and which in our text is indicated by the four prostrations, was, Pauthier alleges, not properly a Chinese form, but only introduced by the Mongols. Baber indeed speaks of it as the *Kornish*, a Moghul ceremony, in which originally " the person who performed it kneeled nine times and touched the earth with his brow each time." He describes it as performed very elaborately (nine times *twice*) by his younger uncle in visiting the elder. But in its essentials the ceremony must have been of old date at the Chinese Court ; for the Annals of the Thang Dynasty, in a passage cited by M. Pauthier himself,* mention that ambassadors from the famous Hárún ar Rashíd in 798 had to perform the " ceremony of kneeling and striking the forehead against the ground." And M. Pauthier can scarcely be right in saying that the practice was disused by the Ming Dynasty and only reintroduced by the Manchus ; for in the story of Shah Rukh's embassy the performance of the K'o-tow occurs repeatedly.

[" It is interesting to note," writes Mr. Rockhill (*Rubruck*, p. 22), "that in A.D. 981 the Chinese Envoy, Wang Yen-tê, sent to the Uigur Prince of Kao-chang, refused to make genuflexions (*pai*) to him, as being contrary to the established usages as regards envoys. The prince and his family, however, on receiving the envoy, all faced eastward (towards Peking) and made an obeisance (*pai*) on receiving the imperial presents (*shou-tzǔ*)." (*Ma Twan-lin*, Bk 336, 13.)—H. C.]

(*Gaubil*, 142 ; *Van Braam*, I. 20-21 ; *Baber*, 106 ; *N. et E.* XIV. Pt. I. 405, 407, 418.)

The enumeration of *four* prostrations in the text is, I fancy, quite correct. There are several indications that this number was used instead of the three times three of later days. Thus Carpini, when introduced to the Great Kaan, "bent the left knee four times." And in the Chinese bridal ceremony of " Worshipping the Tablets," the genuflexion is made four times. At the court of Shah Abbas an obeisance evidently identical was repeated four times. (*Carp.* 759 ; *Doolittle*, p. 60 ; *P. Della Valle*, I. 646.)

---

* *Gaubil*, cited in *Pauthier's Hist. des Relations Politiques de la Chine*, etc., p. 226.

CHAPTER XVI.

CONCERNING THE TWELVE THOUSAND BARONS WHO RECEIVE ROBES
OF CLOTH OF GOLD FROM THE EMPEROR ON THE GREAT
FESTIVALS, THIRTEEN CHANGES A-PIECE.

Now you must know that the Great Kaan hath set
apart 12,000 of his men who are distinguished by the
name of *Keshican*, as I have told you before; and on
each of these 12,000 Barons he bestows thirteen changes
of raiment, which are all different from one another: I
mean that in one set the 12,000 are all of one colour;
the next 12,000 of another colour, and so on; so that
they are of thirteen different colours. These robes are
garnished with gems and pearls and other precious
things in a very rich and costly manner.[1] And along
with each of these changes of raiment, *i.e.* 13 times in
the year, he bestows on each of those 12,000 Barons a
fine golden girdle of great richness and value, and likewise a pair of boots of *Camut*, that is to say of *Borgal*,
curiously wrought with silver thread; insomuch that
when they are clothed in these dresses every man of
them looks like a king![2] And there is an established
order as to which dress is to be worn at each of those
thirteen feasts. The Emperor himself also has his
thirteen suits corresponding to those of his Barons; in
colour, I mean (though his are grander, richer, and
costlier), so that he is always arrayed in the same colour
as his Barons, who are, as it were, his comrades. And
you may see that all this costs an amount which it is
scarcely possible to calculate.

Now I have told you of the thirteen changes of
raiment received from the Prince by those 12,000 Barons,
amounting in all to 156,000 suits of so great cost and
value, to say nothing of the girdles and the boots which

are also worth a great sum of money.   All this the Great
Lord hath ordered, that he may attach the more of
grandeur and dignity to his festivals.

And now I must mention another thing that I had
forgotten, but which you will be astonished to learn from
this Book.   You must know that on the Feast Day a
great Lion is led to the Emperor's presence, and as soon
as it sees him it lies down before him with every sign
of the greatest veneration, as if it acknowledged him for
its lord; and it remains there lying before him, and entirely unchained.   Truly this must seem a strange story
to those who have not seen the thing! [3]

Note I.—On the *Keshican*, see note 1 to chap. xii., and on the changes of
raiment note 3 to chap. xiv., and the remarks there as to the number of distributions.   I confess that the stress laid upon the number 13 in this chapter makes the
supposition of error more difficult.   But there is something odd and unintelligible
about the whole of the chapter except the last paragraph.   For the 12,000 *Keshican*
are here all elevated to *Barons;* and at the same time the statement about their
changes of raiment seems to be merely that already made in chapter xiv.   This
repetition occurs only in the French MSS., but as it is in all these we cannot reject
it.

Note 2.—The words *Camut* and *Borgal* appear both to be used here for what we
call *Russia-Leather*.   The latter word in one form or another, *Bolghár, Borgháli,* or
Bulkál, is the term applied to that material to this day nearly all over Asia.   Ibn
Batuta says that in travelling during winter from Constantinople to the Wolga he
had to put on three pairs of boots, one of wool (which we should call stockings), a
second of wadded linen, and a third of *Borgháli,* "*i.e.* of horse-leather lined with
wolf-skin."   Horse-leather seems to be still the favourite material for boots among all
the Tartar nations.   The name was undoubtedly taken from *Bolghar* on the Wolga,
the people of which are traditionally said to have invented the art of preparing skins
in that manner.   This manufacture is still one of the staple trades of Kazan, the city
which in position and importance is the nearest representative of Bolghar now.

Camut is explained by Klaproth to be "leather made from the back-skin of a
camel."   It appears in Johnson's Persian Dictionary as *Kámú,* but I do not know
from what language it originally comes.   The word is in the Latin column of the
Petrarchian Vocabulary with the Persian rendering *Sagri*.   This shows us what is
meant, for *Saghrí* is just our word *Shagreen,* and is applied to a fine leather granulated
in that way, which is much used for boots and the like by the people of Central Asia.
[In Turkish *ṣāghri* or *saghri* is the name both for the buttocks of a horse and the
leather called *shagreen* prepared with them.   (See *Devic, Dict. Étym.*)—H. C.]   In
the commercial lists of our Indian north-west frontier we find as synonymous *Saghri*
or *Kímukht,* "Horse or Ass-hide."   No doubt this latter word is a form of *Kámú* or
Camut.   It appears (as *Keimukht,* "a sort of leather") in a detail of imports to
Aden given by *Ibn al Wardi,* a geographer of the 13th century.

Instead of Camut, Ramusio has *Camoscia, i.e.* Chamois, and the same seems to be
in all the editions based on Fra Pipino's version.   It may be a misrendering of

camutum or *camutium ;* or is there any real connexion between the Oriental *Kámú Kímukht*, and the Italian *camoscia?* (*I. B.* II. 445; *Klapr. Mém.* vol. III. ; *Davies's Trade Report*, App. p. ccxx. ; *Vámbéry's Travels*, 423; *Not. et Ext.* II. 43.)

Fraehn (writing in 1832) observes that he knew no use of the word *Bolghár*, in the sense of Russian leather, older than the 17th century. But we see that both Marco and Ibn Batuta use it. (*F. on the Wolga Bulghars*, pp. 8-9.)

Pauthier in a note (p. 285) gives a list of the garments issued to certain officials on these ceremonial occasions under the Mongols, and sure enough this list includes "pairs of boots in red leather." Odoric particularly mentions the broad golden girdles worn at the Kaan's court.

[La Curne, *Dict.*, has *Bulga*, leather bag ; old Gallic word from which are derived *bouge* et *bougete, bourse ;* he adds in a note, "Festus writes : '*Bulgas* galli sacculos scorteos vocant.'"—H. C.]

Note 3.—"Then come mummers leading lions, which they cause to salute the Lord with reverence." (*Odoric*, p. 143.) A lion sent by Mirza Baisangar, one of the Princes of Timur's House, accompanied Shah Rukh's embassy as a present to the Emperor ; and like presents were frequently repeated. (See *Amyot*, XIV. 37, 38.)

CHAPTER XVII.

HOW THE GREAT KAAN ENJOINETH HIS PEOPLE TO SUPPLY HIM WITH GAME.

THE three months of December, January, and February, during which the Emperor resides at his Capital City, are assigned for hunting and fowling, to the extent of some 40 days' journey round the city ; and it is ordained that the larger game taken be sent to the Court. To be more particular : of all the larger beasts of the chase, such as boars, roebucks, bucks, stags, lions, bears, etc., the greater part of what is taken has to be sent, and feathered game likewise. The animals are gutted and despatched to the Court on carts. This is done by all the people within 20 or 30 days' journey, and the quantity so despatched is immense. Those at a greater distance cannot send the game, but they have to send the skins after tanning them, and these are employed in the making of equipments for the Emperor's army.[1]

NOTE 1.—So Magaillans : "Game is so abundant, especially at the capital, that every year during the three winter months you see at different places, intended for despatch thither, besides great piles of every sort of wildfowl, rows of four-footed game of a gunshot or two in length : the animals being all frozen and standing on their feet. Among other species you see three sundry kinds of bears . . . . and great abundance of other animals, as stags and deer of different sorts, boars, elks, hares, rabbits, squirrels, wild-cats, rats, geese, ducks, very fine jungle-fowl, etc., and all so cheap that I never could have believed it" (pp. 177-178). As this writer mentions *wild-cats*, we may presume that the "lions" of Polo also were destined to be eaten.

["Kubilai Khan kept a whole army, 14,000 men, huntsmen, distributed in Peking and other cities in the present province of Chili (*Yuen-shi*). The Khan used to hunt in the Peking plain from the beginning of spring, until his departure to Shang-tu. There are in the Peking department many low and marshy places, stretching often to a considerable extent and abounding in game. In the biography of *Ai-sie* (*Yuen shi*, chap. cxxxiv.), who was a Christian, it is mentioned that Kubilai was hunting also in the department of Pao-ting fu." (*Palladius*, p. 45.)—H. C.]

CHAPTER XVIII.

OF THE LIONS AND LEOPARDS AND WOLVES THAT THE KAAN KEEPS
FOR THE CHASE.

THE Emperor hath numbers of leopards[1] trained to the chase, and hath also a great many lynxes taught in like manner to catch game, and which afford excellent sport.[2] He hath also several great Lions, bigger than those of Babylonia, beasts whose skins are coloured in the most beautiful way, being striped all along the sides with black, red, and white. These are trained to catch boars and wild cattle, bears, wild asses, stags, and other great or fierce beasts. And 'tis a rare sight, I can tell you, to see those lions giving chase to such beasts as I have mentioned! When they are to be so employed the Lions are taken out in a covered cart, and every Lion has a little doggie with him. [They are obliged to approach the game against the wind, otherwise the animals would scent the approach of the Lion and be off.][3]

There are also a great number of eagles, all broken to catch wolves, foxes, deer, and wild goats, and they do catch them in great numbers. But those especially that are trained to wolf-catching are very large and powerful birds, and no wolf is able to get away from them.[4]

Note 1.—The Cheeta or Hunting-Leopard, still kept for the chase by native noblemen in India, is an animal very distinct from the true leopard. It is much more lanky and long-legged than the pure felines, is unable to climb trees, and has claws only partially retractile. Wood calls it a link between the feline and canine races. One thousand Cheetas were attached to Akbar's hunting establishment; and the chief one, called Semend-Manik, was carried to the field in a palankin with a kettledrum beaten before him. Boldensel in the first half of the 14th century speaks of the Cheeta as habitually used in Cyprus; but, indeed, a hundred years before, these animals had been constantly employed by the Emperor Frederic II. in Italy, and accompanied him on all his marches. They were introduced into France in the latter part of the 15th century, and frequently employed by Lewis XI., Charles VIII., and Lewis XII. The leopards were kept in a ditch of the Castle of Amboise, and the name still borne by a gate hard by, *Porte des Lions*, is supposed to be due to that circumstance. The *Moeurs et Usages du Moyen Age* (Lacroix), from which I take the last facts, gives copy of a print by John Stradanus representing a huntsman with the leopard on his horse's crupper, like Kúblái's (*supra*, Bk. I. ch. lxi.); Frederic II. used to say of his Cheetas, "they knew how to ride." This way of taking the Cheeta to the field had been first employed by the Khalif Yazid, son of Moáwiyah. The Cheeta often appears in the pattern of silk damasks of the 13th and 14th centuries, both Asiatic and Italian. (*Ayeen Akbery*, I. 304, etc.; *Boldensel*, in *Canisii Thesaurus*, by *Basnage*, vol. IV. p. 339; *Kington's Fred. II.* I. 472, II. 156; *Bochart*, *Hierozoica*, 797; *Rock's Catalogue*, *passim*.)

[The hunting equipment of the Sultan consisted of about thirty falconers on horse-back who carried each a bird on his fist. These falconers were in front of seven horsemen, who had behind a kind of tamed tiger at times employed by His Highness for hare-hunting, notwithstanding what may be said to the contrary by those who are inclined not to believe the fact. It is a thing known by everybody here, and cannot be doubted except by those who admit that they believe nothing of foreign customs. These tigers were each covered with a brocade cloth—and their peaceful attitude, added to their ferocious and savage looks, caused at the same time astonishment and fear in the soul of those whom they looked upon. (*Journal d'Antoine Galland*, trad. par Ch. Schefer, I. p. 135.) The Cheeta (*Gueparda jubata*) was, according to Sir W. Jones, first employed in hunting antelopes by Hushing, King of Persia, 865 B.C.—H. C.]

Note 2.—The word rendered Lynxes is *Leu cervers* (G. Text), *Louz serviers* of Pauthier's MS. C, though he has adopted from another *Loups* simply, which is certainly wrong. The *Geog. Latin* has "*Linceos i.e. lupos cerverios.*" There is no doubt that the *Loup-cervier* is the Lynx. Thus Brunetto Latini, describing the Loup-cervier, speaks of its remarkable powers of vision, and refers to its agency in the production of the precious stone called *Liguire* (*i.e. Ligurium*), which the ancients fancied to come from *Lync-urium*; the tale is in Theophrastus). Yet the quaint Bestiary of Philip de Thaun, published by Mr. Wright, identifies it with the Greek Hyena :—

> "*Hyena* e Griu num, que nus beste apellum,
> Ceo est *Lucervere*, oler fait et mult est fere."

[The Abbé Armand David writes (*Missions Cathol.* XXI. 1889, p. 227) that there is in China, from the mountains of Manchuria to the mountains of Tibet, a lynx called by the Chinese *T'u-pao* (earth-coloured panther) ; a lynx somewhat similar to the *loup-cervier* is found on the western border of China, and has been named *Lyncus Desgodinsi.*—H. C.]

Hunting Lynxes were used at the Court of Akbar. They are also mentioned by A. Hamilton as so used in Sind at the end of the 17th century. This author calls the animal a *Shoe-goose!* *i.e. Siya-gosh* (Black-ear), the Persian name of the Lynx. It is still occasionally used in the chase by natives of rank in India. (*Brunetto Lat. Tresor,* p. 248 ; *Popular Treatises on Science written during Mid. Ages,* 94 ; *Ayeen Akbery,* u.s. ; *Hamilt. E. Indies,* I. 125 ; *Vigne,* I. 42.)

NOTE 3.—The conception of a Tiger seems almost to have dropped out of the European mind during the Middle Ages. Thus in a mediæval Bestiary, a chapter on the Tiger begins : " *Une Beste est qui est apelée Tigre c'est une manière de* Serpent." Hence Polo can only call the Tigers, whose portrait he draws here not incorrectly, *Lions.* So also nearly 200 years later Barbaro gives a like portrait, and calls the animal *Leonza.* Marsden supposes judiciously that the confusion may have been promoted by the ambiguity of the Persian *Sher.*

The Búrgút Eagle. (After Atkinson.)

"𝕰l a encore aiglies qe sunt afaités à prendre leus et voupes et dain et chavriou, et en prennent assez."

The Chinese pilgrim, Sung-Yun (A.D. 518), saw two young lions at the Court of Gandhára. He remarks that the pictures of these animals common in China, were not at all good likenesses. (*Beal,* p. 200.)

We do not hear in modern times of Tigers trained to the chase, but Chardin says of Persia : " In hunting the larger animals they make use of beasts of prey trained for the purpose, *lions,* leopards, *tigers,* panthers, ounces."

NOTE 4.—This is perfectly correct. In Eastern Turkestan, and among the Kirghiz to this day, eagles termed *Búrgút* (now well known to be the Golden Eagle) are tamed and trained to fly at wolves, foxes, deer, wild goats, etc. A Kirghiz will give a good horse for an eagle in which he recognises capacity for training. Mr. Atkinson gives vivid descriptions and illustrations of this eagle (which he calls "Bear coote"), attacking both deer and wolves. He represents the bird as striking one claw into the neck, and the other into the back of its large prey, and then tearing out the liver with its beak. In justice both to Marco Polo and to Mr. Atkinson, I have pleasure in adding a vivid account of the exploits of this bird, as witnessed by one of my kind correspondents, the Governor-General's late envoy to Kashgar. And I trust Sir Douglas Forsyth will pardon my quoting his own letter just as it stands * :— "Now for a story of the *Burgoot*—Atkinson's 'Bearcoote.' I think I told you it was the Golden Eagle and supposed to attack wolves and even bears. One day we came across a wild hog of enormous size, far bigger than any that gave sport to the Tent Club in Bengal. The Burgoot was immediately let loose, and went straight at the hog, which it kicked, and flapped with its wings, and utterly *flabbergasted*, whilst our Kashgaree companions attacked him with sticks and brought him to the ground. As Friar Odoric would say, I, T. D. F., have seen this with mine own eyes."—Shaw describes the rough treatment with which the Búrgút is tamed. Baber, when in the Bajaur Hills, notices in his memoirs : "This day Búrgút took a deer." (*Timkowski*, I. 414; *Levchine*, p. 77; *Pallas, Voyages*, I. 421; *J. R. A. S.* VII. 305; *Atkinson's Siberia*, 493; and *Amoor*, 146-147; *Shaw*, p. 157; *Baber*, p. 249.)

[The Golden Eagle (*Aquila chrysaetus*) is called at Peking *Hoy tiao* (black eagle). (*David et Oustalet, Oiseaux de la Chine*, p. 8.)—H. C.]

CHAPTER XIX.

CONCERNING THE TWO BROTHERS WHO HAVE CHARGE OF THE KAAN'S HOUNDS.

THE Emperor hath two Barons who are own brothers, one called Baian and the other Mingan; and these two are styled *Chinuchi* (or *Cunichi*), which is as much as to say, "The Keepers of the Mastiff Dogs."[1] Each of these brothers hath 10,000 men under his orders; each body of 10,000 being dressed alike, the one in red and the other in blue, and whenever they accompany the Lord to the chase, they wear this livery, in order to be recognized. Out of each body of 10,000 there are 2000 men who are each in charge of one or more great mastiffs, so that the whole number of these is very large. And when the Prince goes a-hunting, one of those Barons, with his 10,000 men and something like 5000 dogs, goes

---

* Dated Yangi Hissar, 10th April, 1874.

towards the right, whilst the other goes towards the left with his party in like manner. They move along, all abreast of one another, so that the whole line extends over a full day's journey, and no animal can escape them. Truly it is a glorious sight to see the working of the dogs and the huntsmen on such an occasion! And as the Lord rides a-fowling across the plains, you will see these big hounds coming tearing up, one pack after a bear, another pack after a stag, or some other beast, as it may hap, and running the game down now on this side and now on that, so that it is really a most delightful sport and spectacle.

[The Two Brothers I have mentioned are bound by the tenure of their office to supply the Kaan's Court from October to the end of March with 1000 head of game daily, whether of beasts or birds, and not counting quails; and also with fish to the best of their ability, allowing fish enough for three persons to reckon as equal to one head of game.]

Now I have told you of the Masters of the Hounds and all about them, and next will I tell you how the Lord goes off on an expedition for the space of three months.

NOTE 1.—Though this particular Bayan and Mingan are not likely to be mentioned in history, the names are both good Mongol names; *Bayan* that of a great soldier under Kúblái, of whom we shall hear afterwards; and *Mingan* that of one of Chinghiz's generals.

The title of "Master of the Mastiffs" belonged to a high Court official at Constantinople in former days, *Sámsúnji Báshi,* and I have no doubt Marco has given the exact interpretation of the title of the two Barons: though it is difficult to trace its elements. It is read variously *Cunici (i.e. Kunichi)* and *Cinuci (i.e Chinuchi).* It is evidently a word of analogous structure to *Kushchi,* the Master of the Falcons; *Parschi,* the Master of the Leopards. Professor Schiefner thinks it is probably corrupted from *Noghaichi,* which appears in Kovalevski's Mongol Dict. as "*chasseur qui a soins des chiens courants.*" This word occurs, he points out, in Sanang Setzen, where Schmidt translates it *Aufseher über Hunde.* (See *S. S.* p. 39.)

The metathesis of *Noghai*-chi into *Kuni*-chi is the only drawback to this otherwise apt solution. We generally shall find Polo's Oriental words much more accurately expressed than this would imply—as in the next chapter. I have hazarded a suggestion of (Or. Turkish) *Chong-It-chi,* "Keeper of the Big Dogs," which Professor Vámbéry thinks possible. (See "*chong,* big, strong,' in his *Tschagataische Sprachstudien,*

p. 282, and note in *Lord Strangford's Selected Writings*, II. 169.) In East Turkestan they call the Chinese *Chong Káfir*, "The Big Heathen." This would exactly correspond to the rendering of Pipino's Latin translation, "*hoc est canum magnorum Praefecti.*" *Chinuchi* again would be (in Mongol) "Wolf-keepers." It is at least possible that the great dogs which Polo terms mastiffs may have been known by such a name. We apply the term Wolf-dog to several varieties, and in Macbeth's enumeration we have—

> ———— "Hounds, and greyhounds, mongrels, spaniels, curs,
> Shoughs, water rugs, and *Demi-Wolves.*"

Lastly the root-word may be the Chinese *Kiuen*, "dog," as Pauthier says. The mastiffs were probably Tibetan, but may have come through China, and brought a name with them, like *Boule-dogues* in France.

[Palladius (p. 46) says that *Chinuchi* or *Cunici* "have no resemblance with any of the names found in the *Yuen shi*, ch. xcix., article *Ping chi* (military organisation), and relating to the hunting staff of the Khan, viz. : *Si pao ch'i* (falconers), *Ho r ch'i* (archers), and *Ke lien ch'i* (probably those who managed the hounds)."—H. C.]

CHAPTER XX.

HOW THE EMPEROR GOES ON A HUNTING EXPEDITION.

AFTER he has stopped at his capital city those three months that I mentioned, to wit, December, January, February, he starts off on the 1st day of March, and travels southward towards the Ocean Sea, a journey of two days.[1] He takes with him full 10,000 falconers, and some 500 gerfalcons besides peregrines, sakers, and other hawks in great numbers; and goshawks also to fly at the water-fowl.[2] But do not suppose that he keeps all these together by him; they are distributed about, hither and thither, one hundred together, or two hundred at the utmost, as he thinks proper. But they are always fowling as they advance, and the most part of the quarry taken is carried to the Emperor. And let me tell you when he goes thus a-fowling with his gerfalcons and other hawks, he is attended by full 10,000 men who are disposed in couples; and these are called

Toscaol, which is as much as to say, "Watchers." And the name describes their business.[3] They are posted from spot to spot, always in couples, and thus they cover a great deal of ground! Every man of them is provided with a whistle and hood, so as to be able to call in a hawk and hold it in hand. And when the Emperor makes a cast, there is no need that he follow it up, for those men I speak of keep so good a look out that they never lose sight of the birds, and if these have need of help they are ready to render it.

All the Emperor's hawks, and those of the Barons as well, have a little label attached to the leg to mark them, on which is written the names of the owner and the keeper of the bird. And in this way the hawk, when caught, is at once identified and handed over to its owner. But if not, the bird is carried to a certain Baron, who is styled the *Bularguchi,* which is as much as to say "The Keeper of Lost Property." And I tell you that whatever may be found without a known owner, whether it be a horse, or a sword, or a hawk, or what not, it is carried to that Baron straightway, and he takes charge of it. And if the finder neglects to carry his trover to the Baron, the latter punishes him. Likewise the loser of any article goes to the Baron, and if the thing be in his hands it is immediately given up to the owner. Moreover, the said Baron always pitches on the highest spot of the camp, with his banner displayed, in order that those who have lost or found anything may have no difficulty in finding their way to him. Thus nothing can be lost but it shall be incontinently found and restored.[4]

And so the Emperor follows this road that I have mentioned, leading along in the vicinity of the Ocean Sea (which is within two days' journey of his capital city, Cambaluc), and as he goes there is many a fine sight to be seen, and plenty of the very best entertainment in

hawking ; in fact, there is no sport in the world to
equal it !

The Emperor himself is carried upon four elephants in
a fine chamber made of timber, lined inside with plates of
beaten gold, and outside with lions' skins [for he always
travels in this way on his fowling expeditions, because he is
troubled with gout]. He always keeps beside him a dozen
of his choicest gerfalcons, and is attended by several of
his Barons, who ride on horseback alongside. And sometimes, as they may be going along, and the Emperor from
his chamber is holding discourse with the Barons, one
of the latter shall exclaim : " Sire ! Look out for Cranes ! "
Then the Emperor instantly has the top of his chamber
thrown open, and having marked the cranes he casts one
of his gerfalcons, whichever he pleases ; and often the
quarry is struck within his view, so that he has the most
exquisite sport and diversion, there as he sits in his
chamber or lies on his bed ; and all the Barons with him
get the enjoyment of it likewise ! So it is not without
reason I tell you that I do not believe there ever existed
in the world or ever will exist, a man with such sport and
enjoyment as he has, or with such rare opportunities.[5]

And when he has travelled till he reaches a place
called CACHAR MODUN,[6] there he finds his tents pitched,
with the tents of his Sons, and his Barons, and those of
his Ladies and theirs, so that there shall be full 10,000
tents in all, and all fine and rich ones. And I will tell
you how his own quarters are disposed. The tent in
which he holds his courts is large enough to give cover
easily to a thousand souls. It is pitched with its door
to the south, and the Barons and Knights remain in
waiting in it, whilst the Lord abides in another close to
it on the west side. When he wishes to speak with any
one he causes the person to be summoned to that other
tent. Immediately behind the great tent there is a fine

large chamber where the Lord sleeps ; and there are also
many other tents and chambers, but they are not in contact with the Great Tent as these are.   The two
audience-tents and the sleeping-chamber are constructed
in this way.   Each of the audience-tents has three poles,
which are of spice-wood, and are most artfully covered
with lions' skins, striped with black and white and red,
so that they do not suffer from any weather.   All three
apartments are also covered outside with similar skins of
striped lions, a substance that lasts for ever.[7]   And inside
they are all lined with ermine and sable, these two being
the finest and most costly furs in existence.   For a robe
of sable, large enough to line a mantle, is worth 2000
bezants of gold, or 1000 at least, and this kind of skin is
called by the Tartars "The King of Furs."   The beast
itself is about the size of a marten.[8]   These two furs of
which I speak are applied and inlaid so exquisitely, that it
is really something worth seeing.   All the tent-ropes are
of silk.   And in short I may say that those tents, to
wit the two audience-halls and the sleeping-chamber,
are so costly that it is not every king could pay for
them.

Round about these tents are others, also fine ones
and beautifully pitched, in which are the Emperor's ladies,
and the ladies of the other princes and officers.   And then
there are the tents for the hawks and their keepers, so
that altogether the number of tents there on the plain is
something wonderful.   To see the many people that are
thronging to and fro on every side and every day there,
you would take the camp for a good big city.   For you
must reckon the Leeches, and the Astrologers, and the
Falconers, and all the other attendants on so great a
company ; and add that everybody there has his whole
family with him, for such is their custom.

The Lord remains encamped there until the spring,

and all that time he does nothing but go hawking round
about among the canebrakes along the lakes and rivers
that abound in that region, and across fine plains on
which are plenty of cranes and swans, and all sorts of
other fowl. The other gentry of the camp also are
never done with hunting and hawking, and every day
they bring home great store of venison and feathered
game of all sorts. Indeed, without having witnessed it,
you would never believe what quantities of game are
taken, and what marvellous sport and diversion they all
have whilst they are in camp there.

There is another thing I should mention ; to wit, that
for 20 days' journey round the spot nobody is allowed,
be he who he may, to keep hawks or hounds, though
anywhere else whosoever list may keep them. And
furthermore throughout all the Emperor's territories,
nobody however audacious dares to hunt any of these
four animals, to wit, hare, stag, buck, and roe, from the
month of March to the month of October. Anybody
who should do so would rue it bitterly. But those people
are so obedient to their Lord's command, that even if a
man were to find one of those animals asleep by the road-
side he would not touch it for the world! And thus the
game multiplies at such a rate that the whole country
swarms with it, and the Emperor gets as much as he
could desire. Beyond the term I have mentioned, however, to wit that from March to October, everybody may
take these animals as he list.[9]

After the Emperor has tarried in that place, enjoying
his sport as I have related, from March to the middle of
May, he moves with all his people, and returns straight
to his capital city of Cambaluc (which is also the capital of
Cathay, as you have been told), but all the while continuing to take his diversion in hunting and hawking as
he goes along.

NOTE I.—" *Vait vers midi jusques à la Mer Occeane, ou il y a deux journées.*" It is not possible in any way to reconcile this description as it stands with truth, though I do not see much room for doubt as to the direction of the excursion. Peking is 100 miles as the crow flies from the nearest point of the coast, at least six or seven days' march for such a camp, and the direction is south-east, or nearly so. The last circumstance would not be very material as Polo's compass-bearings are not very accurate. We shall find that he makes the general line of bearing from Peking towards Kiangnan, *Sciloc* or S. East, hence his *Midi* ought in consistency to represent *S. West*, an impossible direction for the Ocean. It is remarkable that Ramusio has *Greco* or *N. East*, which would by the same relative correction represent *East*. And other circumstances point to the frontier of Liao-tong as the direction of this excursion. Leaving the *two days* out of question, therefore, I should suppose the " Ocean Sea " to be struck at Shan-hai-kwan near the terminus of the Great Wall, and that the site of the standing hunting-camp is in the country to the north of that point. The Jesuit Verbiest accompanied the Emperor Kanghi on a tour in this direction in 1682, and almost immediately after passing the Wall the Emperor and his party seem to have struck off to the left for sport. Kúblái started on the " 1st of March," probably however the 1st of the second Chinese month. Kanghi started from Peking on the 23rd of March, on the hunting-journey just referred to.

NOTE 2.—We are told that Bajazet had 7000 falconers and 6000 dog-keepers; whilst Sultan Mahomed Tughlak of India in the generation following Polo's, is said to have had 10,000 falconers, and 3000 other attendants as beaters. (*Not. et Ext.* XIII. p. 185.)

The Oriental practice seems to have assigned one man to the attendance on every hawk. This Kaempfer says was the case at the Court of Persia at the beginning of last century. There were about 800 hawks, and each had a special keeper. The same was the case with the Emperor Kanghi's hawking establishment, according to Gerbillon. (*Am. Exot.* p. 83 ; *Gerb.* 1st Journey, in *Duhalde.*)

NOTE 3.—The French MSS. read *Toscaor ;* the reading in the text I take from Ramusio. It is Turki, *Toskáúl,* توسقاول, defined as " Gardien, surveillant de la route ; Wächter, Wache, Wegehüter." (See *Zenker*, and *Pavet de Courteille.*) The word is perhaps also Mongol, for Rémusat has *Tosiyal*=" Veille." (*Mél. As.* I. 231.) Such an example of Polo's correctness both in the form and meaning of a Turki word is worthy of especial note, and shows how little he merits the wild and random treatment which has been often applied to the solution of like phrases in his book.

[Palladius (p. 47) says that he has heard from men well acquainted with the customs of the Mongols, that at the present day in " battues," the leaders of the two flanks which surround the game, are called *toscaul* in Mongol.—H. C.]

NOTE 4. — The remark in the previous note might be repeated here. The *Bularguji* was an officer of the Mongol camp, whose duties are thus described by Mahomed Hindú Shah in a work on the offices of the Perso-Mongol Court. " He is an officer appointed by the Council of State, who; at the time when the camp is struck, goes over the ground with his servants, and collects slaves of either sex, or cattle, such as horses, camels, oxen, and asses, that have been left behind, and retains them until the owners appear and prove their claim to the property, when he makes it over to them. The *Bularguji* sticks up a flag by his tent or hut to enable people to find him, and so recover their lost property." (*Golden Horde*, p. 245.) And in the Appendix to that work (p. 476) there is a copy of a warrant to such a Bularguji or Provost Marshal. The derivation appears therein as from *Bularghu*, " Lost property." Here again it was impossible to give both form and meaning of the word more exactly than Polo has done. Though Hammer writes these terminations in *ji* (*dschi*), I believe *chi* (tschi) is preferable. We have this same word *Bularghu* in a grant of privileges to the Venetians by the Ilkhan Abusaid, 22nd December, 1320, which has been published by M. Mas Latrie : "*Item, se algun cavalo* bolargo *fosse trovado apreso de algun vostro veneciano*," etc.—" If any stray horse shall be found in the possession of a Venetian," etc. (See *Bibl. de l'Ecole des Chartes*, 1870—*tirage à part*, p. 26.)

["There are two Mongol terms, which resemble this word *Bularguchi*, viz. *Balagachi* and *Buluguchi*. But the first was the name used for the door-keeper of the tent of the Khan. By Buluguchi the Mongols understood a hunter and especially sable hunters. No one of these terms can be made consistent with the accounts given by M. Polo regarding the Bularguchi: In the *Kui sin tsa shi*, written by Chow Mi, in the former part of the 14th century, interesting particulars regarding Mongol hunting are found." (*Palladius*, 47.) In chapter 101. *Djan-ch'i*, of the *Yuen-shi*, Falconers are called *Ying fang pu lie*, and a certain class of the Falconers are termed *Bo-lan-ghi*. (*Bretschneider*, *Med. Res.* I. p. 188.)—H. C.]

NOTE 5.—A like description is given by Odoric of the mode in which a successor of Kúblái travelled between Cambaluc and Shangtu, with his falcons also in the chamber beside him. What Kúblái had adopted as an indulgence to his years and gout, his successors probably followed as a precedent without these excuses.

[With regard to the gout of Kúblái Khan, Palladius (p. 48) writes: " In the Corean history allusion is made twice to the Khan's suffering from this disease. Under the year 1267, it is there recorded that in the 9th month, envoys of the Khan with a letter to the King arrived in Corea. Kubilai asked for the skin of the *Akirho munho*, a fish resembling a cow. The envoy was informed that, as the Khan suffered from swollen feet it would be useful for him to wear boots made of the skin of this animal, and in the 10th month, the king of Corea forwarded to the Khan seventeen skins of it. It is further recorded in the Corean history, that in the 8th month of 1292, sorcerers and *Shaman* women from Corea were sent at the request of the Khan to cure him of a disease of the feet and hands. At that time the king of Corea was also in Peking, and the sorcerers and Shaman women were admitted during an audience the King had of the Khan. They took the Khan's hands and feet and began to recite exorcisms, whilst Kubilai was laughing."—H. C.]

NOTE 6.—Marsden and Pauthier identify Cachar Modun with *Tchakiri Mondou*, or *Moudon*, which appears in D'Anville's atlas as the title of a "Levée de terre naturelle," in the extreme east of Manchuria, and in lat. 44°, between the Khinga Lake and the sea. This position is out of the question. It is more than 900 miles, *in a straight line* from Peking, and the mere journey thither and back would have taken Kúblái's camp something like six months. The name *Kachar Modun* is probably Mongol, and as *Katzar* is = "land, region," and *Modun* = "wood" or "tree," a fair interpretation lies on the surface. Such a name indeed has little individuality. But the Jesuit maps have a *Modun Khotan* ("Wood-ville") just about the locality supposed, viz. in the region north of the eastern extremity of the Great Wall.

[Captain Gill writes (*River of Golden Sand*, I. p. 111): "This country around Urh-Chuang is admirably described [in *Marco Polo*, pp. 403, 406], and I should almost imagine that the Kaan must have set off south-east from Peking, and enjoyed some of his hawking not far from here, before he travelled to Cachar Modun, wherever that may have been."

"With respect to Cachar Modun, Marco Polo intends perhaps by this name Ho-si wu, which place, together with Yang-ts'un, were comprised in the general name *Ma t'ou* (perhaps the *Modun* of M. Polo). Ma-t'ou is even now a general term for a jetty in Chinese. Ho-si in the Mongol spelling was Ha-shin. D'Ohsson, in his translation of Rashid-eddin renders *Ho-si* by *Co-shi* (*Hist. des Mongols*, I. p. 95), but Rashid in that case speaks not of Ho-si wu, but of the Tangut Empire, which in Chinese was called Ho-si, meaning west of the (Yellow) River. (See *supra*, p. 205). Ho-si wu, as well as Yang-ts'un, both exist even now as villages on the Pei-ho River, and near the first ancient walls can be seen. Ho-si wu means: 'Custom's barrier west of the (Pei-ho) river.' " (*Palladius*, p. 45.) This identification cannot be accepted on account of the position of Ho-si wu.—H. C.]

Note 7.—I suppose the best accessible illustration of the Kaan's great tent may be that in which the Emperor Kienlung received Lord Macartney in the same region in 1793, of which one view is given in Staunton's plates. Another exists in the Staunton Collection in the B. M., of which I give a reduced sketch.

Kúblái's great tent, after all, was but a fraction of the size of Akbar's audience-tents, the largest of which held 10,000 people, and took 1000 *farráshes* a week's work to pitch it, with machines. But perhaps the manner of *holding* people is differently estimated. (*Aín Akb.* 53.)

In the description of the tent-poles, Pauthier's text has "*trois coulombes de fust* de pieces *moult bien encuierées*," etc. The G. T. has "*de leing* d'especies *mout bien curés*," etc. The Crusca, "*di* spezie *molto belle*," and Ramusio going off at a tangent, "*di legno intagliate con grandissimo artificio e indorate*." I believe the translation in the text to indicate the true reading. It might mean camphor-wood, or the like. The tent-covering of tiger-skins is illustrated by a passage in Sanang Setzen, which speaks of a tent covered with panther-skins, sent to Chinghiz by the Khan of the Solongos (p. 77).

The Tents of the Emperor Kienlung.

[Grenard (pp. 160-162) gives us his experience of Tents in Central Asia (Khotan). "These Tents which we had purchased at Tashkent were the 'tentes-abris' which are used in campaign by Russian military workshops, only we made them larger by a third. They were made of grey Kirghiz felt, which cannot be procured at Khotan. The felt manufactured in this town not having enough consistency or solidity, we took Aksu felt, which is better than this of Khotan, though inferior to the felt of Russian Turkestan. These felt tents are extremely heavy, and, once damp, are dried with difficulty. These drawbacks are not compensated by any important advantage ; it would be an illusion to believe that they preserve from the cold any better than other tents. In fact, I prefer the Manchu tent in use in the Chinese army, which is, perhaps, of all military tents the most practical and comfortable. It is made of a single piece of double cloth of cotton, very strong, waterproof for a long time, white inside, blue outside, and weighs with its three tipped sticks and its wooden poles, 25 kilog. Set up, it forms a ridge roof 7 feet high and shelters fully ten men. It suits servants perfectly well. For the master who wants to work, to write, to draw, occasionally to receive officials, the ideal tent would be one of the same material, but of larger proportions, and comprising two parallel vertical partitions and surmounted by a ridge roof. The round form of Kirghiz and Mongol tents is also very comfortable, but it requires a complicated and inconvenient wooden frame-work, owing to which it takes some considerable time to raise up the tent."—H. C.]

Note 8.—The expressions about the sable run in the G. T., "*et l'apellent les*

Tartarz les roi des pelaines," etc. This has been curiously misunderstood both in versions based on Pipino, and in the Geog. Latin and Crusca Italian. The Geog. Latin gives us "*vocant eas Tartari* Lenoidae Pellonae"; the Crusca, "*chiamanle li Tartari* Leroide Pelame"; Ramusio in a very odd way combines both the genuine and the blundered interpretation: "*E li Tartari la chiamano* Regina delle Pelli; *e gli animali si chiamano* Rondes." Fraehn ingeniously suggested that this *Rondes* (which proves to be merely a misunderstanding of the French words *Roi des*) was a mistake for *Kunduz*, usually meaning a "beaver," but also a "sable." (See *Ibn Foszlan*, p. 57.) *Condux*, no doubt with this meaning, appears coupled with *vair*, in a Venetian Treaty with Egypt (1344), quoted by Heyd. (II. 208.)

Ibn Batuta puts the ermine above the sable. An ermine pelisse, he says, was worth in India 1000 dinárs of that country, whilst a sable one was worth only 400 dinárs. As Ibn Batuta's Indian dinárs are *Rupees*, the estimate of price is greatly lower than Polo's. Some years ago I find the price of a *Sack*, as it is technically called by the Russian traders, or robe of fine sables, stated to be in the Siberian market about 7000 banco rubels, *i.e.* I believe about 350*l.* The same authority mentions that in 1591 the Tzar Theodore Ivanovich made a present of a pelisse valued at the equivalent of 5000 *silver* rubels of modern Russian money, or upwards of 750*l.* Atkinson speaks of a *single* sable skin of the highest quality, for which the trapper demanded 18*l.* The great mart for fine sables is at Olekma on the Lena. (See *I. B.* II. 401-402; *Baer's Beiträge*, VII. 215 *seqq.*; *Upper and Lower Amoor*, 390.)

NOTE 9.—Hawking is still common in North China. Pétis de la Croix the elder, in his account of the *Yasa*, or institutes of Chinghiz, quotes one which lays down that between March and October "no one should take stags, deer, roebucks, hares, wild asses, nor some certain birds," in order that there might be ample sport in winter for the court. This would be just the reverse of Polo's statement, but I suspect it is merely a careless adoption of the latter. There are many such traps in Pétis de la Croix. (Engl. Vers. 1722, p. 82.)

CHAPTER XXI.

REHEARSAL OF THE WAY THE YEAR OF THE GREAT KAAN IS DISTRIBUTED.

ON arriving at his capital of Cambaluc,[1] he stays in his palace there three days and no more; during which time he has great court entertainments and rejoicings, and makes merry with his wives. He then quits his palace at Cambaluc, and proceeds to that city which he has built, as I told you before, and which is called Chandu, where he has that grand park and palace of cane, and where he keeps his gerfalcons in mew. There he spends the summer, to escape the heat, for the situation is a very cool one. After stopping there from the beginning of May to the 28th of August, he takes his departure (that is the time when they sprinkle the white mares' milk as I told you), and returns to his capital Cambaluc. There he stops, as I have told you also, the month of September, to keep his Birthday Feast, and also throughout October, November, December, January, and February, in which last month he keeps the grand feast of the New Year, which they call the White Feast, as you have heard already with all particulars. He then sets out on his march towards the Ocean Sea, hunting and hawking, and continues out from the beginning of March to the middle of May; and then comes back for three days only to the capital, during which he makes merry with his wives, and holds a great court and grand entertainments. In truth, 'tis something astonishing, the magnificence displayed by the Emperor in those three days; and then he starts off again as you know.

Thus his whole year is distributed in the following manner : six months at his chief palace in the royal city of Cambaluc, to wit, *September, October, November, December, January, February;*

Then on the great hunting expedition towards the sea, *March, April, May;*

Then back to his palace at Cambaluc for *three days;*

Then off to the city of Chandu which he has built, and where the Cane Palace is, where he stays *June, July, August;*

Then back again to his capital city of Cambaluc.

So thus the whole year is spent; six months at the capital, three months in hunting, and three months at the Cane Palace to avoid the heat. And in this way he passes his time with the greatest enjoyment; not to mention occasional journeys in this or that direction at his own pleasure.

NOTE 1.—This chapter, with its wearisome and whimsical reiteration, reminding one of a game of forfeits, is peculiar to that class of MSS. which claims to represent the copy given to Thibault de Cepoy by Marco Polo.

Dr. Bushell has kindly sent me a notice of a Chinese document (his translation of which he had unfortunately mislaid), containing a minute contemporary account of the annual migration of the Mongol Court to Shangtu. Having traversed the Kiu Yung Kwan (or Nankau) Pass, where stands the great Mongol archway represented at the end of this volume, they left what is now the Kalgan post-road at Tumuyi, making straight for Chaghan-nor (*supra*, p. 304), and thence to Shangtu. The return journey in autumn followed the same route as far as Chaghan-nor, where some days were spent in fowling on the lakes, and thence by Siuen-hwa fu (*"Sindachu,"* *supra*, p. 295) and the present post-road to Cambaluc

CHAPTER XXII.

CONCERNING THE CITY OF CAMBALUC, AND ITS GREAT TRAFFIC AND POPULATION.

You must know that the city of Cambaluc hath such a multitude of houses, and such a vast population inside the walls and outside, that it seems quite past all possibility. There is a suburb outside each of the gates, which are twelve in number;[1] and these suburbs are so great that they contain more people than the city itself [for the suburb of one gate spreads in width till it meets the suburb of the next, whilst they extend in length some three or four miles]. In those suburbs lodge the foreign merchants and travellers, of whom there are always great numbers who have come to bring presents to the Emperor, or to sell articles at Court, or because the city affords so good a mart to attract traders. [There are in each of the suburbs, to a distance of a mile from the city, numerous fine hostelries[2] for the lodgment of merchants from different parts of the world, and a special hostelry is assigned to each description of people, as if we should say there is one for the Lombards, another for the Germans, and a third for the French-men.] And thus there are as many good houses outside

Plain of Cambaluc ; the City in the distance ; from the Hills on the north-west.

of the city as inside, without counting those that belong to the great lords and barons, which are very numerous.

You must know that it is forbidden to bury any dead body inside the city. If the body be that of an Idolater it is carried out beyond the city and suburbs to a remote place assigned for the purpose, to be burnt. And if it be of one belonging to a religion the custom of which is to bury, such as the Christian, the Saracen, or what not, it is also carried out beyond the suburbs to a distant place assigned for the purpose. And thus the city is preserved in a better and more healthy state.

Moreover, no public woman resides inside the city, but all such abide outside in the suburbs. And 'tis wonderful what a vast number of these there are for the foreigners; it is a certain fact that there are more than 20,000 of them living by prostitution. And that so many can live in this way will show you how vast is the population.

[Guards patrol the city every night in parties of 30 or 40, looking out for any persons who may be abroad at unseasonable hours, *i.e.* after the great bell hath stricken thrice. If they find any such person he is immediately taken to prison, and examined next morning by the proper officers. If these find him guilty of any misdemeanour they order him a proportionate beating with the stick. Under this punishment people sometimes die; but they adopt it in order to eschew bloodshed; for their *Bacsis* say that it is an evil thing to shed man's blood].

To this city also are brought articles of greater cost and rarity, and in greater abundance of all kinds, than to any other city in the world. For people of every description, and from every region, bring things (including all the costly wares of India, as well as the fine and precious goods of Cathay itself with its provinces), some for the sovereign, some for the court, some for the city which is so great, some for the crowds of Barons and Knights, some for the great hosts of the Emperor which are quartered round about; and thus between court and city the quantity brought in is endless.

As a sample, I tell you, no day in the year passes that there do not enter the city 1000 cart-loads of silk alone, from which are made quantities of cloth of silk and gold, and of other goods. And this is not to be wondered at; for in all the countries round about there is no flax, so that everything has to be made of silk. It is true, indeed, that in some parts of the country there is cotton and hemp, but not sufficient for their wants. This, however, is not of much consequence, because silk is so abundant and cheap, and is a more valuable substance than either flax or cotton.

Round about this great city of Cambaluc there are some 200 other cities at various distances, from which traders come to sell their goods and buy others for their lords; and all find means to make their sales and purchases, so that the traffic of the city is passing great.

Note 1.—It would seem to have been usual to reckon *twelve* suburbs to Peking down to modern times. (See *Deguignes*, III. 38.)

Note 2.—The word here used is *Fondaco*, often employed in mediæval Italian in the sense nearly of what we call a *factory*. The word is from the Greek πανδοκεῖον, but through the Arabic *Fandúk*. The latter word is used by Ibn Batuta in speaking of the hostelries at which the Mussulman merchants put up in China.

CHAPTER XXIII.

[Concerning the Oppressions of Achmath the Bailo, and the Plot that was formed against Him.[1]

You will hear further on how that there are twelve persons appointed who have authority to dispose of lands, offices,

and everything else at their discretion. Now one of
these was a certain Saracen named ACHMATH, a shrewd
and able man, who had more power and influence with
the Grand Kaan than any of the others ; and the Kaan
held him in such regard that he could do what he pleased.
The fact was, as came out after his death, that Achmath
had so wrought upon the Kaan with his sorcery, that
the latter had the greatest faith and reliance on everything he said, and in this way did everything that
Achmath wished him to do.

This person disposed of all governments and offices,
and passed sentence on all malefactors ; and whenever
he desired to have any one whom he hated put to death,
whether with justice or without it, he would go to the
Emperor and say : " Such an one deserves death, for he
hath done this or that against your imperial dignity."
Then the Lord would say : " Do as you think right,"
and so he would have the man forthwith executed. · Thus
when people saw how unbounded were his powers, and
how unbounded the reliance placed by the Emperor on
everything that he said, they did not venture to oppose
him in anything. No one was so high in rank or power
as to be free from the dread of him. If any one was
accused by him to the Emperor of a capital offence, and
desired to defend himself, he was unable to bring proofs
in his own exculpation, for no one would stand by him,
as no one dared to oppose Achmath. And thus the
latter caused many to perish unjustly.[2]

Moreover, there was no beautiful woman whom he
might desire, but he got hold of her ; if she were unmarried, forcing her to be his wife, if otherwise, com-
pelling her to consent to his desires. Whenever he
knew of any one who had a pretty daughter, certain
ruffians of his would go to the father, and say : " What
say you ? Here is this pretty daughter of yours ; give

her in marriage to the Bailo Achmath (for they called
him 'the Bailo,' or, as we should say, 'the Vicegerent'),[3]
and we will arrange for his giving you such a government or such an office for three years." And so the man
would surrender his daughter. And Achmath would go
to the Emperor, and say : " Such a government is vacant,
or will be vacant on such a day. So-and-So is a proper
man for the post." And the Emperor would reply :
" Do as you think best ; " and the father of the girl was
immediately appointed to the government. Thus either
through the ambition of the parents, or through fear of
the Minister, all the beautiful women were at his beck,
either as wives or mistresses. Also he had some five-
and-twenty sons who held offices of importance, and
some of these, under the protection of their father's name,
committed scandals like his own, and many other abominable iniquities. This Achmath also had amassed great
treasure, for everybody who wanted office sent him a
heavy bribe.

In such authority did this man continue for two-and-
twenty years. At last the people of the country, to wit
the Cathayans, utterly wearied with the endless outrages
and abominable iniquities which he perpetrated against
them, whether as regarded their wives or their own
persons, conspired to slay him and revolt against the
government. Amongst the rest there was a certain
Cathayan named Chenchu, a commander of a thousand,
whose mother, daughter, and wife had all been dis-
honoured by Achmath. Now this man, full of bitter
resentment, entered into parley regarding the destruction
of the Minister with another Cathayan whose name was
Vanchu, who was a commander of 10,000. They came
to the conclusion that the time to do the business would
be during the Great Kaan's absence from Cambaluc.
For after stopping there three months he used to go to

Chandu and stop there three months; and at the same time his son Chinkin used to go away to his usual haunts, and this Achmath remained in charge of the city; sending to obtain the Kaan's orders from Chandu when any emergency arose.

So Vanchu and Chenchu, having come to this conclusion, proceeded to communicate it to the chief people among the Cathayans, and then by common consent sent word to their friends in many other cities that they had determined on such a day, at the signal given by a beacon, to massacre all the men with beards, and that the other cities should stand ready to do the like on seeing the signal fires. The reason why they spoke of massacring the bearded men was that the Cathayans naturally have no beard, whilst beards are worn by the Tartars, Saracens, and Christians. And you should know that all the Cathayans detested the Grand Kaan's rule because he set over them governors who were Tartars, or still more frequently Saracens, and these they could not endure, for they were treated by them just like slaves. You see the Great Kaan had not succeeded to the dominion of Cathay by hereditary right, but held it by conquest; and thus having no confidence in the natives, he put all authority into the hands of Tartars, Saracens, or Christians who were attached to his household and devoted to his service, and were foreigners in Cathay.

Wherefore, on the day appointed, the aforesaid Vanchu and Chenchu having entered the palace at night, Vanchu sat down and caused a number of lights to be kindled before him. He then sent a messenger to Achmath the Bailo, who lived in the Old City, as if to summon him to the presence of Chinkin, the Great Kaan's son, who (it was pretended) had arrived unexpectedly. When Achmath heard this he was much surprised, but made haste to go, for he feared the Prince greatly.   When he arrived at the gate he met a Tartar called Cogatai, who was Captain of the 12,000 that formed the standing garrison of the City ; and the latter asked him whither he was bound so late ?   " To Chinkin, who is just arrived."   Quoth Cogatai, " How can that be ?   How could he come so privily that I know nought of it ? "   So he followed the Minister with a certain number of his soldiers.   Now the notion of the Cathayans was that, if they could make an end of Achmath, they would have nought else to be afraid of. So as soon as Achmath got inside the palace, and saw all that illumination, he bowed down before Vanchu, supposing him to be Chinkin, and Chenchu who was standing ready with a sword straightway cut his head off.   As soon as Cogatai, who had halted at the entrance, beheld this, he shouted " Treason ! " and instantly dis-charged an arrow at Vanchu and shot him dead as he sat.   At the same time he called his people to seize Chenchu, and sent a proclamation through the city that any one found in the streets would be instantly put to death.   The Cathayans saw that the Tartars had dis-covered the plot, and that they had no longer any leader, since Vanchu was killed and Chenchu was taken.   So they kept still in their houses, and were unable to pass the signal for the rising of the other cities as had been settled.   Cogatai immediately dispatched messengers to the Great Kaan giving an orderly report of the whole affair, and the Kaan sent back orders for him to make a careful investigation, and to punish the guilty as their misdeeds deserved.   In the morning Cogatai examined all the Cathayans, and put to death a number whom he found to be ringleaders in the plot.   The same thing was done in the other cities, when it was found that the plot extended to them also.

After the Great Kaan had returned to Cambaluc he
was very anxious to discover what had led to this affair,
and he then learned all about the endless iniquities of
that accursed Achmath and his sons.   It was proved that
he and seven of his sons (for they were not all bad) had
forced no end of women to be their wives, besides those
whom they had ravished.   The Great Kaan then ordered
all the treasure that Achmath had accumulated in the
Old City to be transferred to his own treasury in the
New City, and it was found to be of enormous amount.
He also ordered the body of Achmath to be dug up and
cast into the streets for the dogs to tear ; and commanded
those of his sons that had followed the father's evil
example to be flayed alive.[4]

These circumstances called the Kaan's attention to
the accursed doctrines of the Sect of the Saracens, which
excuse every crime, yea even murder itself, when committed on such as are not of their religion.   And seeing
that this doctrine had led the accursed Achmath and his
sons to act as they did without any sense of guilt, the
Kaan was led to entertain the greatest disgust and
abomination for it.   So he summoned the Saracens and
prohibited their doing many things which their religion
enjoined.   Thus, he ordered them to regulate their
marriages by the Tartar Law, and prohibited their
cutting the throats of animals killed for food, ordering
them to rip the stomach in the Tartar way.

Now when all this happened Messer Marco was
upon the spot.][5]

NOTE 1.—This narrative is from Ramusio's version, and constitutes one of the
most notable passages peculiar to that version.

The name of the oppressive Minister is printed in Ramusio's Collection *Achmach*.
But the *c* and *t* are so constantly interchanged in MSS. that I think there can be
no question this was a mere clerical error for *Achmath*, and so I write it.   I have
also for consistency changed the spelling of *Xandu*, *Chingis*, etc., to that hitherto
adopted in our text of *Chandu*, *Chinkin*, etc.

NOTE 2.—The remarks of a Chinese historian on Kúblái's administration may be appropriately quoted here : " Hupilai Han must certainly be regarded as one of the greatest princes that ever existed, and as one of the most successful in all that he undertook.   This he owed to his judgment in the selection of his officers, and to his talent for commanding them.   He carried his arms into the most remote countries, and rendered his name so formidable that not a few nations spontaneously submitted to his supremacy.   Nor was there ever an Empire of such vast extent.   He cultivated literature, protected its professors, and even thankfully received their advice.   Yet he never placed a Chinese in his cabinet, and he employed foreigners only as Ministers.   These, however, he chose with discernment, *always excepting the Ministers of Finance.*   He really loved his subjects ; and if they were not always happy under his government, it is because they took care to conceal their sufferings. There were in those days no Public Censors whose duty it is to warn the Sovereign of what is going on : and no one dared to speak out for fear of the resentment of the Ministers, who were the depositaries of the Imperial authority, and the authors of the oppressions under which the people laboured.   Several Chinese, men of letters and of great ability, who lived at Hupilai's court, might have rendered that prince the greatest service in the administration of his dominions, but they never were intrusted with any but subordinate offices, and they were not in a position to make known the malversations of those public blood-suckers." (*De Mailla,* IX. 459-460.)

AHMAD was a native of Fenáket (afterwards Sháh-Rúkhia), near the Jaxartes, and obtained employment under Kúblái through the Empress Jamui Khatun, who had known him before her marriage.   To her Court he was originally attached, but we find him already employed in high financial office in 1264.   Kúblái's demands for money must have been very large, and he eschewed looking too closely into the character of his financial agents or the means by which they raised money for him. Ahmad was very successful in this, and being a man of great talent and address, obtained immense influence over the Emperor, until at last nothing was done save by his direction, though he always *appeared* to be acting under the orders of Kúblái. The Chinese authorities in Gaubil and De Mailla speak strongly of his oppressions, but only in general terms, and without affording such particulars as we derive from the text.

The Hereditary Prince Chingkim was strongly adverse to Ahmad ; and some of the high Chinese officials on various occasions made remonstrance against the Minister's proceedings ; but Kúblái turned a deaf ear to them, and Ahmad succeeded in ruining most of his opponents. (*Gaubil,* 141, 143, 151 ; *De Mailla,* IX. 316-317 ; *D'Ohsson,* II. 468-469.)

[The Rev. W. S. Ament (*Marco Polo in Cambaluc,* 105) writes : "No name is more execrated than that of Ah-ha-ma (called Achmath by Polo), a Persian, who was chosen to manage the finances of the Empire.   He was finally destroyed by a combination against him while the Khan was absent with Crown Prince Chen Chin, on a visit to Shang Tu."   Achmath has his biography under the name of *A-ho-ma* (Ahmed) in the ch. 205 of the *Yuen-shi,* under the rubric "Villanous Ministers." (*Bretschneider, Med. Res.* I. p. 272.)—H. C.]

NOTE 3.—This term *Bailo* was the designation of the representative of Venetian dignity at Constantinople, called *Podestà* during the period of the Latin rule there, and it has endured throughout the Turkish Empire to our own day in the form *Balios* as the designation of a Frank Consul.   [There was also a Venetian *bailo* in Syria.—H. C.] But that term itself could scarcely have been in use at Cambaluc, even among the handful of Franks, to designate the powerful Minister, and it looks as if Marco had confounded the word in his own mind with some Oriental term of like sound, possibly the Arabic *Wáli,* "a Prince, Governor of a Province, ... a chief Magistrate." (*F. Johnson.*)   In the *Roteiro* of the Voyage of Vasco da Gama (2nd ed. Lisbon, 1861, pp. 53-54) it is said that on the arrival of the ships at Calicut the King sent "a man who was called the *Bale,* which is much the same as *Alquaide.*"   And the Editor gives the same explanation that I have suggested.

I observe that according to Pandit Manphúl the native governor of Kashgár, under the Chinese Amban, used to be called the *Baili Beg*. [In this case *Baili* stands for *beiléh*.—H. C.] (*Panjab Trade Report*, App. p. cccxxxvii.)

NOTE 4.—The story, as related in De Mailla and Gaubil, is as follows. It contains much less detail than the text, and it differs as to the manner of the chief conspirator's death, whilst agreeing as to his name and the main facts of the episode.

In the spring of 1282 (Gaubil, 1281) Kúblái and Prince Chingkim had gone off as usual to Shangtu, leaving Ahmad in charge at the Capital. The whole country was at heart in revolt against his oppressions. Kúblái alone knew, or would know, nothing of them.

WANGCHU, a chief officer of the city, resolved to take the opportunity of delivering the Empire from such a curse, and was joined in his enterprise by a certain sorcerer called Kao Hoshang. They sent two Lamas to the Council Board with a message that the Crown Prince was returning to the Capital to take part in certain Buddhist ceremonies, but no credit was given to this. Wangchu then, pretending to have received orders from the Prince, desired an officer called CHANG-Y (perhaps the Chenchu of Polo's narrative) to go in the evening with a guard of honour to receive him. Late at night a message was sent to summon the Ministers, as the Prince (it was pretended) had already arrived. They came in haste with Ahmad at their head, and as he entered the Palace Wangchu struck him heavily with a copper mace and stretched him dead. Wangchu was arrested, or according to one account surrendered, though he might easily have escaped, confident that the Crown Prince would save his life. Intelligence was sent off to Kúblái, who received it at Chaghan-Nor. (See Book I. ch. lx.) He immediately despatched officers to arrest the guilty and bring them to justice. Wangchu, Chang-y, and Kao Hoshang were publicly executed at the Old City ; Wangchu dying like a hero, and maintaining that he had done the Empire an important service which would yet be acknowledged. (*De Mailla*, IX. 412-413 ; *Gaubil*, 193-194 ; *D'Ohsson*, II. 470.) [Cf. *G. Phillips*, in *T'oung-Pao*, I. p. 220.— H. C.]

NOTE 5.—And it is a pleasant fact that Messer Marco's presence, and his upright conduct upon this occasion, have not been forgotten in the Chinese Annals : " The Emperor having returned from Chaghan-Nor to Shangtu, desired POLO, Assessor of the Privy Council, to explain the reasons which had led Wangchu to commit this murder. Polo spoke with boldness of the crimes and oppressions of Ahama (Ahmad), which had rendered him an object of detestation throughout the Empire. The Emperor's eyes were opened, and he praised the courage of Wangchu. He complained that those who surrounded him, in abstaining from admonishing him of what was going on, had thought more of their fear of displeasing the Minister than of the interests of the State." By Kúblái's order, the body of Ahmad was taken up, his head was cut off and publicly exposed, and his body cast to the dogs. His son also was put to death with all his family, and his immense wealth confiscated. 714 persons were punished, one way or other, for their share in Ahmad's malversations. (*De Mailla*, IX. 413-414.)

What is said near the end of this chapter about the Kaan's resentment against the Saracens has some confirmation in circumstances related by Rashiduddin. The refusal of some Mussulman merchants, on a certain occasion at Court, to eat of the dishes sent them by the Emperor, gave great offence, and led to the revival of an order of Chinghiz, which prohibited, under pain of death, the slaughter of animals by cutting their throats. This endured for seven years, and was then removed on the strong representation made to Kúblái of the loss caused by the cessation of the visits of the Mahomedan merchants. On a previous occasion also the Mahomedans had incurred disfavour, owing to the ill-will of certain Christians, who quoted to Kúblái a text of the Koran enjoining the killing of polytheists. The Emperor sent for the Mullahs, and asked them why they did not act on the Divine injunction? All they could say was that the time was not yet come ! Kúblái ordered them for execution, and was only appeased by the intercession of Ahmad, and the introduction of a divine with more tact, who smoothed over obnoxious applications of the text. (*D'Ohsson*, II. 492-493.)

CHAPTER XXIV.

How the Great Kaan causeth the Bark of Trees, made into something like Paper, to pass for Money over all his Country.

Now that I have told you in detail of the splendour of this City of the Emperor's, I shall proceed to tell you of the Mint which he hath in the same city, in the which he hath his money coined and struck, as I shall relate to you.   And in doing so I shall make manifest to you how it is that the Great Lord may well be able to accomplish even much more than I have told you, or am going to tell you, in this Book.   For, tell it how I might, you never would be satisfied that I was keeping within truth and reason!

The Emperor's Mint then is in this same City of Cambaluc, and the way it is wrought is such that you might say he hath the Secret of Alchemy in perfection, and you would be right!   For he makes his money after this fashion.

He makes them take of the bark of a certain tree, in fact of the Mulberry Tree, the leaves of which are the food of the silkworms,—these trees being so numerous that whole districts are full of them.   What they take is a certain fine white bast or skin which lies between the wood of the tree and the thick outer bark, and this they make into something resembling sheets of paper, but black.   When these sheets have been prepared they are cut up into pieces of different sizes.   The smallest of these sizes is worth a half tornesel; the next, a little

larger, one tornesel ; one, a little larger still, is worth
half a silver groat of Venice ; another a whole groat ;
others yet two groats, five groats, and ten groats.
There is also a kind worth one Bezant of gold, and
others of three Bezants, and so up to ten. All these
pieces of paper are [issued with as much solemnity and
authority as if they were of pure gold or silver ; and on
every piece a variety of officials, whose duty it is, have to
write their names, and to put their seals. And when all
is prepared duly, the chief officer deputed by the Kaan
smears the Seal entrusted to him with vermilion, and
impresses it on the paper, so that the form of the Seal
remains printed upon it in red ; the Money is then
authentic. Any one forging it would be punished with
death.] And the Kaan causes every year to be made
such a vast quantity of this money, which costs him
nothing, that it must equal in amount all the treasure
in the world.

  With these pieces of paper, made as I have described,
he causes all payments on his own account to be made ;
and he makes them to pass current universally over all
his kingdoms and provinces and territories, and whithersoever his power and sovereignty extends. And nobody,
however important he may think himself, dares to refuse
them on pain of death. And indeed everybody takes
them readily, for wheresoever a person may go throughout the Great Kaan's dominions he shall find these
pieces of paper current, and shall be able to transact all
sales and purchases of goods by means of them just as
well as if they were coins of pure gold. And all the
while they are so light that ten bezants' worth does not
weigh one golden bezant.

  Furthermore all merchants arriving from India or
other countries, and bringing with them gold or silver
or gems and pearls, are prohibited from selling to any one

but the Emperor. He has twelve experts chosen for this business, men of shrewdness and experience in such affairs; these appraise the articles, and the Emperor then pays a liberal price for them in those pieces of paper. The merchants accept his price readily, for in the first place they would not get so good an one from anybody else, and secondly they are paid without any delay. And with this paper-money they can buy what they like anywhere over the Empire, whilst it is also vastly lighter to carry about on their journeys. And it is a truth that the merchants will several times in the year bring wares to the amount of 400,000 bezants, and the Grand Sire pays for all in that paper. So he buys such a quantity of those precious things every year that his treasure is endless, whilst all the time the money he pays away costs him nothing at all. Moreover, several times in the year proclamation is made through the city that any one who may have gold or silver or gems or pearls, by taking them to the Mint shall get a handsome price for them. And the owners are glad to do this, because they would find no other purchaser give so large a price. Thus the quantity they bring in is marvellous, though these who do not choose to do so may let it alone. Still, in this way, nearly all the valuables in the country come into the Kaan's possession.

When any of those pieces of paper are spoilt—not that they are so very flimsy neither—the owner carries them to the Mint, and by paying three per cent. on the value he gets new pieces in exchange. And if any Baron, or any one else soever, hath need of gold or silver or gems or pearls, in order to make plate, or girdles, or the like, he goes to the Mint and buys as much as he list, paying in this paper-money.[1]

Now you have heard the ways and means whereby the Great Kaan may have, and in fact *has*, more treasure

than all the Kings in the World ; and you know all about
it and the reason why.   And now I will tell you of the
great Dignitaries which act in this city on behalf of the
Emperor.

NOTE I.—It is surprising to find that, nearly two centuries ago, Magaillans, a
missionary who had lived many years in China, and was presumably a Chinese
scholar, should have utterly denied the truth of Polo's statements about the paper-
currency of China.   Yet the fact even then did not rest on Polo's statement only.
The same thing had been alleged in the printed works of Rubruquis, Roger Bacon,
Hayton, Friar Odoric, the Archbishop of Soltania, and Josaphat Barbaro, to say
nothing of other European authorities that remained in manuscript, or of the
numerous Oriental records of the same circumstance.

The issue of paper-money in China is at least as old as the beginning of the 9th
century.   In 1160 the system had gone to such excess that government paper
equivalent in nominal value to 43,600,000 ounces of silver had been issued in six years,
and there were local notes besides ; so that the Empire was flooded with rapidly
depreciating paper.

The *Kin* or "Golden" Dynasty of Northern Invaders who immediately preceded
the Mongols took to paper, in spite of their title, as kindly as the native sovereigns.
Their notes had a course of seven years, after which new notes were issued to the
holders, with a deduction of 15 per cent.

The Mongols commenced their issues of paper-money in 1236, long before they
had transferred the seat of their government to China.   Kúblái made such an issue
in the first year of his reign (1260), and continued to issue notes copiously till the end.
In 1287 he put out a complete new currency, one note of which was to exchange
against *five* of the previous series of equal nominal value !   In both issues the
paper-money was, in official valuation, only equivalent to half its nominal value in
silver ; a circumstance not very easy to understand.   The paper-money was called *Chao*.

The notes of Kúblái's first issue (1260-1287) with which Polo may be supposed
most familiar, were divided into three classes ; (1) *Notes of Tens*, viz. of 10, 20, 30, and
50 *tsien* or cash ; (2) *Notes of Hundreds*, viz. of 100, 200, and 500 *tsien ;* and (3) *Notes
of Strings* or *Thousands* of cash, or in other words of *Liangs* or ounces of silver (otherwise *Tael*), viz. of 1000 and 2000 *tsien*.   There were also notes printed on silk for 1,
2, 3, 5, and 10 ounces each, valued at par in silver, but these would not circulate.
In 1275, it should be mentioned, there had been a supplementary issue of small notes
for 2, 3, and 5 cash each.

Marsden states an equation between Marco's values of the Notes and the actual
Chinese currency, to which Biot seems to assent.   I doubt its correctness, for his
assumed values of the groat or *grosso* and tornesel are surely wrong.   The grosso ran
at that time 18 to the gold ducat or sequin, and allowing for the then higher relative
value of silver, should have contained about 5*d.* of silver.   The ducat was also equivalent to 2 *lire*, and the *tornese* (*Romanin*, III. 343) was 4 deniers.   Now the denier is
always, I believe $\frac{1}{240}$ of the *lira*.   Hence the *tornese* would be $\frac{9}{20}$ of the *grosso*.

But we are not to look for *exact* correspondences, when we see Polo applying
round figures in European coinage to Chinese currency.

His bezant notes, I agree with Marsden, here represent the Chinese notes for one
and more ounces of silver.   And here the correspondence of value is much nearer than
it seems at first sight.   The Chinese *liang* or ounce of silver is valued commonly at
6*s.* 7*d.*, say roundly 80*d.** But the relation of gold and silver in civilized Asia was

---

* Even now there are at least eight different *taels* (or liangs) in extensive use over the Empire, and
varying as much as from 96 to 106 ; and besides these are many local *taels*, with about the same limits
of variation.—(*Williamson's Journeys*, I. 60.)

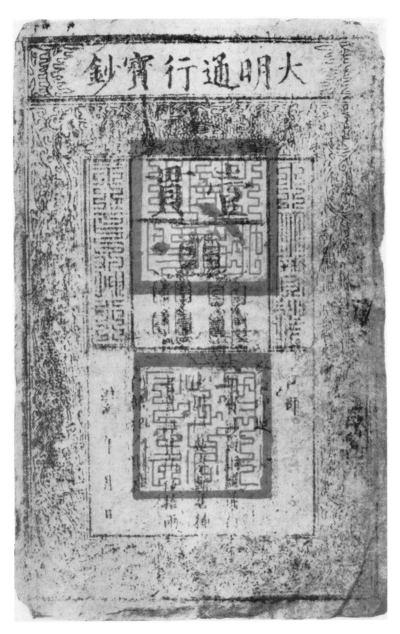

Bank-Note of the Ming Dynasty.   [*To face p.* 426.

then (see ch. I. note 4, and also *Cathay*, pp. ccl. and 442) as 10 to 1, not, as with us now, more than 15 to 1. Wherefore the *liang* in relation to gold would be worth 120*d*. or 10*s*., a little over the Venetian ducat and somewhat less than the bezant or dinár. We shall then find the table of Chinese issues, as compared with Marco's equivalents, to stand thus :—

CHINESE ISSUES, AS RECORDED.	MARCO POLO'S STATEMENT.
For 10 ounces of silver (viz. the Chinese *Ting*) *	10 bezants.
For I ounce of silver, *i.e.* I *liang*, or 1000 *tsien* (cash)	I ,,
For 500 *tsien*	10 groats.
200 ,,	5 ,, (should have been 4).
100 ,,	2 ,,
50 ,,	1 ,,
30 ,,	½ ,, (but the proportionate equivalent of half a groat would be 25 *tsien*).
20 ,,	
10 ,,	I tornesel (but the proportionate equivalent would be 7½ *tsien*).
5 ,,	½ ,, (but prop. equivalent 3¾ *tsien*).

Pauthier has given from the Chinese Annals of the Mongol Dynasty a complete Table of the Issues of Paper-Money during every year of Kúbláï's reign (1260-1294), estimated at their nominal value in *Ting* or tens of silver ounces. The lowest issue was in 1269, of 228,960 *ounces*, which at the rate of 120*d*. to the ounce (see above) =114,480*l*., and the highest was in 1290, viz. 50,002,500 ounces, equivalent at the same estimate to 25,001,250*l*.! whilst the total amount in the 34 years was 249,654,290 ounces or 124,827,144*l*. in nominal value. Well might Marco speak of the vast quantity of such notes that the Great Kaan issued annually !

To complete the history of the Chinese paper-currency so far as we can :

In 1309, a new issue took place with the same provision as in Kúbláï's issue of 1287, *i.e.* each note of the new issue was to exchange against 5 of the old of the same nominal value. And it was at the same time prescribed that the notes should exchange at par with metals, which of course it was beyond the power of Government to enforce, and so the notes were abandoned. Issues continued from time to time to the end of the Mongol Dynasty. The paper-currency is spoken of by Odoric (1320-30), by Pegolotti (1330-40), and by Ibn Batuta (1348), as still the chief, if not sole, currency of the Empire. According to the Chinese authorities, the credit of these issues was constantly diminishing, as it is easy to suppose. But it is odd that all the Western Travellers speak as if the notes were as good as gold. Pegolotti, writing for mercantile men, and from the information (as we may suppose) of mercantile men, says explicitly that there was no depreciation.

The Ming Dynasty for a time carried on the system of paper-money; with the difference that while under the Mongols no other currency had been admitted, their successors made payments in notes, but accepted only hard cash from their people !†
In 1448 the *chao* of 1000 cash was worth but 3. Barbaro still heard talk of the Chinese paper-currency from travellers whom he met at Azov about this time; but after 1455 there is said to be no more mention of it in Chinese history.

I have never heard of the preservation of any note of the Mongols; but some of the Ming survive, and are highly valued as curiosities in China. The late Sir G. T. Staunton appears to have possessed one; Dr. Lockhart formerly had two, of which he gave one to Sir Harry Parkes, and retains the other. The paper is so dark as to

---

* [The Archimandrite Palladius (*l.c.*, p. 50, note) says that "the *ting* of the Mongol time, as well as during the reign of the Kin, was a unit of weight equivalent to fifty *liang*, but not to ten *liang*. Cf. *Ch'u keng lu*, and *Yuen-shi*, ch. xcv. The *Yuen pao*, which as everybody in China knows, is equivalent to fifty *liang* (taels) of silver, is the same as the ancient *ting*, and the character *Yuen* indicates that it dates from the *Yuen* Dynasty."—H. C.]

† This is also, as regards Customs payments, the system of the Government of modern Italy.

explain Marco's description of it as black. By Dr. Lockhart's kindness I am enabled to give a reduced representation of this note, as near a facsimile as we have been able to render it, but with some *restoration*, *e.g.* of the *seals*, of which on the original there is the barest indication remaining.

[Mr. Vissering (*Chinese Currency*, Addenda, I.-III.) gives a facsimile and a description of a Chinese banknote of the Ming Dynasty belonging to the collection of the Asiatic Museum of the Academy of Sciences at St. Petersburg. "In the eighth year of the period *Hung-wu* (1375), the Emperor Tai-tsu issued an order to his minister of finances to make the *Pao-tsao* (precious bills) of the *Ta-Ming* Dynasty, and to employ as raw material for the composition of those bills the fibres of the mulberry tree."—H. C.]

Notwithstanding the disuse of Government issues of paper-money from that time till recent years, there had long been in some of the cities of China a large use of private and local promissory notes as currency. In Fuchau this was especially the case; bullion was almost entirely displaced, and the banking-houses in that city were counted by hundreds. These were under no government control; any individual or company having sufficient capital or credit could establish a bank and issue their bills, which varied in amount from 100 cash to 1000 dollars. Some fifteen years ago the Imperial Government seems to have been induced by the exhausted state of the Treasury, and these large examples of the local use of paper-currency, to consider projects for resuming that system after the disuse of four centuries. A curious report by a Committee of the Imperial Supreme Council, on a project for such a currency, appears among the papers published by the Russian Mission at Peking. It is unfavourable to the particular project, but we gather from other sources that the Government not long afterwards did open banks in the large cities of the Empire for the issue of a new paper-currency, but that it met with bad success. At Fuchau, in 1858, I learn from one notice, the dollar was worth from 18,000 to 20,000 cash in Government Bills. Dr. Rennie, in 1861, speaks of the dollar at Peking as valued at 15,000, and later at 25,000 paper cash. Sushun, the Regent, had issued a vast number of notes through banks of his own in various parts of Peking. These he failed to redeem, causing the failure of all the banks, and great consequent commotion in the city. The Regent had led the Emperor [Hien Fung] systematically into debauched habits which ended in paralysis. On the Emperor's death the Empress caused the arrest and execution of Sushun. His conduct in connection with the bank failures was so bitterly resented that when the poor wretch was led to execution (8th November, 1861), as I learn from an eye-witness, the defrauded creditors lined the streets and cheered. *

The Japanese also had a paper-currency in the 14th century. It is different in form from that of China. That figured by Siebold is a strip of strong paper doubled, 6¼ in. long by 1¾ in. wide, bearing a representation of the tutelary god of riches, with long inscriptions in Chinese characters, seals in black and red, and an indication of value in ancient Japanese characters. I do not learn whether notes of considerable amount are still used in Japan; but Sir R. Alcock speaks of banknotes for small change from 30 to 500 cash and more, as in general use in the interior.

Two notable and disastrous attempts to imitate the Chinese system of currency took place in the Middle Ages; one of them in Persia, apparently in Polo's very presence, the other in India some 36 years later.

The first was initiated in 1294 by the worthless Kaikhatu Khan, when his own and his ministers' extravagance had emptied the Treasury, on the suggestion of a financial officer called 'Izzuddín Muzaffar. The notes were direct copies of Kúblái's, even the Chinese characters being imitated as part of the device upon them.† The

---

* The first edition of this work gave a facsimile of one of this unlucky minister's notes.

† On both sides, however, was the Mahomedan formula, and beneath that the words *Víranjín Túrjí*, a title conferred on the kings of Persia by the Kaan. There was also an inscription to the following effect: that the Emperor in the year 693 (A.H.) had issued these auspicious *chao*, that all who forged or uttered false notes should be summarily punished, with their wives and children, and their property confiscated; and that when these auspicious notes were once in circulation, poverty would vanish, provisions become cheap, and rich and poor be equal (*Cowell*). The use of the term *chao* at Tabriz may be compared with that of *Bánklót*, current in modern India.

Chinese name *Chao* was applied to them, and the Mongol Resident at Tabriz, Pulad Chingsang, was consulted in carrying out the measure.   Expensive preparations were made for this object ; offices called *Chao-Khánahs* were erected in the principal cities of the provinces, and a numerous staff appointed to carry out the details.   Ghazan Khan in Khorasan, however, would have none of it, and refused to allow any of these preparations to be made within his government.   After the constrained use of the Chao for two or three days Tabriz was in an uproar ; the markets were closed ; the people rose and murdered 'Izzuddín ; and the whole project had to be abandoned. Marco was in Persia at this time, or just before, and Sir John Malcolm not unnaturally suggests that he might have had something to do with the scheme ; a suggestion which excites a needless commotion in the breast of M. Pauthier.   We may draw from the story the somewhat notable conclusion that *Block-printing* was practised, at least for this one purpose, at Tabriz in 1294.

The other like enterprise was that of Sultan Mahomed Tughlak of Delhi, in 1330-31.   This also was undertaken for like reasons, and was in professed imitation of the Chao of Cathay.   Mahomed, however, used copper tokens instead of paper ; the copper being made apparently of equal weight to the gold or silver coin which it represented.   The system seems to have had a little more vogue than at Tabriz, but was speedily brought to an end by the ease with which forgeries on an enormous scale were practised.   The Sultan, in hopes of reviving the credit of his currency, ordered that every one bringing copper tokens to the Treasury should have them cashed in gold or silver.   " The people who in despair had flung aside their copper coins like stones and bricks in their houses, all rushed to the Treasury and exchanged them for gold and silver.   In this way the Treasury soon became empty, but the copper coins had as little circulation as ever, and a very grievous blow was given to the State."

An odd issue of currency, not of paper, but of leather, took place in Italy a few years before Polo's birth.   The Emperor Frederic II., at the siege of Faenza in 1241, being in great straits for money, issued pieces of leather stamped with the mark of his mint at the value of his Golden Augustals.   This leather coinage was very popular, especially at Florence, and it was afterwards honourably redeemed by Frederic's Treasury.   Popular tradition in Sicily reproaches William the Bad among his other sins with having issued money of leather, but any stone is good enough to cast at a dog with such a surname.

[Ma Twan-lin mentions that in the fourth year of the period Yuen Show (B.C. 119), a currency of white metal and *deer-skin* was made.   Mr. Vissering (*Chinese Currency*, 38) observes that the skin-tallies " were purely tokens, and have had nothing in common with the leather-money, which was, during a long time, current in Russia. This Russian skin-money had a truly representative character, as the parcels were used instead of the skins from which they were cut ; the skins themselves being too bulky and heavy to be constantly carried backward and forward, only a little piece was cut off, to figure as a token of possession of the whole skin.   The ownership of the skin was proved when the piece fitted in the hole."

Mr Rockhill (*Rubruck*, 201 note) says : " As early as B.C. 118, we find the Chinese using 'leather-money' (*p'i pi*).   These were pieces of white deer-skin, a foot square, with a coloured border.   Each had a value of 40,000 cash. (*Ma Twan-lin*, Bk. 8, 5.) "

Mr Charles F. Keary (*Coins and Medals*, by S. Lane Poole, 128) mentions that " in the reign of Elizabeth there was a very extensive issue of private tokens in lead, tin, latten, and *leather*."—H. C.]

(*Klapr.* in *Mém. Rel. à l'Asie*, I. 375 *seqq.* ; *Biot*, in *J. As.* sér. III. tom. iv. ; *Marsden* and *Pauthier*, in loco ; *Parkes*, in *J. R. A. S.* XIII. 179 ; *Doolittle*, 452 *seqq.* ; *Wylie, J. of Shanghai Lit. and Scient. Soc.* No. I. ; *Arbeiten der kais. russ. Gesandsch. zu Peking*, I. p. 48 ; *Rennie, Peking*, etc., I. 296, 347 ; *Birch*, in *Num. Chron.* XII. 169 ; Information from *Dr. Lockhart* ; *Alcock*, II. 86 ; *D'Ohsson*, IV. 53 ; *Cowell*, in *J. A. S. B.* XXIX. 183 *seqq.* ; *Thomas, Coins of Patan Sovs. of*

Hind. (from *Numism. Chron.* 1852), p. 139 *seqq.*; *Kington's Fred. II.* II. 195; *Amari*, III. 816; *W. Vissering, On Chinese Currency*, Leiden, 1877.)

["Without doubt the Mongols borrowed the bank-note system from the Kin. Up to the present time there is in Si-ngan-fu a block kept, which was used for printing the bank-notes of the Kin Dynasty. I have had the opportunity of seeing a print of those bank-notes, they were of the same size and shape as the bank-notes of the Ming. A reproduction of the text of the Kin bank-notes is found in the *Kin shi ts'ui pien.* This copy has the characters *pao küan* (precious charter) and the years of reign *Chêng Yew*, 1213-1216. The first essay of the Mongols to introduce bank-notes dates from the time of Ogodai Khan (1229-1242), but Chinese history only mentions the fact without giving details. At that time silk in skeins was the only article of a determinate value in the trade and on the project of *Ye lü ch'u ts'ai*, minister of Ogodai, the taxes were also collected in silk delivered by weight. It can therefore be assumed that the name *sze ch'ao* (*i.e.* bank-notes referring to the weight of silk) dates back to the same time. At any rate, at a later time, as, under the reign of Kubilai, the issuing of bank-notes was decreed, silk was taken as the standard to express the value of silver and 1000 *liang* silk was estimated = 50 *liang* (or 1 *ting*) silver. Thus, in consequence of those measures, it gradually became a rule to transfer the taxes and rents originally paid in silk, into silver. The wealth of the Mongol Khans in precious metals was renowned. The accounts regarding their revenues, however, which we meet with occasionally in Chinese history, do not surprise by their vastness. In the year 1298, for instance, the amount of the revenue is stated in the *Siu t'ung Kien* to have been :—

19,000 *liang* of gold = (190,000 *liang* of silver, according to the exchange of that time at the rate of 1 to 10).

60,000 *liang* of silver.

3,600,000 *ting* of silver in bank-notes (*i.e.* 180 millions *liang*); altogether 180,250,000 *liang* of silver.

The number seems indeed very high for that time. But if the exceedingly low exchange of the bank-notes be taken into consideration, the sum will be reduced to a modest amount." (*Palladius*, pp. 50-51.)—H. C.]

[Dr. Bretschneider (*Hist. Bot. Disc.*, I. p. 4) makes the following remark :—" Polo states (I. 409) that the Great Kaan causeth the bark of great Mulberry-trees, made into something like paper, to pass for money." He seems to be mistaken. Paper in China is not made from mulberry-trees but from the *Broussonetia papyrifera*, which latter tree belongs to the same order of Moraceae. The same fibres are used also in some parts of China for making cloth, and Marco Polo alludes probably to the same tree when stating (II. 108) " that in the province of Cuiju (Kwei chau) they manufacture stuff of the bark of certain trees, which form very fine summer clothing."—H. C.]

CHAPTER XXV.

Concerning the Twelve Barons who are set over all the Affairs of the Great Kaan.

You must know that the Great Kaan hath chosen twelve great Barons to whom he hath committed all the necessary affairs of thirty-four great provinces; and now I will tell you particulars about them and their establishments.

You must know that these twelve Barons reside all together in a very rich and handsome palace, which is inside the city of Cambaluc, and consists of a variety of edifices, with many suites of apartments. To every province is assigned a judge and several clerks, and all reside in this palace, where each has his separate quarters. These judges and clerks administer all the affairs of the provinces to which they are attached, under the direction of the twelve Barons. Howbeit, when an affair is of very great importance, the twelve Barons lay in before the Emperor, and he decides as he thinks best. But the power of those twelve Barons is so great that they choose the governors for all those thirty-four great provinces that I have mentioned, and only after they have chosen do they inform the Emperor of their choice. This he confirms, and grants to the person nominated a tablet of gold such as is appropriate to the rank of his government.

Those twelve Barons also have such authority that they can dispose of the movements of the forces, and send them whither, and in such strength, as they please. This is done indeed with the Emperor's cognizance, but still the orders are issued on their authority. They are styled SHIENG, which is as much as to say "The Supreme Court," and the palace where they abide is also called *Shieng*. This body forms the highest authority at the Court of the Great Kaan; and indeed they can favour and advance whom they will. I will not now name the thirty-four provinces to you, because they will be spoken of in detail in the course of this Book.[1]

___

NOTE 1.—Pauthier's extracts from the Chinese Annals of the Dynasty, in illustration of this subject, are interesting. These, as he represents them, show the Council of Ministers usually to have consisted of twelve high officials, viz.: two *Ch'ing-siang* [丞 相] or (chief) ministers of state, one styled, "of the Right," and the other "of the Left"; four called *P'ing-chang ching-ssé*, which seems to mean something like ministers in charge of special departments; four assistant ministers; two Counsellors.

Rashiduddin, however, limits the Council to the first two classes: "Strictly speaking, the Council of State is composed of four Ch'ing-sang (*Ch'ing-siang*) or great officers (*Wazírs* he afterwards terms them), and four Fanchán (*P'ing-chang*) or associated members, taken from the nations of the Tajiks, Cathayans, Ighurs, and Arkaún" (*i.e.* Nestorian Christians). (Compare p. 418, *supra.*)

[A Samarkand man, Seyyd Tadj Eddin Hassan ben el Khallal, quoted in the *Masálak al Absár*, says: "Near the Khan are two amírs who are his ministers; they are called *Djing San* جينكسان (Ch'ing-siang). After them come the two *Bidjan* بجان (P'ing Chang), then the two *Zoudjin* زوجين (Tso Chen), then the two *Yudjin* يوجين (Yu Chen), and at last the *Landjun* لنجون (Lang Chang), head of the scribes, and secretary of the sovereign. The Khan holds a sitting every day in the middle of a large building called *Chen* شن (Sheng), which is very like our Palace of Justice." (*C. Schefer, Cent. Ec. Langues Or.*, pp. 18-19.)—H. C.]

In a later age we find the twelve Barons reappearing in the pages of Mendoza: "The King hath in this city of Tabin (Peking), where he is resident, a royal council of twelve counsellors and a president, chosen men throughout all the kingdom, and such as have had experience in government many years." And also in the early centuries of the Christian era we hear that the Khan of the Turks had his twelve grandees, divided into those of the Right and those of the Left, probably a copy from a Chinese order then also existing.

But to return to Rashiduddin: "As the Kaan generally resides at the capital, he has erected a place for the sittings of the Great Council, called *Sing* . . . . The dignitaries mentioned above are expected to attend daily at the Sing, and to make themselves acquainted with all that passes there."

The *Sing* of Rashid is evidently the Shieng or Sheng (*Scieng*) of Polo. M. Pauthier is on this point somewhat contemptuous towards Neumann, who, he says, confounds Marco Polo's twelve Barons or Ministers of State with the chiefs of the twelve great provincial governments called *Sing*, who had their residence at the chief cities of those governments; whilst in fact Polo's *Scieng* (he asserts) has nothing to do with the *Sing*, but represents the Chinese word *Siang* "a minister," and "the office of a minister." [There was no doubt a confusion between *Siang* 相 and *Sheng* 省.—H. C.]

It is very probable that two different words, *Siang* and *Sing*, got confounded by the non-Chinese attachés of the Imperial Court; but it seems to me quite certain that they applied the same word, Sing or Sheng, to both institutions, viz. to the High Council of State, and to the provincial governments. It also looks as if Marco Polo himself had made that very confusion with which Pauthier charges Neumann. For whilst here he represents the twelve Barons as forming a Council of State at the capital, we find further on, when speaking of the city of Yangchau, he says: "*Et si siet en ceste cité uns des xii Barons du Grant Kaan; car elle est esleue pour un des xii sieges,*" where the last word is probably a mistranscription of *Sciengs*, or *Sings*, and in any case the reference is to a distribution of the empire into twelve governments.

To be convinced that *Sing* was used by foreigners in the double sense that I have said, we have only to proceed with Rashiduddin's account of the administration. After what we have already quoted, he goes on: "The *Sing* of Khanbaligh is the most eminent, and the building is very large. . . . *Sings* do not exist in all the cities, but only in the capitals of great provinces. . . . In the whole empire of the Kaan there are twelve of these Sings; but that of Khanbaligh is the only one which has Ching-sangs amongst its members." Wassáf again, after describing the greatness of Khanzai (Kinsay of Polo) says: "These circumstances characterize the capital itself, but four hundred cities of note, and embracing ample territories, are dependent on its jurisdiction, insomuch that the most inconsiderable of those cities surpasses Baghdad and Shiraz. In the number of these cities are Lankinfu and Zaitun, and Chinkalán; for they call Khanzai a *Shing*, *i.e.* a great city in which the high and mighty Council of Administration holds its meetings." Friar Odoric again says: "This empire hath been divided by the Lord thereof into twelve parts, each one thereof is termed a Singo."

Polo, it seems evident to me, knew nothing of Chinese. His *Shieng* is no direct attempt to represent *any* Chinese word, but simply the term that he had been used to employ in talking Persian or Turki, in the way that Rashiduddin and Wassáf employ it.

I find no light as to the thirty-four provinces into which Polo represents the empire as divided, unless it be an enumeration of the provinces and districts which he describes in the second and third parts of Bk. II., of which it is not difficult to reckon thirty-three or thirty-four, but not worth while to repeat the calculation.

[China was then divided into twelve *Sheng* or provinces: Cheng-Tung, Liao-Yang, Chung-Shu, Shen-Si, Ling-Pe (Karakorum), Kan-Suh, Sze-ch'wan, Ho-Nan Kiang-Pe, Kiang-Ché, Kiang-Si, Hu-Kwang and Yun-Nan. Rashiduddin (*J. As.*, XI. 1883, p. 447) says that of the twelve Sing, Khanbaligh was the only one with *Chin-siang*. We read in *Morrison's Dict.* (Pt. II. vol. i. p. 70): "Chin-seang, a Minister of State, was so called under the Ming Dynasty." According to Mr. E. H. Parker (*China Review*, xxiv. p. 101), *Ching Siang* were abolished in 1395. I imagine that the thirty-four provinces refer to the *Fu* cities, which numbered however *thirty-nine*, according to *Oxenham's Historical Atlas.*—H. C.]

(*Cathay*, 263 *seqq.* and 137; *Mendoza*, I. 96; *Erdmann*, 142; *Hammer's Wassáf*, p. 42, but corrected.)

CHAPTER XXVI.

How the Kaan's Posts and Runners are sped through many Lands and Provinces.

Now you must know that from this city of Cambaluc proceed many roads and highways leading to a variety of provinces, one to one province, another to another; and each road receives the name of the province to which it leads; and it is a very sensible plan.[1] And the messengers of the Emperor in travelling from Cambaluc, be the road whichsoever they will, find at every twenty-five miles of the journey a station which they call *Yamb*,[2] or, as we should say, the "Horse-Post-House." And at each of those stations used by the messengers, there is a large and handsome building for them to put up at, in which they find all the rooms furnished with fine beds and all other necessary articles in rich silk, and where they are provided with everything they can want. If even a king were to arrive at one of these, he would find himself well lodged.

At some of these stations, moreover, there shall be posted some four hundred horses standing ready for the use of the messengers; at others there shall be two hundred, according to the requirements, and to what the Emperor has established in each case. At every twenty-five miles, as I said, or anyhow at every thirty miles, you find one of these stations, on all the principal highways leading to the different provincial governments; and the same is the case throughout all the chief provinces subject to the Great Kaan.[3] Even when the messengers have to pass through a roadless tract where neither house nor hostel exists, still there the station-houses have been established just the same, excepting that the intervals are somewhat greater, and the day's journey is fixed at thirty-five to forty-five miles, instead of twenty-five to thirty. But they are provided with horses and all the other necessaries just like those we have described, so that the Emperor's messengers, come they from what region they may, find everything ready for them.

And in sooth this is a thing done on the greatest scale of magnificence that ever was seen. Never had emperor, king, or lord, such wealth as this manifests! For it is a fact that on all these posts taken together there are more than 300,000 horses kept up, specially for the use of the messengers. And the great buildings that I have mentioned are more than 10,000 in number, all richly furnished, as I told you. The thing is on a scale so wonderful and costly that it is hard to bring oneself to describe it.[4]

But now I will tell you another thing that I had forgotten, but which ought to be told whilst I am on this subject. You must know that by the Great Kaan's orders there has been established between those post-houses, at every interval of three miles, a little fort with some forty houses round about it, in which dwell the people who act as the Emperor's foot-runners. Every one of those runners wears a great wide belt, set all over with bells, so that as they run the three miles from post to post their bells are heard jingling a long way off. And thus on reaching the post the runner finds another man similarly equipt, and all ready to take his place, who instantly takes over whatsoever he has in charge, and with it receives a slip of paper from the clerk, who is always at hand for the purpose ; and so the new man sets off and runs his three miles. At the next station he finds his relief ready in like manner ; and so the post proceeds, with a change at every three miles. And in this way the Emperor, who has an immense number of these runners, receives despatches with news from places ten days' journey off in one day and night ; or, if need be, news from a hundred days off in ten days and nights ; and that is no small matter ! (In fact in the fruit season many a time fruit shall be gathered one morning in Cambaluc, and the evening of the next day it shall reach the Great Kaan at Chandu, a distance of ten days' journey.[5] The clerk at each of the posts notes the time of each courier's arrival and departure ; and there are often other officers whose business it is to make monthly visitations of all the posts, and to punish those runners who have been slack in their work.[6]) The Emperor exempts these men from all tribute, and pays them besides.

Moreover, there are also at those stations other men equipt similarly with girdles hung with bells, who are employed for expresses when there is a call for great haste in sending despatches to any governor of a province, or to give news when any Baron has revolted, or in other such emergencies; and these men travel a good two hundred or two hundred and fifty miles in the day, and as much in the night. I'll tell you how it stands. They take a horse from those at the station which are standing ready saddled, all fresh and in wind, and mount and go at full speed, as hard as they can ride in fact. And when those at the next post hear the bells they get ready another horse and a man equipt in the same way, and he takes over the letter or whatever it be, and is off full-speed to the third station, where again a fresh horse is found all ready, and so the despatch speeds along from post to post, always at full gallop, with regular change of horses. And the speed at which they go is marvellous. (By night, however, they cannot go so fast as by day, because they have to be accompanied by footmen with torches, who could not keep up with them at full speed.)

Those men are highly prized; and they could never do it, did they not bind hard the stomach, chest and head with strong bands. And each of them carries with him a gerfalcon tablet, in sign that he is bound on an urgent express; so that if perchance his horse break down, or he meet with other mishap, whomsoever he may fall in with on the road, he is empowered to make him dismount and give up his horse. Nobody dares refuse in such a case; so that the courier hath always a good fresh nag to carry him.[7]

Now all these numbers of post-horses cost the Emperor nothing at all; and I will tell you the how and the why. Every city, or village, or hamlet, that stands near one of those post-stations, has a fixed demand made on it for as many horses as it can supply, and these it must furnish to the post.   And in this way are provided all the posts of the cities, as well as the towns and villages round about them ; only in uninhabited tracts the horses are furnished at the expense of the Emperor himself.

(Nor do the cities maintain the full number, say of 400 horses, always at their station, but month by month 200 shall be kept at the station, and the other 200 at grass, coming in their turn to relieve the first 200.   And if there chance to be some river or lake to be passed by the runners and horse-posts, the neighbouring cities are bound to keep three or four boats in constant readiness for the purpose.)

And now I will tell you of the great bounty exercised by the Emperor towards his people twice a year.

NOTE I.—The G. Text has " *et ce est mout sçue chouse* " ; Pauthier's Text, " *mais il est moult celé.*"   The latter seems absurd.   I have no doubt that *sçue* is correct, and is an Italianism, *saputo* having sometimes the sense of prudent or judicious.   Thus P. della Valle (II. 26), speaking of Shah Abbas : " *Ma noti V.S. i tiri di questo re, saputo insieme e bizzarro,*" "acute with all his eccentricity."

NOTE 2.—Both Neumann and Pauthier seek Chinese etymologies of this Mongol word, which the Tartars carried with them all over Asia.   It survives in Persian and Turki in the senses both of a post-house and a post-horse, and in Russia, in the former sense, is a relic of the Mongol dominion. The ambassadors of Shah Rukh, on arriving at Sukchu, were lodged in the *Yám-Khána,* or post-house, by the city gate ; and they found ninety-nine such Yams between Sukchu and Khanbaligh, at each of which they were supplied with provisions, servants, beds, night-clothes, etc.   Odoric likewise speaks of the hostelries called *Yam,* and Rubruquis applies the same term to quarters in the imperial camp, which were assigned for the lodgment of ambassadors.   (*Cathay,* ccii. 137 ; *Rubr.* 310.)

[Mr. Rockhill (*Rubruck,* 101, note) says that these post-stations were established by Okkodai in 1234 throughout the Mongol empire.   (*D'Ohsson,* ii. 63.)   Dr. G. Schlegel (*T'oung Pao,* II. 1891, 265, note) observes that *iam* is not, as Pauthier supposed, a contraction of *yi-mà,* horse post-house (*yi-mà* means post-horse, and Pauthier makes a mistake), but represents the Chinese character 站, pronounced at present *chán,* which means in fact a road station, a post.   In Annamite, this character 站 is pronounced *trqm,* and it means, according to *Bonet's Dict. Annamite-Français :* " Relais de poste, station de repos."   (See *Bretschneider, Med. Res.* I. p. 187 note.) —H. C.]

NOTE 3.—Martini and Magaillans, in the 17th century, give nearly the same account of the government hostelries.

NOTE 4.—Here Ramusio has this digression : "Should any one find it difficult to understand how there should be such a population as all this implies, and how they can subsist, the answer is that all the Idolaters, and Saracens as well, take six, eight, or ten wives apiece when they can afford it, and beget an infinity of children. In fact, you shall find many men who have each more than thirty sons who form an armed retinue to their father, and this through the fact of his having so many wives. With us, on the other hand, a man hath but one wife; and if she be barren, still he must abide by her for life, and have no progeny; thus we have not such a population as they have.

"And as regards food, they have abundance; for they generally consume rice, panic, and millet (especially the Tartars, Cathayans, and people of Manzi); and these three crops in those countries render an hundred-fold. Those nations use no bread, but only boil those kinds of grain with milk or meat for their victual. Their wheat, indeed, does not render so much, but this they use only to make vermicelli, and pastes of that description. No spot of arable land is left untilled; and their cattle are infinitely prolific, so that when they take the field every man is followed by six, eight, or more horses for his own use. Thus you may clearly perceive how the population of those parts is so great, and how they have such an abundance of food."

NOTE 5.—The Burmese kings used to have the odoriferous *Durian* transmitted by horse-posts from Tenasserim to Ava. But the most notable example of the rapid transmission of such dainties, and the nearest approach I know of to their despatch by telegraph, was that practised for the benefit of the Fatimite Khalif Aziz (latter part of 10th century), who had a great desire for a dish of cherries of Balbek. The Wazir Yakub ben-Kilis caused six hundred pigeons to be despatched from Balbek to Cairo, each of which carried attached to either leg a small silk bag containing a cherry! (*Quat. Makrizi*, IV. 118.)

NOTE 6.—"Note is taken at every post," says Amyot, in speaking of the Chinese practice of last century, "of the time of the courier's arrival, in order that it may be known at what point delays have occurred." (*Mém.* VIII. 185.)

NOTE 7.—The post-system is described almost exactly as in the text by Friar Odoric and the Archbishop of Soltania, in the generation after Polo, and very much in the same way by Magaillans in the 17th century. Posts had existed in China from an old date. They are spoken of by Mas'udi and the *Relations* of the 9th century. They were also employed under the ancient Persian kings; and they were in use in India, at least in the generation after Polo. The Mongols, too, carried the institution wherever they went.

Polo describes the couriers as changed at short intervals, but more usually in Asiatic posts the same man rides an enormous distance. The express courier in Tibet, as described by "the Pandit," rides from Gartokh to Lhasa, a distance of 800 miles, travelling day and night. The courier's coat is *sealed* upon him, so that he dares not take off his clothes till the seal is officially broken on his arrival at the terminus. These messengers had faces cracked, eyes bloodshot and sunken, and bodies raw with vermin. (*J. R. G. S.* XXXVIII. p. 149.) The modern Turkish post from Constantinople to Baghdad, a distance of 1100 miles, is done in twenty days by four Tartars riding night and day. The changes are at Sivas, Diarbekir, and Mosul. M. Tchihatcheff calculates that the night riding accomplishes only one quarter of the whole. (*Asie Mineure*, 2de Ptie. 632-635.)—See I. p. 352, *paï tze.*

CHAPTER XXVII.

HOW THE EMPEROR BESTOWS HELP ON HIS PEOPLE, WHEN THEY ARE AFFLICTED WITH DEARTH OR MURRAIN.

Now you must know that the Emperor sends his Messengers over all his Lands and Kingdoms and Provinces, to ascertain from his officers if the people are afflicted by any dearth through unfavourable seasons, or storms or locusts, or other like calamity; and from those who have suffered in this way no taxes are exacted for that year; nay more, he causes them to be supplied with corn of his own for food and seed. Now this is undoubtedly a great bounty on his part. And when winter comes, he causes inquiry to be made as to those who have lost their cattle, whether by murrain or other mishap, and such persons not only go scot free, but get presents of cattle. And thus, as I tell you, the Lord every year helps and fosters the people subject to him.

[There is another trait of the Great Kaan I should tell you; and that is, that if a chance shot from his bow strike any herd or flock, whether belonging to one person or to many, and however big the flock may be, he takes no tithe thereof for three years. In like manner, if the arrow strike a boat full of goods, that boat-load pays no duty; for it is thought unlucky that an arrow strike any one's property; and the Great Kaan says it would be an abomination before God, were such property, that has been struck by the divine wrath, to enter into his Treasury.[1]]

NOTE I.—The Chinese author already quoted as to Kúblái's character (Note 2, ch. xxiii. *supra*) says : " This Prince, at the sight of some evil prognostic, or when there was dearth, would remit taxation, and cause grain to be distributed to those who were in destitution. He would often complain that there never lacked informers if balances were due, or if *corvées* had been ordered, but when the necessities of the people required to be reported, not a word was said."

Wassáf tells a long story in illustration of Kúblái's justice and consideration for the peasantry. One of his sons, with a handful of followers, had got separated from the army, and halted at a village in the territory of Bishbaligh, where the people gave them sheep and wine. Next year two of the party came the same way and *demanded* a sheep and a stoup of wine. The people gave it, but went to the Kaan and told the story, saying they feared it might grow into a perpetual exaction. Kúblái sharply rebuked the Prince, and gave the people compensation and an order in their favour. (*De Mailla*, ix. 460; *Hammer's Wassaf*, 38-39.)

CHAPTER XXVIII.

How the Great Kaan causes Trees to be Planted by the Highways.

The Emperor moreover hath taken order that all the highways travelled by his messengers and the people generally should be planted with rows of great trees a few paces apart; and thus these trees are visible a long way off, and no one can miss the way by day or night. Even the roads through uninhabited tracts are thus planted, and it is the greatest possible solace to travellers. And this is done on all the ways, where it can be of service. [The Great Kaan plants these trees all the more readily, because his astrologers and diviners tell him that he who plants trees lives long.[1]

But where the ground is so sandy and desert that trees will not grow, he causes other landmarks, pillars or stones, to be set up to show the way.]

Note 1.—In this Kúblái imitated the great King Asoka, or Priyadarsi, who in his graven edicts (*circa* B.C. 250) on the Delhi Pillar, says: "Along the high roads I have caused fig-trees to be planted, that they may be for shade to animals and men. I have also planted mango-trees; and at every half-coss I have caused wells to be constructed, and resting-places for the night. And how many hostels have been erected by me at various places for the entertainment of man and beast." (*J. A. S. B.* IV. 604.) There are still remains of the fine avenues of Kúblái and his successors in various parts of Northern China. (See *Williamson*, i. 74.)

CHAPTER XXIX.

CONCERNING THE RICE-WINE DRUNK BY THE PEOPLE OF CATHAY.

MOST of the people of Cathay drink wine of the kind that I shall now describe. It is a liquor which they brew of rice with a quantity of excellent spice, in such fashion that it makes better drink than any other kind of wine; it is not only good, but clear and pleasing to the eye.[1] And being very hot stuff, it makes one drunk sooner than any other wine.

NOTE 1.—The mode of making Chinese rice-wine is described in Amyot's *Mémoires*, V. 468 *seqq.* A kind of yeast is employed, with which is often mixed a flour prepared from fragrant herbs, almonds, pine-seeds, dried fruits, etc. Rubruquis says this liquor was not distinguishable, except by smell, from the best wine of Auxerre; a wine so famous in the Middle Ages, that the Historian Friar, Salimbene, went from Lyons to Auxerre on purpose to drink it.[*] Ysbrand Ides compares the rice-wine to Rhenish; John Bell to Canary; a modern traveller quoted by Davis, "in colour, and a little in taste, to Madeira." [Friar Odoric (*Cathay*, i. p. 117) calls this wine *bigni;* Dr. Schlegel (*T'oung Pao*, ii. p. 264) says Odoric's wine was probably made with the date *Mi-yin*, pronounced *Bi-im* in old days. But Marco's wine is made of rice, and is called *shao hsing chiu*. Mr. Rockhill (*Rubruck*, p. 166, note) writes: "There is another stronger liquor distilled from millet, and called *shao chiu:* in Anglo-Chinese, *samshu;* Mongols call it *araka, arrak,* and *arreki.* Ma Twan-lin (Bk. 327) says that the Moho (the early Nu-chên Tartars) drank rice wine (*mi chiu*), but I fancy that they, like the Mongols, got it from the Chinese."

Dr. Emil Bretschneider (*Botanicon Sinicum*, ii. pp. 154-158) gives a most interesting account of the use and fabrication of intoxicating beverages by the Chinese. "The invention of wine or spirits in China," he says, "is generally ascribed to a certain I TI, who lived in the time of the Emperor Yü. According to others, the inventor of wine was TU K'ANG." One may refer also to Dr. Macgowan's paper *On the "Mutton Wine" of the Mongols and Analogous Preparations of the Chinese.* (*Jour. N. China Br. R. As. Soc.*, 1871-1872, pp. 237-240.—H. C.]

* *Kington's Fred. II.* II. 457. So, in a French play of the 13th century, a publican in his *patois* invites custom, with hot bread, hot herrings, and wine of Auxerre in plenty:—
　　　　　　　　　　" Chaiens, fait bon disner chaiens ;
　　　　　　　　　　Chi a caut pain et caus herens,
　　　　　　　　　　Et vin d'Aucheurre à plain tonnel."—
　　　　　　　　　　　　　　　(*Théat. Franç. au Moyen Age*, 168.)

CHAPTER XXX.

CONCERNING THE BLACK STONES THAT ARE DUG IN CATHAY, AND ARE BURNT FOR FUEL.

IT is a fact that all over the country of Cathay there is a kind of black stones existing in beds in the mountains, which they dig out and burn like firewood. If you supply the fire with them at night, and see that they are well kindled, you will find them still alight in the morning; and they make such capital fuel that no other is used throughout the country. It is true that they have plenty of wood also, but they do not burn it, because those stones burn better and cost less.[1]

[Moreover with that vast number of people, and the number of hot baths that they maintain—for every one has such a bath at least three times a week, and in winter if possible every day, whilst every nobleman and man of wealth has a private bath for his own use—the wood would not suffice for the purpose.]

NOTE 1.—There is a great consumption of coal in Northern China, especially in the brick stoves, which are universal, even in poor houses. Coal seems to exist in every one of the eighteen provinces of China, which in this respect is justly pronounced to be one of the most favoured countries in the world. Near the capital coal is mined at Yuen-ming-yuen, and in a variety of isolated deposits among the hills in the direction of the Kalgan road, and in the district round Siuen-hwa-fu. (*Sindachu* of Polo, ante ch. lix.) But the most important coal-fields in relation to the future are those of Shan-tung Hu-nan, Ho-nan, and Shan-si. The last is eminently *the* coal and iron province of China, and its coal-field, as described by Baron Richthofen, combines, in an extraordinary manner, all the advantages that can enhance the value of such a field except (at present) that of facile export; whilst the quantity available is so great that from Southern Shan-si alone he estimates the whole world could be supplied, at the present rate of consumption, for several thousand years. "Adits, miles in length, could be driven within the body of the coal. . . . These extraordinary conditions . . . will eventually give rise to some curious features in mining . . . . if a railroad should ever be built from the plain to this region . . . . branches of it will be constructed within the body of one or other of these beds of anthracite." Baron Richthofen, in the paper which we quote from, indicates the revolution in the deposit of the world's wealth and power, to which such facts, combined with other characteristics of China, point as probable; a revolution so vast that its contemplation seems like that of a planetary catastrophe.

In the coal-fields of Hu-nan " the mines are chiefly opened where the rivers intersect the inclined strata of the coal-measures and allow the coal-beds to be attacked by the miner immediately at their out-croppings."

At the highest point of the Great Kiang, reached by Sarel and Blakiston, they found mines on the cliffs over the river, from which the coal was sent down by long bamboo cables, the loaded baskets drawing up the empty ones.

[Many coal-fields have been explored since ; one of the most important is the coal-field of the Yun-nan province ; the finest deposits are perhaps those found in the bend of the Kiang ; coal is found also at Mong-Tzŭ, Lin-ngan, etc. ; this rich coal region has been explored in 1898 by the French engineer A. Leclère. (See *Congrès int. Géog.*, Paris, 1900, pp. 178-184.)—H. C.]

In various parts of China, as in Che-kiang, Sze-ch'wan, and at Peking, they form powdered coal, mixed with mud, into bricks, somewhat like our " patent fuel." This practice is noticed by Ibn Batuta, as well as the use of coal in making porcelain, though this he seems to have misunderstood. Rashiduddin also mentions the use of coal in China. It was in use, according to citations of Pauthier's, before the Christian era. It is a popular belief in China, that every provincial capital is bound to be established over a coal-field, so as to have a provision in case of siege. It is said that during the British siege of Canton mines were opened to the north of the city.

(*The Distribution of Coal in China*, by Baron Richthofen, in *Ocean Highways*, N.S., I. 311 ; *Macgowan* in *Ch. Repos.* xix. 385-387 ; *Blakiston*, 133, 265 ; *Mid. Kingdom*, I. 73, 78 ; *Amyot*, xi. 334 ; *Cathay*, 261, 478, 482 ; *Notes by Rev. A. Williamson* in *J. N. Ch. Br. R. A. S.*, December, 1867 ; *Hedde and Rondot*, p. 63.)

Æneas Sylvius relates as a miracle that took place before his eyes in Scotland, that poor and almost naked beggars, when *stones* were given them as alms at the church doors, went away quite delighted ; for stones of that kind were imbued either with brimstone or with some oily matter, so that they could be burnt instead of wood, of which the country was destitute. (Quoted by *Jos. Robertson, Statuta Eccles. Scotic.* I. xciii.)

CHAPTER XXXI.

How the Great Kaan causes Stores of Corn to be made, to help his People withal in time of Dearth.

You must know that when the Emperor sees that corn is cheap and abundant, he buys up large quantities, and has it stored in all his provinces in great granaries, where it is so well looked after that it will keep for three or four years.[1]

And this applies, let me tell you, to all kinds of corn, whether wheat, barley, millet, rice, panic, or what not, and when there is any scarcity of a particular kind of corn, he causes that to be issued. And if the price of

the corn is at one bezant the measure, he lets them have
it at a bezant for four measures, or at whatever price will
produce general cheapness ; and every one can have food
in this way.   And by this providence of the Emperor's,
his people can never suffer from dearth.   He does the
same over his whole Empire ; causing these supplies to
be stored everywhere, according to calculation of the
wants and necessities of the people.

Note I.—" *Le fait si bien* estuier *que il dure bien trois ans ou quatre* " (Pauthier) :
" *si bien* estudier " (G. T.).   The word may be *estiver* (It. *stivare*), to stow, but I half
suspect it should be *estuver* in the sense of " kiln-dry," though both the Geog. Latin
and the Crusca render it *gubernare.** Lecomte says : " Rice is always stored in the
public granaries for three or four years in advance.   It keeps long if care be taken to
air it and stir it about ; and although not so good to the taste or look as new rice, it is
said to be more wholesome."

The Archbishop of Soltania (A.D. 1330) speaks of these stores.   "The said
Emperor is very pitiful and compassionate. . . . and so when there is a dearth in the
land he openeth his garners, and giveth forth of his wheat and his rice for half what
others are selling it at."   Kúblái Kaan's measures of this kind, are recorded in the
annals of the Dynasty, as quoted by Pauthier.   The same practice is ascribed to the
sovereigns of the T'ang Dynasty by the old Arab *Relations*.   In later days a missionary
gives in the *Lettres Édifiantes* an unfavourable account of the action of these public
granaries, and of the rascality that occurred in connection with them. (*Lecomte*, II.
101 ; *Cathay*, 240 ; *Relat.* I. 39 ; *Let. Ed.* xxiv. 76.)

[The *Yuen-shi* in ch. 96 contains sections on dispensaries (*Hui min yao kü*),
granary regulations (*Shi ti*), and regulations for a time of dearth (*Chen Sü*). (*Bretschneider, Med. Res.* I. p. 187.)—H. C.]

CHAPTER XXXII.

Of the Charity of the Emperor to the Poor.

I HAVE told you how the Great Kaan provides for the
distribution of necessaries to his people in time of
dearth, by making store in time of cheapness.   Now I
will tell you of his alms and great charity to the poor
of his city of Cambaluc.

---

* Marsden observes incidentally (*Hist. of Sumatra*, 1st edition, p. 71) that he was told in Bengal
they used to dry-kiln the rice for exportation, " owing to which, or to some other process, it will continue good for several years."

You see he causes selection to be made of a number of families in the city which are in a state of indigence, and of such families some may consist of six in the house, some of eight, some of ten, more or fewer in each as it may hap, but the whole number being very great.   And each family he causes annually to be supplied with wheat and other corn sufficient for the whole year.   And this he never fails to do every year. Moreover, all those who choose to go to the daily dole at the Court receive a great loaf apiece, hot from the baking, and nobody is denied; for so the Lord hath ordered.   And so some 30,000 people go for it every day from year's end to year's end.   Now this is a great goodness in the Emperor to take pity of his poor people thus!   And they benefit so much by it that they worship him as he were God.

[He also provides the poor with clothes.   For he lays a tithe upon all wool, silk, hemp, and the like, from which clothing can be made; and he has these woven and laid up in a building set apart for the purpose; and as all artizans are bound to give a day's labour weekly, in this way the Kaan has these stuffs made into clothing for those poor families, suitable for summer or winter, according to the time of year.   He also provides the clothing for his troops, and has woollens woven for them in every city, the material for which is furnished by the tithe aforesaid.   You should know that the Tartars, before they were converted to the religion of the Idolaters, never practised almsgiving.   Indeed, when any poor man begged of them they would tell him, "Go with God's curse, for if He loved you as He loves me, He would have provided for you."   But the sages of the Idolaters, and especially the *Bacsis* mentioned before, told the Great Kaan that it was a good work to provide for the poor, and that his idols would be greatly pleased if he did so. And since then he has taken to do for the poor so much as you have heard.[1]]

NOTE 1.—This is a curious testimony to an ameliorating effect of Buddhism on rude nations. The general establishment of medical aid for men and animals is alluded to in the edicts of Asoka ; * and hospitals for the diseased and destitute were found by Fahian at Palibothra, whilst Hiuen Tsang speaks of the distribution of food and medicine at the *Punyasálás* or "Houses of Beneficence," in the Panjáb. Various examples of a charitable spirit in Chinese Institutions will be found in a letter by Père d'Entrecolles in the XVth Recueil of *Lettres Edifiantes ;* and a similar detail in *Nevius's China and the Chinese,* ch. xv. (See *Prinsep's Essays,* II. 15 ; *Beal's Fahhian,* 107 ; *Pèl. Boudd.* II. 190.) The Tartar sentiment towards the poor survives on the Arctic shores :—" The Yakuts regard the rich as favoured by the gods ; the poor as rejected and cast out by them." (*Billings,* Fr. Tranls. I. 233.)

CHAPTER XXXIII.

[CONCERNING THE ASTROLOGERS IN THE CITY OF CAMBALUC.]

[THERE are in the city of Cambaluc, what with Christians, Saracens, and Cathayans, some five thousand astrologers and soothsayers, whom the Great Kaan provides with annual maintenance and clothing, just as he provides the poor of whom we have spoken, and they are in the constant exercise of their art in this city.

They have a kind of astrolabe on which are inscribed the planetary signs, the hours and critical points of the whole year. And every year these Christian, Saracen, and Cathayan astrologers, each sect apart, investigate by means of this astrolabe the course and character of the whole year, according to the indications of each of its Moons, in order to discover by the natural course and disposition of the planets, and the other circumstances of the heavens, what shall be the nature of the weather, and what peculiarities shall be produced by each Moon

---

* As rendered by J. Prinsep. But I see that Professor H. H. Wilson did not admit the passage to bear that meaning.

of the year ; as, for example, under which Moon there shall be thunderstorms and tempests, under which there shall be disease, murrain, wars, disorders, and treasons, and so on, according to the indications of each ; but always adding that it lies with God to do less or more according to His pleasure.   And they write down the results of their examination in certain little pamphlets for the year, which are called *Tacuin*, and these are sold for a groat to all who desire to know what is coming. Those of the astrologers, of course whose predictions are found to be most exact, are held to be the greatest adepts in their art, and get the greater fame.[1]

And if any one having some great matter in hand, or proposing to make a long journey for traffic or other business, desires to know what will be the upshot, he goes to one of these astrologers and says : " Turn up your books and see what is the present aspect of the heavens, for I am going away on such and such a business."   Then the astrologer will reply that the applicant must also tell the year, month, and hour of his birth ; and when he has got that information he will see how the horoscope of his nativity combines with the indications of the time when the question is put, and then he predicts the result, good or bad, according to the aspect of the heavens.

You must know, too, that the Tartars reckon their years by twelves ; the sign of the first year being the Lion, of the second the Ox, of the third the Dragon, of the fourth the Dog, and so forth up to the twelfth ;[2] so that when one is asked the year of his birth he answers that it was in the year of the Lion (let us say), on such a day or night, at such an hour, and such a moment. And the father of a child always takes care to write these particulars down in a book.   When the twelve yearly symbols have been gone through, then they come

back to the first, and go through with them again in the
same succession.]

NOTE 1.—It is odd that Marsden should have sought a Chinese explanation of the
Arabic word *Takwím*, even with Tavernier before him: " They sell in Persia an
annual almanac called *Tacuim*, which is properly an ephemeris containing the
longitude and latitude of the planets, their conjunctions and oppositions, and other
such matter. The *Tacuim* is full of predictions regarding war, pestilence, and
famine ; it indicates the favourable time for putting on new clothes, for getting bled
or purged, for making a journey, and so forth. They put entire faith in it, and whoever can afford one governs himself in all things by its rules." (Bk. V. ch. xiv.)

The use of the term by Marco may possibly be an illustration of what I have elsewhere propounded, viz. that he was not acquainted with Chinese, but that his intercourse and conversation lay chiefly with the foreigners at the Kaan's Court, and probably
was carried on in the Persian language. But not long after the date of our Book we
find the word used in Italian by Jacopo Alighieri (Dante's son) :—

> " A voler giudicare
> Si conviene adequare
> Inprimo il *Taccuino*,
> Per vedere il cammino
> Come i Pianeti vanno
> Per tutto quanto l'anno."
> —*Rime Antiche Toscane*, III. 10.

Marco does not allude to the fact that almanacs were published by the Government, as they were then and still are. Pauthier (515 *seqq.*) gives some very curious
details on this subject from the Annals of the Yuen. In the accounts of the year
1328, it appears that no less than 3,123,185 copies were printed in three different sizes
at different prices, besides a separate almanac for the *Hwei-Hwei* or Mahomedans.
Had Polo not omitted to touch on the issue of almanacs by Government he could
scarcely have failed to enter on the subject of printing, on which he has kept a silence
so singular and unaccountable.

The Chinese Government still " considers the publication of a Calendar of the
first importance and utility. It must do everything in its power, not only to point
out to its numerous subjects the distribution of the seasons, ... but on account
of the general superstition it must mark in the almanac the lucky and unlucky days,
the best days for being married, for undertaking a journey, for making their
dresses, for buying or building, for presenting petitions to the Emperor, and for many
other cases of ordinary life. By this means the Government keeps the people
within the limits of humble obedience ; it is for this reason that the Emperors of
China established the Academy of Astronomy." (*Timk.* I. 358.) The acceptance of
the Imperial Almanac by a foreign Prince is considered an acknowledgment of
vassalage to the Emperor.

It is a penal offence to issue a pirated or counterfeit edition of the Government
Almanac. No one ventures to be without one, lest he become liable to the greatest
misfortunes by undertaking the important measures on black-balled days.

The price varies now, according to Williams, from 1½d. to 5d. a copy. The price
in 1328 was 1 *tsien* or cash for the cheapest edition, and 1 *liang* or tael of silver for
the *édition de luxe ;* but as these prices were in paper-money it is extremely difficult
to say, in the varying depreciation of that currency, what the price really amounted to.

["The Calendars for the use of the people, published by Imperial command, are of
two kinds. The first, *Wan-nien-shu, the Calendar of Ten Thousand Years,* is an
abridgment of the Calendar, comprising 397 years, viz. from 1624 to 2020. The

Mongol "Compendium Instrument" *Kien-e* in the Observatory Garden.

[*To face p.* 448.

second and more complete Calendar is the *Annual Calendar*, which, under the preceding dynasties, was named *Li-je, Order of Days*, and is now called *Shih-hsien-shu, Book of Constant Conformity (with the Heavens)*.   This name was given by the Emperor *Shun-chih*, in the first year of his reign (1644), on being presented by Father John Schall (*Tang Jo-wang*) with a new Calendar, calculated on the principles of European science.   This *Annual Calendar* gives the following indications : (1°) The cyclical signs of the current year, of the months, and of all the days ; (2°) the *long* and *short* months, as well as the *intercalary* month, as the case may be ; (3°) the designation of each day by the 5 *elements*, the 28 constellations, and the 12 *happy presages* ; (4°) the day and hour of the new moon, of the full moon, and of the two dichotomies, *Shang-hsien* and *Hsia-hsien* ; (5°) the day and hour for the *positions* of the sun in the 24 zodiacal signs, calculated for the various capitals of China as well as for Manchuria, Mongolia, and the tributary Kingdoms ; (6°) the hour of sunrise and sunset and the length of day and night for the principal days of the month in the several capitals ; (7°) various superstitious indications purporting to point out what days and hours are auspicious or not for such or such affairs in different places.   Those superstitious indications are stated to have been introduced into the Calendar under the *Yüan* dynasty." (*P. Hoang, Chinese Calendar*, pp. 2-3.)—H. C.]

We may note that in Polo's time one of the principal officers of the Mathematical Board was *Gaisue*, a native of *Folin* or the Byzantine Empire, who was also in charge of the medical department of the Court.   Regarding the Observatory, see note at p. 378, *supra*.

And I am indebted yet again to the generous zeal of Mr. Wylie of Shanghai, for the principal notes and extracts which will, I trust, satisfy others as well as myself that the instruments in the garden of the Observatory belong to the period of Marco Polo's residence in China.*

The objections to the alleged age of these instruments were entirely based on an inspection of photographs.   The opinion was given very strongly that no instrument of the kind, so perfect in theory and in execution, could have been even imagined in those days, and that nothing of such scientific quality could have been made except by the Jesuits.   In fact it was asserted or implied that these instruments must have been made about the year 1700, and were therefore not earlier in age than those which stand on the terraced roof of the Observatory, and are well known to most of us from the representation in Duhalde and in many popular works.

The only authority that I could lay hand on was Lecomte, and what he says was not conclusive.   I extract the most pertinent passages :

" It was on the terrace of the tower that the Chinese astronomers had set their instruments, and though few in number they occupied the whole area.   But Father Verbiest, the Director of the Observatory, considering them useless for astronomical observation, persuaded the Emperor to let them be removed, to make way for several instruments of his own construction.   The instruments set aside by the European astronomers are still in a hall adjoining the tower, buried in dust and oblivion ; and we saw them only through a grated window.   They appeared to us to be very large and well cast, in form approaching our astronomical circles ; that is all that we could make out.   There was, however, thrown into a back yard by itself, a celestial globe of bronze, of about 3 feet in diameter.   Of this we were able to take a nearer view. Its form was somewhat oval ; the divisions by no means exact, and the whole work coarse enough.

" Besides this in a lower hall they had established a gnomon. . . .   This observatory, not worthy of much consideration for its ancient instruments, much less for its situation, its form, or its construction, is now enriched by several bronze instruments which Father Verbiest has placed there.   These are large, well cast,

---

* Besides the works quoted in the text I have only been able to consult Gaubil's notices, as abstracted in Lalande ; and the Introductory Remarks to Mr. J. Williams's *Observations of Comets* . . . . *extracted from the Chinese Annals*, London, 1871.

adorned in every case with figures of dragons," etc. He then proceeds to describe them :

"(1). Armillary Zodiacal Sphere of 6 feet diameter. This sphere reposes on the heads of four dragons, the bodies of which after various convolutions come to rest upon the extremities of two brazen beams forming a cross, and thus bear the entire weight of the instrument. These dragons . . . . are represented according to the notion the Chinese form of them, enveloped in clouds, covered above the horns with long hair, with a tufted beard on the lower jaw, flaming eyes, long sharp teeth, the gaping throat ever vomiting a torrent of fire. Four lion-cubs of the same material bear the ends of the cross beams, and the heads of these are raised or depressed by means of attached screws, according to what is required. The circles are divided on both exterior and interior surface into 360 degrees ; each degree into 60 minutes by transverse lines, and the minutes into sections of 10 seconds each by the sight-edge* applied to them."

Of Verbiest's other instruments we need give only the names : (2) Equinoxial Sphere, 6 feet diameter. (3) Azimuthal Horizon, same diam. (4) Great Quadrant, of 6 feet radius. (5) Sextant of about 8 feet radius. (6) Celestial Globe of 6 feet diameter.

As Lecomte gives no details of the old instruments which he saw through a grating, and as the description of this zodiacal sphere (No. 1) corresponds in some of its main features with that represented in the photograph, I could not but recognize the *possibility* that this instrument of Verbiest's had for some reason or other been removed from the Terrace, and that the photograph might therefore possibly *not* be a representation of one of the ancient instruments displaced by him.†

The question having been raised it was very desirable to settle it, and I applied to Mr. Wylie for information, as I had received the photographs from him, and knew that he had been Mr. Thomson's companion and helper in the matter.

" Let me assure you," he writes (21st August, 1874), " the Jesuits had nothing to do with the manufacture of the so-called Mongol instruments ; and whoever made them, they were certainly on the Peking Observatory before Loyola was born. They are not made for the astronomical system introduced by the Jesuits, but are altogether conformable to the system introduced by Kúblái's astronomer Kọ Show-king. . . . I will mention one thing which is quite decisive as to the Jesuits. *The circle is divided into* 365¼ *degrees*, each degree into 100 minutes, and each minute into 100 seconds. The Jesuits always used the sexagesimal division. Lecomte speaks of the imperfection of the division on the Jesuit-made instruments ; but *those on the Mongol instruments are immeasurably coarser.*

" I understand it is not the ornamentation your friend objects to?‡ If it is, I would observe that there is no evidence of progress in the decorative and ornamental arts during the Ming Dynasty ; and even in the Jesuit instruments that part of the work is purely Chinese, excepting in one instrument, which I am persuaded must have been made in Europe.

" I have a Chinese work called *Luh-King-t'oo-Kaou*, ' Illustrations and Investigations of the Six Classics.' This was written in A.D. 1131-1162, and revised and

---

* *Pinnula.* The French *pinnule* is properly a sight-vane at the end of a traversing bar. The *transverse lines* imply that minutes were read by the system of our *diagonal scales ;* and these I understand to have been subdivided still further by aid of a divided edge attached to the sight-vane ; qu. a Vernier?

† Verbiest himself speaks of the displaced instruments thus . . . . . " ut nova instrumenta astronomica facienda mihi imponeret, quæ scilicet more Europæo affabre facta, et in specula Astroptica Pekinensi collocata, æternam Imperii Tartarici memoriam apud posteritatem servarent, *prioribus instrumentis Sinicis rudioris Minervæ, quæ jam a* trecentis *proxime* annis *speculam occupabant, inde amotis.* Imperator statim annuit illorum postulatis. ㉑ totius rei curam, publico diplomate mihi imposuit. Ego itaque intra quadriennis spatium sex diversi generis instrumenta confeci." This is from an account of the Observatory written by Verbiest himself, and printed at Peking in 1668 (*Liber Organicus Astronomiæ Europæa apud Sinas Restitutæ,* etc.). My friend Mr. D. Hanbury made the extract from a copy of this rare book in the London Institution Library. An enlarged edition was published in Europe. (Dillingen, 1687.)

‡ On the contrary, he considered the photographs interesting, as showing to how late a period the art of fine casting had endured.

Mongol Armillary Sphere in the Observatory Garden.

[To face p. 450.

printed in 1165-1174. It contains a representation of an armillary sphere, which appears to me to be much the same as the sphere in question. There is a solid horizon fixed to a graduated outer circle. Inside the latter is a meridian circle, at right angles to which is a graduated colure ; then the equator, apparently a double ring, and the ecliptic ; also two diametric bars. The cut is rudely executed, but it certainly shows that some one imagined something more perfect. The instrument stands on a cross frame, with 4 dragon supporters and a prop in the centre.*

"It should be remembered that under the Mongol Dynasty the Chinese had much intercourse with Central Asia ; and among others Yelewchootsae, as confidential minister and astronomer, followed Chinghiz in his Western campaign, held intercourse with the astronomers of Samarkand, and on his return laid some astronomical inventions before the Emperor.

"I append a notice of the Observatory taken from a popular description of Peking, by which it will be seen that the construction of these instruments is attributed to Ko Show-king, one of the most renowned astronomers of China. He was the chief astronomer under Kúblái Kaan " [to whom he was presented in 1262 ; he was born in 1231.—H. C.]

"It must be remembered that there was a special vitality among the Chinese under the Yuen with regard to the arts and sciences, and the Emperor had the choice of artizans and men of science from all countries. From the age of the Yuen till the arrival of the Jesuits, we hear nothing of any new instruments having been made ; and it is well known that astronomy was never in a lower condition than under the Ming."†

Mr. Wylie then draws attention to the account given by Trigault of the instruments that Matteo Ricci saw at Nanking, when he went (in the year 1599) to pay a visit to some of the *literati* of that city. He transcribes the account from the French *Hist. de l'Expédition Chrestienne en la Chine*, 1618. But as I have the Latin, which is the original and is more lucid, by me, I will translate from that.‡

"Not only at Peking, but in this capital also (Nanking) there is a College of Chinese Mathematicians, and this one certainly is more distinguished by the vastness of its buildings than by the skill of its professors. They have little talent and less learning, and do nothing beyond the preparation of the almanacs on the rules of calculation made by the ancients ; and when it chances that events do not agree with their calculation they assert that what they had calculated was the regular course of things, but that the aberrant conduct of the stars was a prognostic from heaven of something going to happen on the earth. This something they make out according to their fancy, and so spread a veil over their own blunders. These gentlemen did not much trust Father Matteo, fearing, no doubt, lest he should put them to shame ; but when at last they were freed from this apprehension they came and amicably visited the Father in hope of learning something from him. And when he went to return their visit he saw something that really was new and beyond his expectation.

"There is a high hill at one side of the city, but still within the walls. On the top of the hill there is an ample terrace, capitally adapted for astronomical observation, and surrounded by magnificent buildings which form the residence of the Professors. . . . On this terrace are to be seen astronomical instruments of cast-metal, well worthy of inspection whether for size or for beauty ; *and we certainly have never seen or read of anything in Europe like them.* For nearly 250 years they have stood thus

---

* This ancient instrument is probably the same that is engraved in Pauthier's *Chine Ancienne* under the title of " The Sphere of the Emperor Shun " (B.C. 2255 !).

† After the death of Kúblái astronomy fell into neglect, and when Hongwu, the first Ming sovereign, took the throne (1368) the subject was almost forgotten. Nor was there any revival till the time of Ching. The latter was a prince who in 1573 associated himself with the astronomer Hing-yun-lu to reform the state of astronomy. (*Gaubil.*) What Ricci has recorded (in Trigautius) of the dense ignorance of the Chinese *literati* in astronomical matters is entirely consistent with the preceding statements.

‡ I had entirely forgotten to look at Trigault till Mr. Wylie sent me the extract. The copy I use (*De Christianâ Expeditione apud Sinas . . . Auct. Nicolao Trigautio*) is of *Lugdun.* 1616. The first edition was published at *August. Vindelicorum* (Augsburg) in 1615 : the French, at Lyons, in 1616.

exposed to the rain, the snow, and all other atmospheric inclemencies, and yet they have lost absolutely nothing of their original lustre.   And lest I should be accused of raising expectations which I do not justify, I will do my best in a digression, probably not unwelcome, to bring them before the eyes of my readers.

" The larger of these instruments were four in number.   First we inspected a great globe [A], graduated with meridians and parallels; we estimated that three men would hardly be able to embrace its girth. . . .   A second instrument was a great sphere [B], not less in diameter than that measure of the outstretched arms which is commonly called a geometric pace.   It had a horizon and poles ; instead of circles it was provided with certain double hoops (armillæ), the void space between the pair serving the purpose of the circles of our spheres.   All these were divided into 365 degrees and some odd minutes.   There was no globe to represent the earth in the centre, but there was a certain tube, bored like a gun-barrel, which could readily be turned about and fixed to any azimuth or any altitude so as to observe any particular star through the tube, just as we do with our vane-sights ; *—not at all a despicable device !   The third machine was a gnomon [C], the height of which was twice the diameter of the former instrument, erected on a very large and long slab of marble, on the northern side of the terrace.   The stone slab had a channel cut round the margin, to be filled with water in order to determine whether the slab was level or not, and the style was set vertical as in hour-dials.†   We may suppose this gnomon to have been erected that by its aid the shadow at the solstices and equinoxes might be precisely noted, for in that view both the slab and the style were graduated.   The fourth and last instrument, and the largest of all, was one consisting as it were of three or four huge astrolabes in juxtaposition [D]; each of them having a diameter of such a geometrical pace as I have specified.   The fiducial line, or Alhidada, as it is called, was not lacking, nor yet the Dioptra.‡   Of these astrolabes, one having a tilted position in the direction of the south, represented the equator ; a second, which stood crosswise on the first, in a north and south plane, the Father took for a meridian ; but it could be turned round on its axis ; a third stood in the meridian plane with its axis perpendicular, and seemed to stand for a vertical circle ; but this also could be turned round so as to show any vertical whatever.   Moreover all these were graduated, and the degrees marked by prominent studs of iron, so that in the night the graduation could be read by the touch without a light.   All this compound astrolabe instrument was erected on a level marble platform with channels round it for levelling. On each of these instruments explanations of everything were given in Chinese characters; and there were also engraved the 24 zodiacal constellations which answer to our 12 signs, 2 to each.§   There was, however, one error common to all the instruments, viz. that, in all, the elevation of the Pole was assumed to be 36°. Now there can be no question about the fact that the city of Nanking lies in lat. 32¼°; whence it would seem probable that these instruments were made for another locality, and had been erected at Nanking, without reference to its position, by some one ill versed in mathematical science.‖

---

* " Pinnulis."   † " Et stilus eo modo quo in horologiis ad perpendiculum collocatus."
‡ The Alidada is the traversing index bar which carries the dioptra, pinnules, or sight-vanes. The word is found in some older English Dictionaries, and in France and Italy is still applied to the traversing index of a plane table or of a sextant.   Littré derives it from (Ar.) 'adád, enumeration ; but it is really from a quite different word, al-idádat العضادة " a door-post," which is found in this sense in an Arabic treatise on the Astrolabe.   (See Dozy and Engelmann, p. 140.)
§ This is an error of Ricci's, as Mr. Wylie observes, or of his reporter.
The Chinese divide their year into 24 portions of 15 days each.   Of these 24 divisions twelve called Kung mark the twelve places in which the sun and moon come into conjunction, and are thus in some degree analogous to our 12 signs of the Zodiac.   The names of these Kung are entirely different from those of our signs, though since the 17th century the Western Zodiac, with paraphrased names, has been introduced in some of their books.   But besides that, they divide the heavens into 28 stellar spaces.   The correspondence of this division to the Hindu system of the 28 Lunar Mansions, called Nakshatras, has given rise to much discussion.   The Chinese sieu or stellar spaces are excessively unequal, varying from 24° in equatorial extent down to 24′.   (Williams, op. cit.)   [See P. Hoang, supra p. 449.]
‖ Mr. Wylie is inclined to distrust the accuracy of this remark, as the only city nearly on the 36th parallel is P'ing-yang fu.

Observatory Terrace.

[*To face p.* 452.

" Some years afterwards Father Matteo saw similar instruments at Peking, or rather the same instruments, so exactly alike were they, insomuch that they had unquestionably been made by the same artist.   And indeed it is known that they were cast at the period when the Tartars were dominant in China ; and we may without rashness conjecture that they were the work of some foreigner acquainted with our studies.   But it is time to have done with these instruments."   (*Lib.* IV. *cap.* 5.)

In this interesting description it will be seen that the Armillary Sphere [B] agrees entirely with that represented in illustration facing p. 450.   And the second of his photographs in my possession, but not, I believe, yet published, answers *perfectly* to the curious description of the 4th instrument [D].   Indeed, I should scarcely have been able to translate that description intelligibly but for the aid of the photograph before me.   It shows the three *astrolabes* or graduated circles with travelling indexes arranged exactly as described, and pivoted on a complex frame of bronze ; (1) circle in the plane of the equator for measuring right ascensions ; (2) circle with its axis vertical to the plane of the last, for measuring declinations ; (3) circle with vertical axis, for zenith distances ?   The Gnomon [A] was seen by Mr. Wylie in one of the lower rooms of the Observatory (see below).   Of the Globe we do not now hear ; and that mentioned by Lecomte among the ancient instruments was inferior to what Ricci describes at Peking.

I now transcribe Mr. Wylie's translation of an extract from a Popular Description of Peking :

" The observatory is on an elevated stage on the city wall, in the south-east corner of the (Tartar) city, and was built in the year (A.D. 1279).   In the centre was the *Tze-wei* * Palace, inside of which were a pair of scrolls, and a cross inscription, by the imperial hand.   Formerly it contained the *Hwan-t'ien-e* [B] ' Armillary Sphere ' ; the *Keen-e* [D ?] ' Transit Instrument' (?) ; the *Tung-kew* [A] ' Brass Globe ' ; and the *Leang-t'ien-ch'ih*, ' Sector,' which were constructed by Ko Show-king under the Yuen Dynasty.

" In (1673) the old instruments having stood the wear of long past years, had become almost useless, and six new instruments were made by imperial authority. These were the *T'ien-t'ee* ' Celestial Globe' (6) ; *Chih-taoue* ' Equinoctial Sphere' (2) ; *Hwang-taoue* ' Zodiacal Sphere' (1) ; *Te-p'ing kinge* ' Azimuthal Horizon' (3) ; *Te-p'ing weie* ' Altitude Instrument' (4) ; *Ke-yene* ' Sextant' (5).   These were placed in the Observatory, and to the present day are respectfully used.   The old instruments were at the same time removed, and deposited at the foot of the stage.   In (1715) the *Te-ping King-wei-e* ' Azimuth and Altitude Instrument' was made ; † and in 1744 the *Ke-hang-foo-chin-e* (literally ' Sphere and Tube instrument for sweeping the heavens').   All these were placed on the Observatory stage.

" There is a wind-index-pole called the ' Fair-wind-pennon,' on which is an iron disk marked out in 28 points, corresponding in number to the 28 constellations." ‡

-¦- Mr. Wylie justly observes that the evidence is all in accord, and it leaves, I think, no reasonable room for doubt that the instruments now in the Observatory

---

But we have noted in regard to this (Polo's Pianfu, vol. ii. p. 17) that a college for the education of Mongol youth was instituted here, by the great minister Yeliu Chutsai, whose devotion to astronomy Mr. Wylie has noticed above.   In fact, two colleges were established by him, one at Yenking, *i.e.* Peking, the other at P'ing-yang ; and astronomy is specified as one of the studies to be pursued at these.   (See *D'Ohsson*, II. 71-72, quoting *De Mailla*.)   It seems highly probable that the two sets of instruments were originally intended for these two institutions, and that one set was carried to Nanking, when the Ming set their capital there in 1368.

* The 28 *sieu* or stellar spaces, above spoken of, do not extend to the Pole ; they are indeed very unequal in extent on the meridian as well as on the equator.   And the area in the northern sky not embraced in them is divided into three large spaces called *Yuen* or enclosures, of which the field of circumpolar stars (or circle of perpetual apparition) forms one which is called *Tze-Wei*.   (*Williams*.)   The southern circumpolar stars form a fourth space, beyond the 28 *sieu*.—*Ibid.*

† " This was obviously made in France.   There is nothing Chinese about it, either in construction or ornament.   It is very different from all the others."   (*Note by Mr. Wylie*.)

‡ " There follows a minute description of the brass clepsydra, and the brass gnomon, which it is unnecessary to translate.   I have seen both these instruments, in two of the lower rooms."—*Id.*

garden at Peking are those which were cast aside by Father Verbiest* in 1673 (or 1668); which Father Ricci saw at Peking at the beginning of the century, and of which he has described the duplicates at Nanking ; and which had come down from the time of the Mongols, or, more precisely, of Kúblái Khan.

Ricci speaks of their age as nearly 250 years in 1599 ; Verbiest as nearly 300 years in 1668. But these estimates evidently point to the *termination* of the Mongol Dynasty (1368), to which the Chinese would naturally refer their oral chronology. We have seen that Kúblái's reign was the era of flourishing astronomy, and that the instruments are referred to his astronomer Ko Shéu-king ; nor does there seem any ground for questioning this. In fact, it being once established that the instruments existed when the Jesuits entered China, all the objections fall to the ground.

We may observe that the *number* of the ancient instruments mentioned in the popular Chinese account agrees with the number of important instruments described by Ricci, and the titles of three at least out of the four seem to indicate the same instruments. The catalogue of the new instruments of 1673 (or 1668) given in the native work also agrees *exactly* with that given by Lecomte.† And in reference to my question as to the *possibility* that one of Verbiest's instruments might have been removed from the terrace to the garden, it is now hardly worth while to repeat Mr. Wylie's assurance that there is no ground whatever for such a supposition. The instruments represented by Lecomte are all still on the terrace, only their positions have been somewhat altered to make room for the two added in last century.

Probably, says Mr. Wylie, more might have been added from Chinese works, especially the biography of Ko Shéu-king. But my kind correspondent was unable to travel beyond the books on his own shelves. Nor was it needful.

It will have been seen that, beautiful as the art and casting of these instruments is, it would be a mistake to suppose that they are entitled to equally high rank in scientific accuracy. Mr. Wylie mentioned the question that had been started to Freiherr von Gumpach, who was for some years Professor of Astronomy in the Peking College. Whilst entirely rejecting the doubts that had been raised as to the age of the Mongol instruments, he said that he had seen those of Tycho Brahe, and the former are quite unworthy to be compared with Tycho's in scientific accuracy.

The doubts expressed have been useful in drawing attention to these remarkable reliques of the era of Kúblái's reign, and of Marco Polo's residence in Cathay, though I fear they are answerable for having added some pages to a work that required no enlargement !

[Mr. Wylie sent a most valuable paper on *The Mongol Astronomical Instruments at Peking* to the Congress of Orientalists held at St. Petersburg, which was reprinted at Shanghai in 1897 in *Chinese Researches*. Some of the astronomical instruments have been removed to Potsdam by the Germans since the siege of the foreign Legations at Peking in 1900.—H. C.]

On these auguries, and on diviners and fortune-tellers, see *Semedo*, p. 118 *seqq. ; Kidd*, p. 313 (also for preceding references, *Mid. Kingdom*, II. 152 ; *Gaubil*, 136).

NOTE 2.—¦-The real cycle of the Mongols, which was also that of the Chinese, runs : 1. Rat ; 2. Ox ; 3. Tiger ; 4. Hare ; 5. Dragon ; 6. Serpent ; 7. Horse ; 8. Sheep ; 9. Ape ; 10. Cock ; 11. Dog ; 12. Swine. But as such a cycle [12 earthly branches, *Ti-chih*] is too short to avoid confusion, it is combined with a co-efficient cycle of *ten* epithets [celestial Stems, *T'ien-kan*] in such wise as to produce a 60-year cycle of compound names before the same shall recur. These co-efficient epithets are found in four different forms : (1) From the Elements : Wood, Fire, Earth, Metal, Water, attaching to each a masculine and feminine attribute so as to make ten epithets. (2) From the Colours : Blue, Red, Yellow, White, Black, similarly treated. (3) By

---

* [Ferdinand Verbiest, S.J., was born at Pitthens, near Courtrai ; he arrived in China in 1659 and died at Peking on the 29th January, 1688.—H. C.]

† We have attached letters A, B, C, to indicate the correspondences of the ancient instruments, and cyphers 1, 2, 3, to indicate the correspondences of the modern instruments.

Observatory Instruments of the Jesuits.

[To face p. 454.

terms without meaning in Mongol, directly adopted or imitated from the Chinese, *Ga*, *Yi*, *Bing*, *Ting*, etc. (4) By the five Cardinal Points: East, South, Middle, West, North. Thus 1864 was the first year of a 60-year cycle :—

1864 = (Masc.) *Wood-Rat* Year	= (Masc.) *Blue-Rat* Year.
1865 = (Fem.) *Wood-Ox* Year	= (Fem.) *Blue-Ox* Year.
1866 = (Masc.) *Fire-Tiger* Year	= (Masc.) *Red-Tiger* Year.
1867 = (Fem.) *Fire-Hare* Year	= (Fem.) *Red-Hare* Year.
1923 = (Fem.) *Water-Swine* Year	= (Fem.) *Black-Swine* Year.

And then a new cycle commences just as before.

This Calendar was carried by the Mongols into all their dominions, and it would appear to have long survived them in Persia. Thus a document issued in favour of Sir John Chardin by the *Shaikh-ul-Islâm* of Ispahan, bears the strange date for a Mahomedan luminary of " The year of the Swine." The Hindus also had a 60-year cycle, but with them each year had an independent name.

The Mongols borrowed their system from the Chinese, who attribute its invention to the Emperor Hwang-ti, and its initiation to the 61st year of his reign, corresponding to B.C. 2637. [" It was Ta-nao, Minister to the Emperor Hwang-ti, who, by command of his Sovereign, devised the sexagenary cycle. Hwang-ti began to reign 2697 B.C., and the 61st year of his reign was taken for the first cyclical sign." *P. Hoang, Chinese Calendar*, p. 11.—H. C.] The characters representing what we have called the ten coefficient epithets are called by the Chinese the " Heavenly Stems " ; those equivalent to the twelve animal symbols are the " Earthly Branches," and they are applied in their combinations not to years only, but to cycles of months, days, and hours, such hours being equal to two of ours. Thus every year, month, day, and hour will have two appropriate characters, and the four pairs belonging to the time of any man's birth constitute what the Chinese call the " Eight Characters " of his age, to which constant reference is made in some of their systems of fortune-telling, and in the selection of propitious days for the transaction of business. To this system the text alludes. A curious account of the principles of prognostication on such a basis will be found in *Doolittle's Social Life of the Chinese* (p. 579 *seqq.* ; on the Calendar, see Schmidt's Preface to *S. Setzen ; Pallas, Sammlungen,* II. 228 *seqq.; Prinsep's Essays, Useful Tables;* 146.)

[" Kubilai Khan established in Peking two astronomical boards and two observatories. One of them was a Chinese Observatory (*sze t'ien t'ai*), the other a Mohammedan Observatory (*hui hui sze t'iĕn t'ai*), each with its particular astronomical and chronological systems, its particular astrology and instruments. The first astronomical and calendar system was compiled for the Mongols by Ye-liu Ch'u-ts'ai, who was in Chingis Khan's service, not only as a high counsellor, but also as an astronomer and astrologer. After having been convinced of the obsoleteness and incorrectness of the astronomical calculations in the *Ta ming li* (the name of the calendar system of the Kin Dynasty), he thought out at the time he was at Samarcand a new system, valid not only for China, but also for the countries conquered by the Mongols in Western Asia, and named it in memory of Chingis Khan's expedition *Si ching keng wu yüan li, i.e.* 'Astronomical Calendar beginning with the year *Keng wu,* compiled during the war in the west.' Keng-wu was the year 1210 of our era. Ye-liu Ch'u-ts'ai chose this year, and the moment of the winter solstice, for the beginning of his period ; because, according to his calculations, it coincided with the beginning of a new astronomical or planetary period. He took also into consideration, that since the year 1211 Chingis Khan's glory had spread over the whole world. Ye-liu Ch'u-ts'ai's calendar was not adopted in China, but the system of it is explained in the *Yuen-shi,* in the section on Astronomy and the Calendar.

" In the year 1267, the Mohammedans presented to Kubilai their astronomical calendar (*wan nien li, i.e.*), the calendar of ten thousand years. By taking this denomination in its literal sense, we may conclude that the Mahommedans brought to China the ancient Persian system, founded on the period of 10,000 years. The compilers of the *Yuen-shi* seem not to have had access to documents relating to this system, for they give no details about it. Finally by order of Kubilai the astronomers *Hui-Heng* and *Ko Show-King* composed a new calculation under the name of *Shou-shi-li*, which came into use from the year 1280. It is thoroughly explained in the *Yuen-shi*. Notwithstanding the fame this system generally enjoyed, its blemishes came soon to light. In the sixth month of 1302 an eclipse of the sun happened, and the calculation of the astronomer proved to be erroneous (it seems the calculation had anticipated the real time). The astronomers of the Ming Dynasty explained the errors in the *Shou-shi-li* by the circumstance, that in that calculation the period for one degree of precession of the equinox was taken too long (eighty-one years). But they were themselves hardly able to overcome these difficulties." (*Palladius*, pp. 51-53.)—H. C.]

CHAPTER XXXIV.

[CONCERNING THE RELIGION OF THE CATHAYANS;[1] THEIR VIEWS AS TO THE SOUL; AND THEIR CUSTOMS.

As we have said before, these people are Idolaters, and as regards their gods, each has a tablet fixed high up on the wall of his chamber, on which is inscribed a name which represents the Most High and Heavenly God; and before this they pay daily worship, offering incense from a thurible, raising their hands aloft, and gnashing their teeth[2] three times, praying Him to grant them health of mind and body; but of Him they ask nought else. And below on the ground there is a figure which they call *Natigai*, which is the god of things terrestrial. To him they give a wife and children, and they worship him in the same manner, with incense, and gnashing of teeth,[2] and lifting up of hands; and of him they ask seasonable weather, and the fruits of the earth, children, and so forth.[3]

Their view of the immortality of the soul is after this fashion. They believe that as soon as a man dies, his soul enters into another body, going from a good to a

better, or from a bad to a worse, according as he hath
conducted himself well or ill.   That is to say, a poor man,
if he have passed through life good and sober, shall be
born again of a gentlewoman, and shall be a gentleman ;
and on a second occasion shall be born of a princess
and shall be a prince, and so on, always rising, till he
be absorbed into the Deity.   But if he have borne himself
ill, he who was the son of a gentleman shall be reborn as
the son of a boor, and from a boor shall become a dog,
always going down lower and lower.

The people have an ornate style of speech ; they
salute each other with a cheerful countenance, and with
great politeness ; they behave like gentlemen, and eat
with great propriety.⁴   They show great respect to their
parents ; and should there be any son who offends his
parents, or fails to minister to their necessities, there is
a public office which has no other charge but that of
punishing unnatural children, who are proved to have
acted with ingratitude towards their parents.⁵

Criminals of sundry kinds who have been imprisoned,
are released at a time fixed by the Great Kaan (which
occurs every three years), but on leaving prison they are
branded on one cheek that they may be recognized.

The Great Kaan hath prohibited all gambling and
sharping, things more prevalent there than in any other
part of the world.   In doing this, he said : " I have conquered you by force of arms, and all that you have is
mine ; if, therefore, you gamble away your property, it
is in fact my property that you are gambling away."
Not that he took anything from them however.

I must not omit to tell you of the orderly way in
which the Kaan's Barons and others conduct themselves
in coming to his presence.   In the first place, within a
half mile of the place where he is, out of reverence for
his exalted majesty, everybody preserves a mien of the

greatest meekness and quiet, so that no noise of shrill voices or loud talk shall be heard. And every one of the chiefs and nobles carries always with him a handsome little vessel to spit in whilst he remain in the Hall of Audience—for no one dares spit on the floor of the hall,—and when he hath spitten he covers it up and puts it aside.[6]   So also they all have certain handsome buskins of white leather, which they carry with them, and, when summoned by the sovereign, on arriving at the entrance to the hall, they put on these white buskins, and give their others in charge to the servants, in order that they may not foul the fine carpets of silk and gold and divers colours.]

NOTE 1.—Ramusio's heading has *Tartars*, but it is manifestly of the Cathayans or Chinese that the author speaks throughout this chapter.

NOTE 2.—"*Sbattendo i denti.*" This is almost certainly, as Marsden has noticed, due to some error of transcription. Probably *Battono i fronti*, or something similar, was the true reading. [See following note, p. 461.—H. C.]

NOTE 3.—The latter part of this passage has, I doubt not, been more or less interpolated, seeing that it introduces again as a *Chinese* divinity the rude object of primitive Tartar worship, of which we have already heard in Bk. I. ch. liii. And regarding the former part of the passage, one cannot but have some doubt whether what was taken for the symbol of the Most High was not the ancestral tablet, which is usually placed in one of the inner rooms of the house, and before which worship is performed at fixed times, and according to certain established forms. Something, too, may have been known of the Emperor's worship of Heaven at the great circular temple at Peking, called *T'ien-t'ân*, or Altar of Heaven (see p. 459), where incensed offerings are made before a tablet, on which is inscribed the name Yuh-Hwang Shang-ti, which some interpret as "The Supreme Ruler of the Imperial Heavens," and regard as the nearest approach to pure Theism of which there is any indication in Chinese worship (See *Doolittle*, pp. 170, 625 ; and *Lockhart* in *J. R. G. S.*, xxxvi. 142). This worship is mentioned by the Mahomedan narrator of Shah Rukh's embassy (1421): "Every year there are some days on which the Emperor eats no animal food. . . . . He spends his time in an apartment which contains no idol, and says that he is worshipping the God of Heaven."* (*Ind. Antiquary*, II. 81.)

The charge of irreligion against the Chinese is an old one, and is made by Hayton in nearly the same terms as it often is by modern missionaries: "And though these people have the acutest intelligence in all matters wherein material things are con-

---

* "In the worship carried on here the Emperor acts as a high priest. HE only worships ; and no subject, however high in rank, can join in the adoration." (*Lockhart.*) The actual temple dates from 1420-1430 ; but the *Institution* is very ancient, and I think there is evidence that such a structure existed under the Mongols, probably only *restored* by the Ming. [It was built during the 18th year of the reign of the third Ming Emperor Yung Loh (1403-1425) ; it was entirely restored during the 18th year of K'ien Lung ; it was struck by lightning and burnt down in 1889 ; it is being re-built.—H. C.]

Great Temple of Heaven, Peking.

cerned, yet you shall never find among them any knowledge or perception of spiritual things." Yet it is a mistake to suppose that this insensibility has been so universal as it is often represented. To say nothing of the considerable numbers who have adhered faithfully to the Roman Catholic Church, the large number of Mahomedans in China, of whom many must have been proselytes, indicates an interest in religion ; and that Buddhism itself was in China once a spiritual power of no small energy will, I think, be plain to any one who reads the very interesting extracts in Schott's essay on Buddhism in Upper Asia and China. (*Berlin Acad. of Sciences*, 1846.) These seem to be so little known that I will translate two or three of them. "In the years *Yuan-yeu* of the Sung (A.D. 1086-1093), a pious matron with her two servants lived entirely to the Land of Enlightenment. One of the maids said one day to her companion : 'To-night I shall pass over to the Realm of Amita.' The same night a balsamic odour filled the house, and the maid died without any preceding illness. On the following day the surviving maid said to the lady : 'Yesterday my deceased companion appeared to me in a dream, and said to me : "Thanks to the persevering exhortations of our mistress, I am become a partaker of Paradise, and my blessedness is past all expression in words."' The matron replied : 'If she will appear to me also then I will believe what you say.' Next night the deceased really appeared to her, and saluted her with respect. The lady asked : 'May I, for once, visit the Land of Enlightenment?' 'Yea,' answered the Blessed Soul, 'thou hast but to follow thy handmaiden.' The lady followed her (in her dream), and soon perceived a lake of immeasurable expanse, overspread with innumerable red and white lotus flowers, of various sizes, some blooming, some fading. She asked what those flowers might signify? The maiden replied : 'These are all human beings on the earth whose thoughts are turned to the Land of Enlightenment. The very first longing after the Paradise of Amita produces a flower in the Celestial Lake, and this becomes daily larger and more glorious, as the self-improvement of the person whom it represents advances ; in the contrary case, it loses in glory and fades away.' * The matron desired to know the name of an enlightened one who reposed on one of the flowers, clad in a waving and wondrously glistening raiment. Her whilom maiden answered : 'That is Yangkie.' Then asked she the name of another, and was answered : 'That is Mahu.' The lady then said : 'At what place shall I hereafter come into existence?' Then the Blessed Soul led her a space further, and showed her a hill that gleamed with gold and azure. 'Here,' said she, 'is your future abode. You will belong to the first order of the blessed.' When the matron awoke she sent to enquire for Yangkie and Mahu. The first was already departed, the other still alive and well. And thus the lady learned that the soul of one who advances in holiness and never turns back, may be already a dweller in the Land of Enlightenment, even though the body still sojourn in this transitory world" (pp. 55-56).

What a singular counterpart the striking conclusion here forms to Dante's tremendous assault on a still living villain,—or enemy !

> ———"che per sua opra
> In anima in Cocito già si bagna,
> Ed in corpo par vivo ancor di sopra."
> —*Infern.* xxxiii. 155.

Again : "I knew a man who during his life had killed many living beings, and was at last struck with an apoplexy. The sorrows in store for his sin-laden soul pained me to the heart ; I visited him, and exhorted him to call on the Amita ; but he obstinately refused, and spoke only of indifferent matters. His illness clouded his understanding ; in consequence of his misdeeds he had become hardened. What was

---

* In 1871 I saw in Bond Street an exhibition of (so-called) "spirit" drawings, *i.e.* drawings alleged to be executed by a "medium" under extraneous and invisible guidance. A number of these extraordinary productions (for extraordinary they were undoubtedly) professed to represent the "Spiritual Flowers" of such and such persons ; and the explanation of this as presented in the catalogue was in substance exactly that given in the text. It is highly improbable that the artist had any cognizance of Schott's Essay, and the coincidence was assuredly very striking.

before such a man when once his eyes were closed? Wherefore let men be converted while there is yet time! In this life the night followeth the day, and the winter followeth the summer; that, all men are aware of. But that life is followed by death, no man will consider. Oh, what blindness and obduracy is this!" (p. 93).

Again: "Hoang-ta-tie, of T'ancheu (Changshu-fu in Honan), who lived under the Sung, followed the craft of a blacksmith. Whenever he was at his work he used to call without intermission on the name of Amita Buddha. One day he handed to his neighbours the following verses of his own composing to be spread about:—

> 'Ding dong! The hammer-strokes fall long and fast,
> Until the Iron turns to steel at last!
> Now shall the long long Day of Rest begin,
> The Land of Bliss Eternal calls me in.'

Thereupon he died. But his verses spread all over Honan, and many learned to call upon Buddha" (103).

Once more: "In my own town there lived a physician by name Chang-yan-ming. He was a man who never took payment for his treatment from any one in poor or indifferent circumstances; nay, he would often make presents to such persons of money or corn to lighten their lot. If a rich man would have his advice and paid him a fee, he never looked to see whether it were much. or little. If a patient lay so dangerously ill that Yanming despaired of his recovery, he would still give him good medicine to comfort his heart, but never took payment for it. I knew this man for many a year, and I never heard the word *Money* pass his lips! One day a fire broke out in the town, and laid the whole of the houses in ashes; only that of the physician was spared. His sons and grandsons reached high dignities" (p. 110).

Of such as this physician the apostle said: "Of a truth I perceive that God is no respecter of persons: But in every nation he that feareth Him, and worketh righteousness, is accepted with Him."

["By the 'Most High and Heavenly God,' worshipped by the Chinese, as Marco Polo reports, evidently the Chinese *T'ien*, 'Heaven' is meant, *Lao t'ien ye* in the common language. Regarding 'the God of things terrestrial,' whose figure the Chinese, according to M. Polo, 'placed below on the ground,' there can also be no doubt that he understands the *T'u-ti*, the local 'Lar' of the Chinese, to which they present sacrifices on the floor, near the wall under the table.

"M. Polo reports, that the Chinese worship their God offering incense, raising their hands aloft, and gnashing their teeth. Of course he means that they placed the hands together, or held kindled joss-stick bundles in their hands, according to the Chinese custom. The statement of M. Polo *sbattendo i denti* is very remarkable. It seems to me, that very few of the Chinese are aware of the fact, that this custom still exists among the Taouists. In the rituals of the Taouists the *K'ow-ch'i* (*Ko'w* = 'to knock against,' *ch'i* = 'teeth') is prescribed as a comminatory and propitiatory act. It is effected by the four upper and lower fore-teeth. The Taouists are obliged before the service begins to perform a certain number of *K'ow-chi*, turning their heads alternately to the left and to the right, in order to drive away mundane thoughts and aggressions of bad spirits. The *K'ow-ch'i* repeated three times is called *ming fa ku* in Chinese, *i.e.* 'to beat the spiritual drum.' The ritual says, that it is heard by the Most High Ruler, who is moved by it to grace.

"M. Polo observed this custom among the lay heathen. Indeed, it appears from a small treatise, written in China more than a hundred years before M. Polo, that at the time the Chinese author wrote, all devout men, entering a temple, used to perform the *K'ow-ch'i*, and considered it an expression of veneration and devotion to the idols. Thus this custom had been preserved to the time of M. Polo, who did not fail to mention this strange peculiarity in the exterior observances of the Chinese. As regards the present time it seems to me, that this custom is not known among the people, and even with respect to the Taouists it is only performed on certain occasions, and not in all Taouist temples." (*Palladius*, pp. 53-54.)—H. C.]

NOTE 4.—"True politeness cannot of course be taught by rules merely, but a great degree of urbanity and kindness is everywhere shown, whether owing to the naturally placable disposition of the people, or to the effects of their early instruction in the forms of politeness." (*Mid. Kingdom*, II. 68.) As regards the "ornate style of speech," a well-bred Chinaman never says *I* or *You*, but for the former "the little person," "the disciple," "the inferior," and so on; and for the latter, "the learned man," "the master," or even "the emperor." These phrases, however, are not confined to China, most of them having exact parallels in Hindustani courtesy. On this subject and the courteous disposition of the Chinese, see *Fontaney*, in *Lett. Edif.* VII. 287 *seqq.*; also XI. 287 *seqq.*; *Semedo*, 36; *Lecomte*, II. 48 *seqq.* There are, however, strong differences of opinion expressed on this subject; there is, apparently, much more genuine courtesy in the north than in the south.

NOTE 5.—"Filial piety is the fundamental principle of the Chinese polity." (*Amiot*, V. 129.) "In cases of extreme unfilial conduct, parents sometimes accuse their children before the magistrate, and demand his official aid in controlling or punishing them; but such instances are comparatively rare. . . . If the parent require his son to be publicly whipped by the command of the magistrate, the latter is obliged to order the infliction of the whipping. . . . If after punishment the son remain undutiful and disobedient, and his parents demand it at the hands of the magistrate, the latter must, with the consent of the maternal uncles of the son, cause him to be taken out to the high wall in front of the yamun, and have him there publicly whipped to death." (*Doolittle*, 102-103.)

NOTE 6.—[Mr. Rockhill writes to me that pocket-spitoons are still used in China.—H. C.]

END OF VOL I.

PRINTED AT THE EDINBURGH PRESS, 9 AND 11 YOUNG STREET.

Archway erected under the Mongol Dynasty, at Kiu-Yung Kwan, N. W. of Peking.*

* On the walls of this archway is engraved the inscription in six characters, of which a representation accompanies ch. xv. of Prologue, note 1.

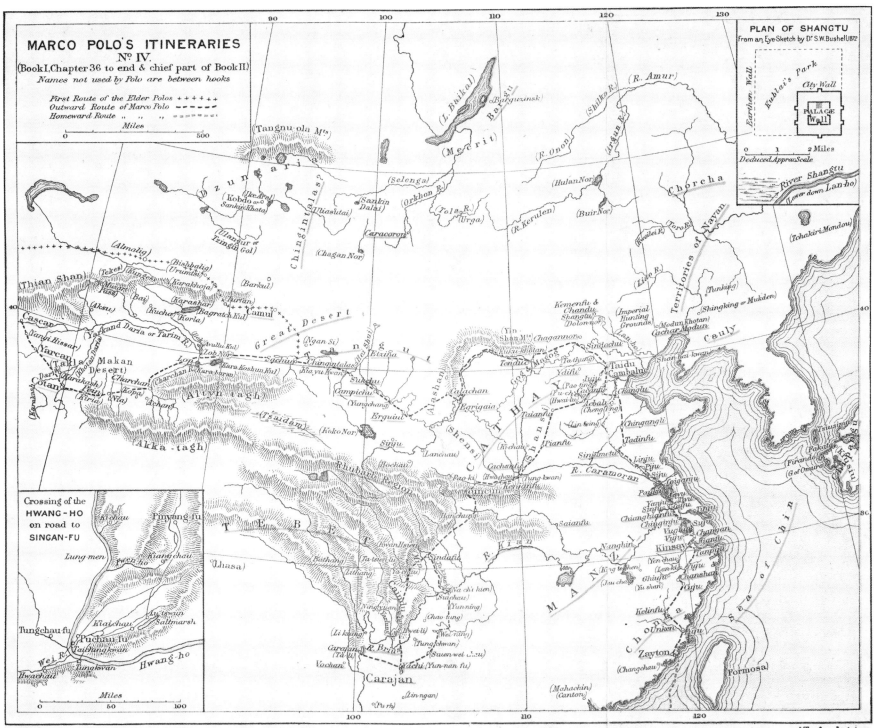

MARCO POLO'S ITINERARIES
Nº IV.
(Book I. Chapter 36 to end & chief part of Book II.)
Names not used by Polo are between hooks

First Route of the Elder Polos +++++
Outward Route of Marco Polo – – –
Homeward Route ,, ,, ,, ◻◻◻◻

Miles
0 ——————— 500

PLAN OF SHANGTU
From an Eye-Sketch by Dr S.W. Bushell 1872

Earthen Wall
Kublai's Park
City Wall
PALACE
Wall

0 1 2 Miles
Deduced Approx Scale

River Shangtu
(Lower down Lan-ho)

Crossing of the HWANG-HO on road to SINGAN-FU

Miles
0 — 50 — 100

{To face last page.